2 DIMENSIONAL QUANTUM FIELD THEORY

Non-perturbative methods in

2 DIMENSIONAL QUANTUM FIELD THEORY

Elcio Abdalla
(Universidade de Sao Paulo)

M. Cristina B. Abdalla
(Instituto de Fisica Teórica – UNESP)

Klaus D. Rothe
(University of Heidelberg)

World Scientific
Singapore • New Jersey • London • Hong Kong

Published by

World Scientific Publishing Co. Pte. Ltd.
5 Toh Tuck Link, Singapore 596224
USA office: 27 Warren Street, Suite 401-402, Hackensack, NJ 07601
UK office: 57 Shelton Street, Covent Garden, London WC2H 9HE

British Library Cataloguing-in-Publication Data
A catalogue record for this book is available from the British Library.

NON-PERTURBATIVE METHODS IN TWO-DIMENSIONAL QUANTUM FIELD THEORY

ISBN-13 978-981-02-0462-4
ISBN-10 981-02-0462-0
ISBN-13 978-981-02-0463-1 (pbk)
ISBN-10 981-02-0463-9 (pbk)

To our parents,
Jose, Ruth, Oswaldo and Julia,
and our children, Filipe
Diana and Leonardo.
Elcio and Cristina

Dedicated to my parents,
Annemarie and Kurt,
to my wife, Neuza,
and to my son, Thomas.
Klaus

FOREWORD

Since the first exactly solvable quantum field theoretical model in two space-time dimensions was proposed by W. Thirring in 1958, the study of two-dimensional quantum field theory has played an important role in the development of a non-perturbative understanding of quantum field theory in general. By the end of the seventies, very comprehensive knowledge had been gathered, revealing an unexpected complexity and richness of the non-perturbative structure of relativistic quantum field theories.

Although interest in such model studies had levelled off by the end of the seventies, this area of research experienced a remarkable comeback in the eighties, when an almost forgotten acquaintance from the seventies – string theory – experienced a dramatic revival as a result of pioneering work due to Green and Schwarz. Even since, two dimensional quantum field theory – which had already found applications in statistical mechanics in the past – has become an important part of elementary particle physics.

While other – more phenomenologically motivated – branches of high energy physics, such as quantum chromodynamics, the electro-weak model of Glashow, Weinberg and Salam, as well as purely group-theoretical studies, have by now been extensively portrayed in a large number of research reports and books the same has not been the case for two-dimensional quantum field theory. This book is intended to fill this gap. Though many of the models to be discussed may not be of immediate relevance to physics in four space-time dimensions, they nevertheless provide a deep structural insight as well as mathematical tools which should prove – and have already proven – useful for the non-perturbative understanding of realistic theories.

This book is intended to cover the most important developments on non-perturbative aspects of two-dimensional quantum field theory since 1958. After two introductory chapters on generalized free fields and their application to the solution of two exactly solvable models, we turn in chapter 4 to a discussion of functional determinants, which shall play an important role in subsequent chapters. This material is to a large extent also relevant to higher dimensions. In chapter 5 we turn to a discussion of the Gross-Neveu model (chiral and non-chiral). This model, though non-renormalizable in higher dimensions, exhibits some interesting properties such as mass transmutation and spontaneous "Kosterlitz-Thouless" type symmetry breakdown. We demonstrate the use of $1/N$ expansion as well as operator methods. Chapters 6 and 7 deal with the classical and quantum aspects of non-linear sigma models, respectively, where very elegant and beautiful results have been obtained. The rather impressive results on the exact S-matrices of a vast number of two-dimensional field-theoretical models could not remain unmentioned as they represent the first examples of non-trivial scattering matrices obeying the principle of minimal analyticity of the sixties. These results are presented in chapter 8. Chapter 9 prepares the ground for non-abelian fermion-boson equivalences (bosonization) which shall play a central role in the remaining part of the book. Chapters 10 to 14 deal with two-dimensional (abelian and non-abelian) gauge theories, including the more recent developments on anomalous chiral gauge theories. In view of the hidden complexity of two-dimensional Quantum Electrodynamics (QED_2), we have thought it appropriate to start off the discussion on gauge theories with the operator solution of this model. This is done in chapter 10, where we discuss, both the case of massless and massive fermions using operator bosonization techniques, with emphasis on delicate questions such as "screening versus confinement" of quarks as well as the existence of charge (soliton) and kink sectors in the bosonized formulation. In chapters 11 to 14 we then choose the functional approach. The ordering of chapters 11/12 and 13/14 (non-abelian versus abelian) may appear unnatural to the readers. The criterion here has not been the degree of complexity, but rather the degree of solvability, the abelian theory being regarded as a special case of the non-abelian one, with the added property of being exactly solvable.

Finally, our survey of two-dimensional field theory would be incomplete without an account of the rapidly developing field of research on two-dimensional conformal field theories. This is left to chapter 15.

This book is intended for students working on their Ph.D. degree, as well as for post-doctoral research workers wishing to become acquainted with non-perturbative aspects in quantum field theory. Although much of the material in this book is special to two dimensions, the techniques used should prove useful also in the development of techniques applicable in higher dimensions. In particular, the last chapter of this book will be of direct interest to researchers who wish to work in the field of conformal field theory and strings.

It is clear that in view of the scope of this book and the tremendous number of published papers on this subject, it is impossible to do full justice to all authors which have contributed to this area of research. For the same reason, a strongly biased selection of the topics, as well as form of presentation had to be made.

When referring to the original literature, much care should be exercised with respect to the conventions and definitions being used. We have therefore summarized the conventions used in this book in appendices A through C. Since extensive use is made of the Dirac algorithm for quantizing constrained systems, we have thought it useful to include a summary of Dirac's method in appendix O.

It is a pleasure to thank Dr Michael Forger for his critical and careful revision of chapter 6. We would also like to thank FAPESP (Fundação de Amparo a Pesquisa do Estado de São Paulo) and DAAD (Deutscher Akademischer Austauschdienst) for their financial support of several exchange visits within the Brazilian-German agreement, which greatly facilitated our collaboration. One of us (K.D.R.) would also like to thank Mrs. M. Steiert and Mrs. G. Einecke for typesetting in TEX one of the chapters, and Dr. W. Wetzel for his helpful assistance concerning computer-related problems. M.C.B.A would like to thank Prof. A. H. Zimerman for critical and helpful discussions during the course "Topics on two dimensional quantum field theory" held at I.F.T. in the 2nd semester of 1987. E. A. wishes to acknowledge the hospitality at the International Center for Theoretical Physics and at SISSA in Trieste, where part of this work was performed. E.A. and M.C.B.A. wish to thank Dr. J. C. Brunelli and Dr. Jorge L. de Lyra for their competent and kind help in the use of TEX software, and computer-related problems.

CONTENTS

Chapter XII Quantum Electrodynamics: Functional Approach

Chapter XIII Non-Abelian Chiral Gauge Theories

Chapter I

INTRODUCTION

The development of Relativistic Quantum Field Theory started in 1932 as a natural extension of Quantum Mechanics to the relativistic domain[1]. The work of Feynman in the late forties provided a powerful tool for the calculation of processes in Quantum Electrodynamics[2]. Second quantization led, however to new conceptual and technical difficulties. Quantum fields had to be regarded as operator-valued distributions, their local products being ill defined as a result of ultraviolet divergencies, which plagued the higher order computations in perturbation theory. This problem was partially mastered via the techniques of renormalization[3,4], and later on completely understood[5,6].

In the early fifties, there appeared a series of papers concerned with the extraction of general non–perturbative properties of quantum field theory from a perturbative setup. Of particular importance for this latter development of quantum field theory were the papers of Lehmann, Symanzik and Zimmermann[7] (and also of Wightman[8], Haag[9], and others [10,11]). The so–called LSZ formalism established the relation between fields and particles, in terms of the asymptotic conditions for the interpolating fields (which are nothing but the interacting fields at finite time). The reduction formula provided the connection between expectation values of fields, and S-matrix elements of particle scattering. From the study of the analytic properties of Feynman diagrams, dispersion relations were derived, which could be used to obtain non–perturbative information[12].

These developments were followed by a new axiomatic approach to Quantum Field Theory (QFT), which became known as constructive QFT. Here functional analysis played an important role. Some of the (non–perturbative) results of the LSZ formalism, which had been based on perturbative studies, could thereby be derived from general principles (if the theory had a non–vanishing mass gap)[13]. An important consequence of this approach was a theorem connecting spin and statistics[14].

Meanwhile, all dynamical calculations in QFT were restricted to perturbation theory. In particular, this rendered calculations involving strong interactions unreliable, and made information about the bound state spectrum only accessible within approximative non-perturbative – and often non-unitary –schemes (e.g. the Bethe-Salpeter equation in the ladder approximation). As a result, QFT had fallen into stagnation, and even discredit, in the late fifties.

These difficulties provided, in particular, the motivation for a new approach to the strong interactions, which became known as S-matrix theory[15], which was to play a dominant role in the sixties. The predictive power of this theory turned out to be very small, since it was entirely based on kinematical principles and analyticity supplemented by the bootstrap idea. An underlying dynamical framework was lacking. Nevertheless, analyticity in the complex angular momentum plane led to the important concept of *duality*, expressing the possibility of representing a given scattering amplitude as a sum over poles in crossed channels.

An explicit realization of these concepts by a remarkable formula proposed by Veneziano[17] led to a new, parallel development in the sixties summarized under the name of *dual models*. However, in the S-matrix (or dual model) approach the high energy behavior was incorrect. Moreover, an analysis of the pole structure of higher order corrections required the introduction of a somewhat mysterious new concept, the *pomeron*[18]. The ever increasing number of parameters which were required to describe experiment within these schemes, and the resulting loss of predictive power, eventually led physicists to abandon them, and to turn again to QFT.

In the meantime, QFT had scored some remarkable successes in the realm of the weak interactions[19]. Moreover, symmetry principles such as those advocated by Gell-Mann and Ne'eman[20] had proven powerful in predicting the masses of strongly interacting particles, as well as the existence of new ones (the Ω^-), without the recourse to dynamical calculations.

This situation led to a revival of QFT in the late sixties and in the seventies, when much attention was given to the non-perturbative aspects. Quantum chromodynamics (QCD) had been proposed as the fundamental theory of the strong interactions[21], but reliable calculations confronting QCD with experimental tests were lacking. The high energy behavior of QFT was investigated by means of the renormalization group and Callan-Symanzik equations[22], which describe the behavior of the theory under finite renormalizations of the parameters. As a result it was possible to relate the zero mass limit, to the high energy behavior. A momentum dependent ("running") coupling constant turns out to characterize the strength of the interaction: depending on the properties of the so called β-function, which appears in the renormalization group equation, the running coupling constant may be sufficiently small for either high or small momenta, to legitimate perturbation theory in one of these regions. In the case of non–abelian gauge theories, perturbation theory turned out to be a good approximation at high energy (asymptotic freedom). Classical solutions of field theory have also played a central role in the (non-perturbative) semiclassical analysis of QFT. Monopole solutions (Minkowski space), and instanton solutions (Euclidean space), have been obtained, showing the importance of the topology of the manifold on which the fields are defined[23]. In this way, more abstract branches of mathematics such as algebraic topology came to play an important role in the unraveling of the structural properties of gauge theories.

Although the above studies have been important for revealing a non trivial and highly interesting structure, exact non-perturbative results are only available for specific models, all in two-dimensional space-time. A complete (non-perturbative) solution of a QFT model means exact knowledge of all its correlation functions. Quantum field theories with such properties have been restricted to 1+1 dimensions. The first such model, describing the current–current interaction of massless fermions was discussed by Thirring[24] in 1958, as an example of a completely soluble quantum field theoretic model obeying the general principles of a QFT[14]. The complete quantum solution was given in a classic paper of Klaiber[25], and was shown to satisfy all Wightman axioms[14]. Up to that time the only models known to obey these axioms had been those describing generalized free fields[26].

Following the above work, Schwinger[11,27] obtained an exact solution of Quantum Electrodynamics in 1+1 dimensions (QED_2). A number of interesting properties, such as the nontrivial vacuum structure of this model, were however only later revealed in the work of Lowenstein and Swieca[28], who explored the consequences of the long range Coulomb force for the charge sectors of the theory. This long range force was interpreted as being responsible for the confinement of quarks[29], that is, their occurrence in the form of permanently bound states of $\bar{q}q$ pairs (baryonic bound states happen to be absent in QED_2). The problem of confinement and the related phenomenon of screening of charge quantum numbers in $d = 2$ has been extensively studied[30], and has served as a basis for sharpening the concepts involved also in higher dimensions.

The surprisingly rich structure of two dimensional quantum electrodynamics was found to describe several important features of non abelian gauge theories, under investigation in the seventies. As a matter of fact, in the late sixties one learned that the short distance singularities of quantum field theory play a key role in the dynamical structure of the theory[31]. The experimental results on lepton-proton scattering at large momentum transfer, required that a realistic theory of the strong interaction be asymptotically free[21,22]. This made QCD the unique candidate for the theory describing strong interactions, since it was shown that no renormalizable theory without non abelian gauge fields can be asymptotically free[32]. The properties of vacuum structure and confinement attributed to QCD_4 were found to be explicitly realized in two-dimensional QED, which made the theory a very interesting "laboratory".

Several further developments in $d = 2$ QFT, of growing importance, followed. Classically exactly integrable models, and the quantization of solitons were extensively studied in two dimensions[33]. Such integrable models were generally characterized by the existence of an infinite number of conservation laws[34]. In the cases where these conservation laws survived quantization, the S-matrices and their associated monodromy matrices could be computed exactly[35,36,37,38,39]. Besides being the first examples of exact S-matrices realizing the ideas of "minimal analyticity" of the sixties, these exact results also play an important role in checking approximation schemes such as the semi-classical approximation, and the $\frac{1}{N}$ expansion, and have important applications in statistical mechanics[40]. Some of the results concerning classical integrability have also been generalized to higher dimensions[41].

In the particular case of the sine-Gordon theory, exact results have also been obtained beyond the S-matrix level[42,37]. Moreover, the S-matrix of fundamental fields was generalized to the full S-matrix describing scattering of bound states and solitons as well[43]. Moreover one finds an unexpected $O(2) \approx U(1)$ symmetry, reflecting the fact that the solitons in the sine-Gordon theory correspond to the fermions of the massive Thirring model. This equivalence, partially conjectured long ago by Skyrme[44], was proven by Coleman[45,46] at the level of Green functions, and later obtained by the use of operatorial methods[47]. In both versions (bosonic or fermionic) the S-matrices could be exactly computed and turn out to be identical[48,42,36].

In the framework of two-dimensional models, the possibility of writing fermions in terms of bosons (bosonization) has been a powerful method for obtaining non-perturbative information. One of the features which should be stressed in this context is that the charge sectors of the fermionic theory corresponds to hidden, charged "soliton" sectors in the purely neutral bosonic theory: dynamical aspects of the fermionic formulation turn into topological properties of the bosonic counterpart. In abelian bosonization the building blocks of the bosonization scheme are the exponentials of the free bosonic fields. The fermion number of this composite operator is directly linked to the infrared behavior of the zero mass scalar fields. This leads to a superselection rule[49], which makes the charged sectors appear in a very natural way.

The $U(1)$ bosonization techniques become cumbersome when applied to non-abelian theories. The reason is twofold: the symmetry transformations of the fermionic theory are non-local with respect to the fundamental Bose fields; these fields lie in non-linear representations of the global symmetry group of the fermions. Significant progress in the direction of non-abelian bosonization was provided by the work of Polyakov and Wiegman[50] on the one hand, and Witten[51] on the other hand. Although these authors discussed the problem in different contexts– chiral QCD_2 and theory of free $O(N)$ Majorana fermions– both arrived at an equivalent bosonic action involving the action of the principal sigma model plus a Wess-Zumino term[52,53]. Hence in $d = 2$ dimensions, fermionic theories were found to exhibit a remarkable "universality" in the bosonic formulation, where the non-linear sigma model and "topological" terms seemed to play a fundamental role.

Non-linear sigma models have a long history. A particularly important role has been played by the class of two-dimensional integrable non-linear sigma models, which have a geometrical origin[54,55,56]. They have been shown to share several properties with Yang-Mills theories in four dimensions[57,58]: at the classical level, both are conformally invariant and present similar geometrical identities as well as non-trivial classical solutions[59,60] (e.g. instantons[61] in the Euclidean formulation). The non linear sigma models for symmetric spaces[54,55], and the Yang-Mills theories for either the self dual sector, or with extended supersymmetry[41], share similar integrability properties. When quantized, non-linear sigma models also exhibit features also believed to be properties of realistic theories, such as confining long range force[56], when the gauge-group is not simple[62], and dynamical mass generation; spontaneous symmetry breakdown presents particular features[63,64]. These properties make them appealing as toy models for the strong interactions[58,65,66]. However, their geometrical origin makes them also very interesting mathematical objects to be studied in their own right.

It turns out that sigma models also play an important role in string theory, where the D-dimensional target manifold is compactified to four dimensional space-time[67]. The action associated with the compactified dimensions is described by a sigma model. The requirement of conformal invariance at the quantum level leads directly to the Einstein equation of general relativity, and predicts its quantum corrections[68,69,70].

Two dimensional space-time has also proven to be an excellent laboratory for the study of gauge-anomalies and the consistency of anomalous chiral gauge theories. The exact solubility of two-dimensional chiral QED[71,73] has played here an important role in opening up a whole new line of developments in the area of chiral gauge theories. In view of the fact that the top quark has not yet been found, the successful quantization of such seemingly inconsistent theories without recourse to an algebraic cancellation of the gauge-anomalies on group-theoretical grounds, is of much interest. An unexpected deep underlying differential geometric significance of such anomalies has thereby been revealed[72,74]. Moreover, one of the most successful uses of sigma models in two dimensions is their relation with respect to four dimensional non abelian gauge theories[66]. Quantum mechanically, sigma models exhibit also desirable features, such as the "secret" long range force[56] generated by quantum fluctuations of the induced gauge field, when the gauge group is not simple[62], unless there is a suitable interaction with fermions (supersymmetric or minimal), liberating the partons[75,76,77,78].

More recently, it was shown that in two dimensional quantum field theories, Poincaré and scale invariance, alone, imply invariance under an infinite dimensional symmetry group[79]. As a result, non trivial correlators can be exactly computed. They are found to be in general related to solutions of hypergeometric differential equations. The parameters labelling these equations, which are regarded as the critical indices, have been classified and characterize the correlators uniquely[79,80]. The conformal algebras are realized in terms of the so called primary fields and their descendents. In Minkowski space, this construction leads naturally to the use of Artin Braids, which relate this problem to the algebraic construction of exact S-matrices, since the star-triangle relations obtained from the infinite local conservation laws, have the same structure as the permutation relations of knot theory[81].

The above ideas may be generalized to include the interaction with conformally invariant gravity[82]. In the light cone gauge the theory simplifies dramatically, due to a new $SL(2,R)$ symmetry[82,83]. The critical indices of the theory may be computed from a very simple equation relating them to the critical indices of the theory in flat space. The results have also been generalized to the supersymmetric case[84].

To summarize, two dimensional models have been an extraordinary laboratory to test ideas in quantum field theory. Thus, the Thirring model provided a realization of an exactly soluble quantum field theory, while the Schwinger and the non linear sigma models were found to exhibit properties of four dimensional non abelian gauge theories. However, two dimensional QFT also plays a direct role in the description of physical reality, having applications in string theories, as well as statistical mechanics. In particular, the methods developed in two-dimensional QFT have been used to extract results concerning the critical behavior of models in statistical mechanics, using conformal invariance alone. An extraordinary amount of physically interesting[85] as well as mathematically elegant concepts[86,87] have emerged from the study of such theories.

Beyond their status as a theoretical laboratory, and their applications in string theories and statistical mechanics, the study of these models has also led to recent developments opening new possibilities for applications of some of the above methods in the study of quantum field theories in higher dimensions. There is a deep relationship between rational conformal invariance in two-dimensional space-time and the Chern-Simons action in three dimensions[88] (which is also equivalent to conformal gravity in three dimensions[89]). The Chern-Simons action proves to be a key element in the generalization of the fermion-boson equivalence to three dimensional space-time[90], and also plays an important role in the discussion of non abelian anomalies of chiral gauge theories in any dimension[74].

The aim of this book is to give a fairly detailed survey of these interesting developments in two-dimensional quantum field theory since the pioneering work of Thirring, without loosing sight of their relevance to the four-dimensional world. Though many of the properties and techniques are peculiar to two-dimensional space-time, the richness of the structure of the models underlies some non-trivial structures to be expected in realistic quantum field theories[91]. Our emphasis therefore lies on the non-perturbative aspects of the two-dimensional models and our tools shall correspondingly be operator and functional methods. We shall nevertheless occasionally make use of Feynman diagram techniques in order to obtain non-perturbative results.

Chapter II

FREE FIELDS

2.1 Introduction

A study of quantum field theory usually starts out with the definition of free fields. In two dimensional space time this is not only the rule, but also shows an interesting feature concerning the zero mass limit of free bosonic fields. This object does not define a (tempered) distribution, due to infrared divergencies, which cannot be tamed in two dimensions[26,92]. The infrared divergencies are shown to cancel, once a superselection rule is defined for correlators involving the exponentials of free fields[26,49]. Therefore, charges may be naturally defined for real bosonic fields in two dimensions, a very important issue that shall be tackled in detail in chapter 3, and also used very often later on. Nonetheless, it is still unclear how to obtain these results from a path integral formulation in the general case, since the equivalence between fermions and bosons is a collective phenomenon (see however appendix D). Moreover, we shall see that topological aspects are the major issue in the case of the bosonization procedure, when we analyse the connection between the massive Thirring model and the sine Gordon theory in chapter 3.

2.2 Bosonic Free Fields

Massive real bosonic fields in two dimensional space-time are defined in terms of the free Lagrangian

$$\mathcal{L} = \frac{1}{2}(\partial_\mu \varphi)^2 - \frac{1}{2}m^2\varphi^2 \quad , \tag{2.1a}$$

with the corresponding Klein-Gordon equation

$$(\Box + m^2)\varphi = 0 \quad . \tag{2.1b}$$

The solution of Eq. (2.1*b*) has the Fourier decomposition:

$$\varphi(x) = \int \frac{dk^1}{\sqrt{2\pi 2k^o}}[a(k)e^{-ikx} + a(k)^\dagger e^{ikx}] = \varphi^{(+)}(x) + \varphi^{(-)}(x) \quad , \tag{2.2}$$

where $k_0 = \sqrt{k_1^2 + m^2}$, and $(+), (-)$ refer to the positive and negative frequency parts.

The momentum conjugate to $\varphi(x)$ is $\Pi(x) = \partial_0\varphi(x)$. Canonical quantization then implies the commutation relations

$$[\varphi(x), \dot\varphi(y)]\Big|_{x^0=y^0} = i\delta(x^1 - y^1) \quad , \tag{2.3}$$

and we have as a consequence the following commutation relation for the operators $a(k)$, and $a(k)^\dagger$:

$$[a(k), a^\dagger(k')] = \delta(k^1 - k'^1) \quad . \tag{2.4}$$

The Fock space is obtained by applying the creation operator $a^\dagger(k)$ successively on states, starting from the vacuum, which is defined by

$$a(k)|0\rangle = 0 \quad . \tag{2.5}$$

The two point function can be easily computed; for the Wightman function, we find

$$\begin{aligned}\Delta^{(+)}(x-y) \equiv \langle 0|\varphi(x)\varphi(y)|0\rangle &= \frac{1}{2\pi}\int \frac{dk^1}{2k^o} e^{-ik(x-y)} = \Delta^{(+)}(x-y; m^2)\\ &= \frac{1}{2\pi} K_0\left(m\sqrt{-(x-y)^2 + i(x^o - y^o)\epsilon}\right)\end{aligned} \tag{2.6}$$

and we have, for the commutator function, the expression

$$i\Delta(x-y) = [\varphi(x), \varphi(y)] = \Delta^{(+)}(x-y) - \Delta^{(+)}(y-x) = -\frac{i}{2} I_0(im\sqrt{x^2})\theta(x^2)\epsilon(x^0) . \tag{2.7}$$

The time ordered product, which is related to the Feynman propagator, is given by

$$\begin{aligned}-i\Delta_F(x-y) = \langle 0|T\varphi(x)\varphi(y)|0\rangle &= \frac{1}{2\pi}\int \frac{dk^1}{2k^0}\Big\{\theta(x^0 - y^0)e^{-ik(x-y)} + \\ \theta(y^0 - x^0)e^{ik(x-y)}\Big\} &= \int \frac{d^2k}{(2\pi)^2}\frac{i}{k^2 - m^2 + i\epsilon} e^{-ik.(x-y)}\end{aligned} \quad . \tag{2.8}$$

The zero mass limit of the above function does not exist, since $\frac{1}{k^2}$ is not a tempered distribution in two dimensions[93,94]. One can nevertheless define a zero mass two-point function by using an infrared (IR) regularization as follows,

$$D^{(+)}(x; \lambda) = \frac{1}{2\pi}\int \frac{dk^1}{2|k^1|}\theta(|k^1| - \lambda)e^{-ikx} \quad , \tag{2.9}$$

where λ is a small IR cutoff.

One finds for this function[1)], for small λ

$$D^{(+)}(x) = -\frac{1}{4\pi}\ln(-\mu^2 x^2 + i x^0 \epsilon) \quad , \tag{2.10}$$

where $\mu = e^\gamma \lambda$, and γ is the Euler constant.

It is clear that in the above the two point function violates the Wightman axiom of positivity[26,92], due to the term that has been subtracted in (2.9) (see footnote 1). Indeed, for test functions with support in $-x^2 > \frac{1}{\mu^2}$, the norm is negative, as we immediately see from (2.10). These arguments however do not preclude the existence of well defined operators obtained from the field $\varphi(x)$. Indeed, the simplest example is the derivative field $\partial_\mu \varphi(x)$, which can be directly defined in terms of the $m \to 0$ limit of the massive scalar field. There is however another operator, the exponencial of the field $\varphi(x)$, which when appropriately defined, plays a fundamental role in the representation of fermionic fields in terms of bosonic ones (bosonization). In fact, the μ independence of these observables as $\mu \to 0$ leads to charge superselection rules, as we next demonstrate.

Consider the two point function given by (2.6) in the zero mass limit; we have

$$\langle 0|\varphi(x)\varphi(0)|0\rangle \underset{m\to 0}{\longrightarrow} -\frac{1}{4\pi}\ln\left[-\frac{m^2 e^{2\gamma}}{4}(x^2 - i x^0 \epsilon)\right] \quad , \tag{2.11}$$

which diverges. In fact, there is no zero mass free scalar field in two dimensions satisfying the Wightman axiom of positivity: consider the distribution satisfying $\Box G(x) = 0$; it has a Fourier transform $\tilde{G}(p)$, which should be Lorentz invariant,

---

1) We calculate (2.9) for $x^0 = 0$ and compare with the zero mass limit of (2.8), at $x^0 = 0$ as well; we have

$$D^{(+)}(x;\lambda) = \frac{1}{2\pi}\lim_{m\to 0}\left\{\int_0^\infty \frac{dk}{\sqrt{k^2+m^2}}\cos kx^1 - \frac{1}{2\pi}\int_0^\lambda \frac{dk}{\sqrt{k^2+m^2}}\cos kx^1\right\} =$$

$$= \frac{1}{2\pi}\lim_{m\to 0}\left\{K_0(mx^1) - \int_0^{\frac{\lambda}{m}} \frac{dy}{\sqrt{y^2+1}}\cos(myx^1)\right\} \quad ;$$

for m and λ small, $\cos(myx) \approx 1$, and using $K_0(mx^1) \approx -\ln\left(\frac{e^\gamma}{2} m x^1\right)$, we thus have

$$D^{(+)}(0,x^1;\lambda) \approx -\frac{1}{2\pi}\ln\frac{e^\gamma}{2} m x^1 - \frac{1}{2\pi}\ln(2\frac{\lambda}{m}) = -\frac{1}{2\pi}\ln(e^\gamma \lambda x^1) \quad .$$

The $i\epsilon$ prescription is chosen in such a way that the integral (2.9) converges. Lorentz invariance implies (2.10).

and concentrated on the positive light cone $p^2 = 0$, and $p^0 \geq 0$, with $p^2\tilde{G}(p) = 0$. The most general distributions satisfying such requirements are[93,95] $\delta(p^0)\delta(p^1)$, and

$$\left\{\theta(p^0+p^1)\left(\frac{1}{p^0+p^1}\right)\right\}_{reg}\delta(p^0-p^1)+\left\{\theta(p^0-p^1)\left(\frac{1}{p^0-p^1}\right)\right\}_{reg}\delta(p^0+p^1) \quad (2.12)$$

where $\{\theta(x)\frac{1}{x}\}_{reg}$ is a regularization of $\theta(x)\frac{1}{x}$.

The first case is trivial; as far as the second one is concerned, it represents a non positive distribution, due to the subtraction implied by the regularization.

We may however construct well defined operators satisfying the Wightman axioms in terms of Wick ordered exponential of zero mass fields[96],

$$:e^{i\alpha\varphi(x)}:=e^{i\alpha\varphi^{(-)}(x)}e^{i\alpha\varphi^{(+)}(x)} \quad ; \qquad (2.13)$$

using the formula

$$e^A e^B = e^B e^A e^{[A,B]} \quad , \qquad (2.14)$$

valid whenever $[A, B]$ is a c-number, we obtain for the vacuum expectation value of the product of exponentials such as (2.13),

$$\left\langle 0\left|e^{i\alpha_1\varphi(x_1)}\dots e^{i\alpha_n\varphi(x_n)}\right|0\right\rangle = e^{-\sum_{i<j}\alpha_i\alpha_j\Delta^{(+)}(x_i-x_j,m)} \quad . \qquad (2.15)$$

It is clear from (2.11) that the zero mass limit of (2.15) is ill defined. Introducing an infrared renormalized field

$$e_\alpha(x) = \left(\frac{m^2e^{2\gamma}}{4}\right)^{\frac{\alpha^2}{8\pi}} :e^{i\alpha\varphi(x)}: \quad , \qquad (2.16)$$

one obtains for the vacuum expectation value of the product $e_{\alpha_1}(x_1)\cdots e_{\alpha_n}(x_n)$,

$$\langle 0\,|e_{\alpha_1}(x_1)\cdots e_{\alpha_n}(x_n)|\,0\rangle = \left(\frac{m^2e^{2\gamma}}{4}\right)^{\sum_{i=1}^{n}\frac{\alpha_i^2}{8\pi}} e^{-\sum_{i<j}\alpha_i\alpha_j\Delta^{(+)}(x_i-x_j;m)} \quad . \qquad (2.17)$$

Using (2.11), we may write (2.17) in the form

$$\langle 0\,|e_{\alpha_1}(x_1)\cdots e_{\alpha_n}(x_n)|\,0\rangle = \left(\frac{m^2e^{2\gamma}}{4}\right)^{\frac{1}{8\pi}\left(\sum_{i=1}^{n}\alpha_i\right)^2}\prod_{i<j}\left[-(x_i-x_j)^2+i\epsilon(x_i^0-x_j^0)\right]^{\frac{\alpha_i\alpha_j}{4\pi}}$$
$$(2.18)$$

In the zero mass limit, the expectation value (2.18) vanishes, unless[49,25,45]

$$\sum_{i=1}^{n} \alpha_i = 0 \quad ; \tag{2.19}$$

equation (2.19) represents a superselection rule. This conclusion is extremely important, implying a natural definition of charge in a theory of real bosonic fields. Note that (2.18) defines a positive definite distribution[26].

In addition to the zero mass field φ it is convenient to define a dual field $\tilde{\varphi}$ via:

$$\partial_\mu \varphi = \epsilon_{\mu\nu} \partial^\nu \tilde{\varphi} \quad . \tag{2.20}$$

The integrability condition of the above equation is obtained by contracting it with either ∂_μ or with $\epsilon_{\mu\rho}\partial^\rho$, yielding:

$$\Box \varphi = 0 = \Box \tilde{\varphi} \quad . \tag{2.21}$$

Therefore $\tilde{\varphi}$ can only be defined by (2.20) in the massless case.

For the two point function involving $\tilde{\varphi}$ one obtains[2)]:

$$[\tilde{\varphi}^{(+)}(x), \tilde{\varphi}^{(-)}(y)] = \langle 0|\tilde{\varphi}(x)\tilde{\varphi}(y)|0\rangle = D^{(+)}(x-y) \tag{2.22}$$

and

$$\left[\varphi^{(+)}(x), \tilde{\varphi}^{(-)}(y)\right] = \langle 0\,|\varphi(x)\tilde{\varphi}(y)|\,0\rangle = \widetilde{D}^{(+)}(x-y) \quad , \tag{2.23}$$

where

$$\widetilde{D}^{(+)}(x) = \frac{1}{4\pi} \ln \frac{ix^+ + \epsilon}{ix^- + \epsilon} \quad , \tag{2.24}$$

and[3)]

$$D^{(+)}(x) = -\frac{1}{4\pi} \ln \left[\mu^2 (i\mu x^+ + \epsilon)(i\mu x^- + \epsilon)\right] \quad . \tag{2.25}$$

The two dimensional Lorentz group is abelian, and takes the simplest form in the light cone variables:

$$\xi^\pm \to \xi^{\pm'} = e^{\pm\chi} \xi^\pm \tag{2.26}$$

with

$$\xi^\pm = \xi^0 \pm \xi^1 \quad , \tag{2.27}$$

---

2) Note that the commutator (2.23) differs from the corresponding one in (A.38) by the additive constant $\frac{i}{4}$.

3) We use $\ln(x + i\epsilon) = \ln|x| + i\pi\theta(-x)$.

and χ is related to the velocity v by $\sinh\chi = \frac{v}{\sqrt{1-v^2}}$. Correspondingly, the Lorentz transformations act on the functions $D^{(+)}(x)$ and $\tilde{D}^{(+)}(x)$ as

$$D^{(+)}(\xi') = D^{(+)}(\xi) \tag{2.28a}$$

$$\widetilde{D}^{(+)}(\xi') = \widetilde{D}^{(+)}(\xi) + \frac{\chi}{2\pi} \quad . \tag{2.28b}$$

As we next show, free Dirac fields may be represented in terms of the exponentials of φ and $\tilde{\varphi}$. Indeed, let us take the two point function

$$\mathcal{F}(\xi) = \left(-\xi^2 + i\epsilon\xi_0\right)^{-\frac{\alpha^2}{4\pi}} \quad ; \tag{2.29}$$

its Fourier transform is given by[92]

$$Z^{(2)}_{\frac{\alpha^2}{2\pi}}(p) = \int d^2\xi e^{ip\xi}\left(-\xi^2 + i\epsilon\xi_0\right)^{-\frac{\alpha^2}{4\pi}} \quad . \tag{2.30}$$

For $\frac{\alpha^2}{2\pi}$ an integer this is the Lorentz invariant distribution

$$Z^{(n)}_l(p) = \frac{1}{2^{l-1}\Gamma\left[\frac{l}{2}\right]\Gamma\left[\frac{l+2-n}{2}\right]}\left(\theta(p_0)\theta(p^2)\sqrt{p^2}\right)^{l-n} \quad , \tag{2.31}$$

with $n = 2$. This expression, valid for $l \geq n$ may be analytically continued to represent (2.29) and (2.30). The Lehmann spectral function for massive fermions in the presence of an exponencial of a massless field has been computed in ref. [92] (page 9, Eqs. 13 and 14), by means of the use of the above distribution for real values of α; for $\alpha \to 0$ one recovers the discrete contribution at the fermion mass.

2.3 Fermionic Free Fields

The massive free spin $\frac{1}{2}$ field in a two dimensional space-time obeys the Dirac equation:

$$(i\,\partial\!\!\!/ - m)\psi = 0 \quad , \tag{2.32}$$

which can be derived from the Lagrangian

$$\mathcal{L} = \overline{\psi}(i\partial\!\!\!/ - m)\psi \quad . \tag{2.33}$$

The Fourier decomposition of the fermionic field is given by

$$\psi(x) = \int \frac{dk^1}{\sqrt{2\pi 2k^0}}\left[b(k)u(k)e^{-ikx} + d^\dagger(k)v(k)e^{ikx}\right] \quad , \tag{2.34}$$

with

$$\{b^\dagger(k), b(p)\} = \{d^\dagger(k), d(p)\} = \delta(k^1 - p^1) \quad , \tag{2.35a}$$

and

$$u(k) = \begin{pmatrix} \sqrt{k^+} \\ \sqrt{k^-} \end{pmatrix} \quad , \quad v(k) = \begin{pmatrix} \sqrt{k^+} \\ -\sqrt{k^-} \end{pmatrix} \quad . \tag{2.35b}$$

The spinors satisfy

$$u_\alpha \overline{u}_\beta = k^\mu \gamma_\mu + m \quad , \tag{2.36a}$$

$$v_\alpha \overline{v}_\beta = k^\mu \gamma_\mu - m \quad , \tag{2.36b}$$

where

$$\gamma_0 = \begin{pmatrix} 0 & 1 \\ 1 & 0 \end{pmatrix} \quad \gamma_1 = \begin{pmatrix} 0 & 1 \\ -1 & 0 \end{pmatrix} \quad . \tag{2.36c}$$

In the zero mass limit we have

$$u(k) = \sqrt{2k^0} \begin{pmatrix} \theta(k^1) \\ \theta(-k^1) \end{pmatrix} \quad , \quad v(k) = \sqrt{2k^0} \begin{pmatrix} \theta(k^1) \\ -\theta(-k^1) \end{pmatrix} \quad . \tag{2.37}$$

Actually, in the massless case, due to the separation of right and left movers, we can rewrite (2.34) in terms of the spinor $u(k)$ alone, by defining an operator $\widetilde{d}(k^1)$ as follows:

$$\widetilde{d}(k^1) = d(k^1) \text{ for } k^1 > 0 \quad , \tag{2.38a}$$

$$\widetilde{d}(k^1) = -d(k^1) \text{ for } k^1 < 0 \quad , \tag{2.38b}$$

Expression (2.34) then takes the form

$$\psi(x) = \int \frac{dk^1}{\sqrt{2\pi 2k^0}} u(k)[b(k)e^{-ikx} + \widetilde{d}^\dagger(k)e^{ikx}] \quad . \tag{2.39}$$

The canonical anticommutator, for the $\psi(x)$ fields (2.34), for arbitrary time reads

$$\{\psi_\alpha(x), \overline{\psi}_\beta(y)\} = iS(x-y)_{\alpha\beta} \quad , \tag{2.40a}$$

where

$$S(x-y) = (i \not\partial + m)\Delta(x-y) \quad , \tag{2.40b}$$

and $\Delta(x-y)$ is given by (2.7). Corresponding to (2.6-8) we also define the two point functions

$$\langle 0|\psi_\alpha(x)\overline{\psi}_\beta(y)|0\rangle = (i \not\partial + m)\Delta^{(+)}(x-y) = S^{(+)}(x-y) \tag{2.41a}$$

$$\langle 0|\overline{\psi}_\beta(x)\psi_\alpha(y)|0\rangle = (i \not\partial - m)_{\alpha\beta}\Delta^{(+)}(x-y) = -S^{(+)}(y-x) \tag{2.41b}$$

$$\langle 0|T\psi_\alpha(x)\overline{\psi}_\beta(y) \mid 0\rangle = (\not\partial - im)\Delta_F(x-y) = iS_F(x-y) \tag{2.41c}$$

These functions are well defined in the zero mass limit.[4] One has for $m = 0$

$$S^{(+)}(x) = -\frac{i}{2\pi}\begin{pmatrix} 0 & \frac{1}{x^- - i\epsilon} \\ \frac{1}{x^+ - i\epsilon} & 0 \end{pmatrix} \quad . \tag{2.42}$$

There exist two conserved local currents: with the usual charge conservation is associated the conserved vector current

$$j_\mu(x) = :\overline{\psi}(x)\gamma_\mu\psi(x): \quad , \tag{2.43}$$

where the double points indicate normal ordering; the chiral symmetry of the Lagrangian (2.33) in the zero mass limit implies in turn the conservation of the axial vector current

$$j^5_\mu(x) = :\overline{\psi}(x)\gamma_\mu\gamma_5\psi(x): \quad = \epsilon_{\mu\nu}j^\nu \quad , \tag{2.44}$$

where we used the relation (see appendix A) $\gamma_\mu\gamma_5 = \epsilon_{\mu\nu}\gamma^\nu$. Conservation of charge ($\partial_\mu j^\mu = 0$) and pseudo-charge ($\partial_\mu j^{\mu 5} = 0$) thus imply

$$\partial_\mu j^\mu = 0 = \epsilon^{\mu\nu}\partial_\mu j_\nu \quad , \tag{2.45}$$

or equivalently

$$\partial_\pm j_\mp = 0 \quad . \tag{2.46}$$

Majorana representation

A second representation for the gamma matrices, useful when dealing with Majorana fermions, is given by

$$\gamma_0 = \begin{pmatrix} 0 & -i \\ i & 0 \end{pmatrix} , \qquad \gamma_1 = \begin{pmatrix} 0 & i \\ i & 0 \end{pmatrix} . \tag{2.47}$$

In terms of this representation, the Dirac equation derived from (2.33) (which retains its form) is given by

$$\partial_+\psi_2 - m\psi_1 = 0 \quad , \tag{2.48a}$$
$$\partial_-\psi_1 + m\psi_2 = 0 \quad . \tag{2.48b}$$

The spinors u and v defined in (2.34) are now given by

$$u(k) = \begin{pmatrix} \sqrt{k_+} \\ i\sqrt{k_-} \end{pmatrix} , \quad v(k) = \begin{pmatrix} \sqrt{k_+} \\ -i\sqrt{k_-} \end{pmatrix} . \tag{2.49}$$

[4] In the zero mass limit, we have $D(x) = -\frac{1}{2}\theta(x^2)\epsilon(x^0)$ and (2.10).

2.4 Bosonization of Fermions

In two dimensions we may always write

$$j_\mu = \partial_\mu \varphi + \epsilon_{\mu\nu}\partial^\nu \tilde{\theta} \quad , \tag{2.50}$$

which corresponds to the decomposition of a vector into a " curl free" and a "divergence free" part. Thus, it follows from (2.45) that φ and $\tilde{\theta}$ are massless fields

$$\Box\varphi = 0 = \Box\tilde{\theta} \quad . \tag{2.51}$$

Hence, we can use (2.20) to express $\tilde{\theta}$ in terms of its dual θ. Defining $\phi = -\sqrt{\pi}(\varphi + \theta)$, (2.43) takes the form

$$j_\mu = -\frac{1}{\sqrt{\pi}}\partial_\mu \phi \quad . \tag{2.52}$$

One may easily convince oneself that j_μ is of the form (2.50), with ϕ a zero mass field, by a direct calculation. Indeed, using (2.34) and (2.43), one has

$$j_\mu(x) =: \overline{\psi}(x)\gamma_\mu\psi(x) := \int \frac{dk^1 dp^1}{2\pi\sqrt{2k^0 2p^0}} \overline{u}(k)\gamma_\mu u(p) \times \{b^\dagger(k)b(p)e^{i(k-p)x} + \\ + b^\dagger(k)d^\dagger(p)e^{i(k+p)x} + d(k)b(p)e^{-i(k+p)x} - d^\dagger(p)d(k)e^{i(k-p)x}\} \,. \tag{2.53}$$

We shall need

$$\frac{1}{2k^0}\sqrt{\frac{k^0}{p^0}}\overline{u}(k)\gamma_\mu u(p) = \frac{k_\mu}{k^0}(\theta(k^1)\theta(p^1) + \theta(-k^1)\theta(-p^1)) = \frac{k_\mu}{k^0}\theta(k^1 p^1) \quad ; \tag{2.54}$$

in order to show that (2.53) can be written in the form (2.52) with ϕ a canonical zero mass field, we factorize the derivative and define $\phi(x)$ by the expression

$$j^\pm(x) = \partial^\pm \phi(x) = \pm i \int \frac{dk^1}{2\pi}\sqrt{k^0}\theta(\pm k^1)\left[c(k^1)e^{ik^1 x^\mp} - c^\dagger(k^1)e^{-ik^1 x^\mp}\right] \\ = \pm i \int \frac{dk^1}{2\pi}\sqrt{k^0}\left[\theta(\pm k^1)c(k^1) - \theta(\mp k^1)c^\dagger(-k^1)\right] e^{ik^1 x^\mp} \tag{2.55}$$

After Fourier transformation, we get

$$\sqrt{k^0}c(k^1) = -i\tilde{j}^+(k)\theta(k^1) + i\tilde{j}^-(k)\theta(-k^1) \quad , \tag{2.56a}$$

$$\sqrt{k^0}c^\dagger(k^1) = i\tilde{j}^+(-k)\theta(k^1) - i\tilde{j}^-(-k)\theta(-k^1) \quad . \tag{2.56b}$$

Using (2.53), (2.54) and (2.56), we obtain[25]

$$c(k^1) = \frac{-i}{\sqrt{k^0}} \int dp^1 \{\theta(p^1 k^1)[b^\dagger(p^1)b(p^1+k^1) - d^\dagger(p^1)d(p^1+k^1)] + \theta((k^1-p^1)p^1)d(k^1-p^1)b(p^1)\} \quad ,(2.57a)$$

$$c^\dagger(k^1) = \frac{i}{\sqrt{k^0}} \int dp^1 \{\theta(p^1 k^1)[b^\dagger(p^1+k^1)b(p^1) - d^\dagger(p^1+k^1)d(p^1)] + \theta((k^1-p^1)p^1)b^\dagger(p^1)d^\dagger(k^1-p^1)\} \quad .(2.57b)$$

We can now compute the commutator of c and $c^\dagger$, in order to verify whether $\phi(x)$ is a canonical field; the q-number terms are found to cancel upon using the support properties of the θ function, and one finds

$$[c(k), c^\dagger(k')] = \delta(k-k') \quad . \tag{2.58}$$

Therefore the c's are the usual annihilation operators (they commute among themselves, as we readily verify using (2.53)) with the vacuum defined by

$$c(k^1)|0\rangle = 0 \quad . \tag{2.59}$$

The bosonic massless fields can be used to build more complex objects having fermionic character. Motivated by the above construction of a canonical free bosonic field out of free fermions, we show now, conversely, how to construct free zero mass fermions out of bosons. The fact that a fermionic field can be written in terms of bosonic fields has been discovered long ago by Jordan and Wigner[97]. Indeed , defining the usual bosonic Fock space by

$$|k\rangle = a^\dagger(k)|0\rangle \quad , \tag{2.60}$$

with the canonical relations

$$[a(k), a^\dagger(k')] = \delta(k^1 - k^{1'}) \quad , \tag{2.61a}$$

$$[a(k), a(k')] = [a^\dagger(k), a^\dagger(k')] = 0 \quad , \tag{2.61b}$$

we define

$$d^\dagger(k) = a^\dagger(k) e^{\{-i\pi \int_{k^1}^{\infty} a^\dagger(k^1)a(k^1)dk^1\}} \quad , \tag{2.62}$$

and verify that it satisfies the anticommutation relations

$$\{d(k), d^\dagger(k^1)\} = \delta(k-k^1) \quad . \tag{2.63}$$

The above construction shows that statistics in two dimensional space-time is a matter of convention. Generalization to higher dimensions of the mapping (2.62) are not Lorentz invariant.

Spin is also a matter of convention in $1+1$ dimensions, since there exists no rotation in 1 space dimension. It is defined as a label of Lorentz transformation properties in $1+1$ dimensions[25], and does not need to be integer or half integer. In fact, for a massless theory, different spin solutions of the Thirring model differing in their "Lorentz spin" correspond to different representations of the conformal group in 2 dimensions[98,99].

Consider the generalized free field

$$\psi^{(\alpha,\beta)}(x) =: e^{i\chi_{\alpha,\beta}(x)} := e^{i\chi^{-}_{\alpha,\beta}(x)} e^{i\chi^{+}_{\alpha,\beta}(x)} \quad , \tag{2.64}$$

where

$$\chi_{\alpha,\beta}(x) = \alpha\varphi(x) + \beta\tilde{\varphi}(x) \quad , \tag{2.65}$$

with φ and $\tilde{\varphi}$ canonically normalized zero mass fields, related by (2.20). Using relations (2.14), (2.23), (2.24) and (2.25) one finds

$$\begin{aligned}\psi^{(\alpha,\beta)}(x)\psi^{(\gamma,\delta)}(y) &= \psi^{(\gamma,\delta)}(y)\psi^{(\alpha,\beta)}(x) e^{\{-[\chi^{-}_{\alpha,\beta}(x),\chi^{+}_{\gamma,\delta}(y)]+[\chi^{-}_{\gamma,\delta}(y),\chi^{+}_{\alpha,\beta}(x)]\}} = \\ &\psi^{(\gamma,\delta)}(y)\psi^{(\alpha,\beta)}(x) e^{-(\alpha\gamma+\beta\delta)(D^{(-)}(x-y)-D^{(-)}(y-x))-(\alpha\delta+\beta\gamma)(\tilde{D}^{(-)}(x-y)-\tilde{D}^{(-)}(y-x)}\end{aligned} \tag{2.66}$$

Using now

$$D^{(-)}(x) - D^{(-)}(-x) = \frac{i}{2}\theta(x^2)\epsilon(x^0) \tag{2.67a}$$

$$\tilde{D}^{(-)}(x) - \tilde{D}^{(-)}(-x) = -\frac{i}{2}\theta(-x^2)\epsilon(x^1) \quad , \tag{2.67b}$$

expression (2.66) reduces to

$$\begin{aligned}&\psi^{(\alpha,\beta)}(x)\psi^{(\gamma,\delta)}(y) \\ &= \psi^{(\gamma,\delta)}(y)\psi^{(\alpha,\beta)}(x) e^{-\frac{1}{2}(\alpha\gamma+\beta\delta)\epsilon(x^0-y^0)\theta((x-y)^2)+\frac{i}{2}(\alpha\delta+\beta\gamma)\epsilon(x^1-y^1)\theta(-(x-y)^2)}\end{aligned} \tag{2.68}$$

From here we see that for space like distances, the fields $\psi^{(\alpha,\beta)}(x)$, $\psi^{(\gamma,\delta)(y)}$

$$\begin{cases} \text{i) commute,} & \text{if} \quad \beta\gamma + \alpha\delta = 4n\pi \\ \text{ii) anticommute,} & \text{if} \quad \beta\gamma + \alpha\delta = 2(2n+1)\pi \end{cases} \tag{2.69}$$

and otherwise neither commute nor anticommute. The case of a field that anticommutes with itself, ($\alpha\beta = (2n+1)\pi$, n integer) will be of particular interest for us.

An arbitrary correlator involving $\psi^{(\alpha,\beta)}(x)$ may be computed by normal ordering with respect to $\phi^{(+)}$ using repeatedly (2.66), as well as the property $\phi\,|0\rangle = 0$; we have

$$\left\langle 0 \middle| \psi^{(\alpha,\beta)}(x_1)\cdots\psi^{(\alpha,\beta)}(x_n)\psi^{\dagger(\alpha,\beta)}(y_1)\cdots\psi^{\dagger(\alpha,\beta)}(y_n) \middle| 0 \right\rangle = e^{F(x_1\cdots x_n, y_1\cdots y_n)} \quad , \tag{2.70}$$

where

$$
\begin{aligned}
F(x_1 \cdots x_n, y_1 \cdots y_n) = \\
= \alpha^2 & \left\{ -\sum_{i<j} \left(D^{(-)}(x_i - x_j) - D^{(-)}(y_i - y_j) \right) + \sum_{i,j} D^{(-)}(x_i - y_j) \right\} \\
+\beta^2 & \left\{ -\sum_{i<j} \left(D^{(-)}(x_i - x_j) - D^{(-)}(y_i - y_j) \right) + \sum_{i,j} D^{(-)}(x_i - y_j) \right\} \\
-2\alpha\beta & \left\{ \sum_{i<j} \left(\tilde{D}^{(-)}(x_i - x_j) + \tilde{D}^{(-)}(y_i - y_j) \right) - \sum_{i,j} \tilde{D}^{-}(x_i - y_j) \right\} . \qquad (2.71)
\end{aligned}
$$

Under Lorentz transformations, $D^{(-)}(x)$ is invariant, while $\tilde{D}^{(-)}(x)$ transforms as stated in (2.28*b*). Thus $F(x,y)$ transforms under Lorentz transformations as

$$F(\Lambda x, \Lambda y) = F(x,y) + n\frac{\alpha\beta}{\pi}\chi \qquad (2.72)$$

We conclude that, the field $\psi^{(\alpha,\beta)}(x)$ must transform as

$$U(\Lambda)\psi^{(\alpha,\beta)}(x)U^{-1}(\Lambda) = \psi^{(\alpha,\beta)}(\Lambda x)e^{\frac{\alpha\beta}{2\pi}\chi} \quad . \qquad (2.73)$$

We define the Lorentz spin s of ψ in terms of the Lorentz transformation property (2.73), that is

$$U(\Lambda)\psi(x)U^{-1}(\Lambda) = e^{s\gamma_5}\psi(\Lambda x) \quad . \qquad (2.74)$$

This implies that the Lorentz-spin of the field $\psi^{(\alpha,\beta)}$ is given by

$$s = \frac{\alpha\beta}{2\pi} \quad , \qquad (2.75)$$

If $\alpha\beta = (2n+1)\pi$, s is a half integer and $\psi^{(\alpha,\beta)}$ anticommutes with itself; if $\alpha\beta = 2n\pi$, s is integer and $\psi^{(\alpha,\beta)}$ commutes with itself (see (2.69)). Of course a true spin does not exist in two dimensions, since there exist no rotations in one spatial dimension.

For massless fields, the "spin" is an arbitrary parameter characterizing the realization of the conformal group in terms of the fields. This is the basis of a boson-fermion duality which we will discuss at length later on. Under dilatations $\widetilde{D}$ is invariant, while D transforms as

$$D^{(-)}(e^{\lambda}x) = D^{(-)}(x) - \frac{\lambda}{2\pi} \quad . \qquad (2.76)$$

Therefore

$$F(e^{\lambda}x, e^{\lambda}y) = F(x,y) + n(\alpha^2+\beta^2)\frac{\lambda}{2\pi} \quad . \tag{2.77}$$

The corresponding transformation laws for ψ reads

$$D^{-1}(\lambda)\psi^{(\alpha,\beta)}(x)D(\lambda) = \psi^{(\alpha,\beta)}(e^{-\lambda}x)e^{\frac{\lambda}{4\pi}(\alpha^2+\beta^2)} \quad ; \tag{2.78}$$

hence, the scale dimension of $\psi^{(\alpha,\beta)}$ is

$$d_{\alpha,\beta} = \frac{\alpha^2+\beta^2}{4\pi} \quad . \tag{2.79}$$

The case $\alpha = \beta = \pm\sqrt{\pi}$ corresponds to a *canonical* spin -$\frac{1}{2}$ field.

For a spinor $\psi(x) = \begin{pmatrix}\psi_1(x)\\ \psi_2(x)\end{pmatrix}$ satisfying the massless Dirac equation, we have in the representation defined in (2.36c) $\psi_1(x) = \psi_1(x^-)$, $\psi_2(x) = \psi_2(x^+)$ so that the corresponding $\chi_{\alpha\beta}$ only depends on x^+ and x^- respectively.

Hence the requirement, that the Dirac equations be satisfied, implies,

$$\left(\frac{\partial}{\partial x^0} + \frac{\partial}{\partial x^1}\right)(\alpha_1\varphi + \beta_1\tilde{\varphi}) = (\alpha_1 - \beta_1)(\partial_0 + \partial_1)\varphi = 0 \quad , \tag{2.80}$$

that is, $\alpha_1 = \beta_1$, for the upper component ψ_1, and

$$(\alpha_2 + \beta_2)(\partial_0 - \partial_1)\varphi = 0 \quad , \tag{2.81}$$

that is, $\alpha_2 = -\beta_2$ for the lower component ψ_2.

Thus, for a canonical, spin $\frac{1}{2}$ fermion satisfying the Dirac equation one has

$$\psi(x) = \sqrt{\frac{\mu}{2\pi}}\begin{pmatrix} :e^{i\sqrt{\pi}(\varphi+\tilde{\varphi})}: \\ :e^{i\sqrt{\pi}(\varphi-\tilde{\varphi})}: \end{pmatrix} \quad , \tag{2.82}$$

with the current

$$j_\mu(x) =:\overline{\psi}\gamma_\mu\psi:= -\frac{1}{\sqrt{\pi}}\partial_\mu\varphi \quad . \tag{2.83}$$

2.5 The RS-Model

As an application of the ideas developed here, we consider the Lagrangian[100]

$$\mathcal{L} = i\bar{\psi}\gamma^\mu\partial_\mu\psi + \frac{1}{2}\partial^\mu\Phi\partial_\mu\Phi - \frac{1}{2}m^2\Phi^2 + g\bar{\psi}\gamma^\mu\gamma_5\psi\partial_\mu\Phi \quad , \tag{2.84}$$

with the equations of motion

$$i\,\partial\!\!\!/\psi + g\gamma^\mu\gamma_5\psi\partial_\mu\Phi = 0 \quad , \tag{2.85a}$$

$$(\Box + m^2)\Phi = g\partial^\mu(\bar{\psi}\gamma^5\gamma_\mu\psi) \quad . \tag{2.85b}$$

The first equation allows a formal solution given by

$$\psi = e^{ig\gamma_5\Phi}\psi^{(0)} \quad , \tag{2.86}$$

where $\psi^{(0)}$ satisfies the free Dirac equation. At the classical level the axial vector current is conserved, and Φ satisfies the free massive Klein Gordon equation.

At the quantum level, Eq. (2.86) is not well defined, and must be normal ordered:

$$\psi_\alpha = e^{ig\gamma^5_{\alpha\beta}\Phi^{(-)}(x)}\psi^{(0)}_\beta(x)e^{ig\gamma^5_{\alpha\beta}\Phi^{(+)}(x)} \quad . \tag{2.87}$$

Since as we show below, Φ continues to be a free field, its decomposition into positive and negative frequency parts is valid at all times. With Eq. (2.87) we are led to replace (2.85a) by the equation of motion

$$i\,\partial\!\!\!/\psi(x) = g\gamma^5\gamma^\mu N[\partial_\mu\Phi\psi](x) \quad , \tag{2.88}$$

with

$$N[\partial_\mu\Phi\psi] \equiv \partial_\mu\Phi^{(-)}\psi + \psi\partial_\mu\Phi^{(+)} \quad . \tag{2.89}$$

The above normal product may be rewritten in terms of the symmetric limit[5)]

$$N[\partial_\mu\Phi\psi] = \lim_{\epsilon\to 0}\frac{1}{2}\{\partial_\mu\Phi(x+\epsilon)\psi(x) + \partial_\mu\Phi(x-\epsilon)\psi(x)\} \tag{2.90}$$

[5)] Eqs.(2.89) and (2.90) differ by a term of order $\mathcal{O}(\epsilon^2)$

$$[\partial_\mu\Phi^{(+)}(x+\epsilon),\psi(x)] + [\partial_\mu\Phi^{(+)}(x-\epsilon),\psi(x)] =$$
$$= \psi(x)\frac{1}{2\pi}\frac{\partial}{\partial\epsilon_\mu}K_0(m\sqrt{-\epsilon^2}) - \psi(x)\frac{1}{2\pi}\frac{\partial}{\partial\epsilon_\mu}K_0(m\sqrt{-\epsilon^2}) = \mathcal{O}(\epsilon^2)$$

The quantum mechanically correct result for the right hand side of Eq. (2.89) may be computed using equation (2.87).

We consider the vector current as defined by the short distance limit[6)]

$$J^\mu_\epsilon(x) = \lim_{\substack{\epsilon\to 0\\ \epsilon^2<0}} f(\epsilon)\left\{\overline{\psi}(x+\epsilon)\gamma^\mu\psi(x)exp\Big(-ig\int_x^{x+\epsilon} dy^\mu \epsilon_{\mu\nu}\partial^\mu\Phi(y)\Big) - \text{v.e.v.}\right\} \quad , \tag{2.91}$$

where v.e.v. stands for "vacuum expectation value" and $f(\epsilon)$ is a suitably chosen renormalization constant.

Using equation (2.87), expanding the exponential up to first order in ϵ, and using (2.14) in order to move $e^{i\gamma_5\Phi^{(+)}(x+\epsilon)}$ to the right of $e^{i\gamma_5\Phi^{(-)}(x)}$, we obtain the following expression

$$J_\mu(x) = \lim_{\substack{\epsilon\to 0\\ \epsilon^2<0}} f(\epsilon)\{\, \overline{\psi}^{(0)}(x+\epsilon)e^{ig\gamma_5\Phi^{(-)}(x+\epsilon)-ig\gamma_5\Phi^{(-)}(x)}\gamma_\mu\times$$
$$e^{-ig\gamma_5\Phi^{(+)}(x+\epsilon)+ig\gamma_5\Phi^{(+)}(x)}\psi^{(0)}(x)e^{-g^2\Delta^{(+)}(\epsilon)}(1-ig\epsilon_\mu\epsilon^{\mu\nu}\partial_\nu\Phi(x))\,\} - \text{v.e.v.} \quad .(2.92)$$

Expanding the exponentials and defining the free vector current

$$j_\mu(x) = :\overline{\psi}^{(0)}(x)\gamma_\mu\psi^{(0)}(x): \quad , \tag{2.93}$$

equation (2.92) takes the form

$$J_\mu(x) = e^{-g^2\Delta^{(+)}(\epsilon)}\times\Bigg\{j_\mu(x)+$$
$$\left\langle 0\middle|\overline{\psi}^{(0)}_\alpha(x+\epsilon)\psi^{(0)}_\beta(x)\middle|0\right\rangle\gamma_\mu\left[1-ig\gamma_5\epsilon^\rho\partial_\rho\Phi - ig\epsilon_\gamma\epsilon^{\gamma\delta}\partial_\delta\Phi\right]\Bigg\}_{\alpha\beta} - \text{v.e.v.} \quad . \tag{2.94}$$

Recalling

$$\left\langle 0\middle|\overline{\psi}^{(0)}(x+\epsilon)_\alpha\psi_\beta(x)\middle|0\right\rangle = -S^{(+)}(-\epsilon) = -\frac{i\not{\epsilon}_{\beta\alpha}}{2\pi\epsilon^2} \quad , \tag{2.95}$$

renormalizing the fermions as

$$\psi(x) = Z_\psi^{\frac{1}{2}}(\epsilon)\psi_R(x) \quad , \quad Z_\psi(\epsilon) = e^{-g^2\Delta^{(+)}(\epsilon)} \quad , \tag{2.96}$$

and using

$$-\operatorname{tr}\frac{\not{\epsilon}}{2\pi\epsilon^2}\gamma_\mu\gamma_5\epsilon^\rho\partial_\rho\Phi = -\frac{1}{\pi}\epsilon_{\mu\nu}\frac{\epsilon^\nu\epsilon^\rho}{\epsilon^2}\partial_\rho\Phi \quad , \tag{2.97a}$$

[6)] The model exhibits a local invariance at the classical level.

$$- \operatorname{tr} \frac{\not{\epsilon}}{2\pi\epsilon^2} \gamma_\mu \epsilon_\alpha \epsilon^{\alpha\beta} \partial_\beta \Phi = -\frac{1}{\pi} \frac{\epsilon_\mu \epsilon_\nu \epsilon^{\nu\rho}}{\epsilon^2} \partial_\rho \Phi \quad , \tag{2.97b}$$

and $(\epsilon_{\mu\nu}\epsilon^\rho \epsilon^\nu - \epsilon^{\rho\nu}\epsilon_\nu \epsilon_\mu) = \epsilon^2 \epsilon^\rho_\mu$, we obtain for the renormalized current, the expression

$$J_\mu(x) = j^f_\mu(x) + \frac{g}{\pi} \epsilon_{\mu\rho} \partial^\rho \Phi \quad . \tag{2.98}$$

The corresponding result for the axial vector current reads

$$N[\overline{\psi}\gamma^\mu \gamma_5 \psi] = j^\mu_{5f}(x) + \frac{g}{\pi} \partial^\mu \Phi \quad . \tag{2.99}$$

Hence, the divergence of the axial current does not vanish, and we get

$$\partial^\mu N[\overline{\psi}\gamma_\mu \gamma_5 \psi] = \frac{g}{\pi} \Box \Phi \quad , \tag{2.100}$$

which results, on account of (2.85b), in the field equation

$$\left[(1 - \frac{g^2}{\pi}) \Box + m^2 \right] \Phi = 0 \quad . \tag{2.101}$$

We notice that the theory breaks down for strong coupling, i.e. when $g \geq \sqrt{\pi}$, due to the appearance of a tachyon. Since there exists a non trivial fermion wave function renormalization constant, the fermion has an anomalous dimension, which can be computed from (2.96), or else be read off the short distance behavior of the fermion two point function. We have

$$\begin{aligned} \langle 0|\psi_\alpha(x)\overline{\psi}_\beta(0)|0\rangle &= \left\langle 0 \middle| :(e^{ig\gamma_5\Phi(x)})_{\alpha\gamma}: \, :(e^{ig\gamma_5\Phi(0)})_{\delta\beta}: \middle| 0 \right\rangle \left\langle 0|\psi^{(0)}_\gamma(x)\overline{\psi}^{(0)}_\delta(0) \mid 0 \right\rangle , \\ &= e^{-g^2(\gamma_5)_{\alpha\gamma}(\gamma_5)_{\delta\beta}\Delta^{(0)}(x)} \frac{-i(\not{x})_{\gamma\delta}}{2\pi x^2} = \frac{1}{2\pi i} \frac{\not{x}}{x^2} e^{g^2\Delta(x)} \quad . \end{aligned} \tag{2.102}$$

For the short distance behavior we find

$$S_F(x) \sim \frac{1}{2\pi i} \frac{\not{x}}{(x^2)^{\frac{g^2}{4\pi}+1}} \quad , \tag{2.103}$$

so that the fermion field scales with an anomalous dimension given by

$$\gamma_\psi = \frac{g^2}{8\pi} \quad . \tag{2.104}$$

Since the coupling constant is not renormalized, the β function vanishes identically.

2.6 Conclusions

This study of free fields in two dimensional space time, reveals unexpected structures, besides being an important departure for formulating a quantum field theory. Although the main applications of the techniques developed here will be subject of the next chapters, such as the complete solution of the Thirring model, the bosonization technique, and others to come through this book, it is important to emphasize some of the points stated in this chapter. The first concerns the relation between the long range behavior of the bosonic fields and the existence of conserved charges in the fermionic theory. We have demonstrated how to construct conserved charges in a theory containing only real bosonic fields, provided the bosonic theory is massless. We will see that the essential point is actually not the masslessness, but the nontrivial behavior at the spatial infinity. In chapter 3 we shall demonstrate the equivalence of the sine Gordon field, which is massive, and the massive Thirring model: the essential point will be that the topological properties of the bosonic theory, are responsible for the charge of the fermionic model.

The free field theory of bosons is not only important in the above context, but it will also provide a nontrivial realization of the conformal algebra, as we shall see in chapter 15. We will see how to construct general correlation functions depending on a character defining the realization of the conformal group. The construction depends only on formula (2.18), modified to include charges at infinity.

As we mentioned in the introduction, we are able, using the techniques developed in this chapter, to construct the exponential fields, which are the basis of all simple soluble models in two dimensional space time. This is still a characteristic result of two dimensional space time, where Wick polynomials of a free massive field are well defined[101]; moreover, while an infinite series of a field does not define a field in the usual sense in more than two space-time dimensions[102], the no go result fails in two dimensions, where entire functions of the elementary fields are well defined objects[26].

The bosonization technique is related to the above results. It has been developed long ago[97], having been undertaken by several authors[103]. However, those authors did not take into consideration the problems related to the massless scalar field in two dimensions, which turns out to be a fundamental piece of the construction.

A very important soluble model, where one uses these ideas, is the Thirring model, which will be thoroughly discussed in the next chapter.

Chapter III

THE THIRRING MODEL

3.1 Introduction

The theory of a massless Dirac field with a quartic interaction was first introduced by Thirring[24]. Up to Fierz rearrangements, the quartic term is unique, and may be described by a current-current interaction. The theory is explicitly soluble, and thus allows an explicit computation of matrix elements, as well as the construction of the eigenstates of the Hamiltonian[24]. Glaser[104] solved the field equations in terms of the free massless Dirac field, but the manipulations were rather formal. Johnson[105] reconsidered the theory, being very careful with the definition of field operator products, and obtained the current in terms of the renormalized operators. By solving the system of equations for the time-ordered functions, he obtained the two-point and four point Green's functions. The operator solution was studied later on[106], as well as the computation of the n-point function[107]. The complete solution of the theory, as well as the study of properties such as Lorentz transformations, infrared behavior, cluster decomposition, and relation with other solutions in the literature were obtained in the classic paper by Klaiber[25].

In this chapter we construct the complete operator solution, as well as the n-point functions, and discuss some properties of the model. Later in the chapter, we also discuss the massive theory, introducing the mass perturbatively. The perturbation theory thus obtained is shown to be equivalent to a corresponding expansion in the sine-Gordon theory[108,44,45]. This equivalence will first be shown for the charge zero sector and then extended to arbitrary charge sectors by generalizing the bosonization formulae of chapter 2 to the interacting case. Both, the massive Thirring and sine-Gordon models are shown to possess an infinite number of conservation laws, which survive quantization. On the classical level these conservation laws imply integrability by the inverse scattering method; at the quantum level they allow one to obtain the exact S matrix of the theory (see chapter 8) as well as some of the form factors.

3.2 The Massless Thirring Model

The Lagrangian defining the Thirring model is that of a massless Dirac fermion with a (quartic) current-current interaction[24]

$$\mathcal{L} = \overline{\psi} i \not\partial \psi + \frac{1}{2} g \overline{\psi} \gamma^\mu \psi \overline{\psi} \gamma_\mu \psi \quad . \tag{3.1}$$

The equation of motion derived from (3.1) reads

$$i\gamma^\mu \partial_\mu \psi + g j^\mu \gamma_\mu \psi = 0 \quad , \tag{3.2}$$

where

$$j^\mu(x) = \overline{\psi} \gamma_\mu \psi \quad . \tag{3.3}$$

The equation of motion (3.2) implies the conservation laws

$$\partial_\mu j^\mu = 0 = \epsilon^{\mu\nu} \partial_\mu j_\nu \quad . \tag{3.4}$$

Hence, at the classical level, the current (3.3) may be written as the gradient of a massless scalar field $\widetilde{\varphi}$:

$$j_\mu = -\frac{1}{\sqrt{\pi}} \partial_\mu \widetilde{\varphi} \quad , \quad \Box \widetilde{\varphi} = 0 \quad . \tag{3.5}$$

In terms of $\widetilde{\varphi}$ the classical equation of motion (3.2) is solved by the Ansatz

$$\psi(x) = e^{-i\frac{g}{\sqrt{\pi}}\widetilde{\varphi}(x)} \psi^{(0)}(x) \quad , \tag{3.6}$$

where $\psi^{(0)}(x)$ is a solution of the Dirac equation. The solution (3.6) is not unique: given a solution ψ of Eq. (3.3), we can construct the one-parameter family of solutions

$$\psi(x) = e^{ic\{\widetilde{\varphi}(x) + \gamma_5 \varphi(x)\}} \psi(x) = e^{i\{\hat{\alpha}\widetilde{\varphi}(x) + \hat{\beta}\gamma_5 \varphi\}} \psi^{(0)}(x) \quad , \tag{3.7}$$

where $\widetilde{\varphi}$ is the dual of φ (see Eq. (2.20)) and classically we have

$$\hat{\alpha} - \hat{\beta} = -\frac{g}{\sqrt{\pi}} \quad . \tag{3.8}$$

If we compute the current with expression (3.7), we verify that classically the current (3.3) is the free current j_μ^f, that is

$$j_\mu(x) = \overline{\psi}^{(0)}(x) \gamma_\mu \psi^{(0)}(x) \equiv j_\mu^f(x) \quad , \tag{3.9}$$

and the solution $\psi(x)$ of the Thirring equation (3.2) can be written entirely in terms of the massless field $\psi^{(0)}(x)$, and the potential $\tilde{\varphi}$. Going over to the quantum case, we choose $\psi^{(0)}$ to be canonical; we may use (2.82), such that expression (3.7) may be rewritten in the form

$$\psi(x) = \sqrt{\frac{\mu}{2\pi}} : e^{i\{\alpha\tilde{\varphi}(x)+\beta\gamma_5\varphi(x)\}} : \begin{pmatrix} 1 \\ 1 \end{pmatrix} \quad , \tag{3.10}$$

where

$$\alpha = \hat{\alpha} + \sqrt{\pi} \quad \text{and} \quad \beta = \hat{\beta} + \sqrt{\pi} \quad , \tag{3.11}$$

and μ is the IR regulator mass associated with φ. The fermion-number and chirality of ψ are contained in the superselection rules implied by the zero mass fields φ and $\tilde{\varphi}$, respectively (see chapter 2). The spin s takes continuous values depending on the parameters α and β:

$$s = \frac{\alpha\beta}{2\pi} = \frac{\hat{\alpha}\hat{\beta} + (\hat{\alpha} + \hat{\beta})\sqrt{\pi}}{2\pi} + \frac{1}{2} \quad . \tag{3.12}$$

Some remarks are in order at this point. The above construction[49] differs from the original solution of Klaiber[25]: the superselection rules are implicit in (3.10), in such a way as to cancel the IR mass parameter μ. The computations are performed using definition (2.16) and the super selection rule (2.19). Klaiber chose to define infrared finite fields at intermediate steps, by working with non-Lorentz-covariant operators involving an IR cutoff; this lack of Lorentz invariance could be compensated by adding (non Lorentz and non translation invariant) terms involving the charge and chiral charge operators in the exponent, in expressions such as (3.10). Lorentz invariance, was only recovered at the end of the computation.

The correlators of the fields can be written as

$$\langle 0 \mid \psi(x_1) \cdots \psi(x_n) \overline{\psi}(y_1) \cdots \overline{\psi}(y_n) \mid 0 \rangle = \left(\frac{\mu}{2\pi}\right)^n e^{F(x_i, y_i)} \tag{3.13}$$

with

$$\begin{aligned} F(x,y) = &- \sum_{j<k} \{(a + b\gamma^5_{x_j}\gamma^5_{x_k}) D(x_j - x_k) + \lambda(\gamma^5_{x_j} + \gamma^5_{x_k})\tilde{D}(x_j - x_k)\} \\ &- \sum_{j<k} \{(a + b\gamma^5_{y_j}\gamma^5_{y_k}) D(y_j - y_k) + \lambda(\gamma^5_{y_j} + \gamma^5_{y_k})\tilde{D}(y_j - y_k)\} \\ &- \sum_{j,k} \{(-a + b\gamma^5_{x_j}\gamma^5_{y_k}) D(x_j - y_k) + \lambda(-\gamma^5_{x_j} + \gamma^5_{y_k})\tilde{D}(x_j - y_k)\} \end{aligned} \tag{3.14}$$

and $a = \hat{\alpha}^2$, $b = \hat{\beta}^2$, $\lambda = \hat{\alpha}\hat{\beta}$. The index in γ^5 indicates where it belongs to, with respect to the Lorentz indices from the left hand side.

We obtained the terms of F using (2.14,15,22,23):

$$e^{i\alpha\varphi^{(+)}(x)}e^{i\alpha\varphi^{(-)}(y)} = e^{i\alpha\varphi^{(-)}(y)}e^{i\alpha\varphi^{(+)}(x)}e^{-\alpha^2 D(x-y)} \tag{3.15a}$$

$$e^{i\beta\gamma^5\tilde{\varphi}^{(+)}(x)}e^{i\beta\gamma^5\tilde{\varphi}^{(-)}(x)} = e^{i\beta\gamma^5\tilde{\varphi}^{(-)}(y)}e^{i\beta\gamma^5\tilde{\varphi}^{(+)}(x)}e^{-\beta^2\gamma_x^5\gamma_y^5 D(x-y)} \tag{3.15b}$$

$$e^{i\alpha\varphi^{(+)}(x)}e^{i\beta\gamma^5\tilde{\varphi}^{(-)}(y)} = e^{i\beta\gamma^5\tilde{\varphi}^{(-)}(y)}e^{i\alpha\tilde{\varphi}^{(+)}(x)}e^{-\alpha\beta\gamma_y^5\tilde{D}(x-y)} \tag{3.15c}$$

$$e^{i\beta\gamma^5\varphi^{(+)}(x)}e^{i\alpha\tilde{\varphi}^{(-)}(y)} = e^{i\alpha\tilde{\varphi}^{(-)}(y)}e^{i\beta\gamma^5\tilde{\varphi}^{(+)}(x)}e^{-\alpha\beta\gamma_x^5\tilde{D}(x-y)} \quad . \tag{3.15d}$$

Using (2.24,25) we obtain from (3.13) the more explicit form

$$\begin{aligned}\langle 0 \mid \psi_1(x_1)\cdots\psi_1(x_n)\psi_1^\dagger(y_1)\cdots\psi_1^\dagger(y_n) \mid 0\rangle &= \Big(\frac{1}{2\pi i}\Big)^2 e^{in\frac{\pi}{2}(\delta_1+\delta_2-1)}\Big(\frac{1}{\mu}\Big)^{n(\delta_1+\delta_2-1)} \\ &\prod_{j<k}(x_j^+ - x_k^+ - i\epsilon)^{\delta_1}(y_j^+ - y_k^+ - i\epsilon)^{\delta_1}\prod_{j,k}(x_j^+ - y_k^+ - i\epsilon)^{-\delta_1} \\ &\prod_{j<k}(x_j^- - x_k^- - i\epsilon)^{\delta_2}(y_j^- - y_k^- - i\epsilon)^{\delta_2}\prod_{j,k}(x_j^- - y_k^- - i\epsilon)^{-\delta_2},\end{aligned} \tag{3.16}$$

where $\delta_1 = \frac{a+b-2\lambda}{4\pi}$, $\delta_2 = \frac{a+b+2\lambda}{4\pi}$.

The corresponding correlators of ψ_2 are obtained by changing the sign of λ and interchanging δ_1 and δ_2, while mixed correlators are computed to be

$$\begin{aligned}&\langle 0 \mid \prod_1^l \psi_1(x_j)\prod_{l+1}^n \psi_2(x_j)\prod_1^l \psi_1^*(y_j)\prod_{l+1}^n \psi_2^*(y_j) \mid 0\rangle = \\ &\frac{\prod_{j=1}^{l}\prod_{k=l+1}^{n}(x_j^+ - x_k^+ - i\epsilon)^\delta(y_j^+ - y_k^+ - i\epsilon)^\delta(x_j^- - x_k^- - i\epsilon)^\delta(y_j^- - y_k^- - i\epsilon)^\delta}{\prod_{j=1}^{l}\prod_{k=l+1}^{n}(x_j^- - y_k^- - i\epsilon)^\delta(x_j^+ - y_k^+ - i\epsilon)^\delta\prod_{l=1}^{n}\prod_{k=1}^{l}(x_j^- - y_k^- - i\epsilon)^\delta(x_j^+ - y_k^+ - i\epsilon)^\delta} \\ &\times\langle 0 \mid \prod_{j=1}^{l}\psi_1(x_j)\psi_1^*(y_j) \mid 0\rangle\langle 0 \mid \prod_{k=l+1}^{n}\psi_2(x_k)\psi_2^*(y_k) \mid 0\rangle\end{aligned} \quad ,(3.17)$$

where $\delta = \frac{a-b}{4\pi}$. From the above we see that the Thirring model satisfies the cluster property for non-vanishing α and β.[1)]

[1)] This means that, for $\eta^2 = -1$ (space like) we have

$$\lim_{\rho\to\infty}\langle 0 \mid \psi(x_1+\rho\eta)\cdots\psi(x_p+\rho\eta)\overline{\psi}(y_1+\rho\eta)\cdots\overline{\psi}(y_q+\rho\eta)\psi(x_{p+1})\cdots\psi(x_n)\overline{\psi}(y_{q+1})\cdots\overline{\psi}(y_n) \mid 0\rangle$$

$$= \langle 0 \mid \psi(x_1)\cdots\psi(x_p)\overline{\psi}(y_1)\cdots\overline{\psi}(y_q) \mid 0\rangle \times \langle 0 \mid \psi(x_{p+1})\cdots\psi(x_n)\overline{\psi}(y_{q+1})\cdots\overline{\psi}(y_n) \mid 0\rangle.$$

In particular one has for the two point function

$$\langle 0 \mid \psi(x)\overline{\psi}(0) \mid 0\rangle = \frac{\mu}{2\pi}e^{-\frac{a+b}{4}} \times \begin{pmatrix} 0 & [\mu(x^- - i\epsilon)]^{-\frac{a+b+2\lambda}{4\pi}}[\mu(x^+ - i\epsilon)]^{-\frac{a+b-2\lambda}{4\pi}} \\ [\mu(x^- - i\epsilon)]^{-\frac{a+b-2\lambda}{4\pi}}[\mu(x^+ - i\epsilon)]^{-\frac{a+b+2\lambda}{4\pi}} & 0 \end{pmatrix} ; \qquad (3.18)$$

the corresponding Fourier transform reads

$$\frac{1}{2\pi}\int dx e^{ipx}\langle 0 \mid \psi(x)\overline{\psi}(0) \mid 0\rangle = \frac{1}{2\mu}\frac{1}{\Gamma\left(\frac{a+b-2\lambda}{4\pi}\right)\Gamma\left(\frac{a+b+2\lambda}{4\pi}\right)}\theta(p^+)\theta(p^-)$$
$$\times \begin{pmatrix} 0 & \left(\frac{p^+}{2\mu}\right)^{\frac{a+b+2\lambda}{4\pi}-1}\left(\frac{p^-}{2\mu}\right)^{\frac{a+b-2\lambda}{4\pi}-1} \\ \left(\frac{p^+}{2\mu}\right)^{\frac{a+b-2\lambda}{4\pi}-1}\left(\frac{p^-}{2\mu}\right)^{\frac{a+b+2\lambda}{4\pi}-1} & 0 \end{pmatrix} \qquad (3.19)$$

The above two point function does not contain any pole. Hence the massless Thirring model does not present, asymptotically, any particle, and has no particle interpretation.

Using formula (3.10), it is possible to arrive at very simple expressions for bilinears involving the Thirring field. The first of them is the current operator

$$j_\mu(x) = \overline{\psi}(x)\gamma_\mu\psi(x) \quad . \qquad (3.20)$$

We treat the simplest case of a canonical spin $\frac{1}{2}$ fermion for which

$$\beta = \frac{\pi}{\alpha} \quad . \qquad (3.21)$$

We define

$$J_\pm(x;\epsilon) = \psi^\dagger(x+\epsilon)(1 \mp \gamma_5)\psi(x) = 2\psi^\dagger_{\substack{2\\1}}(x+\epsilon)\psi_{\substack{2\\1}}(x). \qquad (3.22)$$

Using Eqs. (3.10) and (3.21), one calculates

$$J_\pm(x;\epsilon) = \frac{\mu}{\pi}e^{-i\alpha(\widetilde{\varphi}(x+\epsilon)-\widetilde{\varphi}(x))\pm i\beta(\varphi(x+\epsilon)-\varphi(x))}\left(\frac{\epsilon^\mp}{\epsilon^\pm}\right)^{\frac{\alpha\beta}{2\pi}}(-\mu^2\epsilon^2)^{-\frac{\alpha^2+\beta^2}{4\pi}}$$
$$\approx \left(\frac{\mu}{\pi}\right)(-\mu^2\epsilon^2)^{-\frac{\alpha^2+\beta^2}{4\pi}}\left(\frac{\epsilon^\mp}{\epsilon^\pm}\right)^{1/2}(1 - i\alpha\epsilon^\mu\partial_\mu\widetilde{\varphi} \pm i\beta\epsilon^\mu\partial_\mu\varphi) \quad . \qquad (3.23)$$

Following Mandelstam[47] we define

$$j^\mu(x) = \lim_{\substack{\epsilon=(0,\epsilon^1)\\ \epsilon^1\to 0}} Z(\epsilon)\left(\delta^0_\mu + \frac{\beta^2}{\pi}\delta^1_\mu\right)\{J^\mu(x;\epsilon) - \langle 0 \mid J^\mu(x;\epsilon) \mid 0\rangle\} \quad , \qquad (3.24)$$

where[2)]

$$Z(\varepsilon) = 2(-\mu^2\varepsilon^2)^{\frac{\alpha^2+\beta^2-2\pi}{4\pi}} \quad . \tag{3.25}$$

One finds

$$j_\mu = -\frac{\beta}{\pi}\epsilon_{\mu\nu}\partial^\nu\tilde{\varphi} \quad . \tag{3.26}$$

Other bilinears can also be computed. We have

$$\psi_1^\dagger(x+\epsilon)\psi_2(x) = \frac{\mu}{2\pi}(-\mu^2\epsilon^2)^{-\frac{\alpha^2-\beta^2}{4\pi}} e^{-2i\beta\varphi(x)} \quad . \tag{3.27}$$

Therefore, we identify

$$N[\overline{\psi}\psi](x) = \frac{\mu}{2\pi} {:}\cos\, 2\beta\tilde{\varphi}{:} \tag{3.28a}$$

$$N[\overline{\psi}\gamma_5\psi](x) = \frac{i\mu}{2\pi} {:}\sin\, 2\beta\tilde{\varphi}{:} \quad . \tag{3.28b}$$

Finally, one also has (see appendix E)

$$\overline{\psi}\gamma^\mu\partial_\mu\psi = \frac{\beta^2}{2\pi}(\partial_\mu\tilde{\varphi})^2 \quad . \tag{3.29}$$

3.3 The Massive Thirring Model

3.3.1 Equivalence with sine-Gordon equation

The massive Thirring model is defined by the Lagrangian

$$\mathcal{L} = i\overline{\psi}\gamma^\mu\partial_\mu\psi - m\overline{\psi}\psi + \frac{1}{2}g\overline{\psi}\gamma^\mu\psi\overline{\psi}\gamma_\mu\psi \quad , \tag{3.30}$$

which is nothing but the massive version of Lagrangian (3.1).

The most straightforward way of studying the above theory, is by means of a perturbation expansion in the mass. This requires one to compute the correlator of an arbitrary number of mass insertions in the Thirring model, that is

$$\frac{(-im)^n}{n!}\langle 0 \mid \overline{\psi}\psi(x_1)\cdots\overline{\psi}\psi(x_n) \mid 0\rangle_{Th} = G(x_1\cdots x_n) \quad . \tag{3.31}$$

---

2) Notice that in general, Lorentz invariance is spoiled in intermediate steps. Nevertheless, it is recovered in the end. The normalization of the current is chosen in such a way that the fermion carries one unity of negative charge.

Using equation (3.28), the above correlator is exactly computed in terms of exponentials of free fields. Defining $\tilde{\beta} = 2\beta$ one has[3)]

$$G(x_1 \cdots x_n) \approx \frac{(-im)^n}{n!} \langle 0 \,|{:}\cos\,\tilde{\beta}\varphi(x_1){:}\cdots{:}\cos\,\tilde{\beta}\varphi(x_n){:}\,|\, 0\rangle \quad , \tag{3.32}$$

where : : denotes normal ordering with respect to the field $\varphi(x)$. Eq. (3.32) suggests the equivalence of the fermionic Lagrangian (3.30) with the bosonic Lagrangian

$$\mathcal{L} = \frac{1}{2}(\partial_\mu \varphi)^2 + \frac{m^2}{\tilde{\beta}^2}(\cos\,\tilde{\beta}\varphi - 1) \quad . \tag{3.33}$$

We have thereby identified the fermionic massive Thirring model, with the sine-Gordon theory defined by the Lagrangian (3.33). Note that in the case where the current is given by (3.26), Eq. (3.8) is replaced by

$$\alpha - \beta = -\frac{\beta g}{\pi} \quad , \tag{3.34}$$

and using (3.21), we have

$$\frac{\tilde{\beta}^2}{4\pi} = \frac{1}{1 - \frac{g}{\pi}} \quad . \tag{3.35}$$

The value $\tilde{\beta}^2 = 4\pi$ thus corresponds to a free massive fermion ($g = 0$).

Exponentials of massless fields are rather subtle objects. Recalling results of the previous chapter, we know that they naturally define a superselection rule, once one takes the zero mass limit of the theory. While the limit is not taken, the exponential is IR-cutoff dependent, that is, it will depend on the parameter μ introduced in the previous chapter. A computation of the two point function shows that[45]

$$N_m[e^{-i\tilde{\beta}\varphi}] = \left(\frac{\mu^2}{m^2}\right)^{\frac{\tilde{\beta}^2}{8\pi}} N_\mu e^{-i\tilde{\beta}\varphi} \quad , \tag{3.36}$$

where $N_m(N_\mu)$ stands for "normal ordering" with respect to the mass $m(\mu)$[45]. In particular

$$N_m \cos\,\tilde{\beta}\varphi = \left(\frac{\mu^2}{m^2}\right)^{\frac{\tilde{\beta}^2}{8\pi}} N_\mu \cos\,\tilde{\beta}\varphi \quad . \tag{3.37}$$

Consider the contribution of the above term to the Hamiltonian. For the free Hamiltonian we have

$$\mathcal{H}_0 = \frac{1}{2}\pi^2 + \frac{1}{2}(\partial_1 \varphi)^2 \quad . \tag{3.38}$$

---

3) In the usual treatment of the sine-Gordon equation the parameter β replaces our $\tilde{\beta} = 2\beta$.

Using the usual expansion in terms of the creation and annihilation operators, one obtains

$$\langle H_0^m \rangle = \int \frac{dk}{8\pi} \frac{2k^2 + m^2}{k^0(m)} \quad , \tag{3.39}$$

and

$$\langle \mathcal{H}_0^\mu \rangle = \int \frac{dk}{8\pi} \frac{2k^2 + \mu^2}{k^0(\mu)} \quad , \tag{3.40}$$

with $k^0(\lambda) = \sqrt{(k^1)^2 + \lambda^2}$. Hence

$$N_m H_0 = N_\mu H_0 + \langle H_0^\mu \rangle - \langle H_0^m \rangle \quad = N_\mu H_0 + \frac{1}{8\pi}(\mu^2 - m^2) \quad . \tag{3.41}$$

Adding the mass term in (3.33), Eq. (3.41) is replaced by

$$H = N_m \left[H_0 - \frac{\alpha}{\tilde{\beta}^2} \cos \tilde{\beta}\varphi \right] = N_\mu \left[H_0 - \frac{\alpha}{\tilde{\beta}^2} \left(\frac{\mu^2}{m^2}\right)^{\frac{\tilde{\beta}^2}{8\pi}} \cos \tilde{\beta}\varphi + \frac{1}{8\pi}(\mu^2 - m^2) \right] . \tag{3.42}$$

This expression is unbounded from below as a function of μ^2, if $\tilde{\beta}^2 > 8\pi$. Thus, the sine-Gordon theory will not be well defined in this region. We will come back to the origin of this problem later in this chapter, showing that the theory is superrenormalizable for $0 \le \tilde{\beta}^2 \le 4\pi$, renormalizable for $4\pi \le \tilde{\beta}^2 < 8\pi$, and non renormalizable for $\tilde{\beta}^2 > 8\pi$. For the moment, we discuss some general properties of both theories. We shall come back to the identification between bosonic and fermionic theories (Eqs. (3.20,24,28) and (3.29)) in the next chapter, where we explore Mandelstam's formula non perturbatively in the mass parameter.

3.3.2 Classical conservation laws

The massive Thirring model possesses an infinite number of conservation laws[109]. In order to demonstrate this statement, we consider the field equation in light cone variables

$$i\partial_+ \psi_1 = m\psi_2 - 2g\psi_1 \psi_2^\dagger \psi_2 \quad , \tag{3.43a}$$

$$i\partial_- \psi_2 = m\psi_1 - 2g\psi_2 \psi_1^\dagger \psi_1 \quad , \tag{3.43b}$$

and consider an auxiliary Grassmann valued field $\chi(x;\epsilon)$, depending on a parameter ϵ; χ obeys the following pair of equations

$$\epsilon\partial_- \chi - \chi - 2i\epsilon g \chi \psi_1^\dagger \psi_1 - 4i\epsilon g \psi_1 \chi^\dagger \chi + \psi_1 = 0 \quad , \tag{3.44a}$$

$$\partial_+ \chi + m^2 \epsilon \chi - 2ig\chi\psi_2^\dagger \psi_2 - 4\epsilon g \psi_2 \chi^\dagger \chi + im\psi_2 = 0 \quad . \tag{3.44b}$$

These equations are compatible for arbitrary ϵ, if one takes account of the fermionic character of χ and ψ; this can be seen using the field equations (3.43) above (see appendix F). If we expand χ in powers of ϵ, we see that $\chi(0,x) = \psi_1(x)$, which obeys both equations above, for $\epsilon = 0$; the current $J_\mu(x,\epsilon)$ with light cone components given by

$$J_-(x;\epsilon) = i\psi_1^\dagger\chi - i\chi^\dagger\psi_1 \quad , \tag{3.45a}$$

$$J_+(x;\epsilon) = \epsilon m(\psi_2^\dagger\chi + \chi^\dagger\psi_2) \quad , \tag{3.45b}$$

is conserved (see appendix F):

$$\partial_+ J_-(x;\epsilon) + \partial_- J_+(x;\epsilon) = 0 \quad . \tag{3.46}$$

Equation (3.46) implies the existence of an infinite number of conserved currents $J_\mu^n(x)$ defined in terms of the expansion

$$J_\mu(x;\epsilon) = \sum \epsilon^n J_\mu^n(x) \quad . \tag{3.47}$$

The first few of them are easily calculated considering the expansion

$$\chi(\epsilon) = \sum \epsilon^n \chi^{(n)} \quad . \tag{3.48}$$

Upon substituting (3.48) into (3.44), one finds for the first few terms

$$\chi^{(0)} = \psi_1 \quad , \tag{3.49a}$$

$$\chi^{(1)} = \partial_-\psi_1 \quad , \tag{3.49b}$$

$$\chi^{(2)} = \partial_-^2\psi_1 - 2ig\psi_1^\dagger\psi_1\partial_-\psi_1 \quad , \tag{3.49c}$$

$$\chi^{(3)} = \partial_-^3\psi_1 - 6ig\psi_1\partial_-\psi_1^\dagger\partial_-\psi_1 - 4ig\psi_1^\dagger\psi_1\partial_-^2\psi_1 \quad , \tag{3.49d}$$

which in turn imply that the first non trivial currents are given by

$$J_-^{(1)} = i(\psi_1^\dagger\partial_-\psi_1 - \partial_-\psi_1^\dagger\psi_1) \quad , \tag{3.50a}$$

$$J_+^{(1)} = m(\psi_2^\dagger\psi_1 + \psi_1^\dagger\psi_2) \quad , \tag{3.50b}$$

$$J_-^{(2)} = i(\psi_1^\dagger\partial_-^2\psi_1 - \partial_-^2\psi_1^\dagger\psi_1) \quad , \tag{3.50c}$$

$$J_+^{(2)} = m(\psi_2^\dagger\partial_-\psi_1 + \partial_-\psi_1^\dagger\psi_2) \quad , \tag{3.50d}$$

$$J_-^{(3)} = i(\psi_1^\dagger\partial_-^3\psi_1 - \partial_-^3\psi_1^\dagger\psi_1) + 12g\psi_1^\dagger\psi_1\partial_-\psi_1^\dagger\partial_-\psi_1 \quad , \tag{3.50e}$$

$$J_+^{(3)} = m(\psi_2^\dagger\partial_-^2\psi_1 + \partial_-^2\psi_1^\dagger\psi_2) + 2ig\psi_1^\dagger\psi(\psi_2^\dagger\partial_-\psi_1 - \partial_-\psi_1^\dagger\psi_2) \quad . \tag{3.50f}$$

3.3.3 Quantum conservation laws

In the quantum theory, the conservation laws (3.46) may present anomalous terms. Since the conservation laws above follow from the field equations (3.43), we have to consider the quantum equation of motion. We may use the *BPHZ* renormalization scheme[4,5], in which case, the equations of motion inside composite operators may present the so called anisotropies[110,111]. One defines, in the *BPHZ* scheme, the Lagrangian

$$\mathcal{L} = (1+b)\frac{i}{2}\overline{\psi}\overleftrightarrow{\partial\!\!\!/}\psi - (m-a)\overline{\psi}\psi + \frac{1}{2}(g-c)\overline{\psi}\gamma^\mu\psi\overline{\psi}\gamma_\mu\psi \quad , \tag{3.51}$$

where a, b, c are finite renormalization constants, defined at some renormalization point[5,110].

In fact, this is the case of interest, since the equation of motion are used inside composite operators (see the above form of the currents, which is a polynomial in the basic fields and its derivatives). Inside a normal product, the equations of motion read

$$\begin{aligned}\langle 0 \mid T\, N_{d+\frac{3}{2}}[P(\psi) i\, \partial\!\!\!/\psi](x) X \mid 0\rangle &= \langle 0 \mid T\, N_{d+\frac{3}{2}}[\{P(\psi)\}\frac{m-a}{1+b}\psi](x) X \mid 0\rangle \\ -\langle 0 \mid T\, N_{d+\frac{3}{2}}[P\frac{g-c}{1+b}\gamma_\mu\psi\overline{\psi}\gamma^\mu\psi](x) \mid 0\rangle &+ i\langle 0 \mid T\, N[P](x)\frac{\delta}{m\delta\overline{\psi}(x)} X \mid 0\rangle \quad , \end{aligned} \tag{3.52}$$

where $P[\psi]$ is an arbitrary operator depending on the field ψ and its derivatives, and $N[\{P(\psi)\}]$ denotes the so called anisotropic normal product, which, if one follows the technicality of the *BPHZ* renormalization scheme, has extra subtractions depending on the subdiagrams containing the special vertex, having extra subtractions if subsets of lines growing in $P[\psi]$ are inside the subdiagram (see refs. [4][5][110] and [111] for details). However, the Zimmermann identity[5] relates the anisotropic normal product to the usual one (subtracted according to the general rules of the *BPHZ* scheme)

$$N[\{P[\psi]\}\psi] = N[P\psi] + \sum a_i N[P_i\psi] \quad , \tag{3.53a}$$

where the a_i's are defined in

$$N_d[P] = N_{d+1}[P] + \sum a_i N_{d+1}[P_i] \quad , \tag{3.53b}$$

and where $P_i[\psi]$ are operators with one dimension more than $P[\psi]$. One must show that the anisotropies can all be expressed in terms of total divergencies since they are extra subtractions, and contain external momentum factors, which are derivatives in position space. An explicit computation has been performed for the lowest nontrivial current[112].

Similar objects in the sine-Gordon theory can be correspondingly[34,113,114] obtained. Consider the Lagrangian (3.33) and the following transformation defining the field $\varphi_\gamma(x;\gamma) \equiv \varphi_\gamma$:

$$\partial_+(\varphi_\gamma + \varphi) = \frac{m}{\beta}\gamma^{-1} \sin \beta \frac{\varphi_\gamma - \varphi}{2} \quad , \tag{3.54a}$$

$$\partial_-(\varphi_\gamma - \varphi) = -\gamma 2 \frac{m}{\beta} \sin \beta \frac{\varphi_\gamma + \varphi}{2} \quad . \tag{3.54b}$$

These equations are compatible, since

$$\frac{1}{2}\partial_+\partial_-\varphi_\gamma - \frac{1}{2}\partial_-\partial_+\varphi_\gamma$$
$$= \sin\varphi + \gamma^{-1}\cos\frac{\varphi_\gamma - \varphi}{2}(-\gamma)\sin\frac{\varphi_\gamma + \varphi}{2} + 2\gamma\cos\frac{\varphi_\gamma + \varphi}{2}\sin\frac{\varphi_\gamma - \varphi}{2} = 0 \quad . \tag{3.55}$$

They imply the following conservation law

$$\gamma\partial_+\left\{\cos\beta\frac{\varphi_\gamma + \varphi}{2}\right\} + \gamma^{-1}\partial_-\left\{\cos\beta\frac{\varphi_\gamma - \varphi}{2}\right\} = 0 \quad , \tag{3.56}$$

as can be seen using (3.54).

We can now expand $\varphi(x;\gamma)$ in a power series in γ, around $\gamma = 0$

$$\varphi(x;\gamma) = \sum \varphi^{(n)}\gamma^n \quad , \tag{3.57}$$

with the coefficients $\varphi^{(n)}$ being calculated substituting (3.57) back into (3.56), and equating to zero the terms with equal power in γ. One easily obtains for the first three terms

$$\varphi^{(0)}(x) = \varphi(x) \quad , \tag{3.58a}$$

$$\varphi^{(1)}(x) = -\frac{2}{m^2}\left(\partial_+\varphi(x)\right)^2 \quad , \tag{3.58b}$$

$$\varphi^{(2)}(x) = -\frac{2\beta^2}{m^2}\partial_+^2\varphi(x) \quad . \tag{3.58c}$$

Substituting (3.57) back into (3.56) and equating to zero the coefficient of each power of γ, we obtain an infinite set of conservation laws. The first few of them may be obtained from the explicit expressions (3.58).

The quantum conservation laws may again suffer from the problem of anisotropies. This has been analysed at length in the case of $BPHZ$, and the conservation laws are actually reobtained.

This problem of anisotropies may, in some cases, be tackled by modifying the renormalization scheme such as to avoid anisotropies, including subtractions in the mass, besides those in the external momenta[115]. In the case of the massive Thirring and sine-Gordon models, one could make soft subtractions in the mass[115], around a given (arbitrary) renormalization point $m = \mu$ (non zero, in order to avoid spurious infrared divergences). Similarly to the case of QED_4, where the anisotropies disappear[116], the problem of anisotropies is also solved in this case, and conservation laws depending only on field equations are immediately quantized, preserving the classical symmetries.[4)]

When discussing the problem of computing the exact S-matrix of a certain class of two dimensional models, we shall see that it is essential to have an infinite number of higher conservation laws. Higher local conservation laws imply the conservation of the individual momenta in a scattering process. This result may be obtained by direct computation, using the properties of the model[117], or indirectly, since the local conservation laws must be powers of the external momenta, which consequently are conserved. In these cases, the scattering process, is simple but non trivial, and consists in an exchange of quantum numbers, while particle production is prohibited. This property has been checked up to second order in perturbation theory[117]. A general inductive proof of the absence of pair production at the tree level for any number of external lines has also been obtained[118]. Considering $n \to n$ particle scattering, one can show that the principal value of the tree amplitude vanishes. The proof consists in a direct verification for $3 \to 3$ scattering, and then proceeds by induction. Using crossing symmetry one can relate $n \to n$ scattering to $n-p \to n+p$ scattering. Therefore, the principal value vanishes and the intermediate particles are always on shell. These conditions are very strong in two dimensions, and lead to particle number conservation at each vertex, and therefore also for the whole diagram.

At the one loop level, we use the fact that a general one loop diagram in two dimensions can be exactly computed in terms of the simplest (two propagator) diagrams times a tree amplitude[119].

Thus, consider the loop diagram in figure 3.1, as given by the expression

$$I^{(n)} = \int \frac{d^2k}{(2\pi)^2} \prod_{j=1}^{n} \left[\frac{\delta_j \gamma^\mu (k+p_j)_\mu + m}{(k+p_j)^2 - m^2 + i\epsilon} \right]_{\alpha_j \beta_j} , \tag{3.59}$$

where $\delta_j = 1$ if the momentum goes along with the charge flow, and $\delta_j = -1$ otherwise.

---

4) Let $P\, \not\partial\psi$ be an operator with scale dimension $d + \frac{3}{2}$.Upon using the equation of motion (3.52), this operator will be replaced by $mP(\psi)$, with scale dimension $d + \frac{1}{2}$. Since both operators are subtracted in the same way, this implies an oversubtraction as given by the Zimmermann identity[5,110].

The k_0 integration is easily performed[117] by closing the contour in the lower plane, and using the residue theorem[5)],

$$I^{(n)} = -2\pi i \int dk^1 \sum_{j=1}^{n} \frac{\delta_j \gamma^\mu (k_\mu + p_{\mu_j} + m)}{2\sqrt{(k^1 + p_j^1)^2 + m^2}} \times$$

$$\prod_{\substack{k=1\\k\neq j}}^{4} \frac{\delta_k \gamma^\mu (k_\mu + p_{\mu_j}) + m}{2\sqrt{[(k^1+p_j^1)^2 + m^2](p_k^0 - p_j^0) - 2(k^1 + p_j)(p_k^1 - p_j^1) + (p_k - p_j)^2 + i\epsilon}} \quad (3.60)$$

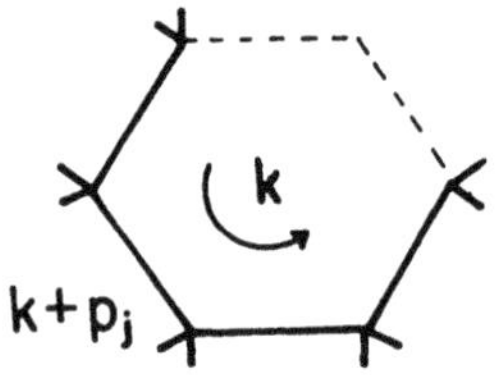

Figure 3.1: One loop contribution to the scattering of 2n particles.

The remaining integral may be performed, by making the change of variable

$$x_j = \frac{k^1 + p_j^1}{\sqrt{(k^1 + p_i^1)^2 + m^2} - m} \quad (3.61)$$

in each term of the sum above, and adding the result.[6)] We have

$$\frac{x_i^2 + 1}{x_i^2 - 1} = \frac{1}{m}\sqrt{(k_i + p_i)^2 + m^2} \quad , \quad (3.62a)$$

$$dx_i = dk_i^1 \frac{m^2}{[\sqrt{(k_i + p_i)^2 + m^2} - m]^2 \sqrt{(k_i + p_i)^2 + m^2}} \quad . \quad (3.62b)$$

Each term in the integrand may thus be expressed as

$$\mathcal{F}_i = (x^2 - 1)^{n-2} (\delta_i \gamma^\mu k_{i\mu} + m) \prod_{\substack{j=1\\j\neq k}}^{n} \frac{\delta_j (\gamma^\mu k_{\mu j} + \gamma^\mu p_{\mu j}) + m}{(p_{kj}^2 + 2mp_{kj}^o)(x_j - x_{kj}^{(-)})(x_j - x_{kj}^{(+)}) + i\epsilon} \quad , \quad (3.63)$$

5) The first term corresponds to the residue at $k_j^0 = \sqrt{(k^1 + p_j)^2 + m^2} - p_j^0$ of the $j^{\underline{th}}$ propagator, and the remaining terms are obtained from the $k \neq j$ propagators, substituting the above expression.

6) When $k^1 \to \pm\infty$, we have $x \to \pm 1$, while $k^1 + p^1 \to 0_\pm$ implies $x \to \pm\infty$. Thus the new integration variable runs in the ranges $[1, \infty)$ and $(-\infty, -1]$.

where x_{kj} are obtained making the $k^{\underline{th}}$ and the $j^{\underline{th}}$ propagators simultaneously zero. We obtain two roots, labeled by $(\pm)$.

The integrand is now rational, and the integral may be performed by elementary means (see also Eqs. (5.36,37)). The poles at $x^2 = 1$ do not contribute, as one can see from a detailed analysis of the residues, or else from the fact that they correspond to poles at $k^1 \to \pm\infty$, which according to the power counting theorem[120], do not contribute for $n \neq 2$ (the case $n = 2$ is trivial in this argument). We now make a partial fraction decomposition

$$\mathcal{F}_j = \sum_{k=1}^{n} \frac{A^{kj}}{x_k - x_{kj}}, \quad A^{kj} = \lim_{x \to x_{kj}} (x - x_{kj})\mathcal{F}(x) \quad ; \tag{3.64}$$

thus, (3.60) may be written as

$$\sum \left(\int_{-\infty}^{-1} + \int_{1}^{\infty} \right) dx_k \sum_{\substack{j=1 \\ j \neq k}}^{n} \left[\frac{1}{x_k - x_{kj}^{+}} - \frac{1}{x_k - x_{kj}^{-}} \right] A_{kj} \quad . \tag{3.65}$$

We perform the elementary integral above, obtaining the result

$$\mathcal{F}(x) = \sum \mathcal{F}_j = \sum_{k=1}^{n} \sum_{j=1}^{n} A_{kj} \ln \frac{1 + x_{kj}^{+}}{1 - x_{kj}^{+}} \frac{1 - x_{kj}^{-}}{1 + x_{kj}^{+}} \quad , \tag{3.66}$$

where we have the following expressions for x_{kj}

$$x_{kj}^{(\pm)} = \frac{2mp_{kj}^{1} \pm p_{kj}^{o} \sqrt{1 - \frac{4m^2}{p_{kj}^2}}}{2mp_{kj}^{o} + p_{kj}^{2}} \quad , \tag{3.67}$$

and where the coefficient A_{kj} in the partial fraction decomposition is given in terms of the product of all propagators with l different from k, j; (we rewrite the numerator of the k$^{\underline{th}}$ and j$^{\underline{th}}$ propagators in terms of the free fermion wave function– to compare with the tree approximation; the factor $\frac{1}{p_{kj}^2 \sqrt{1 - \frac{4m^2}{p_{kj}^2}}}$ arises from the change of variables (3.61)):

$$A_{kj} = \mp 2\pi i \frac{\varpi(k_j)\overline{\varpi}(k_j)}{p_{kj}^2 \sqrt{1 - \frac{4m^2}{p_{kj}^2}}} \prod_{\substack{l=1 \\ l \neq k,j}}^{n} \frac{\delta_l(\gamma \cdot k_l + \gamma \cdot p) + m}{(k_l + p)^2 - m^2 + i\epsilon} \varpi(k_k)\overline{\varpi}(k_k) \equiv \frac{\mp 2\pi i}{p_{kj}^2 \sqrt{1 - \frac{4m^2}{p_{kj}^2}}} T_{kj} \tag{3.68}$$

Here $\varpi(k) = u(k)$ for $\delta = 1$, and $\varpi(k) = iv(k)$ for $\delta = -1$. Substituting $x_{kj}^{(\pm)}$ from (3.67), we have

$$\mathcal{F} = \sum_{k=1}^{n} \sum_{k \neq j=1}^{n} T_{kj} \frac{\mp 2\pi i}{\sqrt{-s_{kj} t_{kj}}} \ln \frac{\sqrt{-s_{kj}} - \sqrt{t_{kj}}}{\sqrt{-s_{kj}} + \sqrt{t_{kj}}} \quad . \tag{3.69}$$

Notice that we have a sum of terms given by the product of the tree diagrams obtained by deleting the lines k and j (T_{kj}),and the loop diagram built from the lines k and j. Our above results concerning the tree diagrams thus tell us that particle production is again forbidden.

We shall use the connection between higher conservation laws and conservation of individual momenta (absence of particle production) as a working hypothesis for calculating the S-matrices of different two dimensional models in chapter 8.

3.4 Bosonization Revisited

We have used the representation of the free Dirac field in terms of bosonic fields, as discussed in chapter 2, in order to demonstrate the equivalence of the (massless) Thirring model to the theory of free zero mass bosons:

$$\mathcal{L}^{Th} = \overline{\psi} i \not\partial \psi + \frac{1}{2} g (\overline{\psi} \gamma^\mu \psi)^2 \widehat{=} \mathcal{L}^{Bos} = \frac{1}{2} \partial^\mu \varphi \partial_\mu \varphi \quad . \tag{3.70}$$

Following Coleman[45] we have similarly demonstrated, in the charge zero sector, the equivalence of a theory of massive fermions with a Thirring like interaction to the sine Gordon theory,

$$\mathcal{L}^{(F)} = \overline{\psi}(i \not\partial - m)\psi + \frac{1}{2} g (\overline{\psi} \gamma^\mu \psi)^2 \widehat{=} \mathcal{L}^{(SG)} = \frac{1}{2} (\partial_\mu \varphi)^2 + \frac{m^2}{\tilde{\beta}^2} (\cos \tilde{\beta} \varphi - 1) \quad , \tag{3.71}$$

provided $\tilde{\beta}$ and g are related by (3.35).

The map from the fermion ψ of the massive Thirring model to the boson φ of the SG theory was based on a perturbation around the free massless theory and only included the charge zero sector.

The sine Gordon theory can be viewed as describing fermion antifermion bound-states in this sector[49]. There exists an extensive literature on this question[121] and we shall not dwell on it further. The inverse map $\varphi \longrightarrow \psi$ was introduced by Mandelstam[47] who was able to write the fundamental fermion of the massive Thirring field, in terms of the bosonic field, thus generalizing the results summarized in Eqs. (3.26-29). As we discuss next, the fermionic field obtained in this way describes the soliton as well as new particle states of the sine Gordon theory, which cannot be interpreted as conventional bound states; there is no element in the local field algebra which connects the vacuum of the sine-Gordon theory to the non trivial soliton states. We shall discover in a later chapter, that the particle scattering of the fundamental field of the sine-Gordon theory can, in turn, be described in terms of the scattering of bound states of the soliton states.

3.4.1 Fermions in terms of bosons

In order to find a fermionization formula, the best is to deal with the dual pair massive Thirring-sine-Gordon, and to express the fermion in terms of the boson, we first note that there is a natural candidate for the fermion in the sine-Gordon theory: the soliton. The soliton operator changes the fundamental field $\psi(x)$ by the constant $2\pi\beta^{-1}$ to the left of its center ("changing the vacuum"), and has no effect to the right of its center. This statement is realized in formulae by[7)]

$$[\varphi(y), \psi(x)] = 2\pi\beta^{-1}\psi(x)\theta(x-y) \quad . \tag{3.72}$$

This commutation relation is realized by the operator[8)]

$$\psi(x) = e^{ic\varphi(x)+i2\pi\beta^{-1}\int_{x^1}^{\infty}\dot{\varphi}(x^0,y^1)dy^1} \quad . \tag{3.73}$$

We can compare (3.73) with (2.82) (or (3.10)), and notice that for massless fields we can identify the line integral in the exponent of (3.73) with $\tilde{\varphi}(x)$, if we carefully rewrite the commutation relations between $\varphi(x)$ and $\tilde{\varphi}(x)$, which will differ after this identification from the previous one by a constant (see footnote 2.2)). For a spin -$\frac{1}{2}$ operator $\psi(x)$ (see 2.69) we find

$$\psi(x) = \sqrt{\frac{\mu}{2\pi}}e^{-\frac{i\pi}{4}\gamma_5}e^{i\frac{\beta}{2}\gamma_5\varphi(x)+i2\pi\beta^{-1}\int_{x^1}^{\infty}dy^1\dot{\varphi}(x^0,y^1)}\begin{pmatrix}1\\1\end{pmatrix} . \tag{3.74}$$

It is now more difficult than before to give a rigorous proof of the formulae such as (3.27,28,29), due to the complicated structure of the interacting theory. It is nevertheless clear that we may still make the identification

$$\hat{N}[\overline{\psi}\psi] = -\frac{\mu}{2\pi}:\cos\beta\varphi: \quad . \tag{3.75}$$

---

7) In the following we drop the tilde on β .

8) Notice the interchange of φ with the line integral in (3.73), as compared with (2.82). This insures the correct statistics for ψ_α and $\psi_\alpha^\dagger$, and requires the introduction of the phase factor $e^{-\frac{i\pi}{4}\gamma_5}$.

The definition of $N[\overline{\psi}\psi]$ in terms of a Wilson expansion requires now an additive and a multiplicative renormalization, as can be checked already for the case of free massive fermions:

$$\overline{\psi}(x+\epsilon)\psi(x) = C(\epsilon) + [f(\epsilon)]N[\overline{\psi}(x)\psi(x)]. \tag{3.76}$$

In order to have more explicit formulae, we write a Källen-Lehmann representation for the 2 point function of $\sin\beta\varphi$[9]

$$\langle 0 \mid \sin\,\beta\varphi(x)\sin\,\beta\varphi(y) \mid 0\rangle = \int_0^\infty (\mu)^2\rho(\mu^2)\Delta^{(+)} + (x-y,\mu^2)d\mu^2. \tag{3.77}$$

For the sine-Gordon field theory, we can use the free field singularity of the correlator, Eqs. (2.11) and (2.18),

$$\langle \sin\,\beta\varphi(\epsilon)\sin\,\beta\varphi(0)\rangle \sim (-\epsilon^2)^{\frac{-\beta^2}{4\pi}} \quad , \tag{3.78}$$

to obtain, from (3.77) the asymptotic behavior

$$\rho(\mu^2) \sim (\mu^2)^{\frac{\beta^2}{4\pi}-3} \quad , \tag{3.79}$$

for large values of μ^2. On the other hand, the equal time commutator $[\varphi(x),\dot{\varphi}(y)]_{e.t.} = i\delta(x^1-y^1)$ will be maintained if and only if

$$\int_0^\infty \rho(\mu^2)d\mu^2 = 1 \quad . \tag{3.80}$$

Using (3.79), the integral in (3.80) is convergent only if

$$\beta^2 < 8\pi \quad . \tag{3.81}$$

This is a condition for the sine-Gordon theory to be well defined. In order to define the mass operator (3.75) we compute its vacuum expectation value; using also the representation $\dot{\varphi}(y) = -i\frac{\delta}{\delta\varphi(y^1)}$, we infer

$$[\sin\,\beta\varphi(x),\dot{\varphi}(y)]_{ET} = -i\beta\cos\,\beta\varphi(x)\delta(x^1-y^1) \quad ; \tag{3.82}$$

---

[9] We start from $\langle\varphi(\epsilon)\varphi(0)\rangle = \int \rho(\mu^2)\Delta(\epsilon;\mu^2)d\mu^2$ and obtain

$$\langle \sin\,\beta\varphi(\epsilon)\sin\,\beta\varphi(0)\rangle = \langle\nabla\varphi\nabla\varphi\rangle = \int (\mu^2)^2\rho(\mu^2)\Delta(\epsilon;\mu^2)d\mu^2 \quad .$$

we have

$$i\beta\langle 0 |: \cos \beta\varphi(x): | 0\rangle \delta(x^1 - y^1) = \langle 0 | [-\Box\varphi(x), \dot{\varphi}(y)]_{ET} | 0\rangle =$$

$$= i \int \rho(\mu^2)\Box\dot{\Delta}(x - y, \mu^2)d\mu^2\Big|_{x^0=y^0} = i \int^{\infty} \mu^2\rho(\mu^2)d\mu^2\delta(x^1 - y^1) \quad . \quad (3.83)$$

Using again (3.79), we see that the integral is finite if $\beta^2 < 4\pi$, and is an infinite constant for $4\pi < \beta^2 < 8\pi$. A logarithmically divergence appears for $\beta^2 = 4\pi$, which corresponds to the case of free fermions. Moreover, using (3.76) and requiring (3.72) to have vanishing vacuum expectation value, we obtain

$$C(\epsilon^1) = \frac{1}{(\epsilon_1^2)^{\frac{\pi}{\beta^2} - (\frac{\beta}{4\pi})^2}} \int s\rho(s)ds \quad . \quad (3.84)$$

We have so far computed the current, the mass term $\overline{\psi}\psi$, and the pseudoscalar $\overline{\psi}\gamma_5\psi$. The kinetic term $i\overline{\psi}\ \partial\!\!\!/\psi$ is again of the form (3.29) (see footnote 7).

As we have witnessed above, the definition of the composite operator $N[\overline{\psi}\psi]$ deserves special consideration for $4\pi < \beta^2 < 8\pi$. It is related with the fact that the exponential field $\psi_\lambda(x) = e^{i\lambda\varphi(x)}$ develops new (non leading)singularities when $\lambda \to 1$, a phenomenon known as cumulative mass effect[122]. Indeed , Coleman's arguments only hold in the region $\dim \overline{\psi}\psi < 1$, where the usual Wick product is enough to define the exponential field . For $1 \leq \dim \overline{\psi}\psi < 2$ the sequence of arguments related to equations (3.76) through (3.84) are in domain of the cumulative mass effect. The case of $\dim \overline{\psi}\psi > 2$ corresponding to $\beta^2 > 8\pi$ in the sine-Gordon model, is non renormalizable, and new terms contribute to the Wilson expansion. The fact that new terms arise can be seen from the mass corrections to the massless theory. A mass correction implies new terms of the order ϵ^2 with respect to previous ones in the definition of composite operators. We refer the interested reader to the literature[121].

3.5 The Soliton as a Disorder Parameter

We have identified the soliton operator of the sine-Gordon theory with the fermion of the massive Thirring model. In its simplest form it is given by (3.73) with $c = 0$:

$$\mu(x) = e^{i\frac{2\pi}{\beta}\int_{x^1}^{\infty} dy^1 \partial_0\varphi(x^0, y^1)} \quad . \quad (3.85)$$

The representation (3.74) for the fermion field operator suggests defining also the operator

$$\sigma(x) = e^{i\frac{\beta}{2}\varphi(x)} \quad . \quad (3.86)$$

In terms of these operators, the upper and lower components of the fermion field may be formally written in the form

$$\psi_1(x) = \sigma(x)\mu(x) \quad , \quad \psi_2(x) = \sigma^*(x)\mu(x) \quad . \tag{3.87}$$

The operators (3.85) and (3.86) satisfy the equal time algebra

$$\mu(x)\sigma(y) = \begin{cases} -\sigma(y)\mu(x) & , \quad y^1 > x^1 \\ \sigma(y)\mu(x) & , \quad y^1 < x^1 \end{cases} \quad , \tag{3.88}$$

This algebra is analogous to the dual algebra of the order and disorder operators in statistical mechanics, as it was first discussed by Kadanoff and Ceva[123]. The nonlocal operator $\mu(x)$ has the property of creating topological excitations which are orthogonal to the vacuum

$$\langle 0 \mid \mu(x) \mid 0 \rangle = 0 \quad . \tag{3.89}$$

This reflects the fact that the operator $\mu(x)$ carries the charge superselection rule of the fermion field (3.87). Similarly one has, in the massive Thirring model

$$\langle 0 \mid \sigma(x) \mid 0 \rangle \neq 0 \quad , \tag{3.90}$$

which reflects the absence of a chiral selection rule for the massive fermion.

The sine-Gordon Lagrangian is invariant under the translation $\varphi \to \varphi + \frac{2\pi}{\beta}$. This corresponds to an invariance under the multiplicative symmetry $\sigma \to -\sigma$. Hence (3.90) shows that this symmetry is spontaneously broken. This is however not the most general situation. In fact, given a dual algebra of the form

$$\mu(x)\phi(y) = \begin{cases} e^{i2\pi/N}\phi(y)\mu(x) & , \quad y^1 > x^1 \\ \mu(y)\phi(x) & , \quad y^1 < x^1 \end{cases} \quad , (x-y)^2 < 0 \quad . \tag{3.91}$$

It has been demonstrated[124] in the framework of axiomatic QFT, that regardless of the form of the Lagrangian $\mathcal{L}(\phi, \partial_\mu \phi)$, the following properties in this case hold:

i)

$$\langle \phi \rangle \langle \mu \rangle = 0 \quad , \tag{3.92}$$

ii)

$$\langle \phi\mu \rangle = 0 \quad , \tag{3.93}$$

iii) whenever

$$\langle \phi \rangle = 0 \quad \text{and} \; \langle \mu \rangle = 0 \quad , \tag{3.94}$$

the mass gap vanishes.

The above theorem expresses a deep relationship between the spectrum and the vacuum expectation values of order and disorder fields. Let us examine how these ideas may be used to obtain the order and disorder operators of a QFT defined by a Lagrangian $\mathcal{L}(\phi, \partial_\mu \phi)$.

It is useful to first review briefly the corresponding situation of the one dimensional Ising model in statistical mechanics. The Hamiltonian reads in this case

$$\mathcal{H} = -J \sum_n \sigma_3(n)\sigma_3(n+1) - \sum_n \sigma_1(n) \quad , \tag{3.95}$$

where $\sigma_1(n)$ are Pauli matrices. The Hamiltonian is invariant under the transformation $\sigma_3(n) \to -\sigma_3(n)$, $\sigma_1(n) \to \sigma_1(n)$. This system is known to posses two phases[125,126], characterized by $\langle\sigma_3\rangle = 0$ (for $J > 1$), and $\langle\sigma_3\rangle \neq 0$ (for $J < 1$). Thus $J = 1$ corresponds to the critical point.

Consider a dual lattice, for which each site n^* is placed between the sites n and $n+1$ of the original lattice, and let $\mu_3(n^*)$ be an operator defined via the dual algebra[123,125]

$$\mu_3(n^*)\sigma_3(m) = \begin{cases} -\sigma_3(m)\mu_3(n^*) & , \quad n > m \\ \sigma_3(m)\mu_3(n^*) & , \quad n \leq m \end{cases} \quad . \tag{3.96}$$

An operator realization of $\mu_3(n^*)$ is[125]

$$\mu_3(n^*) = \prod_{m>n} \sigma_1(m) \quad . \tag{3.97}$$

One finds[123,125] that for $J > 1$, $\langle\mu_3\rangle = 0$, while for $J < 1$, $\langle\mu_3\rangle \neq 0$. Consider for example one of the ground states for $J > 1$ (ordered phase), characterized by all spins oriented up:

$$|\, 0 \,\rangle_{J>1} = |\cdots \uparrow\uparrow \cdots \uparrow \cdots\rangle \quad .$$

The action of $\mu_3(n^*)$ on this state turns it into the state

$$|\cdots \uparrow\uparrow\downarrow\downarrow \cdots\rangle \quad .$$

Hence $\mu_3(n^*)$ has the character of a soliton operator, which creates Bloch walls from the ordered ground state. In the disordered phase, we have $\langle\mu_3\rangle \neq 0$. Hence the phase transition $\langle\mu_3\rangle = 0 \to \langle\mu_3\rangle \neq 0$ at $J = 1$ can be viewed as a soliton condensation.

The above QM description of the one dimensional Ising model is known[125] to have an equivalent description in two dimensional classical statistical mechanics defined by the partition function

$$\mathcal{Z} = \sum_{\{\sigma\}} e^{-\mathcal{H}[\sigma]} \tag{3.98}$$

where $\mathcal{H}$ is the discrete, classical Hamiltonian

$$\mathcal{H} = -J \sum_{n,\mu} \sigma(n)\sigma(n+\hat{r}(\mu)) \quad , \tag{3.99}$$

$\sigma(n) = \pm 1$ are spin variables, n are the lattice sites on a two dimensional lattice, and $\hat{r}(\mu)$, $\mu = 1,2$ are unit vectors associated with the two directions of the lattice; $\mathcal{H}$ is invariant under the transformation $\sigma(n) \to -\sigma(n)$. The dual lattice is composed of sites n^* placed at the center of each plaquette. The disorder variables $\mu(n^*) = \pm 1$ are introduced on the sites of the dual lattice. The disorder correlation function is given by[123]

$$\langle \mu(n_1^*)\mu(n_2^*) \rangle = \sum_{\{\sigma\}} e^{-\mathcal{H}'[\sigma]} \tag{3.100}$$

where $\mathcal{H}'[\sigma]$ differs from (3.99) by the fact that the coupling J is replaced by $-J$ along an arbitrary curve C along the dual lattice sites connecting n_1^* and n_2^*. (see figure 3.2). Notice that the curve C cuts the links of the original lattice, each of which is in turn associated with a pair of spin variables.

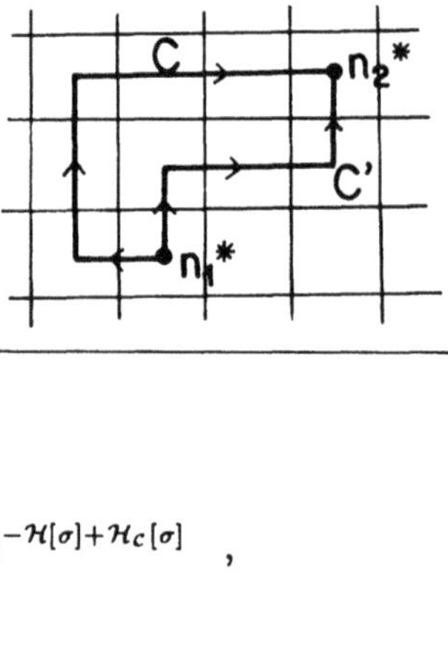

Figure 3.2: Curves C and C' connecting the points n_1^* and n_2^* of the dual lattice.

We may thus evidently write

$$\langle \mu(n_1^*)\mu(n_2^*) \rangle = \sum_{\{\sigma\}} e^{-\mathcal{H}[\sigma]+\mathcal{H}_C[\sigma]} \quad , \tag{3.101a}$$

where

$$\mathcal{H}_C[\sigma] = \begin{cases} 2\mathcal{H}[\sigma] & \text{for } \hat{r}(\mu) \in C \\ 0 & \text{for } \hat{r}(\mu) \notin C \end{cases} \quad . \tag{3.101b}$$

The correlation function (3.101) is actually independent of C. This is easily seen by making the change of variable $\sigma(n) \to \sigma(-n)$ for sites n belonging to an arbitrary region R bounded by the closed curve $\Gamma = C - C'$ (see figure 3.2), which leaves (3.101) invariant. Hence,

$$\langle \mu\mu \rangle_C = \langle \mu\mu \rangle_{C'} \quad . \tag{3.102}$$

This path independence is important in order to have locality properties of the correlation function (3.101), and will play a crucial role in the construction of the disorder operators of QFT.

We first apply the above ideas to the construction of the quantum kinks of theories of the type

$$\mathcal{L} = \partial^\mu \phi^* \partial_\mu \phi - V(\phi, \phi^*) \quad , \tag{3.103}$$

where ϕ is a local, complex scalar field[127]. Let us assume the theory to have a $Z(N)$ symmetry, $\phi(x) \to e^{i\frac{2\pi}{N}}\phi(x)$. As was shown in reference [128], the continuum generalization of the Kadanoff-Ceva formula (3.100) to a kink field operator satisfying the dual algebra (3.96) is given by the "space time" functional

$$\langle \mu(x)\mu^*(y)\rangle = N^{-1}\int \mathcal{D}\phi\mathcal{D}\phi^* e^{-\int d^2z\{\mathcal{L}+\mathcal{L}_C\}} \tag{3.104a}$$

where $\mathcal{L}_C = \int_{x,C}^{y} d\zeta_\mu \delta^2(z-\zeta)\epsilon^{\mu\nu}\Psi_\nu$ is the analogue of $\mathcal{H}_C$ in (3.101), and is to be determined by the requirement that the correlation function (3.104a) be independent of the chosen path C. To this end consider the change of variable $\phi(x) \to e^{i\alpha}\phi(x)$, for x in a region S bounded by $C - C'$. Choose $\alpha = \frac{2\pi}{N}$ corresponding to the $Z(N)$ symmetry of the Lagrangian. Since the integration measure is left invariant by this transformation, we have

$$\begin{aligned}\langle \mu(x)\mu^*(y)\rangle_C &= N^{-1}\int \mathcal{D}\phi\mathcal{D}\phi^* e^{-\int d^2z\{(\mathcal{L}+\mathcal{L}_C)+(\delta\mathcal{L}+\delta\mathcal{L}_C)\}} \\ &= N^{-1}\int \mathcal{D}\phi\mathcal{D}\phi^* e^{-\int d^2z(\mathcal{L}+\mathcal{L}_{C'})} e^{-\int d^2z(\mathcal{L}_C-\mathcal{L}_{C'}+\delta\mathcal{L}+\delta\mathcal{L}_C)}\end{aligned} \tag{3.104b}$$

Hence, independence of the integration path C requires

$$\mathcal{L}_C - \mathcal{L}_{C'} + \delta\mathcal{L} + \delta\mathcal{L}_C = 0 \tag{3.105}$$

Assuming the potential $V(\phi, \phi^*)$ to be invariant under the above transformation, we have,

$$\delta\mathcal{L} = -i\alpha\phi^*\overleftrightarrow{\partial_\mu}\phi\partial_\mu\theta(s) + \alpha^2\phi^*\phi\partial_\mu\theta(s)\partial_\mu\theta(s) \quad , \tag{3.106}$$

where $\theta(s)$ is the two-dimensional Heaviside function with support in S. Now,

$$\partial_\mu\theta(s) = \left(\int_C - \int_{C'}\right)\delta^2(z-\xi)\epsilon_{\mu\nu}d\xi_\nu \quad . \tag{3.107}$$

Making use of this representation, we are thus led to the identification[127]

$$\mathcal{L}_C = -i\alpha\phi^*(z)\overleftrightarrow{\partial_\mu}\phi(z)\int \delta^2(z-\xi)\epsilon_{\mu\nu}d\xi_\nu + \alpha^2\phi^*\phi(z)\int_{x,C}^{y} d\xi_\mu \int_{x,C}^{y} d\eta_\mu \delta^2(z-\xi)\delta^2(z-\eta). \tag{3.108}$$

With this choice,

$$\langle \mu(x)\mu^*(y)\rangle = N^{-1}\int \mathcal{D}\phi\mathcal{D}\phi^* e^{-\int d^2x(\mathcal{L}+\mathcal{L}_C)} \tag{3.109}$$

is independent of the choice of C.

The above result can be cast into a more transparent form, by defining the external field

$$A_\mu(z;C) = \int_{x,C}^{y} \epsilon_{\mu\nu}\delta^2(z-\xi)d\xi_\nu \quad . \tag{3.110}$$

In terms of this field, the functional integral (3.109) takes the form

$$\langle \mu(x)\mu^*(y)\rangle = N^{-1}\int \mathcal{D}\phi\mathcal{D}\phi^* e^{-\int d^2x\{(D_\mu\phi)^*(D_\mu\phi)+V(\phi,\phi^*)\}} \tag{3.111}$$

where D_μ is the covariant derivative $D_\mu = \partial_\mu - i\alpha A_\mu$.

The path independence is now an expression of the invariance of the functional integral under the gauge transformation

$$\begin{aligned} \phi(x) &\to e^{i\alpha\theta(s)}\phi(x) \quad , \\ A_\mu(x;C) &\to A_\mu(x;C) + \partial_\mu\theta(s) \quad . \end{aligned} \tag{3.112}$$

From (3.108,109) we extract the following operator representation, in Minkowski space, for the disorder operator

$$\mu(x) = e^{\alpha\int_{x;C}^{\infty} \epsilon^{\mu\nu}\phi^* \overleftrightarrow{\partial_\nu}\phi d\xi_\nu} \quad . \tag{3.113}$$

It is straightforward to check, that this operator satisfies the dual algebra (3.90), and hence is the QFT generalization of the kink operator $\mu_3(n^*)$ of the Ising model.

We specialize now to the case of a massive ϕ^4 type theory with potential[127]

$$V(\phi,\phi^*) = m^2\phi^*\phi + \frac{\lambda}{4!}\left(\phi^4 + \phi^{*4}\right) + \frac{\mu}{(3!)^2}\left(\phi^*\phi\right)^3 \tag{3.114}$$

This potential possess a $Z(4)$ symmetry $\phi \to e^{i\frac{2\pi}{4}}\phi$. This system can be shown to exist in two phases:

i) symmetric phase.

$$\begin{aligned} &\langle\phi\rangle = 0 \\ &\langle\mu\rangle \neq 0 \quad , \qquad \langle \mu(x)\mu^*(y)\rangle \underset{|x-y|\to\infty}{\longrightarrow} \text{const.} \end{aligned}$$

The phase corresponds to a kink condensate.

ii) partially broken phase.

$$\langle\phi\rangle = 0$$
$$\langle\mu\rangle = 0 \quad , \qquad \langle\mu(x)\mu^*(y)\rangle \underset{|x-y|\to\infty}{\longrightarrow} | x - y |^{-\alpha} \quad ,$$

but $\langle\phi^2\rangle \neq 0$.

Here only the $Z(2)$ symmetry is broken. The μ operator creates massless kinks[128], in agreement with a general theorem[124].

iii) completely broken phase.

$$\langle\phi\rangle \neq 0$$
$$\langle\mu\rangle = 0 \quad , \qquad \langle\mu(x)\mu^*(y)\rangle \underset{|x-y|\to\infty}{\longrightarrow} | x - y |^{-\alpha}\, e^{-m|x-y|} \quad .$$

In this case $\mu(x)$ creates massive kinks with a mass $m \sim \langle\phi\rangle$.

Let us now return to the sine Gordon theory. Repeating the line of reasoning exposed above, we find this time

$$\langle\mu(x)\mu^*(y)\rangle = \int \mathcal{D}\phi e^{-\int d^2z(\mathcal{L}_{SG}+\mathcal{L}_C)} \quad , \tag{3.115}$$

where

$$\mathcal{L}_C(z) = -\frac{2\pi}{\beta}\partial_\mu\phi(z)\int_C \delta^2(z-\xi)\epsilon_{\mu\nu}d\xi_\nu + \frac{1}{2}\left(\frac{2\pi}{\beta}\right)^2 \int_{x;C}^{y} d\xi_\mu \int_{x;C}^{y} d\eta_\mu \delta^2(z-\xi)\delta^2(z-\eta).$$

From (3.115) we extract the following representation for the soliton operator in Minkowski space:

$$\mu(t,x) = e^{i\frac{2\pi}{\beta}\int_{x^1}^{\infty} dz^1 \partial_0\phi(t,z^1)} \tag{3.116}$$

where we have specialized the curve C to lie parallel to the x^1 axis. The exponential (3.116) is nothing but the soliton operator introduced in (3.97)!

The above method of constructing order and disorder operators from a given symmetry of the bosonic Lagrangian in question, can also be used as a tool for constructing bosonic representations of the fermionic operator of other models via the composition (3.87).

3.6 Conclusion

We presented the full quantum solution of the massless Thirring model in terms of free fields. Our building blocks were the exponential fields defined in chapter 2. The correlators have been computed in closed form (Eqs. (3.16,17)), and further properties of the model have been studied. The model is also conformally invariant[129,130] at the quantum level[131]; as it turns out, it is completely equivalent to a free bosonic theory.

The next aim was to modify the model, allowing for a mass for the fundamental fermionic field. Perturbation theory in the mass parameter showed a very important feature at quantum level: the model is equivalent to the bosonic theory, of the sine-Gordon field. The fact that a bosonic theory may describe fermions was suspected long ago[44,108], before the actual proof of this fact, and discussed by several authors. The full quantum equivalence has been proven by Mandelstam (Eqs. (3.73,74)), who has provided the required equivalence in the charged sector, as opposed to Coleman's equivalence, Eqs. (3.28,29), valid only for bilinears in the fermions.

Nevertheless, the aforementioned features are not the only surprising features of the theory; it also exhibits in both the fermionic and bosonic languages, an infinite number of conservation laws, which are a consequence of the Lax pair (3.44), in the case of the Thirring model, and (3.54), in the sine-Gordon case. This is also a feature which is discussed at length in the literature[34,109,113,114,132,133,134]. The consequence at the quantum level, is that one may exactly calculate the full S-matrix of each [109,113,114,132,133,134] of these theories, as well as some form factors[42]. The S-matrix solution will be presented in chapter 8 of this book.

Finally we presented a discussion of the order-disorder formulation of the theory, which provides a deeper physical insight, as well as a method of constructing fermions out of order and disorder operators.

Chapter IV

FUNCTIONAL DETERMINANTS

4.1 Definition and General Remarks

By "functional determinant" we mean in general the determinant of an operator $D[A]$ defined in (continuum) Euclidean space whose value depends on the choice of the field configuration $A(x)$. In the Feynman path-integral approach to quantum field theory one is naturally led to the consideration of determinants of differential operators. Their computation requires regularization of the divergent expressions. The methods used, although seemingly quite different, are all abstracted from the corresponding definition in the finite dimensional case. Among these methods we cite the "ζ-function", "the proper time" and the "Fujikawa method", which we shall discuss in this chapter.

Functional determinants arise in quantum field theory whenever one is dealing with a functional integration of the Gaussian type; that is, whenever the action in question is a second order polynomial in the fields over which we wish to integrate. This may be an exact characteristic of the Lagrangian (fermionic action of QED/QCD); it may be the result of an approximation (semi-classical; effective potential in a $\frac{1}{N}$ expansion, etc.); or it may be the result of rewriting quartic interactions in some fields as a second order polynomial in these fields at the expense of introducing new, auxiliary degrees of freedom ($\mathbb{C}P^{N-1}$, Thirring, Gross-Neveu models).

Conversely, one may represent the determinant of some operator in terms of a Gaussian functional integral over Grassmann-valued fields. This is a very useful technique for dealing with the Faddeev-Popov[135] determinant arising in the process of fixing the gauge in gauge theories.

The Mathews-Salam representation[136] of the Euclidean generating functional of fermionic correlation functions in QED illustrates, how functional determinants make their appearance in physics. It is given by

$$Z[J,\eta,\bar{\eta}] = \int d\mu[A][\mathcal{D}\bar{\psi}][\mathcal{D}\psi]e^{-S_G[A]-S_F[\psi,\bar{\psi},A]}e^{\int(\bar{\eta}\psi+\bar{\psi}\eta)} \qquad (4.1a)$$

$$= \int d\mu[A]e^{-S_G[A]}\mathrm{Det}(i\,\not{\partial}+iM+e\,\not{A})e^{\int\bar{\eta}G(x,y;A)\eta} \quad , \qquad (4.1b)$$

where $d\mu[A]$ is the Faddeev-Popov measure[135] in some suitable gauge, $S_G(S_F)$ is the action associated with the gauge-field (fermions) and $G(x, y; A)$ is the fermionic Greens-function in an external field A^μ:[1)]

$$(i \not\partial + iM + e \not A)G(x, y; A) = \delta^{(d)}(x - y) \quad . \tag{4.1c}$$

The logarithm of a functional determinant has a simple interpretation in terms of Feynman diagrams: It may be represented as an infinite sum of properly weighted one fermion-loop diagrams whose vertices depend on the operator in question. This is easily seen on a formal level. Thus consider an hermitian operator D which we decompose into a "free" (D_0) and an "interacting" part (D_I) as follows:

$$D = D_0 + D_I \quad . \tag{4.2}$$

Borrowing familiar formulas from matrix algebra, we write

$$\omega_D \equiv \ln \det D = \operatorname{tr} \ln(D_0 + D_I) \tag{4.3a}$$

$$\omega_{\frac{D}{D_0}} = \ln \frac{\det D}{\det D_0} = \operatorname{tr} \ln(1 + D_0^{-1} D_I) = \sum_{n=1}^{\infty} \frac{(-1)^{n+1}}{n} \operatorname{tr}\{(\frac{1}{D_0} D_I)^n\} \quad , \tag{4.3b}$$

where $D_0^{-1} D_I$ is to be regarded as a (non-local) operator, and where tr denotes the trace with respect to "internal" indices as well as space-time. The functional ω_D possesses the obvious property, $\omega_{\frac{D}{D_0}} = \omega_D - \omega_{D_0}$.

The sum (4.3) has a simple graphical interpretation: it is given by the sum of (connected) one-loop graphs as shown in figure 4.1

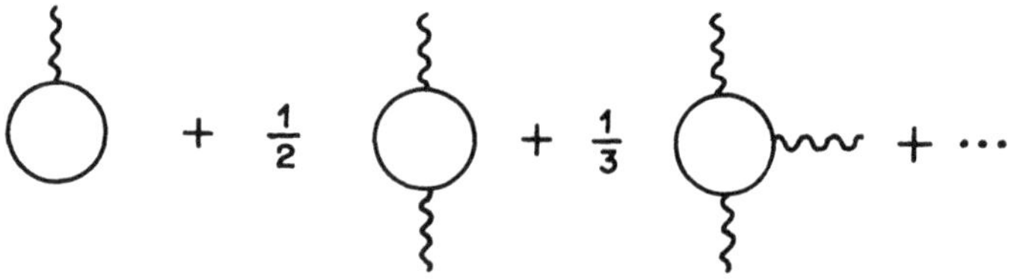

Figure 4.1: Graphical computation of the determinant.

1) The appearance of the "imaginary mass" is peculiar to the Euclidean formulation. A_μ denotes here a hermitian - in general matrix valued - vector field.

with
$-D_I$ for each vertex,
D_0^{-1} for each propagator, a
(-1) overall factor for being a fermion loop, and
$\frac{(n-1)!}{n!} = \frac{1}{n}$ combinatorial factor for a n'th order diagram.

As an example consider the Euclidean Dirac operator of QED,

$$i \not{D} = (i \not{\partial} + e \not{A}) \quad , \tag{4.4}$$

with $A^\mu(x)$ some arbitrarily chosen "external" gauge-field configuration. According to (4.3) we have

$$\omega_{\frac{i\not{D}}{i\not{\partial}}} = \ln\Big[\frac{det(i \not{\partial} + e \not{A})}{det(i \not{\partial})}\Big] = \sum_{n-1}^{\infty} \frac{(-1)^{n+1}}{n} \operatorname{tr}\Big\{\frac{1}{i \not{\partial}} e \not{A}\Big\}^n \tag{4.5a}$$

$$= e \int d^4x S_E(0) \not{A}(x) - \frac{e^2}{2} \int d^4x \int d^4y S_E(x-y) \not{A}(y) S_E(y-x) \not{A}(x) + \cdots \tag{4.5b}$$

where $S_E(x)$ is the inverse of $i \not{\partial}$ (see B.24*b*).

As the first term in (4.5*a*) already shows, there arise ambiguities as a result of divergencies, which will have to be treated by an appropriate regularization technique. In this chapter we shall examine various such techniques which do not rely on perturbation theory.

The exponential of the series (4.3) has a simple representation in terms of a functional integral. Let $\{c^\alpha(x)\}$ be the components of a complex-valued Grassmannian field, with $\{\bar{c}^\alpha(x)\}$ their complex-conjugate. The dimensionality of the index α matches that of the matrix-valued operator D; c and $\bar{c}$ are assumed to obey the Berezin[137] integration rules [C.12]. We then claim, that

$$\frac{\det D}{\det D_0} = \frac{e^{\omega_D}}{e^{\omega_{D_0}}} = \frac{\int \mathcal{D}c\mathcal{D}\bar{c} e^{-\int d^dx \bar{c} D c}}{\int \mathcal{D}c\mathcal{D}\bar{c} e^{-\int d^dx \bar{c} D_0 c}} \quad , \tag{4.6a}$$

where d is the dimensionality of (Euclidean) space-time. The virtue of considering the normalized ratio (4.6) resides in the fact, that thereby the leading divergence is eliminated.

The formal correctness of (4.6) may be established in different ways. One way consists in expanding the integrand of (4.6) in powers of D_I as follows,[2)]

$$\frac{e^{\omega_D}}{e^{\omega_{D_0}}} = \frac{\int \mathcal{D}c\mathcal{D}\bar{c} e^{-\int d^dx \bar{c} D_0 c} \sum_{n=0} \frac{(-1)^n}{n!} [\int (\bar{c} D_I c)]^n}{\int \mathcal{D}c\mathcal{D}\bar{c} e^{-\int d^dx \bar{c} D_0 c}} \tag{4.6b}$$

---

2) D_I may be a local or non-local operator. For compactness of notation we shall frequently refrain from specifying the integration variables.

and then to perform the functional integration. The result is the exponential of the sum of connected one loop diagrams as given by (4.3b).

Another way of formally verifying the correctness of the identification made in (4.6) is to suppose D and D_0 to be operators defined on a compact manifold $\mathcal{M}$, with a discrete eigenvalue spectrum and a complete set of eigenfunctions:

$$Du_n = \lambda_n[D]u_n \quad , \tag{4.7a}$$
$$D_0 v_m = \lambda_m[D_0]v_m \quad . \tag{4.7b}$$

For the time being we shall assume that D and D_0 have no zero eigenvalues.

Expand the fields $c_\alpha(x)$, $\bar{c}_\alpha(x)$ in (4.6) in terms of the above eigenfunctions:

$$c_\alpha(x) = \sum_n a_n u_n(x)_\alpha = \sum_m b_m v_m(x)_\alpha \quad , \tag{4.8a}$$
$$\bar{c}_\alpha(x) = \sum_n \bar{a}_n u_n^\dagger(x)_\alpha = \sum_m \bar{b}_m v_m^\dagger(x)_\alpha \quad , \tag{4.8b}$$

with a_n, $\bar{a}_n$, b_n, $\bar{b}_n$ Grassmann-valued c-number coefficients. Choosing u_n and v_m to be orthonormal with respect to a measure $d\mu(x)$ on $\mathcal{M}$,

$$\int d\mu(x) u_n^\dagger(x) u_{n'}(x) = \delta_{nn'} \quad , \quad \int d\mu(x) v_m^\dagger(x) v_{m'}(x) = \delta_{mm'} \quad , \tag{4.9}$$

we make the change of variable

$$\mathcal{D}\bar{c}\mathcal{D}c \to \prod_n d\bar{a}_n da_n \quad , \tag{4.10a}$$

$$\mathcal{D}\bar{c}\mathcal{D}c \to \prod_m d\bar{b}_m db_m \quad , \tag{4.10b}$$

in the numerator and denominator of (4.6) respectively. Introducing the expansions (4.8) in (4.6) and using (4.7), (4.9), we obtain

$$\frac{e^{\omega_D}}{e^{\omega_{D_0}}} = \frac{\int \prod_n d\bar{a}_n da_n e^{-\sum \lambda_n[D]\bar{a}_n a_n}}{\int \prod_m d\bar{b}_m db_m e^{-\sum \lambda_m[D_0]\bar{b}_m a_m}} \quad . \tag{4.11}$$

Now,

$$e^{-\sum \lambda_n(D)\bar{a}_n a_n} = \prod_n e^{-\lambda_n[D]\bar{a}_n a_n} = \prod_n (1 - \lambda_n[D]\bar{a}_n a_n) \quad . \tag{4.12}$$

The last equality only holds, because of the Grassmann properties $a_n^2 = \bar{a}_n^2 = 0$. Furthermore, according to the Berezin[137] integration rules (C.12),

$$\int \prod_n d\bar{a}_n da_n e^{-\sum \lambda_n[D]\bar{a}_n a_n} = \prod_n \lambda_n[D] \quad , \tag{4.13}$$

with a corresponding result for the denominator in (4.11). Hence we conclude,

$$\frac{e^{\omega_D}}{e^{\omega_{D_0}}} = \frac{\prod_n \lambda_n[D]}{\prod_n \lambda_n[D_0]} \quad . \tag{4.14a}$$

This confirms once more the identification made in (4.6).

If there are zero modes present, then these should be omitted in (4.14a), resulting in

$$\frac{\det' D}{\det' D_0} = \frac{e^{\omega'_D}}{e^{\omega'_{D_0}}} = \frac{\prod{}' \lambda_n[D]}{\prod{}' \lambda_n[D_0]} \quad , \tag{4.14b}$$

where the "prime" indicates omission of zero modes. Hence $\det' D$ is just the subdeterminant corresponding to the non-vanishing eigenvalues of D. The existence of "zero-modes" is not seen in perturbation theory.

The divergencies occurring in the loop-integrations implicit in (4.3b) have their reflection in the non-existence of the product (4.14) because of the (in general unbounded) growth of the eigenvalues with $n \to \infty$. Thus again, a regularization procedure has to be given for defining the determinant via (4.14). Such procedures now will be discussed in detail.

4.2 Determinants and the Generalized Zeta-Function

4.2.1 Elliptic operators

Definition (4.14) can be used to calculate ω_D in terms of an infinite sum involving the eigenvalues $\lambda_n[D]$. In order to get an answer for generic field configurations A, we need to know the eigenvalues as functionals of A. This is generally not possible. Hence a direct calculation starting from (4.14) will in general be feasible for only specific field configurations. For a generic field configuration one has to develop special techniques. All of these are based, in one form or another, on the definition (4.14), and correspond to different regularizations of the in general divergent product (4.13), as we shall see.

Let D be a differential operator of order m defined on a d-dimensional, *compact* manifold $\mathcal{M}$

$$D = \sum_{|\alpha| \leq m} b_{(\alpha)}(x) D^{(\alpha)} \quad . \tag{4.15a}$$

Here $\hat{x} = (\hat{x}_1 \cdots \hat{x}_d) \in \mathcal{M}$ are local coordinates [3], $(\alpha) = (\alpha_1 \cdots \alpha_d)$ is a multiple index, with $|\alpha| \equiv \alpha_1 + \cdots + \alpha_d$, $b_{(\alpha)}(\hat{x}) = b_{\alpha_1 \cdots \alpha_d}(\hat{x}_1 \cdots \hat{x}_d)$ are smooth, $l \times k$ matrix valued functions, and

$$D^{(\alpha)} \equiv (-i)^{(\alpha)} \frac{\partial^{|\alpha|}}{\partial \hat{x}_1^{\alpha_1} \cdots \partial \hat{x}_d^{\alpha_d}} \quad . \tag{4.15b}$$

We associate with (4.15*a*) the "characteristic polynomial"

$$\sum_{|\alpha| \leq m} b_{(\alpha)}(\hat{x}) \xi^{(\alpha)} \equiv \sum_{|\alpha| \leq m} b_{\alpha_1 \cdots \alpha_d}(\hat{x}) \xi_1^{\alpha_1} \cdots \xi_d^{\alpha_d} \quad , \tag{4.16}$$

where $\xi \in \mathcal{M}$. The restriction of the sum to (α): $|\alpha| = m$ leads to the definition of the "principal symbol" associated with D,

$$\sigma_D(x;\xi) = \sum_{|\alpha| = m} b_{(\alpha)}(\hat{x}) \xi^{(\alpha)} \quad , \tag{4.17}$$

The "principal symbol" σ defines a homomorphism from a k to a l dimensional vector space. D is said do be *elliptic* if its principal symbol is invertible for all $\vec{\xi} \neq 0$, i.e., if the homomorphism defined by σ turns into an isomorphism. This requires that $l = k$ and that σ has no zero eigenvalues for $\xi^{(\alpha)} \neq 0$.

Example 1: Euclidean Dirac operator in $\Re^2$, $D = -i\gamma_\mu \partial_\mu$.

In this case $m = 1$ and $b_{10} = \gamma_1$, $b_{01} = \gamma_2$ and $\sigma(x;\xi) = \not{\xi}$. The eigenvalues of σ are determined by the secular equation,

$$\lambda^2 - (\xi_1^2 + \xi_2^2) = 0 \quad .$$

Hence this operator belongs to the class of "elliptic" operators. Notice that the corresponding operator in Minkowski-space does not satisfy the criterion of ellipticity! It is for this reason that we shall always work in this chapter in Euclidean space.

For us, the central importance of elliptic operators resides in the following theorem due to Seeley[138] and Gilkey[139]: Let D be a positive elliptic operator of order m on a compact Riemannian manifold; there then exists an asymptotic expansion for the *diagonal* part of the associated "heat kernel" $\exp(-tD)$, as given by

$$\langle x | exp(-tD) | x \rangle = \frac{1}{(4\pi t)^{\frac{d}{m}}} [a_o(x|D) + a_1(x|D)t + \cdots] \quad . \tag{4.18}$$

---

[3] These coordinates may serve to label only "patches" in $\mathcal{M}$ as it happens if a non-trivial topology is involved. We use a "hat " to distinguish them from the cartesian coordinates x in $\Re^d$

The expansion coefficients $a_k(x \mid D)$ are referred to as "Seeley coefficients"[138]. For elliptic operator of second order, these can all be calculated! Furthermore, we shall see, that for a special class of operators D, the calculation of $\det D$ can be reduced to the calculation of the first few of these coefficients.

We may in fact relax the "positivity condition" on D such as to include *semi-positive* operators. This will be important, in order to be able to discuss the effect of instantons, which turn out to induce zero-modes of the Dirac operator (4.4).

4.2.2 The zeta-function

We shall generally suppose D to be a semi-positive, elliptic, (matrix-valued) differential operator of order m defined on a compact manifold $\mathcal{M}$ of dimension d. Let $d\mu(\hat{x})$ be the measure[4)] on $\mathcal{M}$, and let λ_n, φ_n be the corresponding (real) eigenvalues and eigenfunctions of D:

$$D\varphi_n = \lambda_n \varphi_n \quad . \tag{4.19a}$$

We choose

$$\sum_n \varphi_n(\hat{x})\varphi_n^\dagger(\hat{y}) = \delta(\hat{x},\hat{y}) \tag{4.19b}$$

$$\int d\mu(x)\varphi_n^\dagger \varphi_m = \delta_{mn} \quad . \tag{4.19c}$$

By assumption, all eigenvalues are positive or zero.

We may define a generalized ζ-function[138]

$$\zeta(s|D) = \sum_n{}' \lambda_n^{-s}[D] \quad , \tag{4.20a}$$

where the "prime" on the sum indicates that zero-eigenvalues are to be omitted. We may write (4.20a) in the compact form,

$$\zeta(s|D) = \mathrm{tr}(D^{-s}) \quad , \tag{4.20b}$$

where the trace includes the internal degrees of freedom and is to be taken over the subspace orthogonal to the zero modes. $\zeta(s \mid D)$ can be analytically extended to a meromorphic function of s in the whole complex plane, with $s = 0$ a regular

---

4) If we choose to parametrize $\hat{x}$ by $x = (x^1, \cdots, x^d) \in \Re^d$, then $d\mu(\hat{x}) = d^d x \sqrt{g(x)}$, where $g(x) = \det(g_{\mu\nu}(x))$ with $g_{\mu\nu}(x)$ the induced metric tensor on $\mathcal{M}$. In that case $\delta(\hat{x},\hat{y}) = g^{-1/2}(x) \times \delta^d(x-y)$.

point[138]. Its value at $s = 0$ may be calculated in terms of the Seeley coefficients, as we shall see.

The quantity of immediate interest to us is obtained by differentiating (4.20a) with respect to s:

$$\frac{d\zeta}{ds}(s \mid D) = -\sum_n{}' \frac{\ln \lambda_n}{(\lambda_n)^s} \quad . \tag{4.21}$$

This suggests to define $\det' D$ by

$$[\ln \det{}' D]_\zeta \equiv \lim_{s\to 0}\left[-\frac{d\zeta}{ds}(s \mid D)\right] \equiv -\zeta'(0 \mid D) \quad . \tag{4.22}$$

The "prime" in (4.21) indicates the restriction of the sum the subspace of non-vanishing zero modes where D is invertible.

The calculation of $\det' D$ is of course plagued by the usual ambiguities related to the ultra violet divergencies of fermion-loop graphs. The definition (4.22) by analytic extension to $s = 0$ corresponds to a very *specific* choice of regularization, the so called *ζ-function regularization* . By normalizing $\det' D$ with respect to the "free" operator D_0[5])

$$\ln \frac{\det' D}{\det D_0} = -(\zeta'(0 \mid D) - \zeta'(0 \mid D_0)) \quad , \tag{4.23}$$

we eliminated the ambiguity associated with the leading divergence.

The operators $D\,(D_0)$ of interest in general carry a dimension. This means that the eigenvalues λ_n, and hence the measure (4.10) carries dimensions, so that the corresponding Berezin integration rules must involve a dimensional scale parameter M. In different terms: the eigenvalues λ_n appearing in the definition (4.20a) of the ζ-function should be normalized with respect to that scale. Hence, unless the number of eigenvalues of D and D_0 match (no zero modes of D), there will exist an ambiguity associated with scale transformations (4.10) of the measure in (4.6). It is easily recognized from (4.21) that this ambiguity is of the form $\zeta(0 \mid D)\ln M$. In order to make this ambiguity explicit, we should thus write[140]

$$\omega_{\frac{D}{D_0}} = -[\zeta'(0 \mid D) - \zeta'(0 \mid D_0)] + [\zeta(0 \mid D) - \zeta(0 \mid D_0)] \ln M \quad . \tag{4.24}$$

This "scaling" ambiguity still does not exhaust all the possible ambiguities. We return to a discussion of this point later in section (4.6) .

The meromorphy of $\zeta(s \mid D)$ and its regularity at $s = 0$ are most easily recognized by introducing the (matrix valued) "heat kernel" associated with D, which satisfies the equation

$$\left(\frac{\partial}{\partial t} + D\right) h(t; \hat{x}, \hat{y} \mid D) = 0 \quad , \quad t \geq 0 \quad , \tag{4.25a}$$

5) The spectrum of D_0 will, for the cases of interest, contain no zero modes.

with the "initial" condition

$$h_{\alpha\beta}(0;\hat{x},\hat{y}\mid D)=\delta_{\alpha\beta}\delta^{(d)}(\hat{x},\hat{y}) \quad , \tag{4.25b}$$

where $\delta^{(d)}(\hat{x},\hat{y})$ denotes the δ-function on the compact manifold $\mathcal{M}$. We have,

$$\int d\mu(\hat{x})h(0;x,y\mid D)=1 \quad . \tag{4.26}$$

The solution to (4.25) is evidently given by

$$h_{\alpha\beta}(t;\hat{x},\hat{y}\mid D)=\sum_n e^{-t\lambda_n}\varphi_n(\hat{x})_\alpha\varphi_n^\dagger(\hat{y})_\beta \quad , \tag{4.27}$$

where the sum includes the zero eigenvalues. $h(t;\hat{x},\hat{y}\mid D)$ is the continuous kernel of the "integral trace-class operator" e^{-tD}:

$$h(t;\hat{x},\hat{y}\mid D)=\langle\hat{x}|e^{-tD}|\hat{y}\rangle \quad , \tag{4.28}$$

where $\{|\,\hat{x}\rangle\}$ denotes an orthonormal set of states on $\mathcal{M}$:

$$\int d\mu(x)\mid\hat{x}\rangle\langle\hat{x}\mid=1 \quad , \quad \langle\hat{x}\mid\hat{y}\rangle=\delta^{(d)}(\hat{x},\hat{y}) \quad .$$

To prove the meromorphy of $\zeta(s\mid D)$ we define a generalized (matrix-valued) ζ-function $\zeta(s;\hat{x},\hat{y}\mid D)$ via the Mellin-transform of the heat-kernel (4.23), after subtraction of the contribution from the zero modes, in order to guarantee the convergence of the t-integration at $t=\infty$[141]:

$$\zeta_{\alpha\beta}(s;\hat{x},\hat{y}\mid D)=\frac{1}{\Gamma(s)}\int_0^\infty dt\,t^{s-1}[h_{\alpha\beta}(t;\hat{x},\hat{y}\mid D)-P^{(0)}_{\alpha\beta}(\hat{x},\hat{y})] \quad . \tag{4.29}$$

Here, $P^{(0)}(\hat{x},\hat{y})$ is the (matrix-valued) projector on the zero-modes

$$P^{(0)}_{\alpha\beta}(\hat{x},\hat{y})=\sum_n\varphi_n^{(0)}(\hat{x})_\alpha\varphi_n^{(0)^*}(\hat{y})_\beta \quad . \tag{4.30}$$

Note, that the integrand in (4.29) decays exponentially with $t\to\infty$.

Using (4.27), and performing the t-integration in (4.29) one finds

$$\zeta(s;\hat{x},\hat{y}\mid D)=\sum_n{}'\frac{\varphi_n(\hat{x})\varphi_n^\dagger(\hat{y})}{\lambda_n^s} \quad , \tag{4.31}$$

and in particular

$$\int d\mu(\hat{x}) tr\zeta(s;\hat{x},\hat{x} \mid D) = \zeta(s \mid D) \quad . \tag{4.32a}$$

For D an *invertible* operator (no zero-modes) we recognize in $\zeta(s;\hat{x},\hat{y} \mid D)$ just the meromorphic extension of the kernel of D^{-s}:

$$\zeta(s;\hat{x},\hat{y} \mid D) = \langle \hat{x} \mid D^{-s} \mid \hat{y} \rangle \quad . \tag{4.33}$$

In the mathematical literature[138,142] also the notation $K_{-s}(D;\hat{x},\hat{y})$ is used to denote the r.h.s of (4.33). In this notation (4.32a) takes the form

$$\zeta(s \mid D) = \int d\mu(\hat{x}) tr K_{-s}(D;\hat{x},\hat{x}) \quad . \tag{4.32b}$$

The definition of $K_{-s}(D;\hat{x},\hat{x})$ can be extended to include any (hermitian or non-hermitian) invertible, elliptic differential operator of order m defined on a compact manifold without boundary[138,142].

The meromorphy of $\zeta(s;x,x \mid D)$ is now easily demonstrated. The only singularities in s of (4.29) arise from the lower end-point of the t-integration. Choose ε to lie within a small interval around $t = 0$ for which the Seeley expansion (4.18) converges. Splitting the integration range into $[0,\varepsilon]$ and $[\varepsilon,\infty)$, we have for an operator of *even* order m,

$$\zeta(s;\hat{x},\hat{x} \mid D) = \frac{1}{\Gamma(s)} \frac{1}{(4\pi)^{\frac{d}{m}}} \int_0^{\varepsilon} dt\, t^{s-1-\frac{d}{m}}$$

$$\times \{(a_0 + a_1 t + \cdots + a_r t^r) - (4\pi)^{\frac{d}{m}} t^{\frac{d}{m}} P^{(0)}(\hat{x},\hat{x})\} + \text{“regular”} \quad , \tag{4.34}$$

where “r” is to be determined below and where “*regular*” stands for the remaining holomorphic part. The poles arising from the lower endpoint integration are evidently located at

$$s = \frac{d}{m} - k \quad , \quad k = 0,1,2,\cdots \quad . \tag{4.35}$$

Now, $\frac{1}{\Gamma(s)}$ has simple zeroes at $s = 0,-1,-2,\cdots$ so that $\zeta(s;\hat{x},\hat{x} \mid D)$ has simple poles at

$$s = 1,2,\cdots,\frac{d}{m} \qquad (d \quad even) \quad , \tag{4.36}$$

but not at $s = 0$. This proves the announced meromorphy of $\zeta(s;x,x \mid D)$ in s, and its regularity at $s = 0$. According to (4.36), r is to be chosen $r \geq \frac{d}{m}$. We have[141] for the residue at the pole at $s = \frac{d}{m} - k$

$$\operatorname*{Res}_{s=(\frac{d}{m}-k)} \zeta(s;\hat{x},\hat{x} \mid D) = \frac{1}{\Gamma(\frac{d}{m}-k)} \frac{1}{(4\pi)^{\frac{d}{m}}} a_k(\hat{x}) \quad , \quad (k \neq \frac{d}{m}) \tag{4.37}$$

In particular, since $s = 0$ is a regular point, we obtain from (4.34),

$$\zeta(0;\hat{x},\hat{x} \mid D) = \frac{a_{\frac{d}{m}}(\hat{x})}{(4\pi)^{\frac{d}{m}}} - P^{(0)}(\hat{x},\hat{x}) \quad . \tag{4.38}$$

Now

$$\operatorname{tr}\int d\mu(\hat{x})P^{(0)}(\hat{x},\hat{x}) = \sum \int d\mu(x)\varphi_n^{\dagger(0)}(x)\varphi_n^{(0)}(x) = N \quad , \tag{4.39}$$

with tr being the trace over matrix indices, and N is the number of zero modes. It therefore follows from (4.38) and (4.32a),

$$\zeta(0 \mid D) = tr\frac{\alpha_{\frac{d}{m}}}{(4\pi)^{\frac{d}{m}}} - N \quad , \tag{4.40}$$

where, in general,

$$\alpha_k \equiv \int d\mu(\hat{x})a_k(\hat{x}) \quad . \tag{4.41}$$

The Seeley coefficients play a central role in the computation of functional determinants.

4.3 One Point Compactification

Before we demonstrate, how the Seeley coefficients may be calculated, we shall comment on the manifold $\mathcal{M}$ the operator D is to be defined on.

4.3.1 Stereographic projection of Dirac operator

Theorems about differential operators usually refer to operators defined on a compact manifold, or to boundary value problems.

Under suitable asymptotic conditions one may compactify $\Re^d$. The most familiar method is that of stereographic mapping of $\Re^d$ onto a d-dimensional sphere S^d in $d+1$ dimensions, whereby infinity is mapped either on the north or the south pole[143–146]. In the case of the Dirac operator (4.4) this presupposes a sufficiently good asymptotic behaviour for the gauge field A_μ. Thus, a given gauge field tending to pure gauge at infinity will in general "live" on the stereographic sphere on a patch excluding either the north or the south pole. If the Dirac operator is to exist on the whole sphere, then the winding (Pontryagin) number[6)] associated with A_μ will have to be integer.

---

6) In the mathematical literature it is referred to as the first Chern number[147].

Consider in particular the stereographic projection of $\Re^2$ onto a 2-dimensional sphere of radius R, with infinity being mapped onto its south-pole[143–146].

Let $\vec{r}'$ be the vector from the south pole to the point $(x_1, x_2) \in \Re^2$ (see figure 4.2). If R is the radius of the stereographic sphere, its length r' is given by

$$r' = \frac{2R}{\cos\vartheta} \quad , \tag{4.42}$$

where ϑ is the angle between $\vec{r}'$ and the z-axis. If we denote by φ the corresponding azimuthal angle, we have $x_1 = r' \sin\vartheta\cos\varphi$, $x_2 = r'\sin\vartheta\sin\varphi$.

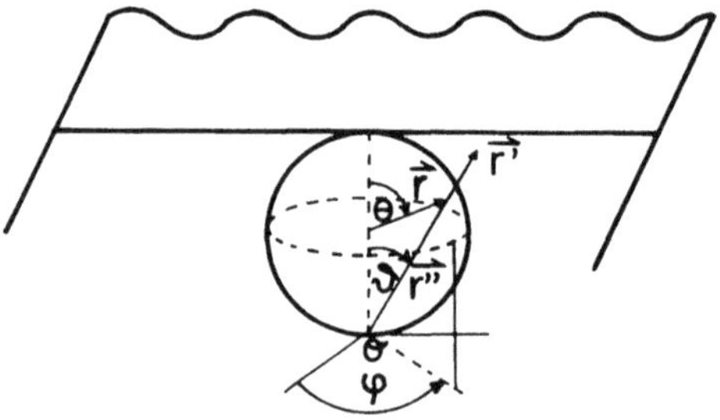

Figure 4.2: Stereographic projection of the real plane onto a 2 dimensional sphere.

If we denote by $\vec{r}''$ the segment of the vector $\vec{r}'$ extending from $\mathcal{O}$ to the intersection point of $\vec{r}'$ with the sphere, we also have

$$r''\sin\vartheta = R\sin\theta \quad , \quad r''\cos\vartheta = R(1+\cos\theta) \quad , \tag{4.43}$$

were θ is the polar angle defined in figure 4.2 and $r'' = \mid \vec{r}'' \mid$. From (4.43) one obtains

$$\tan\vartheta = \frac{\sin\theta}{1+\cos\theta} \quad . \tag{4.44}$$

Combining (4.42,43) we have

$$x_\mu = \frac{2Rr_\mu}{R+r_3} \quad , \quad \mu = 1,2 \tag{4.45}$$

with $\vec{r} = (R\sin\theta\cos\varphi, R\sin\theta\sin\varphi, R\cos\theta)$, and we have the inverse formulae

$$\begin{aligned} r_\mu &= \frac{4R^2}{4R^2+x^2}x_\mu \\ r_3 &= R\frac{4R^2-x^2}{4R^2+x^2} \quad . \end{aligned} \tag{4.46}$$

Let (ξ_μ) be the angular variables [7)]

$$\xi_1 = \theta \quad , \quad \xi_2 = \varphi \quad , \tag{4.47}$$

labelling a point on the sphere, and let ds^2 be the square of Euclidean element of length:

$$ds^2 = \delta_{\mu\nu} dx_\mu dx_\nu \quad . \tag{4.48}$$

In terms of the variables (4.47) on the sphere

$$ds^2 = \hat{g}_{\lambda\rho}(\xi) d\xi_\lambda d\xi_\rho \quad , \tag{4.49}$$

with $\hat{g}_{\lambda\rho}(\xi)$ the induced metric on S^2:

$$\hat{g}_{\lambda\rho}(\xi) = \frac{\partial x_\mu}{\partial \xi_\lambda} \frac{\partial x_\mu}{\partial \xi_\rho} \quad . \tag{4.50}$$

From (4.45) and (4.47) one finds

$$ds^2 = \left(\frac{4R^2 + x^2}{4R^2} \right) d\sigma^2 \quad , \tag{4.51a}$$

with

$$d\sigma^2 = R^2 d\theta^2 + R^2 \sin^2 \theta d\varphi^2 \quad , \tag{4.51b}$$

the square of the element of length on S^2.

Hence the stereographic mapping corresponds to a conformal mapping. In the literature it is common to consider instead the stereographic mapping of $y_\mu = 2x_\mu$ onto S^2, which leads to the slightly modified formulae[143-146]

$$x_\mu = \frac{Rr_\mu}{R + r_3}$$

$$r_\mu = \frac{2R^2}{r^2 + x^2} x_\mu \quad , \quad r_3 = \left(\frac{R^2 - x^2}{R^2 + x^2} \right) R \tag{4.52}$$

$$ds^2 = \Omega_R^{-2} d\sigma^2 \quad , \tag{4.53}$$

with

$$\Omega_R = \frac{2R^2}{R^2 + x^2} \quad . \tag{4.54}$$

From (4.53) we see that

$$d\sigma^2 = g_{\mu\nu}(x) dx_\mu dx_\nu \quad , \tag{4.55a}$$

[7)] ξ_1, ξ_2 correspond to $\hat{x}_1$, $\hat{x}_2$ defined earlier.

with

$$g_{\mu\nu}(x) = \Omega_R^2(x)\delta_{\mu\nu} \quad . \tag{4.55b}$$

The integration measure on the stereographic sphere is correspondingly

$$d^2\sigma = \sqrt{g}d^2x \quad , \tag{4.56a}$$

with

$$g = \det(g_{\mu\nu}) = \Omega_R^4 \quad . \tag{4.56b}$$

The Dirac operator (4.3) will be of central importance in our discussions. We now show how to project it from $\Re^2$ onto the stereographic sphere[143–146].

Proposition 1

$$i\gamma_\mu\partial_\mu = \frac{1}{R}\Omega_R^2(x)V^\dagger(s_{ab}l_{ab}+1)V \quad , \tag{4.57a}$$

where

$$V = \frac{1}{\sqrt{2}}\left(1 + i\frac{\gamma\cdot x}{R}\right) \quad , \tag{4.57b}$$

and where s_{ab}, l_{ab} are the spin and orbital angular momentum operators,

$$s_{ab} = \frac{1}{2}\epsilon_{abc}\sigma_c \quad , \quad l_{ab} = r_a p_b - r_b p_a \quad , \tag{4.57c}$$

with

$$p_a = \frac{1}{i}\frac{\partial}{\partial r_a} \quad . \tag{4.57d}$$

Proof: Using

$$p_\mu = \frac{1}{i}\left(\frac{\partial x_\nu}{\partial r_\mu}\right)\frac{\partial}{\partial x_\nu} \quad ,$$

it is a simple algebraic exercise to show that,

$$\begin{aligned} r_\mu p_\nu &= -ix_\mu\partial_\nu \\ r_\mu p_3 &= \frac{i}{R}x_\mu x_\nu\partial_\nu \\ r_3 p_\mu &= -\frac{iR}{2}\frac{R^2 - x^2}{R^2}\partial_\mu \quad . \end{aligned}$$

As a consequence

$$l_{\mu\nu} = -i(x_\mu\partial_\nu - x_\nu\partial_\mu) \tag{4.58a}$$

$$l_{3\mu} = -\frac{i}{R}x_\mu x_\nu\partial_\nu - \frac{i}{2R}(R^2 - x^2)\partial_\mu \quad . \tag{4.58b}$$

Using these results proposition 1 is verified by explicit algebraic calculation. Relation (4.57a) converts the Dirac operator $i\gamma_\mu\partial_\mu$ into the hyperspherical operator $s_{ab}l_{ab} + 1$.

Proposition 2

$$i\,\not{D} = \frac{1}{R}\Omega_R^2(x)V^\dagger(s_{ab}L_{ab}+1)V \quad , \tag{4.59a}$$

where

$$L_{ab} = l_{ab} + \hat{l}_{ab} \quad . \tag{4.59b}$$

with l_{ab} given by (4.57c), and,

$$\hat{l}_{ab} = -\frac{e}{R}(r_a\hat{A}_b - r_b\hat{A}_a) \quad , \tag{4.59c}$$

where,

$$\hat{A}_\mu = \frac{1}{2}\frac{(R^2+x^2)}{R^2}A_\mu + \frac{x_\mu x_\nu}{R^2}A_\nu \quad , \tag{4.59d}$$

$$\hat{A}_3 = -\frac{x_\mu}{R}A_\mu \quad .$$

Proof: First of all we check, that $\hat{A}_\mu$ lives in the tangent space of the stereographic sphere:

$$r_a\hat{A}_a = 0 \quad . \tag{4.60}$$

The proof proceeds again algebraically. One finds after some calculation,

$$\hat{l} \equiv s_{ab}\hat{l}_{ab} = -e\gamma_5\epsilon_{\mu\nu}x_\mu A_\nu + \frac{e}{2R}(R^2-x^2)(\gamma\cdot A) + \frac{e}{R}(x\cdot A)(\gamma\cdot x) \quad .$$

Using the identities (B.18) and (B.19) as well as

$$\gamma_\mu\gamma_\nu\gamma_\lambda = \delta_{\mu\nu}\gamma_\lambda - \delta_{\mu\lambda}\gamma_\nu + \delta_{\nu\lambda}\gamma_\mu \quad , \tag{4.61}$$

one finds after some further calculation,

$$\frac{\Omega_R^2}{2R}\Big(1 - i\frac{\gamma\cdot x}{R}\Big)\hat{l}\Big(1 + i\frac{\gamma\cdot x}{R}\Big) \equiv e\,\not{A} \quad . \tag{4.62}$$

Combining this result with our previous result for $i\,\not{\partial}$ one obtains[146]

$$(i\,\not{\partial} + \not{A}) = \frac{1}{R}\Omega_R^2V^\dagger D_{s^2}V \quad , \tag{4.63}$$

with

$$D_{s^2} = 1 + s_{ab}L_{ab} \tag{4.64}$$

the Dirac operator on the stereographic sphere.

4.3.2 The associated Dirac operator

The eigenvalue spectrum of the (dimensionless) operator D_{s^2} is evidently discrete[8])

$$D_{s^2}\tilde{u}_n(\Omega) = \tilde{\lambda}_n \tilde{u}_n(\Omega) \quad , \tag{4.65a}$$

where Ω stands for the variables (4.47) defined on the sphere. The eigenfunctions are chosen to be orthonormal

$$\int d\Omega \tilde{u}_n^\dagger(\Omega)\tilde{u}_m(\Omega) = \delta_{nm} \quad , \tag{4.65b}$$

and are complete:

$$\sum_n \tilde{u}_n(\Omega)_\alpha \tilde{u}_n^\dagger(\Omega')_\beta = \delta_{\alpha\beta}\delta^2(\Omega - \Omega') \quad . \tag{4.65c}$$

In order to establish the link with the usual Dirac operator, we consider instead of the eigenfunctions $\tilde{u}_n$, the functions u_n defined by

$$\tilde{u}_n = RVu_n \quad . \tag{4.66}$$

From (4.63) and (4.65) it then immediately follows that

$$i \not{D} u_n = \lambda_n \Omega_R u_n \quad , \tag{4.67}$$

where $\lambda_n = \tilde{\lambda}_n / R$. Since

$$V^\dagger V = \frac{R^2 + x^2}{2R^2} = \Omega_R^{-1}$$

$$d\Omega = \frac{1}{R^2} d\sigma^2 = \frac{\Omega_R^2}{R^2} dx^2 \quad , \tag{4.68}$$

it follows from (4.65) that,

$$\int d^2x \Omega_R(x) u_n^\dagger(x) u_m(y) = \delta_{nm} \tag{4.69a}$$

$$\sum u_n(x)_\alpha u_n^\dagger(y)_\beta = \Omega_R^{-1}\delta^2(x - y) \quad . \tag{4.69b}$$

Hence, solving the eigenvalue problem (4.65) is equivalent to solving the associated eigenvalue problem (4.67) with (4.69).

---

8) We denote eigenfunctions referring specifically to the Dirac operator by u_n rather than φ_n.

Since $\sqrt{g} = \Omega_R^2$, expressions (4.69) are not yet of the "standard" form. To get them in standard form, define

$$\hat{u}_n \equiv \Omega_R^{-\frac{1}{2}} u_n \tag{4.70a}$$

$$i\,\hat{\not{D}} \equiv \Omega_R^{-\frac{3}{2}} i\,\not{D}\,\Omega_R^{\frac{1}{2}} \quad . \tag{4.70b}$$

Then the eigenvalue problem reads

$$i\,\hat{\not{D}}\hat{u}_n = \lambda_i \hat{u}_n \quad , \tag{4.71}$$

with

$$\int d^2x \sqrt{g}\,\hat{u}_n^\dagger(x)\hat{u}_m(x) = \delta_{nm}$$

$$\sum_n \hat{u}_n(x)_\alpha \hat{u}_n(y)_\beta = \frac{1}{\sqrt{g}}\delta_{\alpha\beta}\delta^{(2)}(x-y) \quad . \tag{4.72}$$

One easily verifies, that the operator $i\,\hat{\not{D}}$ is hermitian with respect to the measure $d^2x\sqrt{g}$.

The usefulness of the stereographic projection resides in the following observation. Consider the functional $(d = 2)$

$$\frac{e^{\omega_{i\not{D}}}}{e^{\omega_{i\not{\partial}}}} = \frac{\int \mathcal{D}\bar{\psi}\mathcal{D}\psi e^{-\int d^2x \bar{\psi} i \not{D}\psi}}{\int \mathcal{D}\bar{\psi}\mathcal{D}\psi e^{-\int d^2x \bar{\psi} i \not{\partial}\psi}} = \frac{\int \mathcal{D}\bar{\hat{\psi}}\mathcal{D}\hat{\psi} e^{-\int d^2x \sqrt{g}\bar{\hat{\psi}} i \hat{\not{D}}\hat{\psi}}}{\int \mathcal{D}\bar{\hat{\psi}}\mathcal{D}\hat{\psi} e^{-\int d^2x \sqrt{g}\bar{\hat{\psi}} i \hat{\not{\partial}}\hat{\psi}}} \quad , \tag{4.73}$$

where we have assumed the Jacobian of the transformation $\psi \to \hat{\psi}$ to cancel in the ratio. (This we expect to be true if the dimension of the space spanned by ψ and $\hat{\psi}$ is the same, i.e., if there are no zero modes). Expand now $\hat{\psi}$ in terms of the eigenfunctions of $i\,\hat{\not{D}}(i\,\hat{\not{\partial}})$

$$\hat{\psi} = \sum_r b_n \hat{u}_n \quad , \quad etc.$$

and make the change of variable $\int \mathcal{D}\bar{\hat{\psi}}\mathcal{D}\hat{\psi} \to \int \prod_n d\bar{b}_n db_n \quad ,$ *etc.* We then obtain (the last manipulation is formal)

$$e^{(\omega_{i\not{D}} - \omega_{i\not{\partial}})} = \frac{\prod \lambda_n[A]}{\prod \lambda_n[0]} = \frac{\det i\,\hat{\not{D}}}{\det i\,\hat{\not{\partial}}} = \frac{\det[\Omega_R^{-\frac{3}{2}} i\,\not{D}\Omega_R^{\frac{1}{2}}]}{\det[\Omega_R^{-\frac{3}{2}} i\,\not{\partial}\Omega_R^{\frac{1}{2}}]} = \frac{\det i\,\not{D}}{\det i\,\not{\partial}} \quad . \tag{4.74}$$

Hence, as a result of the underlying conformal invariance, the dependence on R formally drops out, if there are no zero modes. (See section 6 for a more detailed discussion of this point).

4.3.3 Vortex on the stereographic sphere

Let us emphasize that the one-point compactification of the Dirac operator discussed above presupposes that the gauge field A^a_μ "lives" on the corresponding compact manifold $\mathcal{M}$. This will be the case if A^a_μ tends to a pure gauge-field configuration at infinity, i.e.

$$eA_\mu(x) \underset{|x|\to\infty}{\longrightarrow} g^{-1}(x)\partial_\mu g(x) \quad , \quad g(x) \in SU(N) \quad . \tag{4.75}$$

A_μ then exists on coordinate patches of $\mathcal{M}$, as we now illustrate.

Example: Let A_μ be given by the $U(1)$ vortex-field

$$A_\mu(x) = -\frac{n}{e}\epsilon_{\mu\lambda}\frac{x_\lambda}{R^2+x^2} \quad ; \tag{4.76}$$

we have

$$A_\mu(x) \to -\frac{n}{e}\epsilon_{\mu\lambda}\frac{x_\lambda}{x^2} = \frac{1}{e}\partial_\mu\Lambda \quad , \tag{4.77}$$

with $\Lambda(x) = n \operatorname{arc\,ctn}\frac{x_2}{x_1}$.

Hence the vortex-field (4.76) has the property (4.75) of being pure gauge at infinity. As we now show, it admits, as a consequence, a one-point compactification if n is an integer[141].

Identifying Euclidean infinity with a circle of radius $R \to \infty$, and parametrizing $x_{1,2}$ on this circle by R and an angle θ, $x_1 = R\sin\theta$, $x_2 = R\cos\theta$, we have

$$\Lambda(R,\theta) = n\theta \quad . \tag{4.78}$$

Hence (4.76) corresponds to the Pontryagin number n:

$$e\int\frac{d^2x}{4\pi}\epsilon_{\mu\nu}F_{\mu\nu}(x) = n\int\frac{d\Omega}{4\pi} = n \quad .$$

According to (4.59d) the field corresponding to A_μ on the stereographic sphere of radius R is given by

$$\hat{A}_\mu = -\frac{n}{2e}\epsilon_{\mu\nu}\frac{x_\nu}{R^2} \quad , \quad \mu = 1,2 \quad , \quad \hat{A}_3 = 0 \quad . \tag{4.79}$$

Let us write $\hat{A}_\alpha$, $\alpha = 1,2,3$, in terms of the coordinates of that sphere, with the south pole identified with Euclidean infinity. From (4.52) and (4.46b) we have

$$x_1 = \frac{R\sin\theta\cos\varphi}{1+\cos\theta} \quad , \quad x_2 = \frac{R\sin\theta\sin\varphi}{1+\cos\theta} \quad , \tag{4.80}$$

Defining a unit vector tangent to the stereographic sphere, pointing in the azimuthal direction

$$\hat{e}_\varphi = (-\sin\varphi, \cos\varphi) \tag{4.81}$$

we have[141]

$$\hat{\mathcal{A}} = -\frac{n}{2eR}\frac{\sin\theta}{1+\cos\theta}\hat{e}_\varphi \quad , \quad \hat{A}_3 = 0 \quad . \tag{4.82}$$

We see that $\hat{\mathcal{A}}$ does not exist globally on the stereographic sphere, but only on a patch which excludes the south pole. We could have considered however equally well a stereographic mapping where the north pole is identified with infinity. In that case one obtains

$$\hat{\mathcal{A}}' = \frac{n}{2eR}\frac{\sin\theta}{1-\cos\theta}\hat{e}_\varphi \quad , \quad \hat{A}'_3 = 0 \quad . \tag{4.83}$$

Thus $\hat{\mathcal{A}}$ and $\hat{\mathcal{A}}'$ live on two different, but *overlapping* patches defined by

$$(I) \quad -\pi < \theta < \pi \quad , \quad (II) \quad 0 < \theta < 2\pi$$

respectively. In the region of overlap, both $\hat{\mathcal{A}}$ and $\hat{\mathcal{A}}'$ are defined and are in fact related by

$$\hat{\mathcal{A}}' = \hat{\mathcal{A}} + \frac{n}{eR\sin\theta}\hat{e}_\varphi \quad . \tag{4.84a}$$

Recalling that

$$\underset{\sim}{\nabla} = \underset{\sim}{e}_r\frac{\partial}{\partial r} + \underset{\sim}{e}_\theta\frac{1}{r}\frac{\partial}{\partial\theta} + \underset{\sim}{e}_\varphi\frac{1}{r\sin\theta}\frac{\partial}{\partial\varphi} \quad , \tag{4.84b}$$

we therefore see that $\hat{\mathcal{A}}'$ is just the gauge transformation of $\hat{\mathcal{A}}$,

$$\hat{\mathcal{A}}' = \hat{\mathcal{A}} + \frac{1}{e}\underset{\sim}{\nabla}\Lambda \quad , \tag{4.84c}$$

with Λ given by (4.78). $\hat{\mathcal{A}}'$ and $\hat{\mathcal{A}}$ thus differ by a bonafide gauge transformation if n is integer. Correspondingly *gauge invariant* quantities exist in this case *globally* on the stereographic sphere.

4.4 Calculation of Seeley Coefficients

As an example we present here the calculation of the first few Seeley coefficients for an important class of operators. Although Gilkey's method[139,148] is completely general and rigorous, we shall follow the more pragmatic "coincidence limit" method of de Witt[149,150] which is more easily understandable for physicists. In this book we shall need only the first non-trivial Seeley coefficient. The result is contained in the following

Theorem: Let A be the hermitian order-two operator defined on $\Re^d$,

$$A = D_\mu^\dagger D_\mu + X \quad , \quad X^\dagger = X \quad , \tag{4.85a}$$

with $D_\mu = \partial_\mu - iA_\mu$, $D_\mu^\dagger = -\partial_\mu + iA_\mu^\dagger$, where A_μ and X are matrix valued fields. Then

$$\langle x \mid e^{-\epsilon A} \mid x \rangle = \frac{1}{(4\pi\epsilon)^{\frac{d}{2}}}[1 + a_1\epsilon + \cdots] \quad , \tag{4.86a}$$

with

$$a_1 = -X - \frac{1}{4}\{(A_\mu A_\mu^\dagger + A_\mu^\dagger A_\mu) - (A_\mu A_\mu + A_\mu^\dagger A_\mu^\dagger)\} - \frac{1}{2}\{i[(\partial_\mu A_\mu - \partial_\mu A_\mu^\dagger)] + [A_\mu^\dagger A_\mu - A_\mu A_\mu^\dagger]\} \quad . \tag{4.86b}$$

Proof: We seek solutions to the heat equation [9)]

$$\left\{\frac{\partial}{\partial t} + (D_\mu^\dagger D_\mu + X)\right\} h(t; x, y) = 0 \quad , \tag{4.87a}$$

with the initial condition,

$$h(0; x, y) = \delta^{(d)}(x - y) \quad . \tag{4.87b}$$

The singular behaviour (4.87b) in x, y suggests making the Ansatz

$$h(t; x, y) = f(t; x, y) h_0(t; x, y) \quad , \tag{4.88a}$$

with

$$h_0(t; x, y) = \frac{1}{(4\pi t)^{\frac{d}{2}}} e^{-\frac{\sigma}{2t}} \quad , \tag{4.88b}$$

9) We omit specification of the operator A in the argument of the heat-kernel and zeta function.

and σ the geodesic interval between x and y

$$\sigma = \frac{1}{2}(x-y)^2 \quad . \tag{4.89}$$

The function h_0 satisfies the equation

$$\left(\frac{\partial}{\partial t} - \nabla\right) h_0(t;x,y) = 0 \quad , \tag{4.90a}$$

with the initial condition

$$h_0(0;x,y) = \delta^{(d)}(x-y) \quad . \tag{4.90b}$$

Now,

$$D_\mu^\dagger D_\mu = -\partial_\mu\partial_\mu + i(\partial_\mu A_\mu) + iA_\mu\partial_\mu + iA_\mu^\dagger\partial_\mu + A_\mu^\dagger A_\mu$$

$$\partial_\mu h_0 = -\frac{\partial_\mu\sigma}{2t} h_0 \quad . \tag{4.91}$$

A simple calculation yields, using (4.90a)

$$D_\mu^\dagger D_\mu f h_0 = h_0(D_\mu^\dagger D_\mu f) - h_0\left(\frac{\partial_\mu\sigma}{2t}\right)(D_\mu^\dagger - D_\mu)f - f\partial_t h_0 \quad .$$

Substitution of the Ansatz (4.88a) into the above equation then leads to a differential equation for f itself:

$$Af - \frac{\partial_\mu\sigma}{2t}(D_\mu^\dagger - D_\mu)f + \partial_t f = 0 \quad . \tag{4.92}$$

Following de Witt[150,141] we make the Ansatz

$$f(t;x,y) = b_0(x,y) + b_1(x,y)t + b_2(x,y)t^2 + \cdots \quad . \tag{4.93}$$

Substituting this Ansatz into Eq. (4.92), and comparing powers in t, we obtain

$$t^{-1}: \quad \partial_\mu\sigma(D_\mu^\dagger - D_\mu)b_0 = 0 \tag{4.94a}$$

$$t^0: \quad Ab_0 - \frac{1}{2}\partial_\mu\sigma(D_\mu^\dagger - D_\mu)b_1 + b_1 = 0 \tag{4.94b}$$

$$t^1: \quad Ab_1 - \frac{1}{2}\partial_\mu\sigma(D_\mu^\dagger - D_\mu)b_2 + 2b_2 = 0 \tag{4.94c}$$

$$t^l: \quad Ab_l - \frac{1}{2}\partial_\mu\sigma(D_\mu^\dagger - D_\mu)b_{l+1} + (l+1)b_{l+1} = 0 \tag{4.94d}$$

Let us denote the coincidence limit $y \to x$ by

$$b_l(x,x) = [b_l](x) = a_l(x) \quad . \tag{4.95}$$

From the initial condition (4.87b) one has

$$a_0 \equiv [b_0] = 1 \quad . \tag{4.96a}$$

In general we have for $l > 0$,

$$a_l = -\frac{1}{l}[Ab_{l-1}] \quad , \tag{4.96b}$$

where the "bracket" denotes the coincidence limit $y \to x$.

Calculation of a_1

From (4.94*b*) we have

$$a_1 = [Ab_0] \quad . \tag{4.97}$$

To compute a_1 it is convenient to rewrite A in (4.85) as follows

$$A = \frac{1}{2}(D_\mu^\dagger D_\mu + D_\mu D_\mu^\dagger) + Z \quad , \tag{4.98}$$

with

$$Z = X + \frac{1}{2}[i(\partial_\mu A_\mu - \partial_\mu A_\mu^\dagger) + (A_\mu^\dagger A_\mu - A_\mu A_\mu^\dagger)] \quad . \tag{4.99}$$

Hence, using (4.98) we have

$$[Ab_0] = \frac{1}{2}[(D_\mu^\dagger D_\mu + D_\mu D_\mu^\dagger)b_0](x) + Z(x) \quad . \tag{4.100}$$

Now, by applying the operators $D_\mu D_\mu$ and $D_\mu^\dagger D_\mu^\dagger$ to the "parallel-transport" equation (4.94*a*) one finds, respectively

$$2(D_\mu D_\mu^\dagger - D_\mu{}^2)b_0 + \partial_\mu\sigma(D^2 D_\mu^\dagger - D^2 D_\mu)b_0 = 0$$

$$2(D_\mu^\dagger D_\mu - D_\mu^\dagger D_\mu^\dagger)b_0 + \partial_\mu\sigma(D^{\dagger 2} D_\mu^\dagger - D^{\dagger 2} D_\mu)b_0 = 0 \quad .$$

Taking the coincidence limit of these equations one finds

$$[(D_\mu D_\mu^\dagger - D_\mu D_\mu)b_0] = 0 \quad , \quad [(D_\mu^\dagger D_\mu - D_\mu^\dagger D_\mu^\dagger)b_0] = 0 \quad .$$

Hence, adding the two results:

$$[(D_\mu^\dagger D_\mu + D_\mu D_\mu^\dagger)b_0] - [(D_\mu D_\mu + D_\mu^\dagger D_\mu^\dagger)b_0] = 0 \quad . \tag{4.101}$$

Now,

$$(D_\mu D_\mu + D_\mu^\dagger D_\mu^\dagger) = -(D_\mu D_\mu^\dagger + D_\mu^\dagger D_\mu) = (A_\mu A_\mu^\dagger + A_\mu^\dagger A_\mu) - (A_\mu A_\mu + A_\mu^\dagger A_\mu^\dagger) \quad ,$$

or, substituting this expression into (4.101), and remembering (4.100), we finally obtain

$$a_1 = -[Ab_0] = -\frac{1}{4}[(A_\mu A_\mu^\dagger + A_\mu^\dagger A_\mu) - (A_\mu A_\mu + A_\mu^\dagger A_\mu^\dagger)] - Z \quad . \tag{4.102}$$

The calculation of the higher Seeley coefficients by the de Witt method becomes increasingly involved. We do not dwell on calculating them.

Example: The Dirac operator of QED_2/QCD_2

As we have seen, the Seeley coefficients are defined for compact, semi-positive operators. Hence the above technique is not directly applicable to the Dirac operator (4.4). We can however circumvent this problem by making the following observation: From (4.31) we have

$$\zeta(s;x,x \mid (i\,\hat{\not{D}})^2) = \sum_n{}' \frac{\hat{u}_n^\dagger(x)\hat{u}_n(x)}{(\lambda_n^2)^s} \quad . \tag{4.103a}$$

Here λ_n and $\hat{u}_n$ are the eigenvalues and eigenfunctions of the eigenvalue problem (4.71), and the sum extends over the positive and negative eigenvalues λ_n. Using (4.70) and (4.56b) one has

$$\sqrt{g}\zeta(s;x,x \mid (i\,\hat{\not{D}})^2) = \Omega_R(x)\zeta(s;x,x \mid (i\,\not{D})^2) \quad , \tag{4.103b}$$

with

$$\mathrm{tr}\,\zeta(s;x,x \mid (i\,\not{D})^2) = \sum{}' \frac{u_n^\dagger(x)u_n(x)}{(\lambda_n^2)^s} \quad . \tag{4.103c}$$

Furthermore, defining

$$\zeta(s;x,x \mid (i\,\not{D}) = \sum_{\lambda_n>0} \frac{u_n(x)u_n^\dagger(x)}{(\lambda_n)^s} + \sum_{\lambda_n<0} \frac{u_n(x)u_n^\dagger(x)}{(-\lambda_n)^s} \quad , \tag{4.104}$$

and using the fact, that for every eigenvalue $\lambda > 0$ of $i\,\not{D}$ with eigenfunction u_n there exists an eigenvalue $-\lambda < 0$ with eigenfunction $\gamma_5 u_n$, one has

$$\mathrm{tr}\,\zeta(0;x,x \mid i\,\not{D}) = tr\zeta(0;x,x \mid (i\,\not{D})^2) \tag{4.105a}$$

$$tr\zeta'(0;x,x \mid i\,\not{D}) = \frac{1}{2} tr\zeta'(0;x,x \mid (i\,\not{D})^2) \quad . \tag{4.105b}$$

It thus suffices to calculate the Seeley coefficients a_k of $D = (i\,\not{D})^2$, with

$$i\,\not{D} = i\,\not{\partial} + e\,\not{A} \quad , \tag{4.106a}$$

and A^μ a hermitian Lie algebra valued field. They are related to the Seeley coefficients $\hat{a}_k$ of $(i\,\hat{\not{D}})^2$ by $\sqrt{g}\hat{a}_k = \Omega_R a_k$. We have

$$(i\,\not{D})^2 = D_\mu^\dagger D^\mu + X \quad , \tag{4.106b}$$

with

$$X = \frac{ie}{4}[\gamma_\mu, \gamma_\nu]F_{\mu\nu} \quad , \tag{4.106c}$$

and

$$F_{\mu\nu} = \partial_\mu A_\nu - \partial_\nu A_\mu - i[A_\mu, A_\nu] \quad . \tag{4.106d}$$

Hence it follows from (4.105),

$$a_1 = -\frac{ie}{4}[\gamma_\mu, \gamma_\nu]F_{\mu\nu} \quad , \tag{4.106e}$$

or in two dimensions[10)]

$$a_1 = \frac{e}{2}\gamma_5 \epsilon_{\mu\nu} F_{\mu\nu} \quad . \tag{4.106f}$$

4.5 Methods for Computing Determinants

Although $\zeta(0 \mid D)$ may generally be calculated exactly in terms of the Seeley coefficients, a similar computation of

$$\omega_D \equiv \ln \det{}' D \quad , \tag{4.107}$$

will only be possible for a restricted class of operators. All methods are based on the behaviour of det D under an infinitesimal change $D \to D + \delta D$. They however correspond to different regularization procedures.

4.5.1 ζ-function regularization[140,148,149,151,152]

Here one departs from the definition

$$\omega_D = -\zeta'(0 \mid D) \quad , \tag{4.108}$$

and considers

$$\begin{aligned} \delta\omega_D &= \underset{s\to 0}{\text{a.c.}} \left[-\frac{d}{ds}\left(\zeta(s \mid D + \delta D) - \zeta(s \mid D)\right)\right] \\ &= \underset{s\to 0}{\text{a.c.}} \frac{d}{ds}\left[s\,\text{tr}'(D^{-s-1}\delta D)\right] \quad , \end{aligned} \tag{4.109}$$

where a.c. stands for "analytic continuation" and tr$'$ denotes the trace over the subspace orthogonal to the zero-modes of D. In the case where $\text{tr}'(D^{-s-1}\delta D)$ admits an analytic extension to $s = 0$, one evidently has

$$\delta\omega_D = [\text{tr}'(D^{-s-1}\delta D)]_{s=0} \quad . \tag{4.110}$$

10) In $d = 2$ our conventions are $i\gamma_\mu\gamma_5 = \epsilon_{\mu\nu}\gamma_\nu$ and $\frac{1}{2}[\gamma_\mu, \gamma_\nu] = i\epsilon_{\mu\nu}\gamma_5$.

Let us compare this result with

$$\ln \frac{\det(D+\delta D)}{\det D} \approx \mathrm{tr}(D^{-1}\delta D) \quad , \tag{4.111}$$

as obtained by using the familiar property

$$\ln \det D = \mathrm{tr} \ln D \quad , \tag{4.112}$$

of determinants. We see that (4.110) defines a particular regularization of the in general divergent trace (4.111).

More explicitly: if δD is of the form

$$\delta D = \sum_n D^{1-\gamma_n} \hat{\rho}_n D^{\gamma_n} \tag{4.113a}$$

with γ_n natural numbers and $\{\hat{\rho}_n\}$ local operators with the property

$$\langle x \mid \hat{\rho}_n \mid y \rangle = \rho_n(x)\delta(x-y) \quad , \tag{4.113b}$$

then, using the cyclicity of the trace, we find [11)]

$$\begin{aligned} \delta\omega_D &= \{\mathrm{tr}[D^{-s}D^{-1}\delta D]\} = \Big\{\sum_n \int d\mu(x) tr[\langle x \mid D^{-s} \mid x\rangle \rho_n(x)]\Big\}_{s=0} \\ &= \sum_n \int d\mu(x) tr[\zeta(0;x,x \mid D)\rho(x)] \quad , \end{aligned} \tag{4.114}$$

where

$$\rho = \sum_n \rho_n \quad . \tag{4.115}$$

Hence for δD of the form (4.113), $\delta\omega_D$ is computable in terms of the Seeley coefficients $a_k(x \mid D)$.

Although the above considerations seem to depend on the invertibility of D, the result in the form (4.114) is also valid in presence of zero modes, provided that δD has the property

$$\int d\mu(x)(u_n^\dagger \delta D u_n) = \lambda_n \int d\mu(x)(u_n^\dagger(x)\rho(x)u_n(x)) \quad . \tag{4.116}$$

In some cases the infinitesimal result (4.114) can be integrated to give ω_D itself. We demonstrate this for the case of QED_2.

---

11) We assume here that $\Re^d$ is compactifiable; x are then the cartesian coordinates in $\Re^d$.

Example

Consider the case of QED_2, with $i\not{D}$ given by the Dirac-operator (4.106). Consider the infinitesimal (Euclidean) chiral $U(1)$ transformation

$$i\not{D} \to e^{\gamma_5\delta\theta(x)} i\not{D} e^{\gamma_5\delta\theta(x)} \qquad (4.117)$$

which implies

$$\delta(i\not{D}) = -i\gamma_5(\not{\partial}\delta\theta) = i\not{D}\gamma_5\delta\theta + \gamma_5\delta\theta i\not{D} \quad . \qquad (4.118)$$

Defining $D = (i\not{D})^2$, we have correspondingly

$$\delta D = \gamma_5\delta\theta D + 2\sqrt{D}\gamma_5\theta\sqrt{D} + D\gamma_5\theta \quad , \qquad (4.119)$$

where $\sqrt{D} \equiv i\not{D}$.

The result (4.119) falls into the category of variations (4.113). Using the cyclicity of the trace we have from (4.114), using (4.38),

$$\delta\omega_{(i\not{D})} = \frac{1}{2}\delta\omega_{(i\not{D})^2} = \frac{1}{2}\int d^2x 4tr\left[\left(\frac{a_1(x)}{4\pi} - P^{(0)}(x,x)\right)\gamma_5\delta\theta(x)\right] \quad . \qquad (4.120)$$

In the absence of zero modes we thus have, using (4.106c),

$$\delta\omega_{(i\not{D})} = \frac{e}{2\pi}\int d^2x(\delta\theta(x)\epsilon_{\mu\nu}F_{\mu\nu}(x)) \quad . \qquad (4.121)$$

This equation is easily integrated to yield $\omega_{(i\not{D})}$ itself. Identifying $\delta\omega_{(i\not{D})}$ with

$$\delta\omega_{(i\not{D})} = \ln\frac{\det[i\not{\partial} + \gamma^\mu(eA_\mu - \tilde{\partial}_\mu\delta\theta)]}{\det(i\not{\partial} + e\gamma^\mu A_\mu)} \quad , \qquad (4.122)$$

we integrate this expression between $\theta = 0$ and θ defined by

$$\theta = \frac{e}{\nabla}\tilde{\partial}_\nu A_\nu = -\frac{e}{2}\nabla^{-1}\epsilon_{\lambda\rho}\partial_\lambda A_\rho = -\frac{e}{4}\nabla^{-1}\epsilon_{\lambda\rho}F_{\lambda\rho} \quad . \qquad (4.123)$$

We thus obtain,

$$\begin{aligned}\omega_{\frac{\not{\partial}}{\not{D}}} &= \ln\frac{\det i\not{\partial}}{\det(i\not{\partial} + e\not{A})} \\ &= \frac{e^2}{8\pi}\int d^2x\int d^2y\epsilon_{\lambda\rho}F_{\lambda\rho}(x)D_E(x-y)\epsilon_{\mu\nu}F_{\mu\nu}(y) \quad ,\end{aligned} \qquad (4.124)$$

where $\Delta D_E(x) = -\delta^2(x)$, and we have used Eqs. (B.7,24a,29a). Using further Eq. (B.5), the result (4.114) takes the more familiar gauge-invariant form ($\omega_{\frac{\not{\partial}}{\not{D}}} = -\omega_{\frac{\not{D}}{\not{\partial}}}$)

$$\omega_{\frac{\not{D}}{\not{\partial}}} = -\frac{e^2}{2\pi}\int d^2x\int d^2y A_\mu\left(\delta_{\mu\nu} - \frac{\partial_\mu\partial_\nu}{\Delta}\right)A_\nu \quad . \qquad (4.125)$$

In the Lorentz gauge, (4.125) reduces to the familiar Schwinger result [153].

4.5.2 Proper-time regularization

Consider representation (4.29) for the ζ-function. We have

$$\zeta'(s;x,y \mid D) = -\frac{\Gamma'(s)}{(\Gamma(s))^2}\int_0^\infty \frac{dt}{t}t^s[h(t;x,y \mid D) - P_0(x,y)] + \frac{1}{\Gamma(s)}\int_0^\infty \frac{dt}{t}t^s \ln t[h(t;x,y \mid D) - P_0(x,y)] \quad . \tag{4.126}$$

For $s > 0$ the integral converges, and we may replace $\int_0^\infty$ by $\int_\varepsilon^\infty$. Keeping $\varepsilon > 0$, we may now go over to the limit $s \to 0$, obtaining the alternative (proper-time) representation[154]

$$\omega_D = [\ln \det D]_\tau = -\int_\varepsilon^\infty \frac{dt}{t}\operatorname{tr}(e^{tD} - P_0) \quad . \tag{4.127}$$

Note that the projector P_0 insures that the integral converges at $t = \infty$. We now derive the analog of Eq. (4.110) in the proper-time regularization scheme.

Consider again the case where zero modes are absent. Then

$$\omega_{D+\delta D} - \omega_D = -\int_\varepsilon^\infty \frac{dt}{t}\operatorname{tr}(e^{-t(D+\delta D)} - e^{-tD}) \quad . \tag{4.128}$$

Now

$$e^{-t(D+\delta D)} - e^{-tD} = -\int_0^1 d\lambda e^{-\lambda t D} t\delta D e^{(\lambda-1)tD} \tag{4.129}$$

or

$$\operatorname{tr}(e^{-t(D+\delta D)} - e^{-tD}) = -t \operatorname{tr} e^{-tD}\delta D \quad . \tag{4.130}$$

Hence

$$\delta\omega = \omega_{D+\delta D} - \omega_D = \int_\varepsilon^\infty dt \operatorname{tr} e^{-tD}\delta D = \operatorname{tr}(e^{-\varepsilon D}D^{-1}\delta D) \quad . \tag{4.131}$$

Comparing this result with (4.110) we see that (4.131) and (4.110) represent the "proper-time" and ζ-function regularization of the formal expression (4.111), respectively.

If $D^{-1}\delta D$ is a local operator, we need to know only the small ε expansion of the *diagonal* elements of $e^{-\varepsilon D}$, as given by the Seeley expansion (4.18), i.e.

$$\langle x \mid e^{-\varepsilon D} \mid x\rangle = \frac{1}{(4\pi\varepsilon)^{\frac{d}{m}}}[a_0 + \varepsilon a_1 + \cdots] \quad . \tag{4.132}$$

Starting from the proper-time representation (4.127) for the determinant of an operator, we now prove a theorem about a class of operators, which will play an important role in our discussion of two-dimensional gauge-theories.

Theorem:[155] Let K_r be an invertible operator parametrized by r having the "polar decomposition"

$$K_r = \Omega_r M \overline{\Omega}_r \quad , \tag{4.133}$$

where Ω_r and $\overline{\Omega}_r$ are local operators in d dimensions,

$$\langle x \mid \Omega_r \mid y \rangle = \Omega_r(x)\delta^d(x-y) \tag{4.134}$$

$$\langle x \mid \overline{\Omega}_r \mid y \rangle = \overline{\Omega}_r(x)\delta^d(x-y) \quad , \tag{4.135}$$

and M does not depend on r. Define

$$\omega(r) := \ln \det K_r^\dagger K_r \quad . \tag{4.136}$$

Then

$$\begin{aligned} \frac{d}{dr}\omega(r) &= \frac{1}{4\pi\varepsilon)^{\frac{d}{m}}} \int d^2x tr(W_r(x) + \overline{W}_r(x)) \\ &+ \frac{\varepsilon}{(4\pi\varepsilon)^{\frac{d}{m}}} \int d^2x tr\{W_r(x)a_r^{(1)}(x) + \overline{W}_r(x)b_r^{(1)}(x)\} + \mathcal{O}(\varepsilon^{2-\frac{d}{m}}) \quad , \end{aligned} \tag{4.137}$$

where $W_r(x)$ and $\overline{W}_r(x)$ are the hermitian matrix valued functions (the "dot" denotes differentiation with respect to r)

$$W_r(x) = (\dot{\Omega}_r(x)\Omega_r^{-1}(x)) + (\dot{\Omega}_r(x)\Omega_r^{-1}(x))^\dagger \tag{4.138a}$$

$$\overline{W}_r(x) = (\overline{\Omega}_r^{-1}(x)\dot{\overline{\Omega}}_r(x)) + (\overline{\Omega}_r^{-1}(x)\dot{\overline{\Omega}}_r(x))^\dagger \quad , \tag{4.138b}$$

and where $a_r^{(1)}$ and $b_r^{(1)}$ are the coincidence limits of the first-non-trivial Seeley coefficients in the small ε expansion of the heat kernel associated with $K_r K_r^\dagger$ and $K_r^\dagger K_r$ respectively:

$$\langle x \mid e^{-\varepsilon K_r K_r^\dagger} \mid x \rangle = \frac{1}{(4\pi\varepsilon)^{\frac{d}{m}}}[1 + \varepsilon a_r^{(1)}(x) + \mathcal{O}(\varepsilon^2)] \tag{4.139a}$$

$$\langle x \mid e^{-\varepsilon K_r^\dagger K_r} \mid x \rangle = \frac{1}{(4\pi\varepsilon)^{\frac{d}{m}}}[1 + \varepsilon b_r^{(1)}(x) + \mathcal{O}(\varepsilon^2)] \quad . \tag{4.139b}$$

Proof: To begin with notice that the theorem makes a statement about a very special variation $\delta_r \omega_D$ of a particular class of positive-definite operators parametrized by r. We have (for the time being we exclude the possibility of zero modes)

$$\omega(r) = -\int_\varepsilon^\infty \frac{dt}{t} \operatorname{tr} e^{-tK_r^\dagger K_r} \quad . \tag{4.140}$$

Differentiating with respect to r we have,

$$\dot{\omega}(r) = \int_{\epsilon}^{\infty} dt\, \mathrm{tr}\{(\dot{K}_r^\dagger K_r + K_r^\dagger \dot{K}_r)e^{-tK_r^\dagger K_r}\} \quad . \tag{4.141}$$

Now,

$$\dot{K}_r^\dagger K_r = K_r^\dagger(\dot{\Omega}_r \Omega_r^{-1})^\dagger K_r + (\overline{\Omega}_r^{-1}\dot{\overline{\Omega}}_r)^\dagger K_r^\dagger K_r \tag{4.142a}$$

$$K_r^\dagger \dot{K}_r = K_r^\dagger(\dot{\Omega}_r \Omega_r^{-1})K_r + K_r^\dagger K_r(\overline{\Omega}_r^{-1}\dot{\overline{\Omega}}_r) \quad . \tag{4.142b}$$

Introducing these expressions in (4.141), using the cyclicity of the trace and furthermore the property

$$e^{-tK_r^\dagger K_r}K_r^\dagger = K_r^\dagger e^{-tK_r K_r^\dagger} \tag{4.143}$$

we find

$$\mathrm{tr}\{(\dot{K}_r^\dagger K_r + K_r^\dagger \dot{K}_r)e^{-tK_r^\dagger K_r}\} = \mathrm{tr}\{W_r K_r K_r^\dagger e^{-tK_r K_r^\dagger} + \overline{W}_r e^{-tK_r^\dagger K_r}\}$$

$$= -\frac{d}{dt}\,\mathrm{tr}\{W_r e^{-tK_r K_r^\dagger} + \overline{W}_r e^{-tK_r^\dagger K_r}\} \quad , \tag{4.144}$$

where W_r and $\overline{W}_r$ have been defined in (4.138). Upon substituting (4.144) into (4.141), the t-integration may now be trivially performed, to yield

$$\dot{\omega}(r) = \mathrm{tr}[W_r e^{-\epsilon K_r K_r^\dagger} + \overline{W}_r e^{-\epsilon K_r^\dagger K_r}] \quad . \tag{4.145}$$

Making use of (4.139), we obtain the announced result (4.137)

Note that

$$\delta_r K_r = [(\dot{\Omega}_r \Omega_r^{-1})K_r + K_r(\overline{\Omega}_r^{-1}\dot{\overline{\Omega}}_r)]dr$$
$$\delta_r K_r^\dagger = K_r^\dagger[(\dot{\Omega}_r \Omega_r^{-1}) + (\overline{\Omega}_r^{-1}\dot{\overline{\Omega}}_r)K_r^\dagger]dr \quad ,$$

so that

$$\delta_r(K_r^\dagger K_r) = \delta_r K_r^\dagger K_r + K_r^\dagger \delta_r K_r$$

$$= \{2K_r^\dagger(\dot{\Omega}_r \Omega_r^{-1})K_r + (\overline{\Omega}_r^{-1}\dot{\overline{\Omega}}_r)K_r^\dagger K_r + K_r^\dagger K_r(\overline{\Omega}_r^{-1}\dot{\overline{\Omega}}_r)\}dr \quad .$$

Hence this variation is in general not of the form (4.113), unless $K_r^\dagger \neq K_r$.

4.5.3 Fujikawa method

The functional $\delta\omega_D$ can also be viewed as the logarithm of the Jacobian of a transformation $\mathcal{D}c\mathcal{D}\bar{c} \to \mathcal{D}c'\mathcal{D}\bar{c}'$, in (4.6), as we now show. This is the point of view taken by Fujikawa[156] who was the first one to develop a method for studying chiral anomalies in the path-integral framework without resorting to perturbation theory. Fujikawa's observation of the non-invariance of the fermionic measure $\mathcal{D}\psi\mathcal{D}\bar{\psi}$ in QED under chiral transformations provided an important break-through in the conceptual understanding of chiral anomalies from the path-integral point of view. In particular, it provides a systematic method for a transparent derivation of "anomalous" chiral Ward-Takahashi identities within a purely path integral framework, without the need of explicitly adding anomalous terms-calculated from Feynman graphs - to the action. In fact, it is understood by now that all known anomalies can be viewed as being a result of the non-invariance of the functional measure under some "symmetry" transformation[157,158]. Nevertheless, as we now demonstrate, Fujikawa's approach is just another variant of what was said above.

In the Fujikawa approach the definition of $\frac{\det(D+\delta D)}{\det D}$ in terms of the path-integral (4.6) represents the starting point. He considers the situation where the transition from D to $D+\delta D$ can be seen as a change of variable of integration. Thus consider

$$\frac{\det[D+\delta D]}{\det[D]} = \frac{\int \mathcal{D}\bar{c}\mathcal{D}ce^{-S[D+\delta D]}}{\int \mathcal{D}\bar{c}\mathcal{D}ce^{-S[D]}} \quad , \tag{4.146a}$$

with

$$S[D] = \int d^dx\bar{c}Dc \quad . \tag{4.146b}$$

Now,

$$S[D+\delta D] = \int \{\bar{c}(1+\frac{1}{2}\delta DD^{-1})D(1+\frac{1}{2}D^{-1}\delta D)c\} + \mathcal{O}((\delta D)^2 \quad .$$

Hence, redefining the fields as follows,

$$c \to c' = (1+\frac{1}{2}D^{-1}\delta D)c \quad , \quad \bar{c} \to \bar{c}' = \bar{c}(1+\frac{1}{2}\delta DD^{-1}) \quad , \tag{4.147}$$

we have $S[D+\delta D] = \int d^dx\bar{c}'Dc' + \mathcal{O}((\delta D)^2)$. Hence the Jacobian J of the transformation (4.147),

$$\mathcal{D}\bar{c}\mathcal{D}c = J\mathcal{D}\bar{c}'\mathcal{D}c' \quad , \tag{4.148}$$

is just the *l.h.s.* of Eq. (4.146*a*), that is

$$J = \frac{\det(D+\delta D)}{\det D} \quad . \tag{4.149}$$

To evaluate the Jacobian factor we expand $c(\bar{c})$ in a complete set of orthonormal functions φ_n:

$$c(x) = \sum a_n \varphi_n(x) \quad , \quad \bar{c}(x) = \sum \bar{a}_n \overline{\varphi}_n(x)$$

$$\int d\mu(x) \overline{\varphi}_n \varphi_m = \delta_{nm} \quad , \tag{4.150}$$

where $a_n(\bar{a}_n)$ are elements of an infinite Grassmann algebra. We have a corresponding expansion for c' and $\bar{c}'$ with coefficients a' and $\bar{a}'$. The connection between the two sets of coefficients is given in terms of a unitary transformation,

$$a' = Ua \quad , \quad \bar{a}' = \bar{a}'\overline{U} \quad , \tag{4.151a}$$

where (we assume again $D^{-1}\delta D$ to be local operators),

$$\begin{aligned} U_{mn} &= \delta_{mn} + \int d\mu(x) \overline{\varphi}_m(x) \frac{1}{2} D^{-1} \delta D \varphi_n(x) \quad , \\ \overline{U}_{mn} &= \delta_{mn} + \int d\mu(x) \overline{\varphi}_m(x) \frac{1}{2} \delta D D^{-1} \varphi_n(x) \quad . \end{aligned} \tag{4.151b}$$

Now, using the property (4.112) of determinants, we have

$$\prod_n d\bar{a}'_n da'_n = \prod_n d\bar{a}_n da_n e^{-tr\, \ln(\overline{U}U)} \quad . \tag{4.152a}$$

Note that due to the Grassmann character of the a_n's, the Jacobi determinant is just the inverse of what one has in the usual (bosonic) case[159]. With the usual replacement

$$\mathcal{D}\bar{c}\mathcal{D}c \to \prod_n d\bar{a}_n \prod_n da_n \quad ,$$

one therefore has

$$\mathcal{D}\bar{c}\mathcal{D}c = e^{\frac{1}{2}\sum_n \int d\mu(x) \overline{\varphi}_n(x)(\delta D D^{-1} + D^{-1}\delta D)\varphi_n(x)} \mathcal{D}\bar{c}'\mathcal{D}c' \quad . \tag{4.152b}$$

Since $\sum_n \varphi_n(x)\overline{\varphi}_n(x) = \delta(0)$, the above expression is divergent and requires in general regularization. Formally we have however

$$\mathcal{D}\bar{c}\mathcal{D}c = e^{\mathrm{tr}\, D^{-1}\delta D} \mathcal{D}\bar{c}'\mathcal{D}c' \quad , \tag{4.152c}$$

or, with the definition of J given by (4.148),

$$\ln J = \mathrm{tr}\, D^{-1}\delta D = \delta\omega_D \quad , \tag{4.153}$$

in agreement with (4.149) and (4.111). Note that the change in sign in $\ln J$ relative to the bosonic case was essential for this agreement.

Although the calculation of (4.153) cannot depend on the choice of basis functions, it will depend on the choice of regulator. The choice of regulator depends on the symmetry one wishes to preserve. It should however be chosen such as to make the trace in (4.153) well defined. Fujikawa considered[156] the case of D being the *square* of the Dirac operator $i\not{D}$, and δD as corresponding to the chiral transformation (4.119)[12)]. For this case ($\delta\hat{\theta}$ is to be regarded as an operator)

$$\delta\omega_{(i\not{D})^2} = \lim_{\varepsilon\to 0} 4\,\mathrm{tr}[e^{-\varepsilon D}\gamma_5\delta\hat{\theta}] \quad , \tag{4.154}$$

where $\exp(-\varepsilon D)$ is a "damping" factor introduced in order to regulate the trace. Fujikawa chose a manifestly gauge invariant regularization with $D = (i\not{D})^2$. His evaluation of the trace (4.153) did not make use of Seeleys expansion, but followed the procedure to be outlined next.

Write (4.154) in the form

$$\delta\omega_{i\not{D}} = \int d^dx \mathcal{A}_{reg}(x)\delta\theta(x) \tag{4.155a}$$

with

$$\mathcal{A}_{reg}(x) = 2\lim_{\varepsilon\to 0}\sum_n \varphi_n^\dagger(x)\gamma_5 e^{-\varepsilon(i\not{D})^2}\varphi_n(x) \quad . \tag{4.155b}$$

Note that $\mathcal{A}_{reg}$ corresponds to a regularization of $\sum_n \varphi^\dagger(x)\gamma_5\varphi(x)$ which Fujikawa has referred to as "primary" definition of "anomaly"[156].

Although the choice of φ_n as eigenfunctions of $i\not{D}$ seems the most natural one, the result should not depend on the choice of basis. Fujikawa chose the plane waves as basis:

$$\mathcal{A}_{reg}(x) = 2\lim_{\varepsilon\to 0_+}\mathrm{tr}\Big\{\gamma_5\int\frac{d^dp}{(2\pi)^2}e^{-ipx}e^{-\varepsilon(i\not{D})^2}e^{ipx}\Big\} \quad . \tag{4.155c}$$

Now, using (4.106c), we have

$$(i\not{D})^2 e^{ipx} = (-D_\mu D_\mu + X)e^{ipx}$$
$$[-D_\mu D^\mu + X]e^{ipx} = e^{ipx}[(p_\mu - iD_\mu)^2 + X] \quad ,$$

12) Actually Fujikawa considered the "active" point of view,of calculating directly the Jacobian J of the transformation (4.117).

where the factor in the r.h.s. is understood to operate on the constant function. Hence

$$\mathcal{A}_{reg}(x) = 2 \lim_{\varepsilon \to 0_+} \mathrm{tr}\Big\{\gamma_5 \int \frac{d^d p}{(2\pi)^2} e^{-\varepsilon[p^2 - 2ip.D - D^2 + X]}\Big\} \quad . \tag{4.155d}$$

Making the change of variable $\sqrt{\varepsilon}p = k$ and expanding the resulting integrand in powers of ε one sets

$$\mathcal{A}_{reg}(x) = 2 \lim_{\varepsilon \to 0_+} \frac{1}{\varepsilon^{\frac{d}{2}}}\Big\{\gamma_5 \int \frac{d^2 k}{(2\pi)^2} e^{-k^2}[1 + 2i\sqrt{\varepsilon}kD + \varepsilon\mathcal{D}^2 - \varepsilon X + \mathcal{O}(\varepsilon^{\frac{3}{2}})]\Big\} \quad . \tag{4.155e}$$

For $d = 2$ we need not compute higher order terms in ε. Using $\int \frac{d^2k}{(2\pi)^2} e^{-k^2} = \frac{1}{4\pi}$, and $\mathrm{tr}\gamma_5 = 0$, we have in $d = 2$,

$$\mathcal{A}_{reg} = -\frac{1}{2\pi}\mathrm{tr}\{\gamma_5 X\} = \frac{e}{2\pi}\epsilon_{\mu\nu}F_{\mu\nu} \quad , \tag{4.156}$$

which, when substituted into (4.155a), yields (4.121).

The fact that no further subtraction is needed is a consequence of working in $d = 2$ dimensions.

The result (4.156) agrees with our earlier calculation, and demonstrates explicitly the independence of the result on the choice of basis. The same result is obtained by working with the eigenfunctions of $i\,\not{\partial}$ [154]. On the other hand, working with the regulator

$$1 = \lim_{\varepsilon \to 0} e^{-\varepsilon(i\not{\partial})^2} \quad , \tag{4.157}$$

one finds $\mathcal{A}_{reg} = 0$, i.e. a trivial Jacobian[160]. (This is evident from the above calculation leading to (4.156)). $\mathcal{A}_{reg} = 0$ corresponds, as we shall see, to the absence of an anomaly in the divergence of the axial vector current, although at the expense of gauge invariance. Indeed it is well known, that there exists no regularization respecting both gauge and chiral invariance.

Basically one is confronted with defining what has been referred to by Fujikawa[156] as the "primary" anomaly, i.e. $\mathcal{A} = 2\sum_n \varphi_n^\dagger(x)\gamma_5\varphi_n(x)$. Definition (4.155$b$) amounts to computing

$$\mathcal{A}_{reg} = 2 \lim_{\varepsilon \to 0} \sum_n \varphi_n^\dagger(x)\gamma_5 e^{-\varepsilon\lambda_n^2}\varphi_n(x) \quad ,$$

where λ_n are the eigenvalues of $i\,\not{D}$. As was shown by Fujikawa[156], the replacement

$$e^{-\varepsilon\lambda_n^2} \to f(\varepsilon\lambda_n^2) \quad ,$$

where $f(z)$ is any smooth function which rapidly converges to zero as $z \to \infty$, leads to the same result for $\mathcal{A}_{reg}$, thus indicating that the sum representing the primary anomaly is conditionally convergent. Formally this sum states that $\mathcal{A}_{reg} = 2\,\mathrm{tr}\,\gamma_5$. The fact that it does not vanish - contrary to what one would naively expect - has its origin[161] in the spectral asymmetry of the operator $\not{D}$, which in fact conforms to the Atiyah-Singer theorem[162] (see section 7). This is exemplified by the choice (4.157) for the regulator, which leads to a vanishing anomaly.

It is interesting that conventional perturbation theory is in agreement with this chiral asymmetry, if a gauge invariant, Pauli-Villars regularization is employed. In the path-integral framework, the anomalous Ward identity for the divergence of the axial vector current now emerges as follows[163]: The Pauli-Villars regularization of the fermion-loop diagrams is equivalent to the addition of $\mathcal{L}_{PV} = \bar{\phi} i \not{D}\phi - M\bar{\phi}\phi$ to the Lagrangian, where ϕ is a spinor field obeying *Bose* statistics. Thus there exists an additional contribution to the functional measure transforming under a chiral transformation with just the inverse of the Jacobian (4.149),

$$\mathcal{D}\bar{\phi}\mathcal{D}\phi = J^{-1}\mathcal{D}\bar{\phi}'\mathcal{D}\phi' \quad ,$$

showing the invariance of $\mathcal{D}\bar{\psi}\mathcal{D}\psi\mathcal{D}\bar{\phi}\mathcal{D}\phi$ for finite Pauli-Villars mass. The corresponding Ward identity now follows in the "normal " way[164] from the transformation property of the modified Lagrangian under the simultaneous local chiral transformations $\psi(x) \to e^{i\alpha(x)\gamma_5\psi(x)}$ and $\phi(x) \to e^{i\alpha(x)\gamma_5\phi(x)}$. The anomaly is recovered in the limit $M \to \infty$.

Finally let us remark, that we have supposed in our analysis that D was a positive semi-definite operator. This is true if $D = (i\not{D})^2$, but is no longer true, if $i\not{D}$ is replaced by the (in Euclidean space non hermitian) "Dirac" operator with axial coupling of the gauge field, $i\tilde{\not{D}} = i\not{\partial} + \not{A}\gamma^5$. Various methods have been proposed to circumvent this problem[164,158,165,166].

One obvious way consists in considering the operator $i\not{\partial} + i\not{A}\gamma^5$ obtained by analytical continuation from A_μ to iA_μ, and then to continue back at the end of the calculation. It is not surprising that the final result is found to agree with the result obtained by naive application of the ζ-function method to the non-hermitian operator $i\tilde{\not{D}}$.

Another approach was considered by Fujikawa[164]. It consists, in considering the eigenfunctions of the hermitian, positive semi-definite operators $\tilde{\not{D}}^\dagger\tilde{\not{D}}$ and $\tilde{\not{D}}\tilde{\not{D}}^\dagger$:

$$\tilde{\not{D}}^\dagger\tilde{\not{D}}\varphi_n(x) = \lambda_n^2\varphi_n(x) \quad ; \quad \tilde{\not{D}}\tilde{\not{D}}^\dagger\phi_n(x) = \lambda_n^2\phi_n(x) \quad , \quad \lambda_n \in \Re \quad .$$

For a suitable choice of phase this corresponds to

$$\tilde{\not{D}}\varphi_n = \lambda_n\phi_n \quad ; \quad \tilde{\not{D}}^\dagger\phi_n = \lambda_n\varphi_n \quad .$$

The Fujikawa analysis is then repeated by expanding $\psi(\overline{\psi})$ in terms of the complete set of eigenfunctions $\varphi_n(\phi_n)$. Since a gauge transformation induces a similarity transformation on $\tilde{\not{D}}^\dagger\tilde{\not{D}}$ and $\tilde{\not{D}}\tilde{\not{D}}^\dagger$, the eigenvalues λ_n are gauge invariant. Hence the outcome of the calculation will be gauge invariant and thus cannot coincide with the result obtained by analytic continuation in A_μ. Indeed, one is actually calculating $\det(\tilde{\not{D}}\tilde{\not{D}}^\dagger)^{1/2}$, in this way[164,166].

4.5.4 Determinant from current-anomaly

So far we have seen how the Jacobian J associated with an infinitesimal transformation of the fermionic measure, or equivalently, the ratio (4.149) may be calculated in terms of the Seeley coefficients. Consider now in particular the infinitesimal change induced in $i\not{D}$ by a gauge transformation parametrized by $\delta\alpha_a(x)$,

$$A_\mu^a(x) \to A_\mu^a(x) + D_\mu^{ab}\delta\alpha^b(x) \quad . \tag{4.158}$$

Define the effective action associated with the Dirac operator (4.106a) by,

$$-W[A] \equiv \ln\frac{\det i\not{D}[A]}{\det i\not{\partial}} \quad . \tag{4.159}$$

From (4.159) and the definition (4.149) of J follows directly,

$$\begin{aligned}\ln J &= -\int d^dx \frac{\delta W[A]}{\delta A_a^\mu(x)}\mathcal{D}^\mu_{ab}\delta\alpha^b(x) \\ &= e\int dx\delta\alpha^b(x)\mathcal{D}^\mu_{ab}J^b_\mu \quad ,\end{aligned} \tag{4.160a}$$

where,

$$eJ^a_\mu(x\mid A) \equiv \frac{\delta W[A]}{\delta A_a^\mu(x)} \quad , \tag{4.160b}$$

is the "external" field current and J is the Jacobian associated with the gauge transformation of the fermionic measure.

As seen from (4.160a), if $\ln J \neq 0$, then the covariant divergence of the current is anomalous. On the other hand we may calculate (under suitable conditions) $\ln J$ directly using e.g. the ζ-function method. Following the notation of (4.155a), we have from (4.153),

$$\ln J = \int d^dx \mathcal{A}_V^a \delta\alpha^a(x) \quad , \tag{4.161}$$

where $\mathcal{A}_V^a$ is the anomaly if (4.158) corresponds to a symmetry transformation of the classical Lagrangian. Comparing (4.161) with (4.160) we have,

$$D_\mu^{ab} J_\mu^b = \mathcal{A}_V^a \quad . \tag{4.162}$$

Note that the current in question is that defined by (4.160), i.e. the *vector* current in a "vector-gauge" theory. For a gauge invariant regularization, this current has no anomaly. Correspondingly, the axial vector current is however expected to have an anomaly. In the following we shall demonstrate this explicitly, using the ζ-function formalism. The procedure[146] parallels the use of the ζ-function in refs.[167].

Axial anomaly equation

The Dirac operator $i \not\partial + e \not A$ anti-commutes with γ_5. Hence there exists for every eigenvalue λ_n a corresponding eigenvalue of opposite sign. Hence, with λ_n the eigenvalues of the associated Dirac operator (4.70*b*), we have [9)]

$$\zeta(s \mid (i \hat{\not D})^2) = \sum_{\lambda_n \neq 0} \frac{1}{(\lambda_n^2)^s} \quad . \tag{4.163}$$

Introduce the short-hand notation

$$\zeta(s \mid (i \hat{\not D})^2)\Big|_0^A \equiv \zeta(s \mid (i \hat{\not D})^2) - \zeta(s \mid (i \hat{\not\partial})^2) \quad ;$$

thus, in the ζ-function regularization

$$\begin{aligned} W[A] &\equiv \zeta'(0 \mid i \hat{\not D})\Big|_0^A = \frac{1}{2}\zeta'(0 \mid (i \hat{\not D})^2)\Big|_0^A \\ &= \lim_{s \to 0}\left\{ -\frac{1}{2} \sum_{\lambda_n \neq 0} \frac{\ln \lambda_n^2[A]}{(\lambda_n^2[A])^s} \right\}\Big|_0^A \quad , \end{aligned} \tag{4.164}$$

where the summation extends over all (non-vanishing) eigenvalues. Taking the variational derivative with respect to A_μ^a, one has

$$\frac{\delta W[A]}{\delta A_\mu^a(x)} = -\lim_{s \to 0} \left\{ \sum_{\lambda_n \neq 0} \frac{\lambda_n \frac{\delta \lambda_n}{\delta A_\mu^a(x)}}{(\lambda_n^2)^{s+1}} + \mathcal{O}(s) \right\} \quad . \tag{4.165}$$

---

9) The spectrum of $i\hat{\not\partial}$ consists of all integers different from zero , each eigenvalue being 2n-fold degenerate. This degeneracy (c-conjugation) is partially split in the presence of a gauge field A_μ.

Now, from the eigenvalue equation (4.67) and the normalization (4.69a) follows that

$$\int d^2x u_n^\dagger i \not{D} u_n = \lambda_n \quad , \tag{4.166}$$

or

$$\frac{\delta\lambda_n}{\delta A_\mu^a(x)} = \lambda_n \frac{\delta}{\delta A_\mu^a(x)} \int d^2x \Omega_R u_n^\dagger u_n + u_n^\dagger(x) e\gamma_\mu t^a u_n(x) \quad . \tag{4.167}$$

The first term on the r.h.s. vanishes, since the u_n's are normalized. Hence

$$\frac{\delta W}{\delta A_\mu^a(x)} = -\underset{s\to 0}{\text{a.c.}} \sum_n{}' \frac{1}{(\lambda_n^2)^s} \frac{(u_n^\dagger(x)e\gamma_\mu t^a u_n(x))}{\lambda_n} \quad . \tag{4.168}$$

Note that the Dirac-Greens function satisfying

$$i \not{D} G(x, y \mid A) = \delta^{(2)}(x - y) \quad , \tag{4.169}$$

has the spectral representation

$$G(x, y \mid A)_{\alpha\beta} = \sum \frac{u_n(x)_\alpha u_n^*(y)_\beta}{\lambda_n} \quad , \tag{4.170}$$

in the *absence* of zero modes. Hence the $s \to 0$ limit in (4.168) can be regarded as a regularization of the familiar (divergent) expression

$$J_\mu^a(x \mid A) = -\text{tr}(t^a \gamma_\mu G(x, x \mid A)) \quad , \tag{4.171}$$

for the case that zero modes are *present*. Noting that

$$\mathcal{D}_\mu^{ab}(u_n^\dagger t^a \gamma_\mu u_n) = 0 \quad , \tag{4.172}$$

we obtain from (4.168) and (4.160) our first result:

$$\mathcal{D}_\mu^{ab} J_\mu^b(x \mid A) = 0 \quad . \tag{4.173}$$

As for the covariant divergence of the axial vector current, we expect to have, in analogy to (4.168)

$$J_\mu^{5,b}(x \mid A) = -\underset{s\to 0}{\text{a.c.}} \sum_n{}' \frac{[u_n^\dagger(x) i\gamma_\mu \gamma_5 t^a u_n(x)]}{\lambda_n^{2s+1}} \quad . \tag{4.174}$$

To see this, it is convenient to take

$$W[A, \tilde{A}] = -\ln\det[i \not{\partial} + e \not{A} + ie \not{\tilde{A}}\gamma_5] \quad , \tag{4.175a}$$

as a starting point, with $\tilde{u}_n$, $\tilde{\lambda}_n$ the eigenfunctions corresponding to this new (still hermitian!) Dirac operator, and to compute

$$\frac{1}{e}\frac{\delta W[A,\tilde{A}]}{\delta \tilde{A}^a_\mu(x)}\Big|_{\tilde{A}=0} = -\underset{s\to 0}{\text{a.c.}}\sum_n \frac{\tilde{\lambda}_n \frac{\delta \tilde{\lambda}_n}{\delta \tilde{A}^a_\mu(x)}}{(\lambda_n^2)^{s+1}}\Big|_{\tilde{A}=0} \quad . \tag{4.175b}$$

From $\int d^2x \Omega_R(x)\tilde{u}^\dagger_n(x)\tilde{u}_n(x) = 1$ and $\int d^2x \tilde{u}^\dagger_n(i\,\not\partial + e\,\not A + ie\tilde{\not A}\gamma_5)\tilde{u}_n = \tilde{\lambda}_n$, we find in analogy to (4.167),

$$\left(\frac{\delta\tilde{\lambda}_n}{\delta \tilde{A}^a_\mu(x)}\right)_{\tilde{A}=0} = e u^\dagger_n(x) i\gamma_\mu\gamma_5 t^a u_n(x) \quad . \tag{4.176}$$

One therefore obtains from here the expected result, Eq. (4.174).

Now,

$$\mathcal{D}^{ab}_\mu(\overline{u}_n i\gamma_\mu\gamma_5 t^b u_n) = -2\lambda_n\Omega_R u^\dagger_n\gamma_5 t^a u_n \quad ; \tag{4.177}$$

hence it follows from (4.174) and (4.103b) that

$$\mathcal{D}^{ab}_\mu J^{5,b}_\mu(x \mid A) = 2\Omega_R(x)\mathrm{tr}\,t^a\gamma_5\zeta(0;x,x \mid (i\,\not D)^2) \quad . \tag{4.178}$$

The *r.h.s.* of (4.178) represents the axial anomaly. It can be calculated using the Seeley expansion (4.18). Moreover, in two space-time dimensions the combined equations (4.173) and (4.178) may be integrated to give $W[A]$ itself, as we now illustrate for the case of QED_2

Example: Effective action of QED_2

For QED_2 (without zero modes) we have (see (4.106d))

$$\Omega_R(x)\zeta(0;x,x \mid (i\,\not D)^2) = \frac{a_1(x)}{(4\pi)} = \frac{e}{8\pi}\gamma_5\epsilon_{\mu\nu}F_{\mu\nu} \quad . \tag{4.179}$$

Recalling further $i\gamma_\mu\gamma_5 = \epsilon_{\mu\nu}\gamma_\nu$, we have

$$J^5_\mu = \epsilon_{\mu\nu}J_\nu \quad , \tag{4.180}$$

or the equations (4.173) and (4.178) now read

$$\begin{aligned} \partial_\mu J_\mu &= 0 \\ \tilde{\partial}_\mu J_\mu &= -\frac{e}{2\pi}\epsilon_{\mu\nu}F^{\mu\nu} \quad . \end{aligned} \tag{4.181}$$

This anomaly equation has the solution

$$J_\mu(x \mid eA) = \frac{e}{2\pi}\int d^2y \tilde{\partial}^x_\mu D_E(x-y)\epsilon_{\mu\nu}F_{\mu\nu} + \partial_\mu \mathcal{F} \quad , \tag{4.182}$$

where $\mathcal{F}$ is an up to now arbitrary function, which is however restricted by Eq. (4.181) to satisfy $\triangle \mathcal{F} = 0$. If J_μ is to be normalizable, then $\mathcal{F}$ can only be a constant. Thus J_μ is determined, and it remains only to integrate equation

$$\frac{1}{e}\frac{\delta W[A]}{\delta A_\mu(x)} = J_\mu(x \mid A) \quad , \tag{4.183}$$

with $J_\mu(x \mid A)$ given by (4.182). In the present example this is easily done, and the result is just (4.124).

This illustrates how in $d = 2$ the effective action itself may be obtained by the *joint* integration of the equations for the divergence of the vector and axial vector current. In chapter 13 we shall illustrate this also for the non-abelian case. Note that this is a non-trivial property of two dimensions. Although the anomaly is also computable in four dimensions, the same is not true for the effective action.

In the preceding example, $W[A]$ was obtained by integrating Eq. (4.183). This was easily done, since $J_\mu(x \mid A)$ was found to be a linear functional of A_μ. In the general case this integration is formally also possible, as we next show.

4.5.5 Pauli-representation of determinant

The functionals $W[A]$ and $J_\mu(x \mid A)$ in (4.160) depend on A_μ and the charge e only via the product eA_μ. Hence

$$\frac{\partial W[A]}{\partial e} = \int d^dx \frac{\delta W[A]}{\delta(eA^a_\mu(x))}A^a_\mu(x) = \int d^2x J^a_\mu(x \mid eA)A^a_\mu(x) \quad , \tag{4.184}$$

where we have exhibited explicitly the dependence on e. Integrating (4.184) with respect to e, we have

$$W[A] = \int_0^e de' \int d^dx J^a_\mu(x \mid e'A)A^a_\mu(x) \quad . \tag{4.185}$$

An alternative proof proceeds as follows:

$$W[A] = -\ln\left(\frac{\det(i \not\partial + e \not{A})}{\det i \not\partial}\right) = -\ln\det(1 + eX) \quad , \tag{4.186}$$

with $X = \frac{1}{i\partial\!\!\!/} A\!\!\!/$. Hence

$$
\begin{aligned}
\frac{\partial W[A]}{\partial e} &= -\frac{d}{de}\ln\det(1+eX) \\
&= -\lim_{\delta e\to 0}\frac{1}{\det(1+eX)}\left\{\frac{\det(1+eX+\delta eX)-\det(1+eX)}{\delta e}\right\} \\
&= -\lim_{\delta e\to 0}\frac{\det(1+\delta e\frac{1}{1+eX}X)-1}{\delta e} = -\operatorname{tr}\frac{1}{1+eX}X = -\operatorname{tr}\Big(\frac{1}{i\,\partial\!\!\!/ + e\,A\!\!\!/}\,A\!\!\!/\Big) \\
&= -\int d^2x \quad tr[\gamma^\mu t^a G(x,x\mid eA)]A^e_\mu(x) \quad ,
\end{aligned}
\tag{4.187}
$$

which is just (4.184).

The representation of the effective action in terms of the parametric integral (4.185) is known as the "Pauli-representation". Since $J^a_\mu(x \mid A)$ is generally calculable only in perturbation theory, it only provides a formal solution to the problem. It nevertheless will prove useful in our discussion of QCD_2 in chapter 11.

4.6 Ambiguities in the Calculation of the Determinant

4.6.1 Ambiguities in the regularization

We have considered so far two ways of regularizing a functional determinant: the "ζ-function" and "proper-time" regularization. The proper-time regularization requires in general the introduction of counterterms to cancel the singular terms in Seeley's expansion (4.132). It was a peculiarity of QED_2 and QCD_2 that these, a priori singular, contributions vanished identically for purely kinematical reasons. This however does not mean that there exists no ambiguity in the calculation of the QED_2/QCD_2 determinant. Thus, to begin with let us examine the relation between the determinants obtained via the above two regularization schemes. Splitting the integral in (4.126) into the intervals $(0,\varepsilon)$, (ε,∞) and taking the limit $s \to 0$ one finds

$$[\ln\det{}'D]_\zeta = [\ln\det{}'D]_\tau + \zeta(0\mid D)\ln\varepsilon \quad . \tag{4.188}$$

Note that the ambiguity is of the same form as that associated with a change in scale of the eigenvalues (see Eq. (4.24)). Now, if we consider ratios of determinants such as

$$\omega_{\frac{D}{D_0}} = \ln\frac{\det D}{\det D_0} \quad , \tag{4.189}$$

then such a scale should not matter, provided that the dimension of the space on which D and D_0 act is the same. This excludes the case where D, but not D_0, has zero-modes. Let us check this idea for the case that D is the Dirac operator

(4.106a). If the gauge-field A_μ has non-zero winding (Pontryagin) number n, then $i\not{D}$ will have zero modes. In two dimensions the number N of zero modes will be just n itself, (see "Atiyah-Singer Theorem" and "vanishing Theorem" in section 7 of this chapter). From (4.40) and (4.106f) we thus have for d=2,

$$\zeta(0 \mid i\hat{\not{D}}) - \zeta(0 \mid i\hat{\not{\partial}}) = -N \quad , \tag{4.190}$$

which indeed vanishes if there are no zero modes.

There exists another method of regularization, which we refer to as the "Pauli-Villars regularization" of a determinant[168,141]. It is given by

$$[\ln \det D]_{p.v.} = -\left[\frac{d}{ds}\zeta_{p.v.}(s \mid A)\right]_{s=0} \quad , \tag{4.191a}$$

where

$$\zeta_{p.v.}(s \mid D) = \sum{}' \left\{ \frac{1}{\lambda_n^s} + \sum_{k=1}^{\nu} c_k \frac{1}{(\lambda_n + M_k)^2} \right\} \quad .$$

Here $M_1, \cdots M_\nu$ are dimensionless Pauli-Villars regulator masses to be taken to ∞ after setting $s = 0$. The number ν is related to the polynomial bound and spectral density of the asymptotic spectral values. The "minimal" value of ν is that, for which $\zeta_{p.v.}(0 \mid D)$ and $\zeta'_{p.v.}(0 \mid D)$ exist *without* the need of analytic continuation. This places restrictions on M_k and ν which are most conveniently determined by relating the $s \to 0$ behaviour to the small t-behaviour in the inverse Mellin transform of (4.191a). Thus, writing

$$\zeta_{p.v.}(s \mid D) = \frac{1}{\Gamma(s)} \int_0^\infty dt \quad t^{s-1} \left(1 + \sum_{k=1}^{\nu} c_k e^{-M_k t}\right) \times \int d\mu(x)[h(t; x, x \mid D) - P_0(x, x \mid D)] \quad , \tag{4.191b}$$

we see that $\zeta_{p.v.}(0 \mid D)$ exists in the ordinary sense, if

$$\sum_{k=1}^{\nu} c_k = -1 \quad , \quad \sum_{k=1}^{\nu} c_k M_k^l = 0 \quad , \tag{4.191c}$$

and $\nu = \frac{d}{m}$.

A simple computation yields[141,167]

$$[\ln \det D]_{p.v.} = -\zeta'_{p.v.}(0 \mid D) = [\ln \det D]_\zeta + \zeta(0 \mid D) \ln M + \\ + \sum_{l=1}^{\frac{d}{m}} \frac{(-1)^l}{l!} \frac{\alpha_{\frac{d}{m}-l}}{(4\pi)^{\frac{d}{m}}} \sum_k c_k M_k^l \ln M_k + \mathcal{O}\left(\frac{1}{M_k^2} \ln M_k\right) \quad , \tag{4.192a}$$

where $\zeta(0 \mid A)$ is the value of $\zeta(s \mid A)$ obtained by analytic continuation,

$$\ln M = \sum_k c_k \ln M_k \quad , \tag{4.192b}$$

and α_r are the coefficients defined in (4.41).

For QED_2, $\frac{d}{m} = 1$ and $\alpha_0 = 0$, so that the ambiguity is again of the form (4.188), that is

$$[\ln \det{}' A]_\zeta = [\ln \det{}' A]_r + \zeta(0 \mid A) \ln \varepsilon \tag{4.193a}$$

$$= [\ln \det A]_{p.v.} - \zeta(0 \mid A) \ln M \quad . \tag{4.193b}$$

4.6.2 Dependence on R

Strictly speaking, our discussion of the QED_2/QCD_2 determinant has concerned the operator $i\,\hat{\not{D}}$ which depends on the radius of the stereographic sphere. The difference (4.190) is seen not to depend on R. This is in fact a general property, which is however not shared by $\zeta'(0 \mid i\,\hat{\not{D}}) - \zeta'(0 \mid i\,\hat{\not{\partial}})$. Hence the "normalized" determinant will in general depend on R. It is in fact easy to show that it will be independent of R, if and only if, the trace of the energy momentum $\theta_{\mu\nu}$ tensor vanishes, that is, in the absence of a trace anomaly. Such an anomaly will always occur if zero modes are present. This follows from the fact, that independent of the dimension d of space[151,140],

$$\int d^2x g_{\mu\nu} \theta_{\mu\nu}(x)|_0^A = -[\zeta(0 \mid (i\,\hat{\not{D}})^2) - \zeta(0 \mid (i\,\hat{\not{\partial}})^2)] \quad . \tag{4.194}$$

In 4 dimensions the trace $\theta_{\mu\mu} \neq 0$ even in the absence of zero modes. In $d = 2$, where $\mathrm{tr} a_1 = 0$, the dependence on R will only enter through the zero modes. However, as we shall see in chapter 12 this dependence on R cancels in the correlation functions. In $d = 4$ this is not the case, and the limit $R \to \infty$ has to be taken at the end, after proper renormalization.

To demonstrate (4.194), we consider as starting point (4.73), with g the determinant of the induced metric $g_{\mu\nu}$ defined in (4.55b). It is convenient to write the action in (4.73) in the form

$$S_F = \int d^2x g_{\mu\nu}(x) \overline{\hat{\psi}}(x) i\gamma_\mu \hat{D}_\nu \hat{\psi}(x) \quad . \tag{4.195}$$

This allows us to define the energy-momentum tensor associated with the Dirac field by

$$\theta_{\mu\nu}(x) = \frac{1}{2} \langle \overline{\hat{\psi}}(x) \gamma_\mu \overleftrightarrow{\hat{D}}_\nu \hat{\psi}(x) \rangle_A = \frac{\delta W}{\delta g_{\mu\nu}(x)} \quad . \tag{4.196}$$

In particular

$$g_{\mu\nu}\theta_{\mu\nu} = \frac{1}{2}\langle\bar{\psi}(x)i\overleftrightarrow{\not{D}}\psi(x)\rangle_A = g_{\mu\nu}(x)\frac{\delta W}{\delta g_{\mu\nu}(x)} \quad . \tag{4.197}$$

The left hand side is just the trace of $\theta_{\mu\nu}$. We now want to calculate the r.h.s. We have

$$\frac{\delta W}{\delta g_{\mu\nu}(x)} = \frac{1}{2}\frac{\delta\zeta'(0 \mid (i\hat{\not{D}})^2)}{\delta g_{\mu\nu}(x)} = -\sum_{\lambda_n{}'\neq 0}\frac{\lambda_n\frac{\delta\lambda_n}{\delta g_{\mu\nu}(x)}}{(\lambda_n^2)^{s+1}} + \mathcal{O}(s) \quad . \tag{4.198}$$

$\delta\lambda_n/\delta g_{\mu\nu}(x)$ is easily calculated; from the Dirac equation (4.71), i.e.

$$\int d^2x g_{\mu\nu}\hat{u}_n^\dagger(x)i\gamma_\mu\hat{D}_\nu\hat{u}_n(x) = \lambda_n \quad , \tag{4.199}$$

one obtains

$$\frac{\delta\lambda_n}{\delta g_{\mu\nu}(x)} = \hat{u}_n^\dagger(x)i\gamma_\mu\hat{D}_\nu\hat{u}_n(x) \quad . \tag{4.200}$$

Hence, substituting this result into (4.198) one finds, upon using once more (4.199),

$$g_{\mu\nu}(x)\frac{\delta W}{\delta g_{\mu\nu}(x)} = -\sqrt{g}tr\zeta(0;x,x \mid (i\,\hat{\not{D}})^2) \tag{4.201a}$$

or

$$\int d^2x g_{\mu\nu}(x)\frac{\delta W}{\delta g_{\mu\nu}(x)}\bigg|_0^A = -[\zeta(0 \mid (i\,\hat{\not{D}})^2) - \zeta(0 \mid (i\,\hat{\not{\partial}})^2)] \quad , \tag{4.201b}$$

which establishes (4.194).

Let us now demonstrate the relation between the R-dependence of $det(i\,\hat{\not{D}}/i\,\hat{\not{\partial}})$ and the trace-anomaly. We have from (4.67) and (4.69a),

$$\int d^2x u_n^\dagger i\,\not{D}u_n = \lambda_n \quad . \tag{4.202}$$

The R-dependence of λ_n only enters through u_n. From the above equation one obtains

$$\frac{1}{\lambda_n}\frac{\delta\lambda_n}{\delta R} = \int d^2x\Omega_R\Big(\frac{\delta u_n^\dagger}{\delta R}u_n + u_n^\dagger\frac{\delta u_n}{\delta R}\Big) \quad .$$

From the normalization condition (4.69a), we have

$$\int d^2x\Omega_R\Big(\frac{\delta u_n^\dagger}{\delta R}u_n + u_n^\dagger\frac{\delta u_n}{\delta R}\Big) = -\int d^2x\frac{\delta\Omega_R}{\delta R}u_n^\dagger u_n \quad ;$$

or

$$\frac{R}{\lambda_n}\frac{\delta\lambda_n}{\delta R} = -\int d^2x\Omega_R^2(x)\frac{x^2}{R^2}u_n^\dagger(x)u_n(x) \quad . \tag{4.203}$$

Correspondingly, we obtain

$$\begin{aligned} R\frac{\delta}{\delta R}W[A] &= -\lim_{s\to 0}\left\{\sum_{\lambda_n\neq 0}\int d^2x\Omega_R^2(x)\frac{x^2}{R^2}\frac{u_n^\dagger(x)u_n(x)}{(\lambda_n^2)^s}\right\}_{s=0} \\ &= -\int d^2x\Big(\Omega_R(x)\frac{x^2}{R^2}\Big)\sqrt{g}\zeta(0;x,x\mid(i\,\hat{\not{D}})^2) \quad ; \end{aligned} \tag{4.204}$$

or because of (4.194),

$$R\frac{\delta W[A]}{\delta R} = \int d^2x\Big(\Omega_R(x)\frac{x^2}{R^2}\Big)g_{\mu\nu}\theta_{\mu\nu}(x) \quad . \tag{4.205}$$

This proves that $W[A]$ will not depend on R, if and only if there is no trace-anomaly.

Proposition: $\zeta(0\mid(i\,\hat{\not{D}})^2)$ is independent of R.

Indeed,

$$\begin{aligned} R\frac{\delta\zeta(0\mid(i\,\hat{\not{D}})^2)}{\delta R} &= \sum_{\lambda_n\neq 0}\frac{-2s}{(\lambda_n^2)^{s+1}}R\lambda_n\frac{\delta\lambda_n}{\delta R}\Big|_{s=0} \\ &= \Big(-2s\int d^2x\sqrt{g}\Big(\Omega_R(x)\frac{x^2}{R^2}\Big)\sum{}'\frac{\hat{u}_n^\dagger(x)\hat{u}_n(x)}{(\lambda_n^2)^s}\Big)_{s=0} \\ &= -2\int d^2x\sqrt{g}\Big(\Omega_R\frac{x^2}{R^2}\Big)[s\hat{\zeta}(s;x,x\mid(i\,\hat{\not{D}})^2)]_{s=0} = 0 \end{aligned} \tag{4.206}$$

Proposition:

$$R\frac{\delta W}{\delta R}\underset{R\to\infty}{\longrightarrow}\int d^2x\Big(\frac{2x^2}{R^2+x^2}\Big)g_{\mu\nu}\theta_{\mu\nu} = 2\int d^2x g_{\mu\nu}\theta_{\mu\nu} - \int d^2x\Omega_R g_{\mu\nu}\theta_{\mu\nu}\underset{R\to\infty}{\longrightarrow} 0 \quad . \tag{4.207}$$

This shows that the effective action approaches a R independent limit, as $R\to\infty$.

4.7 Atiyah-Singer Index Theorem

The Atiyah-Singer theorem[162] is the outgrowth of a long development in algebraic geometry. It relates a topological invariant to the index $\gamma(D)$ of a bounded operator on $\mathcal{H}$, defined by

$$\gamma(D) = \text{index}\, D = \dim\ker D - \dim\ker D^\dagger \quad , \tag{4.208a}$$

where

$$\ker D := \{\psi \in \mathcal{H}; D\psi = 0\} \quad , \tag{4.208b}$$

is the null space of D, and $D^\dagger$ is the adjoint of D. D is called a Fredholm operator if the dimensions of the null-spaces of D and $D^\dagger$ are finite: $\dim\ker D < \infty, \dim\ker D^\dagger < \infty$. Most operators a physicist becomes confronted with have a vanishing index. This is true for operators (defined on a compact manifold which are hermitian or are of the form $1+K$, with K compact. Whereas the triviality of the index for hermitian operators follows directly from its definition (4.205), the latter statement can be understood in terms of the equivalence of the vanishing index with the validity of the *Fredholm alternative*: "Either the inhomogeneous equation $(1+K)\psi = \chi$ has a unique solution for every χ, or the homogeneous equation has a finite number of linearly independent solutions". In the last case the adjoint equation has the same number of solutions.

As we now show, the winding number n of a gauge field A_μ, in d=2 dimensions, is just the index of the operator

$$D = -\partial_1 + i\partial_2 + ie(A_1 + iA_2) \quad . \tag{4.209}$$

Noting that

$$i\not{D} = \begin{pmatrix} 0 & D \\ D^\dagger & 0 \end{pmatrix} \quad , \tag{4.210}$$

we define

$$\text{Index}\, i\not{D} = \ker i\not{D}|_{\gamma^5=1} - \ker i\not{D}|_{\gamma^5=-1} = N_+ - N_- \quad , \tag{4.211}$$

where $N_+(N_-)$ is the number of zero modes of $i\not{D}$ of positive (negative) chirality.

Proposition:

$$N_+ - N_- = n \tag{4.212}$$

The determination of the zero mode asymmetry from the axial anomaly relation was first suggested by Coleman[169] and further elaborated on by a number of authors[170–174]. We shall provide here a physicists proof[146].

Our starting point is the $U(1)$- axial vector current in the presence of an (abelian or non-abelian) external gauge field A_μ. Taking account of the possible existence of zero modes, it is defined in a way analogous to that given by Eq. (4.174):

$$J'_\mu(x \mid A) = -\underset{s\to 0}{\text{a.c.}} \sum_n{}' \frac{u_n^\dagger(x) i\gamma_\mu\gamma_5 u_n(x)}{\lambda_n^{2s+1}} \quad . \tag{4.213}$$

Correspondingly we have[146,174]

$$\partial_\mu {J'}^5_\mu(x \mid A) = 2\Omega_R(x)\mathrm{tr}\gamma_5\zeta(0; x, x \mid (i\not{D})^2) \quad , \tag{4.214}$$

which replaces (4.178). Taking account of possible zero modes, and recalling (4.38), (4.106d), this becomes

$$\partial_\mu {J'}^5_\mu(x \mid A) = \frac{e}{2\pi} tr\epsilon_{\mu\nu}F_{\mu\nu}(x) - 2\Omega_R(x)\sum u_n^{(0)\dagger}(x)\gamma^5 u_n^{(0)}(x) \quad . \tag{4.215}$$

Notice that the first term on the right hand side is just twice the Pontryagin density.

Since the definition of J^5_μ excludes the zero modes, it is natural to suppose that J^5_μ tends to zero at infinity. Hence integrating (4.215) over $\Re^2$, and dropping the surface term at infinity, we obtain

$$\mathrm{tr}\, e\int \frac{d^2x}{4\pi}\epsilon_{\mu\nu}F_{\mu\nu} - (N_+ - N_-) = 0 \quad , \tag{4.216}$$

where tr refers to the trace in group space. For an $SU(N)$ Lie algebra valued $F_{\mu\nu}$, this trace vanishes, $(\pi_1(SU(N)) \sim 0)$ and $N_+ = N_-$. For QED_2, on the other hand, the first term is just equal to the winding number of the gauge-field $(\pi_1(U(1)) = Z)$, and we obtain the result (4.212).

Let us explicitly illustrate the above result for the case of the abelian vortex field (4.76). It is convenient to define the (complex) "light-cone" variables

$$\overline{z} = x_1 - ix_2 \quad , \quad z = x_1 + ix_2 \quad . \tag{4.217}$$

In terms of these variables the associated Dirac eigenvalue problem (4.67) in the external vortex field (4.76) on the unit sphere ($R = 1$) takes the form

$$\begin{pmatrix} 0 & -(2\partial_z - n\frac{\overline{z}}{1+\overline{z}z}) \\ 2\partial_{\overline{z}} + n\frac{z}{1+\overline{z}z} & 0 \end{pmatrix}\begin{pmatrix} \overline{z}_1 \\ \overline{z}_2 \end{pmatrix} = \frac{2\lambda}{1+\overline{z}z}\begin{pmatrix} \overline{z}_1 \\ \overline{z}_2 \end{pmatrix} \quad . \tag{4.218}$$

We only consider the solutions to the zero-eigenvalue problem. Making the ansatz

$$\begin{pmatrix} \overline{z}_1^{(0)} \\ \overline{z}_2^{(0)} \end{pmatrix} = \begin{pmatrix} (1+\overline{z}z)^{\frac{-n}{2}}\varphi_1^{(0)} \\ (1+\overline{z}z)^{\frac{n}{2}}\varphi_2^{(0)} \end{pmatrix} \quad , \tag{4.219}$$

one finds that $\varphi_1^{(0)}$ and $\varphi_2^{(0)}$ are solutions to

$$\partial_{\bar{z}}\varphi_1^{(0)} = 0 \quad , \quad \partial_z\varphi_2^{(0)} = 0 \quad , \tag{4.220}$$

that is $\varphi_1^{(0)} \alpha\, z^{l_+}$, $\varphi_2^{(0)} \alpha\, \bar{z}^{l_-}$. Requiring these null-modes to be normalized with respect to the measure $d^2x \frac{2}{1+\bar{z}z}$, we see that, depending on the sign of n, all the admissible zero modes have the same chirality ("vanishing theorem" for QED_2[13)]). With

$$\chi_\pm = \frac{1 \pm \gamma_5}{2}\begin{pmatrix}1\\1\end{pmatrix} \quad , \tag{4.221}$$

the Pauli spinors of positive and negative chirality, one has

$$\psi_{l_\pm}^{(0)}(x) = \left(\frac{\mu}{2\pi}\right)^{\frac{1}{2}}[\mu(x_1 \pm ix_2)]^{l-1}\frac{1}{[\mu(1+x^2)]^{\frac{|n|}{2}}}\chi_\pm \quad , \quad n \neq 0 \quad . \tag{4.222}$$

Here μ is just a "mass" parameter to give $\psi_{l_\pm}^{(0)}$ the dimension $\frac{1}{2}$. There is no need to orthonormalize these solutions. The condition of normalizability restricts $l_\pm$ to take the values $l = 1, 2, \cdots |n|$. Hence there are exactly $N_\pm = | n |$ zero modes, with

$$\begin{aligned} N_+ &= n \quad , \quad N_- = 0 \quad \text{for} \quad n > 0 \\ N_- &= n \quad , \quad N_+ = 0 \quad \text{for} \quad n < 0 \quad , \end{aligned} \tag{4.223}$$

so that the Atiyah-Singer index theorem (4.212) is satisfied for the vortex (4.76), with the additional property peculiar to QED_2, that the zero modes associated with a vortex configuration of a given "winding" are of one chirality, only ("vanishing theorem"). As we shall explicitly show in chapter 12, this is also a property of arbitrary gauge-field configurations in d=2.

4.8 Conclusion

The discussion in this chapter was geared to provide the necessary foundation for the chapters to follow. It has thus emphasized within a general setting, special aspects of quantum field theory in two space-time dimensions. There exists a very extensive mathematical literature on this subject, which evidently is outside the scope of this book, and which the mathematically inclined reader is asked to consult. An extensive list of references to this literature can be found in the article on "topological methods for gauge theories"[170] by B. Schroer, as well as in refs. [141,176]. Gilkey's method for calculating the Seeley coefficients, including the case of "non-invertible"

13) For QED_4 this "vanishing theorem " has been proven [175] to hold only for instanton configurations.

operators, has been extensively discussed in a "digestible" way for physicists by Gamboa-Saravi et al in refs. [142]. The "Fujikawa" approach has similarly been discussed and applied to a variety of problems by Fujikawa himself[156,157,158,161]. As we have tried to exhibit, all of these approaches are equivalent if due attention is paid to the ambiguities, and we shall have ample opportunity to apply them in the chapters to follow.

Chapter V

FERMIONIC MODELS WITH SELF-INTERACTION

5.1 Introduction

In this chapter we shall consider self interactions of a multiplet of fermionic fields. The symmetry group will be taken as being either $O(N)$ or $U(N)$. These models have a long story, being introduced by Nambu and Jona Lasinio in four dimensions, in order to describe the phenomenology of elementary particles[177] . In the former case we will take the self interaction as being the four fermion scalar scalar one, and we refer to the model as the Gross-Neveu model[178]. One introduces a continuous chiral $U(N) \otimes \widetilde{U}(N)$ symmetry in the latter case, and the model is referred to as the chiral Gross-Neveu model. The classical models have no mass term, being conformally invariant, but a mass is generated by quantum fluctuations, leading to an asymptotically free theory. The theory is solvable in the $N \to \infty$ limit, and an expansion in powers of $1/N$ may be defined, although being more involved in the chiral model. Both models possess an infinite number of conservation laws[179,180] which survive quantization[180], and as a consequence, the S-matrix of the models may be computed exactly.

In section 2 we discuss the $O(N)$ invariant model, and in section 3 the chiral $SU(N)$ model. The interaction of these fermionic theories with σ models will be discussed in chapters 6, 7 and 8. An important issue concerns the chiral symmetry breaking, in the $SU(N)$ symmetric case. There is a theorem in two dimensional space time, prohibiting the appearance of Goldstone particles[93]. The mass generation in the chiral Gross Neveu model might be taken as a counterexample to this theorem. However, we shall see in the operator solution of the model that this problem is solved by the fact that the chirality carrying field decouples from the physical fields; the latter carry no chirality, and have non vanishing mass.

5.2 The $O(N)$ Invariant Gross-Neveu Model

This model consists of an N-plet of Majorana fermions transforming as a fundamental representation of the orthogonal group $O(N)$, with a quartic Fermi-type self interaction, as first introduced by Nambu and Jona-Lasinio[177]. The Lagrangian is given by

$$\mathcal{L} = \frac{i}{2}\overline{\psi}_i \not{\partial}\psi_i + \frac{1}{4}g(\overline{\psi}_i\psi_i)^2 \quad , \tag{5.1}$$

where the summation over repeated indices is implied. This is the most general $O(N)$ invariant, renormalizable and conformally invariant theory for self interacting Majorana fermions in two dimensions[1)] It presents the discrete symmetry $\psi \to \gamma_5\psi$, to be discussed later on, and the (trivial) symmetry $\psi \to -\psi$.

5.2.1 Classical conservation laws

The classical model is an integrable theory, having an infinite number of conservation laws[2)]. The semiclassical study of the model reveals a very rich structure, and the time independent solutions of the equation of motion can be computed systematically[181]. This is a consequence of the fact that the Noether current associated with $O(N)$ rotations, given by

$$j_\mu^{ij} = i\overline{\psi}^j\gamma_\mu\psi^i \quad , \tag{5.2}$$

satisfies

$$\partial^\mu j_\mu^{ij} = 0 \tag{5.3}$$

as a consequence of the $O(N)$ symmetry, and using the equation of motion of the Lagrangian (5.1) as well as a Fierz transformation, it also satisfies[3)]

$$\partial_\mu j_\nu^{ij} - \partial_\nu j_\mu^{ij} + 2g[j_\mu, j_\nu]^{ij} = 0 \quad . \tag{5.4}$$

---

1) We have, for two dimensional Majorana (anticommuting) spinors, the identities $\overline{\psi}\gamma_5\psi = 0 = \overline{\psi}\gamma_\mu\psi$, which follow from the anticommutation of the components, and from the definition $\overline{\psi} = \psi^\dagger\gamma^0 = \psi\gamma^0$ for a Majorana field, with a charge conjugation matrix being unity; in the Weyl representation, the charge conjugation matrix is $C = \gamma_5$; we use the Majorana representation, (A.17), in this chapter. We have also the identity $\delta_{\alpha\beta}\delta_{\gamma\delta} = \frac{1}{2}(\delta_{\alpha\delta}\delta_{\gamma\beta} + (\gamma^\mu)_{\alpha\delta}(\gamma^\mu)_{\gamma\beta} + (\gamma^5)_{\alpha\delta}(\gamma^5)_{\gamma\beta})$ which implies, for Majorana fields the Fierz identity $\overline{\psi}_i\psi_j\overline{\psi}_j\psi_i = -\frac{1}{2}(\overline{\psi}_i\psi_i)^2$, where sum over repeated indices is implicit. In the particular case of $O(2)$ symmetry the Lagrangian (5.1) is just that of the Thirring model as one sees by defining a Dirac field $\psi_D = \psi_1 + i\psi_2$, and using $-2(\overline{\psi}_i\psi_i)^2 = \overline{\psi}_D\gamma^\mu\psi_D\overline{\psi}_D\gamma_\mu\psi_D$.

2) We have discussed integrability of the sine-Gordon equation, and will in chapter 6 discuss integrable non linear sigma models. For the moment, integrable means that the model has higher conservation laws, which are not directly associated with a Lie group symmetry of the Lagrangian. It is rather related with the existence of a set of differential equations for a matrix valued field. These differential equations depend on a potential, which is in turn, related to the integrable model in question[179].

3) We use the equation of motion $i\not{\partial}\psi_i = -g\overline{\psi}_k\psi_k\psi_i$, which implies $i\partial_\mu(\overline{\psi}_j\gamma^\mu\gamma_5\psi_i) = +2g\overline{\psi}\psi\overline{\psi}_j\gamma_5\psi_i$, as well as the Fierz identity $\overline{\psi}_j\gamma^\mu\gamma_5\psi_k\overline{\psi}_k\gamma_\mu\psi_i = \overline{\psi}_j\gamma_5\psi_i\overline{\psi}_k\psi_k$. Therefore, using (A.17c), we have $\epsilon^{\mu\nu}\partial_\mu J_\nu = -2g\epsilon^{\mu\nu}\overline{\psi}_i\gamma_\nu\psi_k\overline{\psi}_k\gamma_\mu\psi_j$ from which (5.4) is obtained.

This is a crucial identity. It implies that the theory is integrable, since it assures the compatibility of the differential equations[132,179]

$$\partial_\mu U = 2gUj_\mu \quad . \tag{5.5}$$

Indeed, taking the derivative with respect to x^ν of (5.5) and subtracting the resulting equation with $\mu \leftrightarrow \nu$ we obtain the integrability condition (5.4). Using current conservation and (5.4) it is also possible[182] to establish the following linear system of equations for any value of the parameter λ

$$\partial_\mu U^{(\lambda)} = -gU^{(\lambda)}\left[(1-\cosh 2\lambda)j_\mu + \sinh 2\lambda \epsilon_{\mu\nu}j^\nu\right] \quad . \tag{5.6}$$

As a result of equations (5.3) to (5.6), there exists a whole one parameter family of conserved currents given by the expression

$$j_\mu^{(\lambda)} = U^{(\lambda)}\left\{(\cosh 2\lambda - 2)\, j_\mu - \sinh 2\lambda \epsilon_{\mu\nu}j^\nu\right\} U^{(\lambda)^{-1}} \quad . \tag{5.7}$$

The current (5.7) can be shown to be conserved. Expanding $j_\mu^{(\lambda)}$ in powers of λ, and proceeding as in chapter 3 one obtains an infinite number of conserved currents and charges (for a more detailed discussion see chapter 6), the first non trivial charge being given by

$$Q^{(2)} = \int dy_1 dy_2 \epsilon(y_1 - y_2) j_0(t,y_1) j_0(t,y_2) + \frac{1}{g}\int dy j_1(t,y) \quad . \tag{5.8}$$

One may, in fact, construct an infinite number of conserved currents in terms of any conserved current j^μ satisfying the zero curvature condition

$$[D_\mu, D_\nu] = \partial_\mu j_\nu - \partial_\nu j_\mu + 2g[j_\mu, j_\nu] = 0 \quad , \tag{5.9}$$

where D_μ denotes the covariant derivative as given by $D_\mu = \partial_\mu + 2gj_\mu$. The algorithm is very simple[183]. We start out with $j_\mu^{(1)} = j_\mu$. Consider a conserved current $j_\mu^{(n)}$, and the corresponding potential, that is

$$j_\mu^{(n)} = \epsilon_{\mu\nu}\partial^\nu j^{(n)} \quad . \tag{5.10a}$$

Construct $j_\mu^{(n+1)}$ as

$$j_\mu^{(n+1)} = D_\mu j^{(n)} \quad . \tag{5.10b}$$

This new element is also conserved. Indeed, using the conservation of j^μ, one has

$$\partial^\mu j_\mu^{(n+1)} = \partial^\mu D_\mu j^{(n)} = D^\mu \partial_\mu j^{(n)} \quad , \tag{5.11}$$

or recalling definition (5.10a) of $j_\mu^{(n)}$, one finds

$$\partial^\mu j_\mu^{(n+1)} = \epsilon_{\mu\nu} D^\mu j^{(n)\nu} = \epsilon_{\mu\nu} D^\mu D^\nu j^{(n-1)} = 0 \quad , \tag{5.12}$$

where use has been made of Eq.(5.9). Therefore we have an infinite number of conserved charges in the classical theory.

5.2.2 Effective potential and β-function in a $\frac{1}{N}$ expansion

We study now the quantization of the Gross-Neveu model. The Lagrangian (5.1) may be rewritten in terms of an auxiliary field $\sigma(x)$ as

$$\mathcal{L} = \frac{i}{2}\bar{\psi}\, \partial\!\!\!/ \psi - \frac{1}{4g}\sigma^2 + \frac{1}{2}\sigma\bar{\psi}\psi \quad . \tag{5.13}$$

Since the dependence on the auxiliary field σ is quadratic, and does not involve derivatives, the "equation of motion" for σ is given by the constraint

$$\sigma = g\bar{\psi}\psi \quad , \tag{5.14}$$

which, when substituted into (5.13), reproduces (5.1)[4)]

The idea of introducing the auxiliary field, is to enable one to first integrate over the fermions, which is equivalent to computing the functional determinant $\det(i\, \partial\!\!\!/ + \sigma)$. The effective potential $V^{(1)}(\sigma)$ is defined as ($-\frac{i}{2}$ times) the logarithm of the determinant for constant σ:

$$V^{(1)}(\sigma) = -\frac{i}{2}\ln\, \det(i\, \partial\!\!\!/ + \sigma) \quad , \quad \sigma = \text{constant} \quad . \tag{5.15a}$$

Its computation corresponds to summing the one loop fermion diagrams at zero momentum[178] (see figure 5.1); we have the result

$$V^{(1)}_{eff}(\sigma) = iN\sum_{n=1}^{\infty}\frac{(-\sigma)^n}{2n}\,\mathrm{tr}\int\frac{d^2k}{(2\pi)^2}\left(\frac{1}{k\!\!\!/}\right)^n = iN\sum_{n=1}^{\infty}\frac{(\sigma)^{2n}}{2n}\int\frac{d^2k}{(2\pi)^2}\frac{1}{(k^2)^n} \quad . \tag{5.15b}$$

The sum (5.15b) corresponds to the expansion of the logarithm

$$V^{(1)}_{eff}(\sigma) = i\frac{N}{2}\int\frac{d^2k}{(2\pi)^2}\ln\left(1-\frac{\sigma^2}{k^2}\right) = \frac{N}{8\pi}\sigma^2\left[\ln\frac{\sigma^2}{\Lambda^2}-1\right] \quad , \tag{5.16}$$

where a cut-off has been introduced, after Wick rotation, at $k^2 = \Lambda^2$ in order that the integral be well defined.

---

4) Alternatively, we can write

$$-\frac{1}{4g}\sigma^2 + \frac{1}{2}\sigma\bar{\psi}\psi = -\frac{1}{4g}(\sigma - g\bar{\psi}\psi)^2 + \frac{1}{4}g(\bar{\psi}\psi)^2 \quad .$$

The first term may be completely eliminated integrating over σ, leaving no trace of that field.

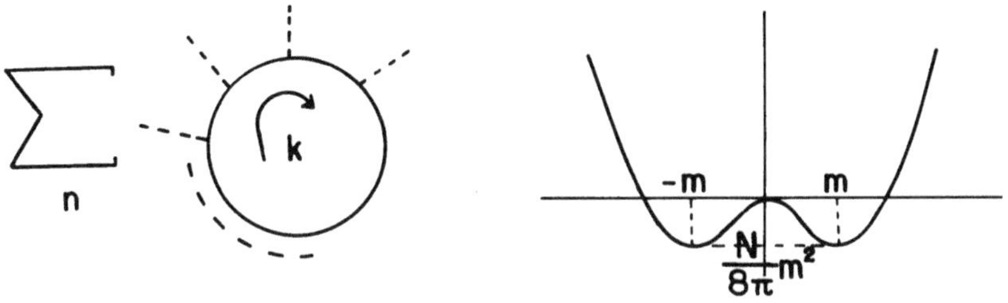

Figure 5.1: Calculation and form of the effective potential.

Taking into account the tree contribution as well, which can be read from (5.13), we finally obtain

$$V_{eff}(\sigma) = \frac{1}{4g}\sigma^2 + \frac{N}{8\pi}\sigma^2 \left[\ln\frac{\sigma^2}{\Lambda^2} - 1\right] \quad . \tag{5.17}$$

This effective potential, whose form is given in figure 5.1 has a maximum at the origin, and a minimum at the point

$$\sigma^2 = m^2 = \Lambda^2 e^{-\frac{2\pi}{Ng}} \quad . \tag{5.18}$$

The maximum at the origin has several consequences. If we try to quantize the theory around $\sigma = 0$ in the large N limit, we obtain a tachyonic contribution to the effective four fermion interaction[178]. On the other hand, quantization around one of the minima at $\sigma = \pm m$ (Eq. 5.18) means that the discrete symmetry

$$\psi \to \gamma_5 \psi \tag{5.19}$$

of the original Lagrangian (5.1), is spontaneously broken by a mass term $\frac{1}{2}m\overline{\psi}\psi$. Hence, for this choice of the vacuum, a mass is spontaneously generated for the fermionic field. The interpretation of m as the mass of the fermionic field requires that it be renormalization group invariant, i.e.

$$\left(\Lambda^2\frac{\partial}{\partial\Lambda^2} + \beta(g)\frac{\partial}{\partial g}\right)m^2 = 0 \quad . \tag{5.20}$$

This permits the computation of the β function of the model, which turns out to be

$$\beta(g) = -\frac{Ng^2}{2\pi} \quad . \tag{5.21}$$

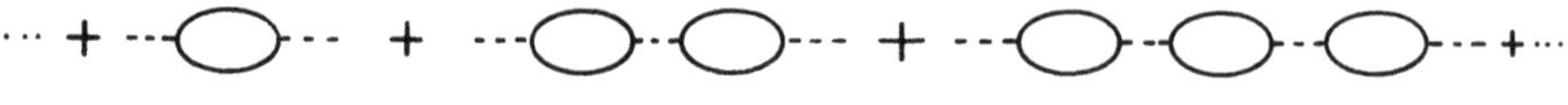

Figure 5.2: Summation of bubble diagrams contributing to fermion interaction in the large N limit.

The sign of the β function shows that the theory is asymptotically free. Moreover, the cut off Λ and the bare coupling constant $g(\Lambda)$ can both be eliminated from the theory, in favor of the generated mass parameter alone. This "mass transmutation" simplifies considerably the discussion of the theory. The "mass transmutation" phenomenon might provide a dynamical mechanism for generating masses in theories which are massless "ab initio", worthwhile to understand[184].

In the limit $N \to \infty$ with gN fixed, the bubble diagrams of figure 5.2 have to be summed (this is a geometric series) and the theory may be solved. We are going to look for corrections in powers of $1/N$. To this end, we rewrite the theory in a way as to allow for the systematic calculation of the corrections.

The generating functional is given by

$$\mathcal{Z}[\eta] = \int \mathcal{D}\psi \mathcal{D}\sigma e^{\frac{i}{2}\int d^2x\left\{\overline{\psi}(i\partial\!\!\!/+\sigma)\psi - \frac{1}{2g}\sigma^2 + 2\overline{\eta}\psi\right\}} \quad ; \tag{5.22}$$

expanding σ around the vacuum expectation value $\langle\sigma\rangle = -m$,

$$\sigma = -m + \frac{\tilde{\sigma}}{\sqrt{N}} \quad , \quad \langle\tilde{\sigma}\rangle = 0 \quad , \tag{5.23}$$

and defining $\Delta\!\!\!/[\tilde{\sigma}] = i\,\partial\!\!\!/ - m + \frac{\tilde{\sigma}}{\sqrt{N}}$, we can write the functional integral in the form

$$\mathcal{Z}\,[\eta] = \int \mathcal{D}\psi \mathcal{D}\tilde{\sigma} e^{\frac{i}{2}\int d^2x\left\{(\overline{\psi}+\overline{\eta}\Delta\!\!\!/[\tilde{\sigma}]^{-1})\Delta_F[\tilde{\sigma}](\psi+\Delta\!\!\!/[\tilde{\sigma}]^{-1}\eta) - \overline{\eta}\Delta\!\!\!/^{-1}\eta - \frac{1}{2Ng}\tilde{\sigma}^2 - \frac{1}{g\sqrt{N}}m\tilde{\sigma}\right\}} \, . \tag{5.24}$$

We integrate out the fermions, writing the determinant thus obtained as

$$\det \Delta\!\!\!/[\tilde{\sigma}] = \mathrm{tr}\ \ln \Delta\!\!\!/[\tilde{\sigma}] \quad .$$

Substituting this expression in (5.24) we have the result

$$\mathcal{Z}\,[\eta] = \int \mathcal{D}\tilde{\sigma} e^{i\int d^2x\left\{-\frac{1}{2}\overline{\eta}\Delta\!\!\!/[\tilde{\sigma}]^{-1}\eta - \frac{1}{4Ng}\tilde{\sigma}^2 + \frac{1}{2g\sqrt{N}}m\tilde{\sigma}\right\} + \frac{1}{2}\mathrm{tr}\left(i\partial\!\!\!/ - m + \frac{\tilde{\sigma}}{\sqrt{N}}\right)} \tag{5.25}$$

The fact that $\tilde{\sigma}$ has vanishing vacuum expectation value, implies that all tadpoles cancel against the linear term in $\tilde{\sigma}$, that is

$$\frac{i}{2g} m \int d^2x \, \tilde{\sigma} + \frac{N}{2} \int d^2x \, \tilde{\sigma} \, \mathrm{tr} \int \frac{d^2k}{(2\pi)^2} \frac{\not{k} + m}{k^2 - m^2} = 0 \quad . \tag{5.26}$$

Using a cutoff Λ^2 for the momentum integral[5)], we obtain Eq.(5.18) for the mass.

The β function has been computed in (5.20,21). A comment concerning this procedure is in order at this point, if one aims at computing higher order corrections. In that case, the β function may depend on the renormalization procedure used to calculate the fermion four point function. Thus the two loop β function computed by normalizing the fermion four point function as in the symmetric theory ($m = 0$) gives again (5.21)[6)]. If instead, the renormalization constants are computed using the non symmetric theory, a different β function is obtained[186,187,188]

$$\beta(g) = -\frac{Ng^2}{2\pi} + 2Ng \frac{a^2}{\mu^2} \frac{1}{1 + 4a^2/\mu^2} \quad , \tag{5.27}$$

with a^2/μ^2 related to g by,

$$\sqrt{1 + 4a^2/\mu^2} \ln \frac{\sqrt{1 + 4a^2/\mu^2} - 1}{\sqrt{1 + 4a^2/\mu^2} + 1} = -\frac{4\pi}{Ng} \quad , \tag{5.28}$$

where the second expression is a modification of the gap equation (5.26)[188]. The $1/N$ correction to the above result has been also discussed in the literature[186].

The two apparently conflicting results are seen to be correct, with different interpretations for the respective β-functions[188]. Indeed we can rewrite the generated mass (5.18) in terms of renormalized parameters. The renormalized coupling constant $g(\mu)$ is defined, at one loop level by

$$\frac{1}{g(\mu)} = \frac{1}{g(\Lambda)} - \frac{N}{\pi} \ln \frac{\Lambda}{\mu} \quad , \tag{5.29}$$

and we obtain from (5.18),

$$m^2 = \mu^2 e^{-\frac{2\pi}{Ng(\mu)}} \quad . \tag{5.30}$$

5) We perform a Wick rotation in the momentum integral, obtaining

$$-\frac{1}{2g} m + \frac{mN}{4\pi} \int_0^{\Lambda^2/m^2} \frac{dx^2}{x^2 + 1} = m \left(-\frac{1}{2g} + \frac{N}{4\pi} \ln \frac{\Lambda^2}{m^2} \right) = 0.$$

6) This procedure is correct. See, with this respect, ref. [185].

In general we can write this result in an explicitly renormalization group invariant way in terms of the β function as

$$m^2 = \mu^2 e^{-\int^g \frac{dg}{\beta(g)}} \quad . \tag{5.31}$$

5.2.3 $\frac{1}{N}$ expansion: Feynman rules

From (5.25) we are able to derive systematically new Feynman rules corresponding to the $1/N$ expansion of the model. The leading order contribution in $1/N$ to the $1PI$ two point function of the field σ is given by the second term in the expansion of the the logarithm in (5.25), plus the tree graph contribution (see figure 5.3).

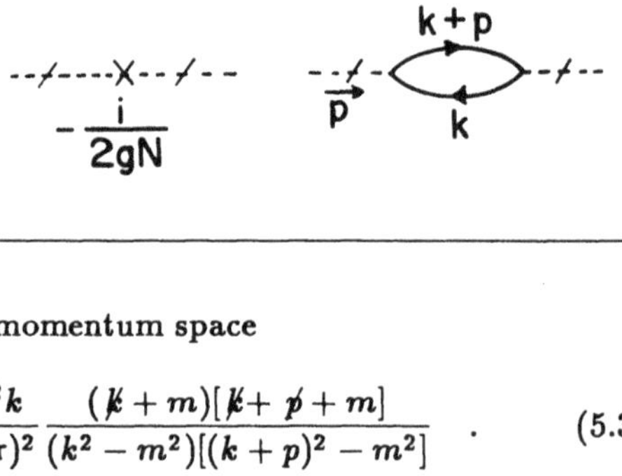

Figure 5.3: Tree and one loop diagrams contributing to the $\tilde{\sigma}$ two point function.

The corresponding result reads is momentum space

$$\tilde{\Gamma}(p) = -\frac{i}{2gN} - \frac{1}{2}\,\mathrm{tr}\int \frac{d^2k}{(2\pi)^2}\frac{(\not{k}+m)[\not{k}+\not{p}+m]}{(k^2-m^2)[(k+p)^2-m^2]} \quad . \tag{5.32}$$

We perform a Wick rotation, and use the fact that the logarithmically divergent part of (5.32) cancels the cut-off dependent constant $-\frac{1}{2g}$ upon use of (5.18,29,30), leaving a finite result. Indeed, we will see that the renormalization of the model in the $1/N$ expansion is very simple due to the mass transmutation. Performing the elementary integral in (5.32) one obtains (in Minkowski space), after cancellation of the divergent term,[7)]

$$\tilde{\Gamma}(p) = \frac{i}{4\pi}\sqrt{\frac{-p^2+4m^2}{-p^2}}\ln\frac{\sqrt{-p^2+4m^2}-\sqrt{-p^2}}{\sqrt{-p^2+4m^2}+\sqrt{-p^2}} \quad . \tag{5.33}$$

---

7) We have $\int \frac{d^2k}{(2\pi)^2}\mathrm{tr}\frac{(\not{k}+m)[\not{k}+\not{p}+m]}{(k^2-m^2)[(k+p)^2-m^2]} = \frac{2}{(2\pi)^2}\int_0^1 dx \int d^2k \frac{k^2-p^2x(1-x)-m^2}{[k^2-m^2+p^2x(1-x)]^2}$, which can be trivially integrated in the angle variables; the radial integration can be performed and results $\frac{1}{4\pi}\int_0^1 dx\left\{i\ln\Lambda^2 - i\ln\left(-m^2+p^2x(1-x)\right) + \frac{2ip^2x(1-x)}{-m^2+p^2x(1-x)}\right\}$. This integral may be performed by elementary means[189].

Table 5.1			
$\tilde{\sigma}$ propagator	······	$\langle 0 \vert \tilde{\sigma}(p)\tilde{\sigma}(-p) \vert 0\rangle$	$(\tilde{\Gamma}(p))^{-1}=4\pi i\frac{\tanh\theta/2}{\theta}$
ψ propagator	———	$\langle 0 \vert \psi_\alpha(p)\overline{\psi}_\beta(-p) \vert 0\rangle$	$\Delta_F(p)^{-1}=\frac{i}{\not p - m}$
$\overline{\psi}\psi\tilde{\sigma}$ vertex		$\sigma(p)\psi_\alpha(q)\overline{\psi}_\beta(-p-q)$	$\frac{i}{\sqrt{N}}$

The $\tilde{\sigma}$ propagator is given by minus the inverse of $\tilde{\Gamma}(p)$[8)]

$$\langle 0 \mid \tilde{\sigma}(p)\tilde{\sigma}(-p) \mid 0\rangle = -(\tilde{\Gamma}(p))^{-1} \quad . \tag{5.34}$$

The Feynman rules can now be read from (5.25), and are summarized in table 1 where we have used $p^2 = 2m^2(1-\cosh\theta) = -4m^2\sinh^2\frac{\theta}{2}$.

In the $1/N$ expansion, the renormalization of the model is very simple. By the power counting theorem[120], the only divergent diagrams are those for which the superficial degree of divergence, given by

$$\delta(\Gamma) = 2 - \frac{1}{2}N_\psi - N_\sigma \quad , \tag{5.35}$$

is non negative, where Γ denotes the diagram, N_ψ the number of external fermion lines, and N_σ the number of external $\tilde{\sigma}$ lines. Thus, we have divergencies in the 2– and 4–fermion correlation functions, in the 2–fermion–one $\tilde{\sigma}$ correlation function, and in the inverse $\tilde{\sigma}$ propagator. The last divergence, and the divergence of the fermion 2 point function, can be absorbed in the fermion and $\tilde{\sigma}$ wave function renormalization, and in the mass renormalization (or equivalently, due to mass transmutation, in the coupling constant renormalization). The superficial degrees of divergence of the remaining correlation functions are actually negative in the $1/N$ expansion, if we consider an appropriate sum of diagrams of the same order. The cancellation mechanism occurs between the diagrams displayed either in figure 5.4 or in figure 5.5: consider a diagram with the external lines ψ_i and $\overline{\psi}_i$; then form a new diagram, joining those lines into a $\tilde{\sigma}$ line, which in turn reproduces the pair ψ^i and $\overline{\psi}^i$ again (see the diagram of figure 5.4), i.e. the above mechanism was used in

8) Indeed,

$$\int \mathcal{D}\sigma\sigma(x)\sigma(y)e^{\frac{1}{2}\int\sigma\Gamma\sigma} = \frac{\delta^2}{\delta J(x)\delta J(y)}\int\mathcal{D}\sigma e^{\frac{1}{2}\int(\sigma\Gamma\sigma-2J\sigma)}\Bigg|_{J=0}$$

$$= \frac{\delta^2}{\delta J(x)\delta J(y)}\int\mathcal{D}\sigma e^{\frac{1}{2}\int\{(\sigma-J\Gamma^{-1})\Gamma(\sigma-\Gamma^{-1}J)-J\Gamma^{-1}J\}}\Bigg|_{J=0} = -\Gamma^{-1}\,.$$

Figure 5.4: Cancellation mechanism of lowest divergencies.

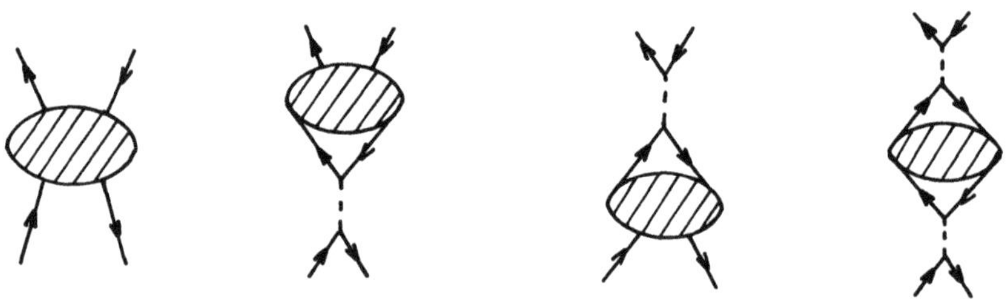

Figure 5.5: Cancellation mechanism of higher divergencies.

the first two, and in the last two diagrams of figure 5.5). It is easy to verify that the diagrams are of the same order in $1/N$, due to a factor of N arising from the fermion loop.

The cancellation is due to the fact that the blob is logarithmically divergent, which is thus a (cut-off dependent) constant when shrunk to a point. In this case, due to the definition (5.34), the divergent pieces cancel as shown in figures 5.6 and 5.7 (the $\tilde{\sigma}$ propagator is the inverse of the renormalized fermion loop).

5.2.4 Leading order S-matrix elements

The $1/N$ expansion of the model is very useful to check several results concerning the S-matrix, which will be computed exactly for the Gross-Neveu model in chapter 9. The essential property allowing for the computation of the exact S-matrix is its factorization in terms of 2 body S matrices. In turn, this property is equivalent to the statement that the probability of pair production is zero [9].

[9] In a factorizable amplitude, the intermediate states (after 2 body collisions) are on shell. Consider the amplitude of pair production, and the 3 body amplitude in figures 5.8 and 5.9, related by crossing. In figure 5.8, $r = -p_1 + p_2 + p_3$, which is on shell if either $p_1 = p_2$ or $p_1 = -p_3$, which are correct configurations. However, in figure 5.9, $r = p_1^* + p_2 + p_3$, consequently $r^2 \geq 3m^2$, by the two dimensional kinematics. Therefore, the total amplitude for pair production (or annihilation) must be zero, in a factorizable amplitude, when all diagrams are added.

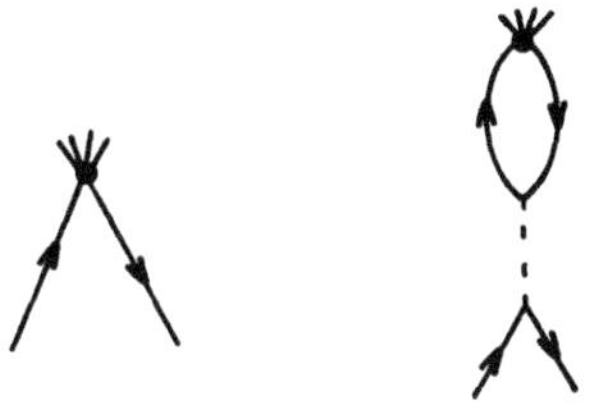

Figure 5.6: Cancellation mechanism concerning diagrams of figure 5.4: the same divergent constant appears in both diagrams, differing by a minus sign.

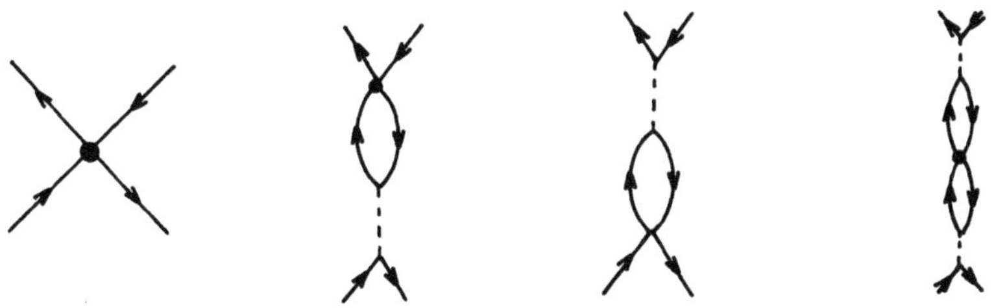

Figure 5.7: Cancellation mechanism concerning diagrams of figure 5.5

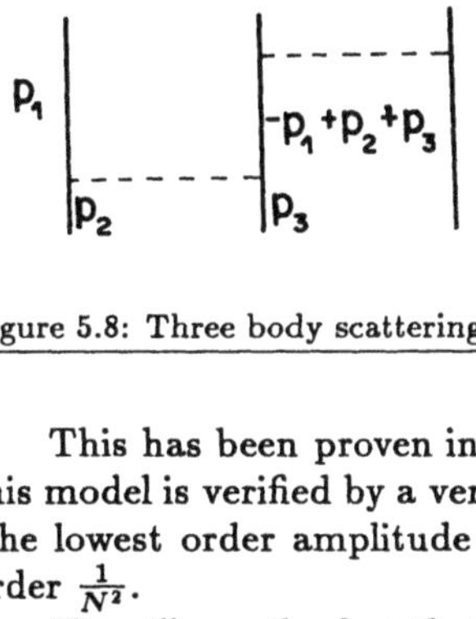

Figure 5.8: Three body scattering amplitude.

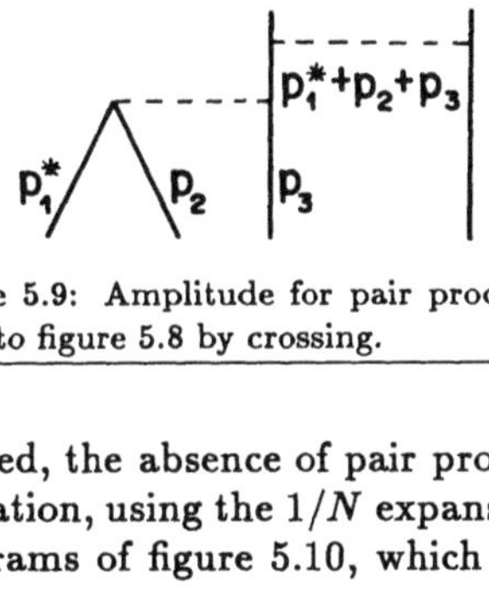

Figure 5.9: Amplitude for pair production related to figure 5.8 by crossing.

This has been proven in general[190]. Indeed, the absence of pair production in this model is verified by a very simple computation, using the $1/N$ expansion[180,191]. The lowest order amplitude is given by diagrams of figure 5.10, which are of the order $\frac{1}{N^2}$.

We will use the fact that in two dimensional space-time, one loop diagrams are exactly computable in terms of tree diagrams times a two-propagator loop (see[117] chapter 3 and figure 5.11[192]). This is due to the fact that the result of performing the k^0 integration using the theorem of residues is a rational function, after the change of variables (see the detailed computation for the massive Thirring model in chapter 3)

$$x_i = \frac{k^1 + p_i^1}{\sqrt{(k^1 + p_i^1)^2 + m^2} - m} \quad . \tag{5.36}$$

The rational function thus obtained may be decomposed into partial fractions, with coefficients calculated at the respective residua. This leaves us with a coefficient given in terms of a sum, where each term of the sum is given by the product of all but two propagators and vertices. The momentum integration only receives a

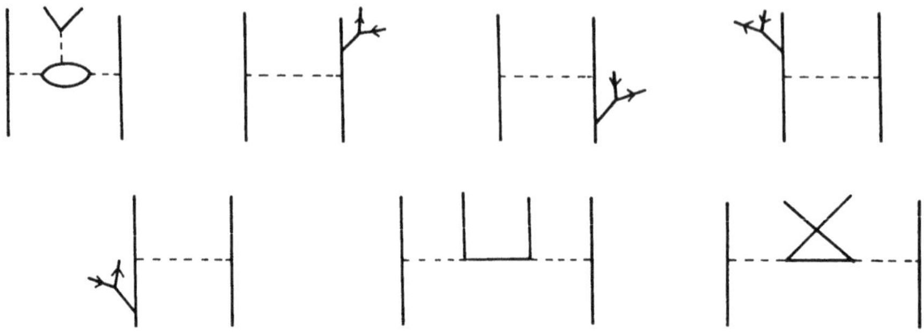

Figure 5.10: Diagrams of order $\frac{1}{N^2}$ contributing to the amplitude of pair production.

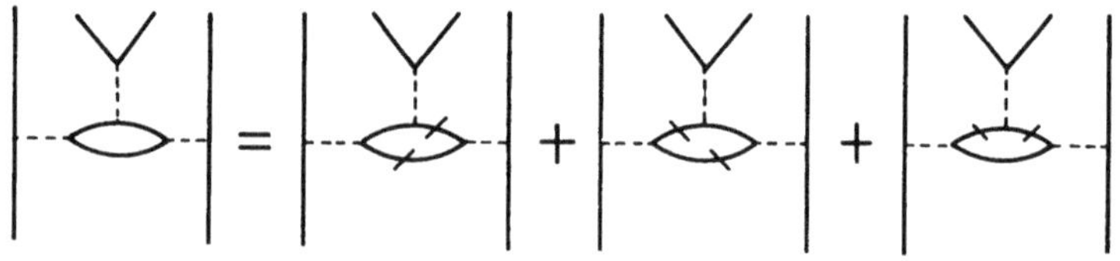

Figure 5.11: Explicit computation of the first diagram of figure 5.10. The dashes in the intermediate figure mean use of the cutting rule (chapter 3), where the momenta are such that the dashed lines are on shell; each diagram here cancels two corresponding diagrams in figure 5.10.

contribution from points where the poles of the two selected propagators coincide. The remaining integral is trivial, and its calculation turns out to be equivalent to

$$\begin{aligned} A(p) &= \int \frac{d^2k}{(2\pi)^2} \frac{1}{(k+p_i)^2 - m^2} \frac{1}{(k+p_j)^2 - m^2} \\ &= -\frac{i}{2\pi\sqrt{p^2(p^2-4m^2)}} \ln \frac{\sqrt{-p^2+4m^2} - \sqrt{-p^2}}{\sqrt{-p^2+4m^2}+\sqrt{-p^2}} = \frac{i\theta}{4\pi m^2 \sinh\theta} \quad . \end{aligned} \tag{5.37}$$

with $p = p_i - p_j$, $p^2 = -4m^2 \sin h^2\theta/2$, $p_i = m(\cosh\theta_i, \sinh\theta_i)$, $\theta = \theta_i - \theta_j$. Here we have used

$$\int \frac{d^2k}{(2\pi)^2} \frac{1}{[k^2 + \alpha(1-\alpha)p^2 - m^2 + i\varepsilon]^2} = \frac{i}{4\pi} \frac{1}{m^2 - \alpha(1-\alpha)p^2}, \tag{5.38a}$$

as well as,

$$\begin{aligned} \int_0^1 d\alpha \frac{\alpha(1-\alpha)}{m^2 - \alpha(1-\alpha)p^2} &= \frac{1}{\sqrt{p^2(p^2-4m^2)}} \ln \left(\frac{\sqrt{p^2(p^2-4m^2)} - p^2}{\sqrt{p^2(p^2-4m^2)}+p^2} \right)^2 \\ &= \frac{4\pi}{p^2} \tilde{\Gamma}_\pi(p), \end{aligned} \tag{5.38b}$$

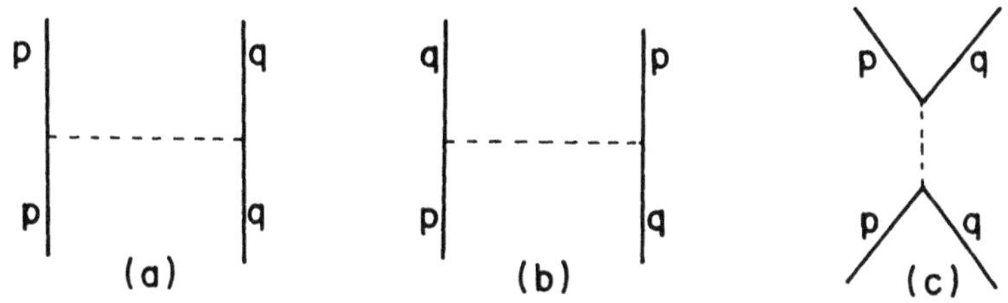

Figure 5.12: Lowest order contribution to 40*a*, 40*b* and 40*c*.

The above procedure permits to compute the diagram 5.10.1 exactly in terms of the tree diagrams 5.10.2-7, and shows that they add up to zero. Using the $1/N$ expansion we are able to obtain the S-matrix up to second order. The two particle scattering of fermions with isospin i, j and rapidities θ_1, θ_2, has the general form[10)]

$$\begin{aligned}\langle \theta_1' k, \theta_2' l \mid \theta_1 i, \theta_2 j\rangle &= \left[\sigma_1(\theta)\delta^{ik}\delta^{jl} + \sigma_2(\theta)\delta^{il}\delta^{kj} + \sigma_3(\theta)\delta^{ij}\delta^{kl}\right]\delta(\theta_1 - \theta_1')\delta(\theta_2 - \theta_2') \\ &- \left[\sigma_1(\theta)\delta^{il}\delta^{jk} + \sigma_2(\theta)\delta^{ik}\delta^{jl} + \sigma_3(\theta)\delta^{ij}\delta^{kl}\right]\delta(\theta_1 - \theta_2')\delta(\theta_2 - \theta_1')\ , \quad (5.39)\end{aligned}$$

where $\theta = \theta_1 - \theta_2$. In lowest order perturbation in $\frac{1}{N}$, we have the contribution due to exchange of the auxiliary $\tilde{\sigma}$ particle (figure 5.12). Using the Feynman rules in table 1 one finds[180]

$$\sigma_1(\theta) = 1 + \frac{2\pi}{N \sinh\theta} \qquad (5.40a)$$

$$\sigma_2(\theta) = \frac{2\pi i}{N\theta} \qquad (5.40b)$$

$$\sigma_3(\theta) = \frac{2\pi i}{N(i\pi - \theta)} \quad . \qquad (5.40c)$$

The computation can also be extended to one loop order. A perturbative expansion of the S-matrix, may also be obtained in this case (see also ref. [36] for further details).

5.2.5 Quantization of the non-local charge

Our next aim is to quantize the non local charge (5.8) in order to obtain non perturbative results from the model. Divergencies in this non local charge arise

10) Notice that instead of the usual momentum delta function, we use deltas in the rapidity. To compare with the usual perturbative calculations, we need to include a factor $\frac{4}{m^2 \sinh\theta}$, since $\delta(p_0 - p_0')\delta(p_1 - p_1') = \frac{4}{m^2 \sinh(\theta - \theta')}\delta(\theta_1 - \theta_1')\delta(\theta_2 - \theta_2')$.

from the short distance singularities of the product of two currents in the first term. We therefore, we have to study the Wilson expansion

$$[j_\mu(x+\varepsilon), j_\nu(x)] = \sum C_{\mu\nu}^{\{\alpha\}}(\varepsilon)\mathcal{O}_{\{\alpha\}}(x) \quad , \tag{5.41}$$

where $\mathcal{O}_{\{\alpha\}}(x)$ are local field operators and $C_{\mu\nu}^{\{\alpha\}}(\varepsilon)$ are c-number functions. The dimension of the operators $\mathcal{O}_{\{\alpha\}}(x)$ can be classified making use of the asymptotic freedom of the theory[178,31]. Since the left hand side of (5.41) has dimension 2, there are only contributions to the right hand side coming from operators of dimension 1 and 2. Moreover, they must be $O(N)$ algebra valued fields. Therefore one finds[35,193]

$$[j_\mu(x,\varepsilon), j_\nu(x)] = C_{\mu\nu}^{\rho}(\varepsilon) j_\rho(x) + D_{\mu\nu}^{\sigma\rho}(\varepsilon)\partial_\sigma j_\rho(x) \quad . \tag{5.42}$$

There are constraints from PT and CP invariance, as well as current conservation, restricting the coefficients $C_{\mu\nu}^{\rho}(\varepsilon)$ and $D_{\mu\nu}^{\sigma\rho}(\varepsilon)$. The first two imply

$$C_{\mu\nu}^{\rho}(\varepsilon) = -C_{\nu\mu}^{\rho}(-\varepsilon) = C_{\nu\mu}^{\rho}(\varepsilon) \tag{5.43a}$$

$$D_{\mu\nu}^{\sigma\rho}(\varepsilon) = -D_{\nu\mu}^{\sigma\rho}(-\varepsilon) - [C_{\nu\mu}^{\rho}(-\varepsilon)\varepsilon^\sigma - \frac{1}{2}g^{\rho\sigma}C_{\nu\mu}^{\lambda}(-\varepsilon)\varepsilon_\lambda] \tag{5.43b}$$

$$D_{\mu\nu}^{\sigma\rho}(-\varepsilon) = D_{\mu\nu}^{\sigma\rho}(\varepsilon) \quad . \tag{5.43c}$$

Using Lorentz invariance, we may thus obtain the general structure of the coefficients, which read

$$C_{\mu\nu}^{\rho}(\varepsilon) = C_1\frac{g_{\mu\nu}\varepsilon^\rho}{\varepsilon^2} + C_2\frac{\varepsilon_\mu\delta_\nu^\rho + \varepsilon_\nu\delta_\mu^\rho}{\varepsilon^2} + C_3\frac{\varepsilon_\mu\varepsilon_\nu\varepsilon^\rho}{(\varepsilon^2)^2} \tag{5.44a}$$

$$\begin{aligned} D_{\mu\nu}^{\sigma\rho}(\varepsilon) &= D_1\frac{(\varepsilon_\mu\delta_\nu^\sigma - \varepsilon_\nu\delta_\mu^\sigma)\varepsilon^\rho}{\varepsilon^2} + D_2\frac{(\varepsilon_\mu\delta_\nu^\rho - \varepsilon_\nu\delta_\mu^\rho)\varepsilon^\sigma}{\varepsilon^2} \\ &+ D_3(\delta_\mu^\rho\delta_\nu^\sigma - \delta_\nu^\rho\delta_\mu^\sigma) + \frac{1}{2}C_2\frac{\varepsilon_\mu\delta_\nu^\rho + \varepsilon_\nu\delta_\mu^\rho}{\varepsilon^2}\varepsilon^\sigma + \frac{1}{2}C_3\frac{\varepsilon_\mu\varepsilon_\nu\varepsilon^\rho\varepsilon^\sigma}{(\varepsilon^2)^2} \quad . \end{aligned} \tag{5.44b}$$

Asymptotic freedom implies that[11)]

$$C_i = \mathcal{O}(|\varepsilon|^{-0}) \quad , \quad D_i = \mathcal{O}(|\varepsilon|^{-0}) \quad . \tag{5.45}$$

Current conservation implies several constraints. One of them is

$$C_1 = -4\varepsilon^2\frac{d}{d\varepsilon^2}\left(D_1(-x^2) - D_3(-x^2)\right) \tag{5.46a}$$

[11)] In ref. [35], one supposes that the fundamental fields have only logarithmic singularities; therefore, a smearing in the space coordinate would be enough to define correlators. In this case it is possible to compute the C_i's exactly. We do not need them, however[194]. See appendix G.

or, for $\varepsilon^o = 0$, $\varepsilon^1 = x$:

$$C_1(-x^2) = -2x\frac{d}{dx}\left(D_1(-x^2) - D_3(-x^2)\right) \tag{5.46b}$$

In order that the cut-off charge

$$Q_\delta = \int_{|y_1-y_2|>\delta} dy_1 dy_2 \epsilon(y_1 - y_2) j_0(t,y_1) j_0(t,y_2) - Z_\delta \int dy j_1(t,y) \quad , \tag{5.47}$$

be finite in the limit $\delta \to 0$, we have [12)]

$$Z_\delta = 2(D_1(-\delta^2) - D_3(-\delta^2)) \quad . \tag{5.48}$$

From the fact that the expansion (5.42) contains no further term, we also have conservation of the non local quantum charge, which follows from

$$\begin{aligned} \frac{dQ_\delta}{dt} &= \int dy \left\{ [j_0(t,y), j_1(t,y+\delta)] - (\delta \to -\delta) \right\} \\ &\quad - 2(D_1(-\delta^2) - D_3(-\delta^2)) \int dy \dot{j}_1(t,y) \to 0 \quad \text{as} \quad \delta \to 0 \quad . \end{aligned} \tag{5.49}$$

We may thus define the quantum non local charge

$$Q^{ab} = \lim_{\delta \to 0} Q_\delta^{ab} \quad , \tag{5.50}$$

which is well defined and conserved. We examine now its action on asymptotic states. We could follow reference [35] and compute $Q^{in/out}$ in terms of asymptotic states. Although the results turn out to be correct the procedure is not rigorous. We shall thus follow [194] and [195], computing first the action on one particle state and then on two particle states, which is our aim. Details not shown in the text are presented in appendix G. The Lorentz transformation (G.6) acts on the non local charge as

$$[T, Q^{ab}] = -\frac{N-2}{\pi} J^{ab} \quad , \tag{5.51}$$

---

12) The divergent piece arises from the first, linearly divergent term in the Wilson expansion (5.42). We have

$$Q_\delta = \text{finite} + \int_\delta^{\mu^{-1}} dx \left\{ -\frac{C_1(x^2)}{x} \right\} \int dy\, j_1(t,y) - Z_\delta \int dy\, j_1(t,y)$$

Therefore we find (5.48) after using (5.46).

where J^{ab} is the isospin operator, acting on a $O(N)$ vector $\mid \theta c\rangle$ as

$$J^{ab}|\theta c\rangle = i\delta^{ac}|\theta b\rangle - i\delta^{bc}|\theta a\rangle \equiv (J^{ab})^{cd}|\theta d\rangle \quad . \tag{5.52}$$

Since the nonlocal charge commutes with translations, it must annihilate the vacuum:

$$Q^{ab}|0\rangle = 0 \quad . \tag{5.53}$$

Moreover, it is a second order antisymmetric tensor and, on a one particle state, must satisfy

$$Q^{ab}|\theta c\rangle = g(\theta)\{\delta^{bc}|\theta a\rangle - \delta^{ac}|\theta b\rangle\} \quad . \tag{5.54}$$

On the other hand, the generator of Lorentz transformations acts on a one particle state as

$$T|\theta a\rangle = \frac{\partial}{\partial\theta}|\theta a\rangle \quad . \tag{5.55}$$

Applying (5.51) on a one particle state, we find

$$\frac{\partial g}{\partial\theta} = -i\frac{N-2}{\pi} \quad . \tag{5.56}$$

Since Q^{ab} is odd under parity, we have

$$g(\theta) = -i\frac{N-2}{\pi}\theta \quad . \tag{5.57}$$

Therefore

$$Q^{ab}|\theta c\rangle = \frac{N-2}{i\pi}\theta\{\delta^{bc}|\theta a\rangle - \delta^{ac}|\theta b\rangle\} \quad . \tag{5.58}$$

The action of the charge on two particle states is described in appendix G. For a two particle state with different momenta we have (see G.15):

$$\left\langle\phi_3 \otimes \phi_4|Q^{\,\substack{out\\ in}}|\phi_1 \otimes \phi_2\right\rangle = \lim_{t\to\pm\infty}\int dxdy\epsilon(x-y)\langle\phi_3|j_0(t,x)|\phi_1\rangle\langle\phi_4|j_0(t,y)|\phi_2\rangle + \langle\phi_1|Q|\phi_1\rangle\langle\phi_2|\phi_2\rangle + (1\leftrightarrow 2) \tag{5.59}$$

where ϕ_i is an asymptotic state with given momentum p_i. The second term has been computed in (5.58). As for the first term we make the following observations: If $p_1 > p_2$, in the limit $t\to\infty$, we have only a contribution from $x < y$ (in the far past, there was no interaction, and the faster particle 1 is to the left of 2). There is a change in sign if either $p_1 < p_2$, or $t\to+\infty$. Therefore, performing the x, y integrals, we may summarize the action of the non local charge on asymptotic states by

$$Q^{ab}\mid\theta_1 i;\theta_2 j\rangle = \mid\theta_1 k;\theta_2 l\rangle\Big\{-\Big[(I^{ac})^{ik}(I^{cb})^{jl} - (I^{ac})^{jl}(I^{cb})^{ik}\Big] + \frac{N-2}{i\pi}\theta_1(I^{ab})^{ik}\delta^{jl} + \frac{N-2}{i\pi}\theta_2(I^{ab})^{jl}\delta^{ik}\Big\} \quad , \tag{5.60a}$$

and

$$\langle\theta_1 i;\theta_2 j \mid Q^{ab} = \langle\theta_1 k;\theta_2 l \mid \Big\{\Big[(I^{ac})^{ik}(I^{cb})^{jl} - (I^{ac})^{jl}(I^{cb})^{ik}\Big] + \frac{N-2}{i\pi}\theta_1 (I^{ab})^{ik}\delta^{jl} + \frac{N-2}{i\pi}\theta_2 (I^{ab})^{jl}\delta^{ik}\Big\} \quad . \qquad (5.60b)$$

The constraints on S-matrix elements[180] implied by the action of the non local charge on asymptotic states are discussed in chapter 8 and appendix G, where we also show that as a result of these constraints, there is no pair production.

5.3 Chiral Gross-Neveu Model

This model is a generalization of the previous one, promoting the discrete symmetry (5.19) to a global continuous chiral $\tilde{U}(1)$ symmetry[178]

$$\psi \rightarrow e^{i\theta\gamma_5}\psi \quad , \qquad (5.61)$$

and the trivial symmetry $\psi \rightarrow -\psi$ to a global $U(1)$ symmetry

$$\psi \rightarrow e^{i\alpha}\psi \quad . \qquad (5.62)$$

In fact, ψ now is a complex Dirac field (see footnote 1)), so that the symmetry is now $SU(N)$ rather than $O(N)$.[13)] The Lagrangian is given by

$$\mathcal{L} = i\overline{\psi}_i \,\partial\!\!\!/ \psi_i + \frac{1}{2}g[(\overline{\psi}_i\psi_i)^2 - (\overline{\psi}_i\gamma_5\psi_i)^2] \quad . \qquad (5.63)$$

where the summation over the $SU(N)$ index i is understood. The Lagrangian (5.63) again defines an integrable model[179,182]. The Noether current associated with the $SU(N)$ symmetry is given by

$$j_{\mu_{ij}} = i\overline{\psi}_j\gamma_\mu\psi_i \quad , \qquad (5.64)$$

Use of the equation of motion and Fierz transformation, shows that it satisfies Eq. (5.3), that is

$$\partial_\mu j_\nu - \partial_\nu j_\mu + 2g[j_\mu, j_\nu] = 0 \quad . \qquad (5.65)$$

This again implies the existence of a non local conserved charge of the form (5.3)

$$Q = \int dy_1 dy_2\, \epsilon(y_1 - y_2) j_0(t,y_1) j_0(t,y_2) + \frac{1}{g}\int dy\, j_1(t,y) \quad , \qquad (5.66)$$

as one readily verifies using current conservation and Eq. (5.65).

Hence equations (5.9) to (5.12) are valid in this case as well, which demonstrates the integrability of the chiral Gross-Neveu model.

---

13) In fact, the complete classical symmetry is $SU(N) \otimes U(1) \otimes U(1)_5$.

5.3.1 Cancellation of infrared singularities

In order to obtain the $1/N$ expansion of this model, we have to reformulate it. As in the previous case, the theory can be reduced to a quadratic form in ψ at the expense of two auxiliary fields,

$$\mathcal{L} = i\overline{\psi}\ \partial\!\!\!/ \psi - \frac{1}{2g}(\sigma^2 + \pi^2) + \overline{\psi}(\sigma + i\pi\gamma_5)\psi \quad . \tag{5.67}$$

However, the $1/N$ expansion of the model using the above Lagrangian can not be performed, due to serious IR problems[196]. Indeed, if we try to do it, we find a massless pole in the π propagator, which plays the role of a massless ("would be") Goldstone boson. Its role is going to be discussed later on. A Goldstone boson in two dimensions is a very problematic object, since two dimensional massless fields present severe IR problems (see chapter 2). In order to deal with this problem, it is common practice to write the fields in terms of $\sigma + i\pi = \rho e^{i\phi}$, such that the Lagrangian reads

$$\mathcal{L} = i\overline{\psi}\ \partial\!\!\!/ \psi - \frac{1}{2g}\rho^2 + \rho\overline{\psi}e^{i\phi\gamma_5}\psi \quad . \tag{5.68}$$

We are going to discuss the quantum theory associated with the above classical Lagrangian in three different ways, obtaining similar results. The more pedestrian approach consists in the extensive use of the bosonization formulae (3.26,28,29). The fermionic fields are bosonized in terms of an N-plet φ_i. The situation is analogous to the massive Thirring model, and one obtains the equivalent bosonic Lagrangian

$$\mathcal{L} = \frac{1}{2}\sum_{i=1}^{N}(\partial_\mu\varphi_i)^2 - \frac{1}{2g}\rho^2 + \frac{\mu}{2\pi}\rho\sum_{i=1}^{N}\cos(\phi + \varphi_i\sqrt{4\pi}) \quad . \tag{5.69}$$

It is now convenient to use the "fermionization" formulae in order to rewrite (5.69) in terms of a new fermion field $\tilde{\psi}_i$[197,198], via the identifications

$$i\overline{\tilde{\psi}}_i\ \partial\!\!\!/ \tilde{\psi}_i = \frac{1}{2}\left[\partial_\mu\left(\varphi_i + \frac{\phi}{\sqrt{4\pi}}\right)\right]^2 \tag{5.70a}$$

$$\overline{\tilde{\psi}}_i\tilde{\psi}_i = \frac{\mu}{2\pi}\cos(\phi + \varphi_i\sqrt{4\pi}) \tag{5.70b}$$

$$\overline{\tilde{\psi}}_i\gamma_\mu\tilde{\psi}_i = -\frac{1}{\sqrt{\pi}}\epsilon_{\mu\nu}\partial^\nu\left(\varphi_i + \frac{\phi}{\sqrt{4\pi}}\right) \quad . \tag{5.70c}$$

The Lagrangian (5.69) then takes the form[198]

$$\mathcal{L} = \overline{\tilde{\psi}}_i i\ \partial\!\!\!/ \tilde{\psi}_i - \frac{1}{2g}\rho^2 + \rho\overline{\tilde{\psi}}_i\tilde{\psi}_i + \frac{1}{2}\partial_\mu\phi\overline{\tilde{\psi}}_i\gamma^\mu\gamma_5\tilde{\psi}_i + \frac{N}{8\pi}(\partial_\mu\phi)^2 \quad . \tag{5.71}$$

The important point is that the massless field ϕ interacts only via its derivative, thus implying the absence of IR problems in the correlation functions of $\tilde{\psi}$. There may be doubts about the validity of the above bosonization formulae for the interpolating fields[199]. However, we may obtain the same Lagrangian (5.71) by computing the fermionic determinant associated with the Lagrangian (5.68). Defining

$$\rho_\pm = \rho e^{\pm i\phi} \quad , \tag{5.72}$$

the Lagrangian (5.68) takes the form

$$\mathcal{L} = \bar{\psi}\left(i\,\partial\!\!\!/ + \rho_+\frac{1+\gamma_5}{2} + \rho_-\frac{1-\gamma_5}{2}\right)\psi - \frac{1}{2g}\rho_+\rho_- \quad . \tag{5.73}$$

It is convenient to consider an auxiliary Lagrangian $\mathcal{L}'$ obtained from (5.73) by the substitution $\partial_\mu \to D_\mu = \partial_\mu - iA_\mu$, with A_μ an "external" $U(1)$-field:

$$\mathcal{L} = \bar{\psi}\left(i\,D\!\!\!\!/ + \rho_+\frac{1+\gamma_5}{2} + \rho_-\frac{1-\gamma_5}{2}\right)\psi - \frac{1}{2}\rho_+\rho_- \quad . \tag{5.74}$$

The corresponding effective action is given by

$$iW = \ln\,\det\left(i\,D\!\!\!\!/ + \rho_+\frac{1+\gamma_5}{2} + \rho_-\frac{1-\gamma_5}{2}\right) \quad . \tag{5.75}$$

The auxiliary Lagrangian (5.75) is invariant under the local transformation

$$\psi \to e^{i\varphi\gamma_5}\psi \tag{5.76a}$$

$$A_\mu \to A_\mu + \epsilon_{\mu\nu}\partial^\nu\varphi \tag{5.76b}$$

$$\rho_\pm \to e^{\mp i\varphi}\rho_\pm \quad , \tag{5.76c}$$

the generators of the corresponding infinitesimal transformation being

$$T = \int d^2x\left\{\epsilon_{\mu\nu}\partial^\nu\varphi\frac{\delta}{\delta A_\mu} + i\varphi\left[\rho_-\frac{\delta}{\delta\rho_-} - \rho_+\frac{\delta}{\delta\rho_+}\right]\right\} \quad . \tag{5.77}$$

The effective action (5.74) however breaks this local invariance. Indeed, the external field axial vector current

$$\tilde{J}_\mu(x;A) = \epsilon_{\mu\nu}J^\nu(x,A) = \epsilon_{\mu\nu}\frac{\delta W}{\delta A_\nu} \quad , \tag{5.78}$$

exhibits the well known Adler-Bell-Jackiw anomaly[74,200,201], which in 1+1 dimensions takes the form (see chapter 4)

$$\partial^\mu\tilde{J}_\mu = \frac{N}{2\pi}\epsilon^{\mu\nu}F_{\mu\nu} \quad . \tag{5.79}$$

Using (5.78) and (5.79) one finds[50,202]

$$TW = \frac{N}{2\pi}\int d^2x\,\varphi\epsilon^{\mu\nu}F_{\mu\nu} \quad . \tag{5.80}$$

Hence, $W[A]$ is not gauge invariant as a result of the anomaly (5.79).

From here it immediately follows, using (5.77), that

$$T^2W = -\frac{N}{\pi}\int d^2x\,(\partial^\mu\varphi)^2 \tag{5.81a}$$

$$T^nW = 0 \quad , \quad n > 2. \quad . \tag{5.81b}$$

Since finite transformation are obtained by exponentiating T, the corresponding series therefore truncates when e^T is applied to W:

$$e^TW = W + \frac{N}{2\pi}\int d^2x\,\varphi\epsilon^{\mu\nu}F_{\mu\nu} - \frac{N}{2\pi}\int d^2x\,(\partial^\mu\varphi)^2 \quad . \tag{5.82a}$$

Therefore

$$W(A_\mu - \epsilon_{\mu\nu}\partial^\nu\varphi, e^{-i\varphi}\rho_+, e^{i\varphi}\rho_-) = \\ W(A_\mu,\rho_+,\rho_-) + \frac{N}{2\pi}\int d^2x\,\epsilon^{\mu\nu}F_{\mu\nu}\varphi - \frac{N}{2\pi}\int d^2x\,(\partial^\mu\varphi)^2 \quad . \tag{5.82b}$$

In particular, we obtain for the actual case of interest, $A_\mu = 0$,

$$W(0,\rho_+,\rho_-) = W(\epsilon_{\mu\nu}\partial^\nu\varphi, e^{-i\varphi}\rho_+, e^{i\varphi}\rho_-) + \frac{N}{2\pi}\int d^2x\,(\partial^\mu\varphi)^2 \quad . \tag{5.83}$$

The left hand side is just the effective action associated with (5.73), while the first term on the right hand side is the effective action associated with (5.71), if we make the identification $\varphi = \frac{\phi}{2}$. This proves the equivalence of Lagrangians (5.71) and (5.73).

5.3.2 The $\frac{1}{N}$ expansion

The large N expansion of the model defined by the Lagrangian (5.71) can be explicitly performed[198], using the method described in Eqs. (5.22) to (5.34). The propagator of the ρ field is exactly the same as that obtained for the $\tilde{\sigma}$ field in the $O(N)$ case [14)].

[14)] We first redefine $\rho = -m + \frac{1}{\sqrt{N}}\tilde{\rho}$; the condition that the field $\tilde{\rho}$ has zero vacuum expectation value, implies that the mass parameter m be given by the expression (5.18).

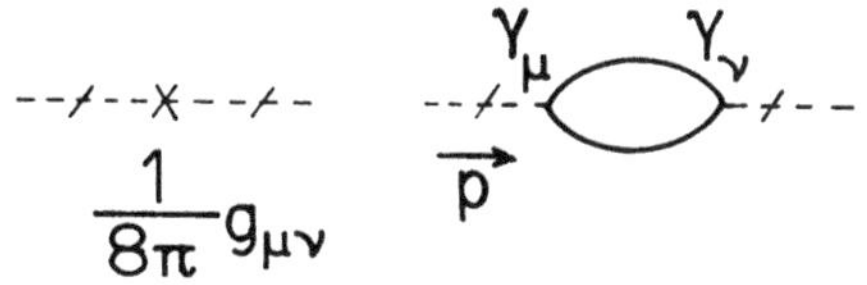

Figure 5.13: Diagrams contributing to the two point function of the field $A_\mu = \sqrt{N}\epsilon_{\mu\nu}\partial^\nu\phi$. The first diagram is simply the tree contribution given in the Lagrangian (5.83), which arouse from (5.81a). The second term is the one loop contribution.

The zeroth order contribution to the ρ-propagator is thus given by expression (5.33), with the diagrams depicted in figure 5.3.

From (5.32) and (5.33) we obtain for $\tilde{\Gamma}(p)$

$$\tilde{\Gamma}(p) = -\frac{i}{2\pi}\frac{\theta}{\tanh\frac{\theta}{2}} \quad , \tag{5.84}$$

where θ is defined by $p^2 = -4m^2\sinh^2\frac{\theta}{2}$.

However, we have now the scalar field ϕ. Since only $\partial_\mu\phi$ occurs in (5.71), only the two point function of $A_\mu = \sqrt{N}\epsilon_{\mu\nu}\partial^\nu\phi$ is needed. We thus have to compute the diagrams given in figure 5.13, given by[15]

$$\tilde{\Gamma}_{\mu\nu}(p) = +\frac{1}{8\pi}g_{\mu\nu} - \frac{1}{4}\operatorname{tr}\int\frac{d^2k}{(2\pi)^2}\gamma_\mu\frac{(\not{k}+m)\gamma_\nu(\not{k}+\not{p}+m)}{(k^2-m^2)[(k+p)^2-m]} \quad . \tag{5.85}$$

The result of performing the integral is

$$\begin{aligned}\tilde{\Gamma}_{\mu\nu}(p) &= \frac{1}{2\pi}\sqrt{\frac{-p^2+4m^2}{-p^2}}\ln\frac{\sqrt{-p^2+4m^2}-\sqrt{-p^2}}{\sqrt{-p^2+4m^2}+\sqrt{-p^2}}(g_{\mu\nu}p^2 - p_\mu p_\nu)\\ &= \frac{1}{2\pi}\theta\tanh\frac{\theta}{2}(g_{\mu\nu}p^2 - p_\mu p_\nu) \quad ,\end{aligned} \tag{5.86}$$

[15] This integral is convergent if a reasonable regularization is used. Indeed, the only divergent piece comes from

$$\int d^2k\,\operatorname{tr}\gamma_\mu\gamma_\rho\gamma_\nu\gamma_\sigma k_\rho k_\sigma f(k^2) \sim \frac{1}{2}g^{\rho\sigma}\operatorname{tr}\gamma_\mu\gamma_\rho\gamma_\nu\gamma_\sigma\int d^2k\,k^2 f(k^2) \quad ,$$

which vanishes due to the identity $\gamma^\rho\gamma_\mu\gamma_\rho = 0$, in 2 dimensions. Therefore, the integral may be performed using the theorem of residua. The anomaly cancels among the two terms, as it should, since the second term corresponds to the effective action.

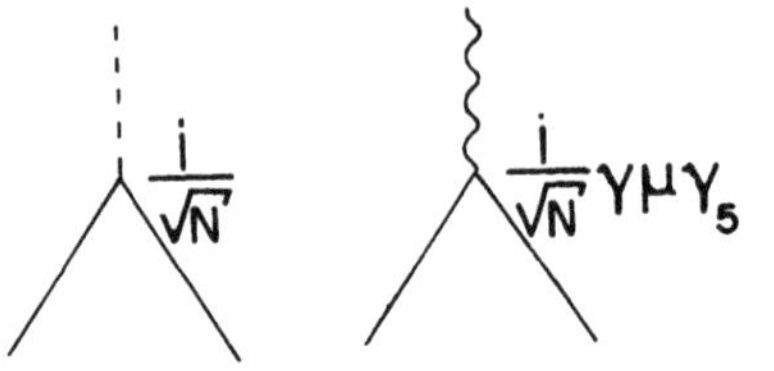

Figure 5.14: Interaction vertices corresponding to the Lagrange density (5.71).

where $p^2 = -4m^2 \sinh^2 \frac{\theta}{2}$.

The vertices are shown in figure 5.14. The amplitudes for particle scattering are thus all free from IR divergencies, and may be computed without difficulty. We shall compute the two particle scattering amplitude in lowest order. The S-matrix elements have the general structure

$$\langle \theta_1' k \theta_2' l \mid \theta_1 i \theta_2 j \rangle = (u_1(\theta)\delta^{ik}\delta^{jl} + u_2(\theta)\delta^{il}\delta^{kj})\delta(\theta_1 - \theta_1')\delta(\theta_2 - \theta_2') + (u_1(\theta)\delta^{il}\delta^{kj} + u_2(\theta)\delta^{ik}\delta^{jl})\delta(\theta_1 - \theta_2')\delta(\theta_2 - \theta_1') \quad . \quad (5.87)$$

The lowest order contributions to $u_1(\theta)$ are given in figure 5.14. The ρ exchange diagram contributes a term $\frac{\pi i}{N \sinh\theta}$ (the ρ-propagator at zero momentum is a constant), while the A_μ exchange contributes (see footnote 10)

$$\frac{(-\frac{i}{2})^2}{4m \sinh\theta} \bar{u}\gamma^\mu u \bar{u}'\gamma_\mu u' \Big(-\frac{4\pi i}{N}\Big) = \frac{\pi i}{N} \coth\theta \quad . \quad (5.88)$$

The resulting amplitude is given by

$$u_1(\theta) = 1 + \frac{i\pi}{N} \coth\left(\frac{\theta}{2}\right) \quad . \quad (5.89)$$

We shall verify in chapter 8 that this is indeed the lowest order approximation of the exact solution. Particle antiparticle backward scattering vanishes. This may also be verified by a perturbative calculation, using an IR cut-off[203].

5.3.3 Operator formulation

This model may also be studied in the operator formalism, which provides, if correctly characterized, an $1/N$ expansion, and a correct understanding of the relation between the "would be" Goldstone boson, and chiral symmetry.

Since the fields ψ_i are fermionic, and define the fundamental representation of $U(N)$, we have the following bosonic representation[199]

$$\psi_i(x) = K_i \left(\frac{\mu}{2\pi}\right)^{\frac{1}{2}} e^{-i\frac{\pi}{4}\gamma_5} \times :e^{-i\sqrt{\frac{\pi}{N}}\left[\gamma_5\chi(x) + \int_{x^1}^\infty dy^1 \dot{\chi}(x^0,y^1)\right]}: \, :e^{-i\sqrt{\pi}\left[\gamma_5\chi_i(x) + \int_{x^1}^\infty dy^1 \dot{\chi}_i(x^0,y^1)\right]}: \quad (5.90)$$

with $i = 1,\dots,N$. Since the $\chi_i's$ are $SU(N)$ valued, they are not independent but satisfy[16)]

$$\sum_{i=1}^{N}\chi_i(x) = 0 \quad . \tag{5.91}$$

The field χ is the potential of the conserved $U(1)$ current. Its zero mass character will ensure that the $U(1)$ symmetry is not spontaneously broken[17)].

In the above, K_i is a Klein factor, necessary to enforce the correct anticommutation relations among different $\psi_i's$. Due to the $U(1)\times\widetilde{U}(1)$ symmetry, the divergence and the curl of the $U(1)$ current both vanish, so that the field $\chi(x)$ is massless. Therefore the fermion fields contain the so called infraparticles[92], and we need to extract them in order to arrive at the physical fields of the theory. They are given by

$$\hat{\psi}_i(x) = K_i\sqrt{\frac{\mu}{2\pi}}e^{-i\sqrt{\pi}\left\{\gamma_5\chi_i(x)+\int_{x^1}^{\infty}dy^1\dot{\chi}_i(x^0,y^1)\right\}} \quad . \tag{5.92}$$

The ψ fields (5.92) will be found to correspond to the field $\widetilde{\psi}_i$ in (5.71). These fields no longer carry $U(1)\times\widetilde{U}(1)$ charge, and transform as a representation of $SU(N)$. The constraint (5.91) implies

$$\hat{\psi}_i^\dagger \sim \frac{1}{(n-1)!}\epsilon_{ii_1\cdots i_{n-1}}\hat{\psi}_{i_1}\cdots\hat{\psi}_{i_{N-1}} \quad , \tag{5.93}$$

where on the right hand side a suitable redefinition of the Klein factor and the normal product prescription is required. Eq.(5.93) states that the antifermions of the chiral Gross-Neveu model can be viewed as a bound state of $N-1$ fermions. We shall use this to determine the S-matrix, and its pole structure.

Asymptotically one expects $\hat{\psi}$ to describe massive particles, so that one should have[199]

$$\hat{\psi}(vt,t) \to \frac{1}{\sqrt{|t|}}\{e^{-im\gamma^{-1}t}\hat{a}(m\gamma v) + e^{im\gamma^{-1}t}\hat{b}^\dagger(m\gamma v)\} \quad , \tag{5.94}$$

where $\gamma = \frac{1}{\sqrt{1-v^2}}$.

The fields $\hat{\psi}_i$ carry spin $s = \frac{1}{2}(1-1/N)$, as can be seen from Eq. (2.70b). They have the property (see 2.62).

$$\hat{\psi}(x,t)\hat{\psi}(y,t) = e^{2\pi is\epsilon(x-y)}\hat{\psi}(y,t)\hat{\psi}(x,t) \quad , \tag{5.95}$$

16) These fields are given by $\chi_f = \sum t^{i_D}_{ff}\phi^{i_D}$ where t^{i_D} are the generators of the "torus" of $SU(N)$, and ϕ^{i_D} the potentials of the corresponding $SU(N)$ currents.

17) In other words, it is the pseudo potential of the $U(1)$ current, while χ_i, are pseudo potentials of the $SU(N)$ currents.

implying an unusual "statistics" for the creation and annihilation operators defined in (5.94)

$$\hat{a}^\dagger(p)\hat{a}^\dagger(p') = e^{2\pi i s\epsilon(p-p')}\hat{a}^\dagger(p')\hat{a}^\dagger(p) \quad . \tag{5.96}$$

Since no scattering theory is known for particles with the above statistics, it is necessary to replace the field $\hat{\psi}$ by another field ψ' with a well defined statistics. This is achieved by introducing in (5.90) free massless scalar and pseudoscalar fields B and A, quantized with metric opposite to that of $\chi(x)$, in such a way that the divergent infrared behavior of ψ induced by $\chi(x)$ is compensated, without affecting the statistics. We define[199]

$$\psi_i'(x) = e^{i\sqrt{\frac{\pi}{N}}[\gamma^5 A(x)+B(x)]}\psi_i(x) \quad . \tag{5.97}$$

Correspondingly the creation and annihilation operators $a^\dagger, a, b^\dagger, b$ are related to $\hat{a}^\dagger, \hat{a}, \hat{b}^\dagger, \hat{b}$ by

$$\hat{a}_{in}^\dagger(p) = a_{in}^\dagger(p)e^{2\pi i(s-\frac{1}{2})\int_p^\infty N_{in}(p')dp'} \tag{5.98}$$

$$\hat{a}_{out}^\dagger(p) = a_{out}^\dagger(p)e^{2\pi i(s-\frac{1}{2})\int_\infty^p N_{out}(p')dp'} \quad . \tag{5.99}$$

where $N_{\substack{in\\out}}$ are the corresponding particle number operators.

Since we expect $\psi_i'(x)$ to be a local field describing massive degrees of freedom, we should have in the far past and future[204]

$$\psi'(vt,t) \underset{t\to\pm\infty}{\longrightarrow} \frac{1}{\sqrt{|t|}}[e^{-im\gamma^{-1}t}a_{\substack{out\\in}}(m\gamma v) + e^{im\gamma^{-1}t}b^\dagger_{\substack{out\\in}}(m\gamma v)] \quad . \tag{5.100}$$

Substitution of ψ in terms of ψ' in (5.63) leads formally to the Lagrangian

$$\mathcal{L} = i\overline{\psi}_i' \,\partial\!\!\!/ \psi_i' + \frac{1}{2}g[(\overline{\psi}_i'\psi_i')^2 - (\overline{\psi}_i'\gamma_5\psi_i')^2] - \frac{1}{2}(\partial_\mu A)^2 - \frac{1}{2}(\partial_\mu B)^2 +$$
$$\frac{\alpha}{\sqrt{N}}\overline{\psi}'\gamma^5\gamma^\mu\psi'\partial_\mu A - \frac{\beta}{\sqrt{N}}\overline{\psi}'\gamma^\mu\psi'\partial_\mu B \quad , \tag{5.101}$$

where we allowed for general couplings α and β, which after renormalization should reduce to $\sqrt{\pi}$ as the renormalized value. We will came back to this point after obtaining the $\frac{1}{N}$ expansion, which we consider next.

The effective action obtained from the Lagrangian (5.101) after introduction of the auxiliary fields σ and π (compare with (5.67)) is given by

$$S_{eff} = -iN \operatorname{tr}\ln\left\{i\,\partial\!\!\!/ + \sigma + i\pi\gamma_5 + \frac{\alpha}{\sqrt{N}}\gamma^5\,\partial\!\!\!/ A - \frac{\beta}{\sqrt{\pi}}\,\partial\!\!\!/ B\right\}$$
$$-\frac{1}{2g}\int d^2x(\sigma^2+\pi^2) - \frac{1}{2}\int d^2x[(\partial_\mu A)^2 + (\partial_\mu B)^2] \quad . \tag{5.102}$$

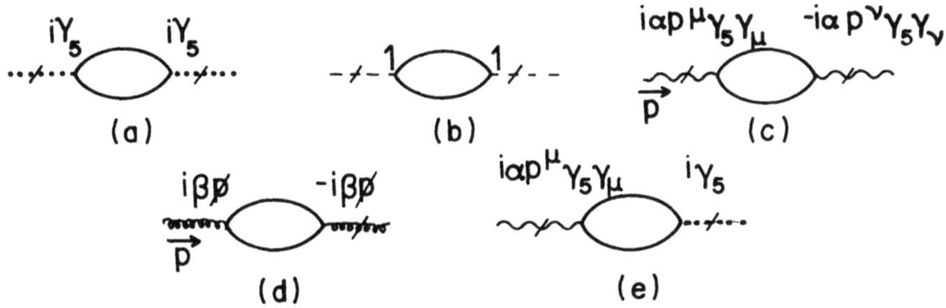

Figure 5.15: Diagrams contributing to the $\pi(a)$,$\sigma(b)$,$A_\mu(c)$, and $B(d)$ proper functions, and for the $A_\mu\pi(e)$ mixed function.

The field σ is found to have a non-vanishing vacuum expectation value $\langle\sigma\rangle = -m$, so that it is convenient to write

$$\sigma = -m + \frac{\tilde{\sigma}}{\sqrt{N}} \quad , \quad \text{and} \quad \pi = \frac{\tilde{\pi}}{\sqrt{N}} \quad . \tag{5.103}$$

The second order contribution to the effective action is given by

$$S^{(2)}_{eff} =$$

$$\frac{i}{2}\,\mathrm{tr}\int d^2x d^2y\,(i\,\partial\!\!\!/ - m)^{-1}\Big[\frac{1}{\sqrt{N}}\tilde{\sigma}(y) + \frac{i}{\sqrt{N}}\tilde{\pi}(y)\gamma_5 + \frac{\alpha}{\sqrt{N}}\gamma_5\;\partial\!\!\!/ A(y) - \frac{\beta}{\sqrt{N}}\;\partial\!\!\!/ B(y)\Big]$$

$$\times(i\,\partial\!\!\!/ - m)^{-1}\Big[\frac{1}{\sqrt{N}}\tilde{\sigma}(x) + \frac{i}{\sqrt{N}}\tilde{\pi}(x)\gamma_5 + \frac{\alpha}{\sqrt{N}}\gamma_5\;\partial\!\!\!/ A(x) - \frac{\beta}{\sqrt{N}}\;\partial\!\!\!/ B(x)\Big]$$

$$-\frac{1}{2gN}\int d^2x\,(\tilde{\sigma}^2 + \tilde{\pi}^2) + \frac{1}{2}\int d^2x\,(A\Box A + B\Box B) \quad .(5.104)$$

The proper two point function are given in terms of the one loop diagrams tabulated in figure 5.15. There are mixed functions as well. We summarize the results. For the σ two point function

$$\tilde{\Gamma}_\sigma(p) = -\,\mathrm{tr}\int \frac{d^2k}{(2\pi)^2}\frac{k\!\!\!/ + m}{k^2 - m^2}\frac{k\!\!\!/ + p\!\!\!/ + m}{(k+p)^2 - m^2} - \frac{i}{gN}$$

$$= \frac{i}{2\pi}\sqrt{\frac{-p^2 + 4m^2}{-p^2}}\ln\frac{\sqrt{-p^2 + 4m^2} - \sqrt{-p^2}}{\sqrt{-p^2 + 4m^2} + \sqrt{-p^2}} = \frac{-i}{2\pi}\frac{\theta}{\tanh\frac{\theta}{2}} \quad , \tag{5.105}$$

which is the same as (5.84). For the π two point function we have

$$\tilde{\Gamma}_{\pi\pi}(p) = \mathrm{tr}\int \frac{d^2k}{(2\pi)^2}\gamma_5\frac{k\!\!\!/ + m}{k^2 - m^2}\gamma_5\frac{k\!\!\!/ + p\!\!\!/ + m}{(k+p)^2 - m^2} - \frac{i}{gN}$$

$$= \frac{i}{2\pi}\sqrt{\frac{-p^2}{-p^2 + 4m^2}}\ln\frac{\sqrt{-p^2 + 4m^2} - \sqrt{-p^2}}{\sqrt{-p^2 + 4m^2} + \sqrt{-p^2}} = -\frac{i}{2\pi}\frac{\theta}{\coth\frac{\theta}{2}} \tag{5.106}$$

For the 1PI two-point function involving A and B one finds

$$\begin{aligned}\tilde{\Gamma}_{AA}(p) &= \alpha^2 \,\mathrm{tr} \int \frac{d^2k}{(2\pi)^2} \frac{\gamma^\mu\gamma_5(\not k + m)\gamma^\nu\gamma_5(\not k + \not p + m)}{(k^2-m^2)[(k+p)^2-m^2]} p_\mu p_\nu - ip^2 \\ &= -i\frac{\alpha^2}{\pi}p^2 - ip^2 + (2m\alpha)^2\tilde{\Gamma}_{\pi\pi} \\ &= 4im^2\sinh^2\frac{\theta}{2}\left\{1 + \frac{\alpha^2}{\pi}\left(1 - \frac{\theta}{\sinh\theta}\right)\right\} \end{aligned} \tag{5.107}$$

$$\begin{aligned}\tilde{\Gamma}_{BB}(p) &= \beta^2 \,\mathrm{tr} \int \frac{d^2k}{(2\pi)^2} \frac{\gamma^\mu(\not k + m)\gamma^\nu(\not k + \not p + m)}{(k^2-m^2)[(k+p)^2-m^2]} p_\mu p_\nu - ip^2 \\ &= -ip^2 = 4im^2\sinh^2\frac{\theta}{2} \quad , \end{aligned} \tag{5.108}$$

where we used, in (5.107) the contribution of the Adler anomaly, and in (5.108) "current conservation". We also have the mixed term

$$\begin{aligned}\tilde{\Gamma}_{\pi A}(p) &= -\alpha \,\mathrm{tr} \int \frac{d^2k}{(2\pi)^2}\gamma^\mu\gamma^5 \frac{\not k + m}{k^2-m^2}\gamma_5 \frac{\not k + \not p + m}{(k+p)^2 - m^2} p_\mu \\ &= -2m\alpha\tilde{\Gamma}_{\pi\pi}(p) = -\frac{m\alpha}{\pi i}\left(\frac{\coth\frac{\theta}{2}}{\theta}\right)^{-1} \quad . \end{aligned} \tag{5.109}$$

All other 1PI two-point functions involving b are found to vanish. In order to obtain the σ and B propagators we just invert (5.105) and (5.108) respectively

$$D_B(p) = \frac{1}{p^2} \quad . \tag{5.110}$$

The computation of the A and π propagators requires on the other hand, the inversion of the 2×2 matrix

$$\Gamma = \begin{pmatrix} \tilde{\Gamma}_{\pi\pi} & \tilde{\Gamma}_{\pi A} \\ \tilde{\Gamma}_{\pi A} & \tilde{\Gamma}_{AA} \end{pmatrix} \quad , \tag{5.111}$$

whose determinant is given by (using Eqs. (5.106,107,109))

$$\det\Gamma = -ip^2(1+\alpha^2/\pi)\tilde{\Gamma}_{\pi\pi} = 4im^2\sinh^2\frac{\theta}{2}\left(1+\frac{\alpha^2}{\pi}\right)\frac{1}{2\pi i}\frac{\theta}{\coth\frac{\theta}{2}} \quad . \tag{5.112}$$

We find

$$D(p^2) = -i\Gamma^{-1} = \begin{pmatrix} D_A(p^2) & D_{A\pi}(p^2) \\ D_{A\pi}(p^2) & D_\pi(p^2) \end{pmatrix} \tag{5.113}$$

with

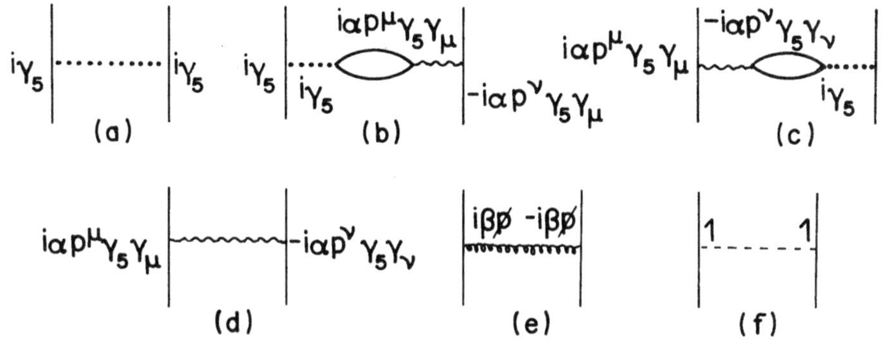

Figure 5.16: Contribution to fermion fermion scattering at order $\frac{1}{N}$.

$$D_A(p^2) = \frac{1}{p^2}\frac{1}{1+\alpha^2/\pi} \tag{5.114}$$

$$D_\pi(p^2) = 2\pi\frac{\coth\frac{\theta}{2}}{\theta}\frac{1+\frac{\alpha^2}{\pi}\left(1-\frac{\theta}{\sinh\theta}\right)}{1+\frac{\alpha^2}{\pi}} \tag{5.115}$$

$$D_{\pi A}(p^2) = \frac{1}{p^2}\frac{2m\alpha}{1+\alpha^2/\pi} \quad . \tag{5.116}$$

We now fix the parameters α and β in (5.101) by requiring that the IR divergencies cancel. We expect to obtain for the non renormalized values, $\alpha = \infty$ (corresponding to $\alpha_{ren} = \sqrt{\pi}$ and $\beta = \beta_{ren} = \sqrt{\pi}$, since B couples to a conserved current). As a matter of fact, if we consider the fermion-fermion interaction to lowest order in the $\frac{1}{N}$ expansion (see figure 5.16) we verify that for $\alpha \to \infty$, the mixed propagator $D_{\pi A}$ gives the contribution (see figure 5.16b, c)

$$\Delta_{16b+c} = -\frac{i}{p^2}\frac{2m\alpha^2}{1+\frac{\alpha^2}{\pi}}\times$$
$$\{p_\mu\overline{u}(p_1)\gamma_5 u(p_1)\overline{u}(p_2)\gamma_5\gamma^\mu u(p_2) + p_\mu\overline{u}(p_1')\gamma_5\gamma^\mu u(p_1)\overline{u}(p_2')\gamma_5 u(p_2)\} \quad . \tag{5.117}$$

We use the equation of motion for the free spinor

$$\overline{u}(p')\gamma_5(\not p' - \not p)u(p) = -2m\overline{u}(p')\gamma_5 u(p) \quad , \tag{5.118}$$

to reduce Eq. (5.117) to the form

$$\Delta_{16b+c} = 2\frac{\alpha^2}{1+\frac{\alpha^2}{\pi}}\frac{i}{p^2}\overline{u}\gamma^5\ \not p u\overline{u}'\gamma^5\ \not p u' = 2\frac{\alpha^2}{1+\frac{\alpha^2}{\pi}}i\frac{p_\mu p_\nu - g_{\mu\nu}p^2}{p^2}\overline{u}\gamma^\mu u\overline{u}'\gamma^\mu u' \quad , \tag{5.119}$$

where we used the identity $\epsilon^{\mu\rho}p_\rho\epsilon^{\nu\sigma}p_\sigma = p^\mu p^\nu - g^{\mu\nu}p^2$. The exchange of a π gives a vanishing contribution for zero momentum transfer. Indeed,

$$\overline{u}(p)\gamma_5 u(q)\overline{u}(q)\gamma_5 u(p) = m^2 \operatorname{tr}\gamma_5 \left(\cosh\frac{\theta-\chi}{2} + \gamma_5 \sinh\frac{\theta-\chi}{2} + \right. \\ \left. i\gamma_0 \sinh\frac{\theta+\chi}{2} + i\gamma^1 \sinh\frac{\theta+\chi}{2} \right) \quad (5.120)$$

vanishes in the limit $q \to p$ because of the trace, so that

$$\Delta_{16a} = 0 \quad . \quad (5.121)$$

Finally, the B exchange (diagram 5.16e) gives the contribution

$$\Delta_{17d} = -\frac{i}{p^2}\beta^2 \overline{u}\gamma^\mu u(p)\overline{u}'\gamma^\nu u \frac{p_\mu p_\nu}{p^2} \quad , \quad (5.122)$$

which partially cancels (5.119), leaving us for $\alpha \to \infty$ and $\beta = \pi$, with

$$\Delta_{16} = \Delta_{16f} - 2i\overline{u}\gamma^\mu u\overline{u}'\gamma_\mu u' \quad , \quad (5.123)$$

which leads again to the result (5.89).

We have thus verified in the $N \to \infty$ limit that both Lagrangians (5.71) and (5.101) lead to the same result for $U(1)$, as they should. In the limit $\alpha \to \infty$, the renormalized coupling α_{ren} indeed turns out to be $\sqrt{\pi}$, as one reads off from the four point function, and the pole in the π-propagator vanishes. To summarize, we conclude that part of the field (5.90) which carries chirality decouples[197,198,199,205] from the physical spectrum and the remaining part describes an $SU(N)$ multiplet with a well defined factorizable S-matrix (see chapter 8). As we shall show in chapter 15 there exists however a critical point in the coupling constant, where these conclusions are not realized, and the theory is conformally invariant. We postpone this question to that chapter.

5.3.4 Quantization of non-local charge

The discussion of the existence and conservation of a non local charge in the quantum chiral Gross-Neveu model follows exactly the same pattern as in the $O(N)$ invariant model. No anomaly exists in this case, since (5.42) gives all possible candidates to the Wilson expansion. As for the asymptotic result, there exist however differences due to the different symmetry group in question; thus, in (5.51) we have to replace $N-2$ by N, and in (5.58) only the first term is present. It is not difficult to see that the action of Q^{ab} on asymptotic states is given in this case by

$$
\begin{aligned}
Q^{ab}|\theta_1 i;\theta_2 j\rangle &= |\theta_1 k;\theta_2 l\rangle\left[-(J^{ac})^{ik}(J^{cb})^{jl} + \frac{N}{i\pi}\theta_1(J^{ab})^{ik}\delta^{jl} + \frac{N}{i\pi}\theta_2(J^{ab})^{jl}\delta^{ik}\right]\\
\langle\theta_1 i;\theta_2 j|Q^{ab} &= \langle\theta_1 k;\theta_2 l|\left[-(J^{ac})^{ki}(J^{cb})^{lj} + \frac{N}{i\pi}\theta_1(J^{ab})^{ki}\delta^{lj} + \frac{N}{i\pi}\theta_2(J^{ab})^{lj}\delta^{ik}\right]\\
Q^{ab}|\theta_1 i;\overline{\theta}_2 j\rangle &= |\theta_1 k;\overline{\theta}_2 l\rangle\left[-(J^{ac})^{ik}(J^{cb})^{jl} + \frac{N}{i\pi}\theta_1(J^{ab})^{ik}\delta^{jl} - \frac{N}{i\pi}\overline{\theta}_2(J^{ab})^{lj}\delta^{ik}\right]\\
\langle\theta_1 i;\overline{\theta}_2 j|Q^{ab} &= \langle\theta_1 k;\overline{\theta}_2 l|\left[-(J^{ac})^{ki}(J^{cb})^{lj} + \frac{N}{i\pi}\theta_1(J^{ab})^{ki}\delta^{lj} - \frac{N}{i\pi}\overline{\theta}_2(J^{ab})^{jl}\delta^{ik}\right]
\end{aligned}
\tag{5.124a, b, c, d}
$$

where J^{ab} is the $SU(N)$ generators (see G.11). These formulae will be used in chapter 8 to fix the exact S-matrix of this model as well.

5.4 Conclusions and Physical Interpretation

The Gross Neveu model is a simple but physically rich model. Its semi classical analysis, both in the $O(N)$ and in the $SU(N)$ symmetric cases reveals that the model has a rich bound state structure[180]. We shall see in chapter 8 that the complete S-matrix of the model can be computed, permitting a reasonable understanding of the $O(N)$ symmetric model.

In the specific case of the chiral Gross Neveu model, a more interesting physical picture arises, due to the chiral symmetry breaking issue. Since as we saw in section 5.3 a mass term is dynamically generated for the fermion, one could be led to conclude that the chiral symmetry is broken. However, this is not possible in two dimensional space time[93]. This problem has been discussed at length by several authors[197,198,205]. The interesting outcome is that the chirality carrying field decouples from the theory (Eq. (5.97)). In the operator language, it is represented by the auxiliary fields A and B. The physical fermion, either (5.92) or (5.97) though exhibiting a non-vanishing mass gap, are chiral singlets. This physical picture is carried over to the supersymmetric $\mathbb{C}P^{N-1}$ model, and lies behind the fact that antiparticles are bound states of particles, in both, the Gross Neveu model, and in the supersymmetric $\mathbb{C}P^{N-1}$ model. This fact permits the computation of the S-matrix for these two cases[206].

The chiral Gross Neveu model will also be used in order to study the $U(1)$ problem in the context of QED_2, in chapter 12, where a nice physical interpretation is obtained.

Chapter VI

NON-LINEAR SIGMA MODELS — CLASSICAL ASPECTS

6.1 Introduction

6.1.1 Historical development

Non-linear sigma models have widespread applications in statistical mechanics as well as in quantum field theory. Historically they have been introduced by Schwinger[207] and used in field theory by Gell-Mann and Lévy in the sixties to describe the phenomenology of pions[208], realizing the expected current algebra[209,210] in terms of a quantum theoretic model. The main drawback was the lack of renormalizability[211]. Moreover, the great success of non-abelian gauge theories, which have become a cornerstone in the field theoretic description of elementary particles, has left the sigma models at a secondary level[21]. Their revival is due to various developments which we will describe in this and the following chapters.

First of all, there have been activities on the statistical mechanics of sigma models, with applications to spin dynamics[212]. Also, it became clear in the late seventies that sigma models are, in several ways, closely related to Yang-Mills theories[57,58,65,66]; they were therefore used as a field theoretic laboratory for understanding the theoretical basis of more realistic theories. More recently, sigma models have appeared in string theory, whose Lagrangian is of the sigma model type[68,69]. These models also appear naturally in the low-energy description of supergravity and of string (or superstring) theory, mainly in the framework of Kaluza-Klein schemes[213]. In addition, they provide realizations of certain infinite-dimensional algebras, such as the Kac-Moody algebras[51], which are very important in the context of string theory and conformal field theory[129]. Finally, sigma models appear quite naturally in several other field-theoretic problems that we shall discuss: integrable models[66,38], confinement[65], $\frac{1}{N}$-expansion[65,75], Wess-Zumino term[51], non-abelian generalization of the bosonization technique[51,53], and others.

6.1.2 Sigma models and current algebra

The sigma model was originally introduced in four dimensions and used to describe the phenomenology of P.C.A.C. (partially conserved axial current). It was not a purely geometrical model, but rather given by the Lagrangian[208,209]

$$\mathcal{L} = \mathcal{L}_0 + \mathcal{L}_1 \tag{6.1}$$

which is the sum of two terms:

$$\mathcal{L}_0 = \overline{\psi}\left[\frac{i}{2}\overleftrightarrow{\not\partial} - g(\sigma + i\vec{\pi}\cdot\vec{\tau}\gamma_5)\right]\psi + \frac{1}{2}\left((\partial\sigma)^2 + (\partial\vec{\pi})^2\right) - \frac{1}{2}\mu^2(\sigma^2 + \vec{\pi}^2) - \frac{\lambda}{4}(\sigma^2 + \vec{\pi}^2)^2 \ , \tag{6.2a}$$

and $\mathcal{L}_1$ is the perturbation

$$\mathcal{L}_1 = c\sigma \ . \tag{6.2b}$$

$\mathcal{L}_0$ has a symmetry under the group $SO(4)$, or equivalently, under the chiral group, $SU(2) \times \widetilde{SU(2)}$, which is realized as

$$\psi \longrightarrow \psi + \frac{i}{2}\vec{\alpha}\cdot\vec{\tau}\psi - \frac{i}{2}\vec{\beta}\cdot\vec{\tau}\gamma_5\psi \quad , \tag{6.3a}$$

$$\vec{\pi} \longrightarrow \vec{\pi} - \vec{\alpha}\wedge\vec{\pi} + \vec{\beta}\sigma \quad , \tag{6.3b}$$

$$\sigma \longrightarrow \sigma - \vec{\beta}\cdot\vec{\pi} \quad . \tag{6.3c}$$

The Noether currents associated with these symmetry transformations are the vector current

$$V_\mu^i = \frac{1}{2}\overline{\psi}\gamma_\mu\tau^i\psi + \varepsilon^{ijk}\pi^j\partial_\mu\pi^k \quad , \tag{6.4a}$$

and the axial vector current

$$A_\mu^i = -\frac{1}{2}\overline{\psi}\gamma_\mu\gamma_5\tau^i\psi + \pi^i\partial_\mu\sigma - \sigma\partial_\mu\pi^i \quad , \tag{6.4b}$$

which are associated with the transformations labelled by the parameters α^i and β^i, respectively. $\mathcal{L}_1$ breaks this symmetry; more precisely, it respects the vector part but breaks the axial part. Therefore, the vector current remains conserved (CVC hypothesis) but the axial current develops a non-vanishing divergence which is, in fact, proportional to the pion field (PCAC hypothesis)[208,209]:

$$\partial^\mu A_\mu^i = c\pi^i \quad . \tag{6.5}$$

This is in accordance with the fact that the operators π^i interpolate between the vacuum and the one-pion states[209]:

$$< 0|\pi^i(0)|\pi^j(q) > = Z_\pi^{\frac{1}{2}}\delta^{ij} \quad . \tag{6.6}$$

We therefore have the following relation for the pion decay constant:

$$f_\pi m_\pi^2 = c Z_\pi^{\frac{1}{2}} \quad . \tag{6.7}$$

A more detailed look at the quantum theory associated with the Lagrangian (6.1) reveals that $\mathcal{L}_1$ induces a non-vanishing expectation value v for σ. We thus define a new field s with vanishing vacuum expectation value by

$$\sigma = s + v \quad , \quad <s> = 0 \quad . \tag{6.8}$$

In terms of s (rather than σ), the Lagrangian is of the form $\mathcal{L} = \mathcal{L}_a + \mathcal{L}_b$, where $\mathcal{L}_b$ contains the piece linear in s:

$$\begin{aligned}\mathcal{L}_a =&\overline{\psi}[\frac{i}{2}\overleftrightarrow{\not\partial} - m - g(s + i\vec{\pi}\cdot\vec{\tau}\gamma_5)]\psi + \frac{1}{2}((\partial\vec{\pi})^2 - \mu_\pi^2\vec{\pi}^2)\\ &+ \frac{1}{2}((\partial s)^2 - \mu_\sigma^2 s^2) - \lambda v s(s^2 + \vec{\pi}^2) - \frac{\lambda}{4}(s^2 + \vec{\pi}^2)^2 \quad ,\end{aligned} \tag{6.9a}$$

$$\mathcal{L}_b = (c - v\mu_\pi^2)s \quad , \tag{6.9b}$$

and where we have introduced the new mass parameters

$$m = gv \quad ; \quad \mu_\pi^2 = \mu^2 + \lambda v^2 \quad ; \quad \mu_\sigma^2 = \mu^2 + 3\lambda v^2 \quad . \tag{6.10}$$

For $v \neq 0$, the pion fields $\vec{\pi}$ and the sigma field σ are no longer degenerate in mass, as they were in the original Lagrangian $\mathcal{L}_0$. The parameter v is determined from the requirement that $\langle s\rangle = 0$. In lowest order, equating the coefficient in Eq. (6.9b) to zero gives $v = c/\mu_\pi^2$.

Obviously, when $\mathcal{L}_1 \neq 0$, the $\widetilde{SU}(2)$ symmetry is broken explicitly. But even when $\mathcal{L}_1 = 0$, we may have two phases: for $\mu^2 > 0$, we are in the symmetric phase: the minimum of the potential occurs at $\sigma^2 + \vec{\pi}^2 = 0$, i.e., $\sigma = \pi^i = 0$, $v = 0$, and the fermions remain massless. For $\mu^2 < 0$, on the other hand, we are in the so-called Goldstone phase[214,21], where the $\widetilde{SU}(2)$ symmetry is broken spontaneously (rather than explicitly): the minimum of the potential occurs at $\sigma^2 + \vec{\pi}^2 = -\mu^2/\lambda \neq 0$, i.e., $v = -\mu^2/\lambda \neq 0$, and the fermions acquire a non-zero mass, while the pion fields $\vec{\pi}$ describe massless (Goldstone) bosons.

The problem with this model is the particle corresponding to the field σ, which has not been observed experimentally. One possibility to get rid of this particle is to realize the chiral $SU(2)$ symmetry non-linearly, by imposing the constraint

$$\sigma^2 + \vec{\pi}^2 = f_\pi^2 \quad . \tag{6.11}$$

In this case f_π is the only parameter of the theory (formally this corresponds to taking the infinite coupling limit $\lambda \to \infty$). The model is now described entirely by the geometry of the target manifold on which it is defined, that is, the sphere S^3. For example, we can use Eq. (6.11) to eliminate the σ field from the Lagrangian, obtaining

$$\mathcal{L}_0 = \overline{\psi}\big[\frac{i}{2}\overleftrightarrow{\not\partial} + g(\sqrt{f_\pi^2 - \vec{\pi}^2} + i\vec{\pi}\cdot\vec{\gamma}\gamma_5)\big]\psi + \frac{1}{2}(\partial\vec{\pi})^2 + \frac{1}{2}\frac{(\vec{\pi}\cdot\partial\vec{\pi})^2}{f_\pi^2 - \vec{\pi}^2} \quad , \tag{6.12a}$$

$$\mathcal{L}_1 = c\sqrt{f_\pi^2 - \vec{\pi}^2} \quad . \tag{6.12b}$$

In four dimensions, the non-linear sigma model as defined by Eq. (6.12) is unfortunately not renormalizable. A study of its renormalization properties has been performed long ago[211]. In lowest order new counterterms such as $(\partial\vec{\pi}\cdot\partial\vec{\pi})^2$ are required. Therefore, this theory is at best a phenomenological low-energy approximation of some more realistic model.

In two dimensions, the non-linear sigma model as defined by Eq. (6.12) (without the matter interaction) is a special case of the general class of non-linear sigma models that we are going to study from now on. These models have a geometric origin – just like non-abelian gauge theories. In the present chapter we shall, at the purely classical level, define non-linear sigma models, and their interaction with matter fields, from the geometric point of view, and investigate their integrability. In the next chapter we shall discuss quantization of these models, and consider further formulations.

6.1.3 Two-dimensional sigma models: preliminaries

Our aim is to provide a general framework for defining classical non-linear sigma models. We are particularly interested in theories that are integrable, i.e., that exhibit an infinite number of conservation laws, as well as in models relevant to more realistic quantum field theory.

We begin with the simplest series of non-linear sigma models, namely the spherical models, more commonly known as the $O(N)$-invariant non-linear sigma models or, somewhat abusively, the $O(N)$ (sigma) models; these have played an interesting and peculiar role in the recent development of quantum field theory. Classically, the field configurations in this kind of model are maps φ from some base space (usually flat Minkowski space or Euclidean space) to the $(N-1)$-dimensional sphere S^{N-1}, so any such map can be viewed as a multiplet of N real scalar fields φ_i, $i = 1, ..., N$, subject to the quadratic constraint

$$\varphi^2 = \varphi_i\varphi_i = 1 \quad , \tag{6.13}$$

with dynamics governed by the Lagrangian

$$\mathcal{L} = \frac{1}{2}(\partial\varphi)^2 = \frac{1}{2}\,\partial^\mu\varphi_i\,\partial_\mu\varphi_i \quad . \tag{6.14}$$

(Here and in the following, summation over repeated indices is understood, as usual.) This is, of course, just the Lagrangian for the free field, but the presence of the constraint (6.13) implies that the model is *not* a free field theory. In fact, variation of the Lagrangian (6.14) with respect to φ subject to the constraint (6.13) gives the equation of motion

$$\partial^2\varphi + (\partial\varphi)^2\varphi = 0 \quad , \tag{6.15a}$$

or more explicitly

$$\partial^\mu\partial_\mu\varphi_i + (\partial^\mu\varphi_j\,\partial_\mu\varphi_j)\varphi_i = 0 \quad , \tag{6.15b}$$

which is obviously a non-linear partial differential equation. (One way to derive (6.15) is by enforcing the constraint with the help of a Lagrange multiplier term that is added to the Lagrangian, giving

$$\mathcal{L} = \frac{1}{2}(\partial\varphi)^2 + \lambda(\varphi^2 - 1) = \frac{1}{2}\,\partial^\mu\varphi_i\,\partial_\mu\varphi_i + \lambda(\varphi_i\varphi_i - 1) \quad . \tag{6.16}$$

Then varying the Lagrangian (6.16) with respect to the Lagrange multiplier λ yields the constraint (6.13), while varying it with respect to φ and using the constraint again gives the equation of motion (6.15).)

The $O(4)$ model, in particular, was originally introduced in the context of strong interactions – as explained before. But it was only in 1975 that Polyakov[58] and Migdal[57] studied it in two dimensions and found strong analogies with four-dimensional non-abelian gauge theories: an analysis of the recursion relations in the lattice theory shows that – just like non-abelian gauge theories[21,215] – it is asymptotically free. The $O(3)$ model, on the other hand, exhibits instanton solutions, which should be viewed as analogues of the instanton solutions found in Yang-Mills theories.

The $O(3)$ sigma model is closely related with the sine-Gordon theory[34]. Indeed, in light-cone variables, the equation of motion reads

$$\partial_+\partial_-\varphi + (\partial_+\varphi\cdot\partial_-\varphi)\varphi = 0 \quad . \tag{6.15c}$$

Now in the region where both derivatives $\partial_+\varphi$ and $\partial_-\varphi$ are non-zero and in fact linearly independent, so that together with the unit vector φ (to which they are orthogonal) they form a frame in three-dimensional space, we introduce the composite field α by

$$\cos\,\alpha = \frac{\partial_+\varphi\cdot\partial_-\varphi}{|\partial_+\varphi||\partial_-\varphi|} \quad ; \tag{6.17}$$

we can decompose the second derivatives $\partial_+^2\varphi$ and $\partial_-^2\varphi$ of φ in terms of φ, $\partial_+\varphi$ and $\partial_-\varphi$. For simplicity, we assume also that

$$(\partial_+\varphi)^2 = 1 = (\partial_-\varphi)^2 \quad , \tag{6.18}$$

which can be achieved by performing an appropriate conformal transformation[1)]. Then

$$\partial_+^2 \varphi = -\varphi - \frac{\partial_+ \alpha}{\sin\,\alpha}\,\partial_- \varphi + \cot\,\alpha\,\partial_+ \alpha\,\partial_+ \varphi \quad . \tag{6.19a}$$

$$\partial_-^2 \varphi = -\varphi - \frac{\partial_- \alpha}{\sin\,\alpha}\,\partial_+ \varphi + \cot\,\alpha\,\partial_- \alpha\,\partial_- \varphi \quad . \tag{6.19b}$$

Differentiating (6.19a) with respect to x^-, or (6.19b) with respect to x^+, and inserting back the other equation now leads to the following equation of motion for α:

$$\partial_+ \partial_- \alpha + \sin\,\alpha = 0 \quad . \tag{6.20}$$

This is the sine-Gordon equation[132]. An analogous connection can be established between the $O(4)$ sigma model and the complex sine-Gordon theory[34]. Therefore, the results on the sine-Gordon equation described in chapter 3 and the analogous results that hold for the complex sine-Gordon equation[216] suggest that the $O(3)$ and $O(4)$ models are both integrable.

Actually the $O(N)$ sigma model is integrable for any value of N, in the following sense. Consider the Noether current j_μ associated with the $O(N)$-transformations, which is a vector field with values in the real antisymmetric $(N \times N)$-matrices, with matrix entries

$$(j_\mu)_{ij} = \varphi_i \overleftrightarrow{\partial_\mu} \varphi_j = \varphi_i\,\partial_\mu \varphi_j - \partial_\mu \varphi_i\,\varphi_j \quad . \tag{6.21}$$

As usual, this current is conserved due to the equation of motion (6.16):

$$\partial^\mu j_\mu = 0 \quad . \tag{6.22}$$

On the other hand, it follows from the constraint (6.13) that it also satisfies the identity

$$\partial_\mu j_\nu - \partial_\nu j_\mu + 2[j_\mu, j_\nu] = 0 \quad . \tag{6.23}$$

This implies, for example, conservation (i.e., time-independence) not only of the standard Noether charge

$$Q^{(1)} = \int dy\, j_0(t,y) \quad , \tag{6.24}$$

but also of the non-local charge

$$Q^{(2)} = \int dy_1 dy_2\, \epsilon(y_1 - y_2)\, j_0(t,y_1)\, j_0(t,y_2) + \int dy\, j_1(t,y) \quad , \tag{6.25}$$

as can be verified from Eqs. (6.22) and (6.23) by direct calculation. (Note that these charges are – just like the current itself – real antisymmetric $(N \times N)$-matrices,

---

1) This is allowed in the classical theory, which is conformally invariant.

so the commutators and products appearing in Eqs. (6.23) and (6.25) are to be understood in the sense of ($N \times N$)-matrices.) In fact, $Q^{(1)}$ and $Q^{(2)}$ are just the first two members of an infinite series of conserved charges $Q^{(n)}$ which – except for $Q^{(1)}$ – turn out to be non-local and which can be derived by an appropriate expansion in the spectral parameter based on the linear system associated with this problem; a detailed exposition will be given, in a more general setting, later on in this chapter.

The way in which the sigma field in the $O(N)$ sigma model can be coupled to fermions so as to preserve these classical properties can be derived by using supersymmetry. To this end, we extend the multiplet of N real scalar fields φ_i to a multiplet of N real scalar superfields Φ_i on which we impose the same quadratic constraint as before:

$$\Phi_i \Phi_i = 1 \quad . \tag{6.26}$$

The superfield expansion of the fields Φ_i reads, in the Majorana representation for the gamma matrices (see appendix A)

$$\Phi_i(x, \theta) = \varphi_i(x) + \bar{\theta}\psi_i(x) + \frac{1}{2}\bar{\theta}\theta F_i(x) \quad , \tag{6.27}$$

where θ is a two-component Majorana spinor. In terms of the component fields φ_i (the scalar fields), ψ_i (the Majorana spinor fields which are their fermionic partners) and F_i (the auxiliary fields), the constraint (6.26) becomes

$$\varphi_i \varphi_i = 1 \quad , \quad \varphi_i \psi_i = 0 \quad , \quad \varphi_i F_i = \frac{1}{2}\bar{\psi}_i \psi_i \quad . \tag{6.28}$$

The corresponding Lagrangian can be written

$$\mathcal{L} = \frac{1}{8} \int d\bar{\theta} d\theta \overline{D\Phi_i} D\Phi_i \quad , \tag{6.29}$$

where $D\Phi_i$ is the superderivative of Φ_i, with spinor components determined from

$$D_\alpha = \frac{\partial}{\partial \bar{\theta}^\alpha} - i(\partial\!\!\!/ \theta)_\alpha = (\gamma_0)_{\alpha\beta} \frac{\partial}{\partial \theta_\beta} - i(\gamma^\mu \theta)_\alpha \partial_\mu \quad , \tag{6.30}$$

and the rules of Berezin[137] integration are such that

$$\int d\bar{\theta} d\theta \bar{\theta}\theta = 4 \quad . \tag{6.31}$$

Using (6.27) and (6.30) to compute the superfield expansion for the superfields $D\Phi_i$ and carrying out the fermionic integrals according to the standard rule (6.31) of Berezin integration, we obtain

$$\mathcal{L} = \frac{1}{2}\, \partial^\mu \varphi_i\, \partial_\mu \varphi_i + \frac{i}{2} \bar{\psi}_i \overleftrightarrow{\partial\!\!\!/} \psi_i + 2F_i F_i \quad . \tag{6.32}$$

Variation of the Lagrangian (6.32) with respect to the fields φ_i, ψ_i and F_i under the constraints (6.28) (e.g., by means of the Lagrange multiplier method described before) leads to the following equations of motion:

$$\partial^\mu \partial_\mu \varphi_i + (\partial^\mu \varphi_j \, \partial_\mu \varphi_j + (\overline{\psi}_j \psi_j)^2) \, \varphi_i - i \overline{\psi}_j \gamma^\mu \psi_i \, \partial_\mu \varphi_j - 2(\overline{\psi}_j \psi_j) \, F_i = 0 \quad , \tag{6.33a}$$

$$i(\delta_{ij} - \varphi_i \varphi_j) \, \partial\!\!\!/ \psi_j + (\overline{\psi}_j \psi_j) \, \psi_i = 0 \quad , \tag{6.33b}$$

$$F_i = \frac{1}{2} (\overline{\psi}_j \psi_j) \, \varphi_i \quad . \tag{6.33c}$$

Inserting Eq. (6.33c) back into Eqs. (6.32) and (6.33a, b), we can eliminate the auxiliary field both from the Lagrangian and from the remaining equations of motion, arriving at[217]

$$\mathcal{L} = \frac{1}{2} \, \partial^\mu \varphi_i \, \partial_\mu \varphi_i + \frac{i}{2} \overline{\psi}_i \overleftrightarrow{\partial\!\!\!/} \psi_i + \frac{1}{2} (\overline{\psi}_i \psi_i)^2 \tag{6.34}$$

as the Lagrangian,

$$\partial^\mu \partial_\mu \varphi_i + (\partial^\mu \varphi_j \, \partial_\mu \varphi_j) \, \varphi_i - i \overline{\psi}_j \gamma^\mu \psi_i \, \partial_\mu \varphi_j = 0 \tag{6.35a}$$

as the bosonic equation of motion and

$$i(\delta_{ij} - \varphi_i \varphi_j) \, \partial\!\!\!/ \psi_j + (\overline{\psi}_j \psi_j) \, \psi_i = 0 \tag{6.35b}$$

as the fermionic equation of motion. In particular, we see that the bosonic sector is the usual $O(N)$ sigma model, while the fermionic sector is nothing but the Gross-Neveu model discussed in chapter 5. The Noether current J_μ associated with the $O(N)$-transformations is now the sum

$$J_\mu = j_\mu + j_\mu^M \tag{6.36}$$

of a contribution j_μ from the bosonic fields φ_i, with matrix entries as given in (6.21), and a contribution j_μ^M from the fermionic matter fields ψ_i, with matrix entries

$$(j_\mu^M)_{ij} = i \overline{\psi}_j \gamma_\mu \psi_i \quad . \tag{6.37}$$

Again, this current is conserved due to the equations of motion (6.35):

$$\partial^\mu J_\mu = 0 \quad . \tag{6.38}$$

On the other hand, it follows from the constraints (6.28), together with the fermionic equation of motion (6.35b), that it also satisfies the relation

$$\partial_\mu (J_\nu + j_\nu^M) - \partial_\nu (J_\mu + j_\mu^M) + 2[J_\mu, J_\nu] = 0 \quad . \tag{6.39}$$

This implies, for example, conservation (i.e., time-independence) not only of the standard Noether charge

$$Q^{(1)} = \int dy\, J_0(t,y) \quad , \tag{6.40}$$

but also of the non-local charge

$$Q^{(2)} = \int dy_1 dy_2\, \epsilon(y_1 - y_2)\, J_0(t,y_1)\, J_0(t,y_2) + \int dy\, (J_1 + j_1^M)(t,y) \quad , \tag{6.41}$$

as can be verified from Eqs. (6.38) and (6.39) by direct calculation. Again, $Q^{(1)}$ and $Q^{(2)}$ are just the first two members of an infinite series of conserved charges $Q^{(n)}$ which – except for $Q^{(1)}$ – turn out to be non-local and which can be derived by an appropriate expansion in the spectral parameter based on the linear system associated with this problem; a detailed exposition will be given, in a more general setting, later on in this chapter.

A natural question that poses itself at this point is of course whether this intriguing structure is specific to the $O(N)$ sigma models or whether it can be extended to other classes of non-linear sigma models, and if so, to which. For example, considering the sphere S^{N-1} as the quotient space $S^{N-1} \cong O(N)/O(N-1)$ or

$$S^{N-1} \cong SO(N)/SO(N-1) \quad ; \tag{6.42}$$

one may ask what happens if one replaces the orthogonal group $O(N)$ by the unitary group $U(N)$ and hence, the $(N-1)$-dimensional sphere S^{N-1} by some quotient space such as $SU(N)/SU(N-1)$ or $SU(N)/U(N-1)$, the latter being nothing but the complex $(N-1)$-dimensional projective space[218]

$$\mathbb{C}P^{N-1} \cong SU(N)/U(N-1) \quad , \tag{6.43}$$

whose points are, by definition, just the one-dimensional subspaces (rays) in the space $\mathbb{C}^N$.

Thus we are directly led to the extremely interesting series of non-linear sigma models commonly known as the $\mathbb{C}P^{N-1}$ models. These models can be formulated in at least two different ways. One of them uses the idea that a ray in $\mathbb{C}^N$ can be identified with the orthogonal projector onto that ray, so that any classical field configuration can be viewed as a complex $(N \times N)$-matrix field P subject to the constraints[218]

$$P^2 = P \quad , \quad P^\dagger = P \quad , \quad \mathrm{tr}(P) = 1 \quad . \tag{6.44}$$

The other method is based on the fact that a ray in $\mathbb{C}^N$ can be represented by a unit vector in $\mathbb{C}^N$ which is however determined only up to a phase, so that any classical field configuration can be viewed as a multiplet of N complex scalar fields z_i, $i = 1, ..., N$, subject to the quadratic constraint[54]

$$\bar{z}_i z_i = 1 \quad , \tag{6.45}$$

together with a freedom under $U(1)$ gauge transformations

$$z_i(x) \to e^{i\alpha(x)} z_i(x) \quad , \tag{6.46}$$

reflecting the fact that the original fields z_i and all of their gauge transforms are just different parametrizations of the same physical field. In this approach, one must make sure that all physical quantities are gauge invariant, which requires the use of covariant derivatives

$$D_\mu z_i = \partial_\mu z_i - i z_i A_\mu \quad , \tag{6.47}$$

where A_μ is a $U(1)$ gauge potential which transforms according to

$$A_\mu \to A_\mu + \partial_\mu \alpha \tag{6.48}$$

under $U(1)$ gauge transformations. The dynamics of the model is then governed by the Lagrangian[56]

$$\mathcal{L} = \overline{D^\mu z_i} D_\mu z_i \quad . \tag{6.49}$$

Here, the $U(1)$ gauge potential appears as a Lagrange multiplier, so its equation of motion expresses it in terms of the fields z_i:

$$A_\mu = -i \bar{z}_j \partial_\mu z_j \quad . \tag{6.50}$$

One of the most attractive features of the $\mathbb{C}P^{N-1}$ models is that they possess instanton solutions for any N. To see this, we must pass from Minkowski space to compactified Euclidean space, which is the two-sphere S^2, so the classical field configurations are now maps from S^2 to $\mathbb{C}P^{N-1}$. These Euclidean field configurations fall into distinct topological classes, called homotopy classes, and due to the fact that $\pi_2(\mathbb{C}P^{N-1}) \cong \mathbb{Z}$, they can be labelled by an integer Q called their winding number and explicitly given by[56]

$$Q = \frac{i}{2\pi} \int d^2x \, \epsilon_{\mu\nu} \, \overline{D_\mu z_i} \, D_\nu z_i = \frac{1}{2\pi} \int d^2x \, \epsilon_{\mu\nu} \, \partial_\mu A_\nu \quad , \tag{6.51}$$

with A_μ given by Eq. (6.50) and $\epsilon_{12} = 1$. Equivalently, using the boundary conditions on the fields z_i at infinity in Euclidean space which guarantee that they represent a map which extends to the point at infinity in S^2, we can write, for $|x|$ large,

$$z_i(x) \sim g(x) z_i \quad ,$$

where g is a phase function describing a mapping from the unit circle $U(1)$ in x-space to the group $U(1)$; then Q is the winding number of g. Now the Euclidean action $S = \int d^2x\, \mathcal{L}$ obeys the inequality

$$S \geq 2\pi |Q| \quad , \tag{6.52}$$

which follows directly from

$$|(D_\mu \pm i\epsilon_{\mu\nu} D_\nu)\, z|^2 \geq 0 \quad . \tag{6.53}$$

The condition for the action to be a minimum is therefore

$$\epsilon_{\mu\nu} D_\nu z = \pm i\, D_\mu z \quad . \tag{6.54}$$

This is (depending on the sign) called the self-duality or anti-self-duality condition. Thus the solutions of Eq. (6.54) are minima of the action within their topological class, which is fixed by Q. Next, introducing the fields $\omega_i^{(j)}$ given by

$$\omega_i^{(j)}(x) = z_i(x)/z_j(x) \quad , \tag{6.55}$$

which, conversely, reproduce the fields z_i (up to an arbitrary phase) according to

$$z_i(x) = e^{i\alpha(x)} \frac{\omega_i^{(j)}(x)}{|\omega_i^{(j)}(x)|} \quad , \tag{6.56}$$

it is not difficult to verify that Eq. (6.54) is further simplified; it becomes

$$\epsilon_{\mu\nu} \partial_\nu \omega_i^{(j)} = \pm i\, \partial_\mu \omega_i^{(j)} \quad . \tag{6.57}$$

This simply means that the $\omega_i^{(j)}$ must be arbitrary functions of $x_1 \pm ix_2$, i.e., arbitrary holomorphic or antiholomorphic functions.

The $O(N)$ sigma models or S^{N-1} models and the $\mathbb{C}P^{N-1}$ models have one specific model in common, namely the $O(3)$ sigma model, which is nothing but the non-linear sigma model on the manifold

$$S^2 \cong \mathbb{C}P^1 \quad , \quad [\text{i.e. } SO(3) \cong SU(2)\, ,\, SO(2) \cong U(1)] \quad , \tag{6.58}$$

so the two series generalize this one model in different directions. Explicitly, the diffeomorphism in Eq. (6.58) is realized via the identification

$$\varphi_a = \bar{z}_i \sigma^a_{ij} z_j \quad , \tag{6.59}$$

where $i, j = 1, 2$, $a = 1, 2, 3$ and the σ_a denote the Pauli matrices. Under this identification, the two Lagrangians (6.14) and (6.49) become identical.

As in the case of the $O(N)$ sigma model, the $\mathbb{C}P^{N-1}$ model turns out to be integrable for any value of N. Indeed, the Noether current j_μ associated with the $SU(N)$-transformations, which is now a vector field with values in the complex traceless antihermitean $(N \times N)$-matrices, with matrix entries

$$(j_\mu)_{ij} = z_i \overleftrightarrow{D}_\mu \bar{z}_j = z_i \, \overline{D_\mu z_j} - D_\mu z_i \, \bar{z}_j \quad , \tag{6.60}$$

again satisfies the conservation law (6.22) together with the identity (6.23), with the same consequences for integrability[66,38,219] as before. Moreover, the entire series of $\mathbb{C}P^{N-1}$ models present instanton solutions, whereas in the series defined on the sphere, this is valid only for the $O(3)$ case.

Again as in the case of the $O(N)$ sigma model, the way in which the sigma field in the $\mathbb{C}P^{N-1}$ model can be coupled to fermions so as to preserve these classical properties can be derived by using supersymmetry. This is done as before, so we simply state the result, without going into any discussion on its derivation from the superfield formalism. The relevant fields are now a multiplet of N complex scalar fields z_i and a multiplet of N Dirac spinor fields ψ_i subject to the constraints

$$\bar{z}_i z_i = 1 \quad , \quad \bar{z}_i \psi_i = 0 \, , \, \overline{\psi}_i z_i = 0 \tag{6.61}$$

and together with a freedom under $U(1)$ gauge transformations

$$z_i(x) \to e^{i\alpha(x)} \, z_i(x) \quad , \quad \psi_i(x) \to e^{i\alpha(x)} \, \psi_i(x) \tag{6.62}$$

which leads to the introduction of covariant derivatives

$$D_\mu z_i = \partial_\mu z_i - i z_i A_\mu \quad , \quad D_\mu \psi_i = \partial_\mu \psi_i - i \psi_i A_\mu \tag{6.63}$$

with A_μ a $U(1)$ gauge potential which transforms as usual (i.e., according to Eq. (6.48)) under $U(1)$ gauge transformations. The dynamics of the model is now governed by the Lagrangian

$$\mathcal{L} = \overline{D^\mu z_i} D_\mu z_i + \frac{i}{2} \overline{\psi}_i \overleftrightarrow{\not{D}} \psi_i + \frac{1}{4} [(\overline{\psi}_i \psi_i)^2 - (\overline{\psi}_i \gamma_5 \psi_i)^2] \quad . \tag{6.64}$$

Again, the $U(1)$ gauge potential appears as a Lagrange multiplier, so its equation of motion expresses it in terms of the fields z_i and ψ_i:

$$A_\mu = -i \bar{z}_j \partial_\mu z_j - \frac{1}{2} \overline{\psi}_j \gamma_\mu \psi_j \quad . \tag{6.65}$$

In particular, we see that the bosonic sector is the usual $\mathbb{C}P^{N-1}$ model, while the fermionic sector is nothing but the chiral Gross-Neveu model discussed in chapter 5.

Also, the Noether current J_μ associated with the $SU(N)$-transformations is again the sum

$$J_\mu = j_\mu + j_\mu^M \tag{6.66}$$

of a contribution j_μ from the bosonic fields z_i, with matrix entries as given in (6.60), and a contribution j_μ^M from the fermionic matter fields ψ_i, with matrix entries

$$(j_\mu^M)_{ij} = i\,\overline{\psi}_j \gamma_\mu \psi_i \quad . \tag{6.67}$$

Again, it satisfies the conservation law (6.38) together with the relation (6.39), with the same consequences for integrability as before. Moreover, if we drop the constraints (6.61) as well as the fermionic self-interaction terms in the Lagrangian (6.64), we obtain what we shall call the minimal model[77] which, as we shall see, is still integrable.

We now want to discuss how the various non-linear sigma models discussed so far fit into a common geometric framework and, in particular, what are the general conditions under which they define (classical) integrable systems.

6.2 Purely Bosonic Non-linear Sigma Models

In this section we shall consider the general formulation of bosonic non-linear sigma models from a mathematically well defined point of view, and the particular case where the model displays the so called dual symmetry. This formulation of sigma models developed and analysed in great detail by Eichenherr and Forger[55,220,221], exhausts, in particular the problem of integrability in the context of the purely bosonic case. The section ends with an example, where we obtain the Grassmannian (and in particular the $\mathbb{C}P^{N-1}$) models as a particular case of the general set up.

6.2.1 Formal development

Preliminaries

Generally speaking, a non-linear sigma model is a field theory of maps between manifolds. More precisely, the classical field configurations in such a model are (smooth) maps φ from a given base space $\mathcal{X}$ to a given target space M, both of which are (connected) pseudo-Riemannian manifolds. In terms of (local) coordinates x^μ on $\mathcal{X}$ and φ^i on M, the Lagrangian reads

$$\mathcal{L} = \frac{1}{2} g^{\mu\nu}\, g_{ij}\, \partial_\mu \varphi^i\, \partial_\nu \varphi^j \quad , \tag{6.68}$$

leading, upon variation of the corresponding action

$$S = \frac{1}{2}\int d^2x \sqrt{|g|}\, g^{\mu\nu}\, g_{ij}\, \partial_\mu \varphi^i\, \partial_\nu \varphi^j \quad , \tag{6.69}$$

to the equations of motion

$$g^{\mu\nu}\left(\nabla_\mu \partial_\nu \varphi^i + \Gamma^i_{jk}\, \partial_\mu \varphi^j\, \partial_\nu \varphi^k\right) = 0 \quad , \tag{6.70}$$

with

$$\nabla_\mu \partial_\nu \varphi^i = \partial_\mu \partial_\nu \varphi^i - \Gamma^\lambda_{\mu\nu} \partial_\lambda \varphi^i \quad . \tag{6.71}$$

Here, the $g_{\mu\nu}$ and g_{ij} are the components of the given metric tensor on $\mathcal{X}$ with respect to x^μ and on M with respect to φ^i, respectively, while the $\Gamma^\lambda_{\mu\nu}$ and Γ^i_{jk} are the corresponding Christoffel symbols,

$$\Gamma^\lambda_{\mu\nu} = \frac{1}{2}\, g^{\kappa\lambda}\left(\partial_\mu g_{\kappa\nu} + \partial_\nu g_{\kappa\mu} - \partial_\kappa g_{\mu\nu}\right) \quad , \tag{6.72}$$

$$\Gamma^i_{jk} = \frac{1}{2}\, g^{il}\left(\partial_j g_{lk} + \partial_k g_{lj} - \partial_l g_{jk}\right) \quad , \tag{6.73}$$

and $|g| = |\det(g_{\mu\nu})|$. Moreover, the g_{ij}, Γ^i_{jk} etc. are to be considered as functions on $\mathcal{X}$ (or appropriate domains therein) by looking at them as functions on M (or appropriate domains therein) and then composing these with the map φ; this implicit dependence on φ (which by the way is responsible for the appearance of the non-linear term in Eq. (6.70)) has for simplicity been suppressed from the notation. Note also that Eqs. (6.68-70) are strikingly similar to the Lagrangian resp. action resp. equations of motion for a freely falling particle moving in M; in that sense, the non-linear sigma model on M is simply the field-theoretic version of the geodesic motion on M (to which it reduces when $\mathcal{X}$ is one-dimensional)[219].

In all that follows, we shall consider only the case where $\mathcal{X}$ is two-dimensional[2)]. Moreover, we shall for simplicity restrict $\mathcal{X}$ to be either two-dimensional Minkowski space or two-dimensional Euclidean space, although even in two dimensions, more general choices are certainly possible and must in fact be allowed[3)]. The necessary generalizations can however be carried out quite easily and we shall therefore disregard this possibility.

[2)] Non-linear sigma models in higher dimensions are well-defined theories at the classical level, but are problematic at the quantum level because they are no longer renormalizable. Even classically, however, integrability and conformal invariance are no longer valid, and the motivation concerning the relation with four-dimensional gauge theories is lost. Nevertheless, they have recently been considered in the literature, especially in three dimensional space-time[222], and in low-energy phenomenology[208,223].

[3)] When considering instantons, for example, we should take $\mathcal{X}$ to be conformally compactified two-dimensional Euclidean space, which is the Riemann sphere S^2, while in string perturbation theory, $\mathcal{X}$ will in general be a (possibly punctured) compact Riemann surface of arbitrarily high genus[224].

The basic ingredient characterizing a non-linear sigma model is thus the choice made for the target space M. One important restriction that we shall always impose is that M be a Riemannian and not just a pseudo-Riemannian manifold: this is both necessary and sufficient to guarantee positivity of the energy in the corresponding non-linear sigma model. Now we can apply a theorem of Nash[225,226] which asserts that any Riemannian manifold M can be isometrically embedded into a Euclidean vector space E [4)] - provided the dimension of E is large enough (as compared to the dimension of M). Then denoting the scalar product on E by $(.,.)$, we can rewrite the Lagrangian (6.68) in the form

$$\mathcal{L} = \frac{1}{2} g^{\mu\nu} \left(\partial_\mu \varphi, \partial_\nu \varphi \right) \quad , \tag{6.74}$$

to be supplemented by the constraints expressing the fact that the E-valued field φ appearing here must be restricted to lie on the embedded submanifold M: this is exactly the situation we encounter in the $O(N)$ sigma models and also in the $\mathbb{C}P^{N-1}$ models if we employ the formulation in terms of projector fields (cf. Eq. (6.44)). Indeed, we can easily make contact between the two forms (6.68) and (6.74) of the Lagrangian if we reexpress the constrained E-valued field φ in (6.74) in terms of the unconstrained fields φ^i in (6.68): the latter are simply the components of the former with respect to curvilinear (local) coordinates on the submanifold M of E. Thus

$$\partial_\mu \varphi = \frac{\partial \varphi}{\partial \varphi^i} \partial_\mu \varphi^i \quad , \tag{6.75}$$

and so (6.68) and (6.74) become identical, with

$$g_{ij} = \left(\frac{\partial \varphi}{\partial \varphi^i}, \frac{\partial \varphi}{\partial \varphi^j} \right) \quad . \tag{6.76}$$

General mathematical set up

So far, we have merely come to the conclusion that the target space M should be some (connected) Riemannian manifold. This of course still leaves us with an enormous freedom of choice and hence calls for some organizing principle. One such principle – and a natural one if we remember that all the important applications of non-linear sigma models in physics have to do with symmetries and symmetry breaking – comes from group theory: the idea is to classify the target space M according to the "size" of its symmetry group G, which is essentially its isometry group[5)]. The two extreme possibilities here are that M either has no symmetries

4) An isometric embedding of one Riemannian manifold into another is a smooth embedding preserving the respective metrics.

5) Isometries of a Riemannian manifold are invertible smooth transformations of that manifold onto itself with smooth inverse which preserve the metric[225,226].

at all or has sufficiently many symmetries to link any two given points. In the first case, the isometry group of M is trivial (consists of the identity alone) or is at best discrete. This situation is in a certain sense generic, a typical example being provided by the so-called Calabi-Yau spaces[227] which play an important role in the compactification of extraneous space-time dimensions in string theory[228]. In the second case, the isometry group of M acts transitively on M, which means that for any two given points in M, there is an isometry of M taking one into the other. In other words, M must be a Riemannian homogeneous space. All other cases are intermediate between these two because any Riemannian manifold can be uniquely decomposed into the disjoint union of orbits under its isometry group[229]. In any case, however, the symmetry group G is not completely fixed by the target space M alone; in fact there are technical advantages in retaining some flexibility in the choice of G. Therefore, we simply assume we are given some connected Lie group G, with Lie algebra $\mathcal{G}$, which acts transitively on M by isometries; this action of G on M will be written in the form[6)]

$$\begin{aligned} G \times M &\to M \\ (g,m) &\to g \cdot m \end{aligned} \tag{6.77}$$

and induces an action

$$\begin{aligned} G \times TM &\to TM \\ (g,u) &\to g \cdot u \end{aligned} \tag{6.78}$$

of G on the tangent bundle TM of M, as well as a representation

$$\begin{aligned} \mathcal{G} &\to \mathcal{X}_K(M) \\ X &\to X_M \end{aligned} \tag{6.79}$$

of $\mathcal{G}$ in the Lie algebra $\mathcal{X}_K(M)$ of Killing vector fields on M[7)]. Explicitly, the action of an element g in G on a tangent vector u in TM is defined by letting g act on a curve in M having u as its derivative, that is, if $u = \frac{d}{dt}m(t)\Big|_{t=0}$,

$$g \cdot u = g \cdot \Big(\frac{d}{dt}m(t)\Big|_{t=0}\Big) = \frac{d}{dt}\left(g \cdot m(t)\right)\Big|_{t=0} , \tag{6.80}$$

---

6) Any action of G on M by isometries defines a homomorphism of G into the isometry group of M which, however, is in general not an isomorphism: it may have a non-trivial kernel, consisting of elements in G that act trivially on M, and its image is in general some subgroup of the isometry group of M; what is needed is just that this subgroup is sufficiently large so that its action on M is still transitive.

7) Killing vector fields on a Riemannian manifold are vector fields on that manifold along which the Lie derivative of the metric vanishes.

whereas the value of the fundamental vector field X_M on M associated with a generator X in $\mathcal{G}$ at a point m in M is defined by letting the one-parameter group generated from X act on m,

$$X_M(m) = \frac{d}{dt}(exp(tX) \cdot m)\bigg|_{t=0} \quad . \tag{6.81}$$

In the rest of this chapter, we shall only consider non-linear sigma models with enough symmetries to exclude the presence of accidental degeneracies. In mathematical language, this means that we suppose the action (6.77) of G on M to be transitive. We also fix, once and for all, an arbitrary reference point m_0 in M and define H to be its stability group: then H is a closed subgroup of G and M becomes identified with the homogeneous space, or coset space, G/H:

$$M = G/H \quad . \tag{6.82}$$

Of course, this coset space cannot be completely arbitrary because we must take into account that M should be a Riemannian manifold on which G is supposed to act by isometries. As a result, it turns out that the stability group H will be compact, the coset space G/H will be reductive and the G-invariant Riemannian metric on M will be induced from a biinvariant pseudo-Riemannian metric on G.[8)] In particular, the statement that the coset space G/H is reductive means that if $\mathcal{G}$ is the Lie algebra of G as before and $\mathcal{H} \subset \mathcal{G}$ denotes the Lie algebra of $H \subset G$, there exists an H-invariant subspace $\mathcal{M}$ of $\mathcal{G}$ which is complementary to the Lie subalgebra $\mathcal{H}$ of $\mathcal{G}$, so that we have an H-invariant direct decomposition[9)]

$$\mathcal{G} = \mathcal{H} \oplus \mathcal{M} \quad , \tag{6.83}$$

In particular, H-invariance of this decomposition implies the following commutation relations:

$$[\mathcal{H},\mathcal{H}] \subset \mathcal{H} \quad , \quad [\mathcal{H},\mathcal{M}] \subset \mathcal{M} \quad . \tag{6.84}$$

For later use, we mention already at this point that the coset space G/H is called (locally) symmetric if in addition we have the commutation relation

$$[\mathcal{M},\mathcal{M}] \subset \mathcal{H} \quad . \tag{6.85}$$

At the opposite end, we have the possibility that

$$[\mathcal{M},\mathcal{M}] \subset \mathcal{M} \quad . \tag{6.86}$$

---

8) In mathematically rigorous form, such conclusions can only be drawn if we impose certain technical conditions on the choice of G and (hence) of H. In all cases of practical importance, however, these conditions are met, so we shall not go into any details and instead refer to the literature[219,66].

9) Here H-invariance refers to the representation of H on $\mathcal{G}$ or on its subspaces obtained by restriction of the adjoint representation of G on $\mathcal{G}$ to H.

This actually means that $\mathcal{M}$ is an ideal in $\mathcal{G}$, so taking the exponential, we see that M appears as a normal Lie subgroup of G. In any case, $\mathcal{M}$ can be identified with the tangent space $T_{m_0}M$ to M at the reference point m_0 - just as $\mathcal{G}$ (or $\mathcal{H}$) can be identified with the tangent space T_1G to G (or T_1H to H) at the group unit 1. Therefore, the G-invariant Riemannian metrics $(.,.)_M$ on M are in one-to-one correspondence with the H-invariant positive definite scalar products $(.,.)_{\mathcal{M}}$ on $\mathcal{M}$ - just as the biinvariant pseudo-Riemannian metrics $(.,.)_G$ on G are in one-to-one correspondence with the G-invariant (more precisely, $Ad(G)$-invariant) non-degenerate scalar products $(.,.)_{\mathcal{G}}$ on $\mathcal{G}$; then the statement that the former is induced from the latter simply means that the direct decomposition (6.83) is orthogonal with respect to $(.,.)_{\mathcal{G}}$ and that $(.,.)_{\mathcal{G}}$ restricted to $\mathcal{M}$ coincides with $(.,.)_{\mathcal{M}}$. We can therefore afford to drop the indices on the various metrics or scalar products and denote them all by the same symbol $(.,.)$, without danger of confusion.

Non linear sigma models on the symmetric space $M = G/H$

With these technicalities out of the way, we can proceed to the formulation of the non-linear sigma model on $M = G/H$ in which G appears as the global symmetry group while H appears as the gauge group. The idea is simply to represent the field configurations of the model not by maps φ from $\mathcal{X}$ to M but by maps g from $\mathcal{X}$ to G, with

$$\varphi(x) = g(x)\,H \quad . \tag{6.87}$$

Locally – i.e., on sufficiently small domains $\mathcal{U} \subset \mathcal{X}$ – this can always be done, but the price to be paid is that the map g from $\mathcal{U}$ to G is of course not unique: in fact, any other map $g \cdot h$ from $\mathcal{U}$ to G with

$$(g \cdot h)(x) = g(x)h(x) \quad , \tag{6.88}$$

where h is any map from $\mathcal{U}$ to H, represents exactly the same field configuration. Conversely, any two maps from $\mathcal{U}$ to G representing the same field configuration must be related according to Eq. (6.88). In other words, describing the non-linear sigma model on M in terms of G-valued fields g rather than M-valued fields φ amounts to introducing the subgroup H as a gauge group, with gauge transformations acting by multiplication from the right,

$$g \rightarrow g \cdot h = gh \quad , \quad \varphi \rightarrow \varphi \quad , \tag{6.89}$$

whereas in both formulations, the group G is a global symmetry group, with global symmetry transformations acting by multiplication from the left,

$$g \rightarrow g_0 \cdot g = g_0 g \quad , \quad \varphi \rightarrow g_0 \cdot \varphi = g_0 \varphi \quad . \tag{6.90}$$

(The index 0 is supposed to indicate that g_0 should not depend on x.) Since all physical quantities must as usual be gauge invariant, it is important to have an associated gauge potential which can be used to define covariant derivatives. This gauge potential A_μ, as well as the covariant derivative $D_\mu g$ of g itself, can be constructed directly from the left invariant Maurer-Cartan form on G (with the field g inserted),

$$g^{-1}dg = (g^{-1}\partial_\mu g)\, dx^\mu \quad ,$$

by taking the orthogonal projection $(.)_{\mathcal{H}}$ from $\mathcal{G}$ onto $\mathcal{H}$ (which annihilates $\mathcal{M}$) resp. the orthogonal projection $(.)_{\mathcal{M}}$ from $\mathcal{G}$ onto $\mathcal{M}$ (which annihilates $\mathcal{H}$); cf. Eq. (6.83). Note that in contrast to the situation in gauge theories, this gauge potential A_μ is not an independent field, but the corresponding gauge field $F_{\mu\nu}$ is defined as usual. Explicitly,

$$A_\mu = \left(g^{-1}\partial_\mu g\right)_{\mathcal{H}} \quad , \tag{6.91a}$$

$$F_{\mu\nu} = \partial_\mu A_\nu - \partial_\nu A_\mu + [A_\mu, A_\nu] \quad , \tag{6.91b}$$

$$k_\mu = \left(g^{-1}\partial_\mu g\right)_{\mathcal{M}} \quad , \tag{6.91c}$$

$$D_\mu g = g k_\mu = \partial_\mu g - g A_\mu \quad . \tag{6.91d}$$

The notation can be justified by observing that under gauge transformations (6.68), A_μ behaves as a gauge potential while $F_{\mu\nu}$, k_μ and $D_\mu g$ are gauge covariant:

$$A_\mu \;\rightarrow\; A_\mu \cdot h = h^{-1} A_\mu h + h^{-1}\partial_\mu h \quad , \tag{6.92a}$$

$$F_{\mu\nu} \;\rightarrow\; F_{\mu\nu} \cdot h = h^{-1} F_{\mu\nu} h \quad , \tag{6.92b}$$

$$k_\mu \;\rightarrow\; k_\mu \cdot h = h^{-1} k_\mu h \quad , \tag{6.92c}$$

$$D_\mu g \;\rightarrow\; D_\mu g \cdot h = (D_\mu g)\, h \quad . \tag{6.92d}$$

The transformation laws (6.92) dictate how one must define higher covariant derivatives; for example,

$$D_\mu D_\nu g = \partial_\mu D_\nu g - D_\nu g\, A_\mu \quad , \tag{6.93a}$$

$$D_\mu k_\nu = \partial_\mu k_\nu + [A_\mu, k_\nu] \quad . \tag{6.93b}$$

In particular, we have the following identities, which will be of central importance later on:

$$F_{\mu\nu} = -[k_\mu, k_\nu]_{\mathcal{H}} \quad , \tag{6.94a}$$

$$D_\mu k_\nu - D_\nu k_\mu = -[k_\mu, k_\nu]_{\mathcal{M}} \quad . \tag{6.94b}$$

They can be proved quite easily from the Maurer-Cartan structure equation (with the field g inserted):

$$\partial_\mu(g^{-1}\partial_\nu g) - \partial_\nu(g^{-1}\partial_\mu g) + [g^{-1}\partial_\mu g, g^{-1}\partial_\nu g] = 0 \quad . \tag{6.94c}$$

Indeed, this equation holds identically for fields g taking values in G, and as a consequence for $g^{-1}\partial g$ taking values in $\mathcal{G}$; due to the commutation relations (6.84), equation (6.94c) implies Eq. (6.94a) when projected onto $\mathcal{H}$ and gives Eq. (6.94b) when projected onto $\mathcal{M}$.

In this formulation, the Lagrangian of the non-linear sigma model on $M = G/H$ can be written in terms of the field φ (cf. Eqs. (6.68), (6.74) - see also for a more mathematical discussion, appendix H)

$$\mathcal{L} = \frac{1}{2}\left(\partial^\mu \varphi, \partial_\mu \varphi\right) \quad , \tag{6.95a}$$

or in terms of the field g:

$$\mathcal{L} = \frac{1}{2}\left(D^\mu g^{-1}, D^\mu g\right) = -\frac{1}{2}\left(g^{-1} D^\mu g, g^{-1} D_\mu g\right) = -\frac{1}{2}\left(D^\mu g\, g^{-1}, D_\mu g\, g^{-1}\right) \quad . \tag{6.95b}$$

The corresponding equation of motion reads

$$D^\mu D_\mu g - D_\mu g\, g^{-1}\, D^\mu g = 0 \quad , \tag{6.96a}$$

or equivalently,

$$D^\mu k_\mu = 0 \quad . \tag{6.96b}$$

Obviously, the Lagrangian (6.95) is gauge invariant (cf. (6.89)) as well as globally invariant (cf. (6.90)), because the metric (.,.) on G is assumed to be biinvariant. The global G-invariance leads to a $\mathcal{G}$-valued Noether current which is explicitly given by

$$j_\mu = -D_\mu g\, g^{-1} = -g k_\mu g^{-1} \quad . \tag{6.97}$$

Obviously, this current is gauge invariant and is conserved:

$$\partial^\mu j_\mu = 0 \quad . \tag{6.98}$$

In fact, the conservation law (6.98) is not only a consequence of the equation of motion (6.96) but is completely equivalent to it - a rather characteristic feature of non-linear sigma models. Moreover, conjugating the identity (6.94b) by g gives the following identity for the current:

$$\partial_\mu j_\nu - \partial_\nu j_\mu + 2[j_\mu, j_\nu] = g\,[k_\mu, k_\nu]_{\mathcal{M}}\, g^{-1} \quad . \tag{6.99}$$

6.2.2 Dual symmetry and higher conservation laws

Although non-linear sigma models defined on general Riemannian homogeneous spaces are interesting in their own right and are useful in a purely theoretical context (e.g., in the description of the scalar sector in supergravity[213] or superstring theories[91]) as well as in particle physics phenomenology[223], there are particular cases where their dynamical structure is restricted by the existence of infinitely many conservation laws, thus rendering them classically integrable. We investigate now the geometric conditions for that situation to occur[66,38,220,221,229].

The approach that we shall follow is to look for a non-trivial family of $\mathcal{G}$-valued potentials $A_\mu^{(\lambda)}$ depending on a spectral parameter λ and satisfying, for all values of λ, a zero curvature condition

$$F_{\mu\nu}^{(\lambda)} \equiv \partial_\mu A_\nu^{(\lambda)} - \partial_\nu A_\mu^{(\lambda)} + [A_\mu^{(\lambda)}, A_\nu^{(\lambda)}] = 0 \quad , \tag{6.100}$$

so that the following system of first order linear differential equations

$$\partial_\mu U^{(\lambda)} = U^{(\lambda)} A_\mu^{(\lambda)} \quad , \tag{6.101}$$

also called the linear system or the Lax pair associated with the problem, is integrable, for all values of λ; its solution, the so-called monodromy matrix, then serves, in a way to be explained below, as the generating functional for an infinite number of conservation laws. The basic Ansatz is that the components of the potentials $A^{(\lambda)}$ should be linear combinations, with λ-dependent coefficients, of the components of the Noether current j:

$$A_\mu^{(\lambda)} = R(\lambda)_\mu^\kappa \, j_\kappa \quad ,$$

where $R(\lambda)$ is some (2×2)-matrix. In order to maintain Lorentz invariance, however, we must require $R(\lambda)$ to commute with all proper orthochronous Lorentz transformations, i.e., with all (2×2)-matrices of the form $\begin{pmatrix} \cosh\chi & \sinh\chi \\ \sinh\chi & \cosh\chi \end{pmatrix}$. (This guarantees that under proper orthochronous Lorentz transformations, $A_\mu^{(\lambda)}$ transforms like j_μ, i.e. as a vector, and hence $U^{(\lambda)}$ transforms as a scalar.) It follows that

$$R(\lambda) = \begin{pmatrix} \alpha(\lambda) & \beta(\lambda) \\ \beta(\lambda) & \alpha(\lambda) \end{pmatrix} \quad ,$$

implying

$$A_\mu^{(\lambda)} = \alpha(\lambda) \, j_\mu + \beta(\lambda) \, \epsilon_{\mu\nu} j^\nu \quad , \tag{6.102}$$

with functions $\alpha(\lambda)$ and $\beta(\lambda)$ to be determined. It is now straightforward to compute the curvature in Eq. (6.100):

$$\begin{aligned}
\epsilon^{\mu\nu} F_{\mu\nu}^{(\lambda)} &= 2\beta \, \partial^\mu j_\mu + 2\alpha \, \epsilon^{\mu\nu} \, \partial_\mu j_\nu + (\alpha^2 - \beta^2) \, \epsilon^{\mu\nu} \, [j_\mu, j_\nu] \\
&= 2\beta \, \partial^\mu j_\mu + \alpha \, \epsilon^{\mu\nu} g \, [k_\mu, k_\nu]_{\mathcal{M}} \, g^{-1} + (\alpha^2 - \beta^2 - 2\alpha) \, \epsilon^{\mu\nu} [j_\mu, j_\nu] \\
&= 2\beta \, \partial^\mu j_\mu + \epsilon^{\mu\nu} \, g \, \big((\alpha^2 - \beta^2 - 2\alpha) \, [k_\mu, k_\nu] + \alpha \, [k_\mu, k_\nu]_{\mathcal{M}} \big) \, g^{-1} \, .
\end{aligned} \tag{6.103}$$

Hence from (6.98) and (6.99) we see that there are precisely two cases in which there exists a non-trivial solution[55]:

a) either M is a (locally) symmetric space, i.e., $[\mathcal{M},\mathcal{M}] \subset \mathcal{H}$ (cf. (6.85)) and

$$\alpha^2 - \beta^2 - 2\alpha = 0 \quad \text{or} \quad (\alpha - 1)^2 - \beta^2 = 1 \quad ,$$

with the solution

$$\alpha(\lambda) = 1 \mp \cosh\lambda \quad , \quad \beta(\lambda) = -\sinh\lambda \quad , \tag{6.104}$$

(both signs are allowed),

b) or M is a Lie group (in fact a normal Lie subgroup of G), i.e., $[\mathcal{M},\mathcal{M}] \subset \mathcal{M}$ (cf. (6.86)) and

$$\alpha^2 - \beta^2 - \alpha = 0 \quad \text{or} \quad (2\alpha - 1)^2 - 4\beta^2 = 1 \quad ,$$

with the solution

$$\alpha(\lambda) = \frac{1}{2}(1 \mp \cosh\lambda) \quad , \quad \beta(\lambda) = -\frac{1}{2}\sinh\lambda \quad , \tag{6.105}$$

(both signs are allowed).

It can be shown[55] that the second alternative can be reformulated as a special case of the first one, so in what follows we shall concentrate on the latter.

In differential geometry, symmetric spaces play an outstanding role because they constitute a class of Riemannian manifolds[10)] that is particularly well understood: in fact they have been completely classified more than half a century ago by É. Cartan[229]. Geometrically, they can be defined as those (connected) Riemannian manifolds M for which the geodesic reflection in any given point m of M[11)] defines an isometry. To be precise, one must distinguish between locally symmetric spaces, where the geodesic reflection in any point of M is supposed to be an isometry of some open neighborhood of that point onto itself, and globally symmetric spaces, where it is assumed that the geodesic reflection in any point of M extends to an isometry of M onto itself; the former can also be characterized as those (connected) Riemannian manifolds whose curvature tensor R is invariant under parallel translations, or to put it differently, is covariantly constant:

$$\nabla R = 0 \quad . \tag{6.106}$$

---

10) In this book, the term "symmetric space" will always mean "Riemannian symmetric space".

11) The simplest way to visualize the geodesic reflection in m is to use normal coordinates around m, in which the geodesics passing through m appear as the straight lines passing through the origin: then geodesic reflection in m appears simply as ordinary reflection in the origin.

As suggested by the terminology, the two types of symmetric spaces are locally indistinguishable[12)] and differ only in certain global properties, the main such property being completeness[13)]. When studying non-linear sigma models, it is therefore sufficient to consider only globally symmetric spaces which in addition can be assumed to be simply connected whenever that may seem convenient[14)]. These spaces are directly amenable to group-theoretical techniques because they turn out to be homogeneous[15)], so they can be written as in Eq. (6.82), and we have an orthogonal direct decomposition as in Eq. (6.83) with commutation relations (6.84) and (6.85). Now if we look at Eq. (6.83) as a decomposition of $\mathcal{G}$ into H-invariant subspaces, we find that both $\mathcal{H}$ and $\mathcal{M}$ will in general still be reducible under H. But if $\mathcal{M}$ turns out to be irreducible, then the symmetric space $M = G/H$ itself is said to be irreducible. Even in that case, however, $\mathcal{H}$ will in general still be reducible and, being a compact Lie algebra, can be decomposed into the orthogonal direct sum of irreducible subspaces[62],

$$\mathcal{H} = \bigoplus_{i=0}^{r} \mathcal{H}_i \quad , \tag{6.107}$$

with commutation relations

$$[\mathcal{H}_i, \mathcal{H}_j] = \{0\} \quad \text{for} \quad i \neq j \quad . \tag{6.108}$$

Here, $\mathcal{H}_0$ denotes the center of $\mathcal{H}$ (which is irreducible if and only if it is one-dimensional), while the $\mathcal{H}_i$, $i = 1, \ldots, r$, denote the simple ideals in the semisimple part of $\mathcal{H}$. For later use, we also note that the $\mathcal{H}$-valued fields A_μ and $F_{\mu\nu}$ introduced above then decompose accordingly, as follows[62]:

$$A_\mu = A_\mu^{(0)} + A_\mu^{(1)} + \cdots + A_\mu^{(r)} \quad , \tag{6.109}$$

$$F_{\mu\nu} = F_{\mu\nu}^{(0)} + F_{\mu\nu}^{(1)} + \cdots + F_{\mu\nu}^{(r)} \quad . \tag{6.110}$$

These decompositions play an important role in the study of anomalies in the quantum theory.

[12)] A theorem in the theory of symmetric spaces states that every point in a locally symmetric space has an open neighborhood which is isometric to some open neighborhood of some point in some globally symmetric space; see Helgason[229], chapter 4 section 5.

[13)] For Riemannian manifolds, the two notions of completeness – completeness as a metric space (every Cauchy sequence converges) and geodesic completeness (every geodesic can be extended to arbitrary values of its affine parameter) – are equivalent.

[14)] Another theorem in the theory of symmetric spaces, showing that this may be done without much loss of generality, states that a globally symmetric space is complete and that conversely, a complete locally symmetric space is, possibly up to a covering, globally symmetric; in fact, it is globally symmetric provided it is also simply connected; otherwise, one may have to replace it by some covering space – such as the universal one – and use the fact that any covering space of a complete locally symmetric space is once again a complete locally symmetric space.

[15)] The group of isometries of a globally symmetric space acts transitively on that space.

We are now ready to describe the classification of symmetric spaces first given by Cartan. First of all, any simply connected, globally symmetric space M is the direct product

$$M = M_0 \times M_- \times M_+ \tag{6.111}$$

of a flat Euclidean space M_0 and two simply connected, globally symmetric spaces M_- and M_+ which are said to be of the compact type and of the non-compact type, respectively[16)]. Moreover, the latter can be further decomposed into the direct product of simply connected, globally symmetric spaces of the compact type and of the non-compact type which in addition are irreducible:

$$M_- = M_-^{(1)} \times \ldots \times M_-^{(r_-)} \quad , \tag{6.112a}$$

$$M_+ = M_+^{(1)} \times \ldots \times M_+^{(r_+)} \quad . \tag{6.112b}$$

Thus it suffices to classify the simply connected, irreducible, globally symmetric spaces. These fall into four distinct classes: the ones of the compact type are either type I or type II, while the ones of the non-compact type are either type III or type IV. Moreover, the type II spaces are the connected compact simple Lie groups, the type IV spaces are the quotients of the connected complex simple Lie groups modulo their maximal compact subgroups, while the type I spaces and type III spaces are listed in table 6.1.

Not all of these spaces are different: we have the identifications gathered in table 6.2.

Moreover, the cases $p = 3, q = 1$ and $p = q = 2$ in the BDI series and the case $N = 2$ in the $DIII$ series are exceptional because they are precisely the ones where the group G is some real form of $SO(4, \mathbb{C})$ and hence fails to be simple; they have been gathered in table 6.3.

Returning to the non-linear sigma models, we have seen that if $M = G/H$ is a symmetric space, so that the inclusion (6.85) holds, then the linear system

$$\partial_\mu U^{(\lambda)} = U^{(\lambda)} \left((1 \mp \cosh \lambda)\, j_\mu - \sinh \lambda \, \epsilon_{\mu\nu} j^\nu \right) \tag{6.113}$$

is integrable, for all values of λ (and with either sign). In fact, the integrability conditions for this linear system are easily seen to consist of two relations: one of them is current conservation

$$\partial^\mu j_\mu = 0 \tag{6.114}$$

and is therefore precisely the equation of motion of the model, while the other is an identity

$$\partial_\mu j_\nu - \partial_\nu j_\mu + 2[j_\mu, j_\nu] = 0 \tag{6.115}$$

16) The terminology can be partly explained by observing that the spaces of the compact type are topologically non-trivial compact manifolds, while the spaces of the non-compact type are topologically trivial, being diffeomorphic (but of course not isometric) with flat Euclidean space; see Helgason[229], chapter 6.

Table 6.1

type	non-compact	compact	rank	dimension	range
AI	$SL(N,R)/SO(N)$	$SU(N)/SO(N)$	$N-1$	$\frac{1}{2}(N-1)(N+2)$	$N\geq 2$
AII	$SU^*(2N)/Sp(N)$	$SU(2N)/Sp(N)$	$N-1$	$(N-1)(2N+1)$	$N\geq 2$
$AIII$	$SU(p,q)/S(U(p)\times U(q))$	$SU(p+q)/S(U(p)\times U(q))$	p	$2pq$	$q\geq p\geq 1$
BDI	$SO_0(p,q)/SO(p)\times SO(q)$	$SO(p+q)/SO(p)\times SO(q)$	p	pq	$q>p\geq 1$
$DIII$	$SO^*(2N)/U(N)$	$SO(2N)/U(N)$	$[\frac{1}{2}N]$	$N(N-1)$	$N\geq 2$
CI	$Sp(N,R)/U(N)$	$Sp(N)/U(N)$	N	$N(N+1)$	$N\geq 1$
CII	$Sp(p,q)/Sp(p)\times Sp(q)$	$Sp(p+q)/Sp(p)\times Sp(q)$	p	$4pq$	$q\geq p\geq 1$
EI	$(e_{6(6)},sp(4))$	$(e_{6(-78)},sp(4))$	6	42	
EII	$(e_{6(2)},su(6)+su(2))$	$(e_{6(-78)},su(6)+su(2))$	4	40	
$EIII$	$(e_{6(-14)},so(10)+R)$	$(e_{6(-78)},so(10)+R)$	2	32	
EIV	$(e_{6(-26)},f_4)$	$(e_{6(-78)}f_4)$	2	26	
EV	$(e_{7(7)},su(8))$	$(e_{7(-133)},su(8))$	7	70	
EVI	$(e_{7(-5)},so(12)+su(2))$	$(e_{7(-133)},so(12)+su(2))$	4	64	
$EVII$	$(e_{7(-25)},e_6+R)$	$(e_{7(-133)},e_6+R)$	3	54	
$EVIII$	$(e_{8(8)},so(16))$	$(e_{8(-248)},so(16))$	8	128	
EIX	$(e_{8(-24)},e_7+su(2))$	$(e_{8(-248)},e_7+su(2))$	4	112	
FI	$(f_{4(4)},sp(3)+su(2))$	$(f_{4(-52)},sp(3)+su(2))$	4	28	
FII	$(f_{4(-20)},so(9))$	$(f_{4(-52)},so(9))$	1	16	
G	$(g_{2(2)},su(2)+su(2))$	$(g_{2(-14)},su(2)+su(2))$	2	8	

Table 6.2

isomorphisms of spaces	isomorphisms of Lie algebras
$AI_{N=2}=AIII_{p=q=1}=BD_{p=2,q=1}=CI_{N=1}$	$su(2)\sim so(3)\sim sp(1), sl(2,R)\sim su(1,1)\sim so(2,1)\sim sp(1,R)$
$BDI_{p=3,q=2}=CI_{N=2}$	$so(5)\sim sp(2), so(3,2)\sim sp(2,R)$
$BDI_{p=4,q=1}=CII_{p=q=1}$	$so(5)\sim sp(2), so(4)\sim sp(1)\times sp(1), so(4,1)\sim sp(1,1)$
$AI_{N=4}=BDI_{p=q=3}$	$su(4)\sim so(6), so(4)\sim so(3)\times so(3), sl(4,R)\sim so(3,3)$
$AII_{N=2}=BDI_{p=5,q=1}$	$su(4)\sim so(6), sp(2)\sim so(5), su^*(4)\sim so(5,1)$
$AIII_{p=q=2}=BDI_{p=4,q=2}$	$su(4)\sim so(6), su(2,2)\sim so(4,2)$
$AIII_{p=3,q=1}=DIII_{N=3}$	$su(4)\sim so(6), su(3,1)\sim so^*(6)$
$BDI_{p=6,q=2}=DIII_{N=4}$	$su(4)\sim so(6), so^*(8)\sim so(6,2)$

Table 6.3

isomorphisms of spaces	isomorphisms of Lie algebras
$BDI_{p=3,q=1}=A_1$	$so(4)\sim su(2)\times su(2), so(3,1)\sim sl(2,C)$
$BDI_{p=2,q=2}=AI_{N=2}\times AI_{N=2}$	$so(4)\sim su(2)\times su(2), so(2,2)\sim sl(2,R)\times sl(2,R)$
$DIII_{N=2}=AI_{N=2}$	$so(4)\sim su(2)\times su(2), so^*(4)\sim su(2)\times sl(2,R)$

that should be viewed as a characteristic feature of non-linear sigma models on symmetric spaces. For completeness, let us rewrite these equations in terms of light cone variables $x^{\pm} = (x^0 \pm x^1)$ and the corresponding derivatives $\partial_{\pm} = (\partial_0 \pm \partial_1)$, replacing the spectral parameter λ by the new spectral parameter $\gamma = \pm\exp(\mp\lambda)$. The linear system (6.113) reads

$$\partial_+ U^{(\gamma)} = U^{(\gamma)} \left((1-\gamma^{-1}) j_+ \right) \quad , \tag{6.116a}$$

$$\partial_- U^{(\gamma)} = U^{(\gamma)} \left((1-\gamma) j_- \right) \quad , \tag{6.116b}$$

while current conservation (6.114) and the identity (6.115) take the form

$$\partial_+ j_- + \partial_- j_+ = 0 \quad , \tag{6.117}$$

and

$$\partial_+ j_- - \partial_- j_+ + 2[j_+, j_-] = 0 \quad , \tag{6.118}$$

respectively.

A complementary point of view emerges if we apply the prescription (6.90) for global symmetry transformations with constant elements g_0 in G replaced by the G-valued solutions $U^{(\lambda)}$ resp. $U^{(\gamma)}$ of the linear system (6.113) resp. (6.116): this defines the so-called dual symmetry transformation (sometimes also referred to as a Bäcklund transformation)

$$g \to g^{(\lambda)} = U^{(\lambda)} g \quad , \quad \varphi \to \varphi^{(\lambda)} = U^{(\lambda)} \varphi \quad , \tag{6.119}$$

resp.

$$g \to g^{(\gamma)} = U^{(\gamma)} g \quad , \quad \varphi \to \varphi^{(\gamma)} = U^{(\gamma)} \varphi \quad . \tag{6.120}$$

It is an unusual kind of symmetry transformation indeed – neither global nor local, but something in between, because the U's are non-trivial but by no means arbitrary functions of x – and it is dual in the sense that the differential equations controlling the space-time dependence of the U's are based on the idea of using a hyperbolic rotation with hyperbolic angle λ to mix the derivative of the original solution with its dual – the derivative being considered as a one-form and its dual being taken in the sense of the Hodge star operator[17]. This becomes even clearer if we consider the linear system satisfied not by the U's but by the g's; it reads

$$\partial_\mu g^{(\lambda)} = g^{(\lambda)} \left(A_\mu \pm k_\mu \cosh\lambda + \epsilon_{\mu\nu} k^\nu \sinh\lambda \right) \quad , \tag{6.121}$$

resp.

[17] In two-dimensional Minkowski space, the Hodge dual of a one-form α is again a one-form $\star\alpha$, with

$$\alpha = \alpha_\mu dx^\mu \quad , \quad \star\alpha = (\star\alpha)_\mu dx^\mu \quad : \quad (\star\alpha)_\mu = \epsilon_{\mu\nu} \alpha^\nu \quad .$$

$$\partial_+ g^{(\gamma)} = g^{(\gamma)} \left(A_+ + \gamma^{-1} k_+\right) \quad , \tag{6.122a}$$

$$\partial_- g^{(\gamma)} = g^{(\gamma)} \left(A_- + \gamma k_-\right) \quad . \tag{6.122b}$$

The crucial property which allows us to speak of a symmetry is now the fact that from every given solution of the equations of motion, the transformation (6.119) resp. (6.120) generates a whole one-parameter family of new solutions. One way to see this is to consider the Noether currents $j_\mu^{(\lambda)}$ resp. $j_\mu^{(\gamma)}$ associated with the $g^{(\lambda)}$ resp. the $g^{(\gamma)}$, which are given by

$$j_\mu^{(\lambda)} = U^{(\lambda)} \left(\pm \cosh\lambda\, j_\mu + \sinh\lambda\, \epsilon_{\mu\nu} j^\nu\right) U^{(\lambda)^{-1}} \quad , \tag{6.123}$$

resp.

$$j_+^{(\gamma)} = -D_+ g^{(\gamma)} g^{(\gamma)^{-1}} = U^{(\gamma)} \left(\gamma^{-1} j_+\right) U^{(\gamma)^{-1}} \quad , \tag{6.124a}$$

$$j_-^{(\gamma)} = -D_- g^{(\gamma)} g^{(\gamma)^{-1}} = U^{(\gamma)} \left(\gamma j_-\right) U^{(\gamma)^{-1}} \quad , \tag{6.124b}$$

and to check by explicit computation that they are conserved, i.e.

$$\partial^\mu j_\mu^{(\lambda)} = 0 \quad , \tag{6.125}$$

resp.

$$\partial_+ j_-^{(\gamma)} + \partial_- j_+^{(\gamma)} = 0 \quad . \tag{6.126}$$

Expanding these conserved currents around some given value λ_0 resp. γ_0, one can derive different types of (infinite series of) conservation laws, both local ones[34,221] obtained by expanding around $\gamma_0 = 0$ and non-local ones[220,231] obtained by expanding around $\gamma_0 = 1$. Here, we shall discuss only the latter. See also appendix I for a more pedestrian discussion.

One possible derivation starts out from the Ansatz

$$U^{(\gamma)} = \exp\, u^{(\gamma)} \quad , \tag{6.127}$$

together with an expansion of $u^{(\gamma)}$ in powers of the parameter $\epsilon = 1 - \gamma$:

$$u^{(\gamma)} = \sum_{n=1}^{\infty} \epsilon^n u^{(n)} \quad . \tag{6.128}$$

Note that the sum only starts at $n = 1$ because at the value $\gamma = 1$, $U^{(\gamma)}$ is identically 1 and hence $u^{(\gamma)}$ is identically zero. Substituting Eq. (6.127) into the linear system (6.116) and using the formula

$$\exp(-X)\, \partial_\mu \exp(X) = \sum_{n=0}^{\infty} \frac{(-1)^n}{(n+1)!} \operatorname{ad}(X)^n \partial_\mu X \quad , \tag{6.129}$$

where $\mathrm{ad}(X)Y = [X,Y]$ and hence $\mathrm{ad}(X)^n Y = [X,[X,\ldots[X,Y]\ldots]]$ with X appearing n times, we obtain

$$(\epsilon - 1)\sum_{n=0}^{\infty}\frac{(-1)^n}{(n+1)!}\,\mathrm{ad}(u^{(\gamma)})^n\,\partial_+ u^{(\gamma)} - \epsilon j_+ = 0 \quad , \tag{6.130a}$$

$$\sum_{n=0}^{\infty}\frac{(-1)^n}{(n+1)!}\,\mathrm{ad}(u^{(\gamma)})^n\,\partial_- u^{(\gamma)} - \epsilon j_- = 0 \quad . \tag{6.130b}$$

Now inserting the expansion (6.128) and collecting all terms of the same order in ϵ, we arrive at two infinite series of equations from which we can recursively determine the light-cone derivatives of the $u^{(n)}$ in terms of the corresponding current components together with the $u^{(m)}$, $m < n$, and their corresponding light-cone derivatives. Note that the expansions for the $\partial_+ u^{(n)}$ and the $\partial_- u^{(n)}$ have a different structure, so that the integrability conditions

$$\partial_+\partial_- u^{(n)} = \partial_-\partial_+ u^{(n)} \tag{6.131}$$

are truly non-trivial. Explicitly, the first two equations (obtained by looking at the coefficient of ϵ and of ϵ^2, respectively) are

$$\partial_+ u^{(1)} = -j_+ \quad , \tag{6.132a}$$

$$\partial_- u^{(1)} = +j_- \quad , \tag{6.132b}$$

and

$$\partial_+ u^{(2)} = -\frac{1}{2}\,[u^{(1)}, j_+] - j_+ \quad , \tag{6.133a}$$

$$\partial_- u^{(2)} = +\frac{1}{2}\,[u^{(1)}, j_-] \quad , \tag{6.133b}$$

respectively; they can be solved by forming the spatial derivative and integrating[18], with the result

$$u^{(1)}(t,y) = -\frac{1}{2}\int dy'\,\epsilon(y-y')\,j_0(t,y') \quad , \tag{6.134}$$

$$\begin{aligned} u^{(2)}(t,y) = &-\frac{1}{8}\int dy_1 dy_2\,\epsilon(y-y_1)\,\epsilon(y_1-y_2)\,[j_0(t,y_1), j_0(t,y_2)] \\ &-\frac{1}{4}\int dy'\,\epsilon(y-y')\,(j_0+j_1)(t,y') \quad . \end{aligned} \tag{6.135}$$

---

[18] We use antisymmetric boundary conditions at infinity, which means that if f is a function that vanishes sufficiently rapidly at $\pm\infty$ to be integrable, we choose its primitive F so as to satisfy $F(-\infty) = -F(+\infty)$; then

$$F(y) = \frac{1}{2}\int_{-\infty}^{y} dy'\,f(y') - \frac{1}{2}\int_{y}^{\infty} dy'\,f(y') = \frac{1}{2}\int dy'\,\epsilon(y-y')\,f(y') \quad .$$

The desired conservation laws

$$\partial_+ j_-^{(n)} + \partial_- j_+^{(n)} = 0 \tag{6.136}$$

can now be obtained from the integrability conditions (6.131) by simply defining the n-th current according to

$$j_+^{(n)} = -\partial_+ u^{(n)} \quad , \tag{6.137a}$$

$$j_-^{(n)} = +\partial_- u^{(n)} \quad , \tag{6.137b}$$

so that the n-th charge, given by

$$q^{(n)}(t) = \int dy\, j_0^{(n)}(t,y) = u^{(n)}(t,-\infty) - u^{(n)}(t,+\infty) \quad , \tag{6.138}$$

is time-independent. Explicitly, the first two charges obtained in this way are $q^{(1)} = Q^{(1)}$ and $q^{(2)} = \frac{1}{2}(Q^{(2)} + Q^{(1)})$, where $Q^{(1)}$ is the standard Noether charge

$$Q^{(1)} = \int dy\, j_0(t,y) \quad , \tag{6.139}$$

while $Q^{(2)}$ is the first non-local charge

$$Q^{(2)} = \frac{1}{2}\int dy_1 dy_2\, \epsilon(y_1 - y_2)\,[j_0(t,y_1), j_0(t,y_2)] + \int dy\, j_1(t,y) \quad . \tag{6.140}$$

Similarly, one can compute further terms in this expansion. The next charge reads

$$\begin{aligned} Q^{(3)} &= \frac{1}{12}\int dy_1 dy_2 dy_3\, \epsilon(y_1 - y_2)\,\epsilon(y_2 - y_3)\,[j_0(t,y_1),[j_0(t,y_2), j_0(t.y_3)]] \\ &\quad + \frac{1}{4}\int dy_1 dy_2\, \epsilon(y_1 - y_2)\,[j_0(t,y_1), j_1(t,y_2)] \quad . \end{aligned} \tag{6.141}$$

There is an alternative method[183] to compute these conserved non-local currents and charges, starting directly from the given conserved local current j_μ and just using that it is curvature free (modulo a factor of 2). Namely, we view the current j_μ as a gauge potential defining a covariant derivative

$$D_\mu^{2j} = \partial_\mu + 2j_\mu \quad , \tag{6.142}$$

which, according to the identity (6.115), has zero curvature:

$$[D_\mu^{2j}, D_\nu^{2j}] = 0 \quad . \tag{6.143}$$

To start the induction, we represent the conserved current $J_\mu^{(1)} = j_\mu$ in terms of a pseudopotential $\chi^{(1)}$:

$$j_\mu = \epsilon_{\mu\nu}\,\partial^\nu \chi^{(1)} \quad , \quad \partial_\mu \chi^{(1)} = \epsilon_{\mu\nu} j^\nu \quad . \tag{6.144}$$

Now we define a new current $J_\mu^{(2)}$ by

$$J_\mu^{(2)} = \partial_\mu \chi^{(1)} + [j_\mu, \chi^{(1)}] = \epsilon_{\mu\nu} j^\nu + [j_\mu, \chi^{(1)}] \quad . \tag{6.145}$$

This current is again conserved:

$$\begin{aligned}\partial^\mu J_\mu^{(2)} &= \epsilon^{\mu\nu}\,\partial_\mu j_\nu + [\partial^\mu j_\mu, \chi^{(1)}] + [j_\mu, \partial^\mu \chi^{(1)}] \\ &= \epsilon^{\mu\nu}\left(\partial_\mu j_\nu + [j_\mu, j_\nu]\right) = 0 \quad .\end{aligned}$$

Explicitly, the solution of Eq. (6.144) is given by

$$\chi^{(1)}(t,y) \;=\; \frac{1}{2}\int dy'\,\epsilon(y-y')\,j_0(t,y') \quad , \tag{6.146}$$

and therefore,

$$J_\mu^{(2)}(t,y) \;=\; \frac{1}{2}\int dy'\,\epsilon(y-y')[j_\mu(t,y), j_0(t,y')] \;+\; \epsilon_{\mu\nu} j^\nu(t,y) \quad , \tag{6.147}$$

so the integral of $J_0^{(2)}(t,y)$ over y yields the first non-local charge (6.140). Iterating the procedure, suppose now that a conserved current $J_\mu^{(n)}$ of order n, $n \geq 2$, has already been defined, and represent it in terms of a pseudopotential $\chi^{(n)}$:

$$J_\mu^{(n)} = \epsilon_{\mu\nu}\,\partial^\nu \chi^{(n)} \quad , \quad \partial_\mu \chi^{(n)} = \epsilon_{\mu\nu} J^{(n)\nu} \quad . \tag{6.148}$$

Now define a new current $J_\mu^{(n+1)}$ of order $n+1$ by

$$J_\mu^{(n+1)} = D_\mu^{2j} \chi^{(n)} = \partial_\mu \chi^{(n)} + 2[j_\mu, \chi^{(n)}] = \epsilon_{\mu\nu} J^{(n)\nu} + 2[j_\mu, \chi^{(n)}] \quad . \tag{6.149}$$

Then for $n > 2$,

$$\partial^\mu J_\mu^{(n+1)} = \partial^\mu D_\mu^{2j} \chi^{(n)} = D_\mu^{2j} \partial^\mu \chi^{(n)} = \epsilon^{\mu\nu} D_\mu^{2j} J_\nu^{(n)} = \epsilon^{\mu\nu} D_\mu^{2j} D_\nu^{2j} \chi^{(n-1)} = 0 \quad ,$$

where we have applied Eq. (6.149) twice (for n and for $n-1$) and have used current conservation (6.114) in the second step and the zero curvature condition (6.143) in the last step. For $n = 2$, however, the proof is slightly different, because the correct definition of $J_\mu^{(2)}$ is Eq. (6.145), which differs from Eq. (6.149) (for $n = 1$) by a factor of 2 in one term:

$$\begin{aligned}
\partial^\mu J_\mu^{(3)} &= \epsilon^{\mu\nu}\, \partial_\mu J_\nu^{(2)} + 2[\partial^\mu j_\mu, \chi^{(2)}] + 2[j_\mu, \partial^\mu \chi^{(2)}] \\
&= \epsilon^{\mu\nu}\, \partial_\mu \left(\epsilon_{\nu\kappa} j^\kappa + [j_\nu, \chi^{(1)}]\right) + 2\epsilon^{\mu\nu}\, [j_\mu, J_\nu^{(2)}] \\
&= \partial^\mu j_\mu + \epsilon^{\mu\nu}\, [\partial_\mu j_\nu, \chi^{(1)}] + \epsilon^{\mu\nu}\, [j_\nu, \partial_\mu \chi^{(1)}] \\
&\qquad + 2\, \epsilon^{\mu\nu} \epsilon_{\nu\kappa}\, [j_\mu, j^\kappa] + 2\, \epsilon^{\mu\nu}\, [j_\mu, [j_\nu, \chi^{(1)}]] \\
&= \epsilon^{\mu\nu}\, [\partial_\mu j_\nu, \chi^{(1)}] + \epsilon^{\mu\nu} \epsilon_{\mu\kappa}\, [j_\nu, j^\kappa] + 2\, \epsilon^{\mu\nu} \epsilon_{\nu\kappa}\, [j_\mu, j^\kappa] \\
&\quad + \epsilon^{\mu\nu} \left([j_\mu, [j_\nu, \chi^{(1)}]] + [j_\nu, [\chi^{(1)}, j_\mu]]\right) \\
&= \epsilon^{\mu\nu} \left[\partial_\mu j_\nu + [j_\mu, j_\nu], \chi^{(1)}\right] \\
&= 0 \quad .
\end{aligned}$$

6.2.3 An explicit example: the Grassmannians

The formulation of non-linear sigma models given above is not always suitable for specific computations, and in fact, the Lagrangians often used in the literature are written in a different language[232,217,56,75]. This happens, for example, for the symmetric spaces $SU(N)/S(U(p) \times U(q))$ in the $AIII$ series or the symmetric spaces $SO(N)/SO(p) \times SO(q)$ in the BDI series ($N = p+q$), which are commonly known as the complex Grassmannians and the real Grassmannians, respectively; they include the complex projective spaces $\mathbb{C}P^{N-1}$ and the spheres S^{N-1}, respectively, as special cases ($p = 1$). In order to facilitate comparison of the previous results with those appearing in other parts of this book and elsewhere, we will exhibit the most important objects, such as the Lagrangian $\mathcal{L}$ or the Noether current j_μ, in terms of the fields X which generalize the fields z of the $\mathbb{C}P^{N-1}$ models or the fields φ of the $O(N)$ sigma models. For the sake of definiteness, we consider the complex case; the real case is easily obtained by requiring all fields to be real.

For the Lie algebra $su(N)$ of traceless anti-hermitean complex $(N \times N)$-matrices, we use the block notation:

$$N \updownarrow \underbrace{\begin{pmatrix} \cdots \\ \cdots \end{pmatrix}}_{\longleftrightarrow\; N} = \begin{pmatrix} \underbrace{(\cdots)}_{} & \underbrace{(\cdots)}_{} \\ (\cdots) & (\cdots) \end{pmatrix} \begin{matrix} \updownarrow p \\ \updownarrow q \end{matrix} \qquad \begin{matrix} \\ \longleftrightarrow\; p \quad \longleftrightarrow\; q \end{matrix} \quad . \tag{6.150}$$

In this case, the spaces appearing in the orthogonal direct decomposition (6.83) consist of the block diagonal and block off-diagonal matrices in $su(N)$:

$$\mathcal{H} = \left\{ \begin{pmatrix} A & 0 \\ 0 & B \end{pmatrix} \,\middle|\, A^\dagger = A,\ B^\dagger = B, \quad \operatorname{tr} A + \operatorname{tr} B = 0 \right\} \quad , \tag{6.151}$$

$$\mathcal{M} = \left\{ \begin{pmatrix} 0 & -R^\dagger \\ R & 0 \end{pmatrix} \right\} \quad . \tag{6.152}$$

The orthogonal direct decomposition (6.107) of the subalgebra $\mathcal{H}$ then takes the form

$$\mathcal{H} = \mathcal{H}_0 \oplus \mathcal{H}_1 \oplus \mathcal{H}_2 \quad , \tag{6.153}$$

with

$$\mathcal{H}_0 = \left\{ i\lambda \begin{pmatrix} 1/p & 0 \\ 0 & -1/q \end{pmatrix} \middle| \, \lambda \in \mathbb{R} \right\} \cong u(1) \cong \mathbb{R} \quad , \tag{6.154a}$$

$$\mathcal{H}_1 = \left\{ \begin{pmatrix} A & 0 \\ 0 & 0 \end{pmatrix} \middle| \, A^\dagger = A\,, \quad \operatorname{tr} A = 0 \right\} \cong su(p) \quad , \tag{6.154b}$$

$$\mathcal{H}_2 = \left\{ \begin{pmatrix} 0 & 0 \\ 0 & B \end{pmatrix} \middle| \, B^\dagger = B\,, \quad \operatorname{tr} B = 0 \right\} \cong su(q) \quad . \tag{6.154c}$$

Furthermore, the field g can be written in the form

$$g = (X \; Y) \quad , \tag{6.155}$$

where g, X, Y are matrices with N rows and with N, p, q columns, respectively. In terms of the fields X and Y, the constraints

$$g^\dagger g = 1_N \quad \text{and} \quad g g^\dagger = 1_N \tag{6.156a}$$

become

$$X^\dagger X = 1_p\,,\; X^\dagger Y = 0\,,\; Y^\dagger X = 0\,,\; Y^\dagger Y = 1_q \tag{6.156b}$$

and

$$X X^\dagger + Y Y^\dagger = 1_N \tag{6.156c}$$

respectively. Next, we have

$$g^{-1}\partial_\mu g = \begin{pmatrix} X^\dagger \partial_\mu X & X^\dagger \partial_\mu Y \\ Y^\dagger \partial_\mu X & Y^\dagger \partial_\mu Y \end{pmatrix} \quad , \tag{6.157}$$

so from Eqs. (6.151) and (6.152),

$$A_\mu = \begin{pmatrix} A_\mu^X & 0 \\ 0 & A_\mu^Y \end{pmatrix} \qquad \text{with} \qquad \begin{aligned} A_\mu^X &= X^\dagger \partial_\mu X \\ A_\mu^Y &= Y^\dagger \partial_\mu Y \end{aligned} \quad , \tag{6.158}$$

$$F_{\mu\nu} = \begin{pmatrix} F^X_{\mu\nu} & 0 \\ 0 & F^Y_{\mu\nu} \end{pmatrix} \quad \text{with} \quad \begin{aligned} F^X_{\mu\nu} &= D_\mu X^\dagger D_\nu X - D_\nu X^\dagger D_\mu X \\ F^Y_{\mu\nu} &= D_\mu Y^\dagger D_\nu Y - D_\nu Y^\dagger D_\mu Y \end{aligned} \quad , \tag{6.159}$$

and

$$k_\mu = \begin{pmatrix} 0 & X^\dagger D_\mu Y \\ Y^\dagger D_\mu X & 0 \end{pmatrix} \quad , \tag{6.160}$$

with

$$D_\mu k_\nu = \begin{pmatrix} 0 & X^\dagger D_\mu D_\nu Y \\ Y^\dagger D_\mu D_\nu X & 0 \end{pmatrix} \quad , \tag{6.161}$$

where the covariant derivatives of the fields X and Y are defined by

$$D_\mu X = \partial_\mu X - X X^\dagger \partial_\mu X \quad , \qquad D_\mu D_\nu X = \partial_\mu D_\nu X - D_\nu X X^\dagger \partial_\mu X \quad , \tag{6.162a}$$

$$D_\mu Y = \partial_\mu Y - Y Y^\dagger \partial_\mu Y \quad , \qquad D_\mu D_\nu Y = \partial_\mu D_\nu Y - D_\nu Y Y^\dagger \partial_\mu Y \quad , \tag{6.162b}$$

which implies, in particular,

$$X^\dagger D_\mu X = 0 \quad , \quad Y^\dagger D_\mu Y = 0 \quad . \tag{6.163}$$

The invariant scalar product on the Lie algebra $\mathcal{G} = su(N)$ (which is unique up to a constant factor since $\mathcal{G} = su(N)$ is simple) will be normalized according to

$$(X, Y) = -\frac{1}{2} \operatorname{tr}(XY) \quad . \tag{6.164}$$

The Lagrangian (6.95) of the model can now be conveniently rewritten in terms of covariant derivatives of the field X only or of the field Y only; this leads – after multiplication by a total factor of 2 – to the result

$$\mathcal{L} = \operatorname{tr} D^\mu X^\dagger D_\mu X = \operatorname{tr} D^\mu Y^\dagger D_\mu Y \quad . \tag{6.165}$$

Similarly, we have for the Noether current (6.97):

$$j_\mu = D_\mu X \, X^\dagger - X \, D_\mu X^\dagger = D_\mu Y \, Y^\dagger - Y \, D_\mu Y^\dagger \quad . \tag{6.166}$$

It is worth noting that these representations exhibit only a part of the gauge symmetry explicitly, namely the $(SU(p) \times U(1))$-part in the X-representation and the $(SU(q) \times U(1))$-part in the Y-representation.

6.3 Non-linear Sigma Models with Fermions[19)]

6.3.1 Definition and properties

Our next aim is to show how the pure non-linear sigma models discussed in the previous section can be extended so as to include matter fields, that is, various other kinds of fields with which the sigma field may interact[20)]. These non-linear sigma models with matter fields are still geometric in the sense that – apart from possible self-interaction terms in the Lagrangian involving the matter fields alone – they are completely specified in terms of geometric data. Namely, recall that the sigma field sector is fixed by picking a (pseudo-)Riemannian manifold M, the target space; the sigma field is then simply a (smooth) map $\varphi : \mathcal{X} \to M$. Similarly[233,234], the matter field sector is fixed by picking a vector bundle V over M, the target bundle, with a given fibre metric $< .,. >$ and a given compatible linear connection (covariant derivative) D^V on it[21)]; the matter field ψ (along a given sigma field φ) is then a (smooth) section of the vector bundle $S \otimes \varphi^* V$ over $\mathcal{X}$, where the vector bundle $\varphi^* V$ over $\mathcal{X}$ results from pulling the vector bundle V over M back to $\mathcal{X}$ with the map $\varphi : \mathcal{X} \to M$[22)], while S is the appropriate tensor/spinor bundle over $\mathcal{X}$ reflecting the tensorial/spinorial nature of the matter fields[23)]. [Thus S is the trivial (real or complex) line bundle over $\mathcal{X}$ if ψ is supposed to be a (real or complex) scalar field, the tangent bundle of $\mathcal{X}$ if ψ is supposed to be a vector field, the (real or complex) spinor bundle of $\mathcal{X}$ if ψ is supposed to be a (Majorana or Dirac) spinor field, etc..] All these vector bundles over $\mathcal{X}$ – S, $\varphi^* V$ and $S \otimes \varphi^* V$ – carry a fibre metric and a compatible linear connection (covariant derivative): these are the ingredients one needs to define the Lagrangian for the matter fields.

In what follows, we do not present the construction of non-linear sigma models with matter fields in full generality, but make some simplifying assumptions. First of all, we suppose, as before, that $\mathcal{X}$ is two-dimensional and is in fact two-dimensional Minkowski space; the discussion of two-dimensional orientable pseudo-Riemannian manifolds (of arbitrary signature) can be carried out easily. Also, we consider not only a restricted class of target spaces but also a restricted class of target bundles, assuming, as before, that M is a Riemannian homogeneous space G/H and, in addition, that V is an associated vector bundle $G \times_H V_0 \to G/H$ over

---

19) In this section, we use the Majorana representation for the gamma matrices.

20) The term "matter fields" is here used in the same spirit as in general relativity, or in gauge theories, where all fields except the metric tensor itself, or the gauge potentials themselves, are referred to as "matter fields".

21) The vector bundle V over M should be real or complex, and in fact (pseudo-)Riemannian real or (pseudo-)Hermitean complex with respect to the given fibre metric, depending on whether the matter fields are supposed to be real or complex.

22) Pulling back fibre bundles with maps between their base spaces essentially amounts to a relabeling of base points: thus the fibre of $\varphi^* V$ at the point x in $\mathcal{X}$ is by definition the fibre of V at the point $\varphi(x)$ in M.

23) We assume for simplicity that only one (multiplet of) matter field(s) occurs; the general case is trivially obtained by taking direct sums.

G/H, obtained from the principal H-bundle $G \to G/H$ over G/H by means of a given representation of the stability group H on a given vector space V_0:

$$M = G/H \quad , \quad V = G \times_H V_0 \quad . \tag{6.167}$$

Explicitly – according to the definition of an associated bundle[226,235] –points in the total space $V = G \times_H V_0$ are equivalence classes $[g, v]$ of pairs (g, v), with $g \in G$ and $v \in V_0$, where

$$\begin{aligned}[g_1, v_1] = [g_2, v_2] &\Leftrightarrow (g_1, v_1) \sim (g_2, v_2) \\ &\Leftrightarrow \text{there exists } h \in H \text{ such that } g_2 = g_1 h \text{ and } v_2 = h^{-1} \cdot v_1\end{aligned} \tag{6.168}$$

Finally, we shall deal exclusively with matter fields that describe spin $-\frac{1}{2}$ fermions, so depending on whether the fermions are neutral or charged, the matter fields will be Majorana spinors or Dirac spinors, and accordingly V_0 will be a real vector space carrying an orthogonal representation of H or a complex vector space carrying a unitary representation of H. It turns out that the choice of this representation completely determines the whole theory. In particular, the representation of H on $\mathcal{M}$, obtained from the adjoint representation of G on $\mathcal{G}$ by restriction, gives rise to the tangent bundle of M as the corresponding associated bundle ($V_0 = \mathcal{M} \Rightarrow V = TM$) and leads to the supersymmetric non-linear sigma model on $M = G/H$.

The formulation of the model now proceeds by representing its field configurations not by maps φ from $\mathcal{X}$ to M and sections ψ of $S \otimes \varphi^* V$ but by maps g from $\mathcal{X}$ to G and sections χ of $S \otimes V_0$, that is, ordinary V_0-valued spinor fields χ on $\mathcal{X}$, with

$$\varphi(x) = g(x)H \quad , \quad \psi(x) = [g(x), \chi(x)] \quad . \tag{6.169}$$

So once again H plays the role of the gauge group, with gauge transformations acting by

$$g \to gh \quad , \quad \chi \to h^{-1} \cdot \chi \quad , \tag{6.170}$$

whereas G is a global symmetry group, with global symmetry transformations acting by

$$g \to g_0 g \quad , \quad \chi \to \chi \quad , \tag{6.171}$$

Moreover, in addition to the composite vector fields A_μ and k_μ coming from the bosonic sector, we now have further composite fields defined in the fermionic sector by forming all possible expressions that are quadratic in the fermion field χ. (Of course all such combinations will – just like A_μ and k_μ and the fermion field χ itself – automatically be globally invariant under G.) For the explicit definition, note first that within our conventions we have:

$$\overline{\psi}\phi = \psi^\dagger \gamma_0 \phi \quad , \tag{6.172}$$

so that

$$(\overline{\psi}\gamma_\mu\phi)^\dagger = \overline{\phi}\gamma_\mu\psi \quad , \quad (\overline{\psi}\phi)^\dagger = \overline{\phi}\psi \quad , \quad (\overline{\psi}\gamma_5\phi)^\dagger = -\overline{\phi}\gamma_5\psi \quad . \tag{6.173}$$

We define a vector field B_μ, a scalar field B and a pseudoscalar field B_5, all taking values in $\mathcal{H}$, by requiring that their scalar product with any generator $t \in \mathcal{H}$ is given by

$$(B_\mu, t) = -\frac{i}{2}\overline{\chi}\gamma_\mu t \cdot \chi \quad , \tag{6.174a}$$

$$(B, t) = -\frac{i}{2}\overline{\chi} t \cdot \chi \quad , \tag{6.174b}$$

$$(B_5, t) = \frac{1}{2}\overline{\chi}\gamma_5 t \cdot \chi \quad . \tag{6.174c}$$

(The fact that all these fields do take values in $\mathcal{H}$, rather than just its complexification, follows from the observation that the expressions on the rhs of Eqs. (6.174a–c) are always real; this in turn follows directly from Eq. (6.173), together with the fact that the generators $t \in \mathcal{H}$ are represented by antihermitean matrices on V_0.) In terms of a basis of generators $t_j \in \mathcal{H}$, with $(t_j, t_k) = g_{jk}$ and $(g^{jk}) = (g_{jk})^{-1}$, we have

$$B_\mu = B_\mu^j t_j \quad , \quad B_\mu^j = -\frac{i}{2} g^{jk}\, \overline{\chi}\gamma_\mu\, t_k \cdot \chi \quad , \tag{6.175a}$$

$$B = B^j t_j \quad , \quad B^j = -\frac{i}{2} g^{jk}\, \overline{\chi}\, t_k \cdot \chi \quad , \tag{6.175b}$$

$$B_5 = B_5^j t_j \quad , \quad B_5^j = \frac{1}{2} g^{jk}\, \overline{\chi}\gamma_5\, t_k \cdot \chi \quad . \tag{6.175c}$$

Similarly, we define a vector field C_μ, a scalar field C and a pseudoscalar field C_5, all of which are real isoscalars, by

$$C_\mu = \frac{1}{2}\overline{\chi}\gamma_\mu\chi \quad , \quad C = \frac{1}{2}\overline{\chi}\chi \quad , \quad C_5 = \frac{i}{2}\overline{\chi}\gamma_5\chi \quad . \tag{6.176}$$

If the representation of H on V_0 is reducible and

$$V_0 = \bigoplus_r V_0^{(r)} \quad , \tag{6.177}$$

is the orthogonal direct decomposition of V_0 into irreducible invariant subspaces, then we can decompose C_μ, C and C_5 accordingly:

$$C_\mu = \sum_r C_\mu^{(r)} \quad , \quad C = \sum_r C^{(r)} \quad , \quad C_5 = \sum_r C_5^{(r)} \quad , \tag{6.178}$$

$$C_\mu^{(r)} = \frac{1}{2}\overline{\chi}^{(r)}\gamma_\mu\chi^{(r)} \quad , \quad C^{(r)} = \frac{1}{2}\overline{\chi}^{(r)}\chi^{(r)} \quad , \quad C_5^{(r)} = \frac{i}{2}\overline{\chi}^{(r)}\gamma_5\chi^{(r)} \quad , \tag{6.179}$$

where $\chi^{(r)}$ is the orthogonal projection of χ onto $V_0^{(r)}$. (Again, the fact that all these fields are real follows directly from Eq. (6.173), together with the fact that the orthogonal projection operators from V_0 to the $V_0^{(r)}$ are hermitean.) Under gauge transformations (6.170), the B-fields transform covariantly,

$$B_\mu \to B_\mu \cdot h = h^{-1}B_\mu h \quad , \tag{6.180a}$$
$$B \to B \cdot h = h^{-1}Bh \quad , \tag{6.180b}$$
$$B_5 \to B_5 \cdot h = h^{-1}B_5 h \quad , \tag{6.180c}$$

while the C-fields remain invariant. Again, the transformation laws dictate how one must define covariant derivatives; for example,

$$D_\mu\chi = \partial_\mu\chi + A_\mu \cdot \chi \quad , \tag{6.181}$$
$$D_\mu B_\nu = \partial_\mu B_\nu + [A_\mu, B_\nu] \quad , \tag{6.182a}$$
$$D_\mu B = \partial_\mu B + [A_\mu, B] \quad , \tag{6.182b}$$
$$D_\mu B_5 = \partial_\mu B_5 + [A_\mu, B_5] \quad . \tag{6.182c}$$

By conjugation under the bosonic field g, we arrive at the following gauge invariant (and globally covariant) fields:

$$j_\mu^M = gB_\mu g^{-1} \quad , \quad j = gBg^{-1} \quad , \quad j_5^M = gB_5 g^{-1} \quad . \tag{6.183}$$

The most general conformally invariant Lagrangian for the model under consideration reads[23]:

$$\mathcal{L} = \frac{1}{2}(D^\mu g, D_\mu g) + \frac{i}{4}\overline{\chi}\not{D}\chi + \frac{1}{2}\mathcal{L}_F \quad , \tag{6.184}$$

where the fermionic self-interaction term $\mathcal{L}_F$ is given by:

$$\mathcal{L}_F = a(B^\mu, B_\mu) + b(B,B) + b_5(B_5, B_5) + \sum_{r,s}\left\{c_{rs}\, C^{(r)\mu}\, C_\mu^{(s)} + d_{rs}\, C^{(r)}\, C^{(s)} + (d_5)_{rs}\, C_5^{(r)}\, C_5^{(s)}\right\} \quad . \tag{6.185}$$

Obviously, this Lagrangian is gauge invariant (cf. (6.170),(6.180)) as well as globally invariant (cf. (6.171)). Notice also that, due to Fierz identities, the terms appearing in the expression (6.185) for $\mathcal{L}_F$ are not all independent but satisfy certain relations,

[23] The requirement of classical conformal invariance avoids the introduction of new mass parameters, and the interactions remain geometrical.

whose explicit form, however, depends on the group H and on its representation on V_0. In any case, the Lagrangian (6.184-185) leads to

$$D^\mu k_\mu - D^\mu B_\mu + [B^\mu, k_\mu] = 0 \tag{6.186}$$

as the bosonic equation of motion and

$$\begin{aligned} \not{D}\chi &= a\gamma^\mu B_\mu \cdot \chi + bB \cdot \chi + ib_5\gamma_5 B_5 \cdot \chi \\ &+ \sum_{r,s} \left\{ ic_{rs}\gamma^\mu \chi^{(r)} C_\mu^{(s)} + id_{rs}\chi^{(r)} C^{(s)} - (d_5)_{rs}\gamma_5 \chi^{(r)} C_5^{(s)} \right\} \end{aligned} \tag{6.187}$$

as the fermionic equation of motion. As a result, the composite vector field B_μ satisfies the covariant conservation law

$$D^\mu B_\mu = 0 \quad , \tag{6.188}$$

(expressing fermion number conservation), while the covariant curl of B_μ turns out to be

$$\begin{aligned} &\epsilon^{\mu\nu}(D_\mu B_\nu - a[B_\mu, B_\nu], t_j) = \\ &\qquad -\frac{b}{4} g^{kl}(\overline{\chi}\gamma_5 [t_j, t_k]_+ \cdot \chi)(\overline{\chi}\, t_l \cdot \chi) + \frac{b_5}{4} g^{kl}(\overline{\chi}\, [t_j, t_k]_+ \cdot \chi)(\overline{\chi}\gamma_5 t_l \cdot \chi) \\ &-\frac{1}{2}\sum_{r,s} d_{rs}(\overline{\chi}^{(r)}\gamma_5\, t_j \cdot \chi^{(r)})(\overline{\chi}^{(s)}\chi^{(s)}) + \frac{1}{2}\sum_{r,s}(d_5)_{rs}(\overline{\chi}^{(r)}\, t_j \cdot \chi^{(r)})(\overline{\chi}^{(s)}\gamma_5\chi^{(s)}) \end{aligned} \tag{6.189}$$

where $[.,.]_+$ denotes the anticommutator. When $b = b_5 = 0$ and $d_{rs} = (d_5)_{rs} = 0$, the complicated expression on the rhs vanishes, and we simply get

$$D_\mu B_\nu - D_\nu B_\mu - 2a[B_\mu, B_\nu] = 0 \quad . \tag{6.190}$$

The $\mathcal{G}$-valued Noether current corresponding to global G-invariance is now given by

$$J_\mu = j_\mu + j_\mu^M \quad . \tag{6.191}$$

It is obviously gauge invariant and is, of course, conserved:

$$\partial^\mu J_\mu = 0 \quad . \tag{6.192}$$

In fact, the conservation law (6.192) is, once again, not only a consequence of the bosonic equation of motion (6.186) but is completely equivalent to it. Moreover, conjugating Eq. (6.188) by g gives

$$\partial^\mu j_\mu^M + [j^\mu, j_\mu^M] = 0 \quad , \tag{6.193}$$

while combining Eqs. (6.192) and (6.193) gives

$$\partial^\mu j_\mu - [j^\mu, j_\mu^M] = 0 \quad . \tag{6.194}$$

When $b = b_5 = 0$ and $d_{rs} = (d_5)_{rs} = 0$, we can also conjugate Eq. (6.190) by g to obtain

$$\partial_\mu j_\nu^M - \partial_\nu j_\mu^M + [j_\mu, j_\nu^M] - [j_\nu, j_\mu^M] - 2a[j_\mu^M, j_\nu^M] = 0 \quad . \tag{6.195}$$

6.3.2 Dual symmetry and higher conservation laws

We shall now investigate the conditions under which the model defined by the Lagrangian (6.184-185) is integrable[233]. As in the purely bosonic case, we shall look for a non-trivial family of $\mathcal{G}$-valued potentials $A_\mu^{(\lambda)}$ depending on a spectral parameter λ and satisfying, for all values of λ, the zero curvature condition (6.100), so that the corresponding linear system or Lax pair (6.101) is integrable, for all values of λ. In the present case, the Ansatz is that the components of the potentials $A^{(\lambda)}$ should be linear combinations, with λ-dependent coefficients, of the components of the two contributions j and j^M to the Noether current J, i.e.,

$$A_\mu^{(\lambda)} = R(\lambda)_\mu^\kappa \, j_\kappa + S(\lambda)_\mu^\kappa \, j_\kappa^M \quad ,$$

where $R(\lambda)$ and $S(\lambda)$ are (2×2)-matrices which, in order to maintain Lorentz invariance, must be of the form

$$R(\lambda) = \begin{pmatrix} \alpha(\lambda) & \beta(\lambda) \\ \beta(\lambda) & \alpha(\lambda) \end{pmatrix} \quad , \quad S(\lambda) = \begin{pmatrix} \gamma(\lambda) & \delta(\lambda) \\ \delta(\lambda) & \gamma(\lambda) \end{pmatrix} \quad ,$$

implying

$$A_\mu^{(\lambda)} = \alpha(\lambda)\, j_\mu + \beta(\lambda)\, \epsilon_{\mu\nu} j^\nu + \gamma(\lambda)\, j_\mu^M + \delta(\lambda)\, \epsilon_{\mu\nu} j^{M\nu} \quad , \tag{6.196}$$

with functions $\alpha(\lambda)$, $\beta(\lambda)$, $\gamma(\lambda)$ and $\delta(\lambda)$ to be determined. Again, it is straightforward to compute the curvature in Eq. (6.100) for this case:

$$\begin{aligned} \epsilon^{\mu\nu} F_{\mu\nu}^{(\lambda)} &= 2\beta\, \partial^\mu j_\mu + 2\alpha\, \epsilon^{\mu\nu}\, \partial_\mu j_\nu + 2\delta\, \partial^\mu j_\mu^M + 2\gamma\, \epsilon^{\mu\nu}\, \partial_\mu j_\nu^M \\ &\quad + (\alpha^2 - \beta^2)\, \epsilon^{\mu\nu}\, [j_\mu, j_\nu] + (\gamma^2 - \delta^2)\, \epsilon^{\mu\nu}\, [j_\mu^M, j_\nu^M] \\ &\quad + 2(\alpha\gamma - \beta\delta)\, \epsilon^{\mu\nu}\, [j_\mu, j_\nu^M] + 2(\alpha\delta - \beta\gamma)\, [j^\mu, j_\mu^M] \\ &= 2\beta\, (\partial^\mu j_\mu - [j^\mu, j_\mu^M]) + 2\delta\, (\partial^\mu j_\mu^M + [j^\mu, j_\mu^M]) \\ &\quad + 2(\alpha\delta - \beta\gamma + \beta - \delta)\, [j^\mu, j_\mu^M] + 2\alpha\, \epsilon^{\mu\nu}\, \partial_\mu j_\nu \\ &\quad + (\alpha^2 - \beta^2)\, \epsilon^{\mu\nu}\, [j_\mu, j_\nu] + 2\gamma\, \epsilon^{\mu\nu}\, \partial_\mu j_\nu^M \\ &\quad + 2(\alpha\gamma - \beta\delta)\, \epsilon^{\mu\nu}\, [j_\mu, j_\nu^M] + (\gamma^2 - \delta^2)\, \epsilon^{\mu\nu}\, [j_\mu^M, j_\nu^M] \quad . \end{aligned} \tag{6.197}$$

When $b = b_5 = 0$ and $d_{rs} = (d_5)_{rs} = 0$, Eqs. (6.193-195) show that the contributions coming from the fermion field can be made to vanish provided that

$$\alpha\delta - \beta\gamma + \beta - \delta = 0 \quad , \tag{6.198a}$$

$$\gamma = \alpha\gamma - \beta\delta \quad , \tag{6.198b}$$

$$\gamma = -\frac{1}{2a}(\gamma^2 - \delta^2) \quad . \tag{6.198c}$$

As in the purely bosonic case, we infer that there are precisely two cases in which there exists a non-trivial solution:

a) either M is a (locally) symmetric space, i.e., $[\mathcal{M}, \mathcal{M}] \subset \mathcal{H}$ (cf. (6.85)) and

$$\alpha^2 - \beta^2 - 2\alpha = 0 \quad \text{or} \quad (\alpha - 1)^2 - \beta^2 = 1 \quad ,$$

with the solution

$$\alpha(\lambda) = 1 \mp \cosh\lambda \quad , \quad \beta(\lambda) = -\sinh\lambda \quad , \tag{6.199a}$$

and

$$\gamma(\lambda) = \frac{1}{2}(1 - \cosh 2\lambda) \quad , \quad \delta(\lambda) = \mp\frac{1}{2}\sinh 2\lambda \quad , \tag{6.199b}$$

(both signs are allowed),

b) or M is a Lie group (in fact a normal Lie subgroup of G), i.e., $[\mathcal{M}, \mathcal{M}] \subset \mathcal{M}$ (cf. (6.86)) and

$$\alpha^2 - \beta^2 - \alpha = 0 \quad \text{or} \quad (2\alpha - 1)^2 - 4\beta^2 = 1 \quad ,$$

with the solution

$$\alpha(\lambda) = \frac{1}{2}(1 \mp \cosh\lambda) \quad , \quad \beta(\lambda) = -\frac{1}{2}\sinh\lambda \quad , \tag{6.200a}$$

and

$$\gamma(\lambda) = \frac{1}{2}(1 \mp \cosh\lambda) \quad , \quad \delta(\lambda) = -\frac{1}{2}\sinh\lambda \quad , \tag{6.200b}$$

(both signs are allowed).

In both cases, Eq. (6.198c) requires that in addition,

$$a = -\frac{1}{2} \quad . \tag{6.201}$$

Again, the second alternative can be reformulated as a special case of the first one, so in what follows we shall concentrate on the latter.

To summarize, we have shown that the non-linear sigma model on a Riemannian symmetric space $M = G/H$ coupled to fermions that transform under an arbitrary representation of H, defined by the Lagrangian

$$\mathcal{L} = \frac{1}{2}(D^\mu g, D_\mu g) + \frac{i}{4}\bar{\chi}\overleftrightarrow{\not{D}}\chi - \frac{1}{4}(B^\mu, B_\mu) + \sum_{r,s} c_{rs}\, C^{(r)\mu} C^{(s)}_\mu \quad , \tag{6.202}$$

with arbitrary coefficients c_{rs}, is an integrable model in the sense that the linear system

$$\partial_\mu U^{(\lambda)} = U^{(\lambda)} \left((1 \mp \cosh\lambda)\, j_\mu - \sinh\lambda\, \epsilon_{\mu\nu} j^\nu + \frac{1}{2}(1 - \cosh 2\lambda)\, j_\mu^M \mp \frac{1}{2} \sinh 2\lambda\, \epsilon_{\mu\nu} j^{M\nu} \right) \tag{6.203}$$

is integrable, for all values of λ (and with either sign). In light cone variables, it reads

$$\partial_+ U^{(\gamma)} = U^{(\gamma)} \left((1 - \gamma^{-1})\, j_+ + \frac{1}{2}(1 - \gamma^{-2})\, j_+^M \right) \quad , \tag{6.204a}$$

$$\partial_- U^{(\gamma)} = U^{(\gamma)} \left((1 - \gamma)\, j_- + \frac{1}{2}(1 - \gamma^2)\, j_-^M \right) \quad , \tag{6.204b}$$

where, as before, $\gamma = \pm\exp(\mp\lambda)$. Moreover, it can be shown that[236], once again, this linear system can be used to define a dual symmetry transformation

$$g \to g^{(\lambda)} = U^{(\lambda)} g \quad , \quad \chi \to \chi^{(\lambda)} \quad , \tag{6.205}$$

resp.

$$g \to g^{(\gamma)} = U^{(\gamma)} g \quad , \quad \chi \to \chi^{(\gamma)} \quad , \tag{6.206}$$

where

$$\chi^{(\lambda)} = \pm \cosh\lambda\, \chi + \sinh\lambda\, \gamma_5 \chi \quad , \tag{6.207}$$

resp.

$$\chi_+^{(\gamma)} = \gamma^{-1} \chi_+ \quad , \quad \chi_-^{(\gamma)} = \gamma\, \chi_- \quad , \tag{6.208}$$

with ϕ_+ and ϕ_- denoting the chiral components of a spinor ϕ,

$$\phi = \phi_+ + \phi_- \quad , \quad \phi_+ = \frac{1}{2}(1 + \gamma_5)\,\phi \quad , \quad \phi_- = \frac{1}{2}(1 - \gamma_5)\,\phi \quad , \tag{6.209}$$

provided that the coefficients c_{rs} in the Lagrangian (6.202) vanish as well. In any case, the corresponding Noether currents $J_\mu^{(\lambda)}$ resp. $J_\mu^{(\gamma)}$, given by

$$J_\mu^{(\lambda)} = U^{(\lambda)} \left(\pm \cosh\lambda\, j_\mu + \sinh\lambda\, \epsilon_{\mu\nu} j^\nu + \cosh 2\lambda\, j_\mu^M \pm \sinh 2\lambda\, \epsilon_{\mu\nu} j^{M\nu} \right) U^{(\lambda)^{-1}} \quad , \tag{6.210}$$

resp.

$$J_+^{(\gamma)} = U^{(\gamma)} \left(\gamma^{-1} j_+ + \gamma^{-2} j_+^M \right) U^{(\gamma)^{-1}} \quad , \tag{6.211a}$$

$$J_-^{(\gamma)} = U^{(\gamma)} \left(\gamma\, j_- + \gamma^2 j_-^M \right) U^{(\gamma)^{-1}} \quad , \tag{6.211b}$$

are conserved, because the expression

$$
\begin{aligned}
& U^{(\lambda)^{-1}}\, \partial^\mu J^{(\lambda)}_\mu\, U^{(\lambda)} \\
&= \pm \cosh\lambda\, \partial^\mu j_\mu + \sinh\lambda\, \epsilon_{\mu\nu}\, \partial^\mu j^\nu + \cosh 2\lambda\, \partial^\mu j^M_\mu \pm \sinh 2\lambda\, \epsilon_{\mu\nu}\, \partial^\mu j^{M\nu} \\
&\quad + [\, U^{(\lambda)^{-1}}\, \partial^\mu U^{(\lambda)} , U^{(\lambda)^{-1}} J^{(\lambda)}_\mu\, U^{(\lambda)}\,] \\
&= \pm \cosh\lambda\, \partial^\mu j_\mu + \sinh\lambda\, \epsilon^{\mu\nu}\, \partial_\mu j_\nu + \cosh 2\lambda\, \partial^\mu j^M_\mu \pm \sinh 2\lambda\, \epsilon^{\mu\nu}\, \partial_\mu j^M_\nu \\
&\quad + \big[\, (1 \mp \cosh\lambda)\, j^\mu - \sinh\lambda\, \epsilon^{\mu\nu} j_\nu + \frac{1}{2}(1 - \cosh 2\lambda)\, j^{M\mu} \mp \frac{1}{2} \sinh 2\lambda\, \epsilon^{\mu\nu} j^M_\nu\,, \\
&\qquad \pm \cosh\lambda\, j_\mu + \sinh\lambda\, \epsilon_{\mu\nu} j^\nu + \cosh 2\lambda\, j^M_\mu \pm \sinh 2\lambda\, \epsilon_{\mu\nu} j^{M\nu}\, \big] \\
&= \pm \cosh\lambda\, \partial^\mu j_\mu + \sinh\lambda\, \epsilon^{\mu\nu}\, \partial_\mu j_\nu + \cosh 2\lambda\, \partial^\mu j^M_\mu \pm \sinh 2\lambda\, \epsilon^{\mu\nu}\, \partial_\mu j^M_\nu \\
&\quad + \big((1 \mp \cosh\lambda)\, \sinh\lambda \pm \sinh\lambda\, \cosh\lambda\big)\, \epsilon^{\mu\nu}\, [j_\mu, j_\nu] \\
&\quad \pm \frac{1}{2}\big((1 - \cosh 2\lambda)\, \sinh 2\lambda + \sinh 2\lambda\, \cosh 2\lambda\big)\, \epsilon^{\mu\nu}\, [j^M_\mu, j^M_\nu] \\
&\quad + \Big((1 \mp \cosh\lambda)\, \cosh 2\lambda \mp \frac{1}{2}(1 - \cosh 2\lambda)\, \cosh\lambda \\
&\qquad \pm \sinh\lambda\, \sinh 2\lambda \mp \frac{1}{2}\, \sinh 2\lambda\, \sinh\lambda\Big)\, [j^\mu, j^M_\mu] \\
&\quad + \Big(\pm(1 \mp \cosh\lambda)\, \sinh 2\lambda + \frac{1}{2}\, \sinh 2\lambda\, \cosh\lambda \\
&\qquad + \sinh\lambda\, \cosh 2\lambda + \frac{1}{2}(1 - \cosh 2\lambda)\, \sinh\lambda\Big)\, \epsilon^{\mu\nu}\, [j_\mu, j^M_\nu] \\
&= \pm \cosh\lambda \big(\partial^\mu j_\mu - [j^\mu, j^M_\mu]\big) + \cosh 2\lambda \big(\partial^\mu j^M_\mu + [j^\mu, j^M_\mu]\big) \\
&\quad + \sinh\lambda\, \epsilon^{\mu\nu} \big(\partial_\mu j_\nu + [j_\mu, j_\nu]\big) \\
&\quad \pm \sinh 2\lambda\, \epsilon^{\mu\nu} \big(\partial_\mu j^M_\nu + [j_\mu, j^M_\nu] + \frac{1}{2}[j^M_\mu, j^M_\nu]\big)
\end{aligned}
$$

vanishes according to Eqs. (6.115) and (6.193-195). Expanding these conserved currents around $\gamma_0 = 1$, $\lambda_0 = 0$, we obtain an infinite series of conservation laws which (except for the very first) are non-local; their derivation proceeds much as in the purely bosonic case. For example, we can, once again, expand the solution $U^{(\gamma)}$ of the linear system (6.204) in powers of the parameter $\epsilon = 1 - \gamma$ (cf. (6.127-129)) to obtain

$$(\epsilon - 1)^2 \sum_{n=0}^{\infty} \frac{(-1)^n}{(n+1)!}\, \mathrm{ad}(u^{(\gamma)})^n\, \partial_+ u^{(\gamma)} - \epsilon(\epsilon - 1) j_+ - \frac{1}{2}\epsilon(\epsilon - 2) j^M_+ = 0 \qquad (6.212a)$$

$$\sum_{n=0}^{\infty} \frac{(-1)^n}{(n+1)!}\, \mathrm{ad}(u^{(\gamma)})^n\, \partial_- u^{(\gamma)} - \epsilon j_- + \frac{1}{2}\epsilon(\epsilon - 2) j^M_- = 0 \quad , \qquad (6.212b)$$

replacing Eq. (6.130),

$$\partial_+ u^{(1)} = -J_+ \quad , \tag{6.213a}$$

$$\partial_- u^{(1)} = +J_- \quad , \tag{6.213b}$$

$$\partial_+ u^{(2)} = -\frac{1}{2}\,[u^{(1)}, J_+] - J_+ - \frac{1}{2} j_+^M \quad , \tag{6.214a}$$

$$\partial_- u^{(2)} = +\frac{1}{2}\,[u^{(1)}, J_-] - \frac{1}{2} j_-^M \quad , \tag{6.214b}$$

and

$$u^{(1)}(t,y) = -\frac{1}{2}\int dy'\,\epsilon(y-y')\,J_0(t,y') \quad , \tag{6.215}$$

$$\begin{aligned} u^{(2)}(t,y) &= -\frac{1}{8}\int dy_1\,dy_2\,\epsilon(y-y_1)\,\epsilon(y_1-y_2)\,[J_0(t,y_1), J_0(t,y_2)] \\ &\quad - \frac{1}{4}\int dy'\,\epsilon(y-y')\,(J_0 + J_1 + j_1^M)(t,y') \quad , \end{aligned} \tag{6.216}$$

replacing Eqs. (6.132-133) and (6.134-135), respectively, leading to the standard Noether charge

$$Q^{(1)} = \int dy\, J_0(t,y) \quad , \tag{6.217}$$

and the first non-local charge

$$Q^{(2)} = \frac{1}{2}\int dy_1\,dy_2\,\epsilon(y_1-y_2)\,[J_0(t,y_1), J_0(t,y_2)] + \int dy\,(J_1 + j_1^M)(t,y) \quad . \tag{6.218}$$

The next charge reads

$$\begin{aligned} Q^{(3)} &= \frac{1}{12}\int dy_1\,dy_2\,dy_3\,\epsilon(y_1-y_2)\,\epsilon(y_2-y_3)\,[J_0(t,y_1), [J_0(t,y_2), J_0(t.y_3)]] \\ &\quad + \frac{1}{4}\int dy_1\,dy_2\,\epsilon(y_1-y_2)\,[J_0(t,y_1), (J_1 + j_1^M)(t,y_2)] + \frac{1}{2}\int dy\, j_0^M(t,y) \end{aligned} \tag{6.219}$$

On the other hand, the alternative method of computing these higher conserved currents and charges by means of a recursive algorithm is now more complicated, because there is no conserved curvature free current around[237]. (It is easy to prove that $J_\mu = j_\mu + j_\mu^M$ is the only linear combination of j_μ and j_μ^M having zero divergence, while $2j_\mu$ is the only linear combination of j_μ and j_μ^M having zero curvature.) Still, the first step is rather simple: we begin, as before, by representing the conserved current $J_\mu^{(1)} = J_\mu$ in terms of a pseudopotential $\chi^{(1)}$:

$$J_\mu = \epsilon_{\mu\nu}\, \partial^\nu \chi^{(1)} \quad , \quad \partial_\mu \chi^{(1)} = \epsilon_{\mu\nu} J^\nu \quad , \tag{6.220}$$

and we define a new current $J_\mu^{(2)}$ by

$$J_\mu^{(2)} = \partial_\mu \chi^{(1)} + [J_\mu, \chi^{(1)}] + \epsilon_{\mu\nu} j^{M\nu} = \epsilon_{\mu\nu}\,(J^\nu + j^{M\nu}) + [J_\mu, \chi^{(1)}] \quad . \tag{6.221}$$

This current is again conserved:

$$\begin{aligned} \partial^\mu J_\mu^{(2)} &= \epsilon^{\mu\nu}\, \partial_\mu (J_\nu + j_\nu^M) + [\partial^\mu j_\mu, \chi^{(1)}] + [j_\mu, \partial^\mu \chi^{(1)}] \\ &= \epsilon^{\mu\nu} \left(\partial_\mu (J_\nu + j_\nu^M) + [J_\mu, J_\nu] \right) = 0 \quad . \end{aligned}$$

Explicitly, the solution of Eq. (6.220) is given by

$$\chi^{(1)}(t,y) \;=\; \frac{1}{2} \int dy'\, \epsilon(y-y')\, J_0(t,y') \quad , \tag{6.222}$$

and therefore,

$$J_\mu^{(2)}(t,y) \;=\; \frac{1}{2} \int dy'\, \epsilon(y-y')[J_\mu(t,y), J_0(t,y')] \;+\; \epsilon_{\mu\nu}\,(J^\nu + j^{M\nu}) \quad , \tag{6.223}$$

so the integral of $J_0^{(2)}(t,y)$ over y yields the first non-local charge (6.218).

Let us conclude this section with a few remarks concerning the role of Dirac spinors vs. Majorana spinors, the use of commuting or anticommuting variables for the fermions, and the question of how the supersymmetric sigma model on $M = G/H$ fits into the present framework.

So far, we have worked with Dirac spinors transforming under a given unitary representation of H on a given complex vector space V_0. Sometimes, however, we need to deal with Majorana fermions, which we are going to view as Dirac spinors satisfying an additional reality constraint:

$$\chi^* = \chi \quad . \tag{6.224}$$

In this case V_0 is the complexification of some real vector space W_0, and the unitary representation of H on V_0 is the complex extension of an orthogonal representation of H on W_0. The generators of H (i.e, the elements of $\mathcal{H}$) are then represented by real antisymmetric linear transformations on V_0, while the projection operators from V_0 to the $V_0^{(r)}$ are real symmetric linear transformations. Due to the Majorana constraint (6.224) we then have the following relations:

i) for commuting (Majorana) spinors:

$$\overline{\chi}\overleftrightarrow{\partial\!\!\!/}\chi = \frac{1}{2}\partial^{\mu}(\overline{\chi}\gamma_{\mu}\chi) \quad , \quad B_{\mu} = 0 \, , \, B_5 = 0 \, , \, C^{(r)} = 0 \quad , \tag{6.225}$$

ii) for anticommuting (Majorana) spinors

$$B = 0 \, , \, C_{\mu}^{(r)} = 0 \, , \, C_5^{(r)} = 0 \quad . \tag{6.226}$$

This implies that commuting Majorana spinors are dynamically trivial, unless one replaces the H-invariant scalar product by an antisymmetric bilinear form ($Sp(2N, R)$ instead of $SO(2N)$). In that case, however, one would be dealing with a non-compact sigma model. Therefore, we shall work with anticommuting (Dirac or Majorana) spinors only – which is physically the more relevant case anyway.

Consider now the situation where

$$W_0 = \mathcal{M} \quad , \tag{6.227}$$

with the orthogonal representation of H derived from the adjoint representation of G on $\mathcal{G} = \mathcal{H} \oplus \mathcal{M}$ by restriction. In this case we have a supersymmetric model. Indeed, the supersymmetry transformations are given by

$$\delta_{\varepsilon} g \;=\; g\,\overline{\varepsilon}\chi \quad , \tag{6.228a}$$

$$\delta_{\varepsilon}\chi \;=\; -i\not{k}\varepsilon \quad . \tag{6.228b}$$

where ε is an anticommuting (constant) Majorana spinor (for a proof see ref. [233]). Moreover, if we are dealing with a Hermitean symmetric space, there exists a generator I of the centre of the stability algebra $\mathcal{H}$ which induces the complex structure on the manifold; it has the properties

$$[I, X] = 0 \quad \text{for} \quad X \in \mathcal{H} \quad , \quad [I, [I, X]] = -X \quad \text{for} \quad X \in \mathcal{M} \quad . \tag{6.229}$$

We then have an $n = 2$ extended supersymmetry, given by

$$\delta'_{\varepsilon} g \;=\; g\,[I, \overline{\varepsilon}\chi] \tag{6.230a}$$

$$\delta'_{\varepsilon}\chi \;=\; i[I, \not{k}]\varepsilon \quad . \tag{6.230b}$$

Indeed, this is not surprising, since $n = 2$ supersymmetry in two dimensions requires the target space to be a Kähler manifold[238,239]. The corresponding superfield formulation may be obtained by defining

$$\psi = g\chi \quad , \tag{6.231}$$

in terms of which Eq. (6.228) becomes

$$\delta_\varepsilon g = \bar{\varepsilon}\psi \quad , \tag{6.232a}$$

$$\delta_\varepsilon \psi = -i\not{\partial} g\, \varepsilon + (\bar{\varepsilon}\psi)\, g^{-1}\psi \quad . \tag{6.232b}$$

Then the formula

$$g(x,\theta) = g(x) + \bar{\theta}\psi(x) + \frac{1}{2}\bar{\theta}\theta G(x) \tag{6.233}$$

provides the usual superfield expansion of supersymmetric sigma models (with auxiliary field G) described in the literature[239,240]. Higher extended supersymmetries may also be obtained. For $n = 4$ one has to consider hyper-Kähler manifolds[241].

6.3.3 Construction of an explicit example

In the following, we want to show how the general method for introducing fermions into non-linear sigma models discussed so far can be applied to the Grassmannian sigma models presented in Sect. 6.2.3. The fermion fields that we are going to consider here transform under a representation of $H = S(U(p) \times U(q))$ which is derived either from the fundamental representation of $G = SU(N)$ or from the adjoint representation of $G = SU(N)$. In any case, we shall obtain more explicit expressions for the composite fields B_μ and j_μ^M, because according to Eq. (6.151), the $(N \times N)$-matrix field B_μ is block diagonal,

$$B_\mu = \begin{pmatrix} B_\mu^z & 0 \\ 0 & B_\mu^Y \end{pmatrix} \quad , \tag{6.234}$$

so according to Eq. (6.155), j_μ^M takes the form

$$j_\mu^M = zB_\mu^z z^\dagger + YB_\mu^Y Y^\dagger \quad , \tag{6.235a}$$

or more explicitly,

$$(j_\mu^M)_{ij} = z_i^a\,(B_\mu^z)^{ab}\,\bar{z}_j^b + Y_i^c\,(B_\mu^Y)^{cd}\,\overline{Y}_j^d \quad . \tag{6.235b}$$

Here B_μ^z is a $(p\times p)$-matrix field with entries $(B_\mu^z)^{ab}$ and B_μ^Y is a $(q\times q)$-matrix field with entries $(B_\mu^Y)^{cd}$ while j_μ^M is an $(N \times N)$-matrix field with entries $(j_\mu^M)_{ij}$ [25]. For the explicit determination of the matrix elements $(B_\mu^z)^{ab}$ and $(B_\mu^Y)^{cd}$ in terms of the fermion field, we use Eq. (6.174a) together with the definition (6.164) of the scalar product $(.,.)$ on the Lie algebra $\mathcal{G} = su(N)$: thus for any generator $t \in \mathcal{H}$,

$$\operatorname{tr}(B_\mu^z A) + \operatorname{tr}(B_\mu^Y D) = \operatorname{tr}(B_\mu t) = -2(B_\mu, t) = i\,\bar{\chi}\gamma_\mu t\cdot\chi$$
$$\text{for} \quad t = \begin{pmatrix} A & 0 \\ 0 & D \end{pmatrix} \quad . \tag{6.236}$$

---

[25] Indices a and b correspond to upper components and run from 1 to p, indices c and d correspond to lower components and run from 1 to q, while indices i and j correspond to both types of components and run from 1 to N.

Fundamental representation: the minimal model

Consider the fundamental representation of $SU(N)$ on the N-dimensional complex vector space $\mathbb{C}^N$. This is of course an irreducible representation of $G = SU(N)$, but under restriction to the subgroup $H = S(U(p) \times U(q))$ it becomes reducible, splitting into two subrepresentations defined on the two H-invariant subspaces

$$V_0^z = \left\{ \begin{pmatrix} u \\ 0 \end{pmatrix} \right\} \cong \mathbb{C}^p \qquad \text{and} \qquad V_0^Y = \left\{ \begin{pmatrix} 0 \\ v \end{pmatrix} \right\} \cong \mathbb{C}^q \quad . \tag{6.237}$$

(Both of these subrepresentations are then irreducible). Suppose now that the fermion field χ – which in the present case will be denoted by ψ – takes values either in V_0^z, with components ψ^a $(a = 1, ..., p)$, or in V_0^Y, with components ψ^c $(c = 1, ..., q)$: all of these will be Dirac spinor fields since the representations of H in question are complex, and we have either $\psi^c = 0$ (first case) or $\psi^a = 0$ (second case). Covariant derivatives of the fermion field are then given either by

$$D_\mu \psi = \partial_\mu \psi + (A_\mu^z)\psi \quad , \tag{6.238a}$$

or

$$D_\mu \psi = \partial_\mu \psi + (A_\mu^Y)\psi \quad . \tag{6.238b}$$

More explicitly, we have either

$$D_\mu^{ab} \psi^b = (\delta^{ab} \partial_\mu + (A_\mu^z)^{ab})\psi^b \quad , \tag{6.239a}$$

or

$$D_\mu^{cd} \psi^d = (\delta^{cd} \partial_\mu + (A_\mu^Y)^{cd})\psi^d \quad . \tag{6.239b}$$

Thus the kinetic term (plus the minimal coupling term) for the fermions in the Lagrangian (6.202) becomes either

$$\overline{\chi} \overleftrightarrow{\not{D}} \chi = \overline{\psi}^a \overleftrightarrow{\not{D}}^{ab} \psi^b \quad , \tag{6.240a}$$

or

$$\overline{\chi} \overleftrightarrow{\not{D}} \chi = \overline{\psi}^c \overleftrightarrow{\not{D}}^{cd} \psi^d \quad , \tag{6.240b}$$

while the composite field B_μ can be computed from Eq. (6.236), with the result that either

$$(B_\mu^z)^{ab} = -i \overline{\psi}^a \gamma_\mu \psi^b \quad , \quad (B_\mu^Y)^{cd} = 0 \quad , \tag{6.241a}$$

or

$$(B_\mu^z)^{ab} = 0 \quad , \quad (B_\mu^Y)^{cd} = -i \overline{\psi}^c \gamma_\mu \psi^d \quad . \tag{6.241b}$$

Therefore, according to Eq. (6.235),

$$(j_\mu^M)_{ij} = -i z_i^a (\overline{\psi}^a \gamma_\mu \psi^b) \overline{z}_j^b - i Y_i^c (\overline{\psi}^c \gamma_\mu \psi^d) \overline{Y}_j^d \quad . \tag{6.242}$$

(Of course, one of the two contributions always vanishes.) In the Lagrangian (6.202) (with all $c_{rs} = 0$), it is convenient to eliminate the field Y in the first case (when $\psi^c = 0$) and the field z in the second case (when $\psi^a = 0$); this leads – after multiplication by a total factor of 2 – to the result

$$\mathcal{L} = \overline{D^\mu z_i^a}\, D_\mu z_i^a + \frac{i}{2}\,\overline{\psi}^a \overleftrightarrow{\not{D}}^{ab}\,\psi^b - \frac{1}{4}\,(\overline{\psi}^a\gamma^\mu\psi^b)(\overline{\psi}^b\gamma_\mu\psi^a) \quad , \tag{6.243a}$$

or

$$\mathcal{L} = \overline{D^\mu Y_i^c}\, D_\mu Y_i^c + \frac{i}{2}\,\overline{\psi}^c \overleftrightarrow{\not{D}}^{cd}\,\psi^d - \frac{1}{4}\,(\overline{\psi}^c\gamma^\mu\psi^d)(\overline{\psi}^d\gamma_\mu\psi^c) \quad , \tag{6.243b}$$

respectively. This can be rewritten as

$$\mathcal{L} = \overline{\tilde{D}^\mu z_i^a}\, \tilde{D}_\mu z_i^a + \frac{i}{2}\,\overline{\psi}^a \overleftrightarrow{\tilde{\not{D}}}^{ab}\,\psi^b \quad , \tag{6.244a}$$

or

$$\mathcal{L} = \overline{\tilde{D}^\mu Y_i^c}\, \tilde{D}_\mu Y_i^c + \frac{i}{2}\,\overline{\psi}^c \overleftrightarrow{\tilde{\not{D}}}^{cd}\,\psi^d \quad , \tag{6.244b}$$

respectively, where $\tilde{D}_\mu$ denotes the covariant derivative with respect to a new (and a priori independent) gauge potential $\tilde{A}_\mu$, which however appears in this Lagrangian only as a Lagrange multiplier: then variation of $\mathcal{L}$ with respect to $\tilde{A}_\mu$ gives

$$\tilde{A}_\mu^z = A_\mu^z + \frac{1}{2}B_\mu^z \quad , \quad \tilde{A}_\mu^Y = A_\mu^Y + \frac{1}{2}B_\mu^Y \tag{6.245}$$

as the equation of motion for this Lagrange multiplier, and inserting this back into the Lagrangian (6.244) reproduces the Lagrangian (6.243). This argument provides the motivation for denoting the model derived from the fundamental representation as the minimal model[77].

Adjoint representation: the supersymmetric model

Consider now the adjoint representation of $SU(N)$ on its Lie algebra $su(N)$. Again, this is an irreducible representation of $G = SU(N)$, but under restriction to the subgroup $H = S(U(p) \times U(q))$ it becomes reducible, splitting into two subrepresentations defined on the two H-invariant subspaces $\mathcal{H}$ and $\mathcal{M}$; cf. Eqs. (6.151-152). (The latter is then irreducible – expressing the fact that the Grassmannians are irreducible Riemannian symmetric spaces; cf. Sect. 6.2.2 – while the former, which is nothing but the adjoint representation of H, is further reducible[233]; its decomposition into irreducible constituents is given by Eq. (6.154).) We suppose for the following that the fermion field χ takes values in $\mathcal{M}$: it will be a Majorana spinor field since the representation of H on $\mathcal{M}$ is real. In matrix form, the Majorana condition (6.224) reads[233]

$$\chi^\dagger = -\chi \quad , \tag{6.246}$$

because there is the conjugation

$$\begin{array}{ccc} sl(N,\mathbb{C}) & \longrightarrow & sl(N,\mathbb{C}) \\ X & \longmapsto & -X^\dagger \end{array} \tag{6.247}$$

on $sl(N,\mathbb{C})$ whose fixed point set is precisely the compact real form $su(N)$. It is then convenient to introduce a new fermion field ψ by setting

$$\psi = g\chi \quad . \tag{6.248}$$

In contrast to the original fermion field χ, this shifted fermion field ψ transforms exactly as the bosonic field g, both under global symmetry transformations ($g \to g_0 g$, $\psi \to g_0\psi$ but $\chi \to \chi$) and under gauge transformations ($g \to gh$, $\psi \to \psi h$ but $\chi \to h^{-1}\chi h$). Therefore, we shall decompose ψ in exactly the same way as g; cf. Eq. (6.155):

$$\psi = (\psi^z \, , \psi^Y) \quad , \tag{6.249}$$

where ψ, ψ^z, ψ^Y are matrices with N rows and with N, p, q columns, respectively; thus ψ^z has components $\psi^{z\,a}_{\;i}$ $(a = 1, ..., p\, , i = 1, ..., N)$ and ψ^Y has components $\psi^{Y\,c}_{\;i}$ $(c = 1, ..., q\, , i = 1, ..., N)$. Now inverting Eq. (6.248), we get

$$\chi = \begin{pmatrix} z^\dagger\psi^z & z^\dagger\psi^Y \\ Y^\dagger\psi^z & Y^\dagger\psi^Y \end{pmatrix} \quad . \tag{6.250}$$

Thus the requirement that χ takes values in $\mathcal{M}$ can be expressed in terms of the following constraints on ψ^z and ψ^Y:

$$z^\dagger\psi^z = 0 \qquad \text{and} \qquad Y^\dagger\psi^Y = 0 \quad , \tag{6.251}$$

$$\psi^{z\dagger}Y + z^\dagger\psi^Y = 0 \qquad \text{or} \qquad \psi^{Y\dagger}z + Y^\dagger\psi^z = 0 \quad . \tag{6.252}$$

(The first of them says that χ is block off-diagonal, while the second reflects the Majorana condition (6.246).) Covariant derivatives of the two fermion fields χ and ψ are then given by

$$D_\mu\chi = \partial_\mu\chi + [A_\mu, \chi] \quad , \tag{6.253}$$

and

$$D_\mu\psi = \partial_\mu\psi - \psi A_\mu \quad , \tag{6.254}$$

so that

$$D_\mu\psi^z = \partial_\mu\psi^z - \psi^z A^z_\mu \quad , \tag{6.255a}$$

$$D_\mu\psi^Y = \partial_\mu\psi^Y - \psi^Y A^Y_\mu \quad , \tag{6.255b}$$

or even more explicitly,

$$D^{ab}_\mu \, \psi^{z\,b}_{\;i} = \left(\delta^{ab}\partial_\mu - (A^z_\mu)^{ba}\right)\psi^{z\,b}_{\;i} \quad , \tag{6.256a}$$

$$D_\mu^{cd}\,\psi^{Y}{}_i^d = (\delta^{cd}\partial_\mu - (A_\mu^Y)^{dc})\psi^{Y}{}_i^d \quad . \tag{6.256b}$$

Thus the kinetic term (plus the minimal coupling term) for the fermions in the Lagrangian (6.202) becomes

$$\bar{\chi}\overleftrightarrow{\not{D}}\chi = \frac{1}{2}\left(\overline{\psi^z}{}_i^a\overleftrightarrow{\not{D}}^{ab}\,\psi^{Y}{}_i^b + \overline{\psi^Y}{}_i^c\overleftrightarrow{\not{D}}^{cd}\,\psi^{Y}{}_i^d\right) \quad , \tag{6.257}$$

while the composite field B_μ can be computed from Eq. (6.236), with the result that

$$(B_\mu^z)^{ab} = -i\,\overline{\psi^z}{}_i^a\gamma_\mu\psi^{z}{}_i^b \quad , \tag{6.258a}$$

$$(B_\mu^Y)^{cd} = -i\,\overline{\psi^Y}{}_i^c\gamma_\mu\psi^{Y}{}_i^d \quad . \tag{6.258b}$$

Therefore, according to Eq. (6.235),

$$(j_\mu^M)_{ij} = -iz_i^a\,(\overline{\psi^z}{}_k^a\gamma_\mu\psi^{z}{}_k^b)\,\bar{z}_j^b - iY_i^c\,(\overline{\psi^Y}{}_k^c\gamma_\mu\psi^{Y}{}_k^d)\,\overline{Y}_j^d \quad . \tag{6.259}$$

(Of course, both contributions are non-trivial here.) Now we can eliminate, in any of these expressions, the field ψ^Y in favor of the field ψ^z, or conversely the field ψ^z in favor of the field ψ^Y, because the constraints (6.251-252), together with the constraints (6.156) on the bosonic fields z and Y, imply that

$$\psi^Y = zz^\dagger\,\psi^Y = -z(\psi^z)^\dagger Y \quad , \tag{6.260a}$$

$$\psi^z = YY^\dagger\,\psi^z = -Y(\psi^Y)^\dagger z \quad . \tag{6.260b}$$

After some calculation, we see that this makes the two terms on the rhs of Eq. (6.257) equal, and we also obtain an alternative expression for B_μ^z in terms of ψ^Y (rather than ψ^z) and for B_μ^Y in terms of ψ^z (rather than ψ^Y):

$$(B_\mu^z)^{ab} = iz_i^b\,(\overline{\psi^Y}{}_i^c\gamma_\mu\psi^{Y}{}_j^c)\,\bar{z}_j^a \quad , \tag{6.261a}$$

$$(B_\mu^Y)^{cd} = iY_i^d\,(\overline{\psi^z}{}_i^a\gamma_\mu\psi^{Y}{}_j^a)\,\overline{Y}_j^c \quad , \tag{6.261b}$$

leading to

$$(j_\mu^M)_{ij} = -iz_i^a\,(\overline{\psi^z}{}_k^a\gamma_\mu\psi^{z}{}_k^b)\,\bar{z}_j^b + i\overline{\psi^z}{}_j^a\gamma_\mu\psi^{z}{}_i^a \quad , \tag{6.262}$$

or

$$(j_\mu^M)_{ij} = -iY_i^c\,(\overline{\psi^Y}{}_k^c\gamma_\mu\psi^{Y}{}_k^d)\,\overline{Y}_j^d + i\overline{\psi^Y}{}_j^c\gamma_\mu\psi^{Y}{}_i^c \quad . \tag{6.263}$$

We can therefore eliminate, in the Lagrangian (6.202) (with all $c_{rs} = 0$), the fields ψ^Y and Y in favor of the fields ψ^z and z, or conversely, the fields ψ^z and z in favor of the fields ψ^Y and Y; then – after multiplication by a total factor of 2 – we find

$$\begin{aligned}\mathcal{L} &= \overline{D^\mu z_i^a}\,D_\mu z_i^a + \frac{i}{2}\,\overline{\psi^z}{}_i^a\overleftrightarrow{\not{D}}^{ab}\,\psi^{z}{}_i^b \\ &\quad - \frac{1}{4}\left((\overline{\psi^z}{}_i^a\gamma^\mu\psi^{z}{}_i^b)(\overline{\psi^z}{}_j^b\gamma_\mu\psi^{z}{}_j^a) + (\overline{\psi^z}{}_i^a\gamma^\mu\psi^{z}{}_j^a)(\overline{\psi^z}{}_j^b\gamma_\mu\psi^{z}{}_i^b)\right) \quad , \end{aligned}\tag{6.264}$$

or

$$\mathcal{L} = \overline{D^\mu Y}{}^c_i \, D_\mu Y_i^c + \frac{i}{2} \overline{\psi^Y}{}^c_i \overleftrightarrow{\not{D}}{}^{cd} \, \psi^Y{}^d_i$$
$$- \frac{1}{4} \left((\overline{\psi^Y}{}^c_i \gamma^\mu \psi^Y{}^d_i)(\overline{\psi^Y}{}^d_j \gamma_\mu \psi^Y{}^c_j) + (\overline{\psi^Y}{}^c_i \gamma^\mu \psi^Y{}^c_j)(\overline{\psi^Y}{}^d_j \gamma_\mu \psi^Y{}^d_i) \right) \quad .(6.265)$$

This Lagrangian does indeed describe the supersymmetric Grassmannian sigma model, the supersymmetry transformations being given by

$$\delta_\varepsilon z = \bar{\varepsilon} \psi^z \quad , \tag{6.266a}$$

$$\begin{aligned} \delta_\varepsilon \psi^z &= -i \not{D} z \, \varepsilon + (\bar{\varepsilon} \psi^Y) \, Y^\dagger \psi^z = -i \not{D} z \, \varepsilon - z \, (\overline{\psi^z} \varepsilon) \, \psi^z \\ &= -i \not{D} z \varepsilon + \frac{1}{2} z (\overline{\psi^z} \psi^z) \varepsilon + \frac{1}{2} z (\overline{\psi^z} \gamma_5 \psi^z) \gamma_5 \varepsilon + \frac{1}{2} z (\overline{\psi^z} \gamma^\mu \psi^z) \gamma_\mu \varepsilon \quad , \end{aligned} \tag{6.266b}$$

or

$$\delta_\varepsilon Y = \bar{\varepsilon} \psi^Y \quad , \tag{6.267a}$$

$$\begin{aligned} \delta_\varepsilon \psi^Y &= -i \not{D} Y \, \varepsilon + (\bar{\varepsilon} \psi^z) \, z^\dagger \psi^Y = -i \not{D} Y \, \varepsilon - Y \, (\overline{\psi^Y} \varepsilon) \, \psi^Y \\ &= -i \not{D} Y \varepsilon + \frac{1}{2} Y (\overline{\psi^Y} \psi^Y) \varepsilon + \frac{1}{2} Y (\overline{\psi^Y} \gamma_5 \psi^Y) \gamma_5 \varepsilon + \frac{1}{2} Y (\overline{\psi^Y} \gamma^\mu \psi^Y) \gamma_\mu \varepsilon \end{aligned} \tag{6.267b}$$

cf. Eq. (6.232). (In the last equality in Eqs. (6.266*b*) and (6.267*b*), we have performed a Fierz transformation.)

6.4 Analogies with Four Dimensional Gauge Theories

One of the most interesting aspects of non-linear sigma models is their close similarity with non-abelian gauge theories (Yang-Mills theories) – a similarity that becomes striking[57,58] when one compares two-dimensional sigma models with four-dimensional gauge theories[65,66,242]. This suggests considering the former as a "theoretical laboratory" for testing the validity of ideas about the latter. In fact, the desire to draw conclusions about the dynamics of gauge fields in four dimensions from such analogies has been one of the main motivations behind much of the recent work done on two-dimensional sigma models – work that has not only produced a number of exact results, both in classical field theory (as explained in this chapter) and in quantum field theory (a subject to be addressed in the next chapter), but has also led to the further development of many field theoretical techniques which are useful in a more general context.

Our aim in the following is to describe some of the analogies between classical sigma models and classical gauge theories in more detail. (There are also analogies between the corresponding quantum theories, but these will not concern us here in this chapter.) We begin by explaining in what sense both types of theories are of a profoundly geometric nature. One important aspect in this context is that their action is the L^2-norm square of a "field strength" which is formed out of first derivatives of the field configuration itself. Indeed, in a non-linear sigma model with a connected Riemannian manifold M as target space, where the field configurations are maps from $\mathcal{X}$ to M, the field strength $d\varphi$ corresponding to any such map φ is just its derivative (tangent map); it can be viewed as a one-form on $\mathcal{X}$ with coefficients in an appropriate vector bundle over $\mathcal{X}$ (namely the pull-back φ^*TM of the tangent bundle TM of M to $\mathcal{X}$ via φ), and the action is – possibly up to a multiplicative coupling constant – simply the L^2-norm square $\|d\varphi\|^2$ of that one-form. Similarly, in a non-abelian gauge theory with a connected compact Lie group G as structure group, where the field configurations are connections on some given principal G-bundle $\mathcal{P}$ over $\mathcal{X}$, the field strength F corresponding to any such connection A is just its curvature tensor; it can be viewed as a two-form on $\mathcal{X}$ with coefficients in an appropriate vector bundle over $\mathcal{X}$ (namely the associated vector bundle $\mathcal{P} \times_G \mathcal{G}$ obtained from $\mathcal{P}$ by means of the adjoint representation of the structure group G on its Lie algebra $\mathcal{G}$), and the action is – possibly up to a multiplicative coupling constant – simply the L^2-norm square $\|F\|^2$ of that two-form. Thus in both cases, the action functional is quite simple (in fact it looks exactly like that for a free field theory), but still the theory is intrinsically non-linear and non-trivial, as can be seen, e.g., by examining the interaction terms one obtains when performing a perturbative expansion around some given background field. The appearance of such interaction terms can be traced back to the fact that (in contrast with a free field theory) the first order differential operator that must be applied to a field configuration in order to obtain the corresponding field strength is non-linear. Moreover, this non-linearity is due to curvature: in fact we have a linear theory (free field theory) if and only if the target space M resp. the structure group G is flat[26]; in this sense the interaction is, in both cases, generated by curvature. Note also that for both types of theories, the configuration space naturally falls into disjoint topological sectors, consisting of all maps φ from $\mathcal{X}$ to M within a given homotopy class and of all connections A on a fixed principal G-bundle $\mathcal{P}$ over $\mathcal{X}$ with a given topological type, i.e., chosen within a given isomorphism class, respectively. In the simplest cases, these topological sectors can be classified by a single integer, and the topological sector to which a given field configuration belongs can be read off from a simple formula for this integer, which is then called the topological charge (see below).

The geometric nature of sigma models and of gauge theories as discussed so far is, of course, a general feature and in particular it does not depend on the space-time dimension $d = \dim \mathcal{X}$. The main reason why sigma models in $d = 2$ dimensions and gauge theories in $d = 4$ dimensions play a distinguished role is that these are precisely the critical values of d for which, classically, both types of theories become

[26] Recall that a compact Lie group is flat if and only if it is abelian.

scale invariant and, in fact, conformally invariant. This can easily be seen, for example, by noting that the multiplicative coupling constant mentioned before will then be dimensionless, and that there are no other dimensionful parameters, such as masses, or by noting that the trace of the energy-momentum tensor will then vanish.

Another interesting and important subject in which two-dimensional non-linear sigma models and four-dimensional non-abelian gauge theories are closely parallel is the existence of instanton solutions[242], that is, of certain special solutions to the Euclidean equations of motion which are, within a given topological sector, absolute minima of the Euclidean action, rather than just stationary points. (Such solutions are important for the quantum theory because they make a significant contribution to the functional integral in situations where standard perturbation theory is no longer applicable.) Normally, one begins by looking for such solutions over flat Euclidean space, but due to conformal invariance, one may replace this by conformally compactified Euclidean space, which is the sphere: this motivates the choice $\mathcal{X} = S^2$ for sigma models and $\mathcal{X} = S^4$ for gauge theories.[27)] The topological sectors are then labelled by the elements of the second homotopy group $\pi_2(M)$ of M for sigma models and by the elements of the third homotopy group $\pi_3(G)$ of G for gauge theories[28)]. For the sake of definiteness, let us concentrate on the simplest non-trivial situation, which is the following.

a) For sigma models, suppose that the target space M is a connected Kähler manifold with complex structure I (defining multiplication by i in the tangent spaces of M) and with a Hermitean metric $< .,. >$ whose real part is the Riemannian metric g used before while its imaginary part is a symplectic form ω:

$$< u, v >_m = g_m(u,v) + i\omega_m(u,v) \qquad \text{for } m \in M \, , \, u,v \in T_mM \quad . \tag{6.268}$$

Thus we must have

$$g_m(I_m u, I_m v) = g_m(u,v) \qquad \text{for } m \in M \, , \, u,v \in T_mM \quad , \tag{6.269a}$$

$$\omega_m(I_m u, I_m v) = \omega_m(u,v) \qquad \text{for } m \in M \, , \, u,v \in T_mM \quad , \tag{6.269b}$$

and

$$\omega_m(u,v) = g_m(I_m u, v) \qquad \text{for } m \in M \, , \, u,v \in T_mM \quad . \tag{6.270}$$

---

27) It can be shown that any smooth finite action solution of the equations of motion over Euclidean space extends to a smooth solution of the equations of motion over the sphere, and of course, due to conformal invariance, both solutions have the same action. In this sense, the condition that a solution over Euclidean space has finite action – which is a boundary condition on its behavior at infinity – is a geometric boundary condition because it can be incorporated into the geometry of the base space $\mathcal{X}$ by passing to the conformal compactification.

28) Representing S^3 as the equator in S^4 and using G-valued functions on S^3 to glue together a trivial principal G-bundle over the upper hemisphere with a trivial principal G-bundle over the lower hemisphere, one can show that isomorphism classes of principal G-bundles over S^4 are in one-to-one correspondence with homotopy classes of maps from S^3 to G, i.e., with the elements of $\pi_3(G)$.

Suppose moreover that $\pi_2(M) \cong \mathbb{Z}$, and choose the generator of $\pi_2(M)$ to be consistent with the standard orientations of $S^2 \cong \mathbb{C}P^1$ and M as complex manifolds[29]. Then normalize the symplectic form ω on M so that its integral over the generator of $\pi_2(M)$ is 4π[30]; in view of Eq. (6.270), this also fixes the normalization for the metric g on M. (The value 4π arises by looking at the special case $M = \mathbb{C}P^1$: it is the volume of $S^2 \cong \mathbb{C}P^1$ with respect to the standard metric.) Thus $\omega/4\pi$ generates the second deRham cohomology group $H^2(M)$ of M, over the reals as well as over the integers; this provides an explicit realization of the isomorphisms

$$\begin{array}{ccc} H^2(M,\mathbb{R}) & \cong & \mathbb{R} \\ \cup & & \cup \\ H^2(M,\mathbb{Z}) & \cong & \mathbb{Z} \end{array} \quad . \tag{6.271}$$

Now the topological charge Q of a field configuration φ is given by the integral[244]

$$Q = \frac{1}{4\pi}\int_{S^2} \varphi^*\omega = \frac{1}{4\pi}\int_{S^2} d\varphi \wedge_\omega d\varphi = \frac{1}{4\pi}\int_{S^2} d^2x\,\sqrt{|g|}\,\epsilon^{\mu\nu}\,\omega_{ij}\,\partial_\mu\varphi^i\,\partial_\nu\varphi^j \quad . \tag{6.272}$$

Under the isomorphism $H^2(S^2,\mathbb{Z}) \cong \mathbb{Z}$, this charge corresponds to a cohomology class depending only on the homotopy class of φ. Next, the topological charge provides a lower bound for the Euclidean action

$$S = \frac{1}{2}\int_{S^2} d\varphi \wedge_g \star\, d\varphi = \frac{1}{2}\int_{S^2} d^2x\,\sqrt{|g|}\,g^{\mu\nu}\,g_{ij}\,\partial_\mu\varphi^i\,\partial_\nu\varphi^j \quad , \tag{6.273}$$

namely

$$S \geq 2\pi|Q| \quad , \tag{6.274}$$

which follows directly from

$$|(\partial_\mu \pm i\epsilon_{\mu\nu}\partial_\nu)\varphi|^2 \geq 0 \quad . \tag{6.275}$$

Indeed,

$$\begin{aligned} |(\partial_\mu \pm i\epsilon_{\mu\nu}\partial_\nu)\varphi|^2 &= <(\partial_\mu \pm i\epsilon_{\mu\nu}\partial_\nu)\varphi\,,(\partial_\mu \pm i\epsilon_{\mu\kappa}\partial_\kappa)\varphi> \\ &= \left(g_{jk} + i\omega_{jk}\right)\left(\partial_\mu\varphi^j \mp i\epsilon_{\mu\nu}\partial_\nu\varphi^j\right)\left(\partial_\mu\varphi^k \pm i\epsilon_{\mu\kappa}\partial_\kappa\varphi^k\right) \\ &= 2g_{jk}\,\partial_\mu\varphi^j\,\partial_\mu\varphi^k \mp 2\epsilon_{\mu\nu}\,\omega_{jk}\,\partial_\mu\varphi^j\,\partial_\mu\varphi^k \quad . \end{aligned}$$

---

29) The group of integers has exactly two generators, namely +1 and −1, so we need an additional condition to fix the sign.

30) In the mathematical literature, the possibility of normalizing ω in this way – apart from the factor 4π which is purely conventional – defines what is called a Hodge manifold. Moreover, a fundamental theorem of Kodaira states that, for M compact, a Hodge manifold is necessarily a complex algebraic manifold; cf. ref. [229].

The condition for the action to be a minimum is therefore

$$\epsilon_{\mu\nu}\partial_\nu\varphi = \pm i\,\partial_\mu\varphi \quad , \tag{6.276a}$$

or in coordinate-free notation

$$\star d\varphi = \pm i\, d\varphi \quad , \tag{6.276b}$$

which is (depending on the sign) called the self-duality or anti-self-duality condition; it states that φ should be a holomorphic or antiholomorphic map from $S^2 \cong \mathbb{C}P^1$ to M. Thus (anti-)instantons in sigma models are simply (anti-)holomorphic maps. In particular, we have thus generalized the results on (anti-)instanton solutions derived in Sect. 6.1.3 for the $\mathbb{C}P^{N-1}$ models.

b) For gauge theories, suppose that the structure group G is a compact connected simple Lie group with Lie algebra $\mathcal{G}$, and fix a maximal torus T in G with Lie algebra $\mathcal{T}$, as well as a definite ordering on the corresponding root system. Then we have $\pi_3(G) \cong \mathbb{Z}$, and we can choose the generator of $\pi_3(G)$ to map $S^3 \cong SU(2)$ identically onto the $SU(2)$-subgroup of G generated by, say, the highest root. Next, we normalize the G-invariant (more precisely, $Ad(G)$-invariant) positive definite scalar product $(.,.)$ on $\mathcal{G}$, which is essentially unique[31], so that the long roots have length $\sqrt{2}$; thus for the classical groups

$$(z, Y) = -\mathrm{tr}\,(zY) \quad \text{for } z, Y \in \mathcal{G} \quad \text{when} \quad \mathcal{G} = su(N) \quad \text{or} \quad \mathcal{G} = sp(N) \quad , \tag{6.277a}$$

which differs from our previous convention (6.164) by a factor of 2, and

$$(z, Y) = -\frac{1}{2}\,\mathrm{tr}\,(zY) \quad \text{for } z, Y \in \mathcal{G} \quad \text{when} \quad \mathcal{G} = so(N) \quad . \tag{6.277b}$$

The invariant scalar product $(.,.)$ on $\mathcal{G}$ gives rise to an invariant three-form $(.,[.,.])$ on $\mathcal{G}$[32] and hence to a biinvariant three-form α on G whose integral over the generator of $\pi_3(G)$ is $2\pi^2$. (The value $2\pi^2$ arises by looking at the special case $G = SU(2)$: it is the volume of $S^3 \cong SU(2)$ with respect to the standard metric, which coincides with the metric given by Eq. (6.277a).) Thus $\alpha/2\pi^2$ generates the third deRham cohomology group $H^3(G)$ of G, over the reals as well as over the integers; this provides an explicit realization of the isomorphisms

$$\begin{array}{ccc} H^2(M,\mathbb{R}) & \cong & \mathbb{R} \\ \cup & & \cup \\ H^2(M,\mathbb{Z}) & \cong & \mathbb{Z} \end{array} \quad . \tag{6.278}$$

[31] For a simple Lie group, all invariant bilinear forms on its Lie algebra are proportional.
[32] This is the simplest example of a general correspondence, known as transgression, between generators of symmetric invariants and of antisymmetric invariants on simple Lie algebras.

Now the topological charge Q of a field configuration A is given by the integral[23]

$$Q = \frac{1}{8\pi^2} \int_{S^4} (F, \star F) = \frac{1}{32\pi^2} \int_{S^4} d^4x \, \sqrt{|g|} \, \epsilon_{\mu\nu\kappa\lambda} \, (F_{\mu\nu}, F_{\kappa\lambda}) \quad . \tag{6.279}$$

Under the isomorphism $H^4(S^4, \mathbb{Z}) \cong \mathbb{Z}$, this charge corresponds to a cohomology class depending only on the principal G-bundle $\mathcal{P}$ over S^4 on which A is defined, namely the characteristic class of $\mathcal{P}$ obtained from the given invariant scalar product on $\mathcal{G}$, multiplied by a factor $1/8\pi^2$, by applying the Weil homomorphism[226]; or for $G = SU(N)$, this is precisely the second Chern class. Again, the topological charge provides a lower bound for the Euclidean action

$$S = \int_{S^4} (F, \star F) = \frac{1}{2} \int_{S^4} d^4x \, \sqrt{|g|} \, (F_{\mu\nu}, F_{\mu\nu}) \quad , \tag{6.280}$$

namely

$$S \geq 8\pi^2 |Q| \quad , \tag{6.281}$$

which follows directly from

$$|F_{\mu\nu} \mp \frac{1}{2} \epsilon_{\mu\nu\kappa\lambda} F_{\kappa\lambda}|^2 \geq 0 \quad . \tag{6.282}$$

The condition for the action to be a minimum is therefore

$$\frac{1}{2} \epsilon_{\mu\nu\kappa\lambda} F_{\kappa\lambda} = \pm F_{\mu\nu} \quad , \tag{6.283a}$$

or in coordinate-free notation,

$$\star F = \pm F \quad , \tag{6.283b}$$

which is (depending on the sign) the self-duality or anti-self-duality condition. Thus (anti-) instantons in gauge theories are simply connections with (anti-)self-dual curvature tensor.

6.5 Concluding Remarks

In this chapter, we have discussed non-linear sigma models as classical field theories, emphasizing their geometric nature which they share with non-abelian gauge theories. We have concentrated on two-dimensional sigma models and have explained in what sense they define integrable theories, first when considered by themselves and then when extended to include fermionic matter fields, provided that the target manifold is a Riemannian symmetric space. We have also presented the Grassmannian models as rather concrete and typical examples. To discuss these matters, and more generally all questions related to symmetries of the target space, the most suitable formulation of the models is not in terms of local coordinates but in terms of appropriate redundant variables – even though this forces us to put up with the usual gauge theory problems. The local coordinate formulation, on the other hand, is useful in background field computations which arise, e.g., in quantum gravity calculations within the context of string theory (a subject to be commented upon later). Moreover, it is the only formulation available in full generality, i.e., when the target space has no symmetries.

An interesting observation that deserves comment is that in two dimensions, the Lagrangian (6.68) and action (6.69) are not the most general one. Namely, assume that the target space M carries, besides the given metric g, a given two-form ω. Then using (local) coordinates x^μ on $\mathcal{X}$ and u^i on M, we can write down the Lagrangian

$$\mathcal{L} = \frac{1}{2}\, g^{\mu\nu}\, g_{ij}\, \partial_\mu\varphi^i\, \partial_\nu\varphi^j + \epsilon^{\mu\nu}\, \omega_{ij}\, \partial_\mu\varphi^i\, \partial_\nu\varphi^j \quad , \tag{6.284}$$

and obtain for the action

$$S = \int dv\, \mathcal{L} = \frac{1}{2}\int d^2x\, \sqrt{|g|}\, g^{\mu\nu}\, g_{ij}\, \partial_\mu\varphi^i\, \partial_\nu\varphi^j + \int d^2x\, \sqrt{|g|}\, \epsilon^{\mu\nu}\, \omega_{ij}\, \partial_\mu\varphi^i\, \partial_\nu\varphi^j \quad . \tag{6.285}$$

(For simplicity, we have – as before – absorbed possible coupling constants into the normalization of g and ω on M.) The corresponding equations of motion read

$$g^{\mu\nu}\big(\nabla_\mu\partial_\nu\varphi^i + \Gamma^i_{jk}\, \partial_\mu\varphi^j\, \partial_\nu\varphi^k\big) - \epsilon_{\mu\nu}\, g^{il}(d\omega)_{ljk}\, \partial_\mu\varphi^j\, \partial_\nu\varphi^k = 0 \quad . \tag{6.286}$$

Supposing that $\mathcal{X}$ is closed (i.e., compact and without boundary) or else that the boundary conditions on the admissible field configurations φ are such that boundary terms may be dropped, we infer in particular that

- when ω is exact, $\omega = d\theta$, the equations of motion are unchanged, and the additional contribution to the action vanishes because the additional contribution to the Lagrangian is a total divergence:

$$\omega_{ij} = \partial_i\theta_j - \partial_j\theta_i \qquad \Longrightarrow \qquad \epsilon^{\mu\nu}\, \omega_{ij}\, \partial_\mu\varphi^i\, \partial_\nu\varphi^j = \partial_\mu\big(2\,\epsilon^{\mu\nu}\, \theta_j\, \partial_\nu\varphi^j\big) \quad .$$

- when ω is closed, $d\omega = 0$, the equations of motion are unchanged, and the additional contribution to the action is (up to some normalization constant) simply the topological charge discussed in Sect. 6.4.

In the general case, we see that adding an exact form to ω, $\omega \to \omega + d\theta$, does not change anything in the theory. In fact, we need not even start from a two-form ω which is globally defined on M: all we need is a covering of M by coordinate patches U_a and a collection of two-forms ω_a which are defined on U_a and such that the $\omega_a - \omega_b$ are exact over the $U_a \cap U_b$: the additional contribution to the action can then be defined by using a partition of unity on M subordinate to the covering by the U_a. But this means that the additional contribution to the action really depends not on the two-form ω or more generally the collection of two-forms ω_a but on the three-form α given by $\alpha = d\omega$ on M or more generally by $\alpha = d\omega_a$ on U_a. Indeed, it is a simple consequence of Stokes' theorem, plus the fact that the exterior derivative d commutes with the pull-back under maps between manifolds, that the additional contribution to the action can be rewritten as a three-dimensional integral, namely the integral of α over a three-dimensional manifold $\mathcal{B}$ whose boundary $\partial\mathcal{B}$ is $\mathcal{X}$. Thus the action (6.285) must be replaced by

$$S = \frac{1}{2}\int_{\mathcal{X}} d^2x\,\sqrt{|g|}\,g^{\mu\nu}\,g_{ij}\,\partial_\mu\varphi^i\,\partial_\nu\varphi^j + \int_{\mathcal{B}} d^3x\,\sqrt{|g|}\,\epsilon^{\mu\nu\kappa}\,\alpha_{ijk}\,\partial_\mu\varphi^i\partial_\mu\varphi^j\,\partial_\nu\varphi^k \quad , \tag{6.287}$$

because in differential forms notation, the second term in Eqs. (6.285) and (6.287) reads

$$\int_{\mathcal{X}} \varphi^*\omega \qquad \text{and} \qquad \int_{\mathcal{B}} \varphi^*\alpha \quad ,$$

respectively, and if $\alpha = d\omega$, this is the same thing:

$$\alpha = d\omega \quad , \quad \mathcal{X} = \partial\mathcal{B} \qquad \Longrightarrow \qquad \int_{\mathcal{X}} \varphi^*\omega = \int_{\mathcal{B}} d(\varphi^*\omega) = \int_{\mathcal{B}} \varphi^*\alpha \quad .$$

The equation of motion (6.286) must be replaced by

$$g^{\mu\nu}\left(\nabla_\mu\partial_\nu\varphi^i + \Gamma^i_{jk}\,\partial_\mu\varphi^j\,\partial_\nu\varphi^k\right) - \epsilon_{\mu\nu}\,T^i_{jk}\,\partial_\mu\varphi^j\,\partial_\nu\varphi^k = 0 \quad , \tag{6.288}$$

where

$$T^i_{jk} = g^{il}\alpha_{ljk} \quad . \tag{6.289}$$

The tensor T can be interpreted as the torsion of a new metric connection $\tilde{\nabla}$ on M which differs from the Levi-Cività connection ∇ used before and into which the additional structure on M needed to define this modified theory can be coded. We shall come back to this kind of model in chapter 9.

We have begun our discussion of non-linear sigma models in this chapter by describing the sigma model as originally introduced by Schwinger[207] and developed into an effective low-energy theory for pion-nucleon interactions, realizing the correct current algebra relations, by Gell-Mann and Lévy[208]. This is really just a concrete example of a general phenomenon which explains the universal role that non-linear sigma models play in field theory – this time in four-dimensional space-time. Namely, consider any field theory that exhibits spontaneous breakdown of a continuous global symmetry, from some connected Lie group G to some closed subgroup H. Then according to the Goldstone theorem, the theory must contain $\dim G - \dim H$ massless Goldstone bosons, and no matter what the Lagrangian for the original theory may be, the dynamics of these Goldstone bosons is, to lowest order, governed by the non-linear sigma model on the coset space G/H. Unfortunately, this scenario is unrealistic because Goldstone bosons must have spin 0 and should be massless, or at least reasonably light, so the only candidates are the pions, which however are nowadays considered to be bound states of a quark with an antiquark. In fact, the reasonably light particles which (according to our present knowledge) are fundamental, namely the light quarks and leptons, all have spin 1/2. This suggests that we should perhaps change our point of view about the spontaneous symmetry breaking scenario and apply it either directly to supersymmetry, which would lead to Goldstone fermions instead of Goldstone bosons, or to a continuous global symmetry within a supersymmetric model, which would lead to massless fermions as supersymmetric partners of the Goldstone bosons. In the second case, the effective low-energy theory governing the dynamics of these Goldstone particles would be the supersymmetric sigma model on the coset space G/H. This explains in what way supersymmetric sigma models appear as effective low-energy theories in preon models[245,246]. Similarly, they appear in supergravity and superstring theories[213,67,247].

According to a result of Zumino[238], the mere definition of the supersymmetric sigma model in four-dimensional space-time requires the target space M to be a Kähler manifold. Writing, as before, I for the complex structure, g for the Riemannian metric and ω for the symplectic form on M, and introducing complex (local) coordinates z^α on M, we have

$$ds^2 = g_{\alpha\bar{\beta}}\, dz^\alpha\, d\bar{z}^\beta \tag{6.290}$$

for the Riemannian metric (line element) and

$$\omega = i\, g_{\alpha\bar{\beta}}\, dz^\alpha \wedge d\bar{z}^\beta \tag{6.291}$$

for the symplectic form, where the components can be derived from a so-called Kähler potential $K = K(z, \bar{z})$ according to

$$g_{\alpha\bar{\beta}} = \frac{\partial^2 K(z,\bar{z})}{\partial z^\alpha \partial \bar{z}^\beta} \quad . \tag{6.292}$$

Briefly, then, the M-valued ordinary field φ, with complex components φ^α, is extended to an M-valued superfield Φ, with complex components Φ^α, and the action reads

$$S = \int d^4x\, d^2\theta\, d^2\bar{\theta}\; K(\Phi, \overline{\Phi}) \quad . \tag{6.293}$$

After integration over the Grassmann variables θ and $\bar{\theta}$, the contribution involving only the spin 0 part becomes

$$S_0 = \int d^4x\, g_{\alpha\bar{\beta}}\, \partial^\mu \varphi^\alpha\, \partial_\mu \overline{\varphi}^\beta \quad . \tag{6.294}$$

From the point of view of group theory – which is natural in the context of spontaneous symmetry breaking scenarios – the simplest situation is that where M is a homogeneous Kähler manifold, that is, $M = G/H$ where the complex structure I, the Riemannian metric g and the symplectic form ω are all invariant under G. For G semisimple, these manifolds can be completely classified[248,249,250]; in particular, there exists a simple procedure for enumerating them in terms of so-called painted Dynkin diagrams[251,229]. Unfortunately, it is not directly the homogeneous Kähler manifolds which appear when one considers spontaneous symmetry breaking of continuous global symmetries within supersymmetric models[252], but some more complicated, though closely related class of Kähler manifolds, of which examples have been analyzed in detail[246,253], but the general structure remains to be unraveled.

Returning to the two-dimensional case, our next aim will be to pass to the quantum theory and see how much of all the interesting structure of the classical theory can be preserved under quantization. A central question will be whether the higher conservation laws derived before remain intact at the quantum level, or whether they will develop anomalies. This is very important because it turns out that when there is no anomaly, one can compute explicitly the exact S-matrix of the theory – which is about as close to its exact solution as one can presently get.

Chapter VII

NON-LINEAR SIGMA MODELS — QUANTUM ASPECTS

7.1 Introduction

The quantization of non linear sigma models is the next step towards a full understanding of these theories. The use of sigma models as useful tools in quantum field theories only becomes effective, after the quantum theory is understood. The first question arising in the quantization of geometric models concerns their renormalizability. Since we aim at a sensible quantum theory, this property is crucial. Because of the highly non linear interaction in these models, they are in general not renormalizable. It is thus very fortunate that in 2 dimensions, it is possible to prove in great generality that this property holds[254,255,256]. It is a result of the low dimensionality of space-time, as well as of identities following from the geometry of the manifold[232]. This also applies to supersymmetric extensions.

The problem of renormalization has been discussed for a variety of two dimensional models. The technique used ($\frac{1}{N}$ expansion[232,257], coupling constant perturbation theory[256] and background field expansion[258,259,56,239]) depend in general on the specific model under consideration. Such models include those defined by the Lagrangian $\mathcal{L} = \frac{1}{2}\partial_\mu q \partial^\mu q$, with q subjected to a constraint[239,255]; models defined on a symmetric space such[232,256] as $\mathcal{L} = \overline{D_\mu z D_\mu z}$; sigma models defined in terms of local coordinates, with Lagrangian[247,258,259] $\mathcal{L} = \frac{1}{2}g^{ab}(\phi)\partial_\mu\phi^a\partial_\nu\phi^b\eta^{\mu\nu}$; supersymmetric extensions with $N = 1, 2, 4$ supersymmetry, have also been discussed in the literature[241], as well as generalization to non torsion free models[261].

In this chapter we discuss the above problems and results arising from the quantum theory, such as definition of higher conserved charges, anomalies, pair production and exact S-matrices. Our main emphasis will be on models admitting a $\frac{1}{N}$ expansion, where the existence of quantum conservation laws, anomalies, pair production, and the general dynamics will be studied in detail.

In particular, we discuss the asymptotic freedom in these models, as well as the question of anomalies which arise in higher dimensions when coupling a non conserved (quantum) current to the gauge field of the sigma model. We shall also deal with so called parallelizable manifolds, with a coupling to an antisymmetric tensor field, and discuss the conditions under which conformal invariance is achieved, by means of a non trivial zero of the beta function. Finally, we discuss the perturbative renormalization of sigma models using the background field method, with application to string theories. Some general results are also stated.

7.2 Grassmannian Bosonic Models

7.2.1 $\frac{1}{N}$ expansion

There exists an important class of Lagrangian models discussed in the previous chapter, which have as representative a field belonging to a vector representation of the symmetry group-$O(N)$, $Sp(N)$, $SU(N)$ or a subgroup of them. In those cases it is in general not difficult to obtain an expansion, in powers of $\frac{1}{N}$, of any correlator. To begin with, we outline the general procedure, which we apply later on. We consider quadratic Lagrangians, of the type introduced in the previous chapter (see Eqs. (6.47-49):

$$\mathcal{L} = \overline{D_\mu z} D^\mu z \quad , \tag{7.1a}$$

with $D_\mu z = \partial_\mu z - izA_\mu$, and subject to a constraint

$$\Omega(z, \bar{z}) = N/2f \quad , \tag{7.1b}$$

where the choice for Ω depends on the model, and f is a coupling constant. The gauge field A_μ, belonging to a given representation of the Lie algebra, plays the role of a Lagrange multiplier, since we have no kinetic Lagrangian for it. It corresponds to the gauge field A_μ of Eq. (6.157)[1]. The procedure now consists in realizing the constraints in terms of auxiliary fields, and integrating over the fundamental fields z, thus obtaining an effective action for the auxiliary fields. Our starting point is the generating functional [2]

$$Z[J] = \int \overline{\mathcal{D}z}\mathcal{D}z\mathcal{D}A_\mu \delta\left(\Omega(\bar{z}, z) - N/2f\right)\, e^{\left\{i\int d^2x\left[\overline{D^\mu z}D_\mu z + \bar{J}z + J\bar{z}\right]\right\}} \quad . \tag{7.2}$$

The delta function enforces the constraint (7.1b). We may use an exponential representation of the delta function,

$$\delta\left[\mathcal{F}\right] = \int \mathcal{D}\alpha \quad e^{\left\{i\int dx\alpha(x)\mathcal{F}(x)\right\}} \tag{7.3}$$

---

[1] In this chapter, however, we use hermitian gauge fields. Therefore we have a factor of i with respect to Eq. (6.157).

[2] To be rigorous, functional integrals should always be defined in Euclidean space. However we shall continue to work in Minkowsky space, since the manipulations done here are sufficiently simple not to require euclidianization. At the end, the Feynman rules may be easily obtained in Euclidean space as well, using the usual substitution rules, as the interested reader may verify.

which, when inserted in (7.2), yields

$$Z[0J,\overline{J}] = \int \mathcal{D}\overline{z}\mathcal{D}z\mathcal{D}A_\mu \mathcal{D}\alpha \; e^{i\int d^2x\left[\overline{D^\mu z}D_\mu z+\alpha(\Omega(\overline{z},z)-N/2f)+\overline{J}z+J\overline{z}\right]} \quad . \qquad (7.4)$$

If $\Omega(\overline{z},z)$ is quadratic in $(\overline{z},z)$, the Gaussian integration may be performed explicitly. This is the case. Examples are provided by non exceptional symmetry groups. Thus, we write:

$$\Omega(\overline{z},z) = \frac{1}{2}\Omega_{\overline{z}z}\overline{z}z + \Omega_{zz}zz + \Omega_{\overline{z}\overline{z}}\overline{z}\overline{z} + \frac{1}{2}\Omega_{z\overline{z}}z\overline{z} \quad .$$

Defining $Z = \begin{pmatrix} z \\ \overline{z} \end{pmatrix}$ and $\overline{Z} = (\overline{z} \quad z)$, and $\Omega = \overline{Z}\widetilde{\Omega}Z$ we then have for the Lagrangian

$$\begin{aligned} \mathcal{L} = -\frac{1}{2}\overline{Z}\mathcal{D}^\mu\mathcal{D}_\mu Z + \frac{1}{2}\alpha\left[\overline{Z}\widetilde{\Omega}Z - \frac{N}{2f}\right] + \frac{1}{2}\overline{\mathcal{J}}Z + \frac{1}{2}\mathcal{J}\overline{Z} = \\ -\frac{1}{2}\left\{\overline{Z} + \overline{\mathcal{J}}\left(\mathcal{D}^2 - \alpha\widetilde{\Omega}\right)^{-1}\right\}\left(\mathcal{D}^2 - \alpha\widetilde{\Omega}\right)\left\{Z + \left(\mathcal{D}^2 - \alpha\widetilde{\Omega}\right)^{-1}\mathcal{J}\right\} \\ +\frac{1}{2}\overline{\mathcal{J}}\left(\mathcal{D}^2 - \alpha\widetilde{\Omega}\right)^{-1}\mathcal{J} - \frac{N}{4f}\alpha \quad , \end{aligned} \qquad (7.5a)$$

where:

$$\mathcal{D}^\mu = \begin{pmatrix} D^\mu & 0 \\ 0 & D^\mu \end{pmatrix} \qquad \widetilde{\Omega} = \begin{pmatrix} \Omega_{z\overline{z}} & \Omega_{\overline{z}\overline{z}} \\ \Omega_{zz} & \Omega_{\overline{z}z} \end{pmatrix} \quad , \qquad (7.5b)$$

and $\mathcal{J} = \begin{pmatrix} J \\ \overline{J} \end{pmatrix}$.

At this point, it is convenient to specialize to the (still large class of) Grassmannian models, for which the constraint (7.1b) takes the simple form[3)]

$$\overline{z}z = \frac{N}{2f} \quad , \text{ that is,} \qquad \widetilde{\Omega} = \frac{1}{2}\begin{pmatrix} 1 & 0 \\ 0 & 1 \end{pmatrix} \quad . \qquad (7.6)$$

Note that this corresponds to choosing $\Omega_{zz} = \Omega_{\overline{z}\overline{z}} = 0$ in (7.5). Here N is related to the rank of the symmetry group G, which we take to be $SU(N)$. The gauge group is $H = S(U(p) \otimes U(N-p))$. The real Grasmannians are trivially obtained by restricting z to be real. The $SO(N)$ case then follows trivially from the results.

---

3) In fact, Grassmannians contain a large enough class of models, since other groups may be obtained from unitary groups by convenient constraints[225].

We also choose the gauge group to be $U(p)$, $p \leq N$, and the constraint $\bar{z}^i_a z^i_b = \frac{N}{2f}\delta_{ab}$, which defines the Grassmannian σ models $SU(N)/S\left(U(p) \otimes U(N-p)\right)$- see chapter 6, subsection 2.3. The $\mathbb{C}P^{N-1}$ models correspond to $p = 1$.

Thus, the final form of the action is quadratic in the z fields, and integration over $\mathcal{D}\bar{z}\mathcal{D}z$ results in the inverse determinant of the quadratic operator $-D^2 + \alpha$. We thus have for the generating functional (7.4),

$$\mathcal{Z}[J, \bar{J}] = \int \mathcal{D}A_\mu \mathcal{D}\alpha e^{iS_{eff}[A_\mu,\alpha] + i\int d^2x \int d^2y \bar{J}(x)\left(-D^2+\alpha\right)^{-1}(x,y)J(y)} \tag{7.7a}$$

where

$$S_{eff}[A_\mu, \alpha] = i \operatorname{Tr} \ln\left(-D^2 - m^2 + \alpha\right) - \int d^2x \frac{\sqrt{N}}{2f}\alpha(x) \quad . \tag{7.7b}$$

The mass term is completely irrelevant at the moment, due to the constraint (7.6), according to which it becomes a global constant in the partition function. Later in the quantum theory, it will be fixed by the requirement that the field α has zero vacuum expectation value[4)]. The trace tr above includes summation over group and space-time indices. In order to study the limit $N \to \infty$, it is convenient to rescale the fields as follows[232,56]:

$$\alpha \to \frac{1}{\sqrt{N}}\alpha \quad , \quad A_\mu \to \frac{1}{\sqrt{N}}A_\mu \quad . \tag{7.8}$$

The effective action (7.7b) then takes the form

$$S_{eff} = i \operatorname{Tr} \ln\left\{\left(-\Box + \frac{2i}{\sqrt{N}}A^\mu\partial_\mu + \frac{1}{N}A^\mu A_\mu\right) - m^2 + \frac{\alpha}{\sqrt{N}}\right\} - \frac{\sqrt{N}}{2f}\int d^2x\alpha(x) \quad . \tag{7.9}$$

Figure 7.1: Expectation value contribution to the α field, and tadpole contribution.

---

4) We could as well take the point of view that the constant m^2 is the vacuum expectation value of α.

Figure 7.2: Tadpole contributions for the gauge fields. It vanishes due to Lorentz invariance.

The vacuum expectation value of α, will modify the mass term that we have introduced in (7.7b). As we already mentioned, it is irrelevant in the classical theory, corresponding to a change of variable in the delta functional representation (7.3). However, in the quantum theory, the α interaction gives rise to the tadpole diagrams of figure 7.1; in a mechanism similar to the one discussed in the Gross Neveu model in chapter 5, we have a mass generation for the fundamental fields z. We proceed by noticing that the theory is expandable in a series in $\frac{1}{\sqrt{N}}$. In order to carry out the expansion, we consider the effective action written explicitly in terms of the fields

$$-iS_{eff} = i\frac{\sqrt{N}}{2f}\int d^2x\alpha(x)+$$

$$\mathrm{tr}\ln\left(-\Box - m^2\right) + \mathrm{tr}\ln\left\{1 + \frac{1}{-\Box - m^2}\left(\frac{2i}{\sqrt{N}}A^\mu\partial_\mu - \frac{1}{N}A^\mu A_\mu + \frac{1}{\sqrt{N}}\alpha\right)\right\} . \quad (7.10)$$

The second term is a constant which may be incorporated in to the definition of the partition function. We can expand the effective action as

$$S_{eff} = \sum_{\nu=1}^{\infty} N^{1-\frac{\nu}{2}} S^{(\nu)} \quad . \tag{7.11}$$

The linear term $\mathcal{O}(\frac{1}{\sqrt{N}})$ does not contribute for an external gauge line (figure 7.2) because of Lorentz invariance (there is no vector to contract with the gauge field index). The $\mathcal{O}(\frac{1}{\sqrt{N}})$ term contribution of the α field must be enforced to be zero, in order to have a well defined theory in the limit $N \to \infty$ (or else, in order that α has a vanishing expectation value). It is computed from the diagrams of figure 7.1 to be

$$S^{(1)} = \mathrm{tr}\,\widetilde{\alpha}(0)\left(-\frac{1}{2f} + \int \frac{d^2k}{(2\pi)^2}\frac{i}{k^2 - m^2}\right) \quad , \tag{7.12}$$

where $\widetilde{\alpha}(0)$ denotes the Fourier transform of $\alpha(x)$ at zero momentum. The above expression corresponds to the last term in (7.9) plus the tadpole contribution corresponding to the second diagram of figure 7.1. The integral is logarithmically divergent, and can be regularized via the Pauli-Villars procedure. We have:[5)]

$$i \int \frac{d^2k}{(2\pi)^2} \left(\frac{1}{k^2 - m^2} - \frac{1}{k^2 - \Lambda^2} \right) = \frac{1}{4\pi} \ln \frac{\Lambda^2}{m^2} \quad . \tag{7.13}$$

We define the renormalized coupling by

$$\frac{2\pi}{f(\Lambda)} = \ln \frac{\Lambda^2}{\mu^2} + \frac{2\pi}{f_{ren}(\mu)} \quad . \tag{7.14}$$

The requirement of cancellation of the tadpole contribution against the linear contribution from α then leads us to the following expression for $f_{ren}(\mu)$:

$$\frac{2\pi}{f_{ren}(\mu)} = \ln \frac{\mu^2}{m^2} \quad . \tag{7.15}$$

From (7.14) we see that the bare coupling constant tends to zero as the cut off goes to infinity. Hence the model exhibits asymptotic freedom, since the bare coupling vanishes in the limit where the cut off goes to infinity. On the other hand, Eq. (7.15) represents[178,184] a dynamical mass transmutation.[6)]

Asymptotic freedom is a common feature of non abelian gauge theories in four dimensions. The latter property (mass transmutation (7.12a)) is more special. In the present case where the coupling constant only enters via the constraint (7.1c), f may be eliminated completely in favor of the mass to all orders in N. This means that the theory no longer depends on the coupling constant, but only on the physical mass parameter m, related to the cut-off Λ and coupling constant $f(\Lambda)$ in a renormalization group invariant way[22,178].

$$m^2 = \Lambda^2 e^{-\frac{2\pi}{f(\Lambda)}} \quad . \tag{7.16}$$

Since with the choice (7.15), $S^{(1)} = 0$, the first term contributing to the expansion (7.11) will be $S^{(2)}$. It is computed by expanding the effective action (7.7b) up to second order in the fields A_μ and α; we have

$$iS^{(2)} = \frac{1}{2} \int d^2x \int d^2y \left\{ \alpha^{ab}(x) \Gamma^{(\alpha)}(x-y) \alpha^{ba}(y) + A_\mu^{ab}(x) \Gamma_{\mu\nu}(x-y) A_\nu^{ba}(y) \right\}, \tag{7.17}$$

---

5) This is clearly a two dimensional result. In three dimensional space time we have for the regularized tadpole integral the result $\int_{PV-regul.} \frac{d^3k}{(2\pi)^3} \frac{i}{k^2-m^2} = \frac{\Lambda - m}{2\pi}$, where Λ is the Pauli-Villars (PV) cut-off. In this case, two phases arise in the model. For a thorough discussion of the three dimensional case see [222].

6) Actually, asymptotic freedom arises in a more general context, independent of the large N limit, as has been proven in a perturbative calculation for the sigma models elsewhere[255] (see also the end of the chapter).

with the Fourier transform of $\Gamma^{(\alpha)}$ and $\Gamma_{\mu\nu}$ given by

$$\widetilde{\Gamma}_\alpha(p) \equiv A(p) = \int \frac{d^2k}{(2\pi)^2} \frac{1}{(k^2 - m^2)[(k+p)^2 - m^2]} \quad , \tag{7.18}$$

$$\widetilde{\Gamma}_{\mu\nu}(p) = \int \frac{d^2k}{(2\pi)^2} \frac{(2k_\mu + p_\mu)(2k_\nu + p_\nu)}{(k^2 - m^2)[(k+p)^2 - m^2]} - 2g_{\mu\nu} \int \frac{d^2k}{(2\pi)^2} \frac{1}{k^2 - m^2} \tag{7.19}$$

For a gauge invariant regularization, one finds

$$\Gamma_{\mu\nu}(p) \equiv \left(g_{\mu\nu} - \frac{p_\mu p_\nu}{p^2} \right) F(p) \quad , \tag{7.20}$$

with

$$F(p) = (-p^2 + 4m^2)A(p) - \frac{i}{\pi} \quad , \tag{7.21}$$

and where, performing the integration in (7.18) one obtains

$$A(p) = -\frac{i}{2\pi} \frac{1}{\sqrt{-p^2(-p^2 + 4m^2)}} \ln \frac{\sqrt{-p^2 + 4m^2} - \sqrt{-p^2}}{\sqrt{-p^2 + 4m^2} + \sqrt{-p^2}} \tag{7.22}$$

In order to integrate over the gauge fields, we need to fix the gauge, which, following the by now standard Faddeev-Popov procedure, involves the addition of a gauge fixing term S_{gf} and a Faddeev-Popov ghost term S_{gh} to the effective action, $S_{eff} \to S_{eff} + S_{gf} + S_{gh}$.

It turns out to be convenient to introduce a linear gauge condition of the form $\partial'_\mu A^\mu = 0$, where ∂'_μ is a two vector of pseudo-differential operators, as given by[232]:

$$\left(\partial'_\mu f\right)(x) = \int \frac{d^2k}{(2\pi)^2} e^{ikx} ik_\mu L(k) \tilde{f}(k) \quad , \tag{7.23}$$

with $L(k)$ an even function of k, which we leave unspecified for the moment. Because of the formal identity of this gauge condition with the usual Lorentz gauge condition $\partial_\mu A^\mu = 0$, we can formally take over the results obtained for S_{gf} and S_{gh} when quantizing gauge theories in the Lorentz gauge. Hence

$$S_{gf} = \frac{1}{2\gamma} \int d^2x \partial^{\mu\prime} A_\mu^{ab} \partial^{\nu\prime} A_\nu^{ba} \quad , \tag{7.24a}$$

and

$$S_{gh} = \bar{\zeta} \mathcal{M} \zeta = \int d^2x \bar{\zeta}^{ab} \mathcal{M}^{ab,cd} \zeta^{cd} \quad , \tag{7.24b}$$

where γ is a parameter, ζ are $p \times p$ matrices representing the ghost fields, and $\mathcal{M}$ is the Faddeev-Popov matrix

$$\mathcal{M} = -\partial'_\mu D^\mu \quad , \tag{7.25}$$

with D_μ the covariant derivative acting on ζ as

$$D_\mu \zeta = \partial_\mu \zeta - \frac{i}{\sqrt{N}} [A_\mu, \zeta] \tag{7.26}$$

The gauge fixing term can be combined with $S^{(2)}$, Eq. (7.17), by redefining $\Gamma_{\mu\nu}$ as

$$\Gamma_{\mu\nu} \to \Gamma_{\mu\nu} + \gamma^{-1} p_\mu p_\nu L^2(p) \quad . \tag{7.27}$$

One is tempted to choose $\partial'_\mu = \partial_\mu$, and $L(k) = \text{const}$. This would however lead to an improper high momentum behavior in $\frac{1}{N}$ perturbation theory. Thus we find it convenient to choose

$$L^2(k) = \frac{1}{k^2} F(k) \tag{7.28a}$$

in which case

$$\Gamma_{\mu\nu}(p) = \left(g_{\mu\nu} - (1 - \gamma^{-1}) \frac{p_\mu p_\nu}{p^2} \right) F(p) \quad , \tag{7.28b}$$

which resembles the quantization of gauge theories in the so called Lorentz α gauges. The propagators corresponding to (7.18) and (7.28b) are thus

$$D(p) = -iA^{-1}(p) \tag{7.29a}$$

$$D_{\mu\nu}(p) = i \left(g_{\mu\nu} - (1 - \gamma) \frac{p_\mu p_\nu}{p^2} \right) F^{-1}(p) \quad . \tag{7.29b}$$

The full set of Feynman rules is given in table 7.1. These Feynman rules may be specialized to the case $p = 1$, defining the $\mathbb{C}P^{N-1}$ model (care must be taken with a factor of 2 in the $\bar{z}zAA$ vertex). There are also forbidden diagrams, which have already been taken into account (see figures 7.1 and 7.2).

The lowest order two point functions of A_μ and α, may also be computed from a resummation of diagrams in figure 7.3. The results are the auxiliary field propagators given in table 7.1. Notice also that the L factor cancels in the computation of closed ghost loops.

7.2.2 Renormalization

In order to define a renormalization procedure, we may use the BPHZ scheme, making Taylor subtractions of the integrands around zero momenta[4,5]. We suppose that appropriate I.R. cut offs are used to avoid infra red divergencies. These represent an additional problem[262], which will be discussed below.

Table 7.1		
Feynman rules of the Grassmannian models		
$z\bar{z}$ propagator		$\frac{i}{p^2-m^2}\delta^{ab}\delta^{ij}$
α^{ab} propagator		$-A(p)^{-1}\delta^{ad}\delta^{bc}$
A_μ propagator		$-\left(g_{\mu\nu}-(1-\gamma)\frac{p_\mu p_\nu}{p^2}\right)\delta^{ad}\delta^{bc}F(p)^{-1}$
ζ propagator		$\frac{i}{\sqrt{k^2F(p)}}$
$z\bar{z}\alpha$ vertex		$\frac{i}{\sqrt{N}}\delta^{ad}\delta^{bc}\delta_{ij}$
$\zeta\zeta A_\mu$ vertex	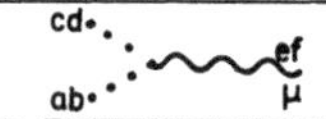	$i\sqrt{\frac{F(p)}{Np^2}}p_\mu(\delta^{ad}\delta^{be}\delta^{fc}-\delta^{af}\delta^{bc}\delta_{de})$
$z\bar{z}A_\mu$ vertex	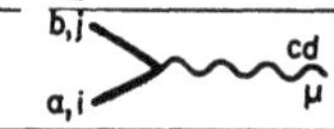	$\frac{i}{\sqrt{N}}(2k_\mu+p_\mu)\delta^{ad}\delta^{bc}$
$z\bar{z}A_\mu A_\nu$ vertex	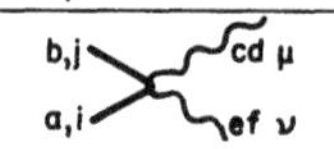	$\frac{i}{N}g_{\mu\nu}\delta_{ij}(\delta^{ad}\delta^{be}\delta^{fc}+\delta^{af}\delta^{bc}\delta^{de})$

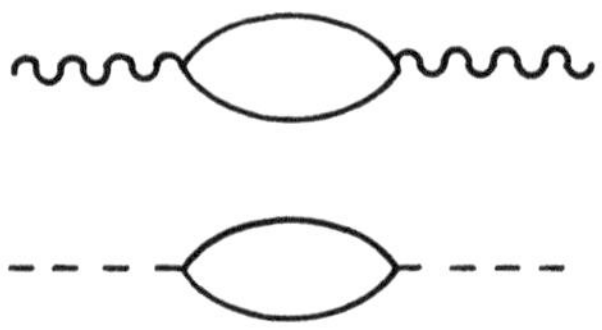

Figure 7.3: Forbidden diagrams; these diagrams have been summed over in the computation of the α and A_μ propagators.

The superficial degree of divergence $\delta(\Gamma)$ of a given proper diagram is obtained analysing the behavior of the propagators at large momenta. The behavior of the functions $A(p)$ and $F(p)$ for large p^2 are:

$$A(p^2) \approx -\frac{i}{2\pi p^2}\ln\left(-\frac{p^2}{m^2}\right) \quad , \tag{7.30a}$$

and

$$F(p^2) \approx \frac{i}{2\pi}\ln\left(-\frac{p^2}{m^2}\right) \quad . \tag{7.30b}$$

Taking into account the large p behavior with a few combinatorial rules counting the lines and vertices[110], it is not very difficult to show that[7)] we arrive at the

7) From the power counting theorem, the superficial degree of divergence is given, in two dimensions, in terms of the number of loops, the number of z lines and the number of momentum dependent vertices, by $\delta(\Gamma) = 2\#loops - 2\#z - lines + \#A\bar{z}z - vertices$; we use the identities $\#loops = \#1 - \#vertices + \#internal\ lines$, and $\sum(\#lines\ ending\ at\ V_i) \times (\#vertices\ V_i) = 2\#internal\ lines + \#external\ lines$.

expression:

$$\delta(\Gamma) = 2 - 2E_\alpha - E_A - E_\zeta, \tag{7.31}$$

where E_x denotes the number of external lines of type x (It is clear that $E_{\overline{z}} = E_z$, and $E_{\overline{\zeta}} = E_\zeta$).

Since $\delta(\Gamma)$ does not depend on E_z, the degree of divergence seems to imply an infinite number of divergent diagrams, differing in the number of external z-legs. However, as discovered by Aref'eva[257], Aref'eva and Azakov[222], and Lowenstein and Speer[254], there exists a cancellation mechanism of divergencies, due to the non-linear constraints of the theory. This cancellation mechanism goes as follows. Let Γ be a connected Feynman diagram with r external $\overline{z}$ and r external z lines, as well as possibly other external lines. Because of charge conservation, we can arrange the r external $\overline{z}$ and z lines in r pairs, such that the diagram is proportional to $\delta_{i_1 j_1} \ldots \delta_{i_r j_r}$ (to see that this is true, follow a line of a given charge through a diagram). Then if for some l, $1 \leq l \leq r$, the lth pair of external $\overline{z} - z$ lines of Γ does not join at the nearest $\overline{z}z\alpha$ or $\overline{z}z\lambda$ vertices, one can define a new connected Feynman diagram Γ_l by joining the $\overline{z}$ line with the z line and connecting to the resulting vertex a new pair of external $\overline{z}z$ lines via an α propagator (see figure 7.4). The procedure can not be repeated, since it would lead to forbidden diagrams. The counterterm of the original diagram is cancelled by the counterterm of the corresponding subdiagram of Γ_l for the following reason. Γ being superficially divergent and proper, it appears as a renormalization part in Γ_l, and the Taylor terms to be subtracted in the process of renormalization of both Γ and Γ_l, are monomials $M(\partial)I_\Gamma|_o$ in partial derivatives of I_Γ, where I_Γ is the unsubtracted Feynman integrand corresponding to a diagram Γ. The remaining piece of Γ_l, on the other hand, just contributes a factor of -1 because of (7.18) and (7.29a) together with the two factors i from the vertices. If $M(\partial)$ has no derivatives, there is a complete cancellation. If $M(\partial)$ contains one derivative, this implies a term $\overline{z}\partial_\mu z$, which gives no contribution due to gauge invariance (as one can prove). If there is a quadratic divergence, so that a term $\overline{D_\mu z} D_\mu z$ appears (two derivatives in $M(\partial)$), then there may or may not be some other pair of lines $\overline{z}/z$; if there is not, this is a wave function renormalization; if there is the attaching procedure can be repeated to cancel the divergence. Thus cancellation occurs for diagrams with more than one pair of external legs. There are also infinities which may be absorbed in a wave function and a mass renormalization.

Figure 7.4: Cancellation mechanism that lowers the superficial degree of divergence of a diagram with external lines $z\overline{z}$.

7.2.3 Infrared divergencies

The infrared problem is rather involved. Processes involving charged fields, such as the correlators $\langle 0|z\bar{z}z\bar{z}|0\rangle$ are not infrared finite. A typical divergent contribution in $\frac{1}{N}$ is given by the diagram of figure 7.5. In the case $p = 1$, corresponding to the $\mathbb{C}P^{N-1}$ model, the problem can be solved if we restrict ourselves to gauge invariant correlators, such as $\langle 0|\mathcal{N}\left[\bar{z}_i(x)z_j(x)\right]\mathcal{N}\left[\bar{z}_k(y)z_l(y)\right]|0\rangle$. In those cases, a cancellation occurs. Indeed, one can prove that[262] *Green functions involving gauge invariant objects are infrared finite.* In the $\mathbb{C}P^{N-1}$ model gauge invariant objects are built from $F_{\mu\nu}$ and its derivatives, as well as covariant derivatives of z and $\bar{z}$. For such objects the cancellation mechanism occurs as follows. We first notice that IR divergencies in the $\frac{1}{N}$ expansion are due to the bad behavior of the gauge field propagator, since for small momentum, we have the following behavior of the function $A(p)$:

$$A(p) \approx \frac{i}{4\pi m^2}\left\{1 + \frac{p^2}{6m^2}\right\} \quad ; \qquad (7.32a)$$

consequently

$$F(p) \approx -\frac{ip^2}{12\pi m^2} \quad . \qquad (7.32b)$$

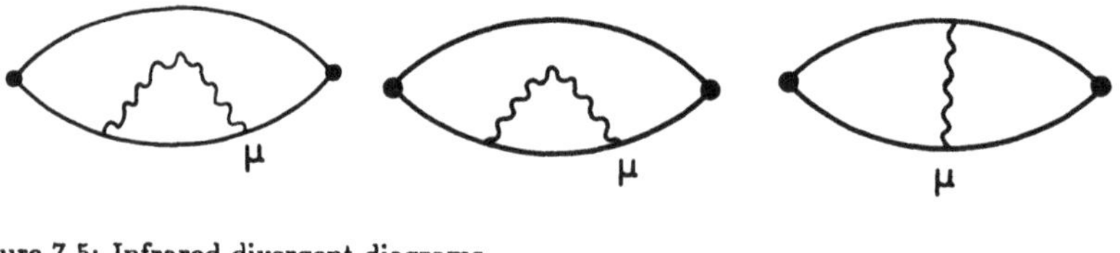

Figure 7.5: Infrared divergent diagrams.

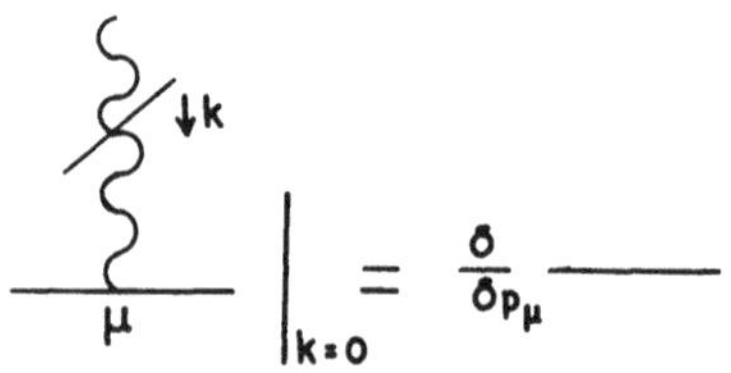

Figure 7.6: Attaching procedure used to combine infrared divergent diagrams, and verify their cancellation.

External gauge field lines are harmless, since we do not integrate over them. The dangerous lines are internal gauge lines, which could produce untamable infrared divergencies.

These internal lines are always connected to a conserved current. Moreover the coupling is such that at zero momentum they correspond to derivatives with respect to the momentum flowing through the lines to which they are connected.[8] As it turns out, adding all possible diagrams, we get for zero gauge field momentum a total derivative. Let us formalize the argument in an explicit one loop example, and afterwards discuss the generalization. Consider the diagram of figure 7.5, contributing to the lowest order of $\langle 0|\mathcal{N}\left[\bar{z}_i(x)z_j(x)\right]\mathcal{N}\left[\bar{z}_k(y)z_l(y)\right]|0\rangle$ in the large N limit. We have the contribution (7.5a) given by

$$D^{5a} = -\delta_{jk}\delta_{il}\int\frac{d^2k}{(2\pi)^2}\int\frac{d^2q}{(2\pi)^2}\left(g_{\mu\nu}-\frac{q_\mu q_\nu}{q^2}\right)F(q)^{-1}\times$$
$$\frac{1}{(p+k)^2-m^2}\frac{[2(k+p)_\mu+q_\mu]}{(p+k+q)^2-m^2}\frac{1}{(k+q)^2-m^2}(2k_\nu+q_\nu)\frac{1}{k^2-m^2} \quad . \quad (7.33)$$

The above integral is IR divergent due to the factor $F(q)^{-1}$. The divergent part is given by the expression

$$D^{5a\,IR} = i12\pi m^2\delta_{jk}\delta_{il}\int\frac{d^2q}{(2\pi)^2q^2}\left(g_{\mu\nu}-\frac{q_\mu q_\nu}{q^2}\right)$$
$$\int\frac{d^2k}{(2\pi)^2}\frac{2(k+p)_\mu}{(p+k)^2-m^2}\frac{1}{(p+k)^2-m^2}\frac{2k_\nu}{(k^2-m^2)^2} \quad , \quad (7.34)$$

which we rewrite as

$$D^{5a\,IR} = i12\pi m^2\delta_{jk}\delta_{il}\int\frac{d^2q}{(2\pi)^2q^2}\left(g_{\mu\nu}-\frac{q_\mu q_\nu}{q^2}\right)\times$$
$$\int\frac{d^2k}{(2\pi)^2}\frac{2(k+p)^\mu}{[(p+k)^2-m^2]^2}\left\{-\frac{\partial}{\partial k_\nu}\right\}\frac{1}{k^2-m^2} \quad . \quad (7.35)$$

We now fix one vertex (the one labelled by μ) and move the other one (the one labelled by ν) cyclically obtaining the expressions for diagrams 7.4b and 7.4c. The divergent part of the three diagrams add up to zero

$$D^{5\,IR} = i12\pi m^2\int\frac{d^2q}{(2\pi)^2}\left(g_{\mu\nu}-\frac{q_\mu q_\nu}{q^2}\right)\times$$
$$\int\frac{d^2k}{(2\pi)^2}\frac{\partial}{\partial k_\nu}\left\{-\frac{1}{k^2-m^2}\frac{2(k_\mu+p_\mu)}{[(k+p)^2-m^2]^2}\right\}=0 \quad , \quad (7.36)$$

where we have used the fact that the second integral contributes a vanishing surface term.

A general internal gauge line can be treated similarly, since it is always coupled to a conserved current, and the interaction occurs always in a matter loop; therefore the generalization is straightforward. The non abelian case is more involved, and there is no general result in the literature, but there exists no a priori reason preventing the same cancellation of divergencies from occurring.

[8] That means that the product of the two lines joining at the given vertex and the vertex itself is equal to the derivative of a single propagator, see figure 7.6.

7.2.4 Physical interpretation of the results

The θ vacuum

The Feynman rules show that the z-particles interact via interchange of scalar (α) and gauge field (A_μ) quanta. The α interaction is short ranged, since $A(p)^{-1}$ has no pole, but the gauge field propagator has a zero mass pole; in two dimensions such a pole generates a "secret" long range force that confines; for a discussion of an analogous problem see chapters 10 and 12. This result can already be foreseen from the fact that infra red divergencies only cancel for gauge invariant objects ("bound states"). Moreover, we can analyse the vacuum properties in more detail, allowing for a θ term in the action of the $\mathbb{C}P^{N-1}$ model ($p = 1$)[56]:

$$S_\theta = S + \theta Q \quad . \tag{7.37}$$

where Q is the topological density,

$$Q = \frac{1}{2\pi\sqrt{N}} \int d^2x \varepsilon^{\mu\nu} \partial_\mu A_\nu \quad . \tag{7.38}$$

This gives rise to new diagrams, with an A_μ-line ending in the vacuum, contributing a factor $-\frac{\theta}{2\pi\sqrt{N}}\varepsilon_{\mu\nu}p^\nu$ to the respective Feynman rule. Correspondingly, the topological charge density $Q(x) = \frac{1}{2\pi}\epsilon^{\mu\nu}\partial_\mu A_\nu$ is found to have a non zero vacuum expectation value,

$$\langle\theta|Q(x)|\theta\rangle = \frac{3m^2}{N\pi}\theta + \mathcal{O}(\frac{1}{N^2}) \quad , \tag{7.39}$$

as computed from diagrams 7.7 representing the annihilation of the gauge field line into the vacuum, with the vertex defined by (7.38). Equation (7.39) is valid for $|\theta| < \pi$, otherwise we have to consider pair production induced by the vacuum energy, which is in that case enough to generate pairs, (see chapter 10 for details). The above discussion shows that the vacuum carries topological charge. The correlators depend on θ, since θ vertices can contribute at zero momentum, due to the massless poles.

figure 7.7: Contribution to the expectation value of the topological charge in a non zero θ vacuum.

7.3 Grassmannian Models Interacting with Fermions

7.3.1 $\frac{1}{N}$ expansion and Feynman rules

We can define a general coupling of Grassmannian models with fermions, introducing quarks with an electric charge e (possibly different from that of the z's) and renormalizable interactions by the Lagrange density[239]

$$\mathcal{L} = \overline{D_\mu z} D^\mu z + \overline{\psi}(i\not{D} - M_B) + \frac{1}{2N}(g_S + e^2 f)\left(\overline{\psi}\gamma_\mu\psi\right)^2 + \frac{g_V}{2N}\left[\left(\overline{\psi}\tau^A\psi\right)^2 - \left(\overline{\psi}\gamma_5\tau^A\psi\right)^2\right] \quad (7.40)$$

with

$$\overline{z}_i^a z_i^b = \frac{N}{2f}\delta^{ab} \quad , \quad (7.41a)$$

$$\overline{z}_i^a \psi_i^b = 0 = \overline{\psi}_i^a z_i^b \quad . \quad (7.41b)$$

and $A = 1, \cdots, F$ a new flavor index. The Lagrangian (7.40) represents a generalization of the theories defined by (7.1) and (7.2). The supersymmetric model is obtained[232,239,233,240] for equal masses and the choice of parameters $e = 1$, $g = 1$, and $F = 1$ (only one flavour). The minimal model is still simpler; it is obtained by deleting the fermionic quartic interactions, and dropping the condition (7.41b). The corresponding Feynman rules will be obtained as a particular case[232,239].

The $\frac{1}{N}$ expansion of the model can be performed following the same lines as used in the purely bosonic case. We have for the generating functional the expression

$$\mathcal{Z}[\overline{J}, J, \overline{\eta}, \eta] = \mathcal{N}^{-1} \int \mathcal{D}\overline{z}\mathcal{D}z\mathcal{D}\overline{\psi}\mathcal{D}\psi \prod \delta\left(\overline{z}z - \frac{N}{2f}\right)\delta(z_i^a\psi_i^b)\delta\left(\overline{\psi}_i^a z_i^b\right) e^{\left[iS + i\int[\overline{J}z + \overline{z}J]\right]} \quad .(7.42)$$

The delta functional can be treated in the same way as in Eq. (7.3a), that is

$$\prod_{x,a,b} \delta\left(\overline{z}_i^a(x) z_i^b(x) - \frac{N}{2f}\delta^{ab}\right)\delta\left(\overline{z}_i^a(x)\psi_i^b(x)\right)\delta\left(\overline{\psi}_i^a(x) z_i^b(x)\right) \sim \int \mathcal{D}\alpha^{ab}\mathcal{D}\overline{c}^{ab}\mathcal{D}c^{ab} e^{i\int d^2x\left[\frac{\alpha^{ba}}{\sqrt{N}}\left(\overline{z}_i^a(x) z_i^b(x) - \frac{N\delta^{ab}}{2f}\right) + \frac{1}{\sqrt{N}}\overline{c}^{ba}\overline{z}_i^a(x)\psi_i^b(x) + \frac{1}{\sqrt{N}}\overline{\psi}_i^a(x) z_i^b(x) c^{ba}\right]} \quad . \quad (7.43)$$

We can treat analogously the quartic interactions, using

$$e^{i\int d^2x\left\{\frac{1}{2gN}B^{ab}B^{ba}\right\}} \sim \int \mathcal{D}\phi^{ab} e^{i\int d^2x\left\{-\frac{1}{2g}\phi^{ab}\phi^{ba} + \frac{1}{\sqrt{N}}B^{ab}\phi^{ba}\right\}} \quad , \quad (7.44a)$$

$$e^{i\int d^2x\left\{\frac{1}{2gN}B_5^{ab}B_5^{ba}\right\}} \sim \int \mathcal{D}\phi_5^{ab} e^{i\int d^2x\left\{-\frac{1}{2g}\phi_5^{ab}\phi_5^{ba} + \frac{1}{\sqrt{N}}B_5^{ab}\phi_5^{ba}\right\}} \quad , \quad (7.44b)$$

where $\sim$ means equality up to normalization constants. We thus arrive at a generating functional which is quadratic in the fundamental fields z and ψ

$$Z\,[J,\bar{J},\eta,\bar{\eta}] =$$

$$N^{-1}\int \mathcal{D}\alpha\mathcal{D}A_\mu\mathcal{D}z\mathcal{D}\bar{z}\mathcal{D}\psi\mathcal{D}\bar{\psi}\mathcal{D}c\mathcal{D}\bar{c}\mathcal{D}\phi\mathcal{D}\phi_5 e^{\left[iS_{fields}+i\int d^2x\{\bar{J}z+J\bar{z}+\bar{\eta}\psi+\bar{\psi}\eta\}\right]} \quad (7.45)$$

where the field action is given by

$$S_{fields}=\int d^2x\,\Big[\overline{D_\mu z}D^\mu z - m^2\bar{z}z + \bar{\psi}(i\not{D} - M_B)\psi +$$

$$\frac{1}{\sqrt{N}}\alpha^{ba}\left(\bar{z}_i^a(x)z_i^b(x) - \frac{N\delta^{ab}}{2f}\right) + \frac{1}{\sqrt{N}}\bar{c}^{ba}\bar{z}_i^a(x)\psi_i^b(x) + \frac{1}{\sqrt{N}}\bar{\psi}_i^a(x)z_i^b(x)c^{ba}$$

$$-\frac{1}{2g}\phi^{abA}\phi^{baA} + \frac{1}{\sqrt{N}}\bar{\psi}^a\tau^A\psi^b\phi_A^{ba} - \frac{1}{2g}\phi_5^{abA}\phi_{5A}^{ba} + i\frac{1}{\sqrt{N}}\bar{\psi}^a\tau^A\gamma_5\psi^b\phi_5^{baA}\Big] = \quad (7.46a)$$

$$\int d^2x[\bar{z}\Delta_B z + \bar{\psi}\,\not{\Delta}\psi + \bar{J}z + \bar{z}J + \bar{\eta}\psi + \bar{\psi}\eta - \frac{\sqrt{N}}{2f}\operatorname{tr}\alpha + \frac{1}{\sqrt{N}}\bar{\psi}zc + \overline{cz}\psi] \quad ,(7.46b)$$

and without loss of generality we have dropped the current-current fermionic interaction in (7.40). The quadratic operators are read off from the previous expressions

$$\Delta_B = -D^\mu D_\mu - m^2 + \frac{\alpha}{\sqrt{N}} \quad , \quad (7.47a)$$

$$\not{\Delta} = i\not{D} - M_B - \frac{1}{\sqrt{N}}(\phi^A + i\phi_5^A\gamma_5)\tau^A \quad , \quad (7.47b)$$

with the covariant derivatives given by

$$D_\mu = \partial_\mu - \frac{ie}{\sqrt{N}}A_\mu \qquad \text{on the fermions} \quad (7.47c)$$

and

$$D_\mu = \partial_\mu - \frac{i}{\sqrt{N}}A_\mu \qquad \text{on the bosons} \quad (7.47d)$$

In order to perform the ψ and z integrations in (7.40), we rewrite the action (including the sources) by first completing the squares in the fields ψ and z, integrating over ψ and then repeating the process for z. To this end we first rewrite S_{fields} as follows

$$\bar{\psi}\,\not{\Delta}\psi + \bar{\eta}\psi + \frac{1}{\sqrt{N}}\bar{\psi}\eta + \frac{1}{\sqrt{N}}\bar{\psi}zc + \overline{cz}\psi =$$

$$\left(\bar{\psi} + (\bar{\eta} + \overline{cz})\,\not{\Delta}^{-1}\right)\not{\Delta}\left(\psi + \not{\Delta}^{-1}(\eta + cz)\right) - (\bar{\eta} + \overline{cz})\,\not{\Delta}^{-1}(\eta + cz) \quad . \quad (7.48)$$

The first term in (7.48) just contributes a determinant to the generating functional (7.45) upon integrating over ψ. The remaining z integration can then be done by again first completing squares using the identity

$$
\begin{aligned}
&\bar{z}\Delta_B z + \bar{J}z + \bar{z}J - (\bar{\eta} + \overline{cz})\,\not{\Delta}^{-1}(\eta + cz) = \\
&\qquad \bar{z}(\Delta_B - \frac{1}{N}\bar{c}\,\not{\Delta}^{-1}c)z + (\bar{J} - \frac{1}{\sqrt{N}}\bar{\eta}\,\not{\Delta}^{-1}c)z + \bar{z}(J - \frac{1}{\sqrt{N}}\bar{c}\,\not{\Delta}^{-1}\eta) - \bar{\eta}\,\not{\Delta}^{-1}\eta \\
&= (\bar{z} + (\bar{J} - \frac{\bar{\eta}\,\not{\Delta}^{-1}c}{\sqrt{N}})(\Delta_B + \frac{\bar{c}\Delta_F^{-1}c}{N})^{-1})(\Delta_B - \frac{\bar{c}\,\not{\Delta}^{-1}c}{N})\times \\
&\qquad\qquad \times(z + (\Delta_B - \frac{\bar{c}\,\not{\Delta}^{-1}c}{N})^{-1}(J - \frac{\bar{c}\,\not{\Delta}^{-1}\eta}{\sqrt{N}})) \\
&\qquad -(\bar{J} - \frac{\bar{\eta}\,\not{\Delta}^{-1}c}{\sqrt{N}})(\Delta_B - \frac{\bar{c}\,\not{\Delta}^{-1}c}{N})^{-1}(J - \bar{c}\,\not{\Delta}^{-1}\eta) \quad . \qquad (7.49)
\end{aligned}
$$

Performing the above mentioned integration over z and ψ we are lead to the following effective action,

$$
\begin{aligned}
S_{eff} = iN\,\mathrm{Tr}\left\{\ln\left(\Delta_B - \frac{1}{N}\bar{c}\,\not{\Delta}^{-1}c\right) - \ln\not{\Delta}\right\} + \\
\int d^2x\left\{-\frac{\sqrt{N}}{2f}\alpha - \frac{1}{2g}\mathrm{tr}(\phi^{A^2} + \phi_5^{A^2})\right\} \quad , \qquad (7.50)
\end{aligned}
$$

with the source term given by the expression

$$
S_{source} = \int \bar{J}'\left(\Delta_B - \frac{1}{N}\int \bar{c}\,\not{\Delta}^{-1}c\right)^{-1} J' + \bar{\eta}\,\not{\Delta}^{-1}\eta \quad , \qquad (7.51)
$$

where $J' = J - \frac{\bar{c}\not{\Delta}^{-1}\eta}{\sqrt{N}}$.

The action (7.50) replaces (7.9). Repeating now the previous analysis, we expand S_{eff} as in (7.10). The dominant term in the $\frac{1}{N}$ expansion is again given by the tadpole contribution; requiring that this term vanishes, we reobtain the dynamically generated mass (7.15) for the boson. The field ϕ also acquires a non-vanishing vacuum expectation value, which we parametrize by writing

$$
\phi^0_{ab} = \varphi^0_{ab} + \sqrt{NF}M_S\delta_{ab} \quad , \qquad (7.52)
$$

where the upper index is the adjoint flavor index. This defines the direction of $U(F)\otimes U(F)$ symmetry breaking. We have the following results for the $\mathcal{O}(\sqrt{N})$ action

$$
S^{(1)} = \tilde{\alpha}(0)\left\{\frac{1}{2f} - \int\frac{d^2k}{(2\pi)^2}\frac{i}{k^2 - m^2}\right\} + \tilde{\varphi}^{(0)}\sqrt{F}\left\{\frac{M_S}{2g} - M\int\frac{d^2k}{(2\pi)^2}\frac{i}{k^2 - m^2}\right\} \qquad (7.53)
$$

where $M = M_B + M_S$. We regularize the integrals with a Pauli Villars cut-off as before (see Eq. (7.12)), obtaining

$$i \int \frac{d^2k}{(2\pi)^2} \frac{1}{k^2 - m^2(M^2)} = \frac{1}{4\pi} \ln \frac{\Lambda^2}{m^2(M^2)} \tag{7.54}$$

where Λ is the ultraviolet PV cut-off. Therefore, imposing that the tadpole contribution $S^{(1)}$ vanishes, we obtain as before the bare couplings as functions of the cut-off and the generated masses:

$$\begin{cases} \frac{1}{f} = \frac{1}{2\pi} \ln \frac{\Lambda^2}{m^2} \\ \frac{1}{g} = \frac{1}{2\pi} \frac{M}{M_S} \ln \frac{\Lambda^2}{M^2} \end{cases} \tag{7.55}$$

Since $S^{(1)} = 0$, the first non-trivial term in the expansion of the action is given by

$$\begin{aligned} iS^{(2)} = \frac{1}{2} \int d^2x d^2y \{ & \alpha^{ab}(x)\Gamma(x-y)\alpha^{ba}(y) + A_\mu^{ab}(x)\Gamma^{\mu\nu}(x-y)A_\mu^{ba}(y) \\ & + 2\bar{c}^{ab}(x)\Gamma_{\bar{c}c}(x-y)c^{ba}(y) + \phi_A^{ab}(x)\Gamma_{AB}^{\phi}(x-y)\phi_B^{ba}(y) + \phi_{5A}^{ab}(x)\Gamma_{\phi_5}^{AB}(x-y)\phi_{5B}^{ba}(y) \\ & + A_\mu^{ab}(x)\Gamma_{\phi_5}^{\mu}(x-y)\phi_{5,0}^{ba}(y) + \phi_{5,0}^{ab}(x)\Gamma_{\phi_5}^{\nu}(x-y)A_\nu^{ba}(y) \} \end{aligned} \tag{7.56}$$

where the individual contributions can be calculated from the diagrams in figure 7.8.

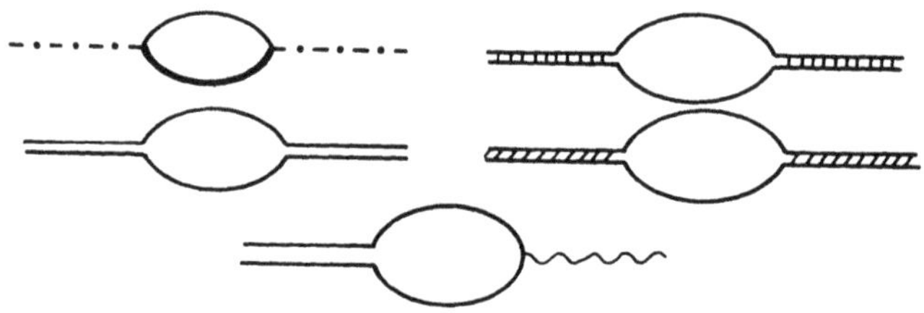

Figure 7.8: Contributions to the auxiliary fields propagators in the Grassmannian model with fermions.

In momentum space the vertex functions read

$$\widetilde{\Gamma}(p) = A(p;m) \quad , \tag{7.57a}$$

$$\widetilde{\Gamma}(p)_{\mu\nu} = \left(g_{\mu\nu} - \frac{p_\mu p_\nu}{p^2}\right) [\, (-p^2 + 4m^2)A(p;m) - 4M^2 F e^2 A(p,M) + (Fe^2 - 1)\frac{i}{\pi} \,]$$

$$= \left(g_{\mu\nu} - \frac{p_\mu p_\nu}{p^2} \right) \tilde{\Gamma}_A(p; m, M) \quad , \tag{7.57b}$$

$$\widetilde{\Gamma}_{\bar{c}c}(p) = \not{p} \left\{ -\frac{i}{4\pi p^2} \ln \frac{M}{m} + \frac{1}{2p^2}(p^2 - m^2 + M^2) A(p; m, M) \right\}$$

$$+ MA(p; m, M) \quad , \tag{7.57c}$$

$$\widetilde{\Gamma}_{\phi}^{AB}(p) = -\delta^{AB} \left\{ i\varepsilon - (p^2 - 4M^2) A(p, M) \right\} \quad , \tag{7.57d}$$

$$\widetilde{\Gamma}_{\phi_5}^{AB}(p) = -\delta^{AB} \left\{ i\varepsilon - p^2 A(p, M) \right\} \quad , \tag{7.57e}$$

$$\widetilde{\Gamma}_{\phi_5}^{\mu}(p) = -\epsilon^{\mu\nu} p_\nu 2e\sqrt{F} M A(p, M) = \epsilon^{\mu\nu} p_\nu \widetilde{\Gamma}_{\phi_5}^{A}(p, M) \quad , \tag{7.57f}$$

where

$$A(p; m, M) = \int \frac{d^2k}{(2\pi)^2} \frac{1}{(k^2 - m^2)((p+k)^2 - M^2)} =$$

$$- \frac{i}{4\pi \sqrt{(-p^2 + M^2 - m^2)^2 - 4m^2 p^2}} \ln \frac{-p^2 + m^2 + M^2 - \sqrt{(-p^2 + M^2 - m^2)^2 - 4m^2 p^2}}{-p^2 + m^2 + M^2 + \sqrt{(-p^2 + M^2 - m^2)^2 - 4m^2 p^2}} \tag{7.58}$$

and $A(p; m) = A(p; m, m)$, given by expression (7.22).

The constant ε is given in terms of the bare constants and the cut-off by

$$\varepsilon = \frac{1}{g} - \frac{1}{2\pi} \ln \frac{\Lambda^2}{m^2} \quad , \tag{7.59}$$

and must be finite, thus defining the scaling property of the mass M_S. The propagators and vertices are given in table 7.2.

In the supersymmetric case, these expressions simplify considerably. We have in this case

$$\varepsilon = 0 \quad , \quad g = f \quad . \tag{7.60}$$

The Feynman rules are immediately obtained, and have been collected in table 7.3. In that table we present the minimally coupled model as well, which corresponds to the Lagrangian of the supersymmetric model without the fermionic self interactions, and with no constraint for the fermionic degrees of freedom, such as (7.41b). We shall return to that model later on.

Table 7.2

Propagators and vertices in the $\frac{1}{N}$ expansion of the Grassmannian model with fermions

$z\bar{z}$ propagator		$\frac{i}{p^2-m^2}\delta^{ab}\delta^{ij}$
$\psi\bar{\psi}$ propagator	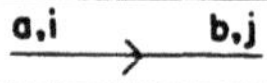	$\frac{i}{\not{p}-M}\delta^{ab}\delta^{ij}\delta_{\alpha\beta}$
α^{ab} propagator	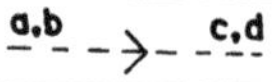	$iA(p)^{-1}\delta^{ad}\delta^{cb}$
A_μ propagator	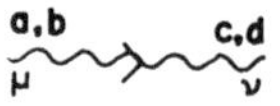	$\frac{-\widetilde{\Gamma}^5}{\widetilde{\Gamma}^A\widetilde{\Gamma}^5+p^2\left(\widetilde{\Gamma}^{A\phi}\right)^2}\left(g_{\mu\nu}-\frac{p_\mu p_\nu}{p^2}\right)\delta^{ad}\delta^{bc}$
ζ propagator	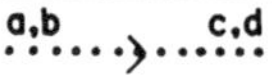	$\frac{i}{\sqrt{k^2F(p)}}\delta^{ad}\delta^{bc}$
ϕ propagator	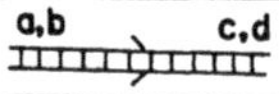	$\left\{-\widetilde{\Gamma}_\phi(p)\right\}^{-1}\delta_{\alpha\beta}\delta^{ad}\delta^{bc}$
ϕ_5 propagator		$\left\{-\widetilde{\Gamma}_{\phi_5}(p)\right\}^{-1}\delta_{\alpha\beta}\delta^{ad}\delta^{bc}$
ϕ_5^0 propagator	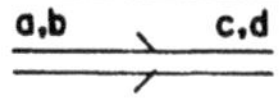	$\frac{-\widetilde{\Gamma}^A}{\widetilde{\Gamma}^A\widetilde{\Gamma}^5+p^2\left(\widetilde{\Gamma}^{A\phi_5}\right)^2}\delta_{\alpha\beta}\delta^{ad}\delta^{bc}$
$A_\mu\phi_5^0$ propagator	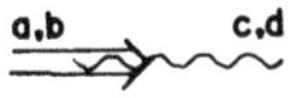	$\frac{-\widetilde{\Gamma}^{A\phi_5}}{\widetilde{\Gamma}^A\widetilde{\Gamma}^5+p^2\left(\widetilde{\Gamma}^{A\phi_5}\right)^2}\epsilon_{\mu\nu}p^\nu$
$\bar{c}c$ propagator	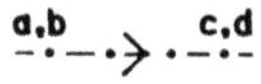	$-(\Gamma_{c\bar{c}})^{-1}\delta^{ad}\delta^{bc}$
$z\bar{z}\alpha$ vertex	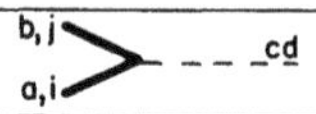	$\frac{i}{\sqrt{N}}\delta^{ad}\delta^{bc}\delta_{ij}$
$\zeta\zeta A_\mu$ vertex		$\sqrt{\frac{F(p)}{Np^2}}ip_\mu(\delta^{ad}\delta^{be}\delta^{fc}-\delta^{af}\delta^{bc}\delta^{de})$
$z\bar{z}A_\mu$ vertex		$\frac{i}{\sqrt{N}}(2k_\mu+p_\mu)\delta^{ad}\delta^{bc}$
$z\bar{z}A_\mu A_\nu$ vertex	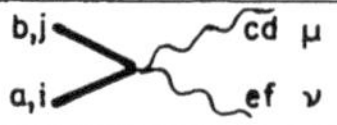	$\frac{i}{N}g_{\mu\nu}\delta_{ij}(\delta^{ad}\delta^{be}\delta^{fc}+\delta^{af}\delta^{bc}\delta^{de})$
$A_\mu\bar{\psi}\psi$ vertex	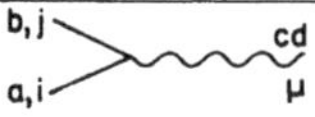	$\frac{i}{\sqrt{N}}\gamma_\mu\delta^{ad}\delta^{bc}\delta_{ij}$
$\phi\bar{\psi}\psi$ vertex	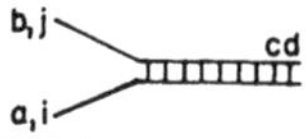	$\frac{i}{\sqrt{N}}\delta^{ad}\delta^{bc}\delta_{ij}\tau^A$
$\phi_5\bar{\psi}\psi$ vertex	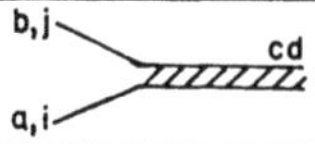	$-\frac{1}{\sqrt{N}}\delta^{ad}\delta^{bc}\delta_{ij}\gamma_5\tau^A$
$\bar{\psi}zc$ vertex		$\frac{i}{\sqrt{N}}\delta^{ad}\delta^{bc}\delta_{ij}$
$\bar{z}\psi\bar{c}$ vertex	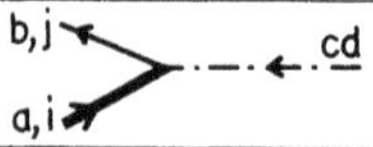	$\frac{i}{\sqrt{N}}\delta^{ad}\delta^{bc}\delta_{ij}$

7.3.2 Physical interpretation of the results

The generalization of the Grassmannian model to include fermions provides a low energy description of the interaction between "quarks" (fermions ψ) and "partons" (bosons z). The low energy effective Lagrangian is obtained from the small momentum approximation of the two point functions (7.57) and vertices, and is given by the expression

$$\mathcal{L}_{eff} = \overline{D^\mu z} D_\mu z - m^2 \bar{z} z + \bar{\psi}(i \not{D} - M)\psi - \frac{M^2 + 2Fe^2m^2}{24\pi m^2 M^2} F_{\mu\nu}^2 + \frac{1}{8\pi M^2}\left(\partial_\mu \varphi_5^A\right)^2 - \frac{1}{2}\varepsilon\left(\varphi_5^A\right)^2 + \frac{ie\sqrt{F}}{4\pi M}\varepsilon^{\mu\nu}F_{\mu\nu}\varphi_5^{(o)} - \frac{1}{\sqrt{N}}\bar{\psi}\tau^A\gamma_5\psi\varphi_5^A \quad . \tag{7.61}$$

The introduction of a θ term of the form (7.37) results in a non-vanishing vacuum expectation value of the gauge invariant field $F = \varepsilon^{\mu\nu}F_{\mu\nu}$ given by

$$\langle\theta|F|\theta\rangle = i\lambda\frac{\theta}{2\pi\sqrt{N}} \quad , \tag{7.62}$$

where

$$\lambda = \varepsilon\frac{12\pi^2 m^2 M^2}{3Fe^2m^2 + \varepsilon\pi(M^2 + 2Fe^2m^2)} \tag{7.63}$$

is the residue of the gauge field propagator, which for $p^2 \approx 0$, behaves as[9)]

$$D_{\mu\nu}^A(p) \approx -\left(g_{\mu\nu} - \frac{p_\mu p_\nu}{p^2}\right)\frac{\lambda}{p^2} \quad . \tag{7.64}$$

From the above, one learns that the propagator decreases with decreasing ε,and vanishes for $\varepsilon = 0$, which is the chiral limit. In this limit, the chiral quark antiquark pairs "screen" the external field. Indeed, we can analogously compute the expectation value of the field $\varphi_5^0 = \frac{ig}{\sqrt{N}}\bar{\psi}\gamma_5\psi$, obtaining

$$\langle\theta|g\bar{\psi}\gamma_5\psi|\theta\rangle = \kappa\frac{\theta}{2\pi} \quad , \tag{7.65}$$

where

$$\kappa = \frac{6\pi\sqrt{F}em^2M}{3Fe^2m^2 + \varepsilon\pi(M^2 + 2Fe^2m^2)} \tag{7.66}$$

is the residuum of the mixed $\langle 0|\varphi_5^{(o)}A_\mu|0\rangle$ propagator. The above expectation is maximal for $\varepsilon = 0$, where the external field is screened. The θ term may be interpreted as external charges at infinity.

In the supersymmetric case, due to the total screening, the physical partons are liberated due to a mechanism similar to the one we saw in the chiral GN model; in that case, antifermions are bound states of fermions and antibosons bound states of bosons and fermions (see chapter 8).

---

9) The infrared behavior is obtained from (7.57).

The $U(1)$ problem in the CP^{N-1} model

In Quantum Chromodynamics with chiral symmetry $U(N)\otimes U(N)$, there is a long standing problem concerning one of the Goldstone bosons[214] associated with the chiral symmetry breaking. The problem consists in the experimental fact that the η' particle associated with the $U(1)_5$ symmetry is not found. Theoretical arguments[263] based on current algebra foresee the mass of that particle to be smaller that $\sqrt{3}m_\pi$, where m_π is the mass of the pions[10)]. 't Hooft[264] has observed that this would be symmetry is broken by the Adler anomaly[201], and one cannot redefine it in a sector with non zero winding number, since the anomaly is a topological density, whose integral is an integer number[23]. However, this solution, based on the topologically non trivial properties of instantons,[23] is not the end of the story: the non conservation of chirality in a $SU(F)\times SU(F)$ symmetric theory, leads to the selection rule[265]

$$\Delta\chi_F = 2F\nu \quad , \tag{7.67}$$

where ν is the winding number of the gauge field[23]. In the framework of the instanton gas approximation, it is not possible to prove the existence of a non zero mass gap for the ninth Goldstone boson[265]. In general, if the chiral current has an anomalous divergence given by

$$\partial_\mu J_5^{\mu F} = \frac{g^2 F}{16\pi^2}F^{\mu\nu}\widetilde{F}_{\mu\nu} + D_F \tag{7.68}$$

where the second term is a soft breaking term given by the quark masses:

$$D_F = 2i\sum_{i=1}^{F} M_i\overline{q}_i\gamma_5 q_i \quad .$$

The mass matrix

$$\not{\Delta} = -\sum_{i=1}^{F} M_i\overline{q}_i q_i$$

has to obey certain Ward identities, if there is no zero mass particle associated with the symmetry breaking. Such Ward identities have been shown to be satisfied if we consider the effective action of the $\mathbb{C}P^{N-1}$ model with interacting fermions, Eq. (7.61)[266]. The $U(1)$ problem can be understood in this model using the $\frac{1}{N}$ expansion, whereas the instanton gas approximation fails to describe the existence of a mass gap[267]. The same situation is found to prevail in four dimensional QCD, thus showing the physical relevance of the $\mathbb{C}P^{N-1}$ theory as a toy model.

A very efficient way of probing the existence of massless fields in a two dimensional gauge theory, is to add the so called θ term to the action. In this case there is a discrete infinity of vacuum states labelled by the winding number, which assumes

---

10) Approximate chiral symmetry is supposed in the calculation.[269]

integral values[268]. The requirement of gauge invariance of the vacuum functional, forces us to take infinite linear combinations of these vacua, obtaining vacua labelled by the θ parameter. As discussed by Jackiw and Rebbi[269], the inclusion of massless fermions drastically changes this infinite vacuum structure. Indeed, although the chiral current is not conserved due to the Adler anomaly, there is a gauge dependent conserved current given by

$$\tilde{J}_5^\mu = J_5^\mu + \frac{1}{4\pi}\epsilon^{\mu\nu\rho\sigma}\,\mathrm{tr}\left(A_\nu\partial_\rho A_\sigma + \frac{2i}{3}A_\nu A_\rho A_\sigma\right) \quad ,$$

defining a corresponding gauge dependent axial charge $\tilde{Q}_5 = \int d^3x \tilde{J}_5^0$. Gauge transformations acting on $\tilde{Q}_5$, shift it by an even integer characterizing the homotopy class. The action of $e^{-\frac{i}{2}\theta'\tilde{Q}_5}$ acting on a vacuum functional labelled by θ, maps it into the vacuum functional labelled by $\theta+\theta'$. Therefore, the parameter θ is physically irrelevant, and all vacua are degenerate in energy. On the other hand, if the fermion is massive, there is no conserved axial charge at all, and the argument fails. In two dimensional gauge theories, the θ parameter corresponds to a background electric field. The above listed properties are common to unbroken as well as to broken gauge theories. However, there are further properties, derived within the instanton gas approximation, that are in general not correct, such as the fact that the θ dependence is exponentially small, and cannot be seen by summing Feynman diagrams. Indeed, the θ dependence of the theory without fermions, in the $\mathbb{C}P^{N-1}$ model is of the order $\frac{1}{N}$, as we have seen in (7.39), and not of order e^{-N}. In fact, the effective theory does not present instantons.

To demonstrate the mechanism envisaged above for the resolution of the $U(1)$ problem, we consider the following effective Lagrangian, motivated by (7.61)

$$\begin{aligned}\mathcal{L} = &(\partial^\mu + iA_\mu)\,\overline{z}\,(\partial^\mu - iA_\mu)\,z - m^2\overline{z}z \\ &-\frac{1}{4}F^{\mu\nu}F_{\mu\nu} + \overline{\psi}\,(i\,\partial\!\!\!/ + A\!\!\!/)\,\psi - M\overline{\psi}\psi + \frac{\theta}{2\pi}\epsilon^{\mu\nu}\partial_\mu A_\nu \quad . \qquad (7.69)\end{aligned}$$

If we bosonize this theory, using the correspondences

$$\overline{\psi} i\,\partial\!\!\!/\psi = \frac{1}{2}(\partial_\mu\vartheta)^2 \quad , \qquad (7.70a)$$

$$\overline{\psi}\gamma^\mu\psi = -\frac{1}{\sqrt{\pi}}\epsilon^{\mu\nu}\partial_\nu\vartheta \quad , \qquad (7.70b)$$

and

$$\overline{\psi}\psi = M\cos\sqrt{4\pi}\vartheta \quad , \qquad (7.70c)$$

we are left with the expression

$$\begin{aligned}\mathcal{L} = &(\partial^\mu + iA_\mu)\,\overline{z}\,(\partial^\mu - iA_\mu)\,z - m^2\overline{z}z \\ &-\frac{1}{4}F^{\mu\nu}F_{\mu\nu} + \frac{1}{2}(\partial_\mu\vartheta)^2 - M^2\cos\sqrt{4\pi}\vartheta - \frac{1}{\sqrt{\pi}}\epsilon^{\mu\nu}\partial_\mu A_\nu\vartheta + \frac{\theta}{2\pi}\epsilon^{\mu\nu}\partial_\mu A_\nu \quad . \qquad (7.71)\end{aligned}$$

We can check as before, that the spectrum of the theory is θ dependent for $M \neq 0$, whereas in the $M = 0$ case, one can eliminate this dependence by the substitution $\vartheta \to \vartheta + \frac{\theta}{2\sqrt{\pi}}$, as predicted by general arguments. For $M = 0$ the $U(1)$ chiral current has an anomalous divergence given by

$$\partial_\mu \left(\overline{\psi}\gamma^\mu\gamma_5\psi\right) = \frac{1}{\pi}\epsilon^{\mu\nu}\partial_\mu A_\nu \quad . \tag{7.72a}$$

In terms of ϑ, this equation takes the form

$$\Box\vartheta = \frac{1}{\sqrt{\pi}}\epsilon^{\mu\nu}\partial_\mu A_\nu \quad . \tag{7.72b}$$

It is easy to see that the right hand side of Eq. (7.72b) induces a mass $\frac{1}{\sqrt{\pi}}$ for the θ field in a way similar to the Kogut Susskind mechanism[270] in the Schwinger model. This is also the case for the general coupling described by Lagrangian (7.40), and resolves the $U(1)$ problem for these models.

Supersymmetric and minimal couplings

In the supersymmetric case, we have $\varepsilon = 0$, and $Fe^2 = 1$ Thus, we see from (7.63), that the pole in the gauge field propagator disappears. The propagators are now well behaved for small momentum, and the long range force disappears. We shall see that the model presents in this case higher conserved quantum charges, implying a factorizable (and calculable) S matrix, with a bound state structure very similar to the one obtained in the chiral Gross Neveu model. This means that there has been a shift to the pole of the gauge field at the expense of the fermion chirality, and only the massive mode excitations appear asymptotically. The shift of the pole is the central feature, and actually does not depend on the supersymmetry, but rather on the existence of an N-plet of fermions coupled minimally to the gauge field.

The minimal models are described by the Lagrangian

$$\mathcal{L} = \overline{D^\mu z} D_\mu z + \overline{\psi} i \not{D}\psi \quad . \tag{7.73}$$

The role of the fermionic fields is to eliminate the pole in the gauge field propagator. Since they are massless, their contribution $\pi^{(\psi)}_{\mu\nu}$ to the gauge field vertex function is given by

$$\pi^{(\psi)}_{\mu\nu} = \frac{i}{\pi}\left(g_{\mu\nu} - \frac{k_\mu k_\nu}{k^2}\right) \quad . \tag{7.74}$$

The gauge propagator is given by the one loop approximation, and reads

$$D_{\mu\nu} = i\left(g_{\mu\nu} - \frac{k_\mu k_\nu}{k^2}\right)\left(F(k) + \frac{i}{\pi}\right)^{-1} \quad . \tag{7.75}$$

Table 7.3

Propagators and vertices in the $\frac{1}{N}$ expansion of the supersymmetric and minimal Grassmannian model

	Supersymmetric model	Minimally coupled model
$z\bar{z}$ propagator	$\delta_{ij}\delta_{ab}\frac{i}{p^2-m^2}$	$\frac{i}{p^2-m^2}\delta^{ab}\delta_{ij}$
$\psi\bar{\psi}$ propagator	$\delta_{ij}\delta_{ab}\frac{i}{\not{p}-m}$	$\delta_{ij}\delta_{ab}\frac{i}{\not{p}-m}$
α^{ab} propagator	$\delta_{ad}\delta_{cb}iA(p)^{-1}$	$\delta_{ad}\delta_{cb}iA(p)^{-1}$
A_μ propagator	$\left(g_{\mu\nu}-\frac{p_\mu p_\nu}{p^2}\right)\frac{-1}{F(p)+i/\pi}\delta^{ad}\delta^{bc}$	$i(g_{\mu\nu}-\frac{p_\mu p_\nu}{k^2})\frac{-1}{F(p)+i/\pi}\delta^{ad}\delta^{bc}$
ζ propagator	$\frac{i}{\sqrt{p^2F(p)}}\delta^{ad}\delta^{bc}$	$\frac{i}{\sqrt{p^2F(p)}}\delta^{ad}\delta^{bc}$
ϕ propagator	$\frac{-1}{F(p)+i/\pi}\delta^{ad}\delta^{bc}$	—
ϕ_5 propagator	$\frac{1}{F(p)+i/\pi}\delta^{ad}\delta^{bc}$	—
$A_\mu\phi_5$ propagator	$\frac{2m}{p^2}\epsilon_{\mu\nu}p^\nu\frac{1}{F(p)+i/\pi}\delta^{ad}\delta^{bc}$	—
$\bar{c}c$ propagator	$\frac{2(\not{p}-2m)}{F(p)+i/\pi}\delta^{ad}\delta^{bc}$	—
$z\bar{z}\alpha$ vertex	$\frac{i}{\sqrt{N}}\delta^{ad}\delta^{bc}\delta_{ij}$	$\frac{i}{\sqrt{N}}\delta^{ad}\delta^{bc}\delta_{ij}$
$\zeta\zeta A_\mu$ vertex	$\sqrt{\frac{F(p)}{Np^2}}ip_\mu(\delta^{ad}\delta^{be}\delta^{fc}-\delta^{af}\delta^{bc}\delta^{de})$	$\sqrt{\frac{F(p)}{Np^2}}ip_\mu(\delta^{ad}\delta^{be}\delta^{fc}-\delta^{af}\delta^{bc}\delta^{de})$
$z\bar{z}A_\mu$ vertex	$\frac{i}{\sqrt{N}}(2k_\mu+p_\mu)\delta^{ad}\delta^{bc}$	$\frac{i}{\sqrt{N}}(2k_\mu+p_\mu)\delta^{ad}\delta^{bc}$
$z\bar{z}A_\mu A_\nu$ vertex	$\frac{i}{N}g_{\mu\nu}\delta_{ij}(\delta^{ad}\delta^{be}\delta^{fc}+\delta^{af}\delta^{bc}\delta^{de})$	$\frac{i}{N}g_{\mu\nu}\delta_{ij}(\delta^{ad}\delta^{be}\delta^{fc}+\delta^{af}\delta^{bc}\delta^{de})$
$A_\mu\bar{\psi}\psi$ vertex	$\frac{i}{\sqrt{N}}\gamma_\mu\delta^{ad}\delta^{bc}\delta_{ij}$	$\frac{i}{\sqrt{N}}\gamma_\mu\delta^{ad}\delta^{bc}\delta_{ij}$
$\phi\bar{\psi}\psi$ vertex	$\frac{i}{\sqrt{N}}\delta^{ad}\delta^{bc}\delta_{ij}$	—
$\phi_5\bar{\psi}\psi$ vertex	$-\frac{1}{\sqrt{N}}\delta^{ad}\delta^{bc}\delta_{ij}\gamma_5$	—

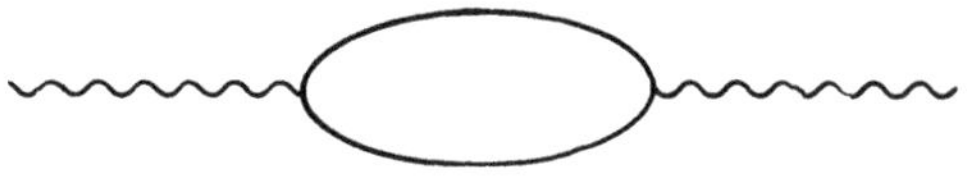

Figure 7.9: Fermionic contributions gauge field two point function.

It has no pole. The fermions remain massless, but the fermionic degrees of freedom decouple, and no asymptotic fermionic state appears in the theory. On the other hand, since the gauge field has no pole, the bosonic particle z is no longer confined. We shall later see that these models may present higher conservation laws in the quantum theory, and their S-matrix for bosonic scattering is computable. We summarize the Feynman rules for the supersymmetric and minimal cases in table 7.3.

The equality of the ϕ and π propagators is a consequence of supersymmetry, since they belong to a common supermultiplet.

The main features of these models are based on the fact that the chirality carrying part of the fermions is "eaten up" by the gauge field, cutting down the long range force, and leaving only the massive fermionic excitations, analogously to the case of the chiral GN model.

Chiral symmetry restoration

We have seen that a spontaneous symmetry breaking is apparently taking place, with N^2-1 Goldstone bosons. However, this is forbidden in two dimensional space time, due to Coleman-Mermin-Wagner's theorem[93]. The resolution[63] is similar to the one found by Witten[197] in the case of the chiral Gross Neveu model, a mechanism that we discussed at length in chapter 5, where the chiral symmetry is realized via the Berezinski-Kosterlitz-Thouless phase[271], with a power law decay for the chiral correlation function.

Consider the field

$$\Phi = \left(\phi^A + i\gamma_5\phi_5^A\right)\tau^A \quad . \qquad (7.76a)$$

This field, appearing in (7.47*b*), may be written in terms of a diagonal matrix χ as follows[63]

$$\Phi = e^{\gamma_5 \vartheta^C \tau^C} e^{-i\beta^B \tau^B} \chi e^{i\beta^D \tau^D} e^{\gamma_5 \vartheta^A \tau^A} , \qquad (7.76b)$$

where ϑ^A and β^A are functions of ϕ and ϕ_5.

The effective action can be computed in terms of the above defined fields, and is quadratic in χ. Since χ is massive, and we are interested here only in the long ranged correlators, we shall neglect its quantum fluctuations, and replace it by its vacuum expectation value, $\sqrt{N}M$. In terms of the remaining fields, we obtain from (7.50) and (7.76), the following effective action (recall the definition (7.47))

$$S_{eff} = iN\mathrm{tr}\ln\Delta_B - iN\mathrm{tr}\ln\left[\not{D} - Me^{2\gamma_5\vartheta^A\tau^A}\right] - \int d^2x \frac{\sqrt{N}}{2f}\alpha \quad . \qquad (7.77)$$

Redefining the fields by $\vartheta^A \to \frac{MN^{-\frac{1}{2}}}{2}\vartheta^A$, we obtain from here the effective action for the fields ϑ^0 and ϑ^A

$$\mathcal{L}_\vartheta = \frac{1}{2}\partial^\mu\vartheta^0\partial^\mu\vartheta^0 + \frac{1}{2}\partial^\mu\vartheta^A\partial_\mu\vartheta^A + \text{ higher order corrections.} \qquad (7.78)$$

The higher order terms go to zero as $N \to \infty$. The chiral correlation function

$$\mathcal{G}(x-y) = \langle 0 \mid \overline{\psi}(x)(1+\gamma_5)\psi(x)\overline{\psi}(y)(1-\gamma_5)\psi(y) \mid 0\rangle \tag{7.79}$$

can be computed using the identification

$$\phi^A = \frac{1}{\sqrt{N}}\overline{\psi}\tau^A\psi \tag{7.80}$$

$$\phi_5^A = \frac{i}{\sqrt{N}}\overline{\psi}\tau^A\gamma_5\psi \quad , \tag{7.81}$$

which follow from the field equations derived from (7.44). Thus we consider

$$\mathcal{G}(x-y) = \langle 0 \mid e^{-\frac{i}{\sqrt{N}}\vartheta^0(x)} e^{\frac{i}{\sqrt{N}}\vartheta^0(y)} \mid 0\rangle \quad , \tag{7.82}$$

which can be computed using the methods of chapter 2, formulae (2.14,15). One obtains

$$\mathcal{G}(x) = \frac{1}{N}|x-y|^{-\frac{1}{N}} \quad . \tag{7.83}$$

We conclude that the correlator decays with a power law. The chiral correlator involving non abelian fields can also be computed for large N, and the same behavior is obtained[63]. The appearance of the power law indicates[197] that the spectral function describes infra particles. We have seen in the case of the chiral Gross Neveu model, that this phenomenon indicates that particles are bound states of antiparticles, as expressed by relation (5.93). In the present case, this interpretation is not very clear due to the confining properties of the model. However, in the supersymmetric theory there is no confinement, and we shall be able to compute the exact S matrix of the model. We shall thereby show that a similar relation between particles and antiparticles exist in this case as well.[78,206]

Asymptotic freedom

As a result of the dynamical mass generation, we obtain asymptotic freedom. This is true because the mass is a renormalization group invariant, and must obey

$$\left\{\Lambda^2\frac{\partial}{\partial\Lambda^2} + \beta\frac{\partial}{\partial f}\right\} m^2 = 0 \quad , \tag{7.84}$$

from which we can read off the β function to first order in $\frac{1}{N}$. Using (7.16), or (7.55), we can read β from (7.84), obtaining

$$\beta(f) = -\frac{f^2}{4\pi N} \quad . \tag{7.85}$$

7.4 Quantization of Higher Conservation Laws

7.4.1 Purely bosonic sigma models and anomalies

It is very important to know, whether the classical higher conservation laws of the non linear sigma models are maintained at the quantum level. The central element in the discussion of the conservation of the nonlocal charge, is the singularity appearing in the first term of the quantum counterpart of Eq. (6.140) which induces, after renormalization, non zero contributions to the time derivative of the non local charge. In that case, the dual symmetry is spoiled by anomalies. Therefore Eq. (6.115), which expresses the integrability of the model, is anomalous in the quantum theory. In order to study this problem, we consider the operator product expansion

$$j_\mu(x+\varepsilon)j_\nu(x) = \sum_{\{\mathcal{O}_{\{\sigma\}}\}} C_{\mu\nu}^{\{\sigma\}}(\varepsilon)\mathcal{O}_{\{\sigma\}}(x) \ , \tag{7.86}$$

where the current $j_\mu(x)$ was defined in (6.97).

We first discuss the purely bosonic theory. Due to asymptotic freedom,[255] we can classify the terms on the right hand side of (7.86) according to their dimension. Therefore, the question of classifying the anomalies is traced back to the classification of operators of dimension 0, 1 and 2, since the left hand side of (7.86) has dimension two. The operators $\mathcal{O}_{\{\sigma\}}(x)$ must take value in the Lie algebra $\mathcal{G}$, and must be globally G-covariant, and locally H-invariant. Any globally covariant and locally invariant operator (notice that G acts from the left, while H acts from the right, see Eq. (6.87-90)) must be of the form

$$L_1 g L_2 g^\dagger \cdots L_{2k-1} g L_{2k} g^\dagger \quad , \tag{7.87}$$

where each L must be either the identity or a product of covariant derivatives such that the total number of them is equal to the dimension of the operator [11)]

The constraints

$$\begin{aligned} g^\dagger g &= 1 \\ g g^\dagger &= 1 \end{aligned} \tag{7.88}$$

permit to rewrite the above operator as[12)]

$$L_1 g g^\dagger \cdots L_k g g^\dagger \quad . \tag{7.89}$$

After decomposing the above operators in irreducible parts, we obtain as potential candidates for operators giving anomalous contributions:[62]

---

11) A G invariant and H covariant operator is given by $L_1 g^\dagger L_2 g \cdots L_{2k-1} g^\dagger L_{2k} g$.

12) The analogous gauge covariant and globally invariant operator is $g^\dagger L_1 g \cdots g^\dagger L_k g$.

i. No operator of dimension zero.
ii. One operator of dimension one, which is the current defined in (6.97), that is $j_\mu = gD_\mu g^\dagger$.
iii. The following operators of dimension two:
  a) the derivative of the current, $\partial_\mu j_\nu = D_\mu g D_\nu g^\dagger + g D_\mu D_\nu g^\dagger$.
  b) the antisymmetric part of the operator $D_\mu D_\nu g g^\dagger$, which turns out to be $gF_{\mu\nu}g^\dagger$, which however, is not irreducible; its irreducible parts are given by $G^{(o)}_{\mu\nu} G^{(1)}_{\mu\nu} \cdots G^{r}_{\mu\nu}$ (see eqs 6.110).

The operators *ii* and *iiia*) are harmless; their effect merely consists in renormalizing the nonlocal charge, as we shall demonstrate in the following. However, the operators in *iiib*) are dangerous. Actually, their sum is equal to the antisymmetric part of the derivative of the current. This follows from the identity

$$\sum_i G^{(i)}_{\mu\nu} = \partial_\mu j_\nu - \partial_\nu j_\mu \tag{7.90}$$

which is nothing but Eq. (6.94*a*). When the group H is not simple, that is, when it has nontrivial ideals, the irreducible components $G^{(i)}_{\mu\nu}$ become candidates for an anomaly. The Wilson expansion for the commutator of two currents can be generally written as:[35]

$$\begin{aligned}[j_\mu(x+\varepsilon), j_\nu(x)] = {} & C^{\rho}_{\mu\nu}(\varepsilon) j_\rho(x) + D^{\sigma\rho}_{\mu\nu}(\varepsilon)\partial_\sigma j_\rho(x) + \\ & \sum_{i=0}^{r} E^{(i)\sigma\rho}_{\ \ \mu\nu}(\varepsilon) G^{(i)}_{\sigma\rho}(x) + \mathcal{N}[j_\mu(x), j_\nu(x)] \quad , \end{aligned} \tag{7.91a}$$

with $\varepsilon^2 < 0$. Because of (7.90), we may require

$$\sum_{i=0}^{r} E^{(i)\sigma\rho}_{\ \ \mu\nu}(\varepsilon) = 0 \quad . \tag{7.91b}$$

The symbol $\mathcal{N}$ means normal product, defining the finite part. We could have equivalently required $D^{\sigma\rho}_{\mu\nu}(\varepsilon)$ to be symmetric with respect to $\rho \leftrightarrow \sigma$. The tensorial nature of the linearly divergent coefficient function $C^{\rho}_{\mu\nu}(\varepsilon)$ and the logarithmically divergent coefficients $E^{(i)\sigma\rho}_{\ \ \mu\nu}(\varepsilon)$ and $D^{\sigma\rho}_{\mu\nu}(\varepsilon)$, can be determined from general principles, such as covariance under the full Poincaré group, current conservation and normalization of the current. Now, under parity-time reversal both the current as well as the derivative operator change sign, while under charge conjugation only the current changes sign. Thus, parity and time reversal imply

$$C^{\rho}_{\mu\nu}(-\varepsilon) = -C^{\rho}_{\mu\nu}(\varepsilon) \tag{7.92}$$

and

$$D^{\sigma\rho}_{\mu\nu}(-\varepsilon) = D^{\sigma\rho}_{\mu\nu}(\varepsilon) \quad , \tag{7.93}$$

while charge conjugation implies

$$C^{\rho}_{\mu\nu}(\varepsilon) = -C^{\rho}_{\nu\mu}(-\varepsilon) \tag{7.94}$$

and

$$D^{\sigma\rho}_{\mu\nu}(\varepsilon) = -D^{\sigma\rho}_{\mu\nu}(-\varepsilon) - [C^{\rho}_{\mu\nu}(-\varepsilon)\varepsilon^{\rho} - \frac{1}{2}\delta^{\rho\sigma}C^{\lambda}_{\nu\mu}(-\varepsilon)] \quad . \tag{7.95}$$

Therefore, using Lorentz invariance we obtain the following general form for the above coefficients

$$C^{\rho}_{\mu\nu}(\varepsilon) = C_1(\varepsilon^2)g_{\mu\nu}\varepsilon^{\rho} + C_2(\varepsilon^2)(\varepsilon_{\mu}\delta^{\rho}_{\nu} + \varepsilon_{\nu}\delta^{\rho}_{\mu}) + C_3(\varepsilon^2)\varepsilon_{\mu}\varepsilon_{\nu}\varepsilon^{\rho} \quad , \tag{7.96}$$

$$\begin{aligned} D^{\sigma\rho}_{\mu\nu}(\varepsilon) = {} & D_1(\varepsilon^2)\varepsilon^{\sigma}(\varepsilon_{\mu}\delta^{\rho}_{\nu} + \varepsilon_{\nu}\delta^{\rho}_{\mu}) + D_2(\varepsilon^2)\varepsilon^{\rho}(\varepsilon_{\mu}\delta^{\sigma}_{\nu} - \varepsilon_{\nu}\delta^{\sigma}_{\mu}) + \\ & D_3(\varepsilon^2)(\delta^{\rho}_{\mu}\delta^{\sigma}_{\nu} - \delta^{\sigma}_{\mu}\delta^{\rho}_{\nu}) + \frac{1}{2}(C^{\sigma}_{\mu\nu}(\varepsilon^2)\varepsilon^{\rho} - \frac{1}{2}g^{\rho\sigma}C^{\lambda}_{\mu\nu}\varepsilon_{\lambda}) \quad , \end{aligned} \tag{7.97}$$

$$E^{\rho\sigma}_{\mu\nu} = E(\varepsilon^2)\varepsilon_{\mu\nu}\varepsilon^{\rho\sigma} \quad . \tag{7.98}$$

Current conservation constrains the coefficients $C_i(\varepsilon^2)$, $D_i(\varepsilon^2)$, and $E(\varepsilon^2)$.[35,75] The first set is defined up to an overall constant, which is fixed by normalization. Current conservation implies the following Ward identity derived from general BPHZ renormalized perturbation theory[5,110,75,272]

$$\begin{aligned} \partial^{x}_{\mu}\langle 0\,|\,T[j_{\mu}(x), j_{\nu}(y)]^{ij}X\,|\,0\rangle = {} & -2c\delta^2(x-y)\langle 0\,|\,Tj^{ij}_{\nu}(y)X\,|\,0\rangle \\ & -\sum_{l}\delta^2(x-x_l)\langle 0\,|\,T[j^{lj}_{\nu}(y)z^{i}(x) - j^{ik}_{\nu}(y)\delta^{jl}z^{k}(x)]X_{\hat{l}}\,|\,0\rangle \\ & +\sum_{m,k}\delta^2(x-y_m)\langle 0\,|\,T[\delta^{im}\bar{z}^{k}(x)j^{kj}_{\nu}(y) - \delta^{km}j^{ik}_{\nu}(y)\bar{z}^{j}(x)]X_{\hat{m}}\,|\,0\rangle \; , \end{aligned} \tag{7.99}$$

where $X = \prod_{l,m} z_l(x_l)\bar{z}_m(y_m)$ and $X_{\hat{l}}$ represents X with the field z_l or $\bar{z}_l$ deleted; the constant c depends on the symmetry group; for $SU(N)$, $c = N$.

The Ward identity together with (7.90) imply several constraints for the coefficients $C^{\rho}_{\mu\nu}(\varepsilon)$, and $D^{\sigma\rho}_{\mu\nu}(\varepsilon)$; we have[75,272]

$$\partial^{\mu}C^{\rho}_{\mu\nu}(\varepsilon) = -2c\delta(\varepsilon)\delta^{\rho}_{\nu} \quad , \tag{7.100}$$

$$\partial^{\mu}D^{\rho\sigma}_{\mu\nu}(\varepsilon) = \partial^{\mu}E^{\rho\sigma}_{\mu\nu} = 0 \quad . \tag{7.101}$$

For $\varepsilon \neq 0$, this leads to the following equations[272]

$$C_1 + 2\varepsilon^2 C_2' + 3C_2 = 0 \quad , \tag{7.102a}$$

$$C_1' + C_2' + \varepsilon^2 C_3' + 2C_3 = 0 \quad , \tag{7.102b}$$

$$2\varepsilon^2 D_1' + 2D_1 - 2D_3' + \frac{1}{2}C_1 = 0 \quad , \tag{7.102c}$$

$$2\varepsilon^2 D_2' + 2D_2 + 2D_3' + \frac{1}{2}C_2 = 0 \quad , \tag{7.102d}$$

$$-2D_1' - 2D_2' + \frac{1}{2}C_3 = 0 \quad , \tag{7.102e}$$

$$D_1 + D_2 + \frac{1}{4}(C_1 + C_2 + C_3) = 0 \quad , \tag{7.102f}$$

$$E' = 0 \quad . \tag{7.102g}$$

The above relations are enough to define a quantum non local charge. Consider

$$Q_\delta = \int_{|y_1 - y_2| > \delta} dy_1 dy_2 \epsilon(y_1 - y_2) j_0(t, y_1) j_0(t, y_2) - Z_\delta \int dy j_1(t, y) \quad , \tag{7.103}$$

with Z_δ a renormalization constant chosen to be

$$Z_\delta = \int_\delta^{\mu^{-1}} dx C_1(-x^2) x \quad , \tag{7.104}$$

where μ^{-1} is a finite number used to avoid spurious divergencies. The divergent part of the double integral is given by the first term in the Wilson expansion (7.76), at equal times:

$$[j_0(t, y + \delta), j_0(t, y)] = C_1(\delta^2)\delta j_1(t, y) \quad . \tag{7.105}$$

Therefore

$$Q_\delta \sim \int dy j_1(t, y) \int_\delta^{\mu^{-1}} C_1(-x^2) x dx - Z_\delta \int dy j_1(t, y) \quad , \tag{7.106}$$

which has a finite limit, for the choice (7.104) of the renormalization constants. The time derivative of the above charge is given by[13)]

$$\begin{aligned} \frac{dQ_\delta}{dt} = \int dy_1 dy_2 \epsilon(y_1 - y_2) \left(\frac{\partial j_1}{\partial y_1}(t, y_1) j_0(t, y_2) + j_0(t, y_1) \frac{\partial j_1}{\partial y_2}(t, y_2) \right) - \\ Z_\delta \int dy j_1(t, y) + \int dy \{ -j_1(t, y + \delta) j_0(t, y) - j_1(t, y - \delta) j_0(t, y) \\ + j_0(t, y) j_1(t, y + \delta) + j_0(t, y) j_1(t, y - \delta) \} - Z_\delta \int dy \frac{\partial j_1}{\partial t}(t, y) \quad . \end{aligned} \tag{7.107}$$

13) We use $\partial^\mu j_\mu = 0$, and perform explicitly the integrations, whenever possible.

We use now the expansion (7.91) and (7.102c,d) to obtain

$$\frac{dQ_\delta}{dt} = \sum_{independent\, G^{(i)}} E^i \int \epsilon^{\mu\nu} G^{(i)}_{\mu\nu}(t,y)dy \quad . \tag{7.108}$$

We thus have conservation of the non local charge if and only if either the gauge group H is simple, in which case there exists no anomaly at all, or the coefficient(s) of the anomaly(ies) is(are) zero.

We can, in some cases, explicitly compute the coefficients as e.g. using a perturbative expansion.

The Eqs. (7.102) do not contain enough information to fix all the coefficients. We use the Wilson expansion in the form

$$\begin{aligned}
&\partial^\nu_x \langle 0|T[j_\mu(y), j_\nu(x)]^{ij} z^l(x')|0\rangle = \\
&\qquad \partial^\nu_x C^\rho_{\mu\nu}(y-x)\langle 0|Tj^{ij}_\rho(x)z^l(x')|0\rangle + C^\rho_{\mu\nu}(y-x)\partial^\nu_x\langle 0|Tj^{ij}_\rho(x)z^l(x')X|0\rangle + \\
&\partial^\nu_x D^{\rho\sigma}_{\mu\nu}(y-x)\langle 0|T\partial_\sigma j^{ij}_\rho(x)z^l(x')X|0\rangle + \partial^\nu_x E^{\rho\sigma}_{\mu\nu}(y-x)\langle 0|Tz_i\bar{z}_j F_{\rho\sigma}(x)z^l(x')X|0\rangle + \\
&\qquad\qquad O(\ln(x-y)^2) \quad .
\end{aligned} \tag{7.109}$$

If we compute the terms involving $\delta(y-x)$ and $\delta(x-x')$, and equate the respective coefficients we obtain[35,75,77,272]

$$C_1(\varepsilon^2) = \frac{c}{2\pi\varepsilon^2} \quad , \tag{7.110a}$$

$$C_2(\varepsilon^2) = -\frac{c}{2\pi\varepsilon^2} \quad , \tag{7.110b}$$

$$C_3(\varepsilon^2) = -\frac{c}{2\pi\varepsilon^2} \quad , \tag{7.110c}$$

$$\varepsilon^2 D_1 - D_3 = -\frac{c}{8\pi}\ln\left(-\mu^2\varepsilon^2\right) \quad , \tag{7.110d}$$

$$\varepsilon^2 D_2 + D_3 = \frac{c}{8\pi}\ln\left(-\mu^2\varepsilon^2\right) \quad . \tag{7.110e}$$

Finally, the function E in (7.98) turns out to be a constant as we explain below. It can be computed only perturbatively. In the cases of $\mathbb{C}P^{N-1}$ and Grasmannians, such a computation has been performed. Some details are given in appendix, together with an explicit perturbative calculation of $C_{\mu\nu}$ and $D^{\sigma\rho}_{\mu\nu}$. We obtain[272,232]

$$E^{CP^{N-1}} = \frac{N}{\pi} \quad , \tag{7.111a}$$

$$E^{Grassmann} = \frac{N}{\pi} \quad . \tag{7.111b}$$

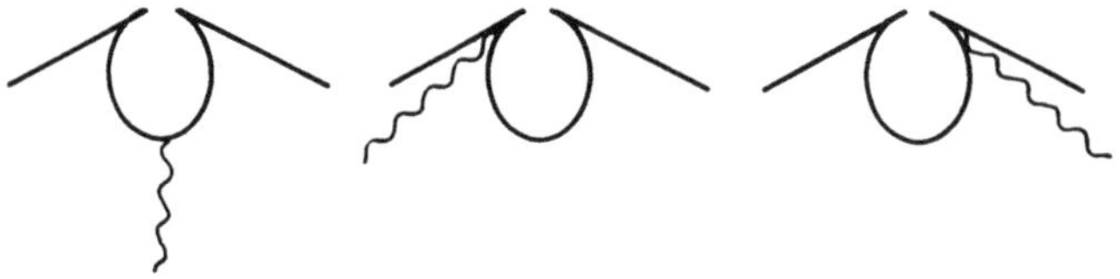

Figure 7.10: Diagrams contributing to the left hand side of Eq. (7.112)

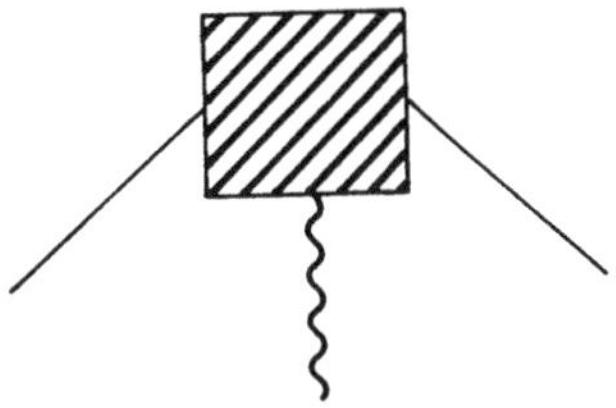

Figure 7.11: Diagrams contributing to $\langle 0|T[J_\mu(\varepsilon), J_\nu(0)] z_l \bar{z}_k A_\lambda|0\rangle$.

In order to obtain non perturbative results later on, we now discuss the question of whether the results for the anomaly receive higher order corrections. From (7.109) and using $C^\rho_{\mu\nu} = C^\rho_{\nu\mu}$ we obtain

$$\int d^2p e^{-i\varepsilon p}\left\{\frac{\partial}{\partial k_\alpha}\langle 0|T[\tilde{j}_\mu(p), j_\nu(0)]\tilde{z}_k(r)\tilde{A}_\lambda(k)|0\rangle^{prop}\Big|_{q=r=k=0} - (\mu \leftrightarrow \nu)\right\}$$

$$= [D^{\rho\sigma}_{\mu\nu}(\varepsilon) - D^{\rho\sigma}_{\nu\mu}(\varepsilon)]\frac{\partial}{\partial k_\alpha}\langle 0|T\partial_\rho j_\sigma(0)\tilde{z}_k(q)\tilde{\bar{z}}_l(r)\tilde{A}_\lambda(k)|0\rangle^{prop}\Big|_{q=r=k=0}$$

$$-4E(\varepsilon^2)\frac{\partial}{\partial k_\alpha}\langle 0|T(z_i\bar{z}_j F_{\mu\nu})(0)\tilde{z}_k\tilde{\bar{z}}A_\lambda(k)|0\rangle^{prop}\Big|_{q=r=k=0} , \qquad (7.112)$$

where the tildes indicate Fourier transform and "prop" stands for "proper". Assuming that our normal products are normalized at zero external momenta, the right-hand side (rhs) of (7.112) turns into

$$2(D^{\alpha\lambda}_{\mu\nu} - D^{\alpha\lambda}_{\nu\mu}) + 4iE(\varepsilon^2)(\delta^\alpha_\mu\delta^\lambda_\nu - \delta^\alpha_\nu\delta^\lambda_\mu) \quad . \qquad (7.113)$$

Notice that $\delta^\alpha_\mu\delta^\lambda_\nu - \delta^\alpha_\nu\delta^\lambda_\mu = -\varepsilon_{\mu\nu}\varepsilon^{\alpha\lambda}$ holds, therefore $E^{\rho\sigma}_{\mu\nu} = E(x^2)\varepsilon_{\mu\nu}\varepsilon^{\rho\sigma}$.

Because of current conservation, $E(\varepsilon^2)$ is a constant which remains to be determined.

Only the graphs in figure 7.10 contribute to the left hand side (lhs) of (7.112). To verify this notice that the diagrams contributing to $\langle 0|T[j_\mu(\varepsilon), j_\nu(0)]z_l\bar{z}_k A_\lambda|0\rangle$ have the general structure shown in figure 7.11. First, it is not difficult to see that A_λ must be attached to the isospin carrying line connected to both vertices, μ and ν. Now, if the derivative $\partial/\partial k_\alpha$ does not act directly on the momentum factors associated with the current vertex, we obtain a result symmetric under the exchange $\mu \leftrightarrow \nu$. Therefore this type of term [figure 7.11a] will not contribute to the lhs of (7.112).

On the other hand, if the derivative acts on the momentum factors associated with the current vertex, the only graphs which contribute are those of figure 7.10. This happens because the insertion of a zero-momentum external gauge field line will produce a result proportional to the derivative with respect to the loop momentum p. Thus after integration we will get zero except for the graphs of figure 7.10, in which case there will be non-vanishing surface terms. But these terms have already been computed in ref[272], so that the result (7.111) holds in every finite order of the $\frac{1}{N}$ expansion.

A note is in order at this point. Perturbation theory alone cannot determine completely the above coefficients. Indeed, the above arguments imply a complete determination of the renormalization constant, which in turn implies an interplay between small and large energy behavior[273]. The renormalization invariant mass is given, in terms of the first two coefficients of the β function, b_0 and b_1 by[194]

$$\mu = \mu_{\overline{MS}} (b_0 f)^{-\frac{b_1}{b_0^2}} e^{-\frac{1}{b_0 f}} \{1 + \mathcal{O}(f)\} \quad , \tag{7.114}$$

where $\overline{MS}$ means a certain renormalization prescription. Thus the coefficient D_1 should have corrections proportional to $\ln f$, which is impossible in perturbation theory. Nevertheless, we should stress the analysis of this chapter contains non perturbative results, which are:

1. The Wilson expansion (7.91) is general, since it is a consequence of asymptotic freedom and the hypothesis that some geometric quantization may be achieved.
2. In order to obtain a well defined non local charge, we did not need to know the explicit form of the coefficients.
3. Although the coefficient of the anomaly is perturbative, its form is not. This fact is enough to arrive at our results.

   See appendix G for further details.

7.4.2 Fermionic interaction and anomaly cancellation

If we have interactions including fermions, the expansion (7.90) involves additional terms. We make the following ansatz for the supersymmetric case

$$\begin{aligned}[J_\mu(x+\varepsilon), J_\nu(x)] = {} & C^\rho_{\mu\nu}(\varepsilon) j_\rho(x) + D^{\sigma\rho}_{\mu\nu}(\varepsilon)\partial_\sigma j_\rho(x) + \sum_{i=0}^{r} D^{(i)}{}^{\sigma\rho}_{\mu\nu}(\varepsilon) G^{(i)}_{\sigma\rho}(x) \\ & + \hat{C}^\rho_{\mu\nu} j^M_\rho + \hat{D}^{\sigma\rho}_{\mu\nu}\partial_\sigma j^M_\rho + \mathcal{N}[J_\mu(x), J_\nu(x)] \quad ,\end{aligned} \tag{7.115}$$

where the current J_μ is the full Noether current, whereas j_μ and j^M_μ correspond to the bosonic and fermionic parts, respectively. Moreover, the coefficients $\hat{C}^\rho_{\mu\nu}$ and $\hat{D}^{\rho\sigma}_{\mu\nu}$ are determined as in the case without fermions, since they obey the same requirements as C and D (see Eqs. (7.92-98)). Using BPHZ perturbation theory they are computed in the same way as C and D. Again, the constant coefficient E can be computed only perturbatively. We find that the anomaly discussed above is cancelled exactly against the Adler Bardeen anomaly[77,272] so that we still have exact conservation of the nonlocal charge, although the gauge group is not simple. Examples are the minimally coupled and supersymmetric $\mathbb{C}P^{N-1}$ models. To illustrate this, we begin by examining the simplest case. We consider $\mathbb{C}P^{N-1}$ models minimally coupled to fermions. In order to understand the mechanism of anomaly cancellation, we verify the conservation of the classical non-local charge. At the classical level they satisfy

$$
\begin{aligned}
[J_\mu, J_\nu]_{ij} &= -\frac{1}{2} z\bar{z}(\partial_\mu J_\nu - \partial_\nu J_\mu)_{ij} + \frac{i}{2}[(z_i\bar{z}_j)\overleftrightarrow{\partial}_\nu(\overline{\psi}\gamma_\mu\psi) - (\nu \leftrightarrow \mu)] \\
&= -\frac{1}{2} z\bar{z}(\partial_\mu J_\nu - \partial_\nu J_\mu)_{ij} - \frac{i}{2}[\partial_\nu(z_i\bar{z}_j\overline{\psi}\gamma_\mu\psi) - (\nu \leftrightarrow \mu)] \quad , \quad (7.116)
\end{aligned}
$$

where, in the last equality, axial-vector current conservation $\epsilon_{\mu\nu}\partial^\mu(\overline{\psi}\gamma^\nu\psi) = 0$ has been used. From (6.128) we see that the classical non local charge

$$
Q^{ij} = \int dy_1 dy_2 \epsilon(y_1 - y_2) J_0^{ik}(y_1, t) J_0^{kj}(y_2, t) - \frac{N}{2f}\int \left(j_i^{ij} - \frac{4if}{N}\overline{\psi}\gamma_1\psi z^i \bar{z}^j \right) dy \quad , \tag{7.117}
$$

is conserved, i.e., $dQ^{ij}/dt = 0$.

In the quantum version of the model all these calculations have to be reexamined. Using the path integral formalism, the fermions can be trivially integrated out; we thus obtain the Feynman rules given in table 7.3, for the $\frac{1}{N}$ expansion.

When comparing with the pure $\mathbb{C}P^{N-1}$ case, we note that the A_μ field has lost its pole at $p^2 = 0$. Heuristically, this means that the partons (the quanta of the z_i fields) are liberated. A factorized S-matrix which agrees in lowest order of $1/N$ with the one obtained by the use of the conservation of (7.117) has been proposed[206,275] (see next chapter). This suggests the existence of conserved quantum non local charges. Taking into account the anomaly entering in the pure $\mathbb{C}P^{N-1}$ case, we conclude that a compensation of this anomaly, arising from the minimal coupling, must exist. To lowest order this can be verified by the study of the short-distance product of the currents, which can be written in the form[5,31]

$$
J_\mu(x+\varepsilon)J_\nu(x) - J_\nu(x)J_\mu(x+\varepsilon) = \mathcal{N}[J_\mu(x+\varepsilon), J_\nu(x)] + \sum_i c_i(\varepsilon)\mathcal{N}[\theta_i](x) \quad , \tag{7.118}
$$

where θ_i corresponds to some formal product of the basic fields and their derivatives. In perturbation theory Eq. (7.118) is an identity. It says that, as $\varepsilon \to 0$, a finite normal product $\mathcal{N}[J_\mu, J_\nu](x)$ can be constructed by subtracting from the formal product $[J_\mu(x+\varepsilon), J_\nu(x)]$ a number of counterterms $\sum_i c_i(\varepsilon)\mathcal{N}[\theta_i]$. Thus, when computing the right-hand-side of (7.118), we can choose any subtraction scheme we prefer. For definiteness we adopt Zimmermann's finite-part prescription[5,110].

In this case, as is well known, the renormalized expression associated with a generic graph Γ is obtained by applying the forest formula, using Taylor operators of minimum degree. According to this procedure Eq. (7.118) becomes

$$[J_\mu(x+\varepsilon), J_\nu(x)] = C^\rho_{\mu\nu} J_\rho + D^{\rho\sigma}_{\mu\nu}\partial_\sigma J_\rho + E^{\rho\sigma}_{\mu\nu} z_i \bar{z}_j F_{\rho\sigma} + \mathcal{N}_2[J_\mu, J_\nu] \quad , \tag{7.119}$$

where the coefficients are found to be given by

$$C^\rho_{\mu\nu} = \frac{N}{2\pi}\left[\frac{g_{\mu\nu}\varepsilon^\rho}{\varepsilon^2} - \left(\frac{\varepsilon_\nu \delta^\rho_\mu}{\varepsilon^2} + \frac{\varepsilon_\mu \delta^\rho_\nu}{\varepsilon^2}\right) - 2\frac{\varepsilon_\mu \varepsilon_\nu \varepsilon^\rho}{(\varepsilon^2)^2}\right] \tag{7.120a}$$

$$D^{\sigma\rho}_{\mu\nu} = \frac{N}{2\pi}[\left(\frac{\gamma}{2} + \frac{1}{4}\ln\frac{-m^2\varepsilon^2}{4}\right)(\delta^\sigma_\mu \delta^\rho_\nu - \delta^\sigma_\nu \delta^\rho_\mu) + \frac{\delta^\sigma_\nu \varepsilon_\mu \varepsilon^\rho}{2\varepsilon^2} - \frac{\delta^\sigma_\mu \varepsilon_\nu \varepsilon^\rho}{2\varepsilon^2}$$

$$+ \frac{\delta^\rho_\mu \varepsilon_\nu \varepsilon^\sigma}{2\varepsilon^2} + \frac{\delta^\rho_\nu \varepsilon_\mu \varepsilon^\sigma}{2\varepsilon^2} - \frac{g_{\mu\nu}\varepsilon^\rho \varepsilon^\sigma}{2\varepsilon^2} + \frac{\varepsilon_\mu \varepsilon^\rho \varepsilon_\nu \varepsilon^\sigma}{(\varepsilon^2)^2}] \tag{7.120b}$$

$$E^{\rho\sigma}_{\mu\nu} = \frac{N}{2\pi}\epsilon_{\mu\nu}\epsilon^{\rho\sigma} \quad . \tag{7.120c}$$

In contrast to the pure $\mathbb{C}P^{N-1}$ case, we find that the normal product in (7.119) gives an additional contribution. For (7.116) we have

$$\mathcal{N}[J_\mu, J_\nu] = \frac{1}{2}\mathcal{N}[(z_i\bar{z}_j)\partial_\nu(z_k\partial_\mu\bar{z}_k + 2\bar{z}_k z_k A_\mu) - (\nu \leftrightarrow \mu)] \quad . \tag{7.121}$$

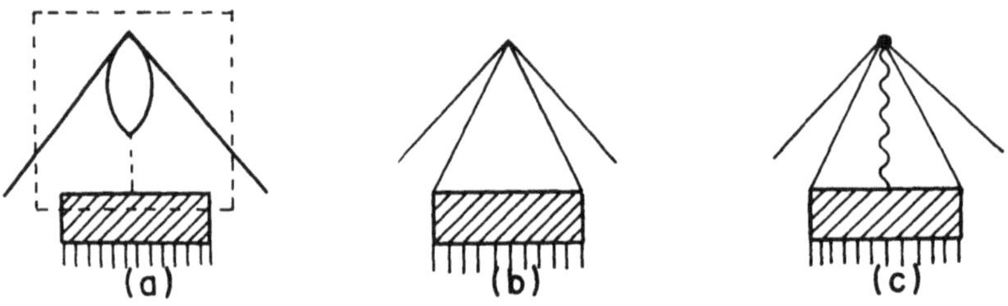

Figure 7.12: Diagrams contributing to the right hand side of (7.121) in the case of minimal coupling.

Now, the graphs contributing to the (rhs) of (7.121) have the structure shown in figure 7.12, where in figure 7.12*b* the detached subgraph of figure 7.12*a* cannot occur. The momentum-space expression for the detached subgraph of figure 7.12*a* is

$$I_{\mu\nu} = -\frac{iN}{2}(2q-p)_\nu\left(g_{\mu\rho} - \frac{q_\mu q_\rho}{q^2}\right)F(q)\left(g^{\rho\sigma} - \frac{q^\rho q^\sigma}{q^2}\right)\frac{i}{F(q)+\frac{i}{\pi}}\tilde{J}_\sigma(q) - (\mu \leftrightarrow \nu) \quad , \tag{7.122}$$

where p is the total momentum entering at the special vertex and $F(q)$ is the function shown in (7.21). Thus

$$I_{\mu\nu} = -\frac{i}{2}(2q-p)_\nu \tilde{J}_\mu(q) - \frac{(2q-p)_\nu}{2\pi}\left(g^{\mu\sigma} - \frac{q^\nu q^\sigma}{q^2}\right)\frac{1}{F(q)+\frac{i}{\pi}}\tilde{J}_\sigma(q) - (\mu \leftrightarrow \nu) \quad , \tag{7.123}$$

where current conservation $q^\mu J_\mu(q) = 0$ has been used.

The first term in (7.123) cancels against the contribution from the graphs of fig 12b. The contribution from the second term, on the other hand is, in coordinate-space, equal to

$$-\frac{N}{\pi} z_i \bar{z}_j(\partial_\nu A_\mu - \partial_\mu A_\nu) - \frac{i\sqrt{N}}{2}\{\partial_\nu[z_i\bar{z}_j(\bar{\psi}\gamma_\mu\psi)] - (\mu \leftrightarrow \nu)\} \quad . \tag{7.124}$$

Thus

$$\mathcal{N}_2[J_\mu, J_\nu] = \frac{N}{\pi} z_i \bar{z}_j F_{\mu\nu} - \frac{i\sqrt{N}}{2}[\partial_\nu(z_i\bar{z}_j\bar{\psi}\gamma_\mu\psi) - (\mu \leftrightarrow \nu)] \quad , \tag{7.125}$$

with $F_{\mu\nu} = \partial_\mu A_\nu - \partial_\nu A_\mu$.

Hence, defining $Q^{ij} = \lim Q^{ij}_\delta$, where

$$Q^{ij}_\delta = \frac{i}{N}\int dy_1 dy_2 \epsilon(y_1 - y_2) J_0^{ik}(y_1,t) J_0^{kj}(y_2,t) - \frac{Z}{N}\int J_1^{ij}(y,t)dy + \frac{i}{\sqrt{N}}\int \bar{\psi}\gamma_1\psi z^i \bar{z}^j dy \tag{7.126}$$

with $Z = \frac{N}{2\pi}\ln(\mu\delta)$, one obtains $dQ^{ij}/dt = 0$ up to order considered.

In the supersymmetric case, we have to deal with a more complicated structure. Nevertheless, the cancellation mechanism works in the same fashion as explained above, and we refer to [77] and [272] for details. The fact that the non local charge is conserved, means that the partons have been liberated by means of the introduction of the fermionic degrees of freedom, similar to the Schwinger mechanism, where the fermions generate a mass for the gauge field, shifting its pole from zero. Therefore, the z quanta are liberated, and confinement does not occur any more. In the supersymmetric case, the physical fermions are similar to the physical fermions of the Gross Neveu model. The action of the non local charge on asymptotic states can be computed, and therefore, we have constraints on the exact S-matrices that determine them up to pole factors, as we shall see in the next chapter. A generalized relation between particles and antiparticles, similar to the one we studied in chapter 5, also exists in the supersymmetric case.

7.5 Perturbative Renormalization and the Background Field Method

7.5.1 Background field method applied to the sigma model

Consider a sigma model defined by the action

$$S = \int d^2x \frac{1}{2} g_{ij}(\phi)\partial^\mu \phi^i \partial_\mu \phi^j \quad . \tag{7.127}$$

The background field method consists in splitting the field into a classical (from now on called background) field and a quantum field. The integration over the quantum fields gives rise to an effective action depending on the background fields, where it is not difficult to read off the necessary counterterms needed to render the theory finite. The advantages of this method, consists in the simplicity of the prescription, and, more important, the preservation of symmetries. The method was introduced by Schwinger[275] and extended to curved space as well as reformulated in covariant geometric language, by B. DeWitt[150]. The method has been extensively used in order to study the renormalization of quantum gravity,[276] non linear sigma models[211], and gauge theories[277]. The method can be explicitly covariantized[278,139,140], and may be generally used to obtain the effective action. The issue of symmetries has also played a rather central role in several discussions[279].

In the case of non linear sigma models, it is very natural to use this method, in view of the difficulty in maintaining the geometry of the system. However, the usual background quantum splitting, consisting in expanding ϕ^i around some classical field configuration φ^i as

$$\phi^i = \varphi^i + \pi^i \quad , \tag{7.128}$$

and treating the quantum fluctuations π^i perturbatively, is not convenient. The reason is that if the field φ^i and $\varphi^i + \pi^i$ are both on the manifold, the difference, π^i, transforms under reparametrization of the manifold in a complicated way, and the calculations suffer from a lack of manifest covariance. Thus we suppose that π^i is a function of a new covariant field ξ^i in terms of which the expansion is defined. We will take ξ^i as the tangent vector to the manifold at φ^i. We define a vector $\lambda^i(s)$, such that $\lambda^i(0) = 0$, $\lambda^i(1) = \pi^i$, and expand $\lambda^i(s)$ in powers of s.[14)] The expansion is given in terms of the vector ξ^i which is a contravariant vector, in the tangent space. This is the method of normal coordinates,[280] which we will not explain in greater detail, but merely state the results. The variable $\lambda^i(s)$ satisfies the differential equation of a geodesic in the manifold $\mathcal{M}$:[281]

$$\ddot{\lambda}^i + \Gamma^i_{jk}\dot{\lambda}^j\dot{\lambda}^k = 0 \quad , \tag{7.129}$$

[14)] For simplicity we suppress the dependence on x.

where the dot means differentiation with respect to the parameter s, and the Christoffel symbol Γ^i_{jk} in defined as usual, in terms of the metric by the expression

$$\Gamma^i_{jk} = g^{il}\left(\frac{\partial g_{lj}}{\partial\varphi^k} + \frac{\partial g_{lk}}{\partial\varphi^j} - \frac{\partial g_{jk}}{\partial\varphi^l}\right) \quad . \tag{7.130}$$

If we successively differentiate (7.129), we obtain expressions of higher derivatives of λ in terms of the lower ones; we can subsequently write a power series for $\lambda(s)$ using (7.130) and its derivatives with respect to φ (we recall that the dependence on s is implicit, that is, only through the fields). After some calculation, we obtain the expansion

$$\lambda^i(s) = \varphi^i + \xi^i s - \frac{1}{2}\Gamma^i_{j_1 j_2}\xi^{j_1}\xi^{j_2}s^2 - \frac{1}{3!}\Gamma^i_{j_1 j_2 j_3}\xi^{j_1}\xi^{j_2}\xi^{j_3}s^3 - \cdots \quad , \tag{7.131}$$

where

$$\xi^i = \dot{\lambda}^i(0) \tag{7.132a}$$

$$\Gamma^i_{j_1\cdots j_n} = \nabla_{j_1}\cdots\nabla_{j_{n-2}}\Gamma^i_{j_{n-1}j_{n-2}} \quad , \tag{7.132b}$$

and where the covariant derivative ∇_j acts only on lower indices

$$\nabla_j\Gamma^i_{kl} = \frac{\partial}{\partial\varphi^j}\Gamma^i_{kl} - \Gamma^n_{jk}\Gamma^i_{nl} - \Gamma^n_{jl}\Gamma^i_{kn} \quad . \tag{7.133}$$

It is clear that ξ^i is a vector, and $\pi^i(\xi) = \lambda^i(1)$ may be expanded in a power series in ξ. Hence we may write the action as a power series in ξ^i, the quantum field, and the coefficients are covariant tensors in the background field φ^i. We may compute the effective action as a function of the background field φ^i, and study the renormalization properties.[259,241] A system of Riemann normal coordinates[280] at the point φ^i is defined by the requirement that the connection coefficients vanish at that point, that is

$$\Gamma^i_{\{j_1\cdots j_n\}}(\varphi^i) = 0 \quad , \tag{7.134}$$

where $\{\}$ stands for symmetrization with respect to the indices, whereas $[]$ means antisymmetrization. This means that geodesics in this system of coordinates are expressed as straight lines (compare with Eq. (7.131)). In this system we may express the derivative of Γ at that point φ^i as a function of the curvature tensor at this point. Indeed, because of (7.134), it follows from

$$R^i_{jkl} = \partial_k\Gamma^i_{jl} - \partial_l\Gamma^i_{jk} + \Gamma^n_{jl}\Gamma^i_{nk} - \Gamma^n_{jk}\Gamma^i_{nl} \quad , \tag{7.135}$$

that

$$\partial_k\Gamma^i_{jl} = \frac{1}{3}(R^i_{jkl} + R^i_{lkj}) \quad ; \tag{7.136}$$

from (7.134) one also obtains

$$\partial_{\{j_1} \cdots \partial_{j_{n-2}} \Gamma^i_{j_{n-1} j_n\}} = 0 \quad , \tag{7.137}$$

which may be used to derive further relations such as

$$\partial_{\{j_1} \partial_{j_2} \Gamma^i_{j_3\} l} = -\frac{1}{2} D_{j_1} R^i_{j_2 l j_3} \quad . \tag{7.138}$$

For a general tensor $T_{j_1 \cdots j_n}(\varphi)$, we have

$$\frac{\partial}{\partial \xi^i} T_{j_1 \cdots j_n}(\varphi) = D_i T_{j_1 \cdots j_n}(\varphi) \tag{7.139}$$

$$\frac{\partial^2}{\partial \xi^i \partial \xi^k} T_{j_1 \cdots j_n}(\varphi) = D_{(i} D_{k)} T_{j_1 \cdots j_n}(\varphi) - \frac{1}{3} \sum_{p=1}^{n} R^l_{i j_p k} T_{j_1 \cdots j_{p-1} l j_{p+1} \cdots j_n}(\varphi) \tag{7.140}$$

Using the above relations, we obtain a general expansion of a tensorial field, at an arbitrary point on the manifold, as given by

$$\begin{aligned} T_{ik}(\varphi + \pi) &= T_{ik}(\varphi) + D_j T_{ik}(\varphi) \xi^j \\ &+ \frac{1}{2} \left\{ D_j D_l T_{ik}(\varphi) - \frac{1}{3} R^n_{jil} T_{nkk}(\varphi) - \frac{1}{3} R^n_{jkl} T_{in}(\varphi) \right\} \xi^j \xi^l + \cdots \quad , \end{aligned} \tag{7.141}$$

with related expansions for higher order tensors, which we do not need now.

For a simple one loop computation, the above results are all we need. We illustrate it for the metric g_{ij}, and for the derivative of the field $\partial_\mu \phi^i$, which in turn is obtained by differentiating (7.131) with respect to s, and putting $s = 1$; we thus obtain

$$\begin{aligned} g_{ij}(\varphi + \pi) &= g_{ij}(\varphi) - \frac{1}{3} R_{i l_1 j l_2} \xi^{l_1} \xi^{l_2} - \frac{1}{3!} D_{l_1} R_{i l_2 j l_3}(\varphi) \xi^{l_1} \xi^{l_2} \xi^{l_3} \\ &+ \frac{1}{5!} (-6 D_{l_1} D_{l_2} R_{i l_3 j l_4} + \frac{16}{3} R_{l_1 j l_2}{}^m R_{l_3 i l_4 m}) \xi^{l_1} \xi^{l_2} \xi^{l_3} \xi^{l_4} + \cdots \end{aligned} \tag{7.142}$$

and

$$\partial_\mu (\varphi^i + \pi^i) = \partial_\mu \varphi^i + D_\mu \xi^i + \frac{1}{3} \partial_\mu \varphi^j R^i_{l_1 l_2 j} \xi^{l_1} \xi^{l_2} + \cdots \quad . \tag{7.143}$$

We arrive at the following background quantum splitting for the action (7.127)

$$\begin{aligned} S(\varphi + \pi) &= S(\varphi) + \int d^2x g_{ij} \partial^\mu \varphi^i D_\mu \xi^j + \frac{1}{2} \int d^2x \{ g_{ij} D^\mu \xi^i D_\mu \xi^j \\ &+ R_{i l_1 l_2 j} \xi^{l_1} \xi^{l_2} \partial^\mu \varphi^i \partial_\mu \varphi^j + \frac{1}{3} D_{l_1} R_{i l_2 l_3 j} \xi^{l_1} \xi^{l_2} \xi^{l_3} \partial^\mu \varphi^i \partial_\mu \varphi^j + \cdots \} \quad . \end{aligned} \tag{7.144}$$

If we use the equation of motion of φ, the linear term vanishes. It contributes to a field redefinition, and does not influence anything else. In order to obtain Feynman rules for the quantum fields, we have to take care of the metric $g_{ij}(\varphi)$ in the kinetic term; we deal with it by redefining ξ^i as

$$\xi^i = e^i_{\ a}\xi^a \quad , \tag{7.145}$$

where

$$e^i_{\ a} e^j_{\ a} = g^{ij} \quad . \tag{7.146}$$

After discarding the linear term the action reads (the procedure is correct, since we do not include terms involving field redefinitions[259])

$$S(\varphi+\pi) = S[\varphi] + \frac{1}{2}\int d^2x \left\{ D^\mu\xi^a D_\mu\xi^a + R_{iabj}\partial^\mu\varphi^i\partial_\mu\varphi^j\xi^a\xi^b \right\} + \cdots , \tag{7.147}$$

where

$$D^\mu\xi^a = \partial^\mu\xi^a + \omega_i^{\ ab}\partial^\mu\varphi^i\xi^b \equiv \partial^\mu\xi^a - iA^{\mu ab}\xi^b \quad , \tag{7.148}$$

and $\omega_i^{\ ab}$ is the spin connection of the manifold.

We are now able to compute the ξ propagator, which is the one of a bosonic field[15])

$$\langle 0 \mid T\xi^a(x)\xi^b(y) \mid 0\rangle = \delta^{ab}\Delta(x-y) \quad . \tag{7.149}$$

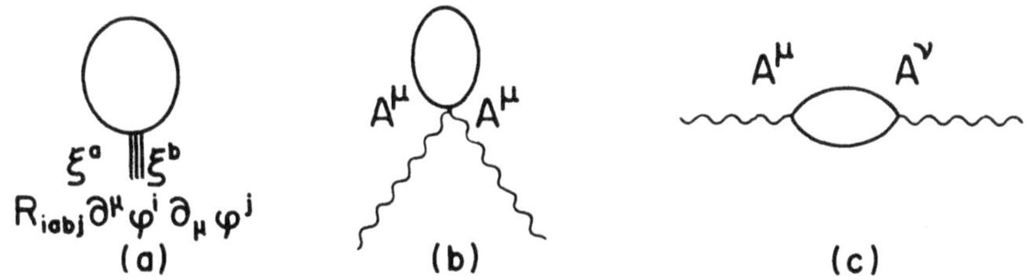

Figure 7.13: Contributions to the one loop counterterm for the bosonic sigma model.

The divergent one loop counterterms arise from the diagrams in figure 7.13a. The contribution to the effective action is given by

$$D_{A1} = R_{iabj}\partial^\mu\varphi^i\partial_\mu\varphi^j\delta^{ab}\int \frac{d^2k}{(2\pi)^2}\frac{i}{k^2-m^2} = R_{iabj}\partial^\mu\varphi^i\partial_\mu\varphi^j\delta^{ab}\frac{1}{2\pi}\ln\frac{\Lambda^2}{m^2} \quad . \tag{7.150}$$

15) We include here a mass to avoid IR divergencies, which are not the issue at this point.

Diagrams 13*b* and 13*c* do not present divergencies if we use a gauge invariant regularization. Thus we obtain a one-loop renormalized effective action

$$S^{(1)}_{eff} = \frac{1}{2}\int d^2x \left(g_{ij} - \frac{1}{2\pi}\ln\frac{\Lambda^2}{m^2}R_{ij}\right)\partial^\mu\varphi^i\partial_\mu\varphi^j \quad , \tag{7.151}$$

where R_{ij} is the Ricci tensor. We can define a β function

$$\beta_{ij} = \frac{\delta g^{ren}_{ij}}{\delta \ln\frac{\lambda^2}{m^2}} = -\frac{1}{2\pi}R_{ij} \quad , \tag{7.152}$$

and arrive at the conclusion that Ricci flat manifolds are one loop finite. Higher loop computations may be performed without new conceptual difficulties (though they may be rather involved!). In the literature one finds, for bosonic models[259],

$$\beta_{ij} = -\frac{1}{2\pi}R_{ij} - \frac{\alpha}{4\pi}R_{iabc}R_j{}^{abc} \tag{7.153}$$

where we introduced a coupling constant α.

For supersymmetric σ models the calculation does not present new difficulties either.[241] The Lagrangian may be written as

$$\mathcal{L} = \frac{1}{2}g_{ij}\partial^\mu\varphi^i\partial_\mu\varphi^j + \frac{i}{2}g_{ij}\overline{\psi}^i\gamma^\mu D_\mu\psi^j + \frac{1}{6}R_{iklj}\overline{\psi}^i\psi^l\overline{\psi}^k\psi^j \quad . \tag{7.154}$$

If we consider the expansions (7.142) to (7.144) above, and the fermionic fields ψ^i to be quantum fields, only few new diagrams are obtained; they correspond to a one loop fermion diagram, with two external "gauge field" lines as defined by the covariant derivative (7.148). See e.g. the second diagram in figure 7.3. We shall use

$$g_{ij}(\varphi)\overline{\psi}^i \not{D}\psi^j = \left(g_{ij}(\varphi) + \frac{1}{3}R_{iabj}\xi^a\xi^b\right)\overline{\psi}^i \not{D}\psi^j \quad , \tag{7.155}$$

where $\psi^a = e^a_i\psi^i$.

To obtain the fermionic contribution at one loop order, we have to deal with

$$S_f = -\frac{1}{2}\int d^2x\overline{\psi}^a\gamma^\mu(\partial_\mu\delta^{ab} + A^{ab}_\mu)\psi^b \, , \tag{7.156}$$

whose contribution is one loop finite due to gauge invariance. When dealing with classical fermionic fields, the second term in (7.155) reproduces the result (7.151) (or 7.152), i.e., renormalizes the metric. Therefore, a Ricci flat manifold is still finite. Higher loop computations are abundant in the literature. We quote the results.

For non supersymmetric non linear σ models, we use the action (7.127) and obtain, computing the diagrams in figure 7.13 the result of Eq. (7.153). If the models are $N = 1$ supersymmetric, it is convenient to use superfields[282]. However they are not needed. The study of the geometry of the manifold reveals that an $N = 2$ supersymmetry requires a Kähler structure[238,241,233], as we have seen in chapter 6. It has been conjectured on the basis of the finiteness of 2 and 3 loop calculations that the $N = 2$ supersymmetry on Ricci flat spaces, or $N = 4$ supersymmetry, imply ultraviolet finite sigma models[282]. Indeed, consider the two loop counterterm to the superfield action

$$S = \int d^2x d^2\theta \frac{1}{4} g_{ij}(\phi) \overline{D}^{\alpha} \phi^i D_{\alpha} \phi^j \quad , \tag{7.157}$$

where

$$\phi^i = \varphi^i + \overline{\theta}\psi^i + \frac{1}{2}\overline{\theta}\theta F^i \quad . \tag{7.158}$$

The background quantum splitting can be made as usual, and we obtain the counterterms using superfield perturbation theory[60]. Up to two loop order, one has to compute the supersymmetrized version of figure 7.13a and the result is summarized by the counterterm

$$\delta S = \frac{1}{32\pi^2} \ln \frac{\Lambda}{m} \int d^2\theta \int d^2x \left(D^k D_k R_{ij} + 2 \left[D^k, D_i \right] R_{kj} \right) \overline{D}\phi^i D\phi^j \quad , \tag{7.159}$$

which vanishes for Ricci flat manifolds, since all terms depend on the Ricci tensor R_{ij}. However,there exists a candidate for a higher loop divergence, whose coefficient is non vanishing. Indeed, if one expands the action up to fourth order, one finds, after a long calculation presented in detail in ref. [260], the result[16)]

$$\beta_{ij} = -\frac{\zeta(3)}{3(4\pi)^4} \tilde{T}_{(ij)} \quad , \tag{7.160}$$

where

$$\begin{aligned} \tilde{T}_{ij} = {} & 2R_{nijm;hk} \left(R^{msrk} R^{n}{}_{(sr)}{}^{h} + R^{msrn} R^{k}{}_{\{sr\}}{}^{h} \right) + 4R_{jkmn;[ih]} R^{m(rk)s} R^{n}_{\{sr\}}{}^{h} \\ & +3 \left(R_{ikht;r} R^{tsrq} R_j^{(hk)}{}_{q;s} + R_{irkt;h} R^{t}{}_{s}{}^{rq} R_j{}^{sh}{}_{q;}{}^{k} + 2R_{ikht;r} R^{t\ rq} R_j{}^{sk}{}_{q;}{}^{h} \right) + \\ & (2R_{rqst;i} - R_{rsqt;i}) R^{t}{}_{h}{}^{rk} R^{hqs}{}_{k;j} - 12 R_{mhki} R_{jrt}{}^{m} (R^{k}{}_{qs}{}^{r} R^{tqsh} + R^{k}{}_{qs}{}^{t} R^{hrsq}), \end{aligned} \tag{7.161}$$

16) There is an ambiguity in (7.160) depending on an arbitrary vector on the manifold, which corresponds to different reparametrizations, and has no physical significance.

where the semi-colon denotes covariant derivative. For the four loop β function of the $N = 2$ supersymmetric sigma model, one obtains the result

$$\beta_{\mu\nu} = -\frac{4\zeta(3)}{3(4\pi)^4}\nabla_\mu\nabla_\nu\Delta K \quad , \tag{7.162}$$

where $\Delta K = R_{\mu\bar{\lambda}\rho\bar{\pi}}R^{\sigma\bar{\lambda}\tau\bar{\pi}}R_{\sigma}{}^{\kappa}{}_{\tau}{}^{\rho} - R_{\mu\bar{\lambda}\rho\bar{\pi}}R^{\rho\bar{\pi}\sigma\bar{\tau}}R_{\sigma\bar{\tau}}{}^{\kappa\bar{\lambda}}$.

The last result, although computed independently from (7.162) in the literature, may be obtained from it by specializing the result to a Kähler manifold, and choosing a different coordinate parametrization in terms of the vector

$$v_j = -\frac{\zeta(3)}{9(4\pi)^4}\nabla_j\Delta K \quad , \tag{7.163}$$

and adding the symmetric derivative of (7.163), $(\nabla_{(i}v_{j)})$ to (7.162).

7.5.2 Parallelizable manifolds, and applications to string theory

It is important, in string theory, to consider a generalized σ model, where there is, in addition to the usual interaction, a coupling to an antisymmetric tensor field (related to a torsion of the manifold), and a scalar coupled to the curvature of the manifold. The Lagrangian is given by[261,68,69]

$$\mathcal{L} = \frac{1}{2\lambda}g_{ij}(\varphi)\partial^\mu\varphi^i\partial_\mu\varphi^j + \frac{1}{2}b_{ij}(\varphi)\epsilon^{\mu\nu}\partial_\mu\varphi^i\partial_\nu\varphi^j + \sqrt{g}R(\varphi)\Phi(\varphi) \quad , \tag{7.164}$$

where we call $\Phi(\varphi)$ the dilaton field, and where $b_{ij} = -b_{ji}$.

Due to the $\epsilon^{\mu\nu}$ factor in the second term, it turns out that its contribution comes with a negative sign with respect to the previous contribution, opening the possibility of obtaining a finite theory, by including torsion and fine tuning the coupling constant. This problem has been analysed by various authors and used largely to compute corrections to an effective string based field theory[68,69].

In general, the antisymmetric tensor field interaction may be obtained from a three dimensional integral over the topological density

$$nS_{WZ} = \frac{n}{12\pi}\int d^3x\epsilon^{\mu\nu\rho}\,\mathrm{tr}\; g^{-1}\partial_\mu g g^{-1}\partial_\nu g g^{-1}\partial_\rho g \quad , \tag{7.165}$$

where the three dimensional space above, has the two dimensional Minkowski space as boundary. The action S_{WZ} is referred to as the Wess Zumino term[52]. When performed over the whole space, it is a topological number. Therefore, different choices of the extension from two to three dimensions gives results which differ by integers. Hence, the coefficient must be quantized. Moreover, writing (7.165) in terms of Lie algebra fields associated with the group valued fields g we obtain

$$S_{WZ} = \frac{1}{12\pi}\int d^3x \epsilon^{\mu\nu\rho} S_{ijk}\partial_\mu\varphi^i\partial_\nu\varphi^j\partial_\rho\varphi^k \quad . \tag{7.166}$$

Under very general conditions[261] (see conclusion of chapter 6) the above torsion S_{ijk} may be derived from a potential b_{ij} as

$$S_{ijk} = \partial_{[i}b_{jk]} \quad , \tag{7.167}$$

in which case we can integrate (7.166) obtaining

$$S_{WZ} = \frac{1}{6\pi}\int d^2x b_{ij}(\varphi)\epsilon^{\mu\nu}\partial_\mu\varphi^i\partial_\nu\varphi^j \quad . \tag{7.168}$$

The form (7.165) of the topological term will be discussed later in chapter 9 in detail. It is not our purpose to consider the background-quantum splitting of (7.164) and its supersymmetric extension in detail here; we merely state the results concerning its renormalization and the computation of the β function. There are conditions under which the topological term does not get renormalized. If the manifold is locally symmetric, that is

$$R_{ijkl;m} = 0 \quad , \tag{7.169}$$

or else, if the torsion $S_{ijk} = \frac{1}{3}\left(b_{ij;k} + b_{jk;i} + b_{ki;j}\right)$ satisfies

$$S_{ijk;l} + S_{ijl;k} = 0 \tag{7.170}$$

the Wess-Zumino term does not get renormalized[261].

Applications to string theory

String theory offers the possibility of describing a unification of all interactions including gravity, based on a minimum input, which is free of the usual ultraviolet divergencies that plague the quantum theory of gravity, and explains the low energy spectrum of chiral matter fields interacting through gauge and gravitational interaction. Nevertheless, it is difficult to extract useful information of such theory; a second quantized version has not been developed, and non perturbative effects, although probably essential for a complete understanding of the theory, are still not available. It is thus very important, in this context, to obtain effective interactions of elementary fields, inspired by string or superstring theory. The propagation of strings in background fields provides such an effective interaction; it is described by sigma models of the type we have been discussing, and it turns out, that one is able to obtain an effective interaction from the requirement of conformal invariance for the sigma model interaction. A bosonic string propagating in a background field is described by the action of the non linear sigma model (7.127,165), or in the language of string theory, by[68,69,91]

$$S = \frac{1}{2\pi\alpha'}\int d^2x \times$$
$$\left\{\frac{1}{2}\sqrt{-g}g^{mn}G_{\mu\nu}\partial_m X^\mu \partial_n X^\nu + \frac{1}{2}\epsilon^{mn}B_{\mu\nu}\partial_m X^\mu \partial_n X^\nu + \frac{1}{2}\alpha'\sqrt{-g}R\phi(X)\right\} , \quad (7.171)$$

where R is the two dimensional scalar curvature. The last term explicitly breaks Weyl invariance. Nevertheless, as one immediately sees, it is of first order in the expansion parameter α'. The field ϕ is called the dilaton field. It is introduced in order to enforce local Weyl invariance, which is broken implicitly by anomalies, and we are entitled to introduce an explicit breaking term. Consistency of string theory, requires the vanishing of the trace of the world sheet energy momentum tensor, which is given by

$$T_n^n = \frac{1}{2\pi}\beta^G_{\mu\nu}\sqrt{-g}g^{mn}\partial_m X^\mu \partial_n X^\nu + \frac{1}{2\pi}\beta^B_{\mu\nu}\epsilon^{mn}\partial_m X^\mu \partial_n X^\nu + \frac{1}{2\pi}\beta^\phi\sqrt{g}R \quad , \quad (7.172)$$

where the β-functions labelled by G, B, ϕ are the β function corresponding to the respective G, B, and ϕ interactions. The requirement that these functions vanish as a consequence of conformal invariance, will lead to constraints on the background fields. We have for the β functions, the expressions

$$\beta^G_{\mu\nu} = R_{\mu\nu} - \frac{1}{4}H^{\lambda\sigma}_\mu + 2\nabla_\mu\nabla_\nu\phi + \mathcal{O}(\alpha') \quad , \quad (7.173a)$$
$$\beta^B_{\mu\nu} = \nabla_\xi H^\xi_{\mu\nu} - 2H^\xi_{\mu\nu}\nabla_\xi\phi + \mathcal{O}(\alpha') \quad , \quad (7.173b)$$
$$\beta^\phi = \frac{D-26}{48\pi^2} + \frac{\alpha'}{16\pi^2}\left[4(\nabla\phi)^2 - 4\nabla^2\phi - R + \frac{1}{12}H^2\right] + \mathcal{O}(\alpha') \quad , (7.173c)$$

where $H_{\mu\nu\rho} = 3\nabla_{[\mu}B_{\nu\rho]}$.

One readily verifies that we recover the Einstein equations from the vanishing of $\beta^G_{\mu\nu}$ in (7.173a), with the action given by the dilaton β function (after a convenient normalization). Consistency implies that the vanishing of the graviton and antisymmetric tensor field β functions, imply the vanishing of the dilaton β^ϕ function.[284] The higher loop contributions will predict corrections to the Einstein field equations, as given by the two loop conformal invariance condition on the graviton field:

$$R_{ij} + \frac{1}{2}\alpha' R_{iabc}R_j{}^{abc} = \nabla_i\nabla_j\phi \quad , \quad (7.174)$$

while for the dilaton we obtain $\nabla^2\phi + (\nabla\phi)^2 = \frac{1}{4}\alpha' R_{ijkl}R^{ijkl}$. where we neglected the antisymmetric tensor field. These effective corrections to the theory may be generalized to the supersymmetric string and to the heterotic string. Several one loop computations have been performed in the literature. The study of corrections to Einstein's equations have been explored using the above results[70].

7.6 Anomalous Non-Linear Sigma Models in Four Dimensions

Supersymmetric four dimensional non linear sigma models are important objects in the context of phenomenological Lagrangians. These models combine in a supermultiplet, the bosonic degrees of freedom characteristic of sigma models, and their fermionic counterparts, allowing for phenomena such as the superhiggs effect, and appearing in the low energy sector of the supergravity and superstring theories. The bosonic degrees of freedom are described by complex scalars, since supersymmetry in four dimensions requires a Kähler manifold. Due to the transformation law $\delta\varphi_i = \bar{\epsilon}\psi$, we know that the fermion must live in the space of tangent vectors TM to the sigma model manifold, M, as we discussed in chapter 6. Counting the degrees of freedom, we see that the fermions must be of the Weyl type [17)]. In terms of local coordinates the Lagrangian is given by

$$\mathcal{L} = g_{i\bar{j}}\partial^\mu\varphi^i\partial_\mu\varphi^j + ig_{i\bar{j}}\bar{\psi}^{\bar{j}}\gamma^\mu\left(\frac{1+\gamma_5}{2}\right)D_\mu\psi^i + \frac{1}{4}R_{i\bar{j}k\bar{l}}\bar{\psi}^{\bar{j}}\psi^i\bar{\psi}^{\bar{k}}\psi^l \quad , \tag{7.175}$$

where $D_\mu\psi^i = \partial_\mu\psi^i + \Gamma^i_{jk}\partial_\mu\varphi^k\psi^j$, and Γ^i_{jk} is the Christoffel symbol computed from the Kähler metric $g_{i\bar{j}}$. An important class of models included in the above class are the Grassmannians, defined by the manifold $\frac{SU(p+q)}{S(U(p)\times U(q))}$, in which case we can write (7.179) in terms of the z fields as defined in chapter 6 for the two dimensional case. As we shall see in chapters 13 and 14, the coupling of Weyl fields to a gauge theory is anomalous, since the gauge field itself is coupled to a non conserved current, as a result of the anomaly. Therefore, the theory is ill defined, if expressed in terms of bosonic background fields. Since low energy dynamics is described by sigma models on homogeneous spaces of the form $\frac{G}{H}$, with G realized non linearly, anomalies prevent the coupling of chiral fermions without spoiling $\frac{G}{H}$ current algebra predictions. However, the condition of absence of anomalies is more delicate for sigma models[285]. Consider the sigma model Lagrangian with fermions given by

$$\mathcal{L} = \mathrm{tr}D^\mu g^{-1}D_\mu g - i\bar{\psi}\,\not{D}\psi \quad . \tag{7.176}$$

---

17) A complex scalar has two degrees of freedom. A Dirac fermion in four dimensions has four degrees of freedom (eight real components, and a factor $\frac{1}{2}$ due to the Dirac equation). A further factor $\frac{1}{2}$ is obtained from the Weyl condition. Moreover, a Majorana fermion in four dimensions is equivalent to a Weyl fermion.

The above Lagrangian is supersymmetric in four dimensions if ψ is a Weyl field. We can parametrize the sigma model field in terms of gauge invariant fields $\phi_i \in \frac{\mathcal{G}}{\mathcal{H}}$, and gauge functions $h(x) \in \mathcal{H}$, so that the Lagrangian turns out to be

$$\mathcal{L} = g_{ij}(\phi)\partial^\mu \phi^i \partial_\mu \phi^j - i\overline{\psi}\gamma^\mu \left[\partial_\mu + h^{-1}\left(\partial_\mu + A_\mu(\phi)\right)h\right]\psi \quad , \tag{7.177}$$

where we use the equation of motion of A_μ following from (7.177), which is, as usually in sigma models, an equation of constraint. From the results of chapter 13, we will learn that the fermionic determinant is in general not gauge invariant, unless we are in an anomaly free representation.[285,286] Thus in general, the Wess Zumino term does not vanish, and the gauge symmetry is broken. In four dimensions, this happens in the $\mathbb{C}P^{n-1}$ and Grassmannian models[287], and for four dimensional N extended supergravity. Therefore, these models must be dealt in the context of anomalous models treated in chapters 13 and 14. In two dimensional space time, supersymmetric models are anomaly free, since we need either Majorana fermions (for $N+1$ supersymmetry), or Dirac fermions (for extended supersymmetry). In two dimensions, the models considered above are anomaly free.

7.7 Conclusion

The study of quantum non linear sigma models in two dimensions has not only revealed a rich structure concerning the formal field theoretic aspects such as renormalizability and quantum integrability, but has also served as a useful playground for gaining insight into the realistic Yang-Mills theories, and the supersymmetric extension, as well as a working device study string theories. These models also play an important role in statistical mechanics, a field of research outside the scope of this book.

Viewing them as a toy model for Yang-Mills theory, we discovered that the (composite) gauge field plays an important role in the dynamics, being responsible for a long range force which confines the partons (elementary fermions), unless the interaction with the fermions is supersymmetric or minimal. Also in the 4-dimensional Yang-Mills theory, supersymmetric coupling plays a special role: if we consider the extended $N=4$ supersymmetric Yang-Mills theory, it is known to be be classically integrable[41]. In two dimensions the Grasmannians have, in general a computable S-matrix for gauge invariant quantities. Their exact integrability allows one to obtain a complete classification of these models (see following chapter). Concepts such as θ vacuum, instantons gas, $U(1)$ problem, can be considered and discussed on a level which is much easier and natural than in the 4 dimensional counterpart.

As we have seen, the non linear σ models can be used to describe the effective interaction of the fundamental fields in string and superstring theory. They include the Einstein's equations, as well as their quantum corrections. This is the first

time since the theory of relativity that this has been achieved starting from fundamental principles. The methods developed so far provide a means for obtaining a perturbative expansion which generates the fundamental interactions.

Chapter VIII

EXACT S-MATRICES OF TWO DIMENSIONAL MODELS

8.1 Introduction

8.1.1 Consequences of higher conservation laws

In this chapter we shall discuss some exact results obtained for integrable theories in 1+1 dimensions, where the existence of an infinite number of conservation laws imposes severe constraints on the dynamics.

In a Poincaré invariant field theory, in $d > 2$ dimensions, the theorem of Coleman and Mandula[288] states that the most general invariance group of a non trivial field theory is the product of the Poincaré group and an internal symmetry group.[1)] The case of graded Lie groups (supersymmetry) has been studied later[289] and leads to similar results.[2)] The basic idea of the proof is that an infinite number of higher conservation laws represented by tensors imply that individual momenta are conserved, and the scattering would consist of an exchange of quantum numbers. This would mean that the S-matrix would not depend analytically on the scattering momenta, that is, the S-matrix would not depend analytically on the scattering angle.

In two dimensional space time the situation is however different. The scattering angle can only be zero or π, and the clash with analyticity no longer exists.

The constraints due to the conservation laws on the scattering process are, in any case, very strong. On the scattering states, the possible conserved charges can only be powers of momenta, due to Lorentz invariance. Therefore, the conservation of an infinite number of local charges implies conservation of the energy, momentum

---

1) One defines a symmetry group of the S-matrix[288], as a group of unitary operators which turn one particle states into one particle states, transform many particle states as if they were tensor products, and commute with the S matrix. The symmetry group G must contain the Poincaré group; moreover there are a finite number of one particle states below any mass parameter M. The elastic scattering amplitudes are analytic functions of the Mandelstam variables in some neighbourhood of the physical region. Under this set of hypothesis, we have either (in $d \neq 2$) a trivial S-matrix, or the symmetry group is isomorphic to the direct product of the Poincaré group and internal symmetry group. If one allows massless particles, the most general symmetry group is the product of the conformal group and an internal symmetry group.

2) If we allow anticommutators as well,[289] the graded algebra is the direct sum of the supersymmetry algebra, and an internal symmetry algebra.

and their powers. Indeed, the higher conserved charges are higher tensors $Q_{\mu_1\cdots\mu_l}$, transforming according to higher representations of the Lorentz group, commuting with one another, and with the momentum[36].

The action of $Q_{\mu_1\cdots\mu_n}$ on asymptotic states is severely restricted by Lorentz invariance. On a one particle state, we have[3)]

$$Q^{\mu_1\cdots\mu_l}|P\rangle = P^{\mu_1}\cdots P^{\mu_l}|P\rangle \quad , \tag{8.1}$$

i.e., the charge is diagonal in the one particle states, and the eigenvalues are integer powers in the momentum. Since $Q_{\mu_1\cdots\mu_l}$ is an integral of a local current, its action on a multiparticle asymptotic state is given by a sum of the action on the one-particle states (since Q is a local operator, we can think of wave packets, which are separated and well localized). Therefore one has

$$Q^{\mu_1\cdots\mu_l}|P_{(1)},\cdots,P_{(m)}\rangle = \sum_{i=1}^{m} P_{(i)}^{\mu_1}\cdots P_{(i)}^{\mu_l}|P_{(1)}\cdots P_{(m)}\rangle \quad . \tag{8.2}$$

We now consider the expectation value of the charge $Q^{\mu_1\cdots\mu_l}$ between two scattering states $\langle P'^{out}_{(1)}\cdots P'^{out}_{(m)}|Q^{\mu_1\cdots\mu_l}|P^{in}_{(1)}\cdots P^{in}_{(n)}\rangle$. Acting with $Q^{\mu_1\cdots\mu_l}$ alternatively on the "out" state and on the "in" state, and equating the results, one has

$$\sum_{i=1}^{n} P_i^{\mu_1}\cdots P_i^{\mu_l} = \sum_{i=1}^{m} P_i'^{\mu_1}\cdots P_i'^{\mu_l} \quad , \tag{8.3}$$

provided the corresponding scattering amplitudes do not vanish. Hence there exists an infinite number of conservation laws that must be obeyed by the external momenta; equations such as (8.3) can only be satisfied if $n = m$, i.e. if there is no particle production, and the individual momenta are conserved. Thus, after a suitable rearrangement, $P_i = P_i'$, and the scattering only consists of time delays and exchange of quantum numbers[36,190].

[3)] We ignore terms such as $g^{\mu\nu}P^\rho|P^\alpha\rangle$ since they are not relevant in the sequel. Also since the mass operator commutes with the charge $Q^{\mu\nu\rho}$ there can be degeneracy; we suppose, for simplicity that the operator $Q^{\mu\nu\rho}$ is diagonal in the mass operator.

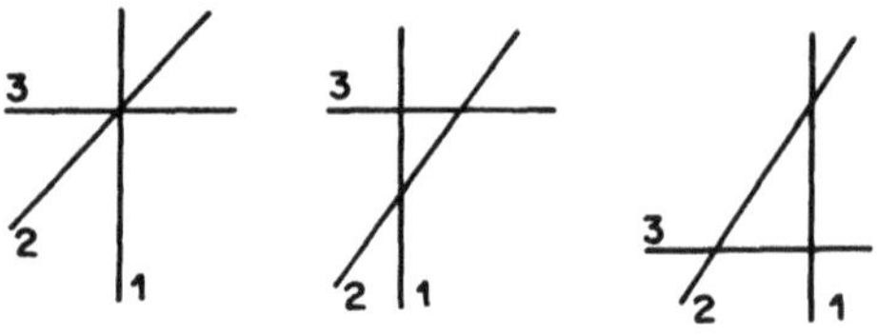

Figures 8.1, 8.2, and 8.3: Action of the operator (8.5) in terms of the space-time diagram.

8.1.2 Factorizable S-matrix

Absence of particle production implies that the S-matrix is of the factorizable type, and the S-matrix for any non trivial scattering is given by the product of all possible two particle scattering amplitudes. This is the content of a theorem using general properties of quantum field theory[190]. This kind of S-matrix is called a factorizable S-matrix. Furthermore the two particle processes are also severely constrained by the so called "factorization relations" (also called "star triangle relations").

We can understand this from the above conservation laws. To this end one further observes that intermediate multiparticle states, with the particles sufficiently separated, should satisfy the same selection rules as described above (see Eq. (8.3)). As a consequence the S-matrix elements for N particle scattering can be expressed as a product of two-particle S-matrices. To show this, we consider again the action of a higher charge $Q^{\mu_1\cdots\mu_n}$ on states. Let us take its pure space component

$$Q^{\overbrace{1\cdots 1}^{n}} \equiv Q^{(n)} \quad , \tag{8.4}$$

and apply[290] the operator $e^{icQ^{(n)}}$ on a state of momentum p

$$e^{icQ^{(n)}}|p\rangle = e^{icp^n}|p\rangle \quad . \tag{8.5}$$

Let the expression

$$\psi(x) = \int dp e^{-a(p-p_0)^2+ip(x-x_0)}$$

be a wave packet centered at the point x_0, and let $e^{icQ^{(n)}}$ act on it; we obtain

$$e^{icQ^{(n)}}\psi(x)e^{-icQ^{(n)}} = \tilde{\psi}(x) = \int dp e^{-a(p-p_0)^2+ip(x-x_0)+icp^n} \quad , \tag{8.6}$$

which is a wave packet now centered at the point $\tilde{x}_0$, given by $\tilde{x}_0 = x_0 - ncp_0^{n-1}$. The shift is proportional to a power of p_0, and hence grows with p_0. Consider now the space-time diagram for the three body amplitude as given in figure 8.1, and act on the states with the exponential of the charge operator, as given by (8.4) and (8.5). This action is such that it moves each line, by an amount which increases with p_0; therefore, in figure 8.1, particle 1 will be much displaced, while the action on particle 3 is very small. Depending on the sign of c in (8.6) we get figures 8.2 and 8.3.

Since $Q^{(n)}$ generates a symmetry, the different scatterings above must be equal, and for c large enough, figures 2 and 3 show that the S-matrix is given by a product of 2 particle S-matrices. Hence the S-matrix is of the factorizable type, which is defined by the fact that intermediate states have on shell momentum, and the intermediate particles are free. The scattering of figure 8.2 is described by the product[36]

$$S_2 = S(p_2, p_3)S(p_1, p_3)S(p_1, p_2) \quad , \tag{8.7a}$$

while figure 8.3 is described by

$$S_2 = S(p_1, p_2)S(p_1, p_3)S(p_2, p_3) \quad , \tag{8.7b}$$

which must be equal. This leads to severe constraints, which are the previously mentioned factorization equations. In (8.7) we have ordered the factors $S(p_i, p_j)$ according to where the particles i and j meet, in temporal order.

The Fock space of the theory is now decomposed irreducibly in subspaces labelled by a definite particle number, that is

$$\mathcal{F} = \bigoplus_{n=0}^{\infty} \int \frac{dP_1^1}{2\omega_1} \cdots \frac{dP_n^1}{2\omega_n} \mathcal{H}_{P_1 \cdots P_n} \quad , \tag{8.8a}$$

with

$$\mathcal{H}_{P_1 \cdots P_n} = \bigoplus_{\alpha_i \in A} (\mathcal{H}_{P_1}^{(\alpha_1)} \otimes \cdots \otimes \mathcal{H}_{P_n}^{(\alpha_n)})^{\Lambda} \quad , \tag{8.8b}$$

where A is the full set of particles, and $\mathcal{H}_P^{(\alpha)}$ is the space of one particle of type α and momentum P. The symbol Λ denotes symmetrization or antisymmetrization, depending on whether we are dealing with bosons or fermions.

A state of $\mathcal{H}_{P_1 \cdots P_n}$ is given by

$$|\alpha_P\rangle = |\alpha_{P_1}, \cdots, \alpha_{P_n}\rangle \quad , \tag{8.9}$$

where $P = \{P_1, \cdots P_n\}$ denotes a given permutation of the first n integers. The above states are normalized to one. Factorization of the S-matrix is expressed by the formula

$$^{out}\langle\alpha'_{P'}|\alpha_P\rangle^{in} = {}^{in}\langle\alpha'_{P'}| \prod_{1 \le i < j \le n} S(P_i, P_j)|\alpha_P\rangle^{in} \quad , \tag{8.10}$$

where $S(P_i, P_j)$ is a representative of the S-matrix of two particles i and j. In general the above S operators might not commute. The expression (8.10) is thus only well defined due to the equality of (8.7a) and (8.7b), which we rewrite as

$$S_{ij}(\theta_{ij})S_{ik}((\theta+\theta')_{ik})S_{jk}(\theta'_{jk}) = S_{jk}(\theta'_{jk})S_{ik}((\theta+\theta')_{ik})S_{ij}(\theta_{ij}) \quad , \tag{8.11}$$

where θ_{ij} is the rapidity defined by $\theta_{ij} = \theta_i - \theta_j$ with

$$p_i = m(\cosh\theta_i, \sinh\theta_i) \quad , \tag{8.12}$$

and where m is the mass of the fundamental particles.

A second, purely algebraic interpretation of equation (8.11) is also possible. We consider the symbols $\{A_i(\theta)\}$ to represent the set of particles. A given n-particle state is defined by the action of a product of these symbols on the vacuum, ordered according to their rapidities: the "in" states are identified with the products in order of decreasing rapidities, while the "out" states are arranged in the order of increasing rapidities. The commutation relations of the A's are defined by the S-matrix, that is

$$A(\theta_1)A'(\theta_2) = S_T(\theta_{12})A'(\theta_2)A(\theta_1) + \cdots \quad , \tag{8.13}$$

where S_T is the transition amplitude for $AA' \to AA'$, and the dots represent other channels. There are different ways to consider the scattering of three particles, and the identification of both ways leads again to equation (8.11).

The two particle S-matrix in a factorizable two dimensional theory is a function of the Mandelstam variable s. It is convenient to write the momenta p_i in terms of the rapidity variable θ_i as defined in (8.12). The two particle S-matrix elements depend on the difference of rapidities. Indeed, they only depend on the variable s, related to $\theta = \theta_i - \theta_j$ by

$$s = (p_i + p_j)^2 = m_i^2 + m_j^2 + 2m_i m_j \cosh\theta \quad . \tag{8.14}$$

For equal masses we have $s = 2m^2(1 + \cosh\theta) = 4m^2(\cosh\frac{\theta}{2})^2$.

In general the two particle amplitudes are analytic functions of s, with cuts along the real axis (see figure 8.4). The scattering amplitude has a cut for $s \leq (m_1 - m_2)^2$ and for $s \geq (m_1 + m_2)^2$. The point $s = (m_1 + m_2)^2$ corresponds to the two particle threshold. The mapping (8.14) transforms the physical sheet in the s plane into a strip $s < \Im m\theta < \pi$, see figure 8.5. The scattering amplitude $S(\theta)$ is real analytic and hence is real on the imaginary θ axis. Moreover, on the real axis $S(-\theta) = S^*(\theta)$.

In the calculation of S-matrices in $D = 2$ dimensions, one first computes the so called minimal S-matrix, which has a minimum number of zeros on the physical sheet, and is such that for large momenta it grows slower than $\exp\frac{p.p'}{m^2}$. At this point, one requires that the S-matrix obeys unitarity and crossing[291]. The first condition turns out to be simply a requirement on the modulus squared of the 2 particle scattering amplitude, since there is no particle production (hence only 2 particle scattering contributes to the total probability)[36].

In a relativistic theory, crossing corresponds to the substitution of an incoming particle of momentum p by an outgoing antiparticle with momentum $-p$. This is

Figure 8.4: Analytic structure of the two particle amplitude in the complex s-plane.

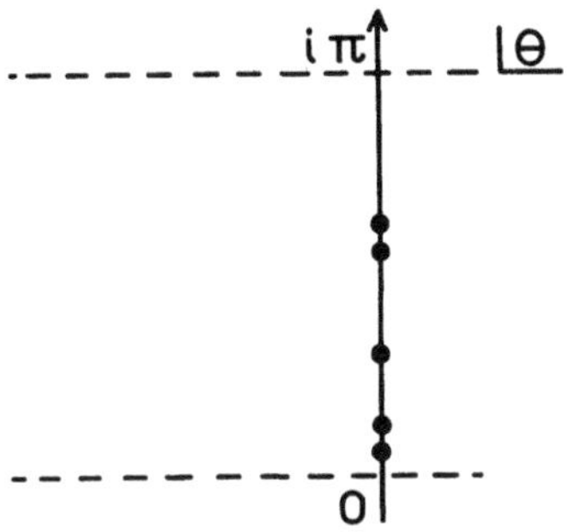

Figure 8.5: Analytic structure of the S-matrix amplitude in the θ plane bound states. The vertical axis denotes Re θ.

equivalent to the substitution $s \to 4m^2 - s$ (*or* $\theta \to i\pi - \theta$). In terms of equations, invariance under crossing implies

$$\langle f_1' f_2' | S(P_1, P_2) | f_1, f_2 \rangle = \langle f_1' \overline{f}_2 | S(P_1, -P_2) | f_1 \overline{f}_2' \rangle \quad . \tag{8.15}$$

8.1.3 Fusion rules

After obtaining the minimal S-matrix, we multiply it by a pole factor (CDD-pole)[292] needed to describe the bound state structure,[292,293] and which in general is of the form

$$P(\theta) = \frac{\sinh \theta + i \sin \lambda}{\sinh \theta - i \sin \lambda} \quad . \tag{8.16}$$

This pole factor will determine a set of bound states, via the so called fusion rule[293] as we now demonstrate.

We start out with a pole such as the one given in (8.16). Suppose we describe a scattering of particles with momenta p_1 and p_2; we take the analytic continuation of these momenta to the complex plane, such that at the bound state pole ($\theta_{12} = i\lambda$) we may parametrize p_1 and p_2 as:

$$p_1 = m\Big(\cosh\Big(\chi + \frac{i\lambda}{2}\Big), \sinh\Big(\chi + \frac{i\lambda}{2}\Big)\Big) \tag{8.17a}$$

$$p_2 = m\Big(\cosh\Big(\chi - \frac{i\lambda}{2}\Big), \sinh\Big(\chi - \frac{i\lambda}{2}\Big)\Big) \quad . \tag{8.17b}$$

In this case, we have a bound state with momentum

$$p_{1+2}^{(2)} = p_1 + p_2 = 2m \cos\frac{\lambda}{2}\Big(\cosh\chi, \sinh\chi\Big) \quad . \tag{8.18}$$

Thus

$$\left[p_{1+2}^{(2)}\right]^2 = (m_2)^2 \quad , \tag{8.19}$$

where

$$m_2 = 2\mu \sin\lambda \quad , \tag{8.20}$$

with $\mu = \frac{m}{2\sin\frac{\lambda}{2}}$.

If we wish to study a 3-particle bound state, we consider the scattering of 3 particles $S(\theta_{12})S(\theta_{13})S(\theta_{23})$, fuse particles 1 and 2 as above, and go simultaneously to the pole of $S(\theta_{23})$ at $\theta_{23} = i\lambda$.

It is convenient to parametrize now p_1, p_2 and p_3 as follows,

$$p_1 = m(\cosh(\chi + i\lambda), \sinh(\chi + i\lambda)) \quad , \tag{8.21a}$$
$$p_2 = m(\cosh\chi, \sinh\chi) \quad , \tag{8.21b}$$
$$p_3 = m(\cosh(\chi - i\lambda), \sinh(\chi - i\lambda)) \quad . \tag{8.21c}$$

We take $\chi = 0$ (center of mass), and obtain m_3 by adding the zero components. Since $p_1 + p_2 = m_2\Big(\cos\frac{\lambda}{2}, i\sin\frac{\lambda}{2}\Big)$ (see 8.18), we have[4)]

$$(p_1 + p_2 + p_3)^0 = m_2 \cos\frac{\lambda}{2} + m\cos\lambda = 2\mu\sin\frac{3\lambda}{2} = m_3 \quad . \tag{8.22}$$

[4)] For $(n+1)$ particles (n odd)

$$p_1 = m\left(\cosh\left(\chi + in\frac{\lambda}{2}\right), \sinh\left(\chi + in\frac{\lambda}{2}\right)\right)$$
$$p_2 = m\left(\cosh\left(\chi - i(n-2)\frac{\lambda}{2}\right), \sinh\left(\chi - i(n-2)\frac{\lambda}{2}\right)\right)$$
$$p_{n+1} = m\left(\cosh\left(\chi - in\frac{\lambda}{2}\right), \sinh\left(\chi - in\frac{\lambda}{2}\right)\right)$$

When $\chi = 0$ we have $p_1 + \cdots + p_{n+1} = m_{n+1}\Big(1, 0\Big)$; we assume

$$p_1 + \cdots + p_n = m_n\left(\cosh\chi^{(n)}, \sinh\chi^{(n)}\right) = m_n\left(\cos\frac{\lambda}{2}, -i\sin\frac{\lambda}{2}\right), \text{by (8.24a)}.$$

In the center of mass: $m_n \sinh\chi^{(n)} + m\sinh\frac{n\lambda}{2} = 0 \Rightarrow \chi^{(n)} = -i\frac{\lambda}{2}$ Indeed:

$$-\sinh\frac{in\lambda}{2}\sinh\frac{\lambda}{2} + \sinh\frac{i\lambda}{2}\sinh\frac{n\lambda}{2} = 0 \quad ; \quad p_{n+1} = m\left(\cos\frac{n\lambda}{2}, -i\sin\frac{n\lambda}{2}\right) \quad .$$

Proceeding inductively we consider the fusion of a n-bound state of mass

$$m_n = 2\mu \sin \frac{n\lambda}{2} \quad , \tag{8.23}$$

and momentum

$$(p_1 + \cdots + p_n) = m_n \left(\cos \frac{\lambda}{2}, i \sin \frac{\lambda}{2} \right) \quad , \tag{8.24a}$$

with the $(n+1)^{th}$ particle of momentum

$$p_1 \cdots + p_{n+1} = m \left(\cos \frac{(n+1)\lambda}{2}, 0 \right) \quad . \tag{8.24b}$$

In this way one finds

$$m_{n+1} = 2\mu \sin \frac{(n+1)\lambda}{2} \quad . \tag{8.24c}$$

We thus proved that in a theory of the integrable type, that is, with a sufficient number of conservation laws such as to prohibit pair creation, the bound state pole of the form (8.16) induces a whole spectrum of bound states obeying a mass formula given by (8.23).

We can summarize the whole program in the following steps[36,37,39]:

1. Set up the factorization equations, either from the local conservation laws, such as in (8.11), or using the non local conservation laws, as below (section 8.3).
2. Impose crossing and unitarity.
3. Compute the minimal S-matrix, that is the one obeying analyticity and having the minimum number of zeros or poles in the physical sheet, and growing asymptotically slower than $\exp |\frac{p_1 p_2}{m^2}|$ for $p_1, p_2 \to \infty$[5)].
4. With some qualitative information about the bound state structure, introduce poles; resonances are supposed to be absent, since unstable particles generally do not exist in a factorizable S-matrix, due to the conservation of the number of particles; (see also footnote 5) below).
5. Check the results by perturbation theory, or any other method available.

The full S-matrix is obtained after consideration of bound state scattering, as well as scattering of more complicated objects, such as kinks and solitons[33]. Although the end result is unique, there are various methods to fulfill the above program, but we will not dwell on detailed explanation of all of them, referring instead the reader to the literature[295].

---

5) The fact that the S-matrix has no redundant poles and zeros in the physical sheet of the transmission amplitude can be proved for one dimensional scattering. In this case all poles in the physical sheet correspond to bound states.

8.1.4 Bound state scattering

We have so far only deduced the spectrum induced by the pole in (8.16). In order to study the bound state scattering amplitudes, we proceed again as above, and obtain the scattering amplitude as the residuum at the bound state pole[296]. Consider for example the scattering shown in figure 8.6. We define

$$S(p_1+p_2,p_3)\bigg|_{\substack{(p_1+p_2)^2=m_2^2\\ p_3^2=m^2}} = \frac{1}{R_2}\operatorname*{Res}_{(p_1+p_2)^2=m_2^2} S_{12}S_{13}S_{23} \quad , \tag{8.25}$$

where

$$R_2 = \operatorname*{Res}_{(p_1+p_2)^2=m_2^2} S_{12} \quad . \tag{8.26}$$

This procedure defines a projection operation, since

$$P_{12} = \frac{1}{R}\operatorname*{Res}_{(p_1+p_2)^2=m_2^2} S(p_1,p_2) \quad , \tag{8.27}$$

acts as a projector. Due to factorization, we have

$$P_{12}S_{13}S_{23} = S_{23}S_{13}P_{12} \quad , \tag{8.28}$$

and as a result no reflexion is possible for the scattering of a bound state and a fundamental particle. This result is extremely important for the determination of the S-matrix of Z_n, chiral Gross-Neveu and supersymmetric $\mathbb{C}P^{n-1}$ models.

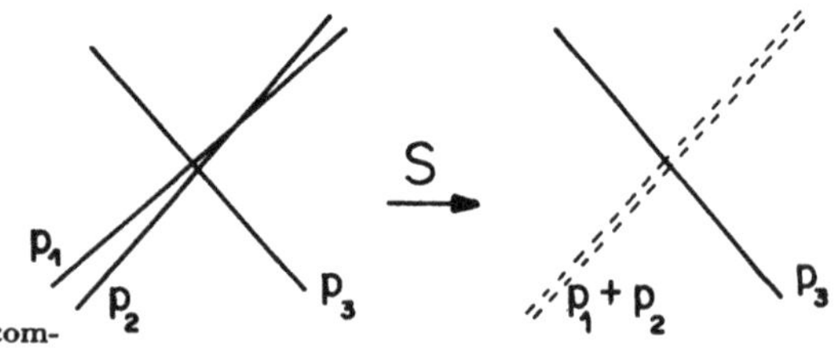

Figure 8.6: Bound state scattering as computed from two particle scattering.

8.2 Classification of Exact S-matrices from the Local Conservation Laws

8.2.1 $SU(N)$ invariant S-matrix

In this section, we shall consider examples of S-matrices associated with the symmetry groups $U(N)$ or $O(N)$, and consider the possible solutions for factorizable S-matrices. The proofs will in general be not complete, since the technical points involved are well explained in the literature[297].

The requirement of $U(N)$ symmetry, implies that the particle-particle and particle-antiparticle scattering amplitudes be of the form

$$\langle P_\gamma(\theta_1')P_\delta(\theta_2')out|P_\alpha(\theta_1)P_\beta(\theta_2)in\rangle = \\ {}_{\alpha\beta}S_{\gamma\delta}(\theta)\delta(\theta_1-\theta_1')\delta(\theta_2-\theta_2') \pm {}_{\alpha\beta}S_{\delta\gamma}(\theta)\delta(\theta_1-\theta_2')\delta(\theta_2-\theta_1') \quad , \tag{8.29a}$$

$$\langle P_\gamma(\theta_1')A_\delta(\theta_2')out|P_\alpha(\theta_1)A_\beta(\theta_2)in\rangle = \\ {}_{\alpha\beta}F_{\gamma\delta}(\theta)\delta(\theta_1-\theta_1')\delta(\theta_2-\theta_2') \pm {}_{\alpha\beta}B_{\delta\gamma}(\theta)\delta(\theta_1-\theta_2')\delta(\theta_2-\theta_1') \quad , \tag{8.29b}$$

respectively, with

$$ {}_{\alpha\beta}S_{\gamma\delta}(\theta) = u_1(\theta)\delta_{\alpha\gamma}\delta_{\beta\delta} + u_2(\theta)\delta_{\alpha\beta}\delta_{\gamma\delta} \quad , \tag{8.30a}$$

$$ {}_{\alpha\beta}F_{\gamma\delta}(\theta) = t_1(\theta)\delta_{\alpha\gamma}\delta_{\beta\delta} + t_2(\theta)\delta_{\alpha\beta}\delta_{\gamma\delta} \quad , \tag{8.30b}$$

$$ {}_{\alpha\beta}B_{\delta\gamma}(\theta) = r_1(\theta)\delta_{\alpha\gamma}\delta_{\beta\delta} + r_2(\theta)\delta_{\alpha\beta}\delta_{\gamma\delta} \quad , \tag{8.30c}$$

where ${}_{\alpha\beta}F_{\gamma\delta}(\theta)$ is called the transmission amplitude, and ${}_{\alpha\beta}B_{\delta\gamma}(\theta)$ the reflection amplitude.

Implementation of the factorization equations (8.11) is now a technical step. They read[297]

$$S.F.F = S^*.F^*.F \quad , \tag{8.31a}$$

$$S.F.B = B^*.B^*.F + F^*.S^*.B \quad , \tag{8.31b}$$

$$F.S.F + B.B.B = F^*.S^*.F + B^*.B^*.B \quad , \tag{8.31c}$$

where

$$A.B.C \sim {}_{\alpha\beta}A_{\delta\epsilon}(\theta_{12})_{\alpha\gamma}B_{\delta\nu}(\theta_{13})_{\beta\delta}C_{\epsilon\lambda}(\theta_{23}) \quad , \tag{8.31d}$$

$$A^*.B^*.C \sim {}_{\delta\epsilon}A_{\kappa\mu}(\theta_{12})_{\alpha\lambda}B_{\delta\nu}(\theta_{13})_{\beta\delta}C_{\epsilon\lambda}(\theta_{23}) \quad , \tag{8.31e}$$

The solutions to (8.31) fall into six classes as given in table 8.1. The function $f(\theta,\lambda)$ is a meromorphic function of θ, for $Re\lambda > 0$; it is uniquely defined by

Table 8.1						
class	$t_1(\theta)$	$u_1(\theta)$	$r_1(\theta)$	$u_2(\theta)$	$r_2(\theta)$	$t_2(\theta)$
I	1	1	0	0	0	0
II	$f(\theta,\lambda)$	$t_1(i\pi-\theta)$	0	$-\lambda\frac{i\pi}{\theta}u_1(\theta)$	0	$-\lambda\frac{i\pi}{i\pi-\theta}t_1(\theta)$
III	$f(\theta,\lambda)f(i\pi-\theta,\lambda)$	$t_1(\theta)$	$-\lambda\frac{i\pi}{\theta}t_1(\theta)$	$r_1(\theta)$	$-\lambda\frac{i\pi}{i\pi-\theta}t_1(\theta)$	$r_2(\theta)$
IV	$f(\theta,\lambda)f(i\pi-\theta,\lambda)$ $\times i\text{th}\frac{1}{2}(\theta+\frac{1}{2}i\pi)$	$-t_1(\theta)$	$-\lambda\frac{i\pi}{\theta}t_1(\theta)$	$r_1(\theta)$	$-\lambda\frac{i\pi}{i\pi-\theta}t_1(\theta)$	$r_2(\theta)$
V	0	0	$\prod_{k=-\infty}^{\infty}\frac{f(\theta,k/2\mu i)}{f(\theta,k/2\mu i+\frac{1}{2})}$	$r_1(\theta)$	$\frac{\sin\mu(i\pi-\theta)}{\sin\mu\theta}r_1(\theta)$	$r_2(\theta)$
VI	0	0	$\prod_{k=-\infty}^{\infty}\frac{f(\theta,k/2\mu i)}{f(\theta,k/2\mu i+\frac{1}{2})}$	$e^{i\mu\theta}r_1(\theta)$	$\frac{\sin\mu(i\pi-\theta)}{\sin\mu\theta}r_1(\theta)$	$e^{i\mu(i\pi-\theta)}r_2(\theta)$

the requirement of being minimal. The only arbitrariness lies in the bound state structure.

From table 8.1, we see for a $U(N)$ symmetry that the solution to the factorization equations is not unique. In the case of $\mathbb{C}P^{N-1}$ and chiral models, the solution will be found to be of class II; to obtain it, we shall use the non local conservation laws$^{6)}$.

The solutions belonging to class III correspond to $O(N)$ symmetry (see explicitly below, the solution for the $O(N)$ invariant sigma and Gross-Neveu models).

8.2.2 S-matrix of the sine-Gordon and massive Thirring model

The sine Gordon theory is defined by the Lagrangian

$$\mathcal{L} = \frac{1}{2}\partial_\mu\varphi\partial^\mu\varphi + \frac{m^2}{\beta}\cos\beta\varphi \quad . \tag{8.32}$$

We have seen that this model has an infinite number of conservation laws, thus prohibiting particle creation. Its dynamics is simple, but non trivial[114] We have only real fields, and the scattering amplitude for forward scattering takes the simple form[298]

$$\langle p_1' p_2' | S | p_1 p_2 \rangle = S(\hat{s})\{\delta(p_1 - p_1')\delta(p_2 - p_2') + \delta(p_1 - p_2')\delta(p_2 - p_1')\} \quad , \tag{8.33}$$

where $\hat{s} = \frac{1}{m^2}p_1 \cdot p_2 = \cosh\theta$, $\theta = \theta_1 - \theta_2$.

Because of the absence of particle production, unitarity takes the form

$$|S(\hat{s})|^2 = 1 \quad , \tag{8.34}$$

$^{6)}$ In fact we have already argued that in the case of the chiral Gross-Neveu model, the particle-antiparticle backward scattering vanishes, due to the fact that the antiparticles are bound states of the particles.

while crossing demands $S(\hat{s}) = S(-\hat{s})$. These requirements imply that $S(\hat{s})$ be of the form

$$S(\hat{s}) = \frac{\sqrt{\hat{s}^2 - 1}\, g(\hat{s}) - i}{\sqrt{\hat{s}^2 - 1}\, g(\hat{s}) + i} \quad , \tag{8.35}$$

with $g(\hat{s})$ some function of $\hat{s}$.

A perturbative calculation in β using (8.32) may be performed without much difficulty, and reveals that $g(\hat{s})$ must be a constant. Thus, the Ansatz (8.35) may be written, using the definition of θ, as

$$S(\theta) = \frac{\sinh\theta - \sin\lambda}{\sinh\theta + \sin\lambda} \quad , \tag{8.36}$$

that is, the minimal S-matrix is unity and we have just one pole[114]; by the fusion method,[293] a whole sequence of charge zero bound states are generated from (8.36). Moreover, the full S-matrix of the model should also describe the scattering of solitons and antisolitons, which are charged fields. We have seen that the sine Gordon is equivalent to the massive Thirring model. Therefore, we expect an $O(2) \approx U(1)$ symmetry at the level of the full S-matrix. This problem, in turn, is tackled as follows. Suppose we have to compute the factorizable S-matrix with $O(2)$ symmetry:

$$\begin{aligned} \langle \theta_1' i\, \theta_2' j\; out | \theta_1 k\, \theta_2 l\; in \rangle = \delta(\theta_1 - \theta_1')\delta(\theta_2 - \theta_2')\big[\delta_{ij}\delta_{kl}s_1(\theta) + \delta_{ik}\delta_{jl}s_2(\theta) + \delta_{il}\delta_{kj}s_3(\theta)\big] \\ + \big(i \longleftrightarrow j \,,\, \theta_1 \longleftrightarrow \theta_2\big) \quad . \end{aligned} \tag{8.37}$$

Equivalently, we can represent as before the particles by fields $A(\theta)$ as we saw in equation (8.13), in terms of which we have the following algebraic relations

$$A_i(\theta_1)A_j(\theta_2) = s_1(\theta)\delta_{ij}A_k(\theta_2)A_k(\theta_1) + s_2(\theta)A_j(\theta_2)A_i(\theta_1) + s_3(\theta)A_i(\theta_2)A_j(\theta_1) \quad . \tag{8.38}$$

We now impose factorization condition on the S-matrix, requiring that (8.38) is an associative product. Note that since the indices i, j, k run from 1 to 2, there are only a few independent tensors; as a result, the equations we shall arrive at, can not be obtained from the $O(N)$ invariant S-matrix, to be computed later, by simply setting $N = 2$. The crucial point is that there is no antisymmetric representation with three indices for $O(2)$, and the factorization relations are less restrictive. Indeed, we shall obtain an S-matrix depending on a free parameter, while in the general $O(N)$ case there are no free parameters left.

Let us discuss the product $A_i(\theta_1)A_j(\theta_2)A_k(\theta_3)$, and use (8.38) successively [7], obtaining

$$\begin{aligned}
&2S_1(\theta)S_3(\theta+\theta')S_1(\theta')+S_1(\theta)S_1(\theta+\theta')S_1(\theta')+S_1(\theta)S_2(\theta+\theta')S_1(\theta')\\
&+S_1(\theta)S_3(\theta+\theta')S_2(\theta')+S_1(\theta)S_3(\theta+\theta')S_3(\theta')+S_2(\theta)S_3(\theta+\theta')S_1(\theta')\\
&+S_2(\theta)S_3(\theta+\theta')S_3(\theta')+S_3(\theta)S_3(\theta+\theta')S_1(\theta')+S_3(\theta)S_3(\theta+\theta')S_2(\theta')\\
&\qquad\qquad=S_3(\theta)S_1(\theta+\theta')S_3(\theta')+S_3(\theta)S_2(\theta+\theta')S_3(\theta'), \qquad (8.39a)
\end{aligned}$$

$$\begin{aligned}
&S_1(\theta)S_1(\theta+\theta')S_2(\theta')+S_1(\theta)S_2(\theta+\theta')S_3(\theta')+S_3(\theta)S_2(\theta+\theta')S_3(\theta')\\
&=S_2(\theta)S_1(\theta+\theta')S_3(\theta')+S_2(\theta)S_3(\theta+\theta')S_3(\theta')+S_3(\theta)S_3(\theta+\theta')S_2(\theta'), \qquad (8.39b)
\end{aligned}$$

where we have

$$\theta=\theta_1-\theta_2 \quad , \quad \theta'=\theta_2-\theta_3 \quad . \qquad (8.40)$$

We define the functions

$$a(\theta)=\frac{S_1(\theta)}{S_3(\theta)} \quad , \quad b(\theta)=\frac{S_2(\theta)}{S_3(\theta)} \quad ; \qquad (8.41)$$

Dividing (8.39) by $S_3(\theta)S_3(\theta+\theta')S_3(\theta')$ we obtain the functional equations

$$\begin{aligned}
&(a(\theta)a(\theta')-1)(1+a(\theta+\theta')+b(\theta+\theta'))\\
&\qquad+(1+a(\theta)+b(\theta))(1+a(\theta')+b(\theta'))-b(\theta)b(\theta')=0 \quad , \qquad (8.42a)
\end{aligned}$$

and

$$a(\theta)a(\theta+\theta')b(\theta')+a(\theta)b(\theta+\theta')+b(\theta+\theta')=b(\theta)a(\theta+\theta')+b(\theta)+b(\theta') \quad .(8.42b)$$

If we set $\theta'=0$ in the second equation, we obtain

$$b(0)(a(\theta)^2-1)=0 \quad , \qquad (8.43)$$

which has the two solutions

$$a(\theta)=\pm 1 \quad . \qquad (8.44)$$

Thus, (8.42b) implies that b is also constant, and we can not satisfy unitarity, unless we take, according (8.43)

$$b(0)=0 \quad . \qquad (8.45)$$

If, on the other hand, we take $\theta=0$ in (8.42b), we learn that if $b(\theta)\neq 0$, then

$$a(0)=0 \quad . \qquad (8.46)$$

---

[7] We can either start commuting $A(\theta_2)$ and $A(\theta_3)$, and afterwards commute $A(\theta_1)$ with the result, we obtain the right hand side, while if we start commuting $A(\theta_1)$ and $A(\theta_2)$ and afterwards commute the result with $A(\theta_3)$, we obtain the left hand side. As an illustration, we obtain the first equation (8.39a) for the coefficient of $A_1A_1A_2$ choosing $i=1=j$, $k=2$, and the second (8.39b) for the coefficient of $A_1A_2A_1$ choosing again $i=j=1$, $k=2$.

We proceed by taking the derivative of (8.42) with respect to θ' at $\theta' = 0$; calling $\alpha = a'(0)$, $\beta = b'(0)$, we obtain

$$a'(\theta) + b'(\theta) = (1 + a(\theta))(\beta(1 + a(\theta)) + \alpha + \beta b(\theta)) \tag{8.47a}$$

$$b'(\theta) = (1 + a(\theta))(\alpha - \beta b(\theta)) \quad . \tag{8.47b}$$

These differential equations are solved by the functions

$$a(\theta) = \tanh \frac{4\pi\theta}{\gamma} \coth \frac{4\pi}{\gamma}(i\delta - \theta) \quad , \tag{8.48a}$$

$$b(\theta) = -i \tan \frac{4\pi\delta}{\gamma} \tanh \frac{4\pi\theta}{\gamma} \quad , \tag{8.48b}$$

with δ and γ being two real parameters, related to α and β. We use crossing symmetry, which in terms of the elementary amplitudes is given by

$$S_2(\theta) = S_2(i\pi - \theta) \quad , \tag{8.49a}$$

$$S_1(\theta) = S_3(i\pi - \theta) \quad , \tag{8.49b}$$

obtaining as a result

$$\delta = \pi \quad . \tag{8.50}$$

On the other hand, unitarity requires

$$S_2(\theta)S_2(-\theta) + S_3(\theta)S_3(-\theta) = 1 \quad . \tag{8.51}$$

We use (8.41) and (8.48) to obtain

$$S_2(\theta)S_2(-\theta) = \frac{\sin^2 \frac{4\pi^2}{\gamma} \sinh^2 \frac{4\pi\theta}{\gamma}}{\sin^2 \frac{4\pi^2}{\gamma} \sinh^2 \frac{4\pi\theta}{\gamma} + \cos^2 \frac{4\pi^2}{\gamma} \cosh^2 \frac{4\pi\theta}{\gamma}} \quad . \tag{8.52}$$

The minimal solution is given by

$$S_2(\theta) = \frac{2}{\pi} \sin \frac{4\pi^2}{\gamma} \sinh \frac{4\pi\theta}{\gamma} \sinh \frac{4\pi(i\pi - \theta)}{\gamma} U(\theta) \quad , \tag{8.53a}$$

where

$$U(\theta) = \Gamma\left(\frac{8\pi}{\gamma}\right)\Gamma\left(1 + \frac{8i\theta}{\gamma}\right)\Gamma\left(1 - \frac{8\pi}{\gamma} - \frac{8i\theta}{\gamma}\right) \prod_{n=1}^{\infty} \frac{R_n(\theta)R_n(i\pi - \theta)}{R_n(0)R_n(i\pi)} \tag{8.53b}$$

and

$$R_n(\theta) = \frac{\Gamma\left(\frac{16\pi n}{\gamma} + \frac{8i\theta}{\gamma}\right)\Gamma\left(1 + \frac{16\pi n}{\gamma} + \frac{8i\theta}{\gamma}\right)}{\Gamma\left(\frac{(2n+1)8\pi}{\gamma} + \frac{8i\theta}{\gamma}\right)\Gamma\left(1 + \frac{(2n-1)8\pi}{\gamma} + \frac{8i\theta}{\gamma}\right)} \quad . \tag{8.53c}$$

In order to describe the scattering of sine Gordon solitons we introduce the charged operators

$$A(\theta) = A_1(\theta) + iA_2(\theta) \quad , \tag{8.54a}$$

$$\overline{A}(\theta) = A_1(\theta) - iA_2(\theta) \quad , \tag{8.54a}$$

with the commutation relations

$$A(\theta)\overline{A}(\theta') = S_T(\theta-\theta')\overline{A}(\theta')A(\theta) + S_R(\theta-\theta')A(\theta')\overline{A}(\theta) \quad , \tag{8.55a}$$

$$A(\theta)A(\theta') = S(\theta-\theta')A(\theta')A(\theta) \quad , \tag{8.55b}$$

where

$$S(\theta) = S_3(\theta) + S_2(\theta) \quad , \tag{8.56a}$$

$$S_T(\theta) = S_1(\theta) + S_2(\theta) \quad , \tag{8.56b}$$

$$S_R(\theta) = S_1(\theta) + S_3(\theta) \quad . \tag{8.56c}$$

Using (8.41,48,53a), one finds

$$S_T(\theta) = -i\frac{sh\frac{8\pi}{\gamma}\theta}{\sin\frac{8\pi^2}{\gamma}}S_R(\theta) \quad , \tag{8.57a}$$

$$S(\theta) = -i\frac{sh\frac{8\pi}{\gamma}(i\pi-\theta)}{\sin\frac{8\pi^2}{\gamma}}S_R(\theta) \quad , \tag{8.57b}$$

$$S_R(\theta) = \frac{1}{\pi}\sin\frac{8\pi^2}{\gamma}U(\theta) \quad . \tag{8.57c}$$

The location of the poles in the transmission and reflexion amplitudes are shown in figure 8.7.

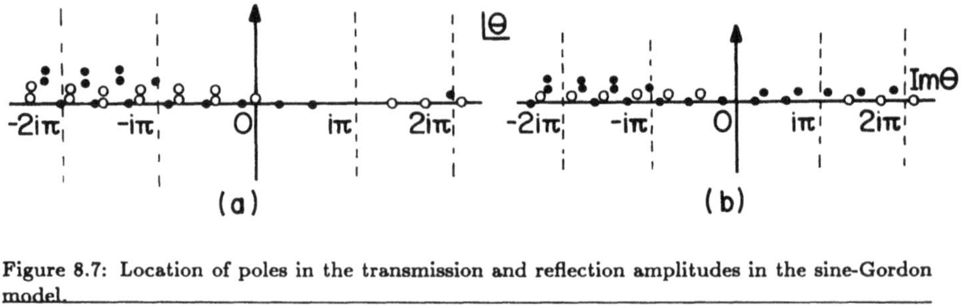

Figure 8.7: Location of poles in the transmission and reflection amplitudes in the sine-Gordon model.

The semi classical spectrum is given by the formula[181]

$$m_n^{semi} = 2m \sin \frac{n\gamma}{16} \quad , \quad n = 1, \cdots < \frac{8\pi}{\gamma} \quad , \tag{8.58}$$

which agrees with the positions of the poles in the above S-matrix. In fact,the semi classical spectrum of the sine-Gordon theory was long ago suspected to be exact[293,181,299]. The semiclassical analysis is thus of extreme importance for this model. If we consider (8.58), we see that, if

$$\gamma = \frac{8\pi}{n} \quad , \tag{8.59}$$

the bound state of order n lies on the boundary of the physical strip. In this case, the semi-classical expression (8.57c) shows vanishing of the particle-antiparticle backward scattering. With the assumption of exactness of the semiclassical spectrum, and vanishing of reflection at (8.59), one derives the expression

$$S_T(\theta) = e^{in\pi} \prod_{k=1}^{n-1} \frac{e^{\theta - i\frac{\pi k}{n}} + 1}{e^{\theta} + e^{-i\frac{\pi k}{n}}} \quad , \tag{8.60}$$

for the transmission amplitude[114,300], which is in complete agreement with the results expressed in equation (8.57a). The semiclassical analysis further shows that

$$\gamma = \frac{\beta^2}{1 - \frac{\beta^2}{8\pi}} \quad . \tag{8.61a}$$

As another check we take γ to be $\gamma = 8\pi$, which, according to (8.59), corresponds to $\beta^2 = 4\pi$, where the sine-Gordon theory describes free massive fermions. In this case, we see that indeed

$$S_T = S = 1 \quad , \quad S_R = 0 \quad .$$

One can further expand the S-matrix in powers of[48]

$$g = \frac{4\pi^2}{\gamma} - \frac{\pi}{2} \quad , \tag{8.61b}$$

and compare with perturbative results of the massive Thirring model.

Finally, one can compute the scattering amplitude of bound states. We define

$$B_n\Big(\frac{\theta_1 + \theta_2}{2}\Big) = \lim_{\theta_1 - \theta_2 \to \frac{in\gamma}{8}} [A(\theta_2)\overline{A}(\theta_1) \pm \overline{A}(\theta_2)A(\theta_1)] \quad , \tag{8.62}$$

where the upper(lower) sign refers to n even(odd) (this corresponds to the correct parity of the bound state (for a discussion see [296])).

The amplitudes are computed according to the procedure described in equations (8.25-28). We have, for the scattering of solitons with the n-bound state

$$A(\theta)B_n(\theta') = S^{(n)}(\theta - \theta')B_n(\theta')A(\theta) \quad . \tag{8.63}$$

This leads to the expression

$$S^{(n)}(\theta) = \frac{\sinh\theta + i\cos\frac{n\gamma}{16}}{\sinh\theta - i\cos\frac{n\gamma}{16}} \prod_{l=1}^{n-1} \frac{\sin^2\left(\frac{n-2l}{32}\gamma - \frac{\pi}{4} + \frac{i\theta}{2}\right)}{\sin^2\left(\frac{n-2l}{32}\gamma - \frac{\pi}{4} + \frac{i\theta}{2}\right)} \quad , \tag{8.64}$$

while for bound state/bound state scattering

$$B_n(\theta)B_m(\theta') = S^{(n,m)}(\theta - \theta')B_m(\theta')B_n(\theta) \quad , \tag{8.65}$$

which leads to

$$S^{(n,m)}(\theta) = \frac{\sinh\theta + i\sin\left(\frac{n+m}{16}\gamma\right)}{\sinh\theta - i\sin\left(\frac{n+m}{16}\gamma\right)} \frac{\sinh\theta + i\sin\left(\frac{n-m}{16}\gamma\right)}{\sinh\theta - i\sin\left(\frac{n-m}{16}\gamma\right)} \times$$
$$\prod_{l=1}^{m-1} \frac{\sin^2\left(\frac{m-n-2l}{32}\gamma + \frac{i\theta}{2}\right)\cos^2\left(\frac{m+n-2l}{32}\gamma + \frac{i\theta}{2}\right)}{\sin^2\left(\frac{m-n-2l}{32}\gamma - \frac{i\theta}{2}\right)\cos^2\left(\frac{m+n-2l}{32}\gamma - \frac{i\theta}{2}\right)} \quad . \tag{8.66}$$

We can interpret B_n as a bound state of $n \geq m$ elementary particles, since we can verify, in the context of poles of the above S-matrix, that

$$B_{n+m}\left(\frac{\theta_1 + \theta_2}{2}\right) = \lim_{\theta_1 - \theta_2 \to i\frac{n+m}{16}\gamma} B_m(\theta_2)B_n(\theta_1) \quad . \tag{8.67}$$

In fact, if we compute $S^{(1,1)}$ we find

$$S^{(1,1)}(\theta) = \frac{\sinh\theta + i\sin\frac{\gamma}{8}}{\sinh\theta - i\sin\frac{\gamma}{8}} \quad , \tag{8.68}$$

which matches the result (8.36).

8.2.3 Exact S-matrix for $O(N)$ symmetry

The Ansatz for a factorizable $O(N)$ invariant S-matrix is still given by expressions (8.37) and (8.38), but as we stated previously, the equations are now more restrictive due to the fact that there are more invariants for $N \geq 3$. We have now the following equations,[180,301]

$$S_2(\theta)S_3(\theta+\theta')S_3(\theta') + S_3(\theta)S_3(\theta+\theta')S_2(\theta') = S_3(\theta)S_2(\theta-\theta')S_3(\theta') \,, \quad (8.69a)$$
$$S_2(\theta)S_1(\theta+\theta')S_1(\theta') + S_3(\theta)S_2(\theta+\theta')S_1(\theta') = S_3(\theta)S_1(\theta+\theta')S_2(\theta') \,, \quad (8.69b)$$
$$NS_1(\theta)S_3(\theta+\theta')S_1(\theta') + S_1(\theta)S_3(\theta+\theta')S_2(\theta') + S_1(\theta)S_2(\theta+\theta')S_1(\theta')$$
$$+S_2(\theta)S_3(\theta+\theta')S_1(\theta') + S_1(\theta)S_1(\theta+\theta')S_1(\theta') = S_3(\theta)S_1(\theta+\theta')S_1(\theta') \,. \quad (8.69c)$$

Defining now

$$a(\theta) = \frac{S_2(\theta)}{S_3(\theta)} \quad , \quad b(\theta) = \frac{S_1(\theta)}{S_2(\theta)} \quad , \tag{8.70}$$

one finds, from (8.69a)

$$a(\theta) + a(\theta') = a(\theta+\theta') \quad , \tag{8.71}$$

which has the solution

$$a(\theta) = i\lambda\theta \quad . \tag{8.72}$$

Substituting this into (8.69b) we obtain

$$b(\theta) = -\frac{i/\lambda}{i\delta - \theta} \quad . \tag{8.73}$$

Equation (8.69c) together with crossing leads to

$$S_3(\theta) = -\frac{2\pi i}{(N-2)\theta}S_2(\theta) \,, \tag{8.74a}$$

$$S_1(\theta) = -\frac{2\pi i}{(N-2)(i\pi-\theta)}S_2(\theta) \quad . \tag{8.74b}$$

We shall come back to these equations latter on, and show that they will lead to the solution of $O(N)$ symmetric σ model, and $O(N)$ invariant Gross-Neveu model.

8.2.4 The Z_N invariant S-matrix

In some cases, the symmetry is enough to entirely fix an S-matrix. An example is provided by a factorizable S-matrix with Z_N symmetry, studied by Köberle and Swieca[302]. No reference to a specific action is needed in this case. The only dynamical input is the realization of the Z_N symmetry by means of the identity

$$Z^{N-1} = \overline{Z} \quad , \tag{8.75}$$

that is, the antiparticle is a bound state of $N-1$ particles (see chapter 5). In those cases, a particle antiparticle scattering at 180° is forbidden (see in this connection the brief discussion following Eq. (8.25)). Thus. the S-matrix is described by

$$\langle p_1' p_2' | S | p_1 p_2 \rangle = u(\theta)[\delta(p_1 - p_1')\delta(p_2 - p_2') + \delta(p_1 - p_2')\delta(p_2 - p_1')] \ , \quad (8.76a)$$

$$\langle \overline{p}_1' p_2' | S | \overline{p}_1 p_2 \rangle = t(\theta)\delta(p_1 - p_1')\delta(p_2 - p_2') \quad . \quad (8.76b)$$

Crossing implies

$$t(\theta) = u(i\pi - \theta) \quad , \quad (8.77)$$

and unitarity requires

$$u(\theta)u(-\theta) = 1 \quad , \quad \text{and} \quad t(\theta)t(-\theta) = 1 \quad . \quad (8.78)$$

The fact that a particle is a bound state of $N-1$ antiparticles in this model implies that $\lambda = \frac{2\pi}{N}$ in Eq. (8.16):

$$P(\theta) = \frac{\sinh \frac{1}{2}\left(\theta + \frac{2\pi i}{N}\right)}{\sinh \frac{1}{2}\left(\theta - \frac{2\pi i}{N}\right)} \quad . \quad (8.79)$$

By fusion, this pole generates the spectrum

$$m_j = m \frac{\sin j\pi/N}{\sin \pi/N} \quad . \quad (8.80)$$

Thus we fix $u(\theta)$ to be[302]

$$u(\theta) = \frac{\sinh \frac{1}{2}\left(\theta + \frac{2\pi i}{N}\right)}{\sinh \frac{1}{2}\left(\theta - \frac{2\pi i}{N}\right)} \quad . \quad (8.81)$$

It is straighforward to compute the amplitude for the scattering of a particle with the bound state of $N-1$ particles. It is given by

$$\prod_{\substack{j=\pm 1, \pm 3 \cdots \pm (N-2),\ N\ odd \\ or\ j=0, \pm 2 \cdots \pm (N-2),\ N\ even}} u\left(\theta + j\frac{\pi i}{N}\right) = u(i\pi - \theta) = t(\theta) \quad , \quad (8.82)$$

which realizes (8.75). For $N = 2$, one finds $S = -1$, which corresponds to the exact solution of the Ising model[303]. The case $N \to \infty$, corresponds to the $O(2)$ symmetric sigma model, which is trivial; indeed, one finds $S = 1$ in this case.

8.3 Quantum Non Local Charges, and Exact S-Matrices

The structure of the Lax pair in integrable theories,[35] implies the existence of a nonlocal conserved charge, as discussed in chapter 6 for the case of σ models, and in chapter 3 for purely fermionic theories. In general, they have, for purely bosonic, or purely fermionic theories, the form (see 6.140, and 5.8)

$$Q^{ij} = \int dy_1 dy_2 \epsilon(y_1 - y_2) j_0(t, y_1)^{ik} j_0(t, y_2)^{kj} + Z \int dy j_1^{ij}(t, y) \quad , \qquad (8.83)$$

where Z is a renormalization constant needed to cancel the (model dependent) divergencies associated with the first term (see chapter 7).

For supersymmetric sigma models, we have the additional contribution due to the fermions,[77,237]

$$Q^{ij}_{susy} = \int dy_1 dy_2 \epsilon(y_1 - y_2)[j_0(t, y_1) + j_0^M(t, y_1)]^{ik}[j_0(t, y_2) + j_0^M(t, y_2)]^{kj} +$$
$$Z_{susy} \int dy [j_1(t, y) + 2 j_1^M(t, y)]^{ij} \quad , \qquad (8.84)$$

where j_μ and j_μ^M have been defined in chapter 6.

We also consider σ models with minimally coupled fermions, for which the conserved charge is given by[77]

$$Q^{ij}_{min} = \int dy_1 dy_2 \epsilon(y_1 - y_2)(j_0(t, y_1) + j_0^M(t, y_1))^{ik}(j_0(t, y_2) + j_0^M(t, y_2))^{kj} +$$
$$Z_{min} \int j_1(t, y)^{ij} dy + \sqrt{N} i \int \overline{\psi} \gamma^1 \psi z^i \overline{z}^j dz \quad . \qquad (8.85)$$

In the case where the quantum non local charge is conserved (see chapter 7, for a discussion of possible anomalies), there exists a strong constraint on the scattering amplitude. This was shown to be true in several important cases. We obtain constraints on the S-matrix elements by studying the action of the nonlocal charge on asymptotic states[35,193]. The naive way of proceeding would be again to write the charges above in terms of asymptotic fields. Despite the fact that this procedure gives the correct results, it is not rigorous,[195] since the asymptotic charges used in the intermediate steps in do not really exist. Here, we shall use the results of chapters 5 and 7, and write down the constraints arising from the conservation of non local charges.

The procedure is completely algebraic, after the action on the asymptotic states is obtained, and the results follow easily after a straightforward (although in general very tedious) algebra.

8.3.1 S-matrices of purely fermionic models

$O(N)$ invariant Gross-Neveu model

We begin with the case of the purely fermionic models, referred to in chapter 5. For the $O(N)$ invariant Gross-Neveu model, we obtain the following action of the nonlocal charge on the asymptotic states (Eq.5.116-119)[193]

$$Q^{ab}|\theta_1 c_1, \theta_2 c_2, \cdots, \theta_l c_l\ in\rangle = |\theta_1 d_1, \theta_2 d_2, \cdots, \theta_l d_l\ in\rangle \left(M^{ab}_{in}\right)_{d_1\cdots d_l, c_1\cdots c_l} \quad (8.86a)$$

$$\langle\theta_1 c_1, \theta_2 c_2, \cdots, \theta_l c_l\ in|Q^{ab} = \left(M^{ab}_{out}\right)_{c_1\cdots c_l, d_1\cdots d_l}\langle\theta_1 d_1, \theta_2 d_2, \cdots, \theta_l d_l\ in| \quad (8.86b)$$

with

$$\left(M^{ab}_{\substack{out\\in}}\right)_{c_1c_2,d_1d_2} =$$

$$\mp\frac{1}{N}\sum_{k<j=1}^{l}\left(I_k^{ac}I_j^{cb} - I_k^{bc}I_j^{ca}\right)\prod_{i\neq j\neq k}\delta_{c_i d_i} + \frac{(N-2)}{iN\pi}\sum_{k=1}^{l}\theta_k I_k^{ab}\prod_{i\neq k}\delta_{c_i d_i} \quad , \quad (8.87a)$$

where I_k^{ab} acts on the isospin of the particle, and its matrix elements are given by

$$I_k^{ab} \equiv \left(I_k^{ab}\right)_{d_k c_k} = \delta^a_{d_k}\delta^b_{c_k} - \delta^b_{d_k}\delta^a_{c_k} \quad (8.87b)$$

The constraint due to the conservation of the non local charge is now immediately obtained considering the expectation value of the charge Q^{ab} between "*in*" and "*out*" states. We have

$$\begin{aligned}\langle\theta_1' c_1' \theta_2' c_2' out|Q^{ab}|\theta_1 c_1, \theta_2 c_2 in\rangle &= \left(M^{ab}_{out}\right)_{c_1'c_2', d_1'd_2'}\langle\theta_1' d_1' \theta_2 d_2 out|\theta_1 c_1, \theta_2 c_2 in\rangle \\ &= \langle\theta_1' c_1' \theta_2' c_2' out|\theta_1 d_1, \theta_2 d_2 in\rangle\left(M^{ab}_{in}\right)_{d_1 d_2, c_1 c_2} \quad (8.88)\end{aligned}$$

The final ingredient is the Ansatz for the factorizable S-matrix, which, due to $O(N)$ symmetry is given by

$$\begin{aligned}&\langle\theta_1' c_1' \theta_2' c_2' out|\theta_1 d_1, \theta_2 d_2 in\rangle = \\ &(4\pi)^2\delta(\theta_1'-\theta_1)\delta(\theta_2'-\theta_2)\left[\sigma_1(\theta)\delta^{c_1'c_2'}\delta^{c_1c_2} + \sigma_2(\theta)\delta^{c_1'c_1}\delta^{c_2'c_2} + \sigma_3(\theta)\delta^{c_1'c_2}\delta^{c_1c_2'}\right] - \\ &(4\pi)^2\delta(\theta_1'-\theta_2)\delta(\theta_2'-\theta_1)\left[\sigma_1(\theta)\delta^{c_1'c_2'}\delta^{c_1c_2} + \sigma_2(\theta)\delta^{c_1'c_2}\delta^{c_2'c_1} + \sigma_3(\theta)\delta^{c_1'c_1}\delta^{c_2c_2'}\right] \quad . (8.89)\end{aligned}$$

Using now the equations (8.88) we arrive at constrains on the amplitudes $\sigma(\theta)$, which read[8)]

$$\sigma_1(\theta) = -\frac{2i\pi}{(N-2)(i\pi-\theta)}\sigma_2(\theta) \ , \quad (8.90a)$$

$$\sigma_3(\theta) = -\frac{2i\pi}{(N-2)\theta}\sigma_2(\theta) \quad . \quad (8.90b)$$

---

8) See appendix J for an explicit computation in the case of sigma models.

Notice that we arrived exactly at the results expressed in (8.74) which followed from the conservation of the local charges.

Unitarity constrains the amplitude $\sigma_2(\theta)$ to satisfy

$$\sigma_2(\theta)\sigma_2(-\theta)+\sigma_3(\theta)\sigma_3(-\theta)=1 \quad , \tag{8.91}$$

and we obtain, using (8.90)

$$\sigma_2(\theta)\sigma_2(-\theta)=\frac{\theta^2}{\left(\frac{2\pi}{N-2}\right)^2+\theta^2} \quad . \tag{8.92}$$

An amplitude obeying this constraint as well as crossing[9]), and containing a minimum number of poles and zeros on the physical sheet, is given by

$$\sigma_2^{(0)}(\theta)=Q(\theta)Q(i\pi-\theta) \quad , \tag{8.93}$$

where

$$Q(\theta)=\frac{\Gamma\left(-\frac{1}{N-2}-i\frac{\theta}{2\pi}\right)\Gamma\left(\frac{1}{2}-i\frac{\theta}{2\pi}\right)}{\Gamma\left(\frac{1}{2}-\frac{1}{N-2}-i\frac{\theta}{2\pi}\right)\Gamma\left(-i\frac{\theta}{2\pi}\right)} \quad . \tag{8.94}$$

Notice that if we change the sign of the parameter $\lambda=\frac{2\pi}{N-2}$, we find another solution of (8.92), which differs from the above by a pole factor; that is, calling the above solution $\sigma_2^{(-)}(\theta)$, we have another solution given by

$$\sigma_2^{(+)}(\theta)=\frac{\sinh\theta-i\sin\frac{2\pi}{N-2}}{\sinh\theta+i\sin\frac{2\pi}{N-2}}\,\sigma_2^{(-)}(\theta) \quad . \tag{8.95}$$

Notice that the multiplicative factor has a zero at $\theta=\frac{2\pi i}{N-2}$.

The bound state spectrum reveals a rich structure[181]. It can be obtained using the fusion method, by starting with a CDD pole at $\lambda=\frac{2\pi}{N-2}$, which in turn generates the spectrum

$$m_j=m\frac{\sin\frac{j\pi}{N-2}}{\sin\frac{\pi}{N-2}} \quad . \tag{8.96}$$

This favors the solution (8.94), since (8.95) turns out to present no bound state pole (the CDD pole cancels against the zero at $\theta=\frac{2\pi i}{N-2}$ referred to above).

---

[9]) Crossing is very simple in the present case, since all particles are real (we are dealing with Majorana spinors), and implies the relation $\sigma_1(\theta)=\sigma_3(i\pi-\theta)$, and $\sigma_2(\theta)=\sigma_2(i\pi-\theta)$

There exist several checks of solution (8.93). The first is the $1/N$ expansion,[180,191] which can be performed up to order $1/N^2$; the results are, up to first order,

$$\sigma_2(\theta) = 1 + \frac{2\pi i}{N \sinh \theta} \tag{8.97a}$$

$$\sigma_3(\theta) = \frac{2\pi i}{N\theta} \tag{8.97b}$$

$$\sigma_1(\theta) = \frac{2\pi i}{N(i\pi - \theta)} \quad , \tag{8.97c}$$

which agree with (5.40). The second is a result of the perturbative expansion of the model[180,304] in the coupling constant, using the relation between the generated mass and the coupling constant[10)]

$$\ln \frac{s}{m^2} = \ln \frac{s}{\mu^2} + \int^{g(\mu)} \frac{dg}{\beta(g)} \quad . \tag{8.98}$$

Up to second order in g the result agrees again with the exact one.

Discussion on the $O(2N)$ symmetric S-matrix

From the above and the factorization equations of the kink-kink[305,306] and fermion-kink scattering amplitudes of the model, we can compute the complete S-matrix for the $O(2N)$ invariant Gross Neveu model[306]. A very rich structure arises. The elementary fermion, turns out to be a bound state of kinks, which are thus particles obeying a generalized statistics. The spin of the kink states are given by $\frac{N}{8}$ and $\frac{N}{8} + \frac{1}{4}$. The situation is very much related to that of the sine Gordon theory, where the Lagrangian field, although presenting a very simple S-matrix structure, is in fact a bound state soliton-antisoliton. The general scattering of bound states is obtained from the rules discussed at the beginning of this chapter (see appendix K). The representations are classified according to the structure of the group $O(2N)$. The cases $N = 2, 3, 4$ may be discussed separately. For $N = 2$, the S-matrix turns out to be a product of two $SU(2)$ invariant S-matrices, as those corresponding to the chiral Gross Neveu model (see later in this chapter). The spin of the kinks are equal to $\frac{1}{4}$, in accordance with the results we presented in chapter 5 for the chiral case, with $N = 2$. The case $N = 3$ is equivalent to the $SU(4)$ chiral Gross Neveu model, and the scattering of two kinks coincides with the scattering of fundamental $SU(4)$ particles. Finally, when $N = 4$ we have the triality symmetry of the Dynkin diagram of $O(8)$. The mass of the fermion is the same as the mass of the kink, as we see looking at the mass formula (8.96), and they all have spin $\frac{1}{2}$. In this case also the different channels of the S-matrix coincide. For further details see appendix K.

---

10) This expansion is valid for $\theta \to \infty$, due to asymptotic freedom.

SU(N) Chiral Gross-Neveu Model

For the chiral Gross-Neveu model, we have the following action of the non-local charge on asymptotic states

$$Q^{ij}|\theta_1 k, \theta_2 l\rangle = \left(M^{ij}_{in}\right)_{kl,mn}|\theta_1 m, \theta_2 n\rangle \ , \tag{8.99a}$$

$$Q^{ij}|\theta_1 k, \overline{\theta}_2 l\rangle = \left(\overline{M}^{ij}_{in}\right)_{kl,mn}|\theta_1 m, \overline{\theta}_2 n\rangle \ , \tag{8.99b}$$

$$\langle\theta_1 k, \theta_2 l|Q^{ij} = \langle\theta_1 m, \theta_2 n|\left(M^{ij}_{out}\right)_{kl,mn} \ , \tag{8.99c}$$

$$\langle\theta_1 k, \overline{\theta}_2 l|Q^{ij} = \langle\theta_1 m, \overline{\theta}_2 n|\left(\overline{M}^{ij}_{out}\right)_{kl,mn} \ , \tag{8.99d}$$

with

$$\left(M^{ij}_{\substack{in\\out}}\right)_{c_1\cdots c_l, d_1\cdots d_l} = \mp\frac{1}{N}\sum_{k<m=1}^{l}\left(I^{il}_k I^{lj}_m - I^{il}_m I^{lj}_k\right)\prod_{i\neq m\neq k}\delta_{c_i d_i} - \sum_{k=1}^{l}\frac{\theta}{i\pi N}{}_k I^{ij}_k\prod_{i\neq j}\delta_{c_i d_i} \ ,$$

where

$$I^{ij}_k \equiv \left(I^{ij}_k\right)_{d_k, c_k} = \begin{cases} \delta^i_{d_k}\delta^j_{c_k} - \frac{1}{N}\delta^{ij}\delta_{c_k d_k} & (k^{th} \ \text{particle}), \\ -\delta^i_{c_k}\delta^j_{d_k} + \frac{1}{N}\delta^{ij}\delta_{c_k d_k} & (k^{th} \ \text{anti-particle}); \end{cases} \tag{8.100a, b}$$

we make the following ansatz for the exact S-matrix

$$\begin{aligned}\langle\theta'_1 c'_1 \theta'_2 c'_2|\theta_1 c_1, \theta_2 c_2\rangle = {} & \delta(\theta'_1 - \theta_1)\delta(\theta'_2 - \theta_2)\left[u_1(\theta)\delta^{c_1 c'_1}\delta^{c_2 c'_2} + u_2(\theta)\delta^{c_1 c'_2}\delta^{c_2 c'_1}\right] \\ & -\delta(\theta'_1 - \theta_2)\delta(\theta'_2 - \theta_1)\left[u_1(\theta)\delta^{c_1 c'_2}\delta^{c_2 c'_1} + u_2(\theta)\delta^{c_1 c'_1}\delta^{c_2 c'_2}\right] \ ,\end{aligned} \tag{8.101a}$$

$$\begin{aligned}\langle\theta'_1 c'_1 \overline{\theta}'_2 c'_2|\theta_1 c_1, \overline{\theta}_2 c_2\rangle = {} & \delta(\theta'_1 - \theta_1)\delta(\theta'_2 - \theta_2)\left[t_1(\theta)\delta^{c_1 c'_1}\delta^{c_2 c'_2} + t_2(\theta)\delta^{c_1 c_2}\delta^{c'_1 c'_2}\right] \\ & +\delta(\theta'_1 - \theta_2)\delta(\theta'_2 - \theta_1)\left[r_1(\theta)\delta^{c_1 c'_1}\delta^{c_2 c'_2} + r_2(\theta)\delta^{c_1 c_2}\delta^{c'_1 c'_2}\right] \ .\end{aligned} \tag{8.101b}$$

Conservation of the non-local charge leads to constraints on the above amplitudes

$$u_2(\theta) = -\frac{2\pi i}{N\theta}u_1(\theta) \ , \tag{8.102a}$$

$$r_1(\theta) = r_2(\theta) = 0 \ , \tag{8.102b}$$

$$t_2(\theta) = -\frac{2\pi i}{N(i\pi - \theta)}t_1(\theta) \ . \tag{8.102c}$$

Moreover, crossing implies

$$t_{1,2}(\theta) = u_{1,2}(i\pi - \theta) \ . \tag{8.102d}$$

This implies that the backward particle-antiparticle scattering vanishes, and the exact S-matrix belongs to class II of table 8.1[307,304,198]. The result can be checked up to first order in a $1/N$ expansion[198,199]. The vanishing of $r(\theta)$ is related to the fact that antiparticles are bound states of particles[34] as discussed in subsection 8.3.1. This result has been checked up to second order in g in perturbation theory.

8.3.2 S-matrices of non-linear sigma models

$O(N)$ invariant model

For the $O(N)$ invariant sigma model we still have a non anomalous non local charge,[62] and the ansatz (8.89) (with a relative plus sign due to boson statistics) is valid[301]. We obtain the same results as for the $O(N)$ invariant Gross-Neveu model, replacing $\frac{1}{N-2}$ by $\frac{-1}{N-2}$ in the minimal solution. The result is given by

$$\sigma_1(\theta) = -\frac{2i\pi}{(N-2)(i\pi-\theta)}\sigma_2(\theta) \tag{8.103a}$$

$$\sigma_3(\theta) = -\frac{2i\pi}{(N-2)\theta}\sigma_2(\theta) \tag{8.103b}$$

$$\sigma_2(\theta) = R(\theta)R(i\pi-\theta) \quad , \tag{8.103c}$$

where

$$R(\theta) = \frac{\Gamma\left(\frac{1}{N-2}-\frac{i\theta}{2\pi}\right)\Gamma\left(\frac{1}{2}-\frac{i\theta}{2\pi}\right)}{\Gamma\left(\frac{1}{2}-\frac{1}{N-2}-\frac{i\theta}{2\pi}\right)\Gamma\left(-\frac{i\theta}{2\pi}\right)} \quad . \tag{8.104}$$

This result shows no bound state pole,[308,309] and has been checked in a $1/N$ expansion up to second order[191]. The usual perturbative expansion can also be checked, as discussed before Eq. (8.98). The model can be coupled to fermions through supersymmetry, and as it turns out, the supersymmetric partner is the Gross Neveu model. The supersymmetric S-matrix is given in detail in reference [290].

$\mathbb{C}P^{N-1}$ models with minimally coupled fermions

For models with $SU(N)$ symmetry we have an anomaly (see discussion in chapter 7) and the purely bosonic models can not be solved exactly. The simplest way of cancelling the anomaly is to introduce minimally coupled fermions. For the $\mathbb{C}P^{N-1}$ model coupled to fermions, with the Lagrangian

$$\mathcal{L} = \overline{D_\mu z} D_\mu z + i\overline{\psi}(\not\partial - \not A)\psi \quad , \tag{8.105}$$

the conserved non local charge is given by the expression[77,78,274]

$$Q^{ij} = \int dy_1 dy_2 \epsilon(y_1-y_2) j_0^{ik}(t,y_1) j_0^{kj}(t,y_2) - Z\int dy j_1 + \int_{-\infty}^{+\infty} z^i \overline{\psi}\gamma_1\psi \overline{z}^j dy \quad , \tag{8.106a}$$

where

$$j_\mu^{ij} = z^i \overleftrightarrow{D}_\mu \overline{z}^j \quad . \tag{8.106b}$$

The conservation of the above charge leads to a number of constraints which suffice to determine the S-matrix[206]. The Ansatz for the $SU(N)$-symmetric S-matrix is given by (8.101), and conservation of the charge leads to (in the case of particle-particle scattering $u(\theta)$ may be obtained by crossing symmetry: $u_{1,2}(\theta) = t_{1,2}(i\pi - \theta)$):

$$
\begin{aligned}
\langle \theta_3 d_1 \theta_4 \bar{d}_2 | Q^{ij} | \theta_1 \bar{c}_1 \theta_2 c_2 \rangle &= \langle \theta_3 d_1 \theta_4 \bar{d}_2 | Q^{ij}_{in} | \theta_1 \bar{c}_1 \theta_2 c_2 \rangle \\
&= \Big(\frac{1}{N} \delta^i_{c'_1} \delta^k_{c_1} \delta^j_{c'_2} \delta^k_{c_2} - \frac{1}{N} \delta^i_{c_2} \delta^k_{c'_2} \delta^j_{c_1} \delta^k_{c'_1} + \frac{\theta_1}{i\pi} \delta^i_{c'_1} \delta^j_{c_1} \delta_{c'_2 c_2} + \frac{\theta - 2}{i\pi} \delta^j_{c'_2} \delta^i_{c_2} \delta_{c'_1 c_1} \Big) \\
&\qquad [(\delta^{c'_1 d_2} \delta^{d_1 c'_2} t_1(\theta) + \delta^{c'_1 c'_2} \delta^{d_1 d_2} t_2(\theta)) \delta(\theta_1 - \theta_3) \delta(\theta_2 - \theta_4) \\
&\quad + (\delta^{c'_1 d_2} \delta^{d_1 c'_2} r_1(\theta) + \delta^{c'_1 c'_2} \delta^{d_1 c_2} r_2(\theta)) \delta(\theta_1 - \theta_4) \delta(\theta_2 - \theta_3)] \quad , \quad (8.107a)
\end{aligned}
$$

for the right action, and

$$
\begin{aligned}
\langle \theta_3 d_1 \theta_4 \bar{d}_2 | Q^{ij} | \theta_1 \bar{c}_1 \theta_2 c_2 \rangle &= \langle \theta_3 d_1 \theta_4 \bar{d}_2 | Q^{ij}_{out} | \theta_1 \bar{c}_1 \theta_2 c_2 \rangle \\
&= \Big(-\frac{1}{N} \delta^i_{d_1} \delta^k_{d'_2} \delta^j_{d_2} \delta^k_{d'_1} + \frac{1}{N} \delta^k_{d_1} \delta^i_{d'_1} \delta^j_{d'_2} \delta^k_{d_2} + \frac{\theta_3}{i\pi} \delta^j_{d'_1} \delta^i_{d_1} \delta_{d_2 d'_2} + \frac{\theta_4}{i\pi} \delta^j_{d_2} \delta^i_{d'_2} \delta_{d_1 d'_1} \Big) \\
&\qquad \times [(\delta^{c_1 d'_2} \delta^{d'_1 c_2} t_1(\theta) + \delta^{c_1 c_2} \delta^{d'_1 d'_2} t_2(\theta)) \delta(\theta_1 - \theta_3) \delta(\theta_2 - \theta_4) + \\
&\qquad (\delta^{c_1 d'_2} \delta^{d'_1 c_2} r_1(\theta) + \delta^{c_1 c_2} \delta^{d'_1 d'_2} r_2(\theta)) \delta(\theta_1 - \theta_4) \delta(\theta_2 - \theta_3)] \quad , \quad (8.107b)
\end{aligned}
$$

for the left action. Equating both right hand sides we obtain

$$ r_1(\theta) = r_2(\theta) = 0 \, , \qquad (8.108a) $$

$$ t_2(\theta) = \frac{-2\pi i}{N(i\pi - \theta)} t_1(\theta) \quad , \qquad (8.108b) $$

with

$$ t_1(\theta) = \frac{\Gamma\left(\frac{1}{2} + \frac{\theta}{2\pi i}\right) \Gamma\left(\frac{1}{2} + \frac{1}{N} - \frac{\theta}{2\pi i}\right)}{\Gamma\left(\frac{1}{2} - \frac{\theta}{2\pi i}\right) \Gamma\left(\frac{1}{2} + \frac{1}{N} + \frac{\theta}{2\pi i}\right)} \quad , \qquad (8.108c) $$

which satisfies

$$ t_1(\theta) t_1(-\theta) = 1 \quad . \qquad (8.108d) $$

In the case of particle-particle scattering, $u(\theta)$ can be obtained from $t(\theta)$ by using crossing symmetry: $u(\theta) = t(i\pi - \theta)$.

The result (8.108) can again be checked using the $1/N$ expansion. There are no bound state poles in the above solution[206,274]. The bosons interact repulsively, and the fermions are screened by the "secret" long range force, while the gauge field looses the zero mass pole.

The same situation holds for the model defined on[310] $\frac{SO(N)}{SO(2) \times SO(N-2)}$, as well as for Grassmannian models with fermions defined on $\frac{SU(N)}{S(U(p) \times U(N-p))}$,[311] if we treat scattering of gauge invariant objects (see appendix J).

The supersymmetric $\mathbb{C}P^{N-1}$ *models*

We now consider the supersymmetric $\mathbb{C}P^{N-1}$ model described by the Lagrangian (6.264), with $p = 1$. We make the Ansatz (the backward particle-antiparticle scattering vanishes as a consequence of the conservation of the non local charges; we have already used this fact in the Ansatz[78])

$$\begin{aligned}\langle B_\beta(\theta_1')B_\delta(\theta_2')|B_\alpha(\theta_1)B_\gamma(\theta_2)\rangle &= [v_1(\theta)\delta_{\alpha\beta}\delta_{\gamma\delta} + v_2(\theta)\delta_{\alpha\delta}\delta_{\gamma\beta}]\delta(\theta_1 - \theta_1')\delta(\theta_2 - \theta_2') \\ &+[v_1(\theta)\delta_{\alpha\delta}\delta_{\gamma\beta} + v_2(\theta)\delta_{\alpha\beta}\delta_{\gamma\delta}]\delta(\theta_1 - \theta_2')\delta(\theta_2 - \theta_1')\ , \quad (8.109a)\end{aligned}$$

$$\begin{aligned}\langle F_\beta(\theta_1')F_\delta(\theta_2')|F_\alpha(\theta_1)F_\gamma(\theta_2)\rangle &= [u_1(\theta)\delta_{\alpha\beta}\delta_{\gamma\delta} + u_2(\theta)\delta_{\alpha\delta}\delta_{\gamma\beta}]\delta(\theta_1 - \theta_1')\delta(\theta_2 - \theta_2') \\ &-[u_1(\theta)\delta_{\alpha\delta}\delta_{\gamma\beta} + u_2(\theta)\delta_{\alpha\beta}\delta_{\gamma\delta}]\delta(\theta_1 - \theta_2')\delta(\theta_2 - \theta_1')\ , \quad (8.109b)\end{aligned}$$

$$\begin{aligned}\langle F_\beta(\theta_1')B_\delta(\theta_2')|F_\alpha(\theta_1)B_\gamma(\theta_2)\rangle &= [c_1(\theta)\delta_{\alpha\beta}\delta_{\gamma\delta} + c_2(\theta)\delta_{\alpha\delta}\delta_{\gamma\beta}]\delta(\theta_1 - \theta_1')\delta(\theta_2 - \theta_2') \\ &+[d_1(\theta)\delta_{\alpha\delta}\delta_{\gamma\beta} + d_2(\theta)\delta_{\alpha\beta}\delta_{\gamma\delta}]\delta(\theta_1 - \theta_2')\delta(\theta_2 - \theta_1')\ , \quad (8.109c)\end{aligned}$$

$$\begin{aligned}&\langle B_\beta(\theta_1')\overline{B}_\delta(\theta_2')|B_\alpha(\theta_1)\overline{B}_\gamma(\theta_2)\rangle = \\ &\qquad [v_1(i\pi - \theta)\delta_{\alpha\beta}\delta_{\gamma\delta} + v_2(i\pi - \theta)\delta_{\alpha\delta}\delta_{\gamma\beta}]\delta(\theta_1 - \theta_1')\delta(\theta_2 - \theta_2')\ , \quad (8.109d)\\ &\langle F_\beta(\theta_1')\overline{F}_\delta(\theta_2')|B_\alpha(\theta_1)\overline{F}_\gamma(\theta_2)\rangle = \\ &\qquad [u_1(i\pi - \theta)\delta_{\alpha\beta}\delta_{\gamma\delta} + u_2(i\pi - \theta)\delta_{\alpha\delta}\delta_{\gamma\beta}]\delta(\theta_1 - \theta_1')\delta(\theta_2 - \theta_2')\ , \quad (8.109e)\\ &\langle B_\beta(\theta_1')\overline{B}_\delta(\theta_2')|F_\alpha(\theta_1)\overline{F}_\gamma(\theta_2)\rangle = \\ &\qquad [d_1(i\pi - \theta)\delta_{\alpha\beta}\delta_{\gamma\delta} + d_2(i\pi - \theta)\delta_{\alpha\delta}\delta_{\gamma\beta}]\delta(\theta_1 - \theta_1')\delta(\theta_2 - \theta_2')\ , \quad (8.109.f)\\ &\langle B_\beta(\theta_1')\overline{F}_\delta(\theta_2')|B_\alpha(\theta_1)\overline{F}_\gamma(\theta_2)\rangle = \\ &\qquad [c_1(i\pi - \theta)\delta_{\alpha\beta}\delta_{\gamma\delta} + c_2(i\pi - \theta)\delta_{\alpha\delta}\delta_{\gamma\beta}]\delta(\theta_1 - \theta_1')\delta(\theta_2 - \theta_2')\ , \quad (8.109.g)\end{aligned}$$

Proceeding as before,one defines the action of the quantum non local charge on asymptotic states, and one arrives at the solution

$$v_1(\theta) = \frac{\sin\left(\frac{\theta}{2i} - \frac{\pi}{N}\right)}{\sin\frac{\theta}{2i}} c_1(\theta)\ , \qquad (8.110a)$$

$$u_1(\theta) = \frac{\sin\left(\frac{\theta}{2i} - \frac{\pi}{N}\right)}{\sin\frac{\theta}{2i}} c_1(\theta)\ , \qquad (8.110b)$$

$$d_1(\theta) = -\frac{\sin\left(\frac{\theta}{2i} - \frac{\pi}{N}\right)}{\sin\frac{\theta}{2i}} c_1(\theta)\ , \qquad (8.110c)$$

and

$$v_2(\theta) = \frac{i\pi}{N\theta} v_1(\theta) , \tag{8.111a}$$

$$u_2(\theta) = \frac{i\pi}{N\theta} u_1(\theta) , \tag{8.111b}$$

$$c_2(\theta) = -\frac{i\pi}{N\theta} c_1(\theta) , \tag{8.111c}$$

$$d_2(\theta) = -\frac{i\pi}{N\theta} d_1(\theta) . \tag{8.111d}$$

Notice that the particle-antiparticle backward scattering vanishes; this is a consequence of the bound state structure, as we have pointed out earlier.

Equations (8.111) are a consequence of the non local conservation laws, while (8.110) are a consequence of factorization of the amplitude for the scattering of the two bosons and one fermion, as well as crossing. Finally, we determine $c_1(\theta)$ from unitarity, which leads to

$$c_1(\theta) = \prod_{l=0}^{\infty} \frac{\Gamma\left(\frac{\theta}{2\pi i} + \frac{1}{N} + l\right)\Gamma\left(1 - \frac{\theta}{2\pi i} + l\right)\Gamma\left(\frac{\theta}{2\pi i} - \frac{1}{N} + l\right)\Gamma\left(1 - \frac{\theta}{2\pi i} + l\right)}{\Gamma\left(1 - \frac{\theta}{2\pi i} + \frac{1}{N} + l\right)\Gamma\left(\frac{\theta}{2\pi i} + l\right)\Gamma\left(1 - \frac{\theta}{2\pi i} - \frac{1}{N} + l\right)\Gamma\left(\frac{\theta}{2\pi i} + l\right)} \tag{8.112}$$

This shows the existence of bound states, with the spectrum given by the fusion rule by

$$m_l = m \frac{\sin \frac{l\pi}{N}}{\sin \frac{\pi}{N}} \quad , \quad l = 1, \cdots, N-1 \quad . \tag{8.113}$$

In this theory, a bound state of $N-1$ fermions is equivalent to an antiboson, i.e

$$\overline{z}_\delta = \frac{1}{(N-1)!} \epsilon_{\delta\alpha_1 \cdots \alpha_{N-1}} f_{\alpha_1} \cdots f_{\alpha_{N-1}} \quad , \tag{8.114a}$$

while the bound state of $N-2$ fermions and a boson is an antifermion

$$\overline{f}_\alpha = \frac{1}{(N-1)!} \epsilon_{\alpha\beta_1 \cdots \beta_{N-1}} f_{\beta_1} \cdots f_{\beta_{N-2}} z_{\beta_{N-1}} \quad . \tag{8.114b}$$

From these formulae it follows that the particle -antiparticle backward scattering amplitudes vanishes identically.

The $O(N)$ supersymmetric sigma model can be similarly handled.

The $\mathbb{C}P^1$ *model*

Another interesting case is that of the $\mathbb{C}P^1$ model, which turns out to be anomaly free. Indeed, the anomaly is given by the expression $z_i\bar{z}_j F^z_{\mu\nu}$, see (7.108); in the $\mathbb{C}P^1$ case, the decomposition (6.155) turns out to be $Z = \begin{pmatrix} z_1 \\ z_2 \end{pmatrix}$ and $Y = \begin{pmatrix} \bar{z}_2 \\ -\bar{z}_1 \end{pmatrix}$. Therefore we have the identity

$$z_i\bar{z}_j F^z_{\mu\nu} = Y_i\overline{Y}_j F^Y_{\mu\nu}$$

which are two versions of the anomaly (7.108). Using equation (6.94) we can show that the anomaly is equal to a total divergence. Therefore, since the model is anomaly free, there is a factorizable S-matrix; indeed, we can calculate the constraints imposed by the conservation of the nonlocal charge[312] using the Ansatz (8.101); we obtain equations (8.108a, b) with $N = 2$, fixing the minimal S matrix. We know that the model is equivalent to the $O(3)$ invariant σ model, if we make the identification

$$\varphi^i = z_\alpha(\tau^i)^{\alpha\beta}\bar{z}_\beta \quad . \tag{8.115a}$$

This identification must hold in terms of the bound state of the $\mathbb{C}P^1$ model[313,312], and can be verified in a numerical analysis[314]. The bound state Ansatz for the S-matrix is given by

$$\begin{aligned}
&\langle b_{c_1c_2}b_{d_1d_2} \mid b_{a_1a_2}b_{b_1b_2}\rangle = \\
&\quad \sigma_1\delta^{a_1b_2}\delta^{a_2b_1}\delta^{c_1d_2}\delta^{c_2d_1} + \sigma_2\delta^{a_1c_1}\delta^{a_2b_1}\delta^{b_2d_2}\delta^{c_2d_1} + \sigma_3\delta^{a_1b_2}\delta^{a_2c_2}\delta^{b_1d_1}\delta^{c_1d_2} \\
&\quad +\sigma_4\delta^{a_1d_1}\delta^{a_2b_1}\delta^{b_2d_2}\delta^{c_1d_2} + \sigma_5\delta^{a_1b_2}\delta^{a_2d_2}\delta^{b_1c_1}\delta^{c_2d_1} + \sigma_6\delta^{a_1d_1}\delta^{a_2d_2}\delta^{b_1d_1}\delta^{b_2d_2} \\
&\quad +\sigma_7\delta^{a_1d_1}\delta^{a_2d_2}\delta^{b_1d_1}\delta^{b_2d_2} + \sigma_8\delta^{a_1c_1}\delta^{a_2d_2}\delta^{b_1d_1}\delta^{b_2c_2} + \sigma_9\delta^{a_1d_1}\delta^{a_2c_2}\delta^{b_1c_1}\delta^{b_2d_2} \, .
\end{aligned} \tag{8.115b}$$

The bound states are defined by means of the introduction of a CDD pole term in $\theta_1 - \theta_2 = i\pi\alpha$, defining the bound state by the operator

$$b_{ab}\left(\frac{1}{2}(\theta_1+\theta_2)\right) = \frac{1}{2}\left(z_a(\theta_1)\bar{z}_b(\theta_2) - z_a(\theta_2)\bar{z}_b(\theta_1)\right) \tag{8.116}$$

The nonlocal charge constraints imply

$$\begin{aligned}
&\sigma_1 = \frac{\sigma_2}{\frac{\theta}{i\pi}+\alpha-1} \quad \sigma_2 = \frac{\sigma_6}{\frac{\theta}{i\pi}-\alpha-1} \quad \sigma_3 = \frac{\sigma_6}{\frac{\theta}{i\pi}+\alpha-1} \\
&\sigma_7 = \frac{\sigma_6}{\left[\frac{\theta}{i\pi}\right]^2} \quad \sigma_8 = \sigma_9 = -\frac{\sigma_6}{\frac{\theta}{i\pi}}\sigma_4 = \sigma_5 = 0 \, .
\end{aligned} \tag{8.117}$$

We rediscover the S-matrix of the $O(3)$ σ model, once we fix $\alpha = 1$, and contract the indices in (8.115b), according to (8.115a). Since one of the values of θ falls off the physical region, we interpret it as corresponding to confinement of the $\mathbb{C}P^1$ particles, while the $O(3)$ particles are seen asymptotically.

Further bosonic anomaly free non linear sigma models

We return to purely bosonic non linear sigma models. This time we deal with more complicated models involving matrix valued fields (principal sigma models), not manageable by means of conventional methods, such as $1/N$ expansion, since they require the summation over all planar diagrams[315]

In order to be free of anomalies, we have to consider symmetric spaces where the gauge subgroup is simple[62] (recall (7.108)) The symmetric spaces of type I with simple gauge groups are:[229]

$$\begin{aligned}
AI: &\quad M = SU(N)/SO(N),\\
AII: &\quad M = SU(2N)/Sp(N),\\
AIII: &\quad M = SU(2)/U(1) = CP^1,\\
BDI: &\quad M = SO(N)/SO(N-1) = S^{N-1},\\
EI: &\quad M = E_6/Sp(4),\\
EIV: &\quad M = E_6/F_4,\\
EV: &\quad M = E_7/SU(8),\\
EVIII: &\quad M = E_8/SO(16),\\
FII: &\quad M = F_4/SO(9),
\end{aligned} \tag{8.118a}$$

We have also the type II symmetric spaces, which are equivalent to group manifolds, thus having a trivial gauge group (principal sigma models).

We shall not discuss the cases involving exceptional groups, and restrict our discussion to the following symmetric spaces

$$\begin{aligned}
BDI &\quad \text{type}I &\quad M &= SO(N)/SO(N-1) = S^{N-1},\\
AIII &\quad \text{type}I &\quad M &= SU(2)/U(1) = CP^1,\\
A &\quad \text{type}II &\quad M &= SU(N),\\
BD &\quad \text{type}II &\quad M &= SO(N),\\
C &\quad \text{type}II &\quad M &= Sp(N),\\
AI &\quad \text{type}I &\quad M &= SU(N)/SO(N),\\
AII &\quad \text{type}I &\quad M &= SU(2N)/Sp(N) \quad .
\end{aligned} \tag{8.118b}$$

The first two cases have been considered previously, and we pass directly to the remaining ones[316,317,318,319,320]. In order to unify our treatment, we employ a gauge invariant formulation, based on the fact that in the above cases, the target manifold is a submanifold of a special unitary group to be defined by an appropriate constraint below. Consider the involutive automorphism of $SU(N)$ given by[319]

$$\sigma_I(g) = g^* \qquad for \quad g \in SU(N) \quad , \tag{8.119}$$

and the involutive automorphism of $SU(2N)$, given by

$$\sigma_{II}(g) = -Jg^*J \qquad for \quad g \in SU(2N) \quad , \tag{8.120}$$

where J is the $2N \times 2N$ matrix

$$J = \begin{pmatrix} 0 & 1_N \\ -1_N & 0 \end{pmatrix} \in SU(2N) \quad . \tag{8.121}$$

Both cases can be resumed in the formula

$$\sigma(g) = Ig^*I^{-1} \quad , \tag{8.122}$$

where

$$g \in SU(N) \quad , \quad I = 1_N \in SU(N) \tag{8.123a}$$

$$g \in SU(2N) \quad , \quad I = J \in SU(2N) \quad . \tag{8.123b}$$

The above involution is used to obtain the respective subspaces as constraints relating g and $\sigma(g)$. These constraints are

$$\sigma(g) = g \Leftrightarrow \begin{cases} g \in SO(N), & I = 1 \quad (\sigma = \sigma_I) \\ g \in Sp(N), & I = J \quad (\sigma = \sigma_{II}) \end{cases} \tag{8.124}$$

which define $SO(N)$ and $Sp(N)$ as subgroups of $SU(N)$ and $SU(2N)$ respectively.

Similarly, we define the cases $\frac{SU(N)}{SO(N)}$ and $\frac{SU(2N)}{Sp(N)}$ as follows

$$\sigma(g) = g^{-1} \Leftrightarrow \begin{cases} \sigma_I(g) = g^{-1} \Leftrightarrow g \in SU(N)/SO(N) \, , \\ \sigma_{II}(g) = g^{-1} \Leftrightarrow g \in SU(2N)/Sp(N) \, . \end{cases} \tag{8.125}$$

The advantage of using the matrix I is that it allows a uniform description of all cases. The formulae will contain a letter A when referring to the case A, BDC when referring to the cases BD or C, and I when referring to the cases AI and AII.

We always have the following constraint, defining either $SU(N)$ or $SU(2N)$ groups:

$$g^{-1} = g^\dagger, \tag{8.126a}$$

$$\det g = 1 \quad . \tag{8.126b}$$

Hence the constraints (8.124) and (8.125) read

$$g = \sigma(g) \equiv Ig^*I^\dagger \qquad \text{or} \qquad g^T = Ig^\dagger I^\dagger, \tag{8.127BDC}$$

$$g^\dagger = \sigma(g) \equiv Ig^*I^\dagger \qquad \text{or} \qquad g^T = IgI^\dagger. \tag{8.127I}$$

The models in question have the following symmetries

$$G = SU(N)_L \times SU(N)_R, \tag{8.128A}$$
$$G = SO(N)_L \times SO(N)_R, \tag{8.128BD}$$
$$G = Sp(N)_L \times Sp(N)_R, \tag{8.128C}$$
$$G = SU(N), \tag{8.128AI}$$
$$G = SU(2N), \tag{8.128AII}$$

which are explicitly summarized by

$$g \longrightarrow g_L g g_R^\dagger \quad , \tag{8.129ABDC}$$

for the first three cases (the so called chiral models) and by

$$g \longrightarrow \sigma(g_o) g g_o^\dagger = I g_o^* I^\dagger g g_o^\dagger \quad , \tag{8.129I}$$

for the type I cases, where we have only a simple symmetry group.

The symmetry (8.129I) is equivalent to

$$g \longrightarrow g_o^{-1} g g_o \quad , \tag{8.129$'I$}$$

where we have used (8.127I).

We also have discrete charge conjugation symmetry

$$g \longrightarrow I g^* I^\dagger \quad , \tag{8.130a}$$

and $g \to g^\dagger$, or equivalently

$$g \longrightarrow I g^\dagger I^\dagger \quad . \tag{8.130b}$$

We can associate with the above continuous symmetries the following Noether currents

$$J_\mu = \left(J_\mu^L, J_\mu^R\right) = \left(-\partial_\mu g g^\dagger, g^\dagger \partial_\mu g\right) \tag{8.131ABDC}$$
$$J_\mu = \frac{1}{2} g^\dagger \partial_\mu g - \frac{1}{2}\sigma\left(\partial_\mu g g^\dagger\right). \tag{8.131I}$$

Due to the constraints (8.127I) we may rewrite (8.131I) as

$$J_\mu = g^\dagger \partial_\mu g \quad . \tag{8.132I}$$

The nonlocal conserved charge is given by the usual expression (see 8.83), and transforms under Lorentz transformations as

$$\left[T, Q^{(2)}\right] = -\frac{N'}{2\pi} Q^{(1)} \quad , \tag{8.133}$$

where $Q^{(2)}$ is the non local charge, $Q^{(1)}$ the ordinary charge and N' has a value which depends on the case in question; one has, respectively

$$N' = N \qquad (8.134A)$$
$$N' = N - 2 \qquad (8.134BD)$$
$$N' = 2N + 2 \qquad (8.134C)$$
$$N' = N \qquad (8.134AI)$$
$$N' = 2N \quad . \qquad (8.134AII)$$

We write for the asymptotic fields [11)]

$$g_{\substack{in \\ out}} = \int d\mu(q)\{b_{\substack{in \\ out}} \exp(-iE(q)t + iqy) + d^*_{\substack{in \\ out}} \exp(+iE(q)t - iqy)\}, \qquad (8.135)$$

where

$$d\mu(q) = \frac{dq}{2\pi E(q)} = \frac{d\theta}{4\pi} \quad ,$$
$$E(q) = \sqrt{q^2 + m^2} = m\cosh\theta \quad ,$$
$$q = m\sinh\theta \quad .$$

with the commutation relations

$$\left[b^{ij}_{\substack{in \\ out}}(q_1), b^{*kl}_{\substack{in \\ out}}(q_2)\right] = \left[d^{ij}_{\substack{in \\ out}}(q_1), d^{*kl}_{\substack{in \\ out}}(q_2)\right] = 2\pi 2E r^{ij,kl}\delta(q_1 - q_2) \qquad (8.136a)$$
$$\left[b^{ij}_{\substack{in \\ out}}(q_1), d^{*kl}_{\substack{in \\ out}}(q_2)\right] = \left[d^{ij}_{\substack{in \\ out}}(q_1), b^{*kl}_{\substack{in \\ out}}(q_2)\right] = 2\pi 2E s^{ij,kl}\delta(q_1 - q_2) \qquad (8.136b)$$

where

$$r^{ij,kl} = \delta^{ik}\delta^{jl} \qquad s^{ij,kl} = 0 \qquad (8.137A)$$
$$r^{ij,kl} = \delta^{ik}\delta^{jl} \qquad s^{ij,kl} = I^{ik}I^{jl} \qquad (8.137BDC)$$
$$r^{ij,kl} = \delta^{ik}\delta^{jl} + I^{il}I^{kj} \qquad s^{ij,kl} = 0 \quad . \qquad (8.137I)$$

Being free fields, $g_{\substack{in \\ out}}$ cannot satisfy the nonlinear constraints (126) as operator conditions, which therefore have to be imposed as conditions on the states[321]. They do however obey the constraints (8.127), which lead to the following identifications between creation (and annihilation) operators:

$$d^{(*)}_{\substack{in \\ out}}(q)I = Ib^{(*)}_{\substack{in \\ out}}(q)$$
$$b^{(*)\,T}_{\substack{in \\ out}}(q)I = Ib^{(*)}_{\substack{in \\ out}}(q) \quad , \quad d^{(*)\,T}_{\substack{in \\ out}}(q)I = Id^{(*)}_{\substack{in \\ out}}(q) \quad . \qquad (8.138)$$

---

11) We use the asymptotic expansion as a working rule to obtain the action of the non local charge on asymptotic states.

Moreover, the asymptotic fields must obey the same type of global symmetries as the interacting field G. Under charge conjugation we have

$$b^{(*)}_{\substack{in\\out}}(q) \to I d^{(*)}_{\substack{in\\out}}(q) I^\dagger \quad , \quad b^{(*)}_{\substack{in\\out}}(q) \to I b^{(*)}_{\substack{in\\out}}(q) I^\dagger \quad . \tag{8.139}$$

We have also the following discrete symmetry

$$b^{(*)}_{\substack{in\\out}}(q) \to I b^{(*)}_{\substack{in\\out}}{}^{T}(q) I^\dagger \quad , \quad d^{(*)}_{\substack{in\\out}}(q) \to I d^{(*)}_{\substack{in\\out}}{}^{T}(q) I^\dagger \quad , \tag{8.140a}$$

or, in terms of states

$$|\,\theta\,, kl\rangle \to \pm\,|\,\theta\,, lk\rangle \quad . \tag{8.140b}$$

As it turns out, the one particle states transform under the following irreducible representations of G:

$$\begin{array}{rlll} N_L^* \times N_R & \text{of} & SU(N)_L \times SU(N)_R & (\text{A} - \text{series}) \\ N_L \times N_R & \text{of} & SO(N)_L \times SO(N)_R & (\text{BD} - \text{series}) \\ N_L \times N_R & \text{of} & Sp(N)_L \times Sp(N)_R & (\text{C} - \text{series}) \\ N \times_s N & \text{of} & SU(N) & (\text{AI} - \text{series}) \\ 2N_L \times_a 2N & \text{of} & SU(2N) & (\text{AII} - \text{series}) \end{array} \tag{8.141}$$

where the subscripts $s(a)$ indicate symmetrization (antisymmetrization) of the tensor product.

We can now write the Ansätze for the S-matrices, impose non local charge conservation, find the minimal solutions and obtain the pole structure. We shall illustrate this in the $SU(N)$ case, and leave the other cases to appendix J. The Ansatz for the $SU(N) \times SU(N)$ invariant solution is

$$\begin{aligned}
&{}_{out}\langle \theta_1' k_1' l_1', \theta_2' k_2' l_2' | \theta_1 k_1 l_1, \theta_2 k_2 l_2 \rangle_{in} \\
&= 16\pi^2 \delta(\theta_1' - \theta_1)\delta(\theta_2' - \theta_2) \times \{ u_1(\theta) \delta^{k_1' k_1} \delta^{k_2' k_2} \delta^{l_1' l_1} \delta^{l_2' l_2} + u_2(\theta) \delta^{k_1' k_2} \delta^{k_2' k_1} \delta^{l_1' l_1} \delta^{l_2' l_2} \\
&\qquad + u_3(\theta) \delta^{k_1' k_1} \delta^{k_2 k_2'} \delta^{l_1' l_2} \delta^{l_2' l_1} + u_4(\theta) \delta^{k_1' k_2} \delta^{k_1 k_2'} \delta^{l_1' l_2} \delta^{l_2' l_1} \} \\
&+ 16\pi^2 \delta(\theta_1' - \theta_2)\delta(\theta_2' - \theta_1) \times \{ u_4(\theta) \delta^{k_1' k_1} \delta^{k_2' k_2} \delta^{l_1' l_1} \delta^{l_2' l_2} + u_3(\theta) \delta^{k_1' k_2} \delta^{k_2' k_1} \delta^{l_1' l_1} \delta^{l_2' l_2} \\
&\qquad + u_2(\theta) \delta^{k_1' k_1} \delta^{k_2 k_2'} \delta^{l_1' l_2} \delta^{l_2' l_1} + u_1(\theta) \delta^{k_1' k_2} \delta^{k_1 k_2'} \delta^{l_1' l_2} \delta^{l_2' l_1} \} \quad ,
\end{aligned} \tag{8.142a}$$

$$\begin{aligned}
&{}_{out}\langle \overline{\theta_1' k_1' l_1'}, \theta_2' k_2' l_2' | \overline{\theta_1 k_1 l_1}, \theta_2 k_2 l_2 \rangle_{in} \\
&= 16\pi^2 \delta(\theta_1' - \theta_1)\delta(\theta_2' - \theta_2) \times \{ t_1(\theta) \delta^{k_1' k_1} \delta^{k_2' k_2} \delta^{l_1' l_1} \delta^{l_2' l_2} + t_2(\theta) \delta^{k_1' k_2} \delta^{k_2' k_1} \delta^{l_1' l_1} \delta^{l_2' l_2} \\
&\qquad + t_3(\theta) \delta^{k_1' k_1} \delta^{k_2 k_2'} \delta^{l_1' l_2} \delta^{l_2' l_1} + t_4(\theta) \delta^{k_1' k_2} \delta^{k_1 k_2'} \delta^{l_1' l_2} \delta^{l_2' l_1} \} \\
&+ 16\pi^2 \delta(\theta_1' - \theta_2)\delta(\theta_2' - \theta_1) \times \{ r_4(\theta) \delta^{k_1' k_1} \delta^{k_2' k_2} \delta^{l_1' l_1} \delta^{l_2' l_2} + r_3(\theta) \delta^{k_1' k_2} \delta^{k_2' k_1} \delta^{l_1' l_1} \delta^{l_2' l_2} \\
&\qquad + r_2(\theta) \delta^{k_1' k_1} \delta^{k_2 k_2'} \delta^{l_1' l_2} \delta^{l_2' l_1} + r_1(\theta) \delta^{k_1' k_2} \delta^{k_1 k_2'} \delta^{l_1' l_2} \delta^{l_2' l_1} \} \, .
\end{aligned} \tag{8.142b}$$

We have to impose conservation of the non local charge. Since there are actually two possible currents (see Eq. (8.131)) we have two non local charges. The computation is now a straightforward, though extremely tedious, exercise. We obtain

$$u_2(\theta) = u_3(\theta) = -\frac{2\pi i}{N\theta} u_1(\theta) \tag{8.144a}$$

$$u_4(\theta) = -\frac{4\pi^2}{N^2\theta^2} u_1(\theta) \tag{8.144b}$$

$$t_i(\theta) = u_i(i\pi - \theta) \tag{8.144c}$$

$$r_i(\theta) = 0 \quad . \tag{8.144d}$$

Unitarity implies

$$u_1(\theta)u_1(-\theta) = \left(\frac{\theta^2}{\theta^2 + (\frac{2\pi}{N})^2}\right)^2 \quad ; \tag{8.145}$$

the minimal solution of (8.145) reads

$$u_1^{min}(\theta) = \left(\frac{\Gamma(1 - \frac{\theta}{2\pi i})\Gamma(\frac{1}{N} + \frac{\theta}{2\pi i})}{\Gamma(\frac{\theta}{2\pi i})\Gamma(1 + \frac{1}{N} - \frac{\theta}{2\pi i})}\right)^2 \quad . \tag{8.146}$$

It remains to determine the CDD pole. This pole can be chosen in such a way that the double poles and double zeros of the minimal solution cancel. This is achieved by the choice

$$P(\theta) = \frac{\sinh \frac{1}{2}(\theta + \frac{2\pi i}{N})}{\sinh \frac{1}{2}(\theta - \frac{2\pi i}{N})} \quad . \tag{8.147}$$

We can interpret the requirement $\det g = 1$ as stating that antiparticles can be obtained as bound states of particles, suggesting the existence of a spectrum of bound states. One finds (8.113).

The solution obtained has also been derived using a nested Bethe Ansatz[317,320].

8.4 Further Developments and Conclusion

Quantum field theoretic problems exhibiting an infinite number of conservation laws can be exactly solved by either using explicitly the conservation laws as done in this chapter, or by the Bethe Ansatz[322] and quantum inverse scattering methods[323].

The basic elements underlying the concept of integrability are the quantum groups or the Yang-Baxter, Zamolodchikov, Faddeev algebras[324].

There is a rather large class of models which can be dealt with using these methods; for a given Lie algebra, we have the class of bosonic models given by the Lagrangian

$$L = \frac{1}{2}(\partial_\mu \vec{\varphi})^2 - m^2 \sum_{\vec{\rho}_j \in \Delta} e^{g\vec{\rho}_j \cdot \vec{\varphi}} \quad , \tag{8.148}$$

where $\vec{\varphi} = (\varphi_1, \cdots, \varphi_n)$, n is the rank of the algebra, and Δ is the set of roots[325]; such models include the sine-Gordon and Liouville theories[324].

Further integrable theories are the complex sine-Gordon[216], the anisotropic O(3) sigma model[326] and anisotropic $SU(2)$ self interaction fermionic model[327], the Bukhvostov-Lipatov model[328], and several integrable models not treated in this book.

It is important to notice that we can associate an statistical mechanics model to a given integrable field theory. To a given vertex, defined by joining together four links we define a statistical weight $P(ij|kl)$, where $ijkl$ are the four directions, such that the partition function is the sum over all configurations, of the product of $P's$, for all vertices

$$Z = \sum_{\text{configs.}} \prod_{\text{vertices}} P(ij|kl)(\theta) \quad , \tag{8.149}$$

where θ is a spectral parameter. If we restrict the above sum to a simple line, we obtain a monodromy matrix[329]

$$T_{kl}(\theta) = \sum_{j_1 \cdots j_{n-1}=1} P^{(1)}_{kj_1}(\theta) \otimes P^{(2)}_{j_1 j_2}(\theta) \cdots P^{(n)}_{j_{n-1} l}(\theta) \quad . \tag{8.150}$$

The trace of T is the transfer matrix. These concepts prove useful in statistical mechanics, and are part of the usual language in that field. The relation concerning integrability in statistical mechanics to that in quantum field theory, is described assuming the existence of a matrix R, acting on the elements $t(\theta) \otimes t(\theta)$ of the tensor product, such that

$$R^{kl}_{ij}(\theta, \theta')F_{kn}(\theta)F_{lm}(\theta') = F_{ik}(\theta')F_{jl}(\theta)R^{ij}_{kl}(\theta, \theta') \quad , \tag{8.151}$$

which implies the existence of a one parameter family of transfer matrices[330].

If one considers the above algebra in the tensor product $T \otimes T \times T$, the equivalence of different paths used to invert the order of factors (or associativity of the algebra) implies a trilinear condition of the type (8.11) for the R-matrix. The operator $T(\theta)$ is called the monodromy operator. It can be computed in terms of the two body physical S-matrix as

$$\langle \theta\alpha | T_{ab}(\lambda) | \theta'\beta \rangle = \delta(\theta - \theta') S_{a\alpha, b\beta}(\theta + \gamma(\lambda)) \quad , \tag{8.152}$$

up to a phase factor, where $\gamma(\lambda)$ is an odd function of λ[39].

From the above expression and the factorization property, it is possible to obtain general matrix elements of $T(\lambda)$ in terms of two body S-matrix elements.

In fact $T(\lambda)$ can be directly related to the associated linear matrix system defining the integrability of the model, as

$$(\partial_\mu A_\mu^{(\lambda)})\phi = 0 \quad , \tag{8.153}$$

where

$$A_\mu^{(\lambda)} = \frac{\lambda}{\lambda^2 - 1}(\lambda j_\mu - \epsilon_{\mu\nu} j^\nu) \quad . \tag{8.154}$$

We have

$$\lim_{x^1 \to \infty} \phi = 1 \quad , \tag{8.155}$$

while

$$\lim_{x^1 \to -\infty} \phi = T \quad , \tag{8.156}$$

which generates, when expanded in a convenient series in λ, the non local conservation laws of the theory.

In this chapter we presented an extremely refined development of quantum field theory. The higher conservation laws obtained in chapters 3, 5 and 6, and quantized versions in chapters 3, 5 and 7 have been used to enforce constraints on the S-matrix elements. The constraints are two fold. First the very existence of higher conservation laws implies that the S-matrix does not admit particle production, and hence is of the factorizable type. The S-matrix is thereby obtained for each model, its actual form being in principle dependent on the model, and also on the particular type of conservation law. In fact, this dependence is superficial, since the requirement of factorizability, together with the use of the particular symmetry involved in the problem, fixes the S-matrix up to an overall factor (which depends on the bound state structure).

Further known S-matrices for several non-linear sigma models have been gathered in appendix J as well.

New integrable models have recently been considered by several authors. These models are obtained from perturbations of conformally invariant field theories by suitably chosen, conformal operators. The S matrices of the theories thus obtained can be computed by the methods presented in this chapter.

Chapter IX

THE WESS-ZUMINO-WITTEN THEORY

9.1 Introduction

In chapters 2 and 3 we have discussed the representation of fermions in terms of bosons and vice-versa. The $U(1)$ symmetry inherent to the Dirac fermions was further generalized in chapter 5, to obtain an operator description of the chiral Gross Neveu model with $SU(N)\otimes\widetilde{U}(1)\otimes U(1)$ symmetry. However, this description does not preserve the structure of the symmetry group. Indeed, let us use the Mandelstam formula (3.73) for the fundamental representation of a given symmetry group G, which for definiteness we take to be $U(N)$,

$$\psi_i(x) = \left(\frac{\mu}{2\pi}\right)^{1/2} K_i e^{-\frac{i\pi}{4}\gamma_5} e^{i\sqrt{\pi}[\gamma_5\varphi_i + \int_{x^1}^{\infty} dy^1 \dot{\varphi}_i(x^0,y^1)]} \begin{pmatrix}1\\1\end{pmatrix} \quad , \tag{9.1}$$

where K_i is a "Klein factor" insuring anticommutativity of the fields for different flavors (see section 3.3 of chapter 5). Here φ_i , $i = 1,\cdots,N$ are taken to be free zero mass fields, so that the corresponding $U(N)$ invariant action is that of N free massless fermions,

$$S = \int d^2x i\overline{\psi}_i \; \partial\!\!\!/ \psi_i \quad . \tag{9.2}$$

Using (9.1) one associates with (9.2) the $O(N)$ invariant bosonic action,

$$S = \int d^2x \frac{1}{2}\partial^\mu\varphi_i\partial_\mu\varphi_i \quad . \tag{9.3}$$

However, the $O(N)$ symmetry of this action does not correspond to a subgroup of the above $U(N)$, since the mapping (9.1) is exponential and non local. Also the $U(N)$ currents are non local when written in terms of φ_i. Therefore, the bosonization technique does not preserve the internal symmetry group in a simple way. On the other hand, setting

$$g(x) = e^{2i\sqrt{\pi}\varphi} \tag{9.4}$$

we know that in the case of a $U(1)$ symmetry, g transforms in the same way as ψ, and the equivalent bosonic action can be written as

$$S = \int d^2x \frac{1}{8\pi} \partial^\mu g^{-1} \partial_\mu g \quad . \tag{9.5}$$

A tentative non abelian generalization of this action consists in replacing (9.5) by the action

$$S = \int d^2x \frac{1}{2\lambda} \operatorname{tr} \partial^\mu g^{-1} \partial_\mu g \quad , \tag{9.6}$$

where g_{ij} are now matrix valued fields, and where we included a coupling constant λ. However, this is no longer a free field theory; the field equations are equivalent to

$$\partial^\mu J_\mu \equiv \partial^\mu (g^{-1} \partial_\mu g) = 0 \quad , \tag{9.7}$$

and equation (9.6) is the action of the principal sigma model previously studied in chapters 6 and 7. It defines an asymptotically free theory with dynamical mass generation; therefore it cannot be equivalent to a theory of free fermions.

Before entering into details about the solution of this problem, we wish to display it in a more general setting[331]. The idea of replacing fermionic degrees of freedom by bosonic ones, plays a fundamental role in the search for effective actions, providing a phenomenological description of the interaction of fermions via a confining potential[332]. In this spirit, the idea of an effective action can be introduced. In $d = 4$ space time dimensions one makes the Ansatz

$$S = S_{P\sigma M} + S_m + S' \tag{9.8}$$

with the chiral action given by

$$S_{P\sigma M} = \frac{f_\pi^2}{8} \int d^2x tr\, \partial^\mu g^{-1} \partial_\mu g \tag{9.9}$$

and the mass term

$$S_m = \frac{f_\pi^2}{8} \int d^2x m^2 tr\, (g + g^{-1}) \tag{9.10}$$

The chiral action describes a sigma model type interaction of the pions, $\pi^a(x)$ which are introduced via the identification,

$$g(x) = e^{-\frac{2i}{f_\pi} \sum T^a \pi^a(x)} \quad . \tag{9.11}$$

The term S_m is a mass term for the pions, and S' an effective self interaction[52], which must be derived from the fundamental theory. The chiral term (as well as the mass term) has a symmetry $g \longleftrightarrow g^{-1}$, which in terms of the pions reads $\pi \longleftrightarrow -\pi$; thus it counts the number of pions modulo 2, which is not a symmetry of the (fundamental) underlying non abelian gauge theory[331]. Thus we need the extra term S' to break this symmetry. The "cure" of this problem gives a clue to the type of term which must be added to the chiral action, without the need of an explicit computation. Indeed, the pions are pseudoscalars in QCD. The parity operation corresponds to $\vec{x} \longrightarrow -\vec{x}$, and $g \longrightarrow g^{-1}$; since this should be a symmetry of the theory, the additional term needed in the equation of motion (9.7) must be parity odd; therefore it must contain the Levi Civita tensor. This suggests the generalization[331]

$$\frac{f_\pi^2}{8}\partial_\mu(g^{-1}\partial^\mu g) + \lambda\epsilon^{\mu\nu\rho\sigma}g^{-1}\partial_\mu g g^{-1}\partial_\nu g g^{-1}\partial_\rho g g^{-1}\partial_\sigma g = 0 \quad , \tag{9.12a}$$

where λ is a coupling constant. However, the second term cannot be derived from a local Lagrangian. Nevertheless, drawing an analogy with a magnetic interaction, Witten[331] was able to obtain the corresponding action as given by

$$S' = nS_{WZ} = -\frac{in}{240\pi^2}\int_0^1 dr \int d^4x\epsilon^{ijklm} g_r^{-1}\partial_i g_r g_r^{-1}\partial_j g_r g_r^{-1}\partial_k g_r g_r^{-1}\partial_l g_r g_r^{-1}\partial_m g_r \; , \tag{9.12b}$$

where $g_r(x)$ is a matrix valued field interpolating between $g(x)$ and the identity:

$$g_0(x) = 1 \quad , \quad g_1(x) = g(x) \quad . \tag{9.13}$$

One refers to S_{WZ} as the Wess-Zumino term (action)[52]. It can be shown to be independent of the choice of (smooth) interpolation.

The parameter n in (9.12b) above must be an integer[51], since S_{WZ} is defined up to a multiple of 2π.

The fact that a bosonic action can describe baryons was foreseen long ago in the papers of Skyrme[44] and Finkelstein and Rubinstein[108] It has been proven that the baryon number of the solution of (9.12a) is non zero[333], and that the action (9.8) describes fermions for odd n[334]. Moreover, Lagrangians similar to (9.8) provide a rather reasonable description of low energy hadron physics[334,335].

In two dimensional space-time, we aim at an equivalent bosonic action for massless fermions, transforming under a non abelian symmetry group. Following the ideas outlined above, one is led to consider the two dimensional zero mass analogue of (9.8) above,

$$S_{WZW} = S_{P\sigma M} + S' \quad , \tag{9.14}$$

where

$$S' = nS_{WZ} = \frac{n}{4\pi}\int_0^1 dr \int d^2x\epsilon^{\mu\nu} tr g_r^{-1}\partial_r g_r g_r^{-1}\partial_\mu g_r g_r^{-1}\partial_\nu g_r \tag{9.15}$$

and

$$S_{P\sigma M} = \frac{1}{8\pi} \int d^2x tr\, \partial^\mu g^{-1} \partial_\mu g \quad . \tag{9.16}$$

We shall refer to S_{WZ}, $S_{P\sigma M}$ and S_{WZW} as the "Wess-Zumino" (WZ), "principal sigma model" ($P\sigma M$), and "Wess-Zumino-Witten" (WZW) action, respectively[51]. They will play an important role in the chapters to follow.

9.2 Existence of a Critical Point

Consider the one parameter family of actions

$$S_{WZNW} = \frac{4\pi}{\lambda^2} S_{P\sigma M} + nS_{WZ} \quad . \tag{9.17}$$

We shall refer to them as the "Wess-Zumino-Novikov-Witten" (WZNW)[52,51,336] action. As we now show, for the choice $\frac{4\pi}{\lambda^2} = n$, S_{WZNW} is the equivalent bosonic action of the (conformally invariant) theory of N non-interacting massless fermions[1)]. To demonstrate this we examine the equations of motions and one-loop β-function following from (9.17).

The equations of motion are obtained by computing the functional variation of S_{WZNW} with respect to g_{ij}. From (9.16) we obtain two alternative forms for $\delta S_{P\sigma M}$:

$$\begin{aligned} \delta S_{P\sigma M} &= \frac{1}{4\pi} \int d^2x\, tr g^{-1} \delta g \partial_\mu (g^{-1} \partial^\mu g) \quad , \\ &= -\frac{1}{4\pi} \int d^2x\, tr \delta g g^{-1} \partial_\mu (g \partial^\mu g^{-1}) \quad . \end{aligned} \tag{9.18}$$

In the case of the principal sigma model ($P\sigma M$) this would lead to the equations of motion $\partial_\mu(g^{-1}\partial^\mu g) = 0$ and $\partial_\mu(g\partial^\mu g^{-1}) = 0$, which are in fact not independent. In the case in question one also needs to compute the variation δS_{WZ}. To this end we note that,[50]

$$\delta(g^{-1}\partial_\mu g) = \nabla_\mu (g^{-1}\delta g) \quad , \tag{9.19}$$

where, ∇_μ acts as follows on Lie-algebra valued functions:

$$\nabla_\mu f = \partial_\mu f + [g^{-1}\partial_\mu g, f] \quad . \tag{9.20}$$

---

1) In the case of a $U(N)$ symmetry, as contemplated here, the fermions are Dirac-fermions. In the case of a $O(N)$ symmetry, as considered in ref. [51], one has to deal with Majorana fermions; the respective actions differ by an overall factor of 1/2.

It has the following useful properties

$$\int d^2x \; \mathrm{tr}[(\nabla_\mu f)h] = -\int d^2x \; \mathrm{tr}[f\nabla_\mu h] \quad , \tag{9.21}$$

$$\nabla^\mu(g^{-1}\partial_\mu g) = \partial^\mu(g^{-1}\partial_\mu g) \quad . \tag{9.22}$$

Making use of these properties, one finds after some calculation[50,51] the following two equivalent results for the variation δS_{WZ}:

$$\delta S_{WZ}[g] = -\frac{1}{4\pi}\int_0^1 dr \int d^2x \frac{\partial}{\partial r}\epsilon^{\mu\nu}\, \mathrm{tr}[g_r^{-1}\delta g_r \partial_\mu(g_r^{-1}\partial_\nu g_r)] \quad , \tag{9.23a}$$

$$= -\frac{1}{4\pi}\int_0^1 dr \int d^2x \frac{\partial}{\partial r}\epsilon^{\mu\nu}\, \mathrm{tr}[\delta g_r g_r^{-1}\partial_\mu(g_r\partial_\nu g_r^{-1})] \quad . \tag{9.23b}$$

Recalling the "boundary conditions" (9.13), one thus finds upon combining (9.23), (9.18) and (9.17)

$$\delta S_{WZNW} = \int d^2x \, \mathrm{tr}\, g^{-1}\delta g \left(\frac{1}{\lambda^2}g^{\mu\nu} - \frac{n}{4\pi}\epsilon^{\mu\nu}\right)\partial_\mu(g^{-1}\partial_\nu g) \quad , \tag{9.24a}$$

$$= \int d^2x tr \delta g g^{-1}\left(-\frac{1}{\lambda^2}g^{\mu\nu} - \frac{n}{4\pi}\epsilon^{\mu\nu}\right)\partial_\mu\left(g\partial_\nu g^{-1}\right) \quad . \tag{9.24b}$$

Setting $\delta S_{WZNW} = 0$, we obtain as equations of motion the conservation laws,

$$\partial_\mu\left(\frac{1}{\lambda^2}g^{\mu\nu} - \frac{n}{4\pi}\epsilon^{\mu\nu}\right)(g^{-1}\partial_\nu g) = 0 \tag{9.25a}$$

$$\partial_\mu\left(\frac{1}{\lambda^2}g^{\mu\nu} + \frac{n}{4\pi}\epsilon^{\mu\nu}\right)(g\partial_\nu g^{-1}) = 0 \quad . \tag{9.25b}$$

For the choice

$$\frac{4\pi}{\lambda^2} = n \quad , \tag{9.26}$$

Eqs. (9.25) just express the conservation of the left (L) and right (R) moving currents

$$j_L^\mu(x) = -i\frac{n}{4\pi}(g^{\mu\nu} - \epsilon^{\mu\nu})(g^{-1}\partial_\nu g) \quad , \tag{9.27a}$$

$$j_R^\mu(x) = -i\frac{n}{4\pi}(g^{\mu\nu} + \epsilon^{\mu\nu})(g\partial_\nu g^{-1}) \quad . \tag{9.27b}$$

This indicates that at the critical point (9.26) the WZNW action is a *candidate* for describing N non-interacting massless fermions.

One-loop β-function

In order to understand the role of the special value (9.26) of λ, we compute the coupling constant renormalization counterterm. To this end we write

$$g(x) = g_0(x)e^{i\lambda\xi(x)} \quad ,$$

where $g_0(x)$ is a classical background, minimizing the action (9.17), and $\xi(x)$ the quantum field. The parameter λ is the same as in (9.17), and has been extracted for convenience. The one loop contributions are obtained by including only terms up to second order in the quantum fields $\xi(x)$. One finds after some algebra

$$S_{WZNW} \approx S_{WZNW}[g_0] + \frac{1}{2}\int d^2x tr g_0^{-1}\partial_\mu g_0 \xi \overleftrightarrow{\partial_\nu} \xi \left(g^{\mu\nu} + \frac{n\lambda^2}{4\pi}\varepsilon^{\mu\nu}\right) + \int d^2x tr \partial_\mu \xi \partial^\mu \xi \; . \quad (9.28)$$

Note that the topological term gives a local contribution (the second term in (9.28))!

The ξ-field propagator is given by $-iD_{ij,kl}(p) = \frac{i\delta_{ik}\delta_{jl}}{p^2}$. Noting that $g_0^{-1}\partial_\mu g_0$ plays the role of an external field, the one-loop effective action is obtained by adding to $S_{WZNW}[g_0]$ the Pauli-Villars regularized contribution of a diagram similar to the one displayed in figure 7.9, with the background fields as external legs, which may be written as

$$tr g_0^{-1}\partial_\mu g_0 \, g_0^{-1}\partial_\nu g_0 \Pi^{\mu\nu}_{reg}(x-y) \quad , \quad (9.29)$$

where $\Pi^{\mu\nu}_{reg}(x)$ is the Fourier transform of $\widetilde{\Pi}^{\mu\nu}_{reg}(p)$ given by

$$\widetilde{\Pi}^{\mu\nu}_{reg}(p) = \frac{-c_V}{4}\int_{\Lambda^2}^{m^2} dm^2 \frac{\partial}{\partial m^2}\int \frac{d^2k}{(2\pi)^2}\frac{(p+2k)_\rho(p+2k)_\sigma}{[(p+k)^2-m^2](k^2-m^2)} \times \left(g^{\mu\rho} + \frac{n\lambda^2\epsilon^{\mu\rho}}{4\pi}\right)\left(g^{\nu\sigma} + \frac{n\lambda^2\epsilon^{\nu\sigma}}{4\pi}\right) \quad , \quad (9.30)$$

where Λ^2 is the Pauli Villars regulator, m^2 is an IR cutoff and c_V is a Casimir eigenvalue associated with the symmetry group in question. For $U(N)$, $c_V = N$, which we take to be the case from now on. One thus obtains for the divergent piece of i times the action

$$-i\, tr\partial^\mu g_0^{-1}\partial_\mu g_0 \frac{N}{8\pi}\left(1 - \left(\frac{n\lambda^2}{4\pi}\right)^2\right)\ln\frac{\Lambda^2}{m^2} \quad . \quad (9.31)$$

Thus, the coupling constant must be renormalized as

$$\frac{1}{\lambda_R^2} = \frac{1}{\lambda^2} - \frac{N}{4\pi}\left(1 - \left(\frac{n\lambda^2}{4\pi}\right)^2\right) \ln \frac{\Lambda^2}{m^2} \quad . \tag{9.32}$$

The corresponding β-function is given by

$$\beta(\lambda_R, n) \equiv \Lambda^2 \left(\frac{\partial \lambda^2}{\partial \Lambda^2}\right)_{\Lambda \to \infty} = -\frac{N}{4\pi}\lambda^4 \left[1 - \left(\frac{n\lambda^2}{4\pi}\right)^2\right] \quad , \tag{9.33}$$

which has a non trivial zero at $\lambda_R^2 = 4\pi/n$ (see 9.26).

9.3 Properties at the Critical Point

9.3.1 The Kac Moody algebra

From the above discussion, we expect the theory to have a conformally invariant fixed point. However, it is clear that the above computation does not prove that the fixed point is exactly at the value (9.26). In order to further investigate the properties of the WZNW action (at the critical point), we proceed with a canonical quantization of the theory.

The WZ action (9.15) only depends linearly on the time derivative of g. Hence it proves useful to write it in the form[337]

$$S'[g] = \frac{n}{4\pi} \int d^2x \, tr A(g) \partial_0 g \quad , \tag{9.34}$$

where we have formally integrated over r, and $A(g)$ is a matrix valued function of g and $\partial_1 g$. We thus obtain for the momentum conjugate to g_{ij}

$$\Pi_{ij} = \frac{n}{4\pi}\partial_0 g_{ji}^{-1} + \frac{n}{4\pi}A_{ji} \equiv \hat{\Pi}_{ij} + \frac{n}{4\pi}A_{ji} \quad . \tag{9.35}$$

Hence there are no constraints.

Taking into account that $A(g)$ is a function of g and its space derivatives (but not of the time derivative of g), we obtain, from the usual Poisson bracket formulation[1)]

$$\left\{\hat{\Pi}_{ij}(x), \hat{\Pi}_{kl}(y)\right\} = -\frac{n}{4\pi} F_{ji;lk}(x) \delta(x^1 - y^1) \tag{9.36a}$$

$$\left\{g_{ij}(x), \hat{\Pi}_{kl}(x)\right\} = \delta_{ik}\delta_{jl}\delta(x^1 - y^1) \quad . \tag{9.36b}$$

[1)] Poisson brackets are always understood to be computed at equal times.

where[337]

$$F_{ij;kl} = \frac{\partial A_{ij}}{\partial g_{lk}} - \frac{\partial A_{kl}}{\partial g_{ji}} = \partial_1 g_{il}^{-1} g_{kj}^{-1} - g_{il}^{-1} \partial_1 g_{kj}^{-1} \quad . \tag{9.36c}$$

At the critical point, the two conserved currents are given by (9.27). In terms of phase-space variables they read

$$j_+(x) = -\frac{in}{2\pi} g^{-1} \partial_+ g = i2\hat{\Pi}^T g - \frac{in}{2\pi} g^{-1} \partial_1 g \quad , \tag{9.37a}$$

$$j_-(x) = -\frac{in}{2\pi} g \partial_- g^{-1} = -i2g\hat{\Pi}^T + \frac{in}{2\pi} g \partial_1 g^{-1} \quad , \tag{9.37b}$$

where $j_+ \equiv \left(j_L^0 - j_L^1\right)$, $j_- \equiv \left(j_R^0 + j_R^1\right)$. The other combinations vanish. Using (9.36a, b), we find the Poisson bracket relations,

$$\left\{j_+^a(x), j_+^b(y)\right\} = 2f^{abc} j_+^c(x) \delta(x^+ - y^+) + \frac{2n}{\pi} \delta^{ab} \delta'(x^+ - y^+) \tag{9.38a}$$

$$\left\{j_-^a(x), j_-^b(y)\right\} = 2f^{abc} j_-^c(x) \delta(x^- - y^-) + \frac{2n}{\pi} \delta^{ab} \delta'(x^- - y^-) \tag{9.38b}$$

$$\left\{j_+^a(x), j_-^b(y)\right\} = 0 \quad . \tag{9.38c}$$

with $j^a = \operatorname{tr} j \tau^a$, and τ^a the $SU(N)$ generators defined in appendix C. Here "prime" on δ represents the derivative with respect to the argument of δ. Thus, we have two Kac-Moody algebras with central extension $k = n$ if (9.33) vanishes.

In the case where the symmetry group is $O(N)$ rather than $SU(N)$, the canonical procedure turns out to be very clumsy, due to the constraints defining the $O(N)$ group, and a new procedure has to be followed. Indeed, following Witten[334],we can straightforwardly obtain the Poisson brackets in that case by rewriting the theory in terms of light cone coordinates, and using light cone quantization. The action so obtained is linear in the "time" derivative, (φ_i now denotes the unconstrained fields parametrizing the group),

$$S = \int d\sigma d\tau A_i(\varphi) \frac{d\varphi_i}{d\tau} \quad , \tag{9.39a}$$

and its variation is given by

$$\begin{aligned} \delta S &= \int d\sigma d\tau \left(\frac{\delta A_i}{\delta \varphi_j} \delta\varphi_j \frac{d\varphi_i}{d\tau} + A_i \frac{d}{d\tau} \delta\varphi_i \right) \\ &= \int d\sigma d\tau \left\{ \frac{\partial A_i}{\partial \varphi_j} - \frac{\partial A_j}{\partial \varphi_i} \right\} \delta\varphi_j \frac{d\varphi_i}{d\tau} \end{aligned} \tag{9.39b}$$

The Poisson bracket may be defined[51] in terms of the inverse of the matrix $F_{ij} = \partial_i A_j - \partial_j A_i$ appearing in (9.39b), that is

$$\{X, Y\}_W = F_{ij}^{-1} \partial_i X \partial_j Y \quad , \tag{9.40a}$$

in which case only derivatives of the function A are required, as already used in (9.36). The variation of the action of the $WZNW$ theory is by now a trivial exercise, and we obtain

$$\delta S = \frac{n}{4\pi}\int d\sigma d\tau \,\mathrm{tr}\, g^{-1}\delta g \frac{d}{d\sigma} g^{-1}\frac{dg}{d\tau} \quad . \tag{9.40b}$$

The inverse of the derivative operator is a step function. We now set $X = j_+^a(x^+)$, and $Y = j_-^a(y^-)$ in (9.40a), to compute

$$\delta X = -\frac{in}{2\pi}\,\mathrm{tr}\, g\tau^a g^{-1}\partial_+\left(\delta g g^{-1}\right) \tag{9.40c}$$

$$\delta Y = \frac{in}{2\pi}\,\mathrm{tr}\, g^{-1}\tau^a g\partial_-\left(g^{-1}\delta g\right) \quad . \tag{9.40d}$$

Using this, we obtain the light-cone algebra

$$\{j_+^a(x), j_+^b(y)\}_W = 4f^{abc}j_+^c(x)\delta(x^+ - y^+) + \frac{2n}{\pi}\delta^{ab}\delta'(x^+ - y^+) \tag{9.41a}$$

$$\{j_-^a(x), j_-^b(y)\}_W = 4f^{abc}j_-^c(x)\delta(x^- - y^-) + \frac{2n}{\pi}\delta^{ab}\delta'(x^- - y^-) \tag{9.41b}$$

$$\{j_-^a(x), j_+^b(y)\}_W = 0 \tag{9.41c}$$

We now show that (9.41) is also the algebra one obtains in a field theory of N free Majorana fermions, with the $O(N)$ symmetric action $S = \frac{i}{2}\int d^2x\overline{\psi}^i\,\partial\!\!\!/\psi^i$. The (hermitian) $O(N)$ currents $j_\mu^{ij} = \overline{\psi}^j\gamma_\mu\psi^i$, $(i,j = 1,\cdots,N)$, may be separated into right and left movers as

$$j_+^{ij} = -2\psi_2^j\psi_2^i \tag{9.42a}$$

$$j_-^{ij} = 2\psi_1^j\psi_1^i \quad . \tag{9.42b}$$

The current $j_+^{ij}(j_-^{ij})$ is a function of $x^+(x^-)$ only, as we see from the equation of motion for ψ.

The above objects satisfy a current algebra relation, which can be computed using the anti-commutation relations for the fermionic fields; indeed, we have for an $O(N)$ invariant Majorana field, the Dirac brackets[2)]

$$\left\{\psi_2^i(x),\psi_2^j(y)\right\}_D = -i\delta^{ij}\delta(x^1-y^1) \Rightarrow \left\{\psi_2^i(x),\psi_2^j(y)\right\}_{AC} = \delta^{ij}\delta(x^1-y^1) \quad . \tag{9.43}$$

[2)] In order to quantize free Majorana fermions consider the canonically conjugate momentum $\Pi_\psi = -\frac{i}{2}\psi$, which in fact is a constraint. From the Dirac formalism[338] we have the (generalized –see appendix O) Dirac bracket $\{\psi(x),\psi(y)\}_D = -i\delta(x^1 - y^1)$, and the corresponding anticommutator $\{\psi(x),\psi(y)\}_{AC} = \delta(x^1 - y^1)$. The original procedure used by Majorana uses the compatibility of Lagrangian and Hamiltonian equations of motion[339]. See footnote 2) in chapter 5 for some properties of Majorana fermions.

From the above we compute[3)]

$$\left[j_+^{ij}(x), j_+^{kl}(y)\right] = 4\left[\psi_2^j(x)\psi_2^i(x), \psi_2^l(y)\psi_2^k(y)\right]$$
$$= 2(\delta^{il} j_+^{kj}(x) + \delta^{jk} j_+^{li}(x) - \delta^{jl} j_+^{ki}(x) - \delta^{ik} j_+^{lj}(x))\delta(x^+ - y^+)$$
$$+ \frac{2i}{\pi}(\delta^{jk}\delta^{il} - \delta^{ik}\delta^{jl})\delta'(x^+ - y^+) \,. \quad (9.44)$$

One checks that the above algebra corresponds exactly to the Poisson algebras (9.38) (the current (9.42*b*) obeys the analogous algebra, and (9.42*a*) commutes with (9.42*b*)).

The algebra (9.44) above is an infinite extension of the $O(N)$ Lie algebra with a central extension given by the Schwinger term, and is called a Kac-Moody algebra, which in general, is defined by the following commutation relation

$$[j^a(x), j^b(y)] = 4if^{abc}j^c(x)\delta(x-y) + \frac{2ik}{\pi}\delta^{ab}\delta'(x-y) \quad . \qquad (9.45)$$

The *level k* of the algebra may differ from one if a higher representation is used for the fermions. An example of an interacting fermionic theory with this property will be discussed in chapter 15.

Since a Kac-Moody algebra defines a conformally invariant theory uniquely for $k = \pm 1$ (unitary irreducible representations are unique), we are thus led to identify the currents (9.37) with $\mathrm{tr}(\tau^a j_\pm)$. In order to complete this identification, we look for the fermionic representation of the bosonic field g_{ij}, as well as the identification of the energy momentum tensor in both theories.

---

3) The central term can be obtained as follows. Consider the most divergent term in the Wick expansion of (see Eq. 2.242)

$$\{\psi_2^j(x)\psi_2^i(x)\psi_2^l(y)\psi_2^k(y)\}_{div} = -\frac{1}{(2\pi)^2}\frac{1}{x^+ - y^+ - i\epsilon}\frac{1}{x^+ - y^+ - i\epsilon}(\delta^{jk}\delta^{il} - \delta^{ik}\delta^{jl})$$

and

$$\{\psi_2^l(y)\psi_2^k(y)\psi_2^j(x)\psi_2^i(x)\}_{div} = -\frac{1}{(2\pi)^2}\frac{1}{x^+ - y^+ + i\epsilon}\frac{1}{x^+ - y^+ + i\epsilon}(\delta^{jk}\delta^{il} - \delta^{ik}\delta^{jl}) \quad .$$

The difference reads

$$\frac{1}{(x^+ - y^+ + i\epsilon)^2} - \frac{1}{(x^+ - y^+ - i\epsilon)^2} = 2i\pi\delta'(x^+ - y^+).$$

9.3.2 The WZW fields in terms of fermions

The commutation relation of g_{ij} with the currents (9.37) is obtained directly from the corresponding Poisson bracket relation (9.36):

$$\left\{j_-^{ij}(x), g^{kl}(y)\right\} = i2g^{il}\delta^{jk}\delta(x^1 - y^1) \quad , \tag{9.46a}$$

$$\left\{j_+^{ij}(x), g^{kl}(y)\right\} = -i2g^{kj}\delta^{il}\delta(x^1 - y^1) \quad , \tag{9.46b}$$

for the $SU(N)$ case, and

$$\left\{j_-^{ij}(x), g_{kl}(y)\right\}_W = 2i\left[g_{il}\delta_{kj} - g_{jl}\delta_{ik})\right]\delta(x^1 - y^1) \quad , \tag{9.46c}$$

$$\left\{j_+^{ij}(x), g_{kl}(y)\right\}_W = -2i\left[(g_{kj}\delta_{il} - g_{ki}\delta_{jl})\right]\delta(x^1 - y^1) \quad , \tag{9.46d}$$

in the $O(N)$ symmetric case. Therefore we conclude that g transforms like the product of fermionic fields. Indeed, using

$$\left\{j_+^{ij}(x), \psi_{2k}(y)\right\} = 2(\psi_{2j}\delta^{ik} - \psi_{2i}\delta^{jk})\delta(x^1 - y^1) \tag{9.47a}$$

and

$$\left\{j_-^{ij}(x), \psi_{1k}(y)\right\} = 2(\psi_{1j}\delta^{ik} - \psi_{1i}\delta^{jk})\delta(x^1 - y^1) \tag{9.47b}$$

in the $O(N)$ symmetric case, we have

$$\left\{j_\pm^{ij}(x), \psi_{1k}\psi_{2l}(y)\right\} = \left\{ \begin{array}{l} -2(\psi_{1k}\psi_{2i}\delta_{lj} - \psi_{1k}\psi_{2j}\delta_{il})\delta(x^1 - y^1) \\ -2(\psi_{1i}\psi_{2l}\delta_{kj} - \psi_{1j}\psi_{2l}\delta_{ik})\delta(x^1 - y^1) \end{array} \right. \tag{9.48}$$

which is to be compared with (9.46). We thus have the identification

$$g_{ij} \rightarrow -\frac{i}{\mu}\psi_{1i}\psi_{2j} \tag{9.49}$$

with an arbitrary renormalization dependent mass parameter μ.

9.3.3 The Sugawara form of the energy momentum tensor

From (9.35) we are able to compute the Hamiltonian of the WZW theory. Using Eqs. (9.18), (9.19) and (9.35), we find

$$H = \int dx^1 \left\{ -\frac{2\pi}{n} tr(\hat{\Pi}^T g)^2 + \frac{n}{8\pi} tr\partial_1 g \partial_1 g^{-1} \right\} \quad . \tag{9.50}$$

The contribution from the Wess-Zumino term (9.15) cancels in the Hamiltonian, since it is linear in the time derivative. We also notice that the Hamiltonian can be entirely written in terms of the current (9.37), since $\hat{\Pi}^T g = \frac{n}{4\pi}\partial_0 g^{-1}g$, which may be substituted in (9.50) to yield

$$H = \int dx^1 \left\{-\frac{n}{8\pi} tr g^{-1}\partial^\mu g g^{-1}\partial_\mu g\right\} = \frac{\pi}{2n}\int dx^1 j^\mu j_\mu \quad . \tag{9.51}$$

This is the so called Sugawara form of the Hamiltonian[340]. A free fermion theory can also be written in this form[341]. Indeed, the fermionic current is known to satisfy the following commutation relations

$$[j_0(t,x), j_0(t,y)] = 0 \quad , \tag{9.52a}$$
$$[j_1(t,x), j_1(t,y)] = 0 \quad , \tag{9.52b}$$
$$[j_0(t,x), j_1(t,y)] = ic\delta'(x^1 - y^1) \quad . \tag{9.52c}$$

We wish to construct an energy momentum tensor, generating translations, that is,

$$[T_{00}(t,x), j_0(t,y)] = ij_1(t,x)\delta'(x-y) \tag{9.53a}$$
$$[T_{00}(t,x), j_1(t,y)] = ij_0(t,x)\delta'(x-y) \tag{9.53b}$$
$$[T_{01}(t,x), j_0(t,y)] = ij_0(t,x)\delta'(x-y) \tag{9.53c}$$
$$[T_{01}(t,x), j_1(t,y)] = ij_1(t,x)\delta'(x-y) \quad . \tag{9.53d}$$

It is not difficult to show, using (9.52), that an energy momentum tensor satisfying the above requirement is given by[341]

$$T_{\mu\nu}(t,x) = \frac{1}{2c}(j_\mu j_\nu + j_\nu j_\mu - g_{\mu\nu} j^\rho j_\rho) \quad . \tag{9.54}$$

The Schwinger term was essential in the construction. Indeed, notice that the coefficient appearing in the expression of the energy-momentum tensor (9.54) is the inverse of the Schwinger term. Therefore, the core of the construction is quantum mechanical.

It is not difficult to see that a canonical short distance expansion of the fermion fields implies that the above energy momentum tensor is equal to the canonical one[342]

$$\theta_{\mu\nu} = \frac{i}{4}(\overline{\psi}\gamma_\mu\partial_\nu\psi + \overline{\psi}\gamma_\nu\partial_\mu\psi - \partial_\mu\overline{\psi}\gamma_\nu\psi - \partial_\nu\overline{\psi}\gamma_\mu\psi) \quad . \tag{9.55}$$

Indeed, defining $T_{\mu\nu}$ in (9.54) in terms of the Wilson short distance expansion for the product $J_\mu(x)J_\nu(x+\epsilon)$, and using

$$\langle 0|\psi(0)\overline{\psi}(\epsilon)|0\rangle \approx -\frac{i}{2\pi}\frac{\gamma^\mu\epsilon_\mu}{\epsilon^2} \quad , \tag{9.56}$$

we find

$$j_\mu(x)j_\nu(x+\epsilon) = -\frac{i}{2\pi}\overline{\psi}(x)\gamma_\mu\gamma_\rho\gamma_\nu\psi(x+\epsilon)\frac{\epsilon^\rho}{\epsilon^2} + \mu \longleftrightarrow \nu \quad . \tag{9.57}$$

Expanding (9.57) in a Laurent series in ϵ, and averaging over ϵ_μ (space like) and $\tilde{\epsilon}_\mu = \epsilon_{\mu\nu}\epsilon^\nu$ (time like)[105] one finds $T_{\mu\nu} = \theta_{\mu\nu}$ for a suitable value of c.

9.3.4 The non-abelian bosonization in the operator language

The above results can also be understood using the methods presented in chapter 3. Although clumsy, they provide a deeper insight into the problem. We start from a theory of massless fermions with a $U(N)$ symmetry group

$$\mathcal{L} = i\overline{\psi}_i \not{\partial}\psi_i \quad , \tag{9.58}$$

and make use of the (abelian) bosonization formula[53]

$$\psi_i(x) = \left[\frac{\mu}{2\pi}\right]^{\frac{1}{2}} \mathcal{N} e^{i\left(\alpha\tilde{\varphi}_i(x)+\beta\gamma_5\varphi_i(x)\right)} \begin{pmatrix} 1 \\ 1 \end{pmatrix} \tag{9.59}$$

or equivalently

$$\psi_{\pm;i}(x) = \left[\frac{\mu}{2\pi}\right]^{\frac{1}{2}} \mathcal{N} e^{i\left(\alpha\tilde{\varphi}_i(x)\mp\beta\varphi_i(x)\right)} \quad , \tag{9.60}$$

where $\mathcal{N}$ is a normal product prescription and where φ and $\tilde{\varphi}$ are the potentials of the (conserved) diagonal currents[4)]; they are related, as usual, by $\partial_\mu\varphi = \tilde{\partial}_\mu\tilde{\varphi}$. We define

$$j_{\pm;ij}(x;\epsilon) = \psi^\dagger_{\pm;j}(x+\epsilon)\psi_{\pm;i}(x) \quad . \tag{9.61}$$

For $i \neq j$ the current (9.61) is well defined for $\epsilon \longrightarrow 0$, since the potentials φ_i and $\tilde{\varphi}_j$ $(i \neq j)$ are independent, while for the diagonal part we need an additive as well as a multiplicative renormalization. Using the abelian bosonization formula (9.60) as well as the commutation relations (2.22-25) for the (free) fields φ and $\tilde{\varphi}$, with equal values for i,j in the product (9.61), we find that

$$j_{\pm\, ij}(x;\epsilon) = \left(i\mu\epsilon^+\right)^{-\frac{(\alpha\pm\beta)^2}{4\pi}\delta_{ij}} \left(i\mu\epsilon^-\right)^{-\frac{(\alpha\mp\beta)^2}{4\pi}\delta_{ij}} e^{-i\alpha\left(\tilde{\varphi}_j(x+\epsilon)-\tilde{\varphi}_i(x)\right)} e^{\pm i\beta(\varphi_j(x+\epsilon)-\varphi_i(x))} \tag{9.62a}$$

Therefore, a finite operator defined from the above is obtained by the procedure

$$j_{\pm\, ij}(x) = \left(i\mu\epsilon^+\right)^{\left(\frac{(\alpha\pm\beta)^2}{4\pi}-1\right)\delta_{ij}} \left(i\mu\epsilon^-\right)^{\left(\frac{(\alpha\mp\beta)^2}{4\pi}-1\right)\delta_{ij}} \times$$
$$[j_{\pm ij}(x;\epsilon) - \delta_{ij} \times \text{divergent constant}] \ . \tag{9.62b}$$

---

4) Notice that $\psi_{\frac{1}{2}} = \psi_{\mp}$. For simplicity we omit the "Klein factor"; see section 3.3 of chapter 5.

Notice that in the canonical case $\alpha^2 = \beta^2 = \pi$, the multiplicative renormalization is very simple.

As a result of this asymmetric renormalization, we verify explicitly that the abelian bosonization prescription does not preserve the symmetry of the theory. We nevertheless proceed using the method as an intermediate step towards the correct results.

We define the bosonic field g_{ij} by means of the short distance expansion of the product

$$g_{ij}(x;\epsilon) = -i\mu^{-1}\psi^{\dagger}_{-;i}(x+\epsilon)\psi_{+;j}(x) \tag{9.63}$$

For $i \neq j$, it is clear again that the above field is finite when $\epsilon \to 0$. It is not difficult to see that it is also finite in this limit, when $i = j$. Indeed

$$g_{ii}(x;\epsilon) = \left[\frac{1}{2\pi}\right]^{\frac{1}{2}} \mathcal{N}e^{-i[\alpha\tilde{\varphi}_i(x+\epsilon)+\beta\varphi_i(x+\epsilon)]}\left[\frac{1}{2\pi}\right]^{\frac{1}{2}}\mathcal{N}e^{i[\alpha\tilde{\varphi}_i(x)-\beta\varphi_i(x)]} =$$
$$\left(-\mu^2\epsilon^2\right)^{\frac{\alpha^2-\beta^2}{4\pi}}\left[\frac{1}{2\pi}\right]\mathcal{N}e^{-i[\alpha\varphi_i(x+\epsilon)+\beta\tilde{\varphi}_i(x+\epsilon)+\alpha\varphi_i(x)-\beta\tilde{\varphi}_i(x)]} \quad . \tag{9.64}$$

The bosonic currents are defined in terms of the operator product expansion of the fields g_{ij} and its derivative; thus consider

$$j^B_{+ij}(x;\epsilon) = ig^{\dagger}_{ki}(x+\epsilon)\partial_+ g_{kj}(x) =$$
$$= i\mu^{-2}\partial^{x'}_+\{:\psi^{\dagger}_{+i}(x+\epsilon)\psi_{-k}(x+\epsilon):\,:\psi^{\dagger}_{-k}(x')\psi_{+j}(x'):\}_{x'=x} \quad . \tag{9.65}$$

Using again the exponential representation of the fermion fields, we have divergent multiplicative factors arising from the commutation of the bosonic creation operators in $\psi^{\dagger}_{-;k}(x+\epsilon)$ and the bosonic annihilation operators in $\psi_{-;k}(x')$, and for $i = j$ analogous factors coming from $\psi_{+;j}(x')$ and $\psi_{+;i}(x+\epsilon)$. Moreover, in the most divergent term, the derivative $\partial^{x'}_+$ acts only on the piece coming from ψ_-, the next term being much smaller as $\epsilon \to 0$. After these algebraic manipulations one obtains an expression for the current operator. Indeed, we have for the most divergent term, the expression

$$j^B_{\pm ij}(x;\epsilon) = -\frac{n(\alpha-\beta)^2}{4\pi}\left(i\mu\epsilon^+\right)^{-\frac{(\alpha-\beta)^2}{4\pi}-1}\left(i\mu\epsilon^-\right)^{-\frac{(\alpha+\beta)^2}{4\pi}-1}:\psi^{\dagger}_{+i}(x+\epsilon)\psi_{+j}(x): \tag{9.66}$$

For $i \neq j$ we need just a multiplicative renormalization from the above procedure, while in the case where $i = j$ we have also to subtract the first term of the Wilson expansion, defining the relation between the bosonic and the fermionic currents[343]. Analogous expressions hold for the right moving components[5)].

[5)] In [53] there are some errors in the intermediate steps of the derivation, although the final results are correct.

9.4 Properties off the Critical Point

9.4.1 Integrability of the Wess-Zumino-Witten action

The equations of motion of the theory may be summarized by the conservation laws (9.25), i.e

$$\partial^{\mu}\left(j_{\mu}-\frac{n\lambda^{2}}{4\pi}\epsilon_{\mu\nu}j^{\nu}\right)=0 \tag{9.68}$$

$$\partial^{\mu}\left(i_{\mu}-\frac{n\lambda^{2}}{4\pi}\epsilon_{\mu\nu}i^{\nu}\right)=0 \quad , \tag{9.69}$$

where

$$j^{\mu}=g^{-1}\partial^{\mu}g \tag{9.70}$$

$$i^{\mu}=\partial^{\mu}gg^{-1} \quad . \tag{9.71}$$

The above currents obey the zero curvature condition

$$\partial_{\mu}j_{\nu}-\partial_{\nu}j_{\mu}+[j_{\mu},j_{\nu}]=0 \tag{9.72}$$

$$\partial_{\mu}i_{\nu}-\partial_{\nu}i_{\mu}-[i_{\mu},i_{\nu}]=0 \quad . \tag{9.73}$$

Thus, using the results of chapter 7, we expect the existence of a conserved non local charge. Indeed,using (9.69), (9.70) and (9.71) one finds that[344,345]

$$Q=\int dy_{1}dy_{2}\epsilon(y_{1}-y_{2})\left[j_{0}(t,y_{1})+\frac{n\lambda^{2}}{4\pi}j_{1}(t,y_{1})\right]\left[j_{0}(t,y_{2})+\frac{n\lambda^{2}}{4\pi}j_{1}(t,y_{2})\right]+ 2\left(1-\left(\frac{n\lambda^{2}}{4\pi}\right)^{2}\right)\int j_{1}(t,y)dy \tag{9.74}$$

and

$$Q'=\int dy_{1}dy_{2}\epsilon(y_{1}-y_{2})\left[i_{0}(t,y_{1})-\frac{n\lambda^{2}}{4\pi}i_{1}(t,y_{1})\right]\left[i_{0}(t,y_{2})-\frac{n\lambda^{2}}{4\pi}i_{1}(t,y_{2})\right]- 2\left(1-\left(\frac{n\lambda^{2}}{4\pi}\right)^{2}\right)\int i_{1}(t,y)dy \tag{9.75}$$

are conserved.

When quantizing the theory, one can use the general methods of section 7.4 to verify that there is no anomalous term in the Wilson expansion of the commutator of the currents:

$$[j_{\mu}(x+\epsilon),j_{\nu}(x)]=C^{\rho}_{\mu\nu}(\epsilon)j_{\rho}(x)+D^{\sigma\rho}_{\mu\nu}(\epsilon)\partial_{\sigma}j_{\rho}(x) \tag{9.76}$$

$$[i_{\mu}(x+\epsilon),i_{\nu}(x)]=C'^{\rho}_{\mu\nu}(\epsilon)i_{\rho}(x)+D'^{\sigma\rho}_{\mu\nu}(\epsilon)\partial_{\sigma}i_{\rho}(x) \quad , \tag{9.77}$$

Hence, we may define finite and conserved charges as follows:

$$Q = \lim_{\delta\to 0} \int_{|y_1-y_2|<\delta} dy_1 dy_2 \epsilon(y_1 - y_2)\left[j_0(t,y_1) + \frac{n\lambda^2}{4\pi} j_1(t,y_1)\right] \times$$
$$\left[j_0(t,y_2) + \frac{n\lambda^2}{4\pi} j_1(t,y_2)\right] + Z_\delta \int \left[j_1(t,y) + \frac{n\lambda^2}{4\pi} j_0(t,y)\right] dy \quad (9.78)$$

$$Q' = \lim_{\delta\to 0} \int_{|y_1-y_2|<\delta} dy_1 dy_2 \epsilon(y_1 - y_2)\left[i_0(t,y_1) - \frac{n\lambda^2}{4\pi} i_1(t,y_1)\right] \times$$
$$\left[i_0(t,y_2) - \frac{n\lambda^2}{4\pi} i_1(t,y_2)\right] + Z_\delta \int \left[i_1(t,y) - \frac{n\lambda^2}{4\pi} i_0(t,y)\right] dy \quad , \quad (9.79)$$

where the relative coefficients have been adjusted by adding a term proportional to the $SU(N)$ charge. The methods of chapter 9 are however difficult to implement, since the present model describes infraparticles, and the asymptotic states can not be described in terms of a usual the Fock space generated by operators creating states of definite mass.

9.4.2 On the solution off the critical point

The equivalence of the chiral model with or without the topological Wess-Zumino term, permits one to construct the solution of the equivalent fermionic model. The latter can be solved by the Bethe Ansatz technique[317]. Since the Bethe Ansatz has been much discussed in the literature, we shall not dwell on this point very much, and proceed quickly to the results, which are worth discussing.

Consider a theory of N families of N_c colored fermions $\psi_{i,f}$, $f = 1, \cdots, N$, $i = 1, \cdots, N_c$, described by the Lagrangian

$$\mathcal{L} = \sum_{f=1}^{N} i\overline{\psi}_f \gamma^\mu \partial_\mu \psi_f - \frac{\lambda}{2} \sum_{f=1}^{N} (\overline{\psi}_f \gamma^\mu \tau^a \psi_f)^2 \quad , \quad (9.80)$$

where τ^a are the generators of $SU(N)$-color, and where summation over color indices is implied. This is easily seen to be equivalent to a theory described by the Lagrangian

$$\mathcal{L} = \overline{\psi}\gamma^\mu (i\partial_\mu + A_\mu)\psi + \frac{1}{2\lambda} \operatorname{tr} A_\mu^2 \quad , \quad (9.81)$$

where $A_\mu = \sum A_\mu^a \tau^a$. Integrating over the fermions in the corresponding partition function yields the equivalent action

$$S_{equiv} = \int d^2x \frac{1}{2\lambda} \operatorname{tr} A_\mu^2 + NW[A_\mu] \quad , \quad (9.82)$$

where $W[A]$ is the effective action $-i \ln \det(i \not\partial + \not A)$, to be calculated in chapter11. In the limit $N \to \infty$ the second term leads to a freezing of the A_μ degrees of freedom, at the value $W[A_\mu] = 0$; since the effective action is gauge invariant, this means that A_μ must be a pure gauge. Writing $A_\mu = ig^{-1}\partial_\mu g$, we thus obtain

$$S_{equiv} = \frac{1}{2\lambda} \operatorname{tr} \int d^2x \partial^\mu g^{-1} \partial_\mu g \quad , \tag{9.83}$$

which shows that in this limit the theory described by (9.80) becomes equivalent to that of the principal sigma model.

The solution of (9.83) using the Bethe Ansatz has been discussed in [317] and leads to the result (8.142); it describes massive fundamental particles, the generated mass being given by the usual formula

$$m^2 = \Lambda^2 e^{-2\pi/\lambda N_c} \quad . \tag{9.84}$$

We now modify the Lagrangian (9.80) by allowing for an unequal number of left and right movers. In this case the spectrum is more complicated, since it turns out that in the infrared region, the system is conformally invariant. The equivalence with a purely fermionic theory continues to be true, although we now have N_L left moving fermions and N_R right moving ones, with $N_R \neq N_L$

$$\begin{aligned}\mathcal{L} = &\sum_{f=1}^{N_L} \overline{\psi}_f \gamma^\mu \frac{(1+\gamma_5)}{2} \partial_\mu \psi_f + \sum_{f=1}^{N_R} \overline{\psi}_f \gamma^\mu \frac{(1-\gamma_5)}{2} \partial_\mu \psi_f \\ &+ \sum_{f=1}^{N_L} \overline{\psi}_f \tau^a \frac{(1+\gamma_5)}{2} \psi_f A_+^a + \sum_{f=1}^{N_R} \overline{\psi}_f \tau^a \frac{(1-\gamma_5)}{2} \psi_f A_-^a + a A_+^a A_-^a \quad ,\end{aligned} \tag{9.85}$$

where a is an arbitrary regulator dependent parameter. In this case, the effective action may be computed as before (see chapter 11). Writing

$$A_+ = ig\partial_+ g^{-1} \quad , \quad A_- = ih\partial_- h^{-1} \quad ,$$

the effective action for a single left-handed or right-handed fermion is given by (see chapter 13)

$$W^{(L)}[A] = -\Gamma[g] \quad , \quad W^{(R)}[A] = -\Gamma[h^{-1}] \quad ,$$

respectively, where Γ is the WZW action (9.14) with $n = 1$.

Using the identity[50] (see chapter 11)

$$\Gamma(gh^{-1}) = \Gamma(g) + \Gamma(h^{-1}) + \operatorname{tr} \int d^2x g^{-1} \partial_+ g h^{-1} \partial_- h \quad , \tag{9.86}$$

the effective action may be written as

$$W[A_+,A_-] = -\frac{1}{2}(N_L+N_R)(\Gamma(g)+\Gamma(h^{-1}))$$
$$-\frac{1}{2}(N_L-N_R)(\Gamma(g)-\Gamma(h^{-1}))+a\int d^2x A_+A_- \quad . \tag{9.87}$$

For A_μ pure gauge ($g=h$),

$$A_\pm = ig^{-1}\partial_\pm g \quad , \tag{9.88}$$

we also have the identity

$$\Gamma(g)-\Gamma(g^{-1}) = \frac{1}{2\pi}\mathrm{tr}\int_0^1 dr\int d^2x\epsilon^{\mu\nu}g^{-1}\dot{g}g^{-1}\partial_\mu g g^{-1}\partial_\nu g \quad . \tag{9.89}$$

Taking $N_L+N_R\to\infty$, freezes again A_μ to a pure gauge excitation (9.88). Substituting (9.89) in (9.87), we obtain for $a=0$ the WZNW action with $n=N_L-N_R$. The solution of the Bethe Ansatz equations for this problem has been obtained in [317]; a mass parameter analogous to the previous one expressing a mass transmutation again appears in the theory,

$$m^2 = \Lambda^2 e^{-2\pi/\lambda N_c} \quad . \tag{9.90}$$

The energy as a function of an external magnetic field H is given by $E(H)=H^2+\cdots$, where the correction terms depend on n. Only after the symmetry $g\to g^{-1}$ is imposed, forcing n to be zero, the theory develops a mass gap.

The supersymmetric theory has not been analysed from this point of view in the literature, but in any case seems to lack the property of integrability[344].

9.4.3 Supersymmetric WZW model[6)]

It is not difficult to obtain the supersymmetric extension of the WZW theory. Consider the purely bosonic action (9.17). Define a superfield

$$G(x,\theta) = g(x)+\bar{\theta}\psi(x)+\frac{1}{2}\bar{\theta}\theta F(x) \quad , \tag{9.91}$$

with the constraint

$$G^\dagger(x,\theta)G(x,\theta)=1 \quad , \tag{9.92}$$

---

6) In this section we use the Majorana representation (A.17) for the gamma matrices.

which is equivalent to

$$g^\dagger g = g g^\dagger = 1 \tag{9.93a}$$

$$\psi^\dagger g + g^\dagger \psi = 0 \tag{9.93b}$$

$$g^\dagger F + F^\dagger g = \overline{\psi}\psi \quad . \tag{9.93c}$$

Define an extension of the superfield $G(x,\theta)$, given by $\tilde{G}(x,\theta,r)$, such that[346]

$$\tilde{G}(x,\theta,1) = G(x,\theta) \tag{9.94}$$

$$\tilde{G}(x,\theta,0) = 1 \quad , \tag{9.95}$$

and the Lagrangian is given by

$$\mathcal{L} = \frac{1}{4\lambda^2}\,\mathrm{tr}\int \frac{d^2\theta}{4}\overline{DG}DG + \frac{n}{8\pi}\,\mathrm{tr}\int_0^1 dr \int \frac{d^2\theta}{4}\tilde{G}^\dagger \dot{\tilde{G}}\overline{D\tilde{G}}\gamma_5 D\tilde{G} \quad . \tag{9.96}$$

where $\dot{\tilde{G}}$ denotes the derivative of $\tilde{G}$ with respect to r, and $D_\alpha = \frac{\partial}{\partial\bar{\theta}_\alpha} - i(\gamma^\mu\theta)_\alpha\partial_\mu$ is the supersymmetric derivative. The terms involving F in the Lagrangian may be easily computed to give

$$\mathcal{L}_F = \frac{1}{4\lambda^2}F^\dagger F - \frac{n}{8\pi}\int_0^1 dr F\frac{d}{dr}\left(\frac{1}{2}\overline{\psi}\psi g^\dagger\right) \quad . \tag{9.97}$$

We use the constraint (9.93c) to eliminate $F^\dagger$. The equation of motion for F is then given by the constraint

$$F = \frac{1}{2}g\overline{\psi}\psi + \frac{n\lambda^2}{8\pi}g\overline{\psi}\gamma_5\psi \quad . \tag{9.98}$$

Substituting (9.98) back into (9.96), and integrating over θ, we have, for the first (chiral) term

$$\mathcal{L}_{ch} = \frac{1}{4\lambda^2}\,\mathrm{tr}\left\{\partial^\mu g^\dagger\partial_\mu g + i\overline{\psi}\,\partial\!\!\!/\psi - \left(\frac{1}{4} + \left(\frac{n\lambda^2}{8\pi}\right)^2\right)(\overline{\psi}\psi)^2\right\} \quad , \tag{9.99}$$

and for the second (Wess-Zumino) term,

$$\mathcal{L}_{WZ} = \frac{n}{8\pi}\int_0^1 dr\left\{\epsilon^{\mu\nu}g^\dagger\dot{g}\partial_\mu g^\dagger\partial_\nu g + \frac{i}{2}\frac{d}{dt}(g^\dagger\partial_\mu g\overline{\psi}\gamma_5\gamma^\mu\psi) - \frac{n\lambda^2}{16\pi}\frac{d}{dr}(\overline{\psi}\psi)^2\right\} \quad , \tag{9.100}$$

Notice the presence of local interactions for the fermions. In arriving at (9.100) we used the following identities

$$\mathrm{tr}(\overline{\psi}\psi)^2 = -\,\mathrm{tr}(\overline{\psi}\gamma_5\psi)^2 = -\frac{1}{2}(\overline{\psi}\gamma^\mu\psi)^2 \quad , \tag{9.101}$$

which are derived by introducing a factor $g^\dagger g = 1$ in the trace, moving g cyclically and Fierz transforming as

$$\mathrm{tr}(\overline{\psi}\psi)^2 = \mathrm{tr}\,\overline{\psi}\psi\overline{\psi}\psi g^\dagger g = \mathrm{tr}(\psi\overline{\psi})(\psi\overline{\psi})gg^\dagger \quad . \tag{9.102}$$

The Fierz transformation

$$\delta_{\alpha\beta}\delta_{\gamma\delta} = \frac{1}{2}[\delta_{\alpha\delta}\delta_{\gamma\beta} + \gamma_{5_{\alpha\delta}}\gamma_{5_{\gamma\beta}} + \gamma_{\mu_{\alpha\delta}}\gamma^\mu_{\gamma\beta}] \tag{9.103}$$

implies

$$\mathrm{tr}(\overline{\psi}\psi)^2 = -\frac{1}{2}\,\mathrm{tr}(\overline{\psi}\psi)^2 - \frac{1}{2}\,\mathrm{tr}(\overline{\psi}\gamma_5\psi)^2 - \frac{1}{2}\,\mathrm{tr}(\overline{\psi}\gamma^\mu\psi)^2 \tag{9.104a}$$

or

$$3\,\mathrm{tr}(\overline{\psi}\psi)^2 = -\,\mathrm{tr}(\overline{\psi}\gamma_5\psi)^2 - \mathrm{tr}(\overline{\psi}\gamma^\mu\psi)^2 \tag{9.104b}$$

We likewise obtain

$$3\,\mathrm{tr}(\overline{\psi}\gamma_5\psi)^2 = -\,\mathrm{tr}(\overline{\psi}\psi)^2 + \mathrm{tr}(\overline{\psi}\gamma^\mu\psi)^2 \quad . \tag{9.105}$$

Identities (9.104) and (9.105) imply (9.101).

Combining (9.99) and (9.100), we obtain the action[346,347]

$$S = S_{WZNW}(g) + \frac{1}{4\lambda^2}\,\mathrm{tr}\int d^2x$$
$$\times\left\{\overline{\psi}i\,\partial\!\!\!/\psi - \frac{1}{4}\left(1-\left(\frac{n\lambda^2}{4\pi}\right)^2\right)(\overline{\psi}\psi)^2 + i\frac{n\lambda^2}{4\pi}g^\dagger\partial^\mu g\overline{\psi}\gamma_5\gamma_\mu\psi\right\} \,. \tag{9.106}$$

It is convenient to work with fields of definite chirality. We define the fields $\chi_\pm$ by

$$\psi = \frac{1-\gamma_5}{2}g\chi_+ + \frac{1+\gamma_5}{2}\chi_- g \quad . \tag{9.107}$$

The field χ corresponds to (6.231) of chapter 6. The constraint (9.93b) implies

$$(\chi_\pm)^\dagger = -\chi_\pm \tag{9.108}$$

and the action reads

$$S = S(g) + \frac{1}{4\lambda^2}\int d^2x\Big\{i(\overline{\chi}_+\,\partial\!\!\!/\chi_+ + \overline{\chi}_-\,\partial\!\!\!/\chi_-)$$
$$-i(1-\alpha)[\overline{\chi}_-\gamma^\mu\partial_\mu gg^\dagger\chi_- + \overline{\chi}_+\gamma^\mu g^\dagger\partial_\mu g\chi_+] - \frac{1}{2}(1-\alpha^2)g\overline{\chi}_+\chi_+g^\dagger\overline{\chi}_-\chi_-\Big\} \,, \tag{9.109}$$

where $\alpha = \frac{n\lambda^2}{4\pi}$, and in the last term we supposed that $\chi_\pm$ are eigenfunctions of γ_5. In the case $\alpha = 1$, the supersymmetric partners of the bosonic WZW fields are fermions in the adjoint representation of the group.

9.5 Conclusion

In this chapter we discussed several properties of the Wess Zumino Witten action at, as well as off criticality. The model turns out to show a rich physics, describing an integrable model for any value of the coupling constant, being asymptotically free for small coupling, and displaying a non trivial zero of the β function. At the critical point the theory is conformally invariant, and the correlators may be computed exactly, a task that shall be accomplished in chapter 15. Off the critical point, integrability implies several properties of the model. It is equivalent to a purely fermionic model with a left-right asymmetry, in the limit where the number of fermions goes to infinity. The Bethe-Ansatz technique may be used to obtain the spectrum of the model. A supersymmetric extension may be obtained as well, and in the critical limit, the supersymmetry relates the Wess Zumino field to a free fermionic theory in the adjoint representation. The superconformal algebra is realised in this case[347].

At the critical point, the theory describes free fermions, and we obtain a non-abelian generalization of the bosonization procedure. This has been obtained in the path integral, as well as in the operator approach, using the abelian bosonization technique as an intermediate step.

We shall come back to this model at the critical point in chapter 15, when discussing the conformal field theory, obtaining the correlation functions.

Chapter X

QUANTUM ELECTRODYNAMICS: OPERATOR APPROACH

10.1 Introduction

Quantum Electrodynamics of massless fermions in 1 + 1 dimensions (massless QED_2) was first studied by J. Schwinger[27,153] in (1962). As it turned out, the solution he obtained using functional methods was incomplete, and missed in fact some of the subtleties that make this model particularly interesting. The same applies to subsequent investigations in the sixties[348]. It was only as late as 1971, that in a now classic paper, paralleling that of Klaiber's paper[25] on the Thirring model, Lowenstein and Swieca[28] presented the complete solution of massless QED_2, using operator methods.

Schwinger's analysis of massless QED_2, which we shall refer to as the (vector) "Schwinger model" showed, that the gauge field becomes massive via a kind of dynamical Higgs mechanism induced by the fermions. The real significance of this only became clear, later, through the work of Lowenstein and Swieca, which revealed a spontaneous breakdown of the global chiral $U(1) \times \widetilde{U}(1)$ symmetry in this model, where $U(1)$ and $\widetilde{U}(1)$ refer to the charge and chirality of the fermion, respectively. As a consequence, the ground state of this model is found to exhibit a doubly infinite degeneracy labelled by fermion number and chirality, the degeneracy with respect to chirality being analogous to that found in four dimensional Quantum Chromodynamics (QCD_4)[268,269]. When going over to an irreducible representation of these vacua, one is naturally led to the θ-vacuum very much familiar from QCD_4.

It has been suggested that the infrared behavior of QCD_4 may be responsible for a confining force permanently binding the quarks (q) and antiquarks ($\bar{q}$) into $q\bar{q}$ pairs. This process has been referred to as "infrared slavery". In QED_2 (and QCD_2) such a confining force is a free bonus of the linear rise of the Coulomb potential characteristic of one space-dimension; hence free asymptotic charges do not exist in the model. This is born out by the fact that charged sectors correspond to states of infinite energy[28,49]. In particular, it means that the charge of the fermions is screened by "quark-antiquark" pairs[1)] created from the vacuum.

This does not yet mean confinement, a property that one expects QCD_4 to have. There could still exist asymptotic states corresponding to screened quarks.

---

1) We shall refer to the fundamental fermions as "quarks", though they carry zero triality.

The search for such states must be restricted to the gauge invariant sector ($\mathcal{H}_{phys}$) of the theory. The asymptotic states in this sector corresponds to eigenstates of the Hamiltonian. In the Schwinger model they are found to be given by the Fock-space of a free pseudoscalar bosonic field $\Sigma(x)$ of mass $e/\sqrt{\pi}$, which makes the confinement of quarks in the Schwinger model manifest.

All states in $\mathcal{H}_{phys}$ can be constructed by applying functionals of $\Sigma(x)$ on the irreducible vacuum state. Care must be taken to distinguish neutral bosonic states from screened fermionic states. Since there is no rotation group in one space dimension, a suitable criterion is provided by "triality", which cannot be screened by the $U(1)$ interaction.

In this chapter, "flavor" replaces "triality". By generalizing the Schwinger model to include flavor, we find[349,350] that one can construct electrically neutral finite energy states carrying the flavor quantum number of the fundamental fermion fields. However, they do not occur asymptotically (confinement).

The confining aspect of QED_2 becomes more interesting in the case of massive fermions. This so called "massive" Schwinger model is no longer exactly soluble. A semi-classical analysis reveals a linear rise of the quark-antiquark $q\bar{q}$ potential up to distances which energetically correspond to the production of a real $q\bar{q}$ pair, and a leveling off for separations exceeding this critical distance[351].

On the non-perturbative level, bosonization techniques analogous to those applied to the massive Thirring model in chapter 3 provide much insight into the problem. These techniques, first used in this context by Coleman, Jackiw and Susskind[352], lead to a modified sine-Gordon (SG) equation, in which the periodic symmetry of the SG equation is explicitly broken by the electromagnetic interaction. One finds that quark states are absent, with the exception for some special θ-worlds, where states with exotic (multiplicative, kink) quantum numbers are liberated[350,351,353,354].

We shall discuss in some detail the properties of massless and massive QED_2, using the bosonization techniques introduced in chapters 2 and 3. It is a remarkable feature of the equivalent bosonic formulation, that the conserved currents can no longer be understood as arising from a Noether symmetry, but rather have a dynamical and topological origin, the $U(1)$ charge being carried by non-local soliton-like states, provided these states exist at all.

The operator approach leads relatively simply to a complete global view of QED_2. This is not true for the functional approach, where the existence of the infinite vacuum degeneracy is linked to a violation of clustering through the existence of "induced" instantons[355], and hence, to the non-simple connectedness of the gauge-field manifold. One is thereby confronted with the appearance of zero modes of the Dirac operator, which complicates considerably the analysis, but at the same time makes it also very interesting. In order to reveal the subtle structure of the model in the functional framework, it is thus extremely helpful, to use the insight gained from the operator approach as a guideline. We therefore present first the operator analysis, although it is likely to be less familiar to the reader, and leave the

Feynman-path-integral approach to chapter 12. For a renormalization-group analysis of massless QED_2 we refer the reader to the work of R. J. Crewther et al[356]. The case of the massless and massive Schwinger model is discussed separately. We shall work in Minkowski space, throughout this chapter.

10.2 The Massless Schwinger Model

10.2.1 Quantum solution

The quantum equations of motion to be solved are

$$i\gamma^\mu \partial_\mu \psi(x) + eN[\gamma^\mu A_\mu(x)\psi(x)] = 0 \tag{10.1a}$$

$$\partial_\mu F^{\mu\nu} + eJ^\nu = 0 \tag{10.1b}$$

with

$$F^{\mu\nu} = \partial^\mu A^\nu - \partial^\nu A^\mu \tag{10.2}$$

the field strength tensor, and J^ν the electromagnetic current

$$J^\nu = N[\bar{\psi}\gamma^\nu\psi] \quad . \tag{10.3}$$

Here $N[\,]$ denotes a suitably defined normal product. Equations (10.1) are formally invariant under global $U(1) \times \tilde{U}(1)$ transformations, as well as local $U(1)$-gauge transformations

$$A^\mu(x) \to A^\mu(x) + \frac{1}{e}\partial^\mu \Lambda(x)$$

$$\psi(x) \to e^{i\Lambda(x)}\psi(x) \quad . \tag{10.4}$$

Note that the coupling constant e carries the dimensions of mass, so that we are dealing with a super-renormalizable theory.

We shall seek solutions in the Lorentz gauge[2)]. As in the case of the usual Gupta-Bleuler formalism one discovers, that the Maxwell equations cannot be satisfied as operator equations, but only hold on a subspace of the Hilbert space, to be defined below[28]. One can nevertheless regularize the vector current such as to be conserved at the operator level. The divergence of the axial vector current $J_5^\mu = \bar{\psi}\gamma^\mu\gamma^5\psi$ is then given by the Adler anomaly[201] (for a perturbative derivation see ref. [360]),

$$\partial_\mu J_5^\mu = \frac{e}{2\pi}\epsilon_{\mu\nu}F^{\mu\nu} \quad . \tag{10.5}$$

[2)] For the discussion of other local covariant gauges see ref. [357]. For a light-cone quantization of this model see, in particular, ref. [358] and [359]. For a quantization in the Coulomb gauge see ref. [28].

We now systematically construct the solution to the equations of motion (10.1). We do this in detail in order to illustrate a method repeatedly used in other sections of this book.

The most general Ansatz for A_μ is given by

$$A_\mu = -\frac{\sqrt{\pi}}{e}(\tilde{\partial}_\mu \Sigma + \partial_\mu \tilde{\eta}) \quad . \tag{10.6}$$

The constant $\frac{\sqrt{\pi}}{e}$ has been introduced for dimensional reasons.

As we already remarked, we shall work in the Lorentz gauge, $\partial_\mu A^\mu = 0$. In this gauge, $\tilde{\eta}$ is just a zero-mass free scalar field:

$$\Box \tilde{\eta} = 0 \quad . \tag{10.7}$$

Hence, using (A.6) we have

$$F_{\mu\nu} = -\frac{\sqrt{\pi}}{e}(\partial_\mu \tilde{\partial}_\nu - \partial_\nu \tilde{\partial}_\mu)\Sigma = \frac{\sqrt{\pi}}{e}\epsilon_{\mu\nu}\Box\Sigma \quad . \tag{10.8}$$

In $d = 1+1$ dimensions, conservation of the current J^μ in (10.1b) and identity $\gamma^\mu\gamma^5 = \epsilon^{\mu\nu}\gamma^\nu$ imply that the vector and axial vector currents can be written as

$$J^\mu \equiv \overline{\psi}\gamma^\mu\psi = -\frac{1}{\sqrt{\pi}}\tilde{\partial}^\mu \Phi \quad , \tag{10.9a}$$

$$J_5^\mu \equiv \overline{\psi}\gamma^\mu\gamma^5\psi = -\frac{1}{\sqrt{\pi}}\partial^\mu \Phi \quad . \tag{10.9b}$$

The anomaly equation (10.5) thus tells us that $\Box\Phi = \Box\Sigma$. Hence $\Phi = \Sigma + h$, where h is a harmonic field,

$$\Box h = 0 \quad . \tag{10.10}$$

In terms of Σ and h the current (10.9a) thus takes the form

$$J_\mu = -\frac{1}{\sqrt{\pi}}\tilde{\partial}_\mu \Sigma + L_\mu \quad , \tag{10.11a}$$

where

$$L_\mu = -\frac{1}{\sqrt{\pi}}\tilde{\partial}_\mu h \quad . \tag{10.11b}$$

Maxwell equations (10.1b) correspondingly read

$$\tilde{\partial}^\nu(\Box + \frac{e^2}{\pi})\Sigma - \frac{e^2}{\sqrt{\pi}}L^\nu = 0 \quad . \tag{10.12}$$

Because of (10.10) we have $\tilde{\partial}_\mu L_\mu = 0$, or $\Box\left(\Box + \frac{e^2}{\pi}\right)\Sigma = 0$, so that Σ is in general a combination of a massive and massless free field.

Without loss of generality we may choose Σ to be a free field of mass $\frac{e^2}{\pi}$:

$$\left(\Box + \frac{e^2}{\pi}\right)\Sigma = 0 \quad , \tag{10.13}$$

its normalization remaining so far unspecified. Eq. (10.13) shows that the gauge field has acquired a mass (dynamical Higgs mechanism) without the loss of gauge invariance!

In view of (10.13), the requirement that (Maxwell's) equations (10.12) be satisfied as operator equations would require L_μ to vanish strongly. As we shall see below, this is not possible. Hence L_μ can only vanish in the weak sense:

$$\langle\psi'|L_\mu(x)|\psi\rangle = 0 \quad . \tag{10.14}$$

This condition defines the physical Hilbert space $\mathcal{H}_{phys}$ of the model.

We define as "observables" all those operators $\mathcal{O}$ which leave $\mathcal{H}_{phys}$ invariant; that is,

$$\mathcal{O}|\psi\rangle \in \mathcal{H}_{phys} \quad \text{if} \quad |\psi\rangle \in \mathcal{H}_{phys} \quad . \tag{10.15}$$

It thus follows from (10.15) that in particular [3)]

$$\langle\psi'|L_\mu(x)\mathcal{O}|\psi\rangle = 0 \quad . \tag{10.16a}$$

Now, the current J_μ as well as $F_{\mu\nu}$ are observables. Thus it follows in particular from (10.16a), that

$$\langle\psi'|L_\mu(x)L_\nu(y)|\psi\rangle = 0 \quad . \tag{10.16b}$$

Hence the operator $L_\mu(x)$, when applied on $|\psi\rangle \in \mathcal{H}_{phys}$, creates zero norm states:

$$\|L[f]|\psi\rangle\| = 0 \quad , \tag{10.17}$$

where $L[f] = \int d^2x f_\mu(x)L^\mu(x)$.

Define the Fock vacuum by

$$\Sigma^{(+)}(x)|0\rangle = 0 \quad , \quad L_\mu^{(+)}(x)|0\rangle = 0 \quad . \tag{10.18}$$

The state $|0\rangle$ evidently belongs to $\mathcal{H}_{phys}$. Choosing $|\psi\rangle = |\psi'\rangle = |0\rangle$ in (10.16), we conclude that

$$\langle 0|[L_\mu^{(+)}(x), L_\nu^{(-)}(y)]|0\rangle = 0 \quad , \quad \textit{for all } x, y \quad . \tag{10.19}$$

We shall return to discuss the implications of (10.19) later, and turn now our attention to an explicit calculation of the Maxwell current (10.11) from its definition (10.3) in terms of the fermionic fields.

---

3) Note that in contrast to h itself, $\tilde{\partial}_\mu h$ is a well defined operator in $d = 1 + 1$ dimensions, free of IR singularities.

10.2.2 The Maxwell current

In order to establish the connection between the harmonic field $h(x)$ and the degrees of freedom characterizing the gauge potential and fermion field ψ, we need to return to Maxwell's equations and calculate the current from its definition (10.3). This evidently requires solving first the Dirac equation (10.1a). It is convenient to define a pseudo-scalar field η "dual" to $\tilde{\eta}$ by

$$\tilde{\partial}_\mu \eta = \partial_\mu \tilde{\eta} \quad , \tag{10.20}$$

which is always possible in $d = 1 + 1$ dimensions, since $\tilde{\eta}$ is a free zero mass field. Inserting (10.6) into (10.1a), the Dirac equation reads

$$[i \not\partial + \sqrt{\pi}\gamma^\mu \gamma_5 \partial_\mu (\Sigma + \eta)]\psi = 0 \quad , \tag{10.21}$$

where we have made use of $\gamma^\mu\gamma_5 = \epsilon^{\mu\nu}\gamma_\nu$.

Eq. (10.21) has the solution,

$$\psi_\alpha(x) =: e^{i\sqrt{\pi}\gamma^5_{\alpha\alpha}(\Sigma(x)+\eta(x))}: \psi^{(0)}_\alpha(x) \quad , \tag{10.22}$$

where the double-dots indicate normal ordering with respect to the Fock space operators Σ and η, and where $\psi^{(0)}_\alpha(x)$ is the (canonically normalized) zero mass free Dirac field. In order to establish the desired relations, we need to compute the fermionic current. We define it in terms of the Wilson short-distance expansion[31]

$$\overline{\psi}(x+\epsilon)\gamma^\mu : e^{ie\int_x^{x+\epsilon} dz^\mu A_\mu(z)} : \psi(x) \approx c^\mu(\epsilon) + \mathcal{Z}(\epsilon) N[\overline{\psi}(x)\gamma^\mu\psi(x)] + \cdots \tag{10.23}$$

where $c^\mu(\epsilon)$ and $\mathcal{Z}(\epsilon)$ are (singular) c-number functions of (the cutoff) ϵ, and where the normal product $N[\overline{\psi}\gamma^\mu\psi](x)$ is taken to define the current. $c^\mu(\epsilon)$ and $\mathcal{Z}(\epsilon)$ play the role of a subtractive and multiplicative renormalization constant, respectively. The double dots denote Wick ordering with respect to the operators Σ and η appearing in A_μ, which have already been established to be free fields.

The Schwinger line integral has been introduced in order to insure formal invariance under the c-number gauge transformation $A_\mu \to A_\mu + \frac{1}{e}\partial_\mu \Lambda$, $\psi \to e^{i\Lambda}\psi$, at each step of the limiting process. As a result, the limit $\epsilon^\mu \to 0$ does not depend on the direction from which it is approached. Moreover it insures that the result of this limit will define a conserved current.

The calculation of the Wilson expansion (10.23) is now straightforward. Proceeding as in the case of the RS model in chapter 2, one obtains, upon using (10.6), (10.20) and (10.22), and keeping only terms up to first order in ϵ,

$$\begin{aligned}&\overline{\psi}(x+\epsilon)\gamma^\mu : e^{ie\int_x^{x+\epsilon} dz^\mu A_\mu(z)} : \psi(x) \\ &\approx \mathcal{Z}(\epsilon)\overline{\psi}^{(0)}(x+\epsilon)\gamma^\mu \left[1 - i\sqrt{\pi}[(\epsilon^\nu\tilde{\partial}_\nu + \gamma^5\epsilon^\nu\partial_\nu)(\Sigma(x)+\eta(x))]\right] : \psi^{(0)}(x) \quad ,\end{aligned} \tag{10.24}$$

where $\mathcal{Z}(\epsilon)$ is a (renormalization) constant resulting from the process of normal ordering the exponentials. It is given by

$$\mathcal{Z}(\epsilon) = e^{f(\epsilon)} = e^{\pi[\Sigma^{(+)}(\epsilon)+\eta^{(+)}(\epsilon),\Sigma^{(-)}(0)+\eta^{(-)}(0)]} \quad . \tag{10.25}$$

Since we have not yet come to fix the normalization the fields Σ and η, we leave the explicit evaluation of $\mathcal{Z}(\epsilon)$ to the end of this section.

We may rewrite the *r.h.s.* of (10.24) as,

$$r.h.s. \approx$$
$$\mathcal{Z}(\epsilon)\overline{\psi}^{(0)}(x+\epsilon)\{\gamma^\mu - i\sqrt{\pi}(\gamma^\mu\epsilon^\nu\tilde{\partial}_\nu + \epsilon^{\mu\lambda}\gamma_\lambda\epsilon^\nu\partial_\nu)(\Sigma(x)+\eta(x))\}\psi^{(0)}(x) \quad . \tag{10.26}$$

Now, using (A.27*a*), as well as $\mathrm{tr}\,\gamma^\mu\gamma^\nu = 2g^{\mu\nu}$, one has

$$\begin{aligned}\overline{\psi}^{(0)}(x+\epsilon)\gamma^\mu\psi^{(0)}(x) &= -\,\mathrm{tr}(\gamma^\mu S^{(+)}(-\epsilon)) - \frac{1}{\sqrt{\pi}}\tilde{\partial}^\mu\varphi \\ &= -\frac{i}{\pi}\frac{\epsilon^\mu}{\epsilon^2} - \frac{1}{\sqrt{\pi}}\tilde{\partial}^\mu\varphi \quad ,\end{aligned} \tag{10.27}$$

where φ is the "potential" of the free fermionic current (see chapter 2),

$$j^\mu(x) =: \psi^{(0)}(x)\gamma^\mu\psi^{(0)}(x) := -\frac{1}{\sqrt{\pi}}\tilde{\partial}^\mu\varphi \quad , \tag{10.28a}$$

with φ satisfying $\Box\varphi = 0$; or because of (A.14), the corresponding expression for the axial vector current reads

$$j_5^\mu(x) =: \psi^{(0)}(x)\gamma^\mu\gamma^5\psi^{(0)}(x) := -\frac{1}{\sqrt{\pi}}\partial^\mu\varphi \quad . \tag{10.28b}$$

For the canonical free fermion field in question, φ is canonically normalized:

$$[\varphi(x), \partial_0\varphi(y)]_{ET} = i\delta(x^1 - y^1) \quad . \tag{10.29}$$

Substituting (10.27) into (10.26) and performing the trace, one finds,

$$r.h.s. \approx \mathcal{Z}(\epsilon)\left\{-\frac{1}{\sqrt{\pi}}\tilde{\partial}^\mu\varphi - \frac{i}{\pi}\left[\frac{\epsilon^\mu}{\epsilon^2} - i\sqrt{\pi}\frac{(\epsilon^\mu\epsilon^\nu - \tilde{\epsilon}^\mu\tilde{\epsilon}^\nu)}{\epsilon^2}\tilde{\partial}_\nu(\Sigma(x)+\eta(x))\right]\right\} \tag{10.30}$$

Recalling the property (A.4), and identifying $c^\mu(z)$ in (10.23) with $-\frac{i}{\pi}\mathcal{Z}(\epsilon)\frac{\epsilon^\mu}{\epsilon^2}$, we obtain for the current

$$N[\overline{\psi}(x)\gamma^\mu\psi(x)] = -\frac{1}{\sqrt{\pi}}\tilde{\partial}^\mu[\Sigma(x) + (\eta(x) + \varphi(x))] \quad . \tag{10.31}$$

Comparing (10.31) with (10.11) it follows that $h = \varphi + \eta$. Condition (10.19) thus implies that the (unphysical) field η is to be quantized with metric opposite to that of φ:

$$[\eta(x), \partial_0 \eta(y)]_{ET} = -i\delta(x^1 - y^1) \quad . \tag{10.32}$$

We shall refer to η as an "indefinite metric" field.

Eq. (10.32) fixes the normalization of η. The normalization of Σ is fixed by requiring the canonical commutations relations

$$[A^1(x), \pi_1(y)]_{ET} = i\delta(x^1 - y^1) \quad . \tag{10.33}$$

Since $\pi_1 = F^{01} = \frac{e}{\sqrt{\pi}}\Sigma$, this means that Σ is canonically normalized, as well:

$$[\Sigma(x), \partial_0 \Sigma(y)]_{ET} = i\delta(x^1 - y^1) \quad . \tag{10.34}$$

The renormalization constant (10.25) is thus seen to be finite.

We have herewith completed the construction of the solution, the Maxwell current being given by

$$J^\mu(x) = -\frac{1}{\sqrt{\pi}}\tilde{\partial}^\mu \Sigma(x) + L^\mu(x) \quad , \tag{10.35a}$$

with

$$L_\mu(x) = -\frac{1}{\sqrt{\pi}}\tilde{\partial}_\mu(\varphi + \eta) \quad . \tag{10.35b}$$

Notice that, unlike the zero mass field $\varphi + \eta$, L_μ is a well defined operator free of IR divergencies !

The corresponding expression for the axial vector current reads,

$$J_5^\mu(x) = -\frac{1}{\sqrt{\pi}}\partial^\mu \Sigma - \frac{1}{\sqrt{\pi}}\partial_\mu(\varphi + \eta) \quad . \tag{10.36a}$$

On $\mathcal{H}_{phys}$,

$$J_5^\mu \hat{=} -\frac{1}{\sqrt{\pi}}\partial^\mu \Sigma \quad . \tag{10.36b}$$

Taking its divergence we verify Eq. (10.5).

In the Lorentz gauge, φ plays the role of a "would be" Goldstone boson. In gauge invariant quantities it appears only in the combination $\varphi + \eta$ with the negative metric zero mass field η, which is a reflection of the $U(1)$ "Goldstone-dipole" of Kogut and Susskind[270] in the Coulomb gauge description. Because of the subsidiary condition (10.14), φ decouples from the physical spectrum, the only trace of it being an infinite degeneracy of the vacuum, as we now demonstrate.

10.2.3 Chiral densities

The computation of the chiral densities defined by

$$\rho_\pm(x) = N\left[\overline{\psi}(x)\frac{1\pm\gamma^5}{2}\psi(x)\right] \tag{10.37}$$

is considerably simpler. Now, that we have established the properties of the fields Σ, φ, and η, it is convenient do depart from the Mandelstam formula[47,361] for the free, canonically normalized, "spin" 1/2 massless fermion field[4)],

$$\psi^{(0)}_\alpha(x) = \left(\frac{\mu}{2\pi}\right)^{1/2} e^{-\frac{i\pi}{4}\gamma^5_{\alpha\alpha}} : e^{i\sqrt{\pi}(\gamma^5_{\alpha\alpha}\varphi(x)+\int_{x^1}^\infty dz^1\partial_0\varphi(x^0,z^1))} : \quad . \tag{10.38}$$

One checks that $\psi^{(0)}_\alpha$ satisfies Fermi's statistics.

$$\psi^\dagger_\alpha(x^0,x^1)\psi_\beta(x^0,y^1) = e^{i\pi(\gamma^5_{\alpha\alpha}\theta(x^1-y^1)-\gamma^5_{\beta\beta}\theta(y^1-x^1))}\psi_\beta(x^0,y^1)\psi^\dagger_\alpha(x^0,x^1) \tag{10.39a}$$
$$\psi_\alpha(x^0,x^1)\psi_\beta(x^0,y^1) = e^{-i\pi(\gamma^5_{\alpha\alpha}\theta(x^1-y^1)-\gamma^5_{\beta\beta}\theta(y^1-x^1))}\psi_\beta(x^0,y^1)\psi_\alpha(x^0,x^1) \tag{10.39b}$$

Note that it is the additive constant $\pm i/4$ in the commutation relations (A.31), which insures the anti-commutativity also for $\alpha \neq \beta$.

Using (10.38), we may write the solution (10.22) to the Dirac equation in the form,

$$\psi_\alpha(x) = \left(\frac{\mu}{2\pi}\right)^{1/2} e^{-\frac{i\pi}{4}\gamma^5_{\alpha\alpha}} : e^{i\sqrt{\pi}\gamma^5_{\alpha\alpha}(\Sigma(x)+\varphi(x)+\eta(x))+i\sqrt{\pi}\int_{x^1}^\infty dz^1\partial_0\varphi(x^0,z^1)} : \quad . \tag{10.40}$$

The $U(1)$ part associated with $\tilde{\varphi}(x)$ cancels out in (10.37) defined via the corresponding short distance expansion

$$\overline{\psi}(x+\epsilon)\frac{1\pm\gamma^5}{2}\psi(x) \approx c(\epsilon) + f(\epsilon)N\left[\overline{\psi}(x)\frac{1\pm\gamma^5}{2}\psi(x)\right] \quad . \tag{10.41}$$

The chirality ± 2 of this density implies $c(\epsilon) = 0$. Reordering again the exponential factors according to their positive and negative frequency parts, we find that $\overline{\psi}(x+\epsilon)(1\pm\gamma_5)\psi(x)$ is finite as $\epsilon \to 0$. Using (A.31), and choosing $f(\epsilon) = 1$, we have[5)]

$$N\left[\overline{\psi}(x)\frac{1\pm\gamma^5}{2}\psi(x)\right] = -\left(\frac{\tilde{\mu}}{2\pi}\right) : e^{\pm 2i\sqrt{\pi}(\Sigma(x)+\varphi(x)+\eta(x))} : \quad , \tag{10.42a}$$

[4)] The exponential of the line integral defining $\tilde{\varphi}$ contains a "Klein" transformation which insures that the anticommutation relations for ψ_α also hold for $\alpha \neq \beta$, as is manifest in (10.39) (see appendix A).

[5)] The extra phase factor in (10.38) combines with that from the additive imaginary constants in (A.31) to yield $e^{i\pi} = -1$. It is needed in order that the mass operator constructed from (10.38) be hermitian.

where $\tilde{\mu} = \mu \left(\frac{\tilde{m}}{\mu}\right)^{1/2}$, $\tilde{m} = \frac{e}{2\sqrt{\pi}} e^{\gamma}$, with γ the Euler constant. We thus have

$$N\left[\overline{\psi}(x)\psi(x)\right] = -\left(\frac{\tilde{\mu}}{\pi}\right) : \cos 2\sqrt{\pi}(\Sigma(x) + \varphi(x) + \eta(x)): \quad , \qquad (10.42b)$$

$$N\left[\overline{\psi}(x)\gamma^5\psi(x)\right] = -i\left(\frac{\tilde{\mu}}{\pi}\right) : \sin 2\sqrt{\pi}(\Sigma(x) + \varphi(x) + \eta(x)): \quad . \qquad (10.42c)$$

We conclude this section with an important remark. We have seen in chapter 2, that the exponential of zero mass scalar fields carries selection rules as a result of the IR behaviour of such fields. In particular, the chiral density (10.37) should carry the chiral selection rule corresponding to the chirality ± 2 of this operator, implying

$$\langle 0|\rho_{\pm}(x)|0\rangle = 0 \quad . \qquad (10.43)$$

According to our discussion in chapter 2, this selection rule is carried by φ. Hence the zero mass, indefinite metric field η should not carry this selection rule. Since η creates (negative norm) states lying in the unphysical part of the Hilbert space we are free to define it with an IR cutoff, without taking this cutoff to infinity. This amounts to define[26,92] its norm in terms of the principal value $\mathcal{P}$,

$$\|\eta^{(-)}[t;f] \mid 0\rangle\|^2 = -\mathcal{P}\frac{1}{2\pi}\int \frac{dp}{2|p|}|\tilde{f}(p)|^2 \quad ,$$

where $\eta^{(-)}[t;f] = \int dy^1 \eta^{(-)}(t,y^1)f(y^1)$, and $\tilde{f}(p)$ denotes the Fourier transform of $f(y^1)$. With this agreement, $e^{i\eta}$ *does not carry a selection rule.*

10.2.4 Vacuum structure

We now show that the QED_2 vacuum is two-fold infinitely degenerate.

The chiral densities (10.41) belong to the class of observables since they commute with the longitudinal part of the current, L_μ. We define the operators[28,349]

$$\sigma_\alpha(x) = e^{i\sqrt{\pi}\gamma^5_{\alpha\alpha}(\varphi(x)+\eta(x))+i\sqrt{\pi}\int_{x^1}^{\infty} dz^1 \partial_0(\varphi(x^0,z^1)+\eta(x^0,z^1))} \quad . \qquad (10.44)$$

Since η is a field quantized with a metric opposite to that of φ, it follows that the σ_α commute with themselves for all space-time points, and have scale dimension zero, so that no normal ordering prescription is needed. They also belong to the class of observables since $[\sigma_\alpha(x), L_\mu(y)] = 0$.

In terms of (10.44), the results (10.42a) for $\rho_\pm$ read

$$\rho_+(x) = -\left(\frac{\tilde{\mu}}{2\pi}\right) : e^{2i\sqrt{\pi}\Sigma(x)} : \sigma_2^+ \sigma_1 \tag{10.45a}$$

$$\rho_-(x) = -\left(\frac{\tilde{\mu}}{2\pi}\right) : e^{-2i\sqrt{\pi}\Sigma(x)} : \sigma_1^+ \sigma_2 \tag{10.45b}$$

It will prove useful to also define the bosonic observables[49]

$$\hat{\psi}_\alpha(x) \equiv \left(\frac{\tilde{\mu}}{2\pi}\right)^{1/2} : e^{i\sqrt{\pi}\gamma^5_{\alpha\alpha}\Sigma(x)} : \sigma_\alpha(x) \quad . \tag{10.45c}$$

Formally these operators are obtained from $\psi_\alpha(x)$ by an operator gauge transformation. By applying $\hat{\psi}_\alpha$ on the Fock-vacuum one obtains what has been referred to as "bleached states"[6)]

We shall refer to σ_α as spurion operators since they carry - as we next show - a bare charge (fermion number) and chiral selection rule, and are otherwise constant operators on $\mathcal{H}_{phys}$.

The free vector and axial vector currents are given by (10.28). Both are conserved. Using (C.13) one has

$$[j_0(x), \sigma_\alpha(y)]_{ET} = -\delta(x^1 - y^1)\sigma_\alpha(x) \tag{10.46a}$$

$$[j_0^5(x), \sigma_\alpha(y)]_{ET} = -\gamma^5_{\alpha\alpha}\delta(x^1 - y^1)\sigma_\alpha(x) \tag{10.46b}$$

Hence the operators σ_α carry a "bare charge" and "bare chirality", where these terms refer to the eigenvalues of[7)] q and $-q_5$ defined by

$$q \equiv \int dx^1 j^0(x) \quad , \quad q_5 \equiv \int dx^1 j_5^0(x) \quad . \tag{10.47}$$

One evidently has

$$[q, \rho_\pm(x)] = 0 \quad , \quad [q_5, \rho_\pm(x)] = \mp 2\rho_\pm(x) \tag{10.48a}$$

$$[q, \hat{\psi}_\alpha(x)] = -\hat{\psi}_\alpha(x) \quad , \quad [q_5, \hat{\psi}_\alpha(x)] = -\gamma^5_{\alpha\alpha}\hat{\psi}_\alpha(x) \tag{10.48b}$$

---

6) The existence of such bleached states in the physical subspace $\mathcal{H}_{phys}$ of QED_2 has been questioned by the authors of ref. [362], who attribute the inclusion of such states to a redundant description of the Lowenstein-Swieca solution. See also ref. [363].

7) "Chirality" is measured by the eigenvalues of γ^5. With our definition (10.28b) of the axial vector current we thus need to define chirality in terms of the negative of the space-integral over j_5^0, in order to conform to this convention.

One readily checks, that the conserved charges q and q_5 generate global $U(1)$ and $\tilde{U}(1)$ transformation on ψ_α [8]. They are however not observables (see following section).

One further checks that the conserved physical charge operator Q, formally given by $Q = \int dx^1 J^0(x)$, does not generate $U(1)$ transformations on ψ_α. On the other hand, the corresponding chiral observable $Q_5 = \int dx^1 J_5^0(x)$ is not conserved, and hence cannot be the generator of chiral $\tilde{U}(1)$ transformations.

These observations represent nothing new. Indeed, j_5^μ is the $d = 2$ analogue of the usual ($d = 4$) gauge variant[9], conserved $U(1)$ axial vector current which one obtains by making use of the fact, that the ABJ anomaly in Eq. (10.5) can be written as a total divergence:

$$j_5^\mu = J_5^\mu - \frac{e}{\pi}\epsilon^{\mu\nu} A_\nu \quad . \tag{10.49}$$

Inspection of expressions (10.28b)and (10.36) for j_5^μ and J_5^μ illustrates a general phenomenon: The Goldstone field φ couples to the gauge variant current j_5^μ, but decouples from the gauge-invariant current J_5^μ on $\mathcal{H}_{phys}$. This is the well known "evasion of the Goldstone Theorem[214]" for the case of gauge theories[365]: spontaneous symmetry breakdown without Goldstone bosons. Hence our terminology, "would be Goldstone boson".

From (10.46) we conclude, that the operator $(\sigma_1\sigma_2)$ only carries fermion number, but no chirality, while the opposite applies to the operator $(\sigma_1^\dagger\sigma_2)$. They are thus the carriers of the superselection rules induced by the infraparticle associated with φ. In fact, on $\mathcal{H}_{phys}$, the operators σ_α do nothing else but to count fermion number (bare charge) and chirality; because of the indefinite metric character of the fields η and $\tilde{\eta}$, they act as constant operators with respect to the observables and thus carry no energy or momentum. This has important consequences to be discussed, now.

The equivalent bosonic Hamiltonian is given by[29,349]

$$H_0 = \frac{1}{2}\int dx^1 \{ :[(\pi_\Sigma)^2 + (\partial_1\Sigma)^2 + \frac{e^2}{\pi}\Sigma^2]: + :[(\pi_\varphi)^2 + (\partial_1\varphi)^2]: - :[(\pi_\eta)^2 + (\partial_1\eta)^2]: \} \tag{10.50}$$

where $\pi_\Sigma = \partial_0\Sigma$, $\pi_\varphi = \partial_0\varphi$ and $\pi_\eta = -\partial_0\eta$.

The mass term in Σ is just a manifestation of the dynamical Higgs mechanism[366], we have referred to before, while the occurrence of the term in η with negative sign is a consequence of its indefinite metric character. As a result, only the first term in

---

8) The action of a "rigid" chiral transformation on φ is given by $[q_5, \varphi] = \frac{i}{\sqrt{\pi}}$, or using (C.13), $e^{-i\alpha\sqrt{\pi}q_5}\varphi e^{i\alpha\sqrt{\pi}q_5} = \varphi + \alpha$.

9) The fact that the gauge variant chiral charge is not a legitimate operator on the physical subspace has been emphasized by 't Hooft[264] in connection with the $U(1)$ problem [363].

(10.50), involving Σ, survives on $\mathcal{H}_{phys}$. Hence the physical spectrum of the Hamiltonian is the Fock-space of (non-interacting) pseudoscalar mesons of mass $e/\sqrt{\pi}$. These bosons can be thought of as asymptotic states of fermion-antifermion pairs permanently bound by the long range Coulomb force (confinement). Furthermore

$$[H, \sigma_\alpha(x)] = 0 \quad . \tag{10.51}$$

Hence the ground state is in fact degenerate, an infinite number of mutually orthogonal ground states being generated by repeated application of σ_α on the Fock-vacuum:

$$|n_1 n_2\rangle = \sigma_1^{n_1}\sigma_2^{n_2}|0\rangle \tag{10.52}$$

with $\langle n_1' n_2' | n_1 n_2\rangle = \delta_{n_1' n_1}\delta_{n_2' n_2}$. One thus obtains from (10.45),

$$\langle n_1' n_2' |\rho_\pm(x)| n_1 n_2\rangle = \left(\frac{-\tilde{\mu}}{2\pi}\right)\delta_{n_1', n_1 \pm 1}\delta_{n_2', n_2 \mp 1} \quad . \tag{10.53a}$$

$$\langle n_1' n_2' |\hat{\psi}_1(x)| n_1 n_2\rangle = \left(\frac{\tilde{\mu}}{2\pi}\right)^{1/2}\delta_{n_1', n_1 + 1}\delta_{n_2' n_2} \quad . \tag{10.53b}$$

Since according to (10.48) $\hat{\psi}_\alpha$ carries chirality $\gamma_{\alpha\alpha}$ and charge -1, its vacuum expectation value would normally vanish. Eq. (10.53b) thus signalizes a spontaneous breakdown of the $U(1) \times \widetilde{U}(1)$ symmetry. This is allowed despite Coleman theorem[93]. Since we are dealing with a gauge theory this spontaneous symmetry breakdown can be realized without the appearance of Goldstone bosons, which have been "eaten" up by the gauge field, thereby giving it a mass $e/\sqrt{\pi}$ (Higgs mechanism).

Since the observables commute with $\sigma_\alpha(x)$ for all space-time points, we can go over to a new set of irreducible ground states with respect to which these observables are diagonal. To this end consider the coherent superposition (in the instanton approximation, a similar situation prevails in QCD_4).

$$|\theta_1\theta_2\rangle = \frac{1}{2\pi}\sum_{n_1,n_2} e^{-in_1\theta_1}e^{-in_2\theta_2}|n_1, n_2\rangle \tag{10.54}$$

of the vacua (10.52), where the summation extends over all integers n_1 and n_2. We have

$$\sigma_\alpha|\theta_1\theta_2\rangle = e^{i\theta_\alpha}|\theta_1\theta_2\rangle \quad . \tag{10.55}$$

Hence, on the gauge-invariant subspace the spurion operators σ_α are reduced to a phase $\exp(i\theta_\alpha)$, where θ_1 and θ_2 label the gauge-invariant vacua. We refer to this process as "spurionization".

In the irreducible basis (10.54), the Hamiltonian (10.50), the chiral densities (10.45) and the operator $\hat{\psi}$ are diagonal. Their diagonal elements are given by

$$H_0 = \frac{1}{2}\int dx^1 : \pi_\Sigma^2 + (\partial_1\Sigma)^2 + \frac{e^2}{\pi}\Sigma^2 : \tag{10.56a}$$

$$\rho_\pm(x) = -\left(\frac{\tilde{\mu}}{2\pi}\right) : e^{\pm 2i\sqrt{\pi}\Sigma(x)} : e^{i(\theta_1-\theta_2)} \tag{10.56b}$$

$$\hat{\psi}_\alpha(x) = \left(\frac{\tilde{\mu}}{2\pi}\right)^{1/2} e^{i\sqrt{\pi}\gamma^5_{\alpha\alpha}\Sigma(x)} e^{i\theta_\alpha} \quad . \tag{10.56c}$$

Hence, in the irreducible basis the vacuum expectation value of $\rho_\pm$ and $\hat{\psi}_\alpha$ no longer vanishes[10)], showing the spontaneous breakdown of the chiral and fermion number symmetry[28,349,29]. Note that the Hamiltonian, (10.56a) is independent of θ_α, so that its spectrum is infinitely degenerate.

With respect to the irreducible vacuum $|\theta_1\theta_2\rangle$, the short distance calculations of the Maxwell current and chiral densities may be conveniently summarized by the following limits[28,49]

$$J^\mu(x) = -\lim_{\epsilon\to 0}\{\mathrm{tr}\,\gamma^0\gamma^\mu T(x, x+\epsilon) - \langle\theta|\,\mathrm{tr}\,\gamma^0\gamma^\mu T(x, x+\epsilon)|\theta\rangle\} \tag{10.57a}$$

$$\rho_\pm = -\lim_{\epsilon\to 0}\mathrm{tr}\,\gamma^0\frac{(1\pm\gamma_5)}{2}T(x, x+\epsilon) \quad , \tag{10.57b}$$

where

$$T_{\alpha\beta}(x,y) = N_\theta(x-y)_{\alpha\beta} : e^{i\sqrt{\pi}[\gamma^5_{\alpha\alpha}\Sigma(x) - \int_x^y dz^\mu \tilde{\partial}_\mu\Sigma(z) - \gamma^5_{\beta\beta}\Sigma(y)]} : \tag{10.58a}$$

with

$$N_\theta(z) = \frac{1}{2\pi}\begin{pmatrix} \frac{i}{z^- - i0} & \tilde{\mu}e^{i(\theta_1-\theta_2)} \\ \tilde{\mu}e^{-i(\theta_1-\theta_2)} & \frac{i}{z^+ - i0} \end{pmatrix} \quad . \tag{10.58b}$$

The minus sign arises from the Fermi statistics obeyed by $\psi_\alpha(x)$. Note that (10.58) is just the finite expression for the (formally) gauge invariant operator

$$T_{\alpha\beta}(x,y) \approx \psi_\alpha(x) e^{ie\int_x^y dz^\mu A_\mu(z)} \psi^\dagger_\beta(y) \quad . \tag{10.59}$$

The importance of the algebra of such observables has been emphasized by Lowenstein and Swieca[28]. Eq. (10.58a) shows that this algebra is isomorphic to a free scalar field of mass $e/\sqrt{\pi}$. In particular, the Hamiltonian only contains free scalar excitations with this mass, thus showing the absence of fermions in its spectrum (confinement of "quarks").

10) The non-vanishing of $\langle\theta \mid \rho_\pm \mid \theta\rangle$ in QCD_4 plays a central role in the QCD sum rules.

10.2.5 Gauge transformations

The situation described above is analogous to the case of a periodic potential in a one dimensional crystal with periodicity a, where semi-classically one has states $|n\rangle$, localized around the nth minimum of the potential, which are related via translations by an integral number of periods a

$$T[ma]|n\rangle = |n+m\rangle \quad , \tag{10.60}$$

with $T[ma] = e^{imaP_{op}}$. However, unlike the case of the Schwinger model, the a priori-degeneracy of these semi-classical ground states in a periodic crystal is lifted by quantum mechanical tunneling, so that the translational invariance is not spontaneously broken in this case. In the case of the (massless) Schwinger model only operators carrying "bare charge" or chirality can induce such a tunneling process. Since H carries neither one of these quantum numbers, the vacuum degeneracy is not lifted. In fact, as we now show[367], the operator with the property analogous to (10.60) is one which realizes topologically non-trivial gauge transformations, so that the different $| n_1 n_2\rangle$ vacua can be viewed as gauge transforms of each other. Since we are working in the Lorentz gauge $\partial_\mu A^\mu = 0$, we can only consider c-number gauge transformations

$$T[\Lambda]\psi(x)T^{-1}[\Lambda] = e^{i\Lambda(x)}\psi(x) \tag{10.61a}$$

$$T[\Lambda]A_\mu(x)T^{-1}[\Lambda] = A_\mu(x) + \frac{1}{e}\partial_\mu\Lambda(x) \quad , \tag{10.61b}$$

with Λ satisfying the Klein-Gordon equation $\Box\Lambda = 0$. The operator $T[\Lambda]$ which induces the gauge transformation (10.61) is given by

$$T[\Lambda] = e^{i\chi[\Lambda]} \quad , \tag{10.62a}$$

with ($\tilde{\varphi}$ and $\tilde{\eta}$ are the fields dual to φ and η, respectively)

$$\chi[\Lambda] = \frac{1}{\sqrt{\pi}}\int_{y_0} dy^1 \left\{(\eta(y)+\varphi(y))\partial_1\Lambda(y) - (\tilde{\eta}(y)+\tilde{\varphi}(y))\partial_0\Lambda(y)\right\} \quad . \tag{10.62b}$$

The time y^0 can be chosen arbitrarily, since $\frac{d}{dt}\chi[\Lambda] = 0$ as a result of Eq. (10.61b). Hence, using

$$[\tilde{\varphi}(x),\varphi(y)]_{ET} = -[\tilde{\eta}(x),\eta(y)]_{ET} = -i\theta(y^1 - x^1) \tag{10.63}$$

one finds

$$[\chi[\Lambda],\sqrt{\pi}\tilde{\eta}(x)] = -i\int_{-\infty}^{x^1} dy^1\partial_1\Lambda \quad , \quad [\chi[\Lambda],\sqrt{\pi}\eta(x)] = -i\int_{-\infty}^{x^1} dy^1\partial_0\Lambda(x) \quad . \tag{10.64}$$

From here, and (C.12) then follow the transformation laws ($\Box\Lambda = 0$).

It was so far implicitly assumed, that $\Lambda(\pm\infty, t) = 0$. In that case $\chi[\Lambda]$ is a well defined operator, and the corresponding operator (10.62) acts like the identity on $\mathcal{H}_{phys}$. If on the other hand $\Lambda(\pm\infty, t) \neq 0$, the operator $T[\Lambda]$ will not act trivially on $\mathcal{H}_{phys}$, but carries a "super selection rule" analogous to σ_α. Indeed, it follows from $\Box\Lambda = 0$, that $\Lambda(x)$ decomposes into a right- and a left-moving wave:

$$\Lambda(x) = f_1(x^1 + x^0) + f_2(x^1 - x^0) \quad . \tag{10.65}$$

Hence we may put (10.62*b*) also in the form

$$\chi[\Lambda] = \frac{1}{\sqrt{\pi}} \sum_{r=1}^{2} \int dy^1 \phi_r(x) \partial_1 f_r \quad , \tag{10.66}$$

where $\phi_r(x) = (\eta(x) + \varphi(x)) + (-1)^r(\tilde{\eta}(x) + \tilde{\varphi}(x))$.

In order to reveal the above mentioned selection rules, we define

$$\Lambda^{[-\frac{1}{2}]}(x) \equiv f_1(x^1 - x^0) \tag{10.67a}$$

$$\Lambda^{[\frac{1}{2}]}(x) \equiv f_2(x^1 + x^0) \tag{10.67b}$$

with the boundary conditions

$$f_r(-\infty) = 0 \quad , \quad f_r(\infty) = (-1)^r \pi \quad . \tag{10.68}$$

The superscript of Λ indicates that regular gauge potentials with these asymptotic properties correspond to a Chern-Pontryagin (winding) number of $\pm\frac{1}{2}$ (see chapter 12).

With the choice of boundary conditions (10.68), we obtain from (10.62), after a partial integration,

$$T[\Lambda^{[+\frac{1}{2}]}] = e^{-i\left\{\int dy^1 (L_0(x) - L_1(x))\Lambda^{[+\frac{1}{2}]}(x)\right\}} \sigma_1(x) \tag{10.69a}$$

$$T[\Lambda^{[-\frac{1}{2}]}] = e^{-i\left\{\int dy^1 (L_0(x) + L_1(x))\Lambda^{[-\frac{1}{2}]}(x)\right\}} \sigma_2(x) \tag{10.69b}$$

where $\sigma_{1,2}$ are the operators defined in (10.44), and L_μ is given by (10.35*b*). Recalling (10.14) we conclude that

$$T[\Lambda^{[\pm\frac{1}{2}]}] \hat{=} \sigma_{\frac{1}{2}}(x) \tag{10.70}$$

on $\mathcal{H}_{phys}$.

Because of (10.53) we thus expect a relation of the form

$$\nu = \frac{\Delta n_1 - \Delta n_2}{2} \quad . \tag{10.71}$$

between the winding number ν of the (Euclidean) field configurations A_μ contributing in a path-integral formulation, and the vacuum to vacuum transition $|n_1 n_2\rangle \to |n_1 + \Delta n_1, n_2 + \Delta n_2\rangle$ (quantum mechanical tunneling).

Since $\Delta(n_1 - n_2)$ represents the change in chirality involved in such a transition, we recognize in (10.71) the Atiyah-Singer theorem discussed in chapter 4.

We shall show in chapter 12, that the above picture of instanton-induced vacuum tunneling is indeed realized in the functional approach. There also exist meron configurations[368] of half-integral winding number, which in two dimensions correspond to configurations with finite action. Their existence is expected on account of the non-vanishing vacuum expectation value of $\langle \theta \mid \hat{\bar{\psi}} \mid \theta \rangle$. In D = 4 dimensions such configurations would correspond to an infinite action, and are thus expected not to contribute.

Finally let us remark, that the currents (10.28), though conserved, are not gauge invariant. Using (10.62) one computes

$$T[\Lambda] j_5^\mu(x) T^{-1}[\Lambda] = j_5^\mu(x) - \frac{1}{\pi}\tilde{\partial}^\mu \Lambda(x) \quad ,$$

in agreement with the corresponding transformation property read off from (10.49). Note that the corresponding charge changes by twice the winding number of the gauge transformation. A similar statement holds for the $U(1)$ charge of QCD_4. Hence q_5 is not invariant under homotopically non-trivial gauge transformations (the boundary conditions (10.67,68) are now replaced by $\Lambda^{[n]}(x) \to 2n\pi$ for $x^1 \to \infty$ and $\Lambda^{[n]} \to 0$, for $x^1 \to -\infty$),

$$T[\Lambda^{[n]}] q_5 T^{-1}[\Lambda^{[n]}] = q_5 + 2n \quad .$$

Hence the operator $T[\Lambda^{[n]}]$ lowers the chirality of a state by $2n$. A similar situation prevails in QCD_4.

10.2.6 Correlation functions and violation of clustering

In a local quantum field theory there exists an intimate relation between the cluster decomposition and the degeneracy of the ground state.

The simplest gauge-invariant correlation functions are those constructed from $\hat{\psi}_\alpha$, $F_{\mu\nu}$, the currents J^μ, J_5^μ, and the chiral densities $\rho_\pm(x)$.

From (10.45a,b) one has

$$\langle n,-n \mid \prod_{i=1}^{n} \rho_+(x_i) \mid 0,0\rangle = \left(-\frac{\tilde{\mu}}{2\pi}\right)^n \langle 0,0| \prod_{i=1}^{n} : e^{2i\sqrt{\pi}\Sigma(x_i)} : |0,0\rangle \quad (10.72a)$$

$$\langle 0,0 \mid \prod_{i=1}^{n} \rho_-(x_i) \mid n,-n\rangle = \left(-\frac{\tilde{\mu}}{2\pi}\right)^n \langle 0,0| \prod_{i=1}^{n} : e^{-2i\sqrt{\pi}\Sigma(x_i)} : |0,0\rangle \quad (10.72b)$$

Using (C.12), the r.h.s. of (10.72) is easily evaluated to give

$$\langle \pm n, \mp n \mid \prod_{i=1}^{n} \rho_\pm(x_i) \mid 0,0\rangle = \left(-\frac{\tilde{\mu}}{2\pi}\right)^n e^{-4\pi \sum_{i<j} \Delta^{(+)}(x_i - x_j; \frac{e^2}{\pi})} \quad . \quad (10.73)$$

This result is also obtained by looking at the clustering properties of the diagonal element

$$\langle 0,0|\rho_-(\xi_1)\cdots\rho_-(\xi_n)\rho_+(x_1)\cdots\rho_+(x_n)|0,0\rangle = (\frac{\mu}{2\pi})^{2n} \times$$
$$e^{-4\pi\sum_{i<j}\Delta^{(+)}(x_i-x_j;\frac{e^2}{\pi})} e^{-4\pi\sum_{i<j}\Delta^{(+)}(\xi_i-\xi_j;\frac{e^2}{\pi})} e^{4\pi\sum_{i,j}\Delta^{(+)}(\xi_i-x_j;\frac{e^2}{\pi})} \quad (10.74)$$

In the limit $\xi_i \to \infty$, $\xi_i - \xi_j$ and x_j fixed, this expression is seen to factorize as follows:

$$\langle 0,0|\rho_-(\xi_1)\cdots\rho_+(x_n)|0,0\rangle \to$$
$$\langle 0,0|\rho_-(\xi_1)\cdots\rho_-(\xi_n)|n,-n\rangle\langle n,-n|\rho_+(x_1)\cdots\rho_+(x_n)|0,0\rangle \quad (10.75)$$

where the r.h.s. is given in terms of expression (10.73).

This violation of clustering is evidently a direct consequence of the vacuum degeneracy and spontaneous breakdown of the chiral symmetry.

Similarly one finds for the observables (10.45c),

$$\langle 0,0|\hat{\psi}_1(\xi_1)\cdots\hat{\psi}_1(\xi_n)\hat{\psi}_1(x_1)\cdots\hat{\psi}_1(x_n)|0,0\rangle \to$$
$$\langle 0,0|\hat{\psi}_1(\xi_1)\cdots\hat{\psi}_1(\xi_n)|n,0\rangle\langle n,0|\hat{\psi}_1(x_1)\cdots\hat{\psi}_1(x_n)|0,0\rangle \quad (10.76)$$

with

$$\langle n,0|\hat{\psi}_1(x_1)\cdots\hat{\psi}_1(x_n)|0,0\rangle = \left(\frac{\tilde{\mu}}{2\pi}\right)^{n/2} e^{-\pi\sum_{i<j}\Delta^{(+)}(x_i-x_j;\frac{e^2}{\pi})} \quad . \quad (10.77)$$

Eqs. (10.75) and (10.77) express the spontaneous symmetry breakdown of $U(1) \times \tilde{U}(1)$. The situation is thus analogous to that of superconductivity, where the vacuum also carries a fermionic quantum number.

10.2.7 Absence of charged states (screening)

Because of the property (10.14), the current (10.35) is given on $\mathcal{H}_{phys}$ by

$$J^\mu = -\frac{1}{\sqrt{\pi}}\tilde{\partial}^\mu \Sigma + L^\mu \hat{=} -\frac{1}{\sqrt{\pi}}\tilde{\partial}^\mu \Sigma \quad . \tag{10.78}$$

The corresponding charge operator Q is thus

$$Q = \frac{1}{\sqrt{\pi}} \int dx^1 \partial_1 \Sigma(t, x^1) \quad . \tag{10.79}$$

Because of the commutation relations (10.33) we have

$$[Q, \hat{\psi}_\alpha(x)] = 0 \quad . \tag{10.80}$$

The operator $\hat{\psi}$ thus carries no charge! It lives in the charge zero sector of the Schwinger model. As we now show, this is in fact the only sector of states having finite energy. To see this, we attempt to construct charged states, by repeated application of the disorder operator

$$\mu_\Sigma(x) =: e^{i\sqrt{\pi}\int_{x^1}^{\infty} dz^1 \partial_0 \Sigma(x^0, z^1)}: \tag{10.81}$$

on the physical vacuum. Using (C.12) one computes

$$[Q, \mu_\Sigma(x)] = -\mu_\Sigma(x) \quad . \tag{10.82}$$

The disorder operator (10.81) is thus a candidate for creating a soliton state with one unit negative charge from the vacuum. We define the order operator by

$$\sigma_\Sigma(x)_\alpha =: e^{i\sqrt{\pi}\gamma^5_{\alpha\alpha}\Sigma(x)}: \tag{10.83}$$

The operators σ_Σ and μ_Σ obey the order disorder algebra (compare with (3.90))

$$\mu_\Sigma(x)\sigma_\Sigma(y) = \begin{cases} -\sigma_\Sigma(y)\mu_\Sigma(x), & y^1 > x^1 \\ \sigma_\Sigma(y)\mu_\Sigma(x), & y^1 < x^1 \end{cases} \tag{10.84}$$

In the irreducible representation we have, after a suitable renormalization

$$N\left[\sigma_\Sigma(x)_\alpha \mu_\Sigma(x)\right] \approx: \psi_\alpha(x) e^{-ie\int_{-\infty}^{x^1} dz^1 A_1(x^0, z^1)}: e^{-i\theta_\alpha} \quad , \tag{10.85}$$

Hence the product of the order and disorder operators is a candidate for the "physical" fermion field[11]. The line integral is an expression of the fact, that a charged particle always "carries with it" an electromagnetic tail.

---

[11] The same is found to be the case for the Ising model[369].

We may view the operator (10.85) as being the result of taking a pair of opposite charges localized at x and y, as represented by the dipole operator[49,370]

$$D_{\alpha\beta}(x,y) = \overline{\psi}_\alpha(x) e^{-ie\int_{x^1}^{y^1} dz^1 A_1(z)} \psi_\beta(y) \quad , \tag{10.86}$$

and placing one of the charges at infinity. Because of the linear rise of the Coulomb potential in one space dimension, we expect this process to require an infinite energy. Indeed, the operator

$$T_{x^1}^{y^1}(t) =: e^{i\sqrt{\pi}\int_{x^1}^{y^1} dz^1 \partial_0 \Sigma(t,z^1)}: \quad , \tag{10.87}$$

just translates $\Sigma(z)$ by $\sqrt{\pi}$ for all $x^1 \leq z^1 \leq y^1$:

$$T_{y^1}^{x^1}(t)\Sigma(t;z^1)T_{x^1}^{y^1}(t) = \Sigma(t,z^1) + \sqrt{\pi}\theta(z^1 - x^1)\theta(y^1 - z^1) \quad . \tag{10.88}$$

Consider the dipole state

$$|D(x^1,y^1;t)\rangle \equiv T_{y^1}^{x^1}(t)|0\rangle \quad . \tag{10.89}$$

Because of (10.88) one obtains from (10.56a)[28]

$$\langle D(x^1,y^1;t)|H_0|D(x,y;t)\rangle - \langle 0|H_0|0\rangle = e^2|y^1 - x^1| \quad , \tag{10.90}$$

where we have dropped the end point contributions to the kinetic energy. Note that (10.90) is the Coulomb potential of two dipoles. As expected, expression (10.90) diverges as $y^1 \to \infty$, thus showing the absence of charged finite energy states from $\mathcal{H}_{phys}$. This is in agreement with Schwinger's intuitive picture of the total "screening" of charge by the electromagnetic interaction[12)].

The total charge of the electric poles at the end of the string is found[359] to oscillate periodically in time with a frequency $e/\sqrt{\pi}$. This is a result of the vacuum polarization effects, whereby the electric energy (10.90) of the dipole is used to create virtual pairs. Since it takes no energy to create a virtual pair of massless fermions, this instability persists for arbitrary separations $|\, x^1 - y^1 \,|$.

The linear rise of the energy $\langle D|H_0|D\rangle$ with $|y^1 - x^1|$ is directly linked to the mass term in (10.56a), which can be traced to the Coulomb potential and is responsible for the short range character of the electromagnetic interaction (intrinsic Higgs mechanism). Indeed we have on $\mathcal{H}_{phys}$, using (10.78)

$$\int dx^1 \frac{e^2}{2\pi}\Sigma^2(t,x^1) \widehat{=} -\frac{e^2}{2}\int dx^1 \int dy^1 J^0(x^1,t)\frac{|x^1 - y^1|}{2}J^0(y^1,t) \tag{10.91}$$

12) For a general discussion concerning the relation between charge screening and the existence of a mass gap, see ref.[371].

where we have used $\partial_1^2 \frac{|x^1-y^1|}{2} = \delta(x^1 - y^1)$.

These properties are also reflected by the asymptotic behavior

$$\left\langle \sigma_\Sigma(x)\sigma_\Sigma^\dagger(y)\right\rangle = e^{-\frac{1}{2}K_0\left(\frac{e}{\sqrt{\pi}}|x-y|\right)} \underset{|x-y|\to\infty}{\longrightarrow} 1 \tag{10.92a}$$

$$\left\langle \mu_\Sigma(x)\mu_\Sigma^\dagger(y)\right\rangle \to 0 \tag{10.92b}$$

which implies $\langle\sigma_\Sigma\rangle \neq 0$ and $\langle\mu_\Sigma\rangle = 0$, corresponding to one of the three possible situations allowed by the order-disorder algebra[124] (see chapter 3).

When turning off the electromagnetic interaction (by setting $e = 0$), Σ turns into a massless field ϕ and the resulting disorder operator

$$\mu_\phi = e^{i\sqrt{\pi}\int_{x^1}^{\infty} dz^1 \partial_0\phi(t,z^1)} \quad , \tag{10.93}$$

now creates finite energy charged sectors from the vacuum, which are the charge sectors of the massless Thirring model. Correspondingly, $:\sigma_\phi(x)\mu_\phi(x):$ is just the fermion of the Thirring model.

10.2.8 The quark-antiquark potential

It is enlightening to examine the screening of the fermionic charge from still another point of view.

Because of the intrinsic Higgs mechanism induced by the vacuum polarization, we expect the potential of an external quark-antiquark ($q\bar{q}$) pair to exhibit a behaviour characteristic of screened charges. This potential can be calculated by either performing a typical Wilson loop[372] calculation (see chapter 12) or by comparing the ground state expectation value of the Hamiltonian (10.56a) in the presence of an external $q\bar{q}$ source, with that obtained in the absence of such a source. Let $x^1 = \frac{L}{2}(-\frac{L}{2})$ be the location of the test charges Q $(-Q)$. The $q\bar{q}$ potential $V(L)$ is then defined by

$$V(L) = \langle\Omega_Q|H_Q|\Omega_Q\rangle - \langle\Omega_0|H_0|\Omega_0\rangle \quad , \tag{10.94}$$

where H_0 (H_Q) and $|\Omega_0\rangle$ ($|\Omega_Q\rangle$) are the Hamiltonian and ground state in the absence (presence) of the test charges, respectively. H_Q is easily obtained, by noting that the charge density in the presence of these test charges is given by

$$\begin{aligned} J^0_{tot}(t,x^1) &= \frac{Q}{e}\left(\delta(x^1 - \frac{L}{2}) - \delta(x^1 + \frac{L}{2})\right) + \frac{1}{\sqrt{\pi}}\partial_1\Sigma(t,x^1) \\ &= \frac{1}{\sqrt{\pi}}\partial_1(\Sigma(t,x^1) - \phi) \quad , \end{aligned} \tag{10.95}$$

with

$$\phi = \sqrt{\pi}\frac{Q}{e}\theta\left(x^1 + \frac{L}{2}\right)\theta\left(\frac{L}{2} - x^1\right) \quad . \tag{10.96}$$

Hence we obtain H_Q from H_0 by replacing J^0 in (10.91) by (10.95), i.e

$$H_Q \equiv H[\phi] = \frac{1}{2}\int dx^1[\pi_\Sigma^2 + (\partial_1\Sigma)^2 + \frac{e^2}{\pi}(\Sigma - \phi)^2] \quad . \tag{10.97}$$

Correspondingly we write

$$|\Omega_Q\rangle = |\Omega[\phi]\rangle \quad , \quad |\Omega_0\rangle = |\Omega[0]\rangle \tag{10.98}$$
$$E[\phi] = \langle\Omega[\phi] \mid H[\phi] \mid \Omega[\phi]\rangle \quad .$$

We work in the Schrödinger representation. Using the completeness of the states $|\Sigma\rangle$, we have

$$E[\phi] = \int_t \mathcal{D}\Sigma \int_t \mathcal{D}\Sigma' \langle\Omega[\phi]|\Sigma\rangle\langle\Sigma|H[\phi]|\Sigma'\rangle\langle\Sigma'|\Omega[\phi]\rangle \quad , \tag{10.99}$$

where $\mathcal{D}\Sigma$ stands for the usual functional measure. Now

$$\langle\Sigma|H[\phi]|\Sigma'\rangle =$$
$$\delta[\Sigma - \Sigma']\frac{1}{2}\int_t dx^1\left\{ - \frac{\delta^2}{\delta\Sigma(x)^2} + \Sigma(x)D\Sigma(x) - \frac{2e^2}{\pi}\phi\Sigma(x) + \frac{e^2}{\pi}\phi^2\right\} \quad , \tag{10.100}$$

where $D = \frac{e^2}{\pi} - \partial_1^2$.

By "diagonalizing" (10.100), we recognize immediately, that the (normalized) ground-state wave-functional $\langle\Sigma|\Omega[\phi]\rangle$ is given by

$$\langle\Sigma|\Omega[\phi]\rangle = \frac{e^{-\frac{1}{2}\int dy^1(\Sigma - \frac{e^2}{\pi}\phi D^{-1})\sqrt{D}(\Sigma - D^{-1}\frac{e^2}{\pi}\phi)}}{\left(\int \mathcal{D}\Sigma e^{-\int dy^1\Sigma\sqrt{D}\Sigma}\right)^{1/2}} \tag{10.101}$$

From (10.100) and (10.101) we conclude that

$$E[\phi] - E[0] = \frac{1}{2}\int [\frac{e^2}{\pi}\phi - (\frac{e^2}{\pi})^2\phi D^{-1}\phi] \quad , \tag{10.102}$$

where the operator D^{-1} has the matrix elements

$$\langle x|D^{-1}|y\rangle = \langle x^1|(\frac{e^2}{\pi} - \partial_1^2)^{-1}|y^1\rangle = \frac{\sqrt{\pi}}{2e}e^{-\frac{e}{\sqrt{\pi}}|x^1 - y^1|} \quad . \tag{10.103}$$

Using (10.96), a simple calculation yields,

$$V(L) \equiv E[\phi] - E[0] = \frac{\sqrt{\pi}}{2}\frac{Q^2}{e}(1 - e^{-\frac{e}{\sqrt{\pi}}L}) \quad . \tag{10.104}$$

As expected, the potential tends to a constant for large separations of the $q\bar{q}$ pair, reflecting the screening of the test-charges by the induced vacuum polarization[29].

In chapter 12 we shall show how this result for the potential is also obtained in the framework of the Feynman path-integral representation of the vacuum expectation value of the Wilson-loop operator $\exp(ie \oint dz^\mu A_\mu(z))$.

10.2.9 Adding flavor

The generalization of the previous results to the case where we have N species of fermions belonging to the fundamental representation of $U(N)$ flavor is straightforward. The fermion fields now carry a flavour index f linked to the free fermionic part. Correspondingly Eq. (10.38) is replaced by[349]

$$\psi^{(0)}_{\alpha,f}(x) = \left(\frac{\mu}{2\pi}\right)^{1/2} e^{-\frac{i\pi}{4}\gamma^5_{\alpha\alpha}} K_f : e^{i\sqrt{\pi}(\gamma^5_{\alpha\alpha}\varphi_f(x) + \int_{x^1}^{\infty} dz^1 \partial_0 \varphi_f(x^0,z^1))} : \quad , \tag{10.105}$$

where φ_f are the "potentials" associated with the N free, conserved currents,

$$j^\mu_f(x) = :\overline{\psi}^{(0)}_f(x)\gamma^\mu\psi^{(0)}_f(x): = -\frac{1}{\sqrt{\pi}}\tilde{\partial}^\mu\varphi_f(x) \quad . \tag{10.106}$$

and K_f is a "Klein factor" insuring that the fields (10.105) also anticommute for different flavor index.

We make an Ansatz for A^μ analogous to (10.6), which we write in the form,

$$A^\mu = -\frac{1}{e}\tilde{\partial}^\mu(\alpha\Sigma(x) + \delta\eta) \quad , \tag{10.107}$$

where we have anticipated the relation (10.20) for a massless field η.

Following the previous line of reasoning, we again expect Σ to be a massive free field. Solving the Dirac equation (10.1a) with A_μ this time given by (10.107), one finds

$$\psi_{\alpha,f}(x) = \left(\frac{\mu}{2\pi}\right)^{1/2} : e^{i\gamma^5_{\alpha\alpha}(\alpha\Sigma(x)+\delta\eta(x))} : \psi^{(0)}_{\alpha,f}(x) \quad , \tag{10.108}$$

A short-distance calculation yields for the Maxwell current

$$J^\mu(x) = -\alpha\frac{N}{\pi}\tilde{\partial}^\mu\Sigma(x) + L^\mu(x) \quad , \tag{10.109}$$

with

$$L^\mu(x) = -\sqrt{\frac{N}{\pi}}\tilde{\partial}^\mu\left(\varphi(x) + \sqrt{\frac{N}{\pi}}\delta\eta(x)\right) \quad , \tag{10.110}$$

where φ is the massless field defined by

$$\varphi = \frac{1}{\sqrt{N}}\sum_{f=1}^{N}\varphi_f \quad . \tag{10.111}$$

The fields φ_f and φ satisfy the canonical commutation relations

$$[\varphi_f(x), \partial_0\varphi_{f'}(y)]_{ET} = i\delta(x^1 - y^1)\delta_{ff'} \quad ,$$

$$[\varphi(x), \partial_0\varphi(y)]_{ET} = i\delta(x^1 - y^1) \quad . \tag{10.112}$$

Again, Maxwell's equations are only satisfied in the subspace $\mathcal{H}_{phys}$, for which

$$\langle\psi'|L^\mu(x)|\psi\rangle = 0 \quad , \quad |\psi'\rangle \, , \, |\psi\rangle \in \mathcal{H}_{phys} \tag{10.113}$$

and provided that

$$\left(\Box + \frac{Ne^2}{\pi}\right)\Sigma = 0 \quad . \tag{10.114}$$

The significance of (10.113) has already been discussed in section 2, and we shall not repeat it here.

Condition (10.113) again implies, that η must be quantized with indefinite metric. Choosing the normalization to be the same as in (10.32), we conclude that $\delta = \sqrt{\frac{\pi}{N}}$. The constant α in (10.107) is then chosen, such that $\psi_{\alpha,f}(x)$ in (10.108) approaches the free fermion field in the limit where the electromagnetic interaction is turned off. This requires $\alpha = \sqrt{\frac{\pi}{N}}$. Note that this choice of α and δ will then imply a canonical short distance behaviour for the interacting fermion field!

In order to arrive at an expression for the physical (gauge invariant) fermion field $\hat{\psi}_{\alpha,f}$ analogous to (10.45c), it is convenient to introduce a new set of $N-1$ potentials ϕ^{i_D} ($i_D = 1, \cdots, N-1$) associated with the $N-1$ free, conserved currents belonging to the torus (maximal abelian subgroup) of $SU(N)_F$:

$$j_\mu^{i_D}(x) =: \overline{\psi}^{(0)}(x)\gamma_\mu\tau^{i_D}\psi^{(0)}(x) := -\frac{1}{\sqrt{\pi}}\tilde{\partial}_\mu\phi^{i_D}(x) \quad , \tag{10.115a}$$

where $\{\tau^{i_D}\}$ are the mutually commuting $N-1$ (hermitian) generators of $SU(N)_F$ with the normalization (C.10). Since the free axial vector currents

$$j_\mu^{5,i_D}(x) =: \overline{\psi}^{(0)}(x)\gamma_\mu\gamma^5\tau^{i_D}\psi^{(0)}(x) := -\frac{1}{\sqrt{\pi}}\partial_\mu\phi^{i_D}(x) \quad , \tag{10.115b}$$

are also conserved[373], it follows that the potentials $\{\phi^{i_D}\}$ are zero mass free fields $\Box\phi^{i_D} = 0$. The relation between these potentials and those in (10.106) and (10.111) is

$$\varphi_f = \frac{1}{\sqrt{N}}\varphi + \sum_{i_D}\tau_{ff}^{i_D}\phi^{i_D} \quad . \tag{10.116}$$

The ϕ^{i_D} are also canonically normalized $[\phi^{i_D}(x), \partial_0\phi^{i_D}(y)]_{ET} = i\delta^{i_D j_D}\delta(x^1 - y^1)$. In terms of φ, the free $U(1)$ and $\tilde{U}(1)$ currents read

$$j^\mu(x) = -\sqrt{\frac{N}{\pi}}\tilde{\partial}_\mu\varphi \quad , \quad j_5^\mu(x) = -\sqrt{\frac{N}{\pi}}\partial_\mu\varphi \quad . \tag{10.117}$$

The expressions (10.45) for the chiral densities, $\hat{\psi}_\alpha$ and the Hamiltonian are now replaced by

$$\rho_+(x)_f = -\left(\frac{\hat{\mu}}{2\pi}\right) :e^{2i\sqrt{\frac{\pi}{N}}\Sigma(x)}: \mathcal{F}^\dagger_{2,f}(x)\mathcal{F}_{1,f}(x): \sigma_2^\dagger\sigma_1 \quad , \tag{10.118a}$$

$$\rho_-(x)_f = -\left(\frac{\hat{\mu}}{2\pi}\right) :e^{-2i\sqrt{\frac{\pi}{N}}\Sigma(x)}: \mathcal{F}^\dagger_{1,f}(x)\mathcal{F}_{2,f}(x): \sigma_1^\dagger\sigma_2 \quad , \tag{10.118b}$$

$$\hat{\psi}_{\alpha,f}(x) = \left(\frac{\hat{\mu}}{2\pi}\right)^{1/2} K_f :e^{i\sqrt{\frac{\pi}{N}}\gamma^5_{\alpha\alpha}\Sigma(x)}: \mathcal{F}_{\alpha,f}(x)\sigma_\alpha \quad , \tag{10.118c}$$

and

$$H_0^{U(N)_F} = H_0 + \sum_{i_D} \frac{1}{2}: ((\pi^{i_D})^2 + (\partial_1\phi^{i_D})^2): \quad , \tag{10.118d}$$

where

$$\rho_\pm(x)_f \equiv N\left[\overline{\psi}_f(x)\frac{1\pm\gamma^5}{2}\psi_f(x)\right] \quad , \tag{10.118d}$$

$$\mathcal{F}_{\alpha,f}(x) = :e^{\left\{i\sqrt{\pi}\gamma^5_{\alpha\alpha}(\varphi_f(x)-\frac{1}{\sqrt{N}}\varphi(x))+i\sqrt{\pi}\int_{x^1}^\infty dz^1\partial_0(\varphi_f(x^0,z^1)-\frac{1}{\sqrt{N}}\varphi(x^0,z^1))\right\}}:$$

$$\sigma_\alpha(x) = e^{\left\{i\sqrt{\frac{\pi}{N}}(\gamma^5_{\alpha\alpha}(\varphi(x)+\eta(x))+\int_{x^1}^\infty dz^1\partial_0(\varphi(x^0,z^1)+\eta(x^0,z^1)))\right\}} \tag{10.119a,b}$$

and H_0 is the Hamiltonian (10.50) of the $U(1)$ Schwinger model. Here $\hat{\mu}=\mu\left(\frac{\tilde{m}}{\mu}\right)^{\frac{1}{2N}}$, $\tilde{m} = \frac{e}{2\sqrt{\pi}}e^\gamma$, and K_f is the Klein factor.

The operator $\sigma_\alpha(x)$ has scale dimension zero, and satisfies the commutation relations (10.46) with q and q_5 now being the $U(1)$ and $\tilde{U}(1)$ charges associated with the currents (10.117). They thus carry the $U(1)$ and $\tilde{U}(1)$ selection rules of the fermion field, as before. Since they commute with the observables for all space-time points, we continue to have the same vacuum degeneracy as in the $U(1)$ case, and in the irreducible basis, the ground states continue to be labelled by two "angles", as in (10.54), with σ_α acting on $|\theta_1\theta_2\rangle$ as in (10.55).
The C-number gauge transformations are now generated by the operator (10.62) with the replacement $\chi[\Lambda] \rightarrow \sqrt{N}\chi[\Lambda]$. In the notation of (10.69) we have, this time

$$T[\Lambda^{[+\frac{1}{2N}]}]\hat{=}\sigma_1 \quad , \quad T[\Lambda^{[-\frac{1}{2N}]}]\hat{=}\sigma_2 \quad \text{on } \mathcal{H}_{phys} \quad . \tag{10.120}$$

Analogously to the $U(1)$ case, the equivalence (10.120) implies the existence of instanton configurations of winding number $\nu = \frac{\Delta n_1 - \Delta n_2}{2N} = \frac{1}{2N}\Delta n_{ch}$ mediating between vacua differing in their chirality by Δn_{ch}.

Unlike the case of the $U(1)$ and $\tilde{U}(1)$ currents (10.35) and (10.36), the $SU(N)$ and $\widetilde{SU}(N)$ torus currents are just equal to the corresponding free fermionic currents[373] (10.115a,b):

$$J_\mu^{i_D}(x) = j_\mu^{i_D}(x) = -\frac{1}{\sqrt{\pi}}\tilde{\partial}_\mu \phi^{i_D}(x)\ , \quad J_\mu^{5,i_D}(x) = j_\mu^{5,i_D}(x) = -\frac{1}{\sqrt{\pi}}\partial_\mu \phi^{i_D}(x) \quad . \tag{10.121a}$$

Denoting the respective charges by Q^{i_D} and $Q_5^{i_D}$ one finds

$$[Q^{i_D}, \mathcal{F}_{\alpha,f}(x)] = -\tau_{ff}^{i_D}\mathcal{F}_{\alpha,f}(x) \tag{10.121b}$$

$$[Q_5^{i_D}, \mathcal{F}_{\alpha,f}(x)] = -\tau_{ff}^{i_D}\gamma_{\alpha\alpha}^5\mathcal{F}_{\alpha,f}(x) \tag{10.121c}$$

With the $U(1)$ currents, $\mathcal{F}_{\alpha,f}$ evidently commutes. This shows that the flavor operator $\mathcal{F}_{\alpha,f}$ carries the $SU(N)_F \times \widetilde{SU}(N)_F$ selection rules implied by the $SU(N)_F \times \widetilde{SU}(N)_F$ symmetry of the Lagrangian, which remains unbroken.

The ground states are flavour neutral. Hence, unlike the $U(1)$ case, one can detect the degeneracy of the $|n_1 n_2\rangle$ vacua via clustering violation, only provided one considers the clustering properties of $SU(N)_L \times SU(N)_R$ singlets. Clustering is then recovered, if we go to the irreducible representation in terms of the vacua $|\theta_1\theta_2\rangle$ constructed as a coherent superposition (10.54) of the $|n_1 n_2\rangle$-vacua.

The above results are easily generalized to include the (gauged) maximal abelian subgroup (Cartan subalgebra) of $SU(N)$-color[349]. In fact it has been argued[374,375], that $SU(N)$ Quantum Chromodynamics in two dimensions may exhibit two phases: the weak coupling ("'t Hooft" phase), described in terms of an $1/N$ expansion, and a strong coupling phase, where the $SU(N)$ gauge group is broken down to the maximal abelian subgroup (torus) of $SU(N)_D$. Although in a gauge-theory with vector coupling there seems to be no reason for such a breakdown to occur, it is nevertheless interesting to examine a generalized QED_2 on $SU(N)_D$. The exact operator solutions can again be constructed for the case of a $SU(N)_D$, $SU(N)_D \times U(1)$ and $SU(N)_D \times U(1) \times SU(M)_F$ symmetry, where $SU(U)_D$ and $U(1)$ are gauge groups, while the flavor symmetry $SU(M)_F$ is a global one. We shall not dwell here on this construction, and refer the reader to the literature[349,350,375]. The essentially new feature of the solution is that the vacuum is now characterized by several chiral and fermionic "angles" whose number is respectively equal to the gauge group in question. In this respect, QCD_4 thus resembles more the $U(1) \times SU(N)_F$ theory, than its extension to the torus of $SU(N)$-color.

10.2.10 Fractional winding number and the $U(1)$ problem

The $U(1)$-problem can be resumed in the following question: how can one avoid in QCD_4 a Nambu-Goldstone realization[214] of the η', which couples to the $U(1)$-axial current? The resolution of this problem lies in the observation that the Adler-Bell-Jackiw (ABJ) anomaly[200,201] occurs in the divergence of the gauge-invariant axial vector current, which however does not generate chiral transformations. As we have witnessed in section (10.2.4), this transformation is generated in our case by q_5.

The ABJ anomaly is closely linked to the existence of topologically non-trivial gauge field configurations via the Atiyah-Singer theorem (see chapter 4). In particular, Crewther[265] has proposed that A_μ-configurations of fractional winding number $1/N$ be responsible for a dynamical generation of the η' mass. To achieve this he needed to break the $\widetilde{SU}(N)_F$ symmetry explicitly, by giving a mass to the fermions. We shall now discuss a model realizing Crewther's ideas[382]. (We refer also to section 3.2 of chapter 7 in order to see how these ideas apply to a more complex problem.)

We have seen above, that the mass operator $\bar{\psi}\psi$ has a non-vanishing θ-vacuum expectation value in the Schwinger model. This, a priori, can be taken to be a signal of a dynamical mass generation for the fermions or as a manifestation of a spontaneous breakdown of the global $\tilde{U}(1)$ symmetry. As we have seen, the latter is the case; in fact, since the fermions of the model do not exist as asymptotic particles, we can not ascribe a "mass" to them. On the other hand, as we have already remarked, a dynamical mass generation via a topological $\tilde{U}(1)$-symmetry breaking induced by instantons with fractional winding number $1/N$ could resolve the so called $\tilde{U}(1)$ problem in QCD_4 with a $U(N)_F$-flavour symmetry.

In the $U(1)$ Schwinger model, only A_μ configurations with integer and half-integer winding numbers contribute (see chapter 12). In particular one finds that configurations with half-integral winding number (merons) are responsible for the non-vanishing θ-vacuum expectation value of operators carrying fermion number.

According to our discussion of the preceding section we may expect instanton configurations of winding number $1/N$ or $1/2N$ to come to play a role, if we endow the fermions with flavor quantum numbers. The resulting Lagrangian now exhibits an $U(N) \times \tilde{U}(N)$ symmetry. Since, as we have seen, the ground state however continues to carry only $U(1) \times \tilde{U}(1)$ quantum numbers, the $SU(N) \times \widetilde{SU}(N)$ symmetry remains unbroken. Hence instanton configurations with winding numbers $1/N$ and $1/2N$ remain undetected as a result of $SU(N) \times \widetilde{SU}(N)$ selection rules which imply the vanishing of correlation functions carrying $SU(N) \times \widetilde{SU}(N)$ quantum numbers. As an example we have[13)] $\langle\theta|\rho_\pm(x)_f|\theta\rangle = 0$, where $\rho_\pm(x)_f$ is given by (10.118a, b).

---

13) We shall frequently denote $|\theta_1\theta_2\rangle$ by just $|\theta\rangle$, and $|0,0\rangle$ by $|0\rangle$, where θ is the chiral angle $\theta_1 - \theta_2$ (see Eq. (10.56b)).

The vanishing of this particular vacuum expectation value is a consequence of the (chiral) $\widetilde{SU}(N)$ selection rules, and is thus also a property of the $SU(N)$ singlet, $\rho_\pm(x) = \frac{1}{N}\sum_f \rho_\pm(x)_f$. This is expressed by the cluster property,

$$\langle\theta|\rho_+(x)\rho_-(x+\xi)|\theta\rangle \underset{|\xi|\to\infty}{\sim} |\xi^2|^{\frac{1}{N}-1} \quad . \tag{10.122}$$

In order to be able to detect instanton configurations with winding number $\frac{1}{N}$, we break the $\widetilde{SU}(N)$ symmetry explicitly by adding to the N flavour Schwinger Lagrangian a fermionic self interaction of the chiral Gross Neveu type[14)] :

$$\mathcal{L} = -\frac{1}{4}F_{\mu\nu}F^{\mu\nu} + \mathcal{L}_{ch}^{GN} + e\bar{\psi}\not{A}\psi \tag{10.123}$$

where

$$\mathcal{L}_{ch}^{GN} = \bar{\psi}i\not{\partial}\bar{\psi} + \frac{g}{2}[(\bar{\psi}\psi)^2 + (\bar{\psi}i\gamma^5\psi)^2] \quad . \tag{10.124}$$

Here the fermions are taken to belong to the fundamental representation of $U(N)_F$. We shall refer to (10.123,124) as the Schwinger-Gross-Neveu (SGN) Lagrangian.

The SGN Lagrangian shares with the QCD_4 Lagrangian of Crewther the property of $SU(N)_F \times U(1) \times \tilde{U}(1)$ invariance. In both cases the chiral $\tilde{U}(1)$ invariance is broken spontaneously via "instanton" configurations with a fractional winding number $\nu = 1/N$. Although there exists no theorem forbidding such a symmetry breakdown in QCD_4, the corresponding phenomenon in $d = 1+1$ dimensions appears at first to contradict Coleman's theorem[93]. One must however keep in mind, that we are dealing with a gauge theory. Since a Lorentz invariant formulation requires the operators to "live" in an indefinite metric Hilbert space, the Coleman theorem does not apply to this case. On the other hand, in the pure chiral Gross-Neveu model (see chapter 5), the global chiral $\tilde{U}(1)$ invariance is de facto not spontaneously broken, even though the dynamical fermion mass generation in this model appears to signalize such a breakdown. The resolution of this paradox is rather subtle, as we have seen in chapter 5.

In order to demonstrate how the addition of a $U(1)$ gauge field interaction to the chiral GN Lagrangian leads to a breakdown of the $U(1) \times \tilde{U}(1)$ symmetry, let us first briefly review the situation for the chiral GN model.

We have seen in chapter 5 that the spontaneous mass generation in the chiral GN model is accompanied by the appearance of zero mode excitations. Since the $U(1)$ vector and axial vector currents are conserved in this model, this is reminiscent of the familiar Nambu-Goldstone mechanism. However, in d = 1+1 dimensions, zero mass scalar particles do not exist. These "would-be" Goldstone-bosons must

---

[14)] Crewther solved the problem by explicitly introducing a mass term for the fermions[265].

therefore exponentiate, if the (fermionic) theory is to respect the Wightman axioms of positivity (see chapter 2). This is contained in the factorization of the fermion field into a $U(1)\times\tilde{U}(1)$ and an $SU(N)$ part, as exhibited by Eq. (5.90). The selection rules carried by the $U(1)\times\tilde{U}(1)$ factor insure that $\langle 0 \mid \mathcal{N}[\overline{\psi}^{GN}\psi^{GN}](x) \mid 0\rangle = 0$, so that there is no spontaneous $\tilde{U}(1)$-symmetry breakdown. How can this be reconciled with a dynamical mass generation?

The resolution of this paradox has been clearly exhibited in a paper of Witten[197]. Let us implement explicitly the constraint (5.91) by introducing the fields ϕ^{i_D} via (10.116). In terms of ϕ^{i_D}, the fields χ_i in (5.90) are given by (f stands for flavour)

$$\chi_f = \sum_{i_D} \tau^{i_D}_{ff}\phi^{i_D} \quad , \tag{10.125}$$

so that equation (5.90) reads

$$\psi_f^{GN}(x) = :e^{i\sqrt{\frac{\pi}{N}}\left(\gamma^5\varphi(x)+\int\limits_{x^1}^{\infty} dy^1\partial_0\varphi\right)}:\hat{\psi}_f^{GN}(x) \tag{10.126a}$$

where $\hat{\psi}_f^{GN}(x)$ represents the $SU(N)$-part

$$\hat{\psi}_f^{GN}(x) = \sqrt{\frac{\mu}{2\pi}}K_f e^{-\frac{i}{4}\pi\gamma^5} e^{i\sqrt{\pi}\sum\limits_{i_D}\tau^{i_D}_{ff}[\gamma^5\phi^{i_D}(x)+\int\limits_{x^1}^{\infty} dy^1\partial_0\phi^{i_D}]} \tag{10.126b}$$

Note that $\hat{\psi}_f^{GN}$ no longer carries $U(1)$-chirality and $U(1)$-charge. As already witnessed in chapter 5, these fields obey a generalized statistics corresponding to a "Lorentz spin" $s = \frac{1}{2}(1-1/N)$.

The fields φ and ϕ^{i_D} are the (canonically normalized) "potentials" of the $U(1)$ and diagonal $SU(N)$ currents, respectively (compare with (10.117) and (10.121a))

$$\overline{\psi}^{GN}(x)\gamma_\mu\psi^{GN}(x) = -\sqrt{\frac{N}{\pi}}\tilde{\partial}_\mu\varphi \tag{10.127}$$

$$\overline{\psi}^{GN}(x)\tau^{i_D}\gamma_\mu\psi^{GN}(x) = -\frac{1}{\sqrt{\pi}}\tilde{\partial}_\mu\phi^{i_D} \quad . \tag{10.128}$$

In terms of these bosonic fields the equivalent chiral GN Lagrangian reads

$$\mathcal{L}_{ch}^{GN} = \mathcal{L}_{U(1)} + \mathcal{L}_{SU(N)} \tag{10.129}$$

where $\mathcal{L}_{U(1)} = \frac{1}{2}\partial^\mu\varphi\partial_\mu\varphi$ and

$$\mathcal{L}_{SU(N)} = \frac{1}{2}\sum_{i_D}\partial^\mu\phi^{i_D}\partial_\mu\phi^{i_D} + \frac{g\mu^2}{2\pi^2}\left\{\Big(\sum_f \cos 2\sqrt{\pi}\sum_{i_D}\tau_{ff}^{i_D}\phi^{i_D}\Big)^2 + \Big(\sum_f \sin 2\sqrt{\pi}\sum_{i_D}\tau_{ff}^{i_D}\phi^{i_D}\Big)^2\right\}$$
$$= \frac{1}{2}\sum_{i_D}\partial^\mu\phi^{i_D}\partial_\mu\phi^{i_D} + \frac{g\mu^2}{\pi}\sum_{f,f'}\cos 2\sqrt{\pi}(\varphi_f - \varphi_{f'}) \quad . \qquad (10.130)$$

In the bosonized version the $U(1)_F$ part of the Lagrangian in thus seen to decouple explicitly from the $SU(N)_F$ part. While $\mathcal{L}_{U(1)}$ is the Lagrangian of a zero mass field, $\mathcal{L}_{SU(N)}$ describes the interaction of $N-1$ massive scalar fields satisfying a coupled set of sine-Gordon equations. As a result, $\hat{\psi}_f^{GN}$ does not carry a chiral selection rule, so that $\langle 0 \mid \mathcal{N}[\overline{\hat{\psi}^{GN}}\hat{\psi}^{GN}](x) \mid 0\rangle \neq 0$ is compatible with no spontaneous chiral $\tilde{U}(1)$ symmetry breakdown. This is Witten's argument[197].

The addition of a $U(1)$ electromagnetic interaction to $\mathcal{L}_{ch}^{GN}$ as in (10.123), amounts to the replacement of $\mathcal{L}_{U(1)}$ in (10.129) by

$$\mathcal{L}_{U(1)} = -\frac{1}{4}F^{\mu\nu}F_{\mu\nu} + \frac{1}{2}\partial^\mu\varphi\partial_\mu\varphi + e\sqrt{\frac{N}{\pi}}\epsilon^{\mu\nu}\partial_\nu\varphi A_\mu + e\frac{\theta}{4\pi}\epsilon^{\mu\nu}F_{\mu\nu} \qquad (10.131)$$

while the $SU(N)$-Lagrangian (10.130) remains unchanged.

The third term in (10.131) represents the interaction of the gauge field with the fermionic $U(1)$ current. Since $\mathcal{L}_{U(1)}$ is the Lagrangian of the Schwinger model, we have added a topological term to describe the θ-sectors of the theory.

By considering the equations of motion following from the bosonic form of the Lagrangian (10.123) one easily verifies, that φ has been turned into a free, massive field by the $U(1)$ interaction, its mass being $m = e\sqrt{\frac{N}{\pi}}$ (Schwinger-Higgs mechanism). Hence, from the point of view of this Lagrangian, φ plays the role of Σ in our previous discussion. The gauge-field has thus "eaten" the would-be Goldstone boson of the GN model and has become "fat". As we already know from our discussion of the Schwinger model, the global chiral $\tilde{U}(1)$ symmetry is now spontaneously broken, thus showing that we have circumvented Coleman's theorem by "embedding" the chiral GN model in a U(1)-gauge theory.

It is interesting that the above Higgs mechanism presents itself in a somewhat different way for the operator (10.126*a*). The addition of a $U(1)$ gauge interaction leads to the replacement of this operator by

$$\psi_f(x)_\alpha =: e^{-i\sqrt{\frac{\pi}{N}}\tilde{\eta}(x)} e^{i\sqrt{\frac{\pi}{N}}\gamma^5_{\alpha\alpha}\Sigma(x)} : \hat{\psi}_f^{GN}(x)\sigma_\alpha \tag{10.132}$$

with

$$A_\mu(x) = -\frac{1}{Ne}\sqrt{\pi}\epsilon_{\mu\nu}\partial^\nu\left(\Sigma(x) + \eta(x)\right) \tag{10.133}$$

where Σ is the massive field satisfying (10.114), and $\tilde{\eta}$ is seen to be a pure gauge degree of freedom. Hence the Lagrangian (10.129) actually summarizes the following process: by adding the electromagnetic interaction, the gauge field A_μ acquires a mass via the familiar Schwinger mechanism, and the field φ, which plays the role of the "would be" η' in the $U(1)$ problem, "spurionizes" on the gauge invariant subspace along the lines discussed in the previous sections.

In chapter 12 we shall compute the vacuum expectation value of the chiral density ρ_+ in the $\frac{1}{N}$ expansion, and show that to leading order, we have $\langle \frac{1}{N} \mid \rho_+(0) \mid 0\rangle = -\frac{m}{2}$, where $\frac{1}{N}$ is the winding number of the gauge-field configuration contributing to this matrix element, and m is identified with the dynamically generated mass of the fermion.

We have thus shown that the Schwinger-Gross-Neveu model described by the Lagrangian (10.129) provides an example of the mechanism envisaged in QCD_4 for explaining the absence of the η' in the experiments. We now proceed to discuss QED_2 for the case, where an explicit mass is given to the fermions.

10.3 The Massive Schwinger Model

10.3.1 Equivalent bosonic formulation

Quantum Electrodynamics with massless fermions (Schwinger Model) in $d = 1 + 1$ dimensions turned out to be exactly solvable at operator level. This is no longer true, if the Dirac equation (10.1a) is replaced by

$$(i\,\partial\!\!\!/ - M + e\,A\!\!\!/)\psi = 0 \quad , \tag{10.134a}$$

describing now the electromagnetic interaction of a massive Dirac fermion. Nevertheless, Coleman's equivalence of the massive Thirring model to the sine-Gordon theory, supplemented by Mandelstam's explicit construction of fermions out of bosons, indicates the way of implementing the bosonization ideas also in the case of massive QED_2. This allows us to obtain a good insight into non-perturbative properties of this model. (For a perturbative treatment in the fermion mass, see ref.[353].)

The Lagrangian $\mathcal{L}$ of massive QED_2 is related to that of massless QED_2, $\mathcal{L}_0$, by the addition of a mass term in the fermions:

$$\mathcal{L} = \mathcal{L}_0 + MN[\overline{\psi}(x)\psi(x)] \quad . \tag{10.134b}$$

The resulting theory still requires no infinite renormalization, except for a redefinition of the zero-point energy.

The mass term explicitly breaks the $\tilde{U}(1)$ symmetry of $\mathcal{L}_0$. The equivalent bosonic form of $\mathcal{L}_0$ is given by (we regard for the moment all variables as c-number valued)

$$\mathcal{L}_0 = \frac{1}{2}(\partial_\mu \Sigma \partial^\mu \Sigma - \frac{e^2}{\pi}\Sigma^2) + \frac{1}{2}\partial_\mu \varphi \partial^\mu \varphi - \frac{1}{2}\partial_\mu \eta \partial^\mu \eta \quad . \tag{10.135}$$

Adding to $\mathcal{L}_0$ the mass term (10.42b), Eq. (10.134b), takes the form (it is convenient to introduce the constants α, β, δ which need no longer take the value $\alpha = \beta = \delta = \sqrt{\pi}$)

$$\mathcal{L} = \mathcal{L}_0 - \frac{\tilde{\mu}M}{\pi} \cos\, 2(\alpha\Sigma(x) + \beta\varphi(x) + \delta\eta(x)) \quad . \tag{10.136}$$

Following Coleman, Jackiw and Susskind[352], we now assume that the Lagrangian (10.136) actually describes the massive Schwinger model. We shall refer to this as the "adiabatic" principle.

The fields Σ , φ and η now are no longer free fields. In fact, from (10.136) we obtain the following (classical) Euler Lagrange equations[376]:

$$\left(\Box + \frac{e^2}{\pi}\right)\Sigma + 2\frac{\alpha\tilde{\mu}M}{\pi} \sin\, [2(\alpha\Sigma + \beta\varphi + \delta\eta)] = 0 \tag{10.137a}$$

$$\Box\eta - 2\frac{\delta\tilde{\mu}M}{\pi} \sin\, [2(\alpha\Sigma + \beta\varphi + \delta\eta)] = 0 \tag{10.137b}$$

$$\Box\varphi + 2\frac{\beta\tilde{\mu}M}{\pi} \sin\, [2(\alpha\Sigma + \beta\varphi + \delta\eta)] = 0 \tag{10.137c}$$

In going over to the quantum theory, the terms involving exponentials have to be understood in terms of an appropriately defined "normal product". In order to be able to define such a product we assume that also in the presence of the mass term, the fields Σ , φ and η continue to have canonical scale dimensions. As we shall see this turns out to be a valid assumption.

We may turn the exponential of a local operator $\chi(x)$ into a well defined, finite operator by applying this operator on a sequence of test functions $h_{x^1}(\eta^1)$ with support around $\eta^1 = x^1$:

$$\chi(x^0, x^1) \to \chi(x^0; h) \equiv \int d\eta^1 h_{x^1}(\eta^1)\chi(x^0, \eta^1) \quad . \tag{10.138}$$

We then define a finite exponential operator of χ by

$$N[e^{\chi(x)}] \equiv \lim_{h\to\delta} \mathcal{Z}(h)e^{\chi(x;h)} \quad , \tag{10.139}$$

where $h \to \delta$ denotes a sequence of test functions $h_{x^1}(\eta^1)$ tending to $\delta(x^1-\eta^1)$, and $\mathcal{Z}(h)$ is a suitably chosen functional of h (renormalization constant). Wick ordering in the case of exponentials of free fields will now be replaced by the "normal product" as defined by the r.h.s. of Eq. (10.139). Under the assumption that $\chi(x)$ is a field with canonical scale dimension, $\mathcal{Z}(h)$ may then be calculated explicitly in the limit $h \to \delta$, by identifying it with the renormalization constant for χ a free field:

$$\lim_{h\to\delta}\left\{\mathcal{Z}(h)e^{\frac{1}{2}[\chi^{(+)}(x;h),\chi^{(-)}(x;h)]}\right\} = 1 \quad . \tag{10.140}$$

To simplify the notation, we shall continue to use the "Wick ordered" notation : : to denote the normal product. Invoking once again the adiabatic principle, we suppose the fermion field satisfying (10.134*a*) to be given by

$$\psi_\alpha(x) = \lim_{h\to\delta} \mathcal{Z}(h)\psi_\alpha(x;h) = \lim_{h\to\delta} \mathcal{Z}(h)\left(\frac{\mu}{2\pi}\right)^{\frac{1}{2}} e^{-\frac{i\pi}{4}\gamma^5_{\alpha\alpha}} e^{i\Phi_\alpha(x;h)} \quad . \tag{10.141a}$$

Here

$$\Phi_\alpha(x;h) \equiv \int_{-\infty}^{\infty} d\eta^1 h_{x^1}(\eta^1)\Phi_\alpha(x) \quad , \tag{10.141b}$$

where

$$\Phi_\alpha(x) = \gamma^5_{\alpha\alpha}(\alpha\Sigma(x) + \delta\eta(x) + \beta\varphi(x)) + \frac{\pi}{\beta}\int_{x^1}^{\infty} dy^1 \partial_0\varphi(x^0,y^1) \quad . \tag{10.141c}$$

The parametrization has been chosen such as to be consistent with that of the mass term in (10.136); note that the dependence on β corresponds to that of a spin $\frac{1}{2}$ free fermion field (see chapter 2).

For a canonical scaling behavior of the fields Σ , φ and η, this Ansatz can be shown to lead to the following results for the current, and the scalar and pseudoscalar densities:

$$N[\overline{\psi}(x)\gamma^\mu\psi(x)] = -\frac{1}{\pi}\tilde{\partial}^\mu\chi(x) = -\frac{\alpha}{\pi}\tilde{\partial}^\mu\Sigma + L_\mu \tag{10.142a}$$

$$N[\overline{\psi}(x)\psi(x)] = -\frac{\tilde{\mu}}{\pi}:\cos\,(2\chi(x)): \tag{10.142b}$$

$$N[\overline{\psi}(x)\gamma^5\psi(x)] = -\frac{i\tilde{\mu}}{\pi}:\sin\,(2\chi(x)): \tag{10.142c}$$

where $L^{\mu} = -\frac{1}{\sqrt{\pi}}\tilde{\partial}_{\mu}(\beta\varphi + \delta\eta)$, and

$$\chi(x) = \alpha\Sigma(x) + \beta\varphi(x) + \delta\eta(x) \quad . \tag{10.143}$$

As previously agreed, the notation : : stands for the limiting process $:e^{i2\chi(x)}:\equiv \lim_{h\to\delta} \mathcal{Z}[h]e^{i2\chi(x;h)}$, with $\mathcal{Z}[h]$ some suitably chosen functional of h to render the limit finite. From the point of view of a Wilson expansion we have,

$$\overline{\psi}(x+\xi)\psi(x) \approx (\bar{\mu}^2\xi^2)^{1-\dim[\overline{\psi}\psi]}\mathcal{M}(x) \quad , \tag{10.144a}$$

where $\mathcal{M}(x)$ is the mass operator

$$\mathcal{M}(x) \equiv -\frac{\tilde{\mu}}{\pi}:\cos\,(2\chi(x)): \tag{10.144b}$$

and $\dim\,[\overline{\psi}\psi]$ is the scale dimension of the mass operator as given by

$$\dim\,[\overline{\psi}\psi] = \frac{\beta^2}{\pi} + \frac{\alpha^2-\delta^2}{\pi} \quad . \tag{10.144c}$$

10.3.2 The quantum Dirac equation[376]

The fermion operator defined by (10.141) should satisfy the massive Dirac equation (10.134a). As we now show, this will be the case if $\beta = \sqrt{\pi}$. To this end we shall make use of the following useful formula. Let $e^{A(x)}$, be a well defined operator. Then

$$\delta e^{A(x)} = \int_0^1 d\lambda e^{\lambda A(x)}\delta A(x) e^{(1-\lambda)A(x)} \quad . \tag{10.145a}$$

For the case that $[A(x), \partial_{\mu}A(x)] = 0$ we recover the naive result $\partial_{\mu}e^{A(x)} = \partial_{\mu}A(x)e^{A(x)}$. In general this commutator does not vanish, however. Making further use of the identity

$$\int_0^1 d\lambda e^{\lambda A} B e^{-(\lambda-1)A} = \frac{1}{2}\{B, e^A\} + \frac{1}{2}\int_0^1 d\lambda \Big[[e^{\lambda A}, B], e^{(1-\lambda)A}\Big] \quad , \tag{10.145b}$$

we obtain

$$\gamma^{\mu}\partial_{\mu}e^{i\Phi_{\alpha}(x;h)} = \frac{i}{2}\gamma^{\mu}\{\partial_{\mu}\Phi(x;h), e^{i\Phi_{\alpha}(x;h)}\}$$
$$+ i\gamma^{\mu}\int_0^1 d\lambda\Big[[e^{i\lambda\Phi_{\alpha}(x;h)}, \partial_{\mu}\Phi_{\alpha}(x;h)], e^{i(1-\lambda)\Phi_{\alpha}(x;h)}\Big] \quad . \tag{10.146}$$

For the case, where Σ , φ and η are free fields, we have $[\Phi_\alpha(x;h), \partial_\mu \Phi_\alpha(x;h)] =$ c-number, so that the second term in (10.146) vanishes. This is not true, if these fields satisfy the equations (10.137). Indeed, although we continue to have $[\Phi_\alpha(x;h), \partial_1 \Phi_\alpha(x;h)] = c -$ number this is no longer true, as far as the time derivative is concerned: from Eqs. (10.137) we obtain

$$\partial_\mu \Phi_\alpha(x;h) = \gamma^5_{\alpha\alpha}\partial_\mu \chi(x;h) + \frac{\pi}{\beta}\tilde{\partial}_\mu \varphi(x;h)$$
$$-2\frac{M\tilde{\mu}}{\pi} g_{\mu 0} \int_{-\infty}^{\infty} d\eta h_x(\eta) \int_\eta^\infty dz^1 \colon \sin\, 2\chi(x^0, z^1) \colon \quad . \qquad (10.147)$$

From here then follows

$$i\gamma^\mu_{\alpha\beta}\partial_\mu e^{i\Phi_\beta(x;h)} = -\frac{1}{2} e \gamma^\mu_{\alpha\beta}\{A_\mu(x;h), e^{i\Phi_\beta(x;h)}\}$$
$$+\frac{\sqrt{\pi}}{2}(\beta - \frac{\pi}{\beta})\gamma^\mu\{j_\mu(x;h), e^{i\Phi_\beta(x;h)}\} + \Delta_\alpha(x;h) \quad (10.148)$$

where

$$A_\mu = -\frac{1}{e}\tilde{\partial}_\mu(\alpha\Sigma + \delta\eta) \quad , \qquad (10.149a)$$
$$j_\mu = -\frac{1}{\sqrt{\pi}}\tilde{\partial}_\mu \varphi \quad , \qquad (10.149b)$$

$$\Delta_\alpha(x;h) = -\frac{2M\tilde{\mu}}{\pi}\gamma^0_{\alpha\beta}\int_{-\infty}^{\infty} d\eta h_x(\eta) \int_\eta^\infty dy^1 \tilde{\Delta}_\beta(x^0; x^1, y^1; h) \quad (10.150a)$$

$$\tilde{\Delta}_\beta(x^0; x^1, y^1; h) = \int_0^1 d\lambda e^{i\lambda\Phi_\beta(x;h)} \colon \sin 2\chi(x^0, y^1) \colon e^{-i\lambda\Phi_\beta(x;h)} e^{i\Phi_\beta(x;h)} \quad (10.150b)$$

Using the fact that $[\Phi(x;h), \chi(y;h)]_{ET} =$ c-number, and that $\Phi(x;h)$ commutes with itself, we can perform, after some manipulations involving repeatedly (C.13), the λ integration, to yield,

$$\tilde{\Delta}(x^0; x^1, y^1; h) = \frac{\left[\colon \cos\, 2\chi(x^0, y^1) \colon, e^{i\Phi_\alpha(x;h)}\right]}{2i[\Phi_\alpha(x;h), \chi(x^0, y^1)]} \quad , \qquad (10.150c)$$

where we have already made the replacement $\chi(x;h) \to \chi(x)$ in the denominator, in order to facilitate the discussion. Now,

$$2i[\Phi_\alpha(x;h), \chi(y)]_{ET} = 2\pi i \int_\infty^\infty d\eta h_x(\eta) \int_\eta^\infty dz^1 [\partial_0 \varphi(x^0, z^1), \varphi(x^0, y^1)]_{ET}$$
$$= 2\pi \int_{-\infty}^{y^1} d\eta h_x(\eta) \quad .$$
$$(10.151)$$

Using furthermore the identity

$$\int_{-\infty}^{\infty} d\eta h_x(\eta) \int_{\eta}^{\infty} dy^1 F(y^1) = \int_{-\infty}^{\infty} dy^1 F(y^1) \int_{-\infty}^{y^1} d\eta h_x(\eta) \quad , \tag{10.152}$$

we can finally cast (10.150a) into the form

$$\Delta_\alpha(x;h) = M\gamma^0_{\alpha\beta} \int_{-\infty}^{\infty} dy^1 \left[-\frac{\tilde{\mu}}{\pi} \colon \cos\, 2\chi(x^0, y^1), e^{i\Phi_\beta(x;h)} \right] \quad . \tag{10.153}$$

Substituting (10.153) into (10.148) and recalling (10.142*b*), we finally obtain,

$$i\, \not{\partial}\psi(x;h) = -e\gamma^\mu \frac{1}{2}\{A_\mu(x;h), \psi(x;h)\} + \frac{\sqrt{\pi}}{2}\Big(\beta - \frac{\pi}{\beta}\Big)\gamma^\mu\{j_\mu(x;h), \psi(x;h)\} +$$
$$M\gamma^0 \int_{-\infty}^{\infty} dy^1 [\mathcal{M}(y), \psi(x;h)]_{ET} \quad , \tag{10.154}$$

with $\mathcal{M}(y)$ the "mass-operator" (10.144*b*).

We recognize in the first term the symmetrized form of the interaction of ψ with the gauge field. The second term corresponds to a Thirring-like interaction. The requirement that it be absent fixes β to be

$$\beta = \sqrt{\pi} \quad . \tag{10.155a}$$

As for the third term, it should yield in the limit $h \to \delta$

$$\gamma^0_{\alpha\beta} \int_{-\infty}^{\infty} dy_1 [\mathcal{M}(y), \psi(x)] = \psi(x) \quad ,$$

provided that the scale dimension of $\mathcal{M}$ is just equal to its canonical ("engineering") dimension. Recalling (10.144*c*), we therefore seek a solution satisfying $\beta^2 + \alpha^2 - \delta^2 = \pi$, or because of (10.155*a*)

$$\alpha^2 = \delta^2 \quad . \tag{10.155b}$$

We have thus shown that our Ansatz (10.141) for the fermion field satisfies the Dirac equation.

With A_μ given by (10.149*a*) and J^μ given by (10.142*a*), Maxwell's equations (10.1*b*) read

$$-\Box\tilde{\partial}^\mu(\alpha\Sigma + \delta\eta) - \frac{e^2}{\pi}\tilde{\partial}^\mu(\alpha\Sigma + \beta\varphi + \delta\eta) = 0$$

or using the equations of motion (10.137*a*, *b*),

$$\tilde{\partial}^\mu \left\{ 2\frac{\tilde{\mu}M}{\sqrt{\pi}}(\alpha^2 - \delta^2) \colon \sin\, 2\chi(x) \colon - \frac{e^2}{\pi}(\beta\varphi + \delta\eta) \right\} = 0 \quad . \tag{10.156}$$

Because of (10.155*b*) this means that Maxwells equations will be satisfied only on the subspace $\mathcal{H}_{phys}$ of $\mathcal{H}$ defined by

$$\langle\Phi|\tilde{\partial}^{\mu}(\beta\varphi+\delta\eta)|\psi\rangle = 0 \quad , \quad |\psi\rangle \, , \, |\Phi\rangle \in \mathcal{H}_{phys} \quad . \tag{10.157}$$

Finally we note that it also follows from (10.137*b*, *c*), that

$$\Box(\beta\varphi+\delta\eta) = \frac{(\beta^2-\delta^2)}{\pi}\frac{2M\bar{\mu}}{\sqrt{\pi}}\sin\, 2\chi \tag{10.158}$$

so that $\tilde{\partial}^{\mu}(\beta\varphi+\delta\eta)$ becomes a purely longitudinal zero mass field, provided $\beta^2 = \delta^2$, In that case (10.157) is a condition on the purely longitudinal part of the current. Together with (10.155*a*) this fixes δ to be

$$\delta = \sqrt{\pi} \quad . \tag{10.155c}$$

We have thus succeeded in demonstrating, that for $\alpha = \beta = \delta = \sqrt{\pi}$ Maxwell's equations are satisfied on the subspace defined by (10.157), the solutions being given by (10.141) and (10.149*a*), with Σ , φ and η solutions to the coupled set of equations (10.137). As before, condition (10.157) implies that η must be canonically quantized with "indefinite" metric. The final expressions for the Maxwell current and field tensor read

$$J^{\mu} = -\frac{1}{\sqrt{\pi}}\tilde{\partial}^{\mu}\left(\Sigma+\varphi+\eta\right) \tag{10.159a}$$

$$F^{\mu\nu} = \frac{e}{\sqrt{\pi}}\epsilon^{\mu\nu}\Sigma \tag{10.159b}$$

and are thus seen to be of the same form as in the $M = 0$ case.

10.3.3 Vacuum structure and all that

The discussion of the ground state structure follows the lines of the massless model. We may formally perform again an operator gauge transformation taking us to the physical fermion and gauge-field operators, $\hat{\psi}$ and $\hat{A}_{\mu}$:

$$\hat{\psi}(x) =: exp\Big[i\sqrt{\pi}\int_{x^1}^{\infty} dy^1\partial_0\eta(x^0,y^1)\Big]\psi(x): \tag{10.160a}$$

$$\hat{A}_{\mu}(x) = A_{\mu}(x) - \frac{\sqrt{\pi}}{e}\partial_{\mu}\int_{x^1}^{\infty} dy^1\partial_0\eta(x^0,y^1) \quad . \tag{10.160b}$$

One checks, however, that $[\mathcal{M}(y), \hat{\psi}(x)] = 0$ so that the above transformation does not leave the Dirac equation invariant. Hence, in the massive Schwinger model, it no longer represents a symmetry transformation of the theory. We can nevertheless introduce the operators (10.160a), which can be written in a form analogous to (10.45c)

$$\hat{\psi}(x) = \left(\frac{\mu}{2\pi}\right)^{\frac{1}{2}} : e^{i\sqrt{\pi}\gamma^5_{\alpha\alpha}\Sigma(x)} : \sigma_\alpha(x) \tag{10.161a}$$

with

$$\sigma_\alpha(x) =: e^{i(\sqrt{\pi}\gamma^5_{\alpha\alpha}(\varphi(x)+\eta(x))+i\sqrt{\pi}\int_{x^1}^{\infty} dy^1 \partial_0(\varphi(x^0,y^1)+\eta(x^0,y^1))} : \quad . \tag{10.161b}$$

Note that the combination $\varphi + \eta$ describes a free, massless field, commuting with itself and its time derivatives at all space-time points. As in the massless case, σ_α carries the chirality and bare charge of the fermion. We may use it to construct again irreducible subspaces of $\mathcal{H}_{phys}$ from a given ground state $|\Omega_0\rangle$ by considering a coherent superposition of the states $\sigma_1^{n_1}\sigma_2^{n_2}|\Omega_0\rangle$:

$$|\theta_1\theta_2\rangle = \sum e^{-in_1\theta_1} e^{-in_2\theta_2} \sigma_1^{n_1}\sigma_2^{n_2}|\Omega_0\rangle \quad . \tag{10.162}$$

Since according to (10.56b) the mass operator (10.144b) depends on the "chiral angle" $\theta = \theta_1 - \theta_2$,

$$\mathcal{M}(x) = -\frac{\tilde{\mu}}{\pi} : \cos\,(2\sqrt{\pi}\Sigma(x) + \theta) : \quad , \tag{10.163}$$

the Hamiltonian, and hence, its spectrum, also depends on θ:

$$H = \int dy^1 \left\{ \frac{1}{2} : (\partial_0\Sigma)^2 + (\partial_1\Sigma)^2 + \frac{e^2}{\pi}\Sigma^2 : -\frac{\tilde{\mu}}{\pi} : \cos\,(2\sqrt{\pi}\Sigma(y) + \theta) : \right\} \quad . \tag{10.164}$$

The presence of the mass term has thus lifted the degeneracy of the ground states of the massless theory with respect to the chiral angle, so that different values of this angle θ now describe different sectors of the theory. The physics of the model is however a periodic function of θ, with period 2π.

10.3.4 Screening versus confinement[351]

We now want to take a closer look at the question of screening and confinement of the fundamental fermions ("quarks") of QED_2.In the case where the fermions are massive, this question no longer admits an exact analysis, and we shall have to resort to a "semi-classical" approximation, which consists in calculating the $q\bar{q}$ potential

$V(L)$ defined in (10.94) by looking for the field configurations which minimize the classical Hamiltonian

$$H = \int dx^1 \left\{ \frac{1}{2}(\pi_\Sigma)^2 + \frac{1}{2}(\partial_1 \Sigma)^2 - v(\Sigma) \right\} \tag{10.165a}$$

with ($m^2 = e^2/\pi$)

$$v(\Sigma; x^1) = -\frac{1}{2}m^2(\Sigma - \phi)^2 - \frac{M^2}{4\pi}[1 - \cos(2\sqrt{\pi}\Sigma + \theta)] \quad , \tag{10.165b}$$

which now replaces (10.97). We have found it convenient to redefine the zero point energy, and to make the replacement $\tilde{\mu}M \to \frac{M^2}{2}$, in order to conform to the notation of ref. [351]. The dependence of v on x^1 arises from that dependence of ϕ defined in (10.96). The reason for choosing the sign of the "potential" as shown, will become apparent in a moment.

Equivalent mechanical problem

Since $v(\Sigma; x^1) \leq 0$, it follows that the configurations Σ that minimize H are time-independent. For such configurations it will be helpful to think of Σ as the location q of a point particle, and of x^1 as the corresponding time-coordinate t. (For a similar approach to the spontaneously broken ϕ^4 theory see S. Coleman's Erice Lectures, ref. [23]). In that case (10.165) becomes the action of a point-particle moving in a time dependent potential $v(q,t)$,

$$H \to S_\phi[q] = \int dt \left(\frac{1}{2}\dot{q}^2 - v(q,t) \right) \quad . \tag{10.166}$$

The classical configuration which we seek, just minimizes this action, i.e., satisfies the corresponding Euler-Lagrange equations

$$\ddot{q}_{cl} + \frac{\partial v}{\partial q_{cl}}(q_{cl}, t) = 0 \tag{10.167}$$

in the respective time intervals, where

$$v(q,t) = \begin{cases} v_0(q) = \frac{-1}{2}m^2 q^2 - \frac{M^2}{4\pi}[1 - \cos(2\sqrt{\pi}q + \theta)] \quad , & |t| < \frac{L}{2} \\ v_{Q/e}(q) = \frac{-1}{2}m^2 (q - \frac{Q}{e}\sqrt{\pi})^2 - \frac{M^2}{4\pi}[1 - \cos(2\sqrt{\pi}q + \theta)] \, , & \frac{-L}{2} < t < \frac{L}{2} \end{cases} \tag{10.168}$$

The potential $V(L)$ in (10.94) is then obtained by calculating

$$V(L) = S_\phi[q_{cl}] - S_0[q_{cl}] \quad . \tag{10.169}$$

In fact, it is easy to see, that $V(L)$ is entirely given in terms of the value of q_{cl} at the times $t = -L/2$ and $t = +L/2$ at which the external source ϕ is turned on and off, respectively, if we assume q_{cl} and $\dot{q}_{cl}$ to be continuous functions of time in the whole interval $-\infty \leq t \leq \infty$. In that case

$$\frac{dV(L)}{dL} = \frac{dS_\phi}{dL}[q_{cl}] = -m^2 \int_{-\infty}^{\infty} dt(q-\phi)\frac{d\phi}{dL}$$

$$= -m^2 \frac{\sqrt{\pi}}{2}\left[q_{cl}\left(\frac{L}{2}\right) + q_{cl}\left(-\frac{L}{2}\right)\right] + m^2 \frac{\pi}{2} L \quad . \tag{10.170}$$

Proposition:

$$q_{cl}(L/2) = q_{cl}(-L/2) \quad . \tag{10.171}$$

Proof: In order that $S_\phi[q_{cl}] < \infty$ (finite energy H) we must have that

$$\dot{q}_{cl}(\pm\infty) = 0 \quad , \quad q_{cl}(\pm\infty) = 0 \quad . \tag{10.172}$$

Now, from the equations of motion (10.167) follows that in the respective time intervals, the energy of our equivalent mechanical system is conserved:

$$\frac{1}{2}\dot{q}_{cl}^2 + v_0(q_{cl}) = \begin{cases} E_- & , \quad |t| < -L/2 \\ E_+ & , \quad |t| > L/2 \end{cases} \tag{10.173a}$$

$$\frac{1}{2}\dot{q}_{cl}^2 + v_{Q/e}(q_{cl}) = E_{Q/e} \qquad -\frac{L}{2} < t < \frac{L}{2} \tag{10.173b}$$

From the asymptotic conditions (10.172) we conclude that

$$E_- = E_+ = 0 \tag{10.174}$$

This means, that

$$\left.\frac{1}{2}\dot{q}_{cl}^2 + v_0(q)\right|_{t=-\frac{L}{2}-0} = \left.\frac{1}{2}\dot{q}_{cl}^2 + v_0(q)\right|_{t=\frac{L}{2}+0} = 0 \tag{10.175a}$$

Moreover from (10.173*b*) it follows that

$$\left.\frac{1}{2}\dot{q}_{cl}^2 + v_{Q/e}(q_{cl})\right|_{t=-\frac{L}{2}+0} = \left.\frac{1}{2}\dot{q}_{cl}^2 + v_{Q/e}(q_{cl})\right|_{t=\frac{L}{2}-0} \tag{10.175b}$$

Continuity of $\dot{q}$ thus tells us, that

$$v_{Q/e}(q_{cl}(-\frac{L}{2})) - v_0(q_{cl}(-\frac{L}{2})) = v_{Q/e}(q_{cl}(\frac{L}{2})) - v_0(q_{cl}(\frac{L}{2})) \tag{10.176}$$

which implies (10.171). This means that the "point particle" must be localized at the same point in the instant that the external source is switched on ($t = -\frac{L}{2}$) and off ($t = \frac{L}{2}$). This will play an important role in our subsequent considerations. Using this property, and integrating Eq. (10.170), we finally obtain

$$V(L) - V(L_0) = \frac{m^2\pi}{4}(L - L_0) - m\sqrt{\pi}\int_{L_0}^{L} dL\, q_{cl}(L/2) \quad . \tag{10.177}$$

We drop from here on the subscript "*cl*" on q_{cl}.

The case $M = 0$

In this case,

$$v(q,t) = -\frac{1}{2}m^2(q-\phi)^2 \quad . \tag{10.178}$$

We choose the external quark source to carry one unit of charge ($\frac{Q}{e} = 1$)[15]. The corresponding potentials $v_0(q)$ and $v_{Q/e}(q)$ defined in (10.168) are represented in figure 10.1 in terms of a slashed and solid line respectively.

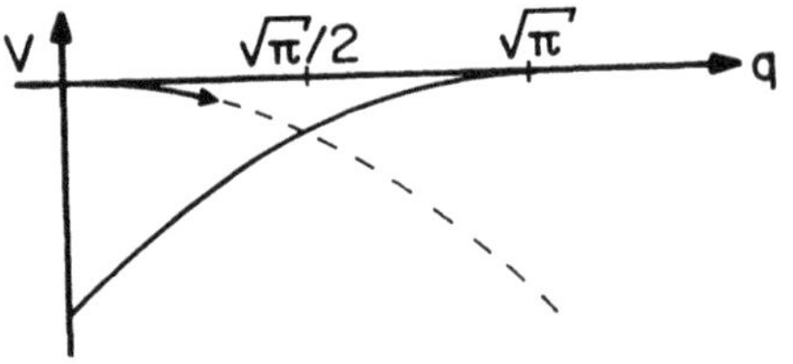

Figure 10.1: Potential v_0 (dashed line) and v_1 (solid line) for M=0. The heavier line with an arrow represents the particle motion

The potential energy seen by the pseudoparticle as it moves during the time interval $[-\infty, \infty]$ is described by the heavy line. Because of the conditions (10.172), the pseudoparticle starts from "rest" and "rolls down" the potential hill represented by v_0; it then experiences a sudden potential drop at $t = -\frac{L}{2}$ without a change of its kinetic energy then "climbs the hill" represented by v_1 until reaching the "turning" point q_0 which it must reach at $t = 0$ for symmetry reason. It then exactly reverses its motion, reaching at $t = \frac{L}{2}$ the same position it had at time $t = -\frac{L}{2}$ (Eq. (10.171)). At this time it is instantly "lifted up" to the potential hill described by v_0, again without change in the kinetic energy, in order to finally return to its original position $q = 0$ at $t = \infty$. In formulae:

$$L = 2\int_{q(-\frac{L}{2})}^{q(0)} \frac{dq}{\sqrt{2E_Q + m^2(q-\sqrt{\pi})^2}} \tag{10.179a}$$

where

$$E_Q = \frac{1}{2}\dot{q}\big(-\frac{L}{2}\big)^2 - \frac{1}{2}m^2\big[q\big(-\frac{L}{2}\big) - \sqrt{\pi}\big]^2 \quad . \tag{10.179b}$$

Now, according to (10.174), $E_0 = \frac{1}{2}\dot{q}\big(-\frac{L}{2}\big)^2 - \frac{1}{2}m^2q^2\big(-\frac{L}{2}\big) = 0$. The turning point $q(0)$ corresponds to the vanishing of the denominator in (10.179). This gives

$$q(0) = \sqrt{\pi}\Big(1 - \sqrt{1 - \frac{2q(-L/2)}{\sqrt{\pi}}}\Big) \quad . \tag{10.180}$$

Integrating (10.179), one thus obtains

[15] If Q is not an integral multiple of e, the charge Q cannot be completely screened by the vacuum polarization, and a long-range confining force will exist, rendering a separation of the external charges impossible without expending infinite energy[352].

$$L = -\frac{2}{m} \ln \sqrt{1 - \frac{2q(-L/2)}{\sqrt{\pi}}} \quad . \tag{10.181}$$

Solving for $q(-L/2)$ and substituting into (10.177) we obtain the result

$$V(L) = \frac{\pi m}{2}(1 - e^{-mL}) \tag{10.182}$$

in agreement with Eq. (10.104).

The potential (10.182) tends to a constant for large L. This reflects the screening of the test-charges induced by the virtual $q\bar{q}$ pairs created from the vacuum, as we pump energy into the system[29]. Since it takes next to no energy to create a real pair of zero mass fermions, the onset of screening is almost immediate, the characteristic length of separation being controlled, by the only dimensional parameter in the theory, $m = e/\sqrt{\pi}$.

The case $M \neq 0$, $\theta = 0$

The situation is now considerably more complicated, figure 10.1 being now replaced by figure 10.2.

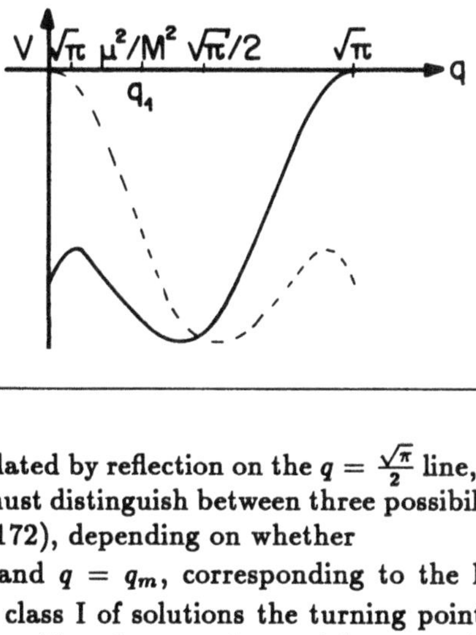

Figure 10.2: Potential v_0 (dashed line) and v_1 (solid line) for $M{>}e$, $\theta{=}0$

The solid and dotted curve are just related by reflection on the $q = \frac{\sqrt{\pi}}{2}$ line, and intersect at this value of q. This time we must distinguish between three possibilities of (finite) energy solutions satisfying (10.172), depending on whether

I) the "particle" moves between $q = 0$ and $q = q_m$, corresponding to the local maximum of the solid curve v_1. For this class I of solutions the turning point q_I^{tp} must obviously approach q_m as $L \to \infty$. For this to happen the particle cannot jump at a later point than a certain critical value q_c smaller than q_m, $0 < q_I(-L/2) < q_c$, $q_c < q_m$.

II) For the second class of solutions, the particle still jumps before reaching q_m, but manages now to overcome the local maximum in a finite time; $q_c < q_{II}(-L/2) < q_m$.

III) For the third class of solutions, $q_m < q(-L/2) < \frac{\sqrt{\pi}}{2}$.

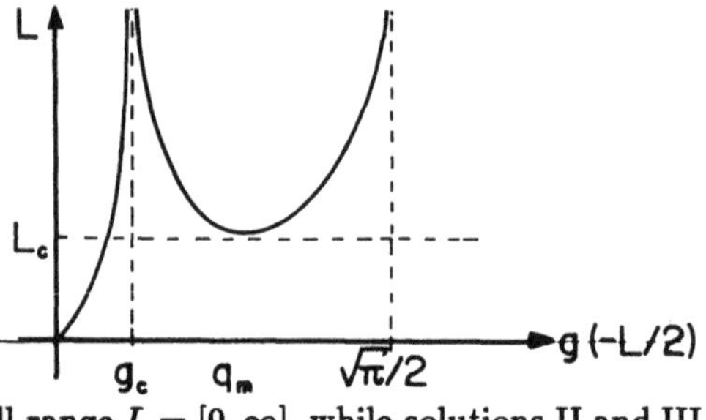

Figure 10.3: Localization of the "particle" as a function of the time, $t=-\frac{L}{2}$ at which the external source is turned on

It is clear that class I covers the full range $L = [0, \infty]$, while solutions II and III only exist for $L \in [L_c, \infty]$, with L_c some "critical" value. The situation is depicted qualitatively in the figure 10.3.

In all three cases, the particle has to "manage" to return to the point $q(-L/2)$ in the time $t = L$ in order to "catch up" again with the potential v_0 when it is "turned on" at $t = L/2$. Only then will it reach $q = 0$ for $t \to \infty$, as required by the "finite energy" condition. This is the meaning of $q(-L/2) = q(L/2)$, which we derived on the basis of energetic considerations. The turning point is obviously reached at $t = 0$ in all three cases. Let us make these statements somewhat more quantitative.

Solutions of Type I and II

To get an estimate of the location of the local maximum of the solid curve in fig. 3 we shall assume that

$$\frac{m}{M} << 1 \quad . \tag{10.183}$$

In that case we may suppose that $q_I << \sqrt{\pi}$ for the actual motion of the pseudoparticle, so that we may replace the potential v_1 in (10.168) by ($\theta = 0!$)

$$v_1(q) \approx -\frac{1}{2}m^2(q+\sqrt{\pi})^2 - \frac{M^2}{2}q^2 \quad . \tag{10.184}$$

From here we find for the location of the maximum

$$q_m \approx \sqrt{\pi}\frac{m^2}{M^2} \quad , \tag{10.185}$$

which conforms with our above assumption. In this approximation we are to solve the equations

$$\ddot{q} - \alpha^2 q = 0 \quad , \quad |t| < \frac{L}{2} \tag{10.186a}$$

$$\ddot{q} - \alpha^2 q = -\sqrt{\pi}m^2 \quad , \quad -\frac{L}{2} < t < \frac{L}{2} \quad , \tag{10.186b}$$

where $\alpha = \sqrt{m^2 + M^2}$.

Imposing the boundary conditions (10.172), as well as continuity of $q(t)$ and $\dot{q}(t)$ at $t = \pm\frac{L}{2}$, we obtain for the type I solution,

$$q_I(t) = \begin{cases} \sqrt{\pi}\frac{m^2}{\alpha^2}e^{-\alpha|t|}\sinh\frac{\alpha L}{2}, & |t| < \frac{L}{2} \\ \sqrt{\pi}\frac{m^2}{\alpha^2}(1 - e^{-\alpha\frac{L}{2}}\cosh\alpha t), & -\frac{L}{2} < t < \frac{L}{2} \end{cases} \tag{10.187}$$

The solution is symmetric with respect to $t \to -t$, as expected. The location of the turning point follows from the requirement that $\dot{q} = 0$ at $t = 0$, which happens for $t = 0$.

$$q_I^{tp} = q_I(0) = \sqrt{\pi}\frac{m^2}{\alpha^2}(1 - e^{-\frac{\alpha L}{2}}) \quad . \tag{10.188}$$

It lies to the left of q_m as expected. Furthermore, it follows from (10.187), that

$$q_I(-\frac{L}{2}) = \sqrt{\pi}\frac{m^2}{\alpha^2}(1 - e^{-\alpha L}) \quad , \tag{10.189}$$

so that substitution of this expression into (10.177) yields

$$V_I(L) = -\frac{\pi}{2}\frac{m^4}{\alpha^3}(1 - e^{-2\alpha L}) + \frac{\pi m^2}{2}\left(1 - \frac{m^2}{\alpha^2}\right)L \quad . \tag{10.190}$$

This potential increases linearly with L, as $L \to \infty$, and therefore cannot correspond to the state of lowest energy for all L! Hence we need to take a look at the other two types of solutions. It is clear, that for a discussion of the solution of type II, we can no longer approximate the cosine in (10.168) by its first two terms in the Taylor expansion. As fig. 10.4 however demonstrates, $q_{II}(-L/2) \to q_I(-L/2)$ as $L \to \infty$, so that this solution will again lead to a linear rise of the potential as $L \to \infty$.

We know however, that "screening" of the test charges, should eventually turn this potential into a constant one as $L \to \infty$. We therefore look at the asymptotic behaviour of the solution of type III.

Asymptotic behaviour of solution of type III

We return to the first integral of the equations of motion, analogous to (10.179), and divide the integration into the two intervals $(q(-L/2), \sqrt{\pi}/2)$ and $(\sqrt{\pi}/2, q(0))$. Since $q(-L/2) \to \frac{\sqrt{\pi}}{2}$, and the turning point lies to the right $\frac{\sqrt{\pi}}{2}$, the dominant contribution for $L \to \infty$ comes from the integration over the second interval:

$$L = \int_{\frac{\sqrt{\pi}}{2}}^{q(0)} dq \frac{1}{\sqrt{2(E_1 - v_1(q))}} + \mathcal{O}(L^{-1}) \quad , \tag{10.191}$$

where E_1 is the total energy (10.173b) of the mechanical system for $-\frac{L}{2} < t < \frac{L}{2}$.

For $L \to \infty$, $q(0) \to \sqrt{\pi}$, and the dominant contribution comes from the region around $\sqrt{\pi}$, where the particle will spend most of its time. Thus we are allowed to make the approximation:

$$\cos 2\sqrt{\pi}q \approx 1 - 2\pi(\sqrt{\pi} - q)^2 \quad . \tag{10.192}$$

In this approximation, the turning point will be given by

$$(\sqrt{\pi} - q(0))^2 = -\frac{2E_1}{\alpha^2} \quad . \tag{10.193}$$

Hence it follows from (10.191) that

$$L \approx -\frac{2}{\alpha} \ln \frac{\sqrt{-\frac{2E_1}{\alpha^2}}}{\frac{\sqrt{\pi}}{2} + \sqrt{\frac{\pi}{4} + \frac{2E_1}{\alpha^2}}} + \mathcal{O}(L^{-1}) \quad . \tag{10.194}$$

From (10.194) we find $E_1 \underset{L \to \alpha}{\longrightarrow} -\frac{\alpha^2 \pi}{2} e^{\alpha L}$, or, using

$$E_1 = \sqrt{\pi} m^2 q(-L/2) - \frac{1}{2}\pi m^2 \quad , \tag{10.195}$$

we obtain

$$q(-L/2) \to \frac{\sqrt{\pi}}{2}(1 - \frac{\alpha^2}{m^2} e^{-\alpha L}) \quad . \tag{10.196}$$

Substituting this expression into (10.177) one finally finds

$$V(L) \to -\frac{\alpha\pi}{2} e^{-\alpha L} + V(\infty) \quad . \tag{10.197}$$

$V(L)$ thus tends exponentially to a constant, as $L \to \infty$, reflecting "screening". The value of this constant may be estimated to $\mathcal{O}(\frac{m}{M})$ by noting that for $m = 0$, Eqs. (10.173) reduce to that of the time independent sine-Gordon theory for $\beta = 2\sqrt{\pi}$ (see Eq. (3.34)). Hence $V(\infty)$ is just the soliton mass M_{sol} for this value of β. Since $M_{sol} = 8M/\beta$, we conclude that

$$V(\infty) \approx M_{sol}(\beta = 2\sqrt{\pi}) = \frac{4M}{\pi} \tag{10.198}$$

and correspondingly we have to $\mathcal{O}(m^2/M^2)$

$$V_{III}(L) \approx \frac{4M}{\pi}(1 - 2e^{-\alpha L}) \quad . \tag{10.199}$$

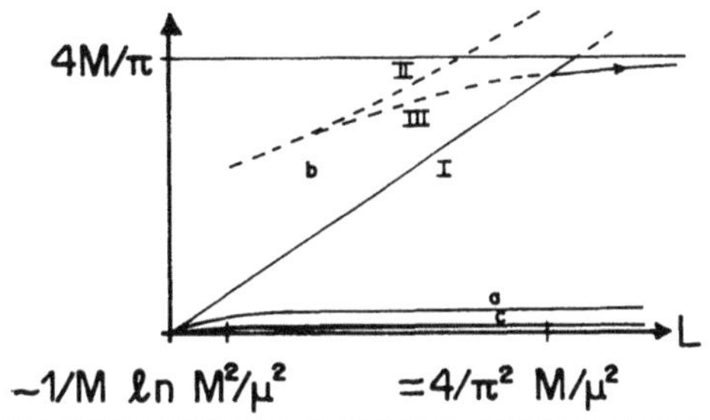

Figure 10.4: Interparticle potential $V(L)$ (solid line) for solutions of type I, II, and III. Dashed lines represent metastable states.

The situation is depicted qualitatively in figure 10.4 below. As is evident from this figure, $V_{II}(L)$ and $V_{III}(L)$ only exist for $L > L_c$. L_c is estimated to be of the order of $\frac{1}{M} \ln \frac{M^2}{m^2}$. From figure 10.5 and (10.177) it is also evident, that $V_{III}(L) \leq V_{II}(L)$, the equality sign holding for $L = L_c$,

It is clear that the true potential $V(L)$ is obtained by going smoothly from branch I to branch II at the intersection point of the curves III and II at $L \approx \frac{8}{\pi^2} \frac{M}{m^2}$, thus ensuring that the interparticle potential corresponds to a configuration of lowest possible energy for every L. It is interesting that the asymptotic value (10.198) corresponds to just barely the value at which real pair production would be possible![16]

So far we have only discussed case $\theta = 0$. For $0 < \theta < \pi$ the situation very much parallels the one described above. The $\theta = \pi$ case is however an exception, as we now discuss.

The case $M = 0$, $\theta = \pi$

For $\theta = \pi$ figure 10.3 is replaced by figure 10.5.

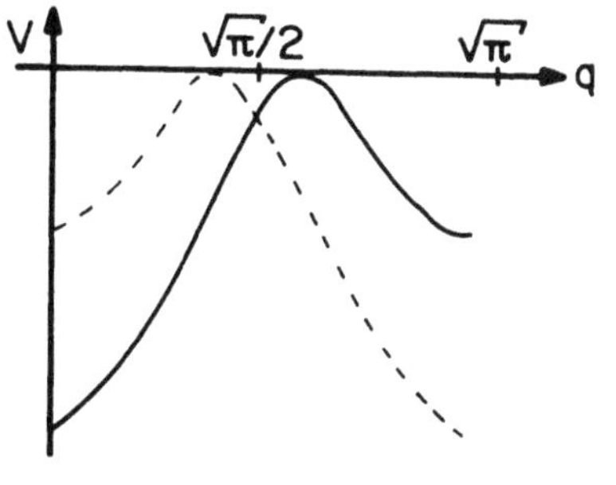

Figure 10.5: Potential v_0 (dashed line) and v_1 (solid line).

[16] This behavior of the $q\bar{q}$ potential has also been confirmed by a Monte-Carlo on the lattice[377].

We now have two degenerate maxima corresponding to the potential in the absence (broken curve) and presence (solid curve) of external charges, respectively.

Since the maxima are separated by a distance of only $\sqrt{\pi}\frac{m^2}{M^2}$, the motion in the equivalent mechanical model minimizing the interparticle potential $V(L)$ will take place between those two maxima, and we are allowed to make the approximation $\cos\left(2\sqrt{\pi}q+\pi\right) \approx 1+2\pi\left(q-\sqrt{\frac{\pi}{2}}\right)^2$. One finds this time[351]

$$V(L) = \frac{\pi}{2}\frac{m^4}{\alpha^3}(1-e^{-ML}) \tag{10.200}$$

Comparing with our previous results we see that there is no linear rise of $V(L)$ over any appreciable range of L, and that the screening effects are even more "violent" than in the massless (Schwinger model) case. We shall now show, that the corresponding states are to be interpreted as "colorless" kink states with multiplicative rather than additive quantum numbers.

10.3.5 Adding flavour

In the preceding section we have witnessed total screening for the "vacuum angle" $\theta = \pi$. We have tentatively identified this phenomenon with a new "phase" corresponding to the appearance of kink-like states. Actually, in a gauge theory in $d = 1+1$ dimensions with dynamical quarks, it is not the perimeter vs. area law that will distinguish between confinement and screening of quarks, since here the vacuum polarization will eventually always lead to screening of "charges" (colors). Thus a possible breakdown of confinement is signalized here by the appearance of (color neutral) states carrying fundamental flavour. This motivates us to reconsider the problem of confinement in the presence of flavor quantum numbers. As we shall see, the appearance of kink states for $\theta = \pi$ witnessed in the previous section is not an artefact of the semiclassical treatment, but can be understood in an entirely non-perturbative context, from the operator point of view[349,350]. This approach has the virtue of revealing some unusual features of our model: In a local $SU(N)$ invariant quantum field theory, it is usually taken for granted that one-particle states belonging to an irreducible representation can be used to construct multiparticles states as direct product states belonging to higher irreducible representations of $SU(N)$. This, in particular, would imply additivity with respect to the quantum numbers associated with the internal symmetry group. Our model discussion will show that this is not necessarily so, and will be the reason for calling these states exotic.

We shall use the following terminology: Sectors characterized by multiplicative (additive) quantum numbers will be referred to as kink (soliton) sectors, in correspondence with the spontaneously broken ϕ^4 (sine-Gordon) theory, where such sectors are known to occur. In that case the existence of such sectors is directly linked to a finite discrete (periodic) symmetry of the Hamiltonian. Thus one would think, that the distinction between kinks and solitons should be an unambiguous one. This, as we shall demonstrate is not always the case. Thus, for the generalized model to be discussed next we occasionally find that states which appear to be kinks are actually exotic solitons. We shall present only the essentials of the new ideas involved. The reader should consult refs. [350] for a more detailed analysis. The generalization of the local $U(1)_G$ symmetry to a $U(1)_G \times SU(N)_F$ symmetry, with $SU(N)_F$ a global flavor symmetry, has already been discussed in the context of the massless ($M = 0$) model.

The transition to the massive case is made in the spirit of the adiabatic principle, by simply adding to $\mathcal{H}$ the appropriate mass terms. We have

$$\mathcal{H}_\theta = \mathcal{H}_{free} + V_\theta \tag{10.201}$$

where

$$\mathcal{H}_{free} = \frac{1}{2}:[(\partial_0\Sigma)^2 + (\partial_1\Sigma)^2]: + \frac{1}{2}\sum_{i_D=1}^{N-1} :[(\partial_0\phi^{i_D})^2 + (\partial_1\phi^{i_D})^2] \quad . \tag{10.202a}$$

and V_θ is the potential

$$V_\theta = \frac{1}{2}\frac{Ne^2}{\pi}\Sigma^2 - \frac{\tilde{\mu}M}{\pi}\sum_f :\cos\left(2\sqrt{\frac{\pi}{N}}\Sigma + 2\sqrt{\pi}\chi_f + \theta\right): \quad , \tag{10.202b}$$

The fields χ_f are related to those in (10.116) by

$$\chi_f \equiv \varphi_f - \frac{1}{\sqrt{N}}\varphi = \sum_{i_D} \tau_{ff}^{i_D}\phi^{i_D} \tag{10.203}$$

and thus satisfy the constraint $\sum_{f=1}^N \chi_f = 0$. The fields Σ and ϕ^{i_D} , $i_D = 1, \cdots N-1$, on the other hand are independent fields; they are the "potentials" of the N diagonal conserved currents of $U(1) \times SU(N)_F$:

$$J_\mu^{i_D}(x) = -\frac{1}{\sqrt{\pi}}\tilde{\partial}_\mu\phi^{i_D}(x) \tag{10.204a}$$

$$J_\mu(x) = -\sqrt{\frac{N}{\pi}}\tilde{\partial}_\mu\Sigma(x) \tag{10.204b}$$

The Hamiltonian density (10.202) has a manifest periodicity in θ with period 2π. The periodicity of the corresponding partition function is however $2\pi/N$. Indeed, we have, if we regard $\mathcal{H}$ as function of Σ and χ_f:

$$\mathcal{H}_{\theta+2\pi/N}(\Sigma, \chi_f) = \mathcal{H}_\theta(\Sigma, \chi'_f) \quad , \tag{10.205}$$

where the shifted fields

$$\chi'_f = \chi_f + \frac{\sqrt{\pi}}{N} \quad , \quad f = 1 \cdots r-1, r+1, \cdots N \tag{10.206a}$$

$$\chi'_r = \chi_r - (N-1)\frac{\sqrt{\pi}}{N} \quad , \tag{10.206b}$$

again satisfy the same constraint as χ_f.

Because of this hidden periodicity it will be sufficient to study the spectrum of H_θ, for θ modulo $2\pi/N$.

Exotic states

For $\theta = 0$ the spectrum of H_θ is known to only contain (finite energy) states interpretable as mesons and baryons. According to Eq. (10.205) the same applies of course also to $H_{\theta=\frac{n2\pi}{N}}$, with $n =$ integer. The corresponding hadronic sectors are created, respectively, by repeated application of the "mesonic" and "barionic" operators (we suppress the dependence on spinor indices)

$$\hat{M} = \mathcal{F}_f \mathcal{F}_f^\dagger \quad , \quad \hat{B} = \prod_{i=1}^{N} \mathcal{F}_{f_i} \quad , \tag{10.207}$$

where $\mathcal{F}_f$ is the "flavor" operator already introduced in (10.119a). Thus using (10.203) (we suppress the spinor index)

$$\mathcal{F}_f(x) = : e^{i\sqrt{\pi}\gamma^5 \chi_f + i\sqrt{\pi}\int_{x^1}^{\infty} dz^1 \partial_0 \chi_f} \tag{10.208}$$

where : : denotes a suitable normal ordering prescription defining $\mathcal{F}_f(x)$ as a unitary operator. This operator continues to satisfy the commutation relation (10.121*b*) and thus carries $SU(N)_F$-quantum numbers. However, since the mass term in (10.134*b*) explicitly breaks the $\tilde{U}(N)$ symmetry, the $\widetilde{SU}(N)$-torus currents of both, the free and interacting fermion field are no longer conserved. Correspondingly the operator (10.208) carries no chiral selection rules. The chiral part in the exponent of (10.208) is nevertheless important in order to have the correct short distance properties with respect to the $SU(N)$ currents. Notice that we are only interested in the sectors, and not specific particle states of the theory. It is thus sufficient to consider the operator (10.208) by itself.

In the zero mass ($M = 0$) case, the operator (10.208) creates a finite energy state belonging to the fundamental representation of $SU(N)_F$. For $M \neq 0$ this is no longer the case, since this operator no longer commutes with the potential (10.202b),

$$[\mathcal{F}_f(x), V_\theta(y)] = -\frac{\tilde{\mu}M}{\pi}\sum_{f'=1}^{N}\Big[:\cos\Big(2\sqrt{\frac{\pi}{N}}\Sigma(y) + 2\sqrt{\pi}\chi_{f'}(y) - \frac{2\pi}{N}\theta(y^1 - x^1) + \theta\Big): \\ -:\cos\Big(2\sqrt{\frac{\pi}{N}}\Sigma(y) + 2\sqrt{\pi}\chi_{f'}(y) + \theta\Big):\Big]\mathcal{F}_f(x) \quad . \tag{10.209}$$

Here we have used $[\chi_f(x), \partial_0\chi_{f'}(y)] = i\left(\delta_{ff'} - \frac{1}{N}\right)$, implying

$$\mathcal{F}_f(t,x^1)\chi_{f'}(t,y^1)\mathcal{F}_f^\dagger(t,x^1) = \chi_{f'}(t,y^1) + \sqrt{\pi}\theta(y^1 - x^1)\left(\delta_{ff'} - \frac{1}{N}\right) \tag{10.210}$$

Note that the translation of $\chi_{f'}$ is a result of the line integral appearing in (10.208). We therefore see, that in general the mass operator is non-local with respect to the "flavor" operator $\mathcal{F}_{f'}$. In fact, except for $N = 1$, the soliton operator $\mathcal{F}_f$ does not create states of finite energy, for any value of θ. However, let us adjoin to it the "kink" operator

$$K = K_\Sigma \prod_{f=1}^{N} K_{\varphi_f} = K_\Sigma \prod_{f=1}^{N} K_{\chi_f} \quad , \tag{10.211}$$

where K_ϕ is a "kink operator"[378] with the property[17)]

$$K_\phi\phi(x)K_\phi^{-1} = \begin{cases} \phi(x) \quad , & x \to -\infty \\ -\phi(x) \quad , & x \to \infty \end{cases} \quad . \tag{10.212a}$$

A formal representation of such a Kink-operator is given by [378]

$$K_\phi = e^{\frac{i}{2}\int dy^1 \phi_{\rm kink}(y)[\phi^2 + (\pi_\phi)^2]} \quad , \tag{10.212b}$$

where $\phi_{\rm kink}$ is any function interpolating between zero at $y^1 = -\infty$ and π at $y^1 = +\infty$. Note that this operator induces a rotation in phase space[18)].

---

17) A construction of such an object can be understood in terms of functional integrals[379].

18) To prove Eq. (10.212a), we note that $B(y^0) = \int_{y^0} dy^1 \phi_{\rm kink}(y)[\phi^2 + \pi_\phi^2]$ has the property

$$[B, \phi(x)]_{ET} = \phi_{\rm kink}(x)\pi(x)$$

$$[B, \pi(x)]_{ET} = -\phi_{\rm kink}(x)\phi(x)$$

We find that $K\mathcal{F}_f$ creates a state of finite energy on the $\theta = \pi/N$ vacuum:

$$K\mathcal{F}_f|\theta = \frac{\pi}{N}\rangle = |K\mathcal{F}_f; \frac{\pi}{N}\rangle \quad . \tag{10.213}$$

A corresponding statement can be made for the operator $\mathcal{F}_f^\dagger K$ acting on $|\theta = \pi/N\rangle$. We identify the state created by $K\mathcal{F}_f$ $(\mathcal{F}_f^\dagger K)$ when acting on the $|\theta = \frac{\pi}{N}\rangle$ vacuum as the quantum field theoretical version of Coleman's quasiclassical "half-asymptotic" particle states[353].

The finite energy state (10.213) has "exotic" transformation properties. As the name kink already suggests, it carries a multiplicative quantum number. Indeed, noting that

$$Q^{i_D} = \frac{1}{\sqrt{\pi}}(\phi^{i_D}(t,\infty) - \phi^{i_D}(t,-\infty)) \quad , \tag{10.214}$$

in the weak sense, one obtains after a simple calculation

$$\langle\frac{\pi}{N}; K\mathcal{F}_{f_1}\cdots K\mathcal{F}_{f_l}|Q^{i_D}|K\mathcal{F}_{f_1}\cdots K\mathcal{F}_{f_l};\frac{\pi}{N}\rangle = \\ -\frac{(1-(-1)^l)}{\sqrt{\pi}}\langle\phi^{i_D}\rangle + \sum_{k=1}^{l}(-1)^{k+1}\tau^{i_D}_{f_k f_k} \quad . \tag{10.215}$$

where we have repeatedly used

$$\mathcal{F}_f^\dagger(x)\phi^{i_D}(y)\mathcal{F}_f(x) = \phi^{i_D}(x) + \sqrt{\pi}\tau^{i_D}_{ff}\theta(y^1 - x^1) \quad , \tag{10.216}$$

as well as translational invariance. Hence the operator $K\mathcal{F}_f$ creates (exotic) sectors with multiplicative "kink" (modulo additive mesonic) quantum numbers, and we refer to it as a kink-soliton operator. In particular we see from (10.216) that repeated application of such a kink-soliton operator to the $|\pi/N\rangle$ vacuum does not lead to states with increasing "exoticity"; for the "exotic" states, flavor is not an additive quantum number, even if $\langle\phi^{i_D}\rangle = 0$. In general $\langle\phi^{i_D}\rangle$ is coupling constant dependent and may be calculated in the semi-classical approximation . Its precise value does

---

For $x^1 \to \infty$, $\phi_{\text{kink}} \to \pi$, and this algebra becomes isomorphic to

$$[\sigma_3, \sigma_i] = 2i\epsilon_{ij}\sigma_j \,, \; \epsilon_{12} = -\epsilon_{21} = 1 \quad ,$$

of the Pauli matrices, with the correspondences $B \to \sigma_3$, $(\pi, \phi) \to (\sigma_1, \sigma_2)$. Hence it is clear that

$$e^{i\pi B}\begin{pmatrix}\phi(x)\\ \pi(x)\end{pmatrix}e^{-i\pi B} \to \begin{pmatrix}-\phi(x)\\ -\pi(x)\end{pmatrix} \text{ for } x^1 \to \infty \quad .$$

not matter, however. These exotic states thus exists for all values of the coupling constant, and there is no evidence for two distinct (kink-soliton and pure soliton) phases.

It is remarkable that $\theta = \pi/N$ is precisely the angle for which the $SU(N)_F$ model becomes P and T invariant. This is easiest seen by performing a $SU(N)_F$ diagonal chiral rotation

$$\chi_f = \chi'_f - \frac{\sqrt{\pi}}{2N} \ , \ f = 1, \cdots, r-1, r+1, \cdots, N \ , \ \chi_r = \chi'_r + (N-1)\frac{\sqrt{\pi}}{2N} \quad ,$$

in the potential (10.202*b*), which removes θ in all but the term involving χ_r, where θ is effectively replaced by π in the sum appearing in V_θ, Eq. (10.202*b*). The resulting potential after this rotation is invariant under the transformation $\Sigma \to -\Sigma \ , \ \chi'_f \to -\chi'_f$. Hence the absence of antiparticle states in the $\theta = \pi/N$ vacuum implies a violation of the CPT theorem! The reason lies in the non-local character of the kink operator, and illustrates one further peculiarity of the theory in question.

Kinks in $SU(2)_F$

$SU(2)_F$ plays an exceptional role in the above discussion. In this case there exists another kink-soliton operator, $K_\Sigma \mathcal{F}_f$, which creates finite energy states as well[354]. This is most readily seen, by noting that for $\theta = \pi/2$ and $N = 2$, the potential (10.202) may be written in the form

$$V_{\frac{\pi}{2}} = \frac{e^2}{\pi}\Sigma^2 + 2\frac{M\bar{\mu}}{\pi}\sin(\sqrt{2\pi}\Sigma)\cos(2\sqrt{\pi}\phi) \quad , \tag{10.217}$$

where we have set $\chi_1 = -\chi_2 = \phi$. For the expectation value of Q^3 and Q, in the state $| K_\Sigma \mathcal{F}_1 \cdots K_\Sigma \mathcal{F}_l; \frac{\pi}{2}\rangle$ one finds in this case

$$\left\langle \frac{\pi}{2}; K_\Sigma \mathcal{F}_1 \cdots K_\Sigma \mathcal{F}_l \right| \begin{pmatrix} Q^3 \\ Q \end{pmatrix} \left| K_\Sigma \mathcal{F}_1 \cdots K_\Sigma \mathcal{F}_l; \frac{\pi}{2} \right\rangle = \begin{pmatrix} \sum\limits_{k=1}^{l} \tau^3_{f_k f_k} \\ -\frac{1}{\sqrt{\pi}}(1-(-1)^l)\langle\Sigma\rangle \end{pmatrix} . \tag{10.218}$$

The flavor quantum number is in this case additive, since K_Σ does not act on χ, whereas the charge quantum number is multiplicative, being just a "relic" of charge. In a semi-classical treatment we identify $\langle\Sigma\rangle$ with the stable (absolute) minima of the potential (10.217). Defining the dimensionless coupling constant $g = \frac{e^2}{2\pi M\bar{\mu}}$, they are given by $2\sqrt{\pi}\phi = n\pi$, and $\xi = 2\sqrt{\pi}\langle\Sigma\rangle$ a solution of $g\xi + (-1)^n \cos\xi = 0$. We conclude that $\langle\Sigma\rangle = (-1)^{n+1}\Sigma_0$, with $\Sigma_0 > 0$ and independent of n. We are thus led to identify the above states with states made up of n quarks, whose charge has been screened by the interaction with the $U(1)$ gauge field.

The U(1) kink revisited

Let us finally return the $U(1)$ case discussed at length in the preceding section. In a semiclassical approximation we had witnessed the appearance of a particular "state" for $\theta = \pi$. Let us recall to this effect, that for the case of the periodic sine-Gordon potential

$$V_{S.G.}(\Sigma) = \frac{M^2}{4\pi}(1 - \cos(2\sqrt{\pi}\Sigma)) \quad , \tag{10.219}$$

with periodicity $\Sigma \to \Sigma + \sqrt{\pi}$ there exist an infinite number of charge sectors with a (conserved) topological charge

$$Q = \frac{1}{\sqrt{\pi}}(\Sigma(t, \infty) - \Sigma(t, -\infty)) \quad , \tag{10.220}$$

which is quantized, since $\Sigma(t, \pm\infty)$ has to coincide with one of the minima of (10.219) for a solution with finite action. Here the operator which takes one from the n'th to the $(n+1)$'th charge sector, is given by the disorder operator

$$\mu_\Sigma(x) = e^{i\sqrt{\pi}\int_{x^1}^{\infty} dz^1 \partial_0 \Sigma(x^0, z^1)} \quad . \tag{10.221}$$

Because of the periodicity of $V_{S.G.}(\Sigma)$, this operator creates "soliton" states of finite energy. This symmetry is however broken in the presence of an electromagnetic interaction via an intrinsic Higgs mechanism generating the mass $\frac{e}{\sqrt{\pi}}$ for Σ. The resulting potential $V(\Sigma) = \frac{e^2}{2\pi}\Sigma^2 + V_{S.G.}(\Sigma)$ is no longer periodic, and as an absolute minimum at $\Sigma = 0$. Hence, for a finite energy solution, the charge can only take the value $Q = 0$.

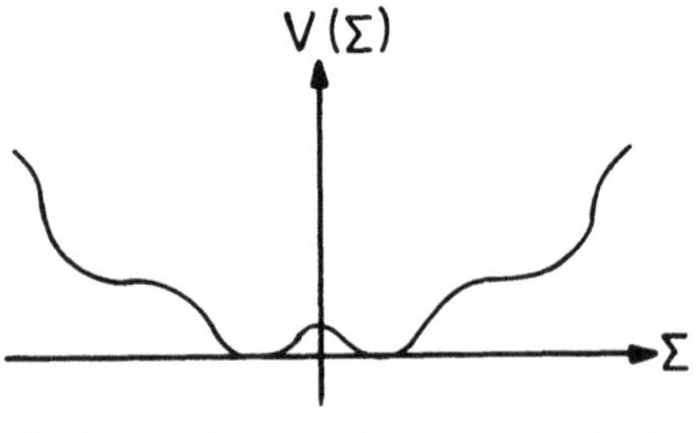

Figure 10.6: $V(\Sigma)$ for $\theta=\pi$ $M^2 >> \frac{e^2}{\pi}$, after suitable "gauging" of minimum.

Let us now shift the argument of the cosine in (10.219) by π. V then takes the form shown in figure 10.6 (after suitable translation in V). Hence the original symmetry $\Sigma \to -\Sigma$ is spontaneously broken; correspondingly the charge (10.220) can now take on two values, showing that Q plays the role of a multiplicative quantum number. We thus identify the state found in our semiclassical discussion for $\theta = \pi$ with a soliton-kink-like state.

It is again a straightforward matter to extend these results to include also the maximal abelian subgroup of $U(1)_{\text{El}} \times SU(N)_{\text{color}}$. We refer the reader to the literature for a discussion of this generalization[350]

10.3.6 Lorentz transformation properties

The bosonic representation (10.141) of the fermion field operator with $\alpha = \beta = \gamma = \sqrt{\pi}$ was shown to be a solution of the massive Dirac equation satisfying the requirement of anticommutativity (spin-statistics) and a canonical short-distance behavior. Although it is expected to transform correctly under Lorentz transformations, this is by no means obvious from (10.141). In fact, were it not for the line integral in the exponent, this expression would transform as a scalar.

Under the infinitesimal Lorentz transformation $x \to \Lambda x$, parametrized by

$$\Lambda^{\mu}_{\ \nu} \approx g^{\mu}_{\ \nu} + \omega^{\mu}_{\ \nu}\,,\ \omega^{\mu\nu} = -\omega^{\nu\mu}\,,\ \omega^{0}_{\ 1} = v \quad , \tag{10.222}$$

the fields (see Eq. (10.143))

$$\chi(x;h) = \int_0^1 d\eta h_x(\eta)\chi(x^0,\eta) \tag{10.223a}$$

$$\tilde{\varphi}(x;h) = \int_0^1 d\eta h_x(\eta^1)\int_\eta dy^1 \partial_0\varphi(x^0,y^1) \tag{10.223b}$$

transform as follows:

$$U[\Lambda]\chi(x;h)U^{-1}[\Lambda] \approx \chi(x;h) + \delta\chi(x;h) \tag{10.224a}$$

$$U[\Lambda]\tilde{\varphi}(x;h)U^{-1}[\Lambda] \approx \tilde{\varphi}(x;h) + \delta\tilde{\varphi}(x;h) \tag{10.224b}$$

where

$$\delta\chi(x;h) = v\int d\eta^1 h_x(\eta^1)(x^0\partial_1 + \eta^1\partial_0)\chi(x^0,\eta^1) \tag{10.224c}$$

and

$$\begin{aligned}\delta\tilde{\varphi}(x;h) &= v\int d\eta^1 h_x(\eta^1)(x^0\partial_1 + \eta^1\partial_0)\tilde{\varphi}(x^0,\eta^1)\\ &- v\frac{2\beta\tilde{\mu}M}{\pi}\int d\eta^1 h_x(\eta^1)\int_{\eta^1}^{\infty} dy^1(y^1-\eta^1){:}\sin 2\chi(x^0,\eta^1){:} \quad .\end{aligned} \tag{10.224d}$$

Here we have used the equation of motion (10.137c) and dropped surface terms arising from partial integrations. As agreed to before, : : stands for a limiting process such as given by (10.140).

For $M = 0$ Eq. (10.224d) also describes the transformation law of a scalar field $\tilde{\varphi}$. This has its roots in the conservation of the axial vector current (10.28b). This property is however completely spoiled by the addition of the mass-term in (10.134b). It is thus remarkable, that on quantum level, the canonical dimension of the mass operator restores the desired Lorentz properties of the fermion field, as we now show.

The transformation properties (10.224) imply for the field $\Phi_\alpha(x;h)$, Eq.(10.142), the transformation

$$\begin{aligned}\delta\Phi_\alpha(x;h) &= v\int d\eta^1 h_x(\eta^1)(x^0\partial_1 + \eta^1\partial_0)\Phi_\alpha(x^0,\eta^1)\\ &\quad - v2\tilde{\mu}M\int d\eta^1 h_x(\eta^1)\int_{\eta^1}^{\infty} dy^1(y^1-\eta^1){:}\sin\,2\chi(x^0,\eta^1){:}\quad .\end{aligned} \tag{10.225}$$

Using once more the equation of motion (10.137c) in the second term, we may rewrite (10.225) as (we use $\partial_1^\eta h_x(\eta^1) = -\partial_1^x h_x(\eta)$)

$$\delta\Phi_\alpha(x;h) = v(x^0\partial_1 + x^1\partial_0)\Phi_\alpha(x;h) + \Delta_1\Phi_\alpha(x;h) + \Delta_2\Phi_\alpha(x;h) \tag{10.226a}$$

where

$$\Delta_1\Phi_\alpha(x;h) \equiv -v\int d\eta^1 h_x(\eta^1)(x^1-\eta^1)\left[\gamma^5_{\alpha\alpha}\partial_0\chi(x^0,\eta^1) - \frac{\pi}{\beta}\partial_1\varphi(x^0,\eta^1)\right]$$

and

$$\Delta_2\Phi_\alpha(x;h) \equiv -v2\tilde{\mu}M\int d\eta^1 h_x(\eta^1)\int_{\eta^1}^{\infty} dy^1(y^1-x^1){:}\sin 2\chi(x^0,\eta^1){:}\quad . \tag{10.226b}$$

Using (10.145a) we then have

$$\delta e^{i\Phi_\alpha(x;h)} = v(x^0\partial_1 + x^1\partial_0)e^{i\Phi_\alpha(x;h)} + \Delta_1 e^{i\Phi_\alpha(x;h)} + \Delta_2 e^{i\Phi_\alpha(x;h)}\quad , \tag{10.227a}$$

where $l = 1, 2$.

$$\Delta_l e^{i\Phi_\alpha(x;h)} := i\int_0^1 d\lambda e^{i\lambda\Phi_\alpha(x;h)}\Delta_l\Phi(x;h)e^{i(1-\lambda)\Phi_\alpha(x;h)}\quad . \tag{10.227b}$$

Using (10.145b) we have for $l = 1$

$$\Delta_1 e^{i\Phi_\alpha(x;h)} = \frac{i}{2}\left\{\Delta_1\Phi_\alpha(x;h), e^{i\Phi_\alpha(x;h)}\right\}\quad , \tag{10.228}$$

since the contribution coming from the double commutator in (10.145b) vanishes identically, because of (C.14).

Recalling (10.141), we thus have

$$\delta\psi_\alpha(x;h) = v(x^0\partial_1 + x^1\partial_0)\psi_\alpha(x;h) + \Delta\psi_\alpha(x;h) \quad , \tag{10.229a}$$

where

$$\Delta\psi_\alpha(x;h) = -\frac{i}{2}\{\Delta_1\Phi(x;h), \psi_\alpha(x;h)\} - iv2M\bar{\mu}\int d\eta^1 h_x(\eta^1)\times$$

$$\int_{\eta^1}^{\infty} dy^1(y^1 - x^1)\int_0^1 d\lambda e^{i\lambda\Phi_\alpha(x;h)}{:}\sin 2\chi(x^0,y^1){:}\, e^{-i\lambda\Phi_\alpha(x;h)}\psi_\alpha(x;h) \quad . \tag{10.229b}$$

Repeating the steps which led from (10.150a) to (10.153), one obtains

$$\Delta\psi_\alpha(x;h) = -\frac{i}{2}v\{S_\alpha(x;h), \psi_\alpha(x;h)\}$$
$$+ivM\pi\int_{-\infty}^{\infty} dy^1(y^1 - x^1)[\mathcal{M}(y), \psi(x;h)]_{ET} \quad , \tag{10.230}$$

where we have set $\Delta_1\Phi_\alpha(x;h) = vS(x;h)$. At this point we choose the canonical values $\alpha = \beta = \delta = \sqrt{\pi}$. In that case the mass operator has scale dimension $d = 1$, so that the last term in (10.230) vanishes due to, locality of the commutator. Hence

$$\Delta\psi_\alpha(x;h) = -\frac{iv}{2}\{S(x;h), \psi_\alpha(x;h)\} \quad . \tag{10.231}$$

The anticommutator may be evaluated using the fact that in the limit $h \to \delta$, only the short-distance behavior is relevant. This allows us to treat the χ and φ as free fields. In this case one has the identity,

$$\frac{1}{2}\{S, e^{i\Phi}\} =: Se^{i\Phi}: + \frac{i}{2}[S^{(+)}, \Phi^{(-)}]{:}\, e^{i\Phi}{:} - \frac{i}{2}[S^{(-)}, \Phi^{(+)}]{:}\, e^{\Phi}{:} \quad ,$$

where : : is just the ordinary Wick-product. Only the last two terms can yield a non-vanishing contribution to (10.230) in the limit $h \to \delta$. Making use of the identity,

$$\lim_{h\to\delta} h_x(\eta^1)\int d\xi^1 h_x(\xi^1)\left(\frac{1}{\eta^1 - \xi^1 + i0} + \frac{1}{\eta^1 - \xi^1 - i0}\right) = -\delta'(\eta^1 - x^1)$$

one finds, after a lengthy, but straightforward, calculation

$$\frac{1}{2}\{S(x;h), \psi_\alpha(x;h)\}_{h\to\delta} = \frac{i}{2}\gamma^5_{\alpha\alpha}\psi_\alpha(x) \quad .$$

Hence (10.229) reduces to

$$\delta\psi_\alpha(x;h) = v(x^0\partial_1 + x^1\partial_0)\psi(x) + \frac{v}{2}\gamma^5_{\alpha\alpha}\psi(x) \quad , \tag{10.232a}$$

corresponding to the infinitesimal form of the transformation law (2.74) of a "spin $\frac{1}{2}$" field in $d = 1 + 1$ dimensions.

10.3.7 The MSM as the limit of a massive vector theory

We have seen, that in QED_2 Maxwell's equations are only satisfied on the (physical) subspace $\mathcal{H}_{phys}$ of the Hilbert space $\mathcal{H}$ defined by (10.157). It is interesting to regard massive QED_2 as the limit of a vector-meson theory (Thirring-Wess model[380]) when the mass m_0 of the vector-meson field B_μ tends to zero. It is clear, that such a limit can only exist for the operators which map into the QED_2 observables in $\mathcal{H}_p$ in the limit $m_0 \to 0$. The "gauge-dependent" operators, on the other hand, are expected to show a pathological behaviour in this limit.

At the classical level the "Thirring-Wess" model is defined by the equations of motion

$$\partial_\mu F^{\mu\nu} + m_0^2 B^\nu + eJ^\nu = 0 \tag{10.233a}$$

$$(i\not\partial - M + e\not B)\psi = 0 \tag{10.233b}$$

where

$$F^{\mu\nu} = \partial^\mu B^\nu - \partial^\nu B^\mu \tag{10.233c}$$

and

$$J^\nu = \overline{\psi}\gamma^\nu\psi \quad . \tag{10.233d}$$

In this model we expect to be able to obtain a well defined local solution of the coupled Proca-Dirac equations (10.233a,b) without having to impose a restriction of the type (10.157). It follows from the conservation of the current (10.233d), that $\partial_\mu B^\mu = 0$.

Guided by the known results[380,381] for the massless ($M = 0$) Thirring-Wess model, we make the following ansatz for the fermion field:

$$\psi_\alpha(x) = \left(\frac{\tilde{\mu}}{2\pi}\right)^{1/2} e^{-i\frac{\pi}{4}\gamma^5_{\alpha\alpha}} : e^{i\Phi_\alpha(x)} : \tag{10.234a}$$

where $\tilde{\mu} = \left(\frac{e}{2\sqrt{\pi}}\right)^\gamma$ and

$$\Phi_\alpha(x) = \gamma^5_{\alpha\alpha}(\alpha\Sigma(x) + \beta\phi(x)) + \frac{\pi}{\beta}\int_{x^1}^{\infty} dy^1 \partial_0\phi(x^0, y^1) \quad . \tag{10.234b}$$

Here : : is to be understood again in the sense of (10.140). The coefficients of ψ and the line integral have again be chosen such that ψ satisfies Fermi statistics.

For the case $e \to 0$ we should recover the free field solution for ψ. Correspondingly we expect in this limit, that $\alpha \to 0$, $\beta \to \sqrt{\pi}$ and that ϕ satisfies the sine-Gordon equation for this value of β:

$$\Box\phi =: \sin\, 2\sqrt{\pi}\phi: \quad . \tag{10.235}$$

For the interacting case, α and β are so far undetermined constants.

In the case $M = 0$, the equivalent Hamiltonian is known to read

$$H_0 = \frac{1}{2}\int :\pi_\Sigma^2 + \pi_\phi^2 + (\partial_1\Sigma)^2 + m^2\Sigma^2 + (\partial_1\phi)^2: \quad , \tag{10.236}$$

with $m^2 = \frac{e^2}{\pi} + m_0^2$, corresponding to a free massless field ϕ and free massive field Σ of mass m. In this case

$$\mathcal{M}(x) \equiv :\overline{\psi}(x)\psi(x): = -\frac{\tilde{\mu}}{\pi}:\cos 2[\alpha\Sigma(x) + \beta\phi(x)]: \tag{10.237}$$

Following the reasoning of section (10.3.1), we are led in the $M \neq 0$ case to the equations of motion

$$(\Box + m^2)\Sigma(x) + 2\alpha\frac{M\tilde{\mu}}{\pi}:\sin 2[\alpha\Sigma(x) + \beta\phi(x)]: = 0 \tag{10.238a}$$

$$\Box\phi(x) + 2\beta\frac{M\tilde{\mu}}{\pi}:\sin 2[\alpha\Sigma(x) + \beta\phi(x)]: = 0 \quad . \tag{10.238b}$$

In order to establish whether our Ansatz (10.234) satisfies the Dirac equation, we compute $i\,\partial\!\!\!/\psi$, using (10.145), and the equation of motion (10.238b). Following the steps of the previous sections one finds

$$i\,\partial\!\!\!/\psi(x;h) = -\frac{e}{2}\gamma^\mu\{B_\mu(x;h), \psi(x;h)\} + M\gamma^0\int_{-\infty}^{\infty} dy^1[\mathcal{M}(x^0,y^1), \psi(x)]_{ET} \tag{10.239a}$$

where

$$B_\mu(x;h) = -\frac{1}{e}\left[\alpha\tilde{\partial}^\mu\Sigma(x;h) + \left(\beta - \frac{\pi}{\beta}\right)\tilde{\partial}^\mu\phi(x;h)\right] \quad . \tag{10.239b}$$

The scale dimension of the mass operator is

$$\dim\,[\overline{\psi}\psi] \equiv d = \frac{\alpha^2 + \beta^2}{\pi} \quad , \tag{10.240}$$

which replaces (10.144c). We expect the commutator appearing in (10.239a) to give an infinite, finite or zero contribution, depending on whether d is greater, equal or smaller than one, the value $d = 1$ being the canonical dimension.

We shall assume $d \leq 1$. It is then simple to check, that the only contribution to the integral comes from the neighbourhood of $y^1 = x^1$. Thus we need to know only the short-distance behaviour of the ϕ-two-point function, in order to compute the integral in (10.239a). We shall assume that the massive Thirring-Wess model has as its short-distance fixed point, the corresponding massless Thirring-Wess model, and that in the terminology of Schroer and Truong[122], there are no cumulative mass effects in the spinor fields. This then implies the following property for the commutator of the fields Σ and ϕ (summarized by φ):

$$[\varphi^{(+)}(x), \varphi^{(-)}(0)]_{ET} \underset{x\to 0}{\approx} -\frac{1}{4\pi}\{\ln[\tilde{\mu}(-x^1 - i0)] + \ln[\tilde{\mu}(-x^1 + i0)]\} \quad . \quad (10.241)$$

Using this property, and taking the limit $h \to \delta$ under the integral sign, which is allowed since we have assumed $d \leq 1$, one obtains,

$$\begin{aligned}
&\int_{-\infty}^{\infty} dy^1 [\mathcal{M}_{\alpha\beta}(y), \psi_\beta(x)]_{ET} = \\
&-\frac{1}{2}\frac{\tilde{\mu}M}{\pi}\left(\frac{\tilde{\mu}}{2\pi}\right)^{1/2} \gamma^0_{\alpha\beta} e^{-\frac{i\pi}{4}\gamma^5_{\beta\beta}} \left(\int_{-\infty}^{\infty} dy^1 {:} e^{2i(\alpha\Sigma(x)+\beta\phi(x))+i\Phi(x)_\beta}{:}\right. \\
&\qquad \times \left\{ \frac{[\mu(x^1 - y^1 - i0)]^{(1+d\gamma^5_{\beta\beta})/2}}{[\mu(x^1 - y^1 - i0)]^{(1-d\gamma^5_{\beta\beta})/2}} - \frac{[\mu(x^1 - y^1 + i0)]^{(1+d\gamma^5_{\beta\beta})/2}}{[\mu(x^1 - y^1 - i0)]^{(1-d\gamma^5_{\beta\beta})/2}} \right\} \\
&- \int_{-\infty}^{\infty} dy^1 {:} e^{-2i(\alpha\Sigma(x)+\beta\phi(x))+i\Phi(x)_\beta}{:} \\
&\qquad \times \left.\left\{ \frac{[\mu(x^1 - y^1 - i0)]^{(1-d\gamma^5_{\beta\beta})/2}}{[\mu(x^1 - y^1 + i0)]^{(1+d\gamma^5_{\beta\beta})/2}} - \frac{[\mu(x^1 - y^1 + i0)]^{(1-d\gamma^5_{\beta\beta})/2}}{[\mu(x^1 - y^1 - i0)]^{(1+d\gamma^5_{\beta\beta})/2}} \right\}\right) \quad (10.242)
\end{aligned}$$

where the summation over the spin index β is understood, and d is given by Eq. (10.240).

Before proceeding, we digress by considering the Proca equation (10.233*a*). Identifying (10.239*b*) with the vector-meson field, we have:

$$F^{\mu\nu} = -\frac{1}{e}\epsilon^{\mu\nu}\left[\alpha\Box\Sigma + \left(\beta - \frac{\pi}{\beta}\right)\Box\phi\right] \quad . \quad (10.243)$$

On the other hand, the current computed from (compare with Eq. (10.58))

$$J^\mu_{TW}(x) = -\lim_{\epsilon\to 0}\{\mathrm{tr}\,[\gamma^0\gamma^\mu T(x+\epsilon, x) - V.E.V.\} \quad (10.244a)$$

with

$$T^{(TW)}_{\alpha\beta}(x,y) = N_0(x-y){:}e^{i[\Phi_\alpha(x) + e\int_x^y dz^\mu B_\mu(z) - \Phi_\beta(y)]}{:} \quad (10.244b)$$

reads

$$J^\mu_{TW}(x) = -\frac{1}{\pi}[\alpha\tilde{\partial}^\mu\Sigma(x) + \beta\tilde{\partial}^\mu\phi(x)] \quad . \quad (10.245)$$

The Proca-equation in thus satisfied, provided that

$$\frac{\alpha^2}{\pi} = \frac{e^2/\pi}{m^2} \quad , \quad \frac{\beta^2}{\pi} = \frac{m_0^2}{m^2} \tag{10.246}$$

implying $d = 1$, consistent with our input. Note that $\alpha \to 0$, $\frac{\beta^2}{\pi} \to 1$ in the limit $e \to 0$ as expected. On the other hand, for $m_0 \to 0$, we have $\alpha \to \sqrt{\pi}$, $\beta \to 0$.

One verifies that

$$[J^0_{TW}(x), \psi(y)]_{ET} = -\delta(x^1 - y^1)\psi(y) \quad , \tag{10.247}$$

with $\psi(y)$ a well defined operator. These commutation relations reflect that unlike in the case of the Schwinger model, there exist non-trivial charge sectors, i.e., no screening! Setting $d = 1$ in (10.242), we find

$$\int_{-\infty}^{\infty} dy^1 [\mathcal{M}(y), \psi(x)]_{ET} = M\psi(x) \quad , \tag{10.248}$$

in accordance with the canonical scale dimension of the mass operator implied by $d = 1$.

We have thus verified that the fermion operator with α and β given by (10.246) satisfies the coupled Proca-Dirac equations (10.233) with

$$J^\mu_{TW}(x) = -\frac{1}{\sqrt{\pi}} \left[\frac{e/\sqrt{\pi}}{m} \tilde{\partial}^\mu \Sigma(x) + \left(\frac{m_0}{m}\right) \tilde{\partial}^\mu \phi(x) \right] \tag{10.249}$$

and

$$B^\mu(x) = -\frac{1}{m} \left[\tilde{\partial}^\mu \Sigma(x) - \frac{e/\sqrt{\pi}}{m_0} \tilde{\partial}^\mu \phi(x) \right] \tag{10.250}$$

The Recovery of the massive Schwinger Model

It is immediately evident from the above results that the limit $m_0 \to 0$ does not exist for the fields B^μ and ψ. This is not surprising since if this limit were to exist, we would obtain a charge-carrying fermion field, as well as operator solutions to Maxwell's equations in contradiction with the previous results. As for the bilocals (10.244*b*) one finds

$$T^{TW}(x^0, x^1; x^0, y^1) \underset{m_0 \to 0}{\longrightarrow} T_{\theta=0}(x^0, x^1; x^0, y^1) \quad , \tag{10.251}$$

where T_θ is given by (10.57), and only involves the Σ field. Hence the ϕ field completely decouples from the observables in this limit! Examples are provided by the mass operator $\mathcal{M}$ and the current J^μ_{TW}. As for $F^{\mu\nu}$, we have

$$F^{\mu\nu} = \frac{1}{m} \epsilon^{\mu\nu} \left[\Box \Sigma - \frac{e\sqrt{\pi}}{m_0} \Box \phi \right] \tag{10.252}$$

or using the equations of motion (10.238) for Σ and ϕ this expression reduces to

$$F^{\mu\nu} = -m\epsilon^{\mu\nu}\Sigma \tag{10.253}$$

showing that $F^{\mu\nu}$ approaches smoothly the corresponding value of the massive Schwinger model in the limit $m_0 \to 0$.

10.4 Conclusion

The Schwinger model has been seen to provide an explicit realization of features believed to be characteristic of QCD_4: spontaneous breakdown of the chiral symmetry without Goldstone bosons, infinite vacuum degeneracy and "infrared slavery" (confinement). In fact, the model served as an excellent laboratory to expose the intimate relation existing between these phenomena: In the Coulomb gauge description, where only physical degrees of freedom are involved, the absence of Goldstone bosons is linked to the correlation of distant regions due to the long range character of the Coulomb force in $d = 2$ dimensions, which results in that a non-rigid chiral transformation cannot be performed without the expense of energy. This has been referred to as the "seizing" of the vacuum[270], meaning that the vacuum cannot support long-wavelength, zero energy excitations. In the Lorentz gauge description adopted in this chapter, the absence of Goldstone bosons is realized via their decoupling from the physical Hilbert space, leaving only massive excitations in $\mathcal{H}_{phys}$. This shows that the long range force has been screened and implies the absence of charged states in the spectrum. In both description the only relique of the would be Goldstone bosons is a chirally non-invariant vacuum. Though "screening" of the quark charge does not yet imply confinement of the quarks, these two properties nevertheless happened to be intimately linked, since the long range of the Coulomb force turned out to be responsible for both, screening and absence of a pole in the gauge invariant fermionic two-point function. The Goldstone bosons only contribute to gauge variant quantities such as j^5_μ, an observation at the heart of the solution of the $U(1)$ problem.

Chapter XI

QUANTUM CHROMODYNAMICS

11.1 Introduction

Unlike the Schwinger model[27], quantum chromodynamics of massless fermions in 1+1 dimensions is no longer exactly solvable. One can nevertheless derive some exact properties, by either using standard methods in perturbation theory, or by obtaining its effective action using heat kernel, or other, methods. In particular, following di Vecchia, Durhuus and Petersen[343], one arrives at an equivalent bosonic action which reduces to that of the Wess-Zumino-Witten[51,52] action (9.14) for $n = 1$ in the limit of vanishing coupling to the gauge field (theory of free fermions with $U(N) \times \tilde{U}(N)$ symmetry.

Quantum chromodynamics in 1+1 dimensions (QCD_2) is a superrenormalizable theory with finite mass and coupling constant renormalizations. The first attempt to obtain its particle spectrum dates back to 1974, and was based on the $1/N$ expansion[315,23]. In 1+1 dimensions the use of axial gauges eliminates the Yang-Mills self-coupling, characteristic of non-abelian gauge theories In the limit where $N \to \infty$ ($e^2 N$ fixed) only planar diagrams without fermion loops are found to contribute. Correspondingly the fermion four-point function is just given by an infinite series of ladder graphs with self-energy insertions for the fermions.

By adopting the light-cone gauge ($A_- = 0$), and formulating the Feynman rules in terms of light-cone coordinates, 't Hooft obtained[383], to leading order in N, the non-linear integral equation

$$\sum(p) = -\frac{ie^2}{2\pi^2}\int \frac{dk_-}{k_-^2}\int dk_+ \frac{(k_- + p_-)}{M^2 + (k+p)^2 + (k_- + p_-)\sum(k+p) - i\epsilon} \quad ,$$

for the fermion self-energy. The "$i\epsilon$-prescription" to be used when performing the dk_+ and dk_- integrations is however ambiguous. Moreover, the severe infrared problems in $d = 2$ dimensions poses serious problems.

The procedure followed by 't Hooft was to perform the contour integration over k_+, using the above $i\epsilon$ prescription, followed by the k_- integration with an infrared cut-off λ, restricting k_- to the integration range $| k_- | \geq \lambda$. Since k_- scales under a boost, this procedure is Lorentz invariant; it also makes the confining aspects of the model manifest, since the quark poles are pushed to infinity as $\lambda \to \infty$. For gauge *invariant* quantities this procedure was shown[384] to be equivalent to interpreting all integrals as principal values, which leads to the solution $\sum(p) = \frac{e^2}{\pi^2}\frac{1}{p_-}$, and correspondingly to the fermion two-point function $S_F(p) = \frac{\not{p}+M+\frac{e^2}{2\pi}\frac{\gamma_-}{p_-}}{p^2-M^2+\frac{e^2}{\pi}}$.

As pointed out by T.T. Wu[385], the use of the principal value prescription is however highly ambiguous due to the non-commutative nature of principal value integrals, as expressed by the Poincaré-Bertrand formula. Moreover, the result for $S_F(p)$ implies a tachyon for $M^2 < e^2/\pi$, and hence in particular for $M = 0$. This has been emphasized by Frishman et al[386].

The simplicity of the above result for $\sum(p)$ reflects the fact, that for 't Hooft's prescription, only the second order self-energy diagram contributes. By reducing the sum over ladder graphs for the fermion four-point function to an integral equation for the corresponding Bethe-Salpeter wave function[387], 't Hooft was able to obtain an approximate formula for the $q\bar{q}$ bound-state spectrum. The masses of these bound states were found to lie asymptotically on a nearly linear "Regge" trajectory. This is unlike the $U(1)$ case, where only one such mesonic state exists.

The ambiguity referred to above, as well as the infrared problems associated with the gluons, which in the $1/N$ approximation remain massless, have made 't Hooft's solution a controversial issue.

One would expect that the ambiguity, associated with the use of Minkowskian light-cone coordenates, would be mastered by going over to the Euclidean formulation via the analytic continuation (Wick rotation), and repeating the calculation in the, now, complex light-cone gauge. This path, followed by T. T. Wu[385], leads, after continuing to Minkowski space to the quite different result

$$\sum(p) = \frac{1}{2p_-}\left\{M^2 - p^2 - \sqrt{(M^2 - p^2) + \frac{4e^2}{\pi}p^2}\right\}^{1/2} .$$

The presence of the anomalous branch cut reflects the fact that, for the "Euclidean" prescription, all planar "rainbow graphs" now contribute to the self-energy in the large N limit. Though this procedure was tested[388] for the case of the Schwinger model, where summation of the perturbation series was shown to lead the correct result for the fermion two-point function, it also remains controversial. Indeed, the implementation of the complex light-cone gauge involves a non-unitary transformation, which does not leave the dynamics invariant[1)]. From this point of view the dramatic discrepancy between the results for $\sum(p)$ obtained by 't Hooft and T. T. Wu may not be surprising. Nevertheless, a discussion of QCD_2 in the axial gauge $n \cdot A = 0$, $n^2 = -1$, reveals[386] that it is inconsistent with the principal value prescription.

[1)] In fact the two-one component of the $M = 0$ Dirac operator becomes holomorphic in the $A_1 - iA_2 = 0$ gauge.

In the $1/N$ limit the gluons remain massless, since fermion loops do not contribute. This is quite unlike the $U(1)$ case, where the photon requires a mass via an intrinsic Higgs mechanism, and has led to speculations[374,375], that QCD_2 may in fact exist in two phases associated with the weak and strong coupling regime. In this picture, the $1/N$ limit would correspond to the weak-coupling limit ('t Hooft's phase), with massless gluons and a mesonic spectrum described by a Regge trajectory. In the strong coupling regime (Higgs phase), on the other hand, the gluons would be massive, and the original $SU(N)_c$-symmetry would be broken down to the maximal abelian subgroup (torus) of $SU(N)_c$.

Calculations attempting to give support to this idea were at first performed by V. Baluni[389], and P. J. Steinhardt[390], for the case of QCD_2 with one flavor. By working in the Weyl basis defined by $B_{ij} = \sum_a t^a_{ij} B^a$, and choosing the gauge

$$
\begin{aligned}
(A^1)_{ij} &= 0 \quad , \quad i = 1, \cdots, N-1 \\
E_{ij} &= 0 \quad , \quad \text{for}\, i \neq j \quad , \quad i,j = 1, \cdots, N
\end{aligned}
$$

where E^a denotes the chromoelectric field, Baluni could express the one flavor, $SU(N)_c \times U(1)_F$ QCD_2 Hamiltonian - after a suitable Fierz transformation - in terms of the chiral densities and (diagonal) $SU(N)_c$-torus currents alone. This allowed him to use conventional (abelian) bosonization techniques, and thus to circumvent the - up to then unresolved - problem of non-abelian bosonization[51]. This is no longer possible if several flavors are involved. Using Witten's non-abelian bosonization techniques[51], D. Gepner extended Baluni's discussion to the case of a $SU(N)_c \times U(N)_F$ symmetry[391].

Alternative approaches have been based on the factorization of the vacuum expectation value of the product of $SU(N)$- symmetric operators in the large N limit. Such a factorization can be understood[392,393,394] in the path-integral framework as being the result of a semi-classical approximation, once the action has been expressed in terms of "isoscalar" bilinears in the fermion fields by a suitable Fierz transformation.

In view of the speculative status of such approximate, though non-perturbative, studies of the bound-state spectrum of (massive) QCD_2, we shall limit ourselves in this chapter to the study of some non-perturbative aspects in QCD_2 of *massless* fermions which can be rigorously proven.

Since the vacuum degeneracy in the Schwinger model complicates substantially the path integral representation of the correlation functions (see following chapter), we have chosen to first first discuss QED_2 from the operator point of view, where this degeneracy is more easily detected. The implementation of such operator techniques in the case of a non-abelian symmetry was first considered by Halpern[121]. They were based on the Mandelstam representation[47] of the form (10.105), but proved cumbersome and inadequate for handling the non-abelian case: Though the baryonic current and $N-1$ "diagonal" torus currents continue to be of the form (10.106,115), being local with respect to the N fundamental Bose fields φ_f, this is no longer true for the "off-diagonal" currents involving the non-commuting generators of $SU(N)$. Moreover, the fields φ_f are required to lie in non-linear representations of the global symmetry group of the fermions, so that the originally manifest symmetry properties on the fermionic level, become spoiled by the conventional (abelian) bosonization procedure. Only if one restricts the color group to the maximal abelian subgroup (Cartan subalgebra), exact solutions may be obtained using conventional operator techniques[349], as we have seen in chapter 10.

Functional techniques, on the other hand, have proven to be a very powerful tool for arriving at equivalent bosonic actions of the fermionic theory, and indicate the way for arriving at non-abelian bosonization formulae without the shortcoming above (see chapter 9). The results are contained in the remarkable paper by E. Witten, ref. [51]. We shall therefore abandon the operator-methods of chapter 10, and base our analysis on the Feynman path-integral approach.

Since the mapping from S^1 to $SU(N)$ is trivial ($\Pi_1\,(SU(N)) = 0$), we expect the Dirac operator for an $SU(N)$ gauge field to have no zero modes of topological origin[2)]; hence we also expect no vacuum degeneracy. This leads us to examine QCD_2 in the framework of conventional perturbation theory. By formally summing the perturbation series, we shall thus arrive at exact representations for the gauge current, the determinant of the Dirac operator and the Dirac Green's functions in an external field, which represent the essential building stones of the Mathews-Salam representation (4.1) of fermionic correlation functions.

As we have already seen in chapter 4, the external field current $J_\mu(x \mid A)$ and fermionic determinant $\det i \not{D}$ of a $U(1)$ gauge theory in two dimensions (QED_2) are exactly calculable. In fact they are, respectively, functionals of first and second degree in A_μ, reflecting the fact, that the perturbation series for $J_\mu(x \mid A)$ and $\ln\det\, i\not{D}[A]$ (connected one-loop graphs) terminates at the first non-trivial order in the coupling constant e. Correspondingly, we could also solve for the exact Dirac-Greens functions (4.169) in this case. Moreover, the structure of both, the determinant and Greens function is such, that all correlation functions of QED_2 can be constructed explicitly via the generating functional (4.1) for the case $M = 0$ (see following chapter). This is to be expected, since massless QED_2 admits exact operator solutions in terms of generalized free fields (see chapter 10).

The same is no longer true for QCD_2, where the perturbation series for $\ln\det i\not{D}$ (effective action) and $J^a_\mu(x \mid A)$ no longer terminates. Nevertheless, the effective action may still be calculated non-perturbatively in closed form by the Fujikawa method[156], the heat-kernel (proper-time) method[154,155,396], or by integrating the anomaly equation[50]. The result is however complicated, and cannot be written explicitly in terms of the gauge potential A^a_μ. An explicit representation in terms of A^a_μ is nevertheless obtained by constructing an equivalent bosonic (Wess-Zumino like) action following the method of di Vecchia et al[343].

[2)] If one restricts the gauge group to the maximal diagonal subgroup of $SU(N)$ (torus), such zero modes are known to occur[395].

In order to appreciate better the difficulties, it is thus instructive to first examine the general case of QCD_2 from the point of view of perturbation theory, and then to discuss some of the results using non-perturbative techniques. This is the path we shall follow. The results will serve as a starting point for our discussion in chapter 12, where we reconsider the case of QED_2, in the Feynman path integral formulation.

We shall work with the Euclidean formulation, our conventions being those stated in Appendix B. The absence of zero modes will be assumed throughout.

11.2 The External Field Current

The theory to be discussed is defined by the Lagrangian

$$\mathcal{L} = -\frac{1}{4}\operatorname{tr} F^a_{\mu\nu} F^{\mu\nu}_a + \overline{\psi}(i\,\partial\!\!\!/ + e\tau^a A\!\!\!/^a)\psi \quad , \tag{11.1a}$$

with the corresponding (classical) Euler-Lagrange equations

$$\mathcal{D}^{ab}_\mu F^{\mu\nu}_b(x) + e\overline{\psi}(x)\gamma^\nu\tau^a\psi(x) = 0 \tag{11.1b}$$

$$i\gamma^\mu\partial_\mu\psi(x) + e\gamma^\mu\tau^a A^a_\mu(x)\psi(x) = 0 \quad , \tag{11.1c}$$

and $F^a_{\mu\nu}$ the chromoelectric field-strength tensor $F^a_{\mu\nu} = \partial_\mu A^a_\nu - \partial_\nu A^a_\mu + ef_{abc}A^b_\mu A^c_\nu$.

We are interested in obtaining exact non-perturbative statements about QCD_2. This does not exclude the use of perturbation theory, provided we sum the perturbation series to all orders. Since we assume the absence of zero modes, this is a possible procedure, which we follow in this section.

From the point of view of perturbation theory, the QCD_2 external field current $J^a_\mu(x \mid A)$ admits an expansion in terms of either tree graphs or one loop fermion graphs. We shall discuss these two representations separately.

11.2.1 Tree graph expansion[395]

For a gauge invariant regularization, the external field current

$$J^a_\mu(x \mid A) = \left\langle \overline{\psi}(x)\tau^a\gamma_\mu\psi(x)\right\rangle_A \tag{11.2}$$

is covariantly conserved (see Eq. (4.173)), and satisfies the anomaly equation (4.215) in the absence of zero modes:

$$\mathcal{D}^{ab}_\mu J^b_\mu(x \mid A) = 0 \tag{11.3a}$$

$$\tilde{\mathcal{D}}^{ab}_\mu J^b_\mu(x \mid A) = -\frac{e}{2\pi}\epsilon_{\mu\nu}F^a_{\mu\nu}(x) \quad . \tag{11.3b}$$

No such anomaly occurs in the "isoscalar" axial current[3]. Defining now the Greens function $\tilde{K}_\mu(x,y \mid A)$ by

$$\mathcal{D}_\mu^{ab}\tilde{K}_\mu^{bc}(x,y \mid A) = 0 \quad . \tag{11.4a}$$

$$\tilde{\mathcal{D}}_\mu^{ab}\tilde{K}_\mu^{bc}(x,y \mid A) = -\delta^{ac}\delta^2(x-y) \quad . \tag{11.4b}$$

we immediately integrate Eq. (11.3), obtaining

$$J_\mu^a(x \mid A) = \frac{e}{2\pi}\int d^2y\tilde{K}_\mu^{ab}(x,y \mid A)\epsilon_{\lambda\rho}F_{\lambda\rho}^b(y) \quad . \tag{11.5}$$

It is convenient to define K_μ^{ab} via $\tilde{K}_\mu^{ab} = \epsilon_{\mu\nu}K_\nu^{ab}$. In terms of light cone coordinates (see appendix B), Eq. (11.5) decomposes as

$$J_\pm^a(x \mid A) = \pm\frac{e}{2\pi}\int d^2yK_\pm^{ab}(x,y \mid A)F_{+-}^b(y) \quad , \tag{11.6}$$

where $K_\pm = K_1 \pm iK_2$, and

$$F_{+-}^a \equiv \partial_+A_-^a - \partial_-A_+^a + gf_{abc}A_+^bA_-^c = -2iF_{12} \quad . \tag{11.7a}$$

$$F_{+-}^a = \partial_+A_-^a - \mathcal{D}_-^{ac}A_+^c = -\partial_-A_+^a + \mathcal{D}_+^{ac}A_-^c \quad . \tag{11.7b}$$

In terms of $K_\pm$, Eqs. (11.4) take the form,

$$\mathcal{D}_\mp^{ab}K_\pm^{bc}(x,y \mid A) = -\delta^{ac}\delta^2(x-y) \quad . \tag{11.8}$$

Using this equation, as well as (11.7b) we obtain from (11.6), after a partial integration,

$$J_\pm^a(x \mid A) = \frac{e}{2\pi}A_\pm^a(x) + \frac{e}{2\pi}\int d^2yK_\pm^{ab}(x,y \mid A)\partial_\pm A_\mp^b(y) \quad . \tag{11.9}$$

It is convenient to introduce the two-point functions

$$D_\pm(x) \equiv \partial_\pm D_E(x) \tag{11.10}$$

where $D_E(x)$ is the Euclidean two point function (B.25a).

---

[3] This is peculiar to two-dimensional space-time. In $d = 4$, the $U(1)$-axial current exhibits the Adler-Bell-Jackiw[200,201] (ABJ) anomaly. It is for this reason that the two-dimensional analogue of QCD_4 is QED_2, rather than QCD_2. It also means that the color-neutral vector and axial-vector currents are respectively given by $\tilde{\partial}_\mu\phi$ and $\partial_\mu\phi$, with ϕ a zero-mass free field. This entails a consistency condition on the bound state wave function in the $1/N$ limit[397].

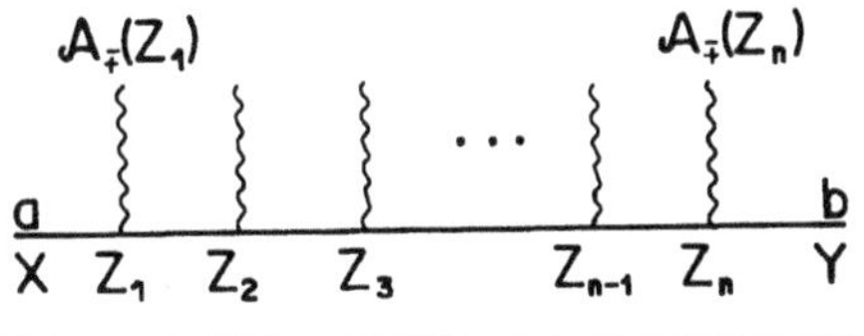

Figure 11.1: n^{th}-order tree-graph contributing to $K_{\pm}^{ab}(x,y|A)$. The corresponding propagators (solid lines) are $D_{\pm}(z_i - z_{i+1})$.

In terms of (11.10), equation (11.8) has the iterative solution

$$K_{\pm}^{ab}(x,y \mid A) = \delta^{ab} D_{\pm}(x-y) + \sum_{n=1}^{\infty} e^n \int d^2z_1 \cdots d^2z_n D_{\pm}(x-z_1)\cdots D_{\pm}(z_n - y)\left[\mathcal{A}_{\mp}(z_1)\cdots\mathcal{A}_{\mp}(z_n)\right]^{ab} \quad (11.11)$$

with $\mathcal{A}_{\pm}$ the (antihermitian) Lie algebra valued fields in the adjoint representation, Eq. (C.5). Expansion (11.11) corresponds to the sum of tree-graphs shown in figure 11.1

For the $U(1)$ case only the first term in (11.11) contributes and we are left with the result (4.182) for the current. This truncation, resulting from f_{acb} being zero for $U(1)$ is also evident upon writing (11.11) in terms of multiple commutations of A_μ in the fundamental representation

$$K_{\pm}^{ab}(x,y \mid A) = \delta^{ab} D_{\pm}(x-y) + \sum_{n=1} (-ie)^n \int d^2z_1 \cdots d^2z_n D_{\pm}(x-z_1)\cdots D_{\pm}(z_n - y) \times \mathrm{tr}\{\tau^a[A_{\mp}(z_1),[A_{\mp}(z_2)\cdots,[A_{\mp}(z_n),\tau^b]\cdots]]\} \quad (11.12)$$

where $A_\mu = \sum A_\mu^c \tau^c$, (see (C.9) and (C.10)).

Introducing (11.12) in (11.9) we obtain

$$J_{\pm}^a(x \mid A) = \frac{e}{2\pi}\left[A_{\pm}^a(x) + \int d^2y \partial_{\pm} D_{\pm}(x-y) A_{\mp}^a(y)\right] + \frac{1}{2\pi}\sum_{n=2}^{\infty} i(-ie)^n \int d^2z_1 \cdots d^2z_n D_{\pm}(x-z_1)\cdots D_{\pm}(z_{n-1}-z_n) \times \mathrm{tr}\left\{\tau^a\left[A_{\mp}(z_1),[A_{\mp}(z_2),\cdots[A_{\mp}(z_{n-1}),\partial_{\pm}A_{\mp}(z_n)]\cdots]\right]\right\} \quad , \quad (11.13)$$

where we have performed the relabellings $n \to n-1$, $y \to z_n$ in order to facilitate comparison with later results.

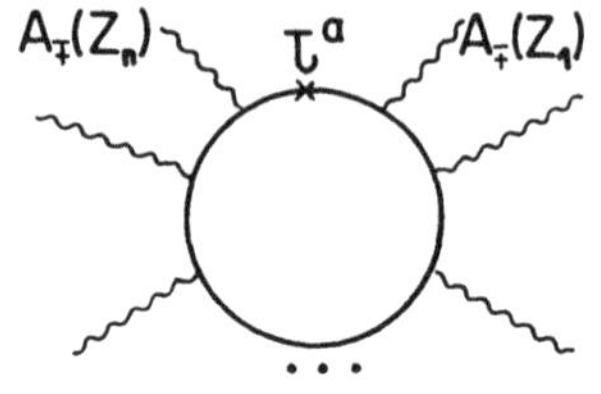

Figure 11.2: n^{th}-order loop contribution to the induced current $J^a_\mp(x|A)$.

11.2.2 Loop expansion

The current defined by (11.2) is evidently given by the sum of one loop graphs shown in figure 11.2.

It is a priori surprising that this expansion should collapse to the sum of tree graph contributions in the adjoint representation, as shown in figure 11.1. Let us examine how this situation comes about. Note first that a similar situation has already been encountered in chapters 3 and 5, for the case of the Thirring and Gross Neveu model, respectively; it is a characteristic feature of two-dimensional space-time.

Because of UV divergencies, the zeroth and first order terms need special care. Hence we calculate them first.

The $\mathcal{O}(e^0)$ contribution vanishes if we agree to set $S_E(0) = 0$. This is suggested by the fact that $S_E(x)$ is an antisymmetric function of x, and corresponds to a symmetric integration in Fourier space. The $n = 1$ contribution must be calculated separately. It is given by (recall Eqs. (4.159), (4.160) and (4.5b))

$$J^a_\mu(x \mid A) = e\int d^2y \mathrm{tr}[S_E(y-x)\gamma_\mu\tau^a S_E(x-y)\,A\!\!\!/(y)] + \mathcal{O}(e^2) \quad . \qquad (11.14)$$

We regularize the integral (11.14) in a gauge-invariant way. Write it as

$$J^a_\mu(x \mid A) = e\int d^2y \Pi^{ab}_{\mu\nu}(y-x)A^b_\nu(y) + \mathcal{O}(e^2) \quad , \qquad (11.15)$$

where $\Pi^{ab}_{\mu\nu}(y-x) = \mathrm{tr}[S_E(y-x)\gamma_\mu\tau^a S_E(x-y)\gamma_\nu\tau^b]$, has the property $\Pi^{ab}_{\mu\nu}(z) = \Pi^{ba}_{\nu\mu}(-z)$.

Gauge covariance demands ($J_\mu \equiv \sum_a \tau^a J^a_\mu$)

$$J_\mu(x \mid A^g) = g^{-1}(x)J_\mu(x \mid A)g(x) \quad , \qquad (11.16)$$

where $A^g_\mu = g^{-1}A_\mu g + \frac{i}{e}g^{-1}\partial_\mu g$. Infinitesimally,

$$A^{a\,g}_\mu(x) = A^a_\mu(x) - \frac{1}{e}\mathcal{D}^{ab}_\mu\Lambda^b(x) \quad . \qquad (11.17)$$

Thus (11.16) requires to $\mathcal{O}(e)$,

$$\partial_\mu \Pi^{ab}_{\mu\nu}(z) = 0 \quad , \tag{11.18}$$

which in momentum space reads

$$k_\mu \tilde{\Pi}^{ab}_{\mu\nu}(k) = \int \frac{d^2p}{(2\pi)^2} \mathrm{tr}\left[\left(\frac{1}{\not p - \not k} - \frac{1}{\not p}\right)\gamma_\nu\right] \quad . \tag{11.19}$$

Hence Eq. (11.18) is satisfied, if we were allowed to make a shift in the integration variable. However, since the integral (11.19) is superficially logarithmically divergent, we need to first regularize it. We adopt the Pauli-Villars regularization,

$$\tilde{\Pi}^{ab}_{\mu\nu}(k)_{reg} = \delta^{ab} \int \frac{d^2p}{(2\pi)^2}[I_{\mu\nu}(k,p,0) - I_{\mu\nu}(k,p,M)] \quad , \tag{11.20}$$

with

$$\begin{aligned} I_{\mu\nu}(k,p;M) &= \mathrm{tr}\left(\frac{1}{\not p - iM}\gamma_\mu \frac{1}{\not p - \not k - iM}\gamma_\nu\right) \\ &= \frac{2[p_\mu(p-k)_\nu + p_\nu(p-k)_\mu - p.(p-k)\delta_{\mu\nu} - M^2\delta_{\mu\nu}]}{(p^2+M^2)((p-k)^2+M^2)} \quad . \end{aligned} \tag{11.21}$$

Going over to Feynman-parameter space one finds

$$\begin{aligned} \tilde{\Pi}^{ab}_{\mu\nu}(k)_{reg} &= 2\delta^{ab}(k^2\delta_{\mu\nu} - k_\mu k_\nu)\times \\ &\int \frac{d^2p}{(2\pi)^2}\int_0^1 d\alpha\alpha(1-\alpha)\left(\frac{1}{[p^2+\alpha(1-\alpha)k^2]^2} - \frac{1}{[p^2+\alpha(1-\alpha)k^2+M^2]^2}\right) \\ -\delta^{ab}\delta_{\mu\nu} &\int \frac{d^2p}{(2\pi)^2}\int_0^1 d\alpha\left(\frac{\alpha(1-\alpha)k^2}{[p^2+\alpha(1-\alpha)k^2]^2} - \frac{\alpha(1-\alpha)k^2+M^2}{[p^2+\alpha(1-\alpha)k^2+M^2]^2}\right) \end{aligned} \tag{11.22}$$

Now, since the integrand only depends on p^2, we can make the replacement $d^2p = \frac{1}{2}dp^2 d\varphi \to \pi dp^2$.

The term proportional to $\delta_{\mu\nu}$ is seen to vanish, after performing the p^2-integ ration. The remaining integration converges for each term individually, and we are left, in the limit $M \to \infty$, with the finite result

$$\tilde{\Pi}^{ab}_{\mu\nu}(k)_{reg} \underset{M\to\infty}{\longrightarrow} \delta^{ab}\frac{1}{\pi}\left(\delta_{\mu\nu} - \frac{k_\mu k_\nu}{k^2}\right) \quad . \tag{11.23}$$

Returning to coordinate space we thus have

$$J^a_\mu(x \mid A) = \frac{e}{\pi}\left(\delta_{\mu\nu} - \frac{\partial_\mu\partial_\nu}{\Delta}\right)A^a_\nu(x) + \mathcal{O}(e^2) \quad . \tag{11.24}$$

In terms of light-cone variables this reads,

$$J^a_\pm(x \mid A) = \frac{e}{2\pi}\left[A^a_\pm(x) + \int d^2y \partial_\pm D_\pm(x-y)A^a_\mp(y)\right] + \mathcal{O}(e^2) \quad . \tag{11.25}$$

Our result (11.25) is seen to agree with the $\mathcal{O}(e)$ term in Eq. (11.13).

Higher order contributions[395]

We now consider the remaining terms in the one loop expansion. It is again convenient to decompose J^a_μ into its light cone components. Using properties (B.17) of the γ matrices, as well as $S_E(x) = i\, \not{\partial} D_E(x) = \frac{i}{2}(\gamma_+ D_- + \gamma_- D_+)$, we obtain, recalling (11.25), the expansion

$$J^a_\pm(x \mid A) = \frac{e}{2\pi}\left[A^a_\pm(x) + \int d^2y D_\pm(x-y)\partial_\pm A^a_\mp(y)\right] + \Delta J^a_\pm(x \mid A) \qquad (11.26a)$$

where

$$\Delta J^a_\pm(x \mid A) = -\sum_{n=2}^{\infty} i(-ie)^n \int d^2z_1 \cdots d^2z_n D_\pm(x-z_1)D_\pm(z_1-z_2)\cdots D_\pm(z_n-x)$$
$$\times\, \mathrm{tr}(\tau^a A_\mp(z_1)\cdots A_\mp(z_n)) \quad . \qquad (11.26b)$$

The manipulations to follow are most easily carried out in momentum space. Define the Fourier transforms $\tilde{A}_\pm(p)$, $\tilde{D}_\pm(p)$ by

$$A_\pm(x) = \int \frac{d^2k}{(2\pi)^2} e^{ik.x} \tilde{A}_\pm(k) \qquad (11.27a)$$

$$D_\pm(x) = \int \frac{d^2p}{(2\pi)^2} e^{ip.x} \tilde{D}_\pm(p) \quad . \qquad (11.27b)$$

From (11.10), one has $\tilde{D}_\pm(p) = \frac{i}{p_\mp}$; we write $\Delta J^a_\pm$ in the form

$$\Delta J^a_\pm(x \mid A) = -\sum_{n=2}^{\infty} i(-ie)^n \int \frac{d^2k_1}{(2\pi)^2}\cdots\frac{d^2k_n}{(2\pi)^2} e^{i(k_1+\cdots+k_n)x}$$
$$J^{(n)}_\pm(k_1\cdots k_n)\mathrm{tr}\{\tau^a \tilde{A}_\mp(k_1)\cdots\tilde{A}_\mp(k_n)\} \quad , \qquad (11.28)$$

where the functions $J^{(n)}_\pm$ are given in terms of the one-loop integrals,

$$J^{(n)}_\pm(k_1\cdots k_n)$$
$$= \int \frac{d^2p}{(2\pi)^2}\tilde{D}_\pm(p)\tilde{D}_\pm(p+k_1+\cdots+k_n)\tilde{D}_\pm(p+k_2+\cdots+k_n)\cdots\tilde{D}_\pm(p+k_n) \quad .(11.29)$$

The individual terms in (11.28) just correspond to the diagram of figure 11.2. We use the algebraic identity,

$$\tilde{D}_\pm(p)\tilde{D}_\pm(p+k) = \tilde{D}_\pm(k)[\tilde{D}_\pm(p) - \tilde{D}_\pm(p+k)] \quad , \qquad (11.30)$$

to derive the following recursion relation:

$$J_{\pm}^{(n)}(k_1 \cdots k_n) = \tilde{D}_{\pm}(k_1 + \cdots + k_n)[J_{\pm}^{(n-1)}(k_2 \cdots k_n) - (-)^n J_{\pm}^{(n-1)}(k_{n-1} \cdots k_1)] \tag{11.31a}$$

These recursion relations can be solved iteratively, subject to the "initial condition"

$$J_{\pm}^{(1)}(k_1) - J_{\pm}^{(1)}(k_2) = \int \frac{d^2p}{(2\pi)^2}[\tilde{D}_{\pm}(p)\tilde{D}_{\pm}(p+k_1) - \tilde{D}_{\pm}(p)\tilde{D}_{\pm}(p+k_2)]. \tag{11.31b}$$

The r.h.s. involves a superficially logarithmically divergent integral of the form:

$$I_{\pm}(k) = -\int \frac{d^2p}{(2\pi)^2} \frac{1}{p_{\mp}} \frac{1}{(p+k)_{\mp}} \quad .$$

Doing first the p_1, and then the p_2 integration yields, using Cauchy's residue theorem, $I_{\pm}(k) = \frac{1}{2\pi}(\frac{k_{\pm}}{k_{\mp}} - 1)$. Performing the integration in the inverse order, yields $I_{\pm}(k) = \frac{1}{2\pi}(\frac{k_{\pm}}{k_{\mp}} + 1)$. We choose the particular regularization given by averaging over these two results:

$$I_{\pm}(k) = \frac{1}{2\pi} \frac{k_{\pm}}{k_{\mp}} \quad . \tag{11.32}$$

We thus obtain for (11.31b),

$$J_{\pm}^{(1)}(k_1) - J_{\pm}^{(1)}(k_2) = -\frac{i}{2\pi}[(k_1)_{\pm}\tilde{D}_{\pm}(k_1) - (k_2)_{\pm}\tilde{D}_{\pm}(k_2)] \quad . \tag{11.33}$$

Looking back at (11.28) we define

$$T_{\pm}^{(n)}(x \mid A) = \int \frac{d^2k_1}{(2\pi)^4} \cdots \frac{d^2k_n}{(2\pi)^4} e^{i(k_1+\cdots+k_n)x}(\tilde{A}_{\mp}(k_1) \cdots \tilde{A}_{\mp}(k_n))J_{\pm}^{(n)}(k_1 \cdots k_n). \tag{11.34}$$

In terms of $T_{\pm}^{(n)}$, the current (11.26) reads

$$J_{\pm}^a(x \mid A)$$
$$= \frac{e}{2\pi}\left[A_{\pm}^a(x) + \int d^2y D_{\pm}(x-y)\partial_{\pm}A_{\mp}^a(y)\right] - \sum_{n=2} i(-ie)^n \operatorname{tr}\left(\tau^a T_{\pm}^n(x \mid A)\right) . \tag{11.35}$$

Proposition[395]*:* For $n \geq 2$, $T_{\pm}^{(n)}(x)$ can be written in the form,

$$T_{\pm}^{(n)}(x \mid A) = -\frac{1}{2\pi}\int \frac{d^2k_1}{(2\pi)^2} \cdots \frac{d^2k_n}{(2\pi)^2} e^{i(k_1+\cdots+k_n)x}\tilde{D}_{\pm}(k_1 + \cdots + k_n) \cdots \tilde{D}_{\pm}(k_n) \times$$
$$\left[\tilde{A}_{\mp}(k_1), \cdots \left[\tilde{A}_{\mp}(k_{k-2}), \left[\tilde{A}_{\mp}(k_{n-1}, \tilde{\phi}_{\mp}(k_n)\right]\right] \cdots\right] \tag{11.36}$$

where $\phi_{\mp}(k) = ik_{\pm}\tilde{A}_{\mp}(k)$.

Proof

Using (11.31) we obtain from (11.34) after a suitable renaming of the integration variables,

$$T_\pm^{(n)}(x \mid A) = \int \frac{d^2k_1}{(2\pi)^2} \cdots \frac{d^2k_n}{(2\pi)^2} e^{i(k_1+\cdots+k_n)x} \tilde{D}_\pm(k_1+\cdots+k_n) J_\pm^{(n-1)}(k_2\cdots k_n) \times$$
$$\{\tilde{A}_\mp(k_1)(\tilde{A}_\mp(k_2)\cdots\tilde{A}_\mp(k_n)) - (-1)^n(\tilde{A}_\mp(k_n)\cdots\tilde{A}_\mp(k_2))\tilde{A}_\mp(k_1)\} \quad (11.37)$$

From (11.31a) we recognize the property

$$J_\pm^{(n-1)}(k_n\cdots k_2) = (-1)^n J_\pm^{(n-1)}(k_2\cdots k_n) \quad . \quad (11.38)$$

Hence we obtain from (11.37), after suitable relabelling

$$T_\pm^{(n)}(x \mid A) = \int \frac{d^2k_1}{(2\pi)^2} \cdots \frac{d^2k_n}{(2\pi)^2} e^{i(k_1+\cdots+k_n)x} \tilde{D}_\pm(k_1+\cdots+k_n) \times$$
$$\times \left[\tilde{A}_\mp(k_1), J_\pm^{(n-1)}(k_2\cdots k_n)\tilde{A}_\mp(k_2)\cdots\tilde{A}_\mp(k_n)\right] \quad (11.39)$$

Using repeatedly (11.31) and (11.39), we find,

$$T_\pm^{(n)}(x \mid A) =$$
$$\int \frac{d^2k_1}{(2\pi)^2} \cdots \frac{d^2k_n}{(2\pi)^2} e^{i(k_1+\cdots+k_n)x} \tilde{D}_\pm(k_1+\cdots+k_n)\cdots\tilde{D}_\pm(k_{n-2}+\cdots+k_n)$$
$$\times \left[\tilde{A}_\mp(k_1), \cdots \left[\tilde{A}_\mp(k_{n-2}), J_\pm^{(2)}(k_{n-1},k_n)\tilde{A}_\mp(k_{n-1})\tilde{A}_\mp(k_n)\right]\cdots\right] \quad .(11.40)$$

Finally, noting that

$$J^{(2)}(k_{n-1},k_n) = \tilde{D}_\pm(k_{n-1}+k_n)(-\frac{1}{2\pi})[(ik_{n-1})_\pm \tilde{D}_\pm(k_{n-1}) - (ik_n)_\pm \tilde{D}_\pm(k_n)] \quad , \quad (11.41)$$

we obtain the claimed result (11.36).

Going over to coordinate space, $T_\pm^{(n)}$ takes the form

$$T_\pm^{(n)}(x \mid A) = -\frac{1}{2\pi}\int d^2z_1\cdots d^2z_n D_\pm(x-z_1)D_\pm(z-z_2)\cdots D_\pm(z_{n-1}-z_n)$$
$$\times [A_\mp(z_1),[A_\mp(z_2)\cdots[A_\mp(z_{n-1}),\partial_\pm A_\mp(z_n)]\cdots]] \quad , \quad (11.42)$$

Introducing this expansion in (11.35), we recover the result (11.13). We thus see that the equivalence of the one loop and tree graph expansion is a genuine two dimensional property with roots in the algebraic identity (11.30).

The corresponding expansion for the effective action is obtained making use of the Pauli-representation (4.185) of the corresponding functional determinant; it takes the form[395,398],

$$W[A] - W[0] = \frac{e^2}{2\pi}\int d^2x \delta^{ab} A^a_\mu(x)\Big(\delta_{\mu\nu} - \frac{\partial_\mu\partial_\nu}{\Delta}\Big)A^b_\nu(x) +$$

$$+\frac{1}{2}\sum_{n=2}^{\infty}\frac{(-ie)^{n+1}}{n+1}\int d^2x \mathrm{tr}[A_-(x)T^{(n)}_+(x \mid A) + A_+(x)T^{(n)}_-(x \mid A)] \quad . \qquad (11.43)$$

The first term represents the exact result for the $U(1)$ case (with the replacement $\delta^{ab} \to 1$), since $T^{(n)}_{\pm} = 0$ for an abelian group. We have thus also shown that the A_μ two-point function computed from (10.6) is the result of summing the geometric series of one-loop diagrams. As (11.43) shows, this is a gauge independent statement!

11.3 QCD_2 Determinant

As we already stated, the implementation of the bosonization techniques in the case of a non-abelian symmetry was first considered by Halpern[121], using operator methods. It turns out that functional methods provide a more adequate approach to this problem, the key role being played by the effective action. We thus turn now to a non-perturbative calculation of the QCD_2 effective action $W[A]$. We shall do this using two different methods.

11.3.1 Heat kernel method

The heat kernel method has been used in one formulation or another to compute the QCD_2 effective action. The calculations have originally been restricted to the so called "decoupling" gauge; The techniques were based on Fujikawa's approach (see chapter 4) and standard heat-kernel methods. Shortcomings of these calculations are that the gauge conditions (Alvarez's "integrability conditions"[154]) are not easily nor globally implementable, except for $SU(2)$.

The following calculation does not rely on such a choice of gauge[155]. Indeed, we may apply theorem (4.137) to a gauge-independent computation of $W[A]$. To this end we notice that in $d = 2$, $A_\pm$ can always be written in the form

$$eA_- = iV^{-1}\partial_- V \quad , \quad eA_+ = iU^{-1}\partial_+ U \quad , \qquad (11.44a)$$

$$eA_\mu = \frac{i}{2}(\delta_{\mu\nu} - i\epsilon_{\mu\nu})V^{-1}\partial_\nu V + \frac{i}{2}(\delta_{\mu\nu} + i\epsilon_{\mu\nu})U^{-1}\partial_\nu U \quad , \qquad (11.44b)$$

with U and V taking values in $SL(N,\mathbb{C})$, the complexification of $SU(N)$.

We now rewrite the Dirac operator using the identity

$$\gamma_\mu A_\mu = \gamma_\nu P_- V^{-1} i\partial_\nu V + \gamma_\nu P_+ U^{-1}\partial_\nu U \quad , \tag{11.45}$$

with $P_{+(-)}$ the projector on right (left) handed fermions, $P_\pm = \frac{1}{2}(1 \pm \gamma_5)$.

Define

$$\Omega = \begin{pmatrix} V^{-1} & 0 \\ 0 & U^{-1} \end{pmatrix} \quad , \quad \overline{\Omega} = \begin{pmatrix} U^{-1} & 0 \\ 0 & V^{-1} \end{pmatrix} \quad . \tag{11.46}$$

Then,

$$\Omega i \not\partial \overline{\Omega} = \begin{pmatrix} 0 & V^{-1}\partial_- V + \partial_- \\ -U^{-1}\partial_+ U - \partial_+ & 0 \end{pmatrix} \quad , \tag{11.47a}$$

$$= i \not\partial + e \not{A} \quad . \tag{11.47b}$$

The Dirac operator thus has the structure of (4.133).

In order to apply the theorem (4.137), we define interpolating fields U_r and V_r depending on a continuous parameter $r \in [0,1]$, subject to the boundary conditions

$$\begin{aligned} U_r(x)&: \quad U_0(x) = 1 \quad , \quad U_1(x) = U(x) \\ V_r(x)&: \quad V_0(x) = 1 \quad , \quad V_1(x) = V(x) \quad , \end{aligned} \tag{11.48}$$

with a corresponding definition of Ω_r, $\overline{\Omega}_r$ and $A_\mu^{(r)}$. Then (11.47) becomes,

$$i \not{D}(A^{(r)}) = \Omega_r i \not\partial \overline{\Omega}_r \quad . \tag{11.49}$$

Now (compare with (4.106))

$$(i \not{D}[A^{(r)}])^2 = -\mathcal{D}_\mu^{(r)}\mathcal{D}_\mu^{(r)} - \frac{e}{2}\gamma_5 \epsilon_{\mu\nu} F_{\mu\nu}^{(r)} \tag{11.50}$$

where, $F_{\mu\nu}^{(r)} = \partial_\mu A_\nu^{(r)} - \partial_\nu A_\mu^{(r)} - ie[A_\mu^{(r)}, A_\nu^{(r)}] = [D_\mu^{(r)}, D_\nu^{(r)}]$. Hence the corresponding first non-trivial Seeley coefficients is given by (see (4.106f))

$$a_1^{(r)}(x) = \frac{e}{2}\gamma_5 \epsilon_{\mu\nu} F_{\mu\nu}^{(r)} \quad . \tag{11.51}$$

Following the notation of (4.138), we furthermore have

$$W_r = \begin{pmatrix} \dot{V}_r^{-1} V_r & 0 \\ 0 & \dot{U}_r^{-1} U_r \end{pmatrix} + hc. \tag{11.52a}$$

$$\overline{W}_r = \begin{pmatrix} U_r^{-1}\dot{U}_r & 0 \\ 0 & V_r^{-1}\dot{V}_r \end{pmatrix} + hc. \tag{11.52b}$$

Now, from $A_\pm^\dagger = A_\mp$ we deduce that

$$U^\dagger = V^{-1} \quad , \quad V^\dagger = U^{-1} \quad , \tag{11.53}$$

so that

$$W_r = (U_r^{-1}\dot{U}_r - V_r^{-1}\dot{V}_r)\gamma_5 \quad . \tag{11.54}$$

Furthermore, $W_r = \overline{W}_r$. Applying formula (4.137) and using (11.51) we thus finally have

$$\dot{\omega}(r) = \frac{e}{2\pi}\int d^2x \mathrm{tr}[(U_r^{-1}\dot{U}_r - V_r^{-1}\dot{V}_r)\epsilon_{\mu\nu}F_{\mu\nu}^{(r)}] \quad . \tag{11.55}$$

We next rewrite the *r.h.s.* in a manifestly gauge invariant form. In terms of (11.44), the gauge transform of A_μ, $A_\mu^g \equiv g^{-1}A_\mu g + \frac{i}{e}g^{-1}\partial_\mu g$, takes the form

$$eA_+^g = i(g^{-1}U^{-1})(\partial_+ Ug) \quad , \tag{11.56a}$$

and

$$eA_-^g = i(g^{-1}V^{-1})(\partial_- Vg) \quad . \tag{11.56b}$$

Hence, under gauge transformation,

$$U \to Ug \quad , \quad V \to Vg \quad . \tag{11.56c}$$

Since the functional determinant should be gauge invariant, we expect it to depend only on the gauge invariant quantity

$$G = UV^{-1} \quad . \tag{11.57}$$

We therefore define

$$G_r = U_r V_r^{-1} \quad . \tag{11.58}$$

We make now a number of observations, which may be verified by explicit calculation[399]:

$$\begin{aligned} &i) \quad \partial_-(G_r^{-1}\partial_+ G_r) = ieV_r F_{+-}^{(r)} V_r^{-1} = ieV_r F_{+-}^{(r)} V_r^{-1} \\ &ii) \quad \partial_+(G_r^{-1}\partial_- G_r) = ieV_r F_{+-}^{(r)} V_r^{-1} - [G_r^{-1}\partial_+ G_r, G_r^{-1}\partial_- G_r] \\ &iii) \quad G_r^{-1}\dot{G}_r = V_r(U_r^{-1}\dot{U}_r - V_r^{-1}\dot{V}_r)V_r^{-1} \quad . \end{aligned} \tag{11.59}$$

Hence, expression (11.55) for $\dot{\omega}$ may be cast into the manifestly gauge invariant form

$$\begin{aligned}\dot{\omega}(r) &= \frac{1}{2\pi}\int d^2x \operatorname{tr}[(V_r^{-1}G_r^{-1}\dot{G}_r V_r)(V_r^{-1}\partial_-(G_r^{-1}\partial_+ G_r)V_r)] \\ &= \frac{1}{2\pi}\int d^2x \operatorname{tr}[G_r^{-1}\dot{G}_r\partial_-(G_r^{-1}\partial_+ G_r)] \\ &= \frac{1}{2\pi}\int d^2x(\delta_{\mu\nu}+i\epsilon_{\mu\nu})\operatorname{tr}[G_r^{-1}\dot{G}_r\partial_\mu(G_r^{-1}\partial_\nu G_r)] \\ &= \frac{1}{2\pi}\int d^2x \operatorname{tr}[G_r^{-1}\dot{G}_r\partial_\mu(G_r^{-1}\partial_\mu G_r)] \\ &\quad -\frac{i}{2\pi}\int d^2x\epsilon_{\mu\nu}\operatorname{tr}[(G_r^{-1}\partial_r G_r)(G_r^{-1}\partial_\mu G_r)(G_r^{-1}\partial_\nu G_r)] \quad . \qquad (11.60)\end{aligned}$$

The first term can be put into a more convenient form. Indeed,

$$\operatorname{tr}[G_r^{-1}\dot{G}_r\partial_\mu(G_r^{-1}\partial_\mu G_r)] = \partial_\mu\operatorname{tr}[G_r^{-1}\dot{G}_r(G_r^{-1}\partial_\mu G_r)] + \frac{1}{2}\frac{d}{dr}\operatorname{tr}[(\partial_\mu G_r^{-1})(\partial_\mu G_r)] \quad .$$

Hence, dropping in (11.60) the contribution from the surface term at x-infinity, and doing the r-integration, we finally obtain, for the (Euclidean) effective action

$$W^{(E)}[A] = -\frac{1}{2}(\omega(1)-\omega(0)) = -\Gamma^{(E)}[UV^{-1}] \qquad (11.61a)$$

where

$$\begin{aligned}\Gamma^{(E)}[G] &= \frac{1}{8\pi}\int d^2x \operatorname{tr}[(\partial_\mu G^{-1})(\partial^\mu G)] \\ &\quad -\frac{i}{4\pi}\int_0^1 dr\int d^2x\epsilon_{\mu\nu}\operatorname{tr}[(G_r^{-1}\partial_r G_r)(G_r^{-1}\partial_\mu G_r)(G_r^{-1}\partial_\nu G_r)] \qquad (11.61b)\end{aligned}$$

is the Euclidean version of the Wess-Zumino-Witten action[4] introduced in chapter 9.

---

[4] Assuming that $G(x)$ tends to the identity (A_μ tends to pure gauge) as $|x| \to \infty$, we may identify the points at infinity. $\Re_2$ may then be viewed as a large two-sphere S^2. The mapping G from S^2 into $SU(N)$ can be extended to a mapping $\overline{G}$ of a solid ball B, with boundary S^2, into $SU(N)$. This allows us to write (11.61b) as ($\epsilon_{123}=1$)

$$\Gamma^{(E)}[G] = \frac{1}{8\pi}\int d^2x \operatorname{tr}[(\partial_\mu G^{-1})(\partial^\mu G)] - \frac{i}{12\pi}\int_B d^3x\epsilon_{ijk}\operatorname{tr}[(\overline{G}^{-1}\partial_i\overline{G})(\overline{G}^{-1}\partial_j\overline{G})(\overline{G}^{-1}\partial_k\overline{G})]$$

This expression is just twice the equivalent bosonic action obtained by Witten for the case of Majorana fermions, where $\overline{G}$ defines the mapping from B into $O(N)$.

The result (11.61) for $W^{(E)}[A]$ is manifestly gauge invariant and has also been obtained from perturbation theory[400]. It is remarkable that, except for an overall sign, it has the same form as the Euclidean version of the Wess-Zumino-Witten action S_{WZW}, expression (9.14), for the choice $n = 1$, which we have denoted here[5)] by $\Gamma^{(E)}$.

Using the Euclidean version of the Polyakov-Wiegman formulae (9.86) (see section (11.3.3) for proof)

$$\Gamma^{(E)}[AB] = \Gamma^{(E)}[A] + \Gamma^{(E)}[B] + \frac{1}{4\pi}\int d^2x \mathrm{tr}[(A^{-1}\partial_+ A)(B\partial_- B^{-1})]$$

and recalling (11.44), we recognize that

$$\det(i \not\partial + e \not A) = e^{-\frac{e^2}{4\pi}\int d^2x \mathrm{tr} A_\mu^2} e^{\Gamma^{(E)}[U]} e^{\Gamma^{(E)}[V^{-1}]} \quad . \tag{11.62}$$

As we shall see in chapter 13, $-\Gamma^{(E)}[U]\,(-\Gamma^{(E)}[V^{-1}])$ are the effective actions of the Dirac fermions with pure right(left)-handed coupling to the gauge field, for a chirally symmetric regularization. The first factor above thus shows that a *gauge invariant* regularization is incompatible with a simple factorization of $\det(i \not\partial + e \not A)$ into a right- and left- handed piece! This observation plays a fundamental role in the Polyakov-Wiegman method, to be discussed next.

11.3.2 Polyakov-Wiegman Method

As promised, we now rederive the result (11.61) in still another way. Polyakov and Wiegman[50] obtained the effective QCD_2 action by first calculating the external field current (11.2) and then functionally integrating the equation (4.160*b*), to obtain $W[A]$. We recapitulate here the procedure followed by those authors. In this subsection we work in Minkowski space to avoid a complex form of the light-cone gauge condition. Correspondingly we drop the superscript (E) on $W^{(E)}$.

The first step consists in integrating Eqs. (11.3). To do this we make use of the gauge invariance of QCD_2, by choosing the gauge $V = 1$, which amounts to the choice

$$A_- = 0 \quad , \quad A_+ = iU^{-1}\partial_+ U \quad , \tag{11.63}$$

Writing Eqs. (11.3) in the form

$$\partial_\mu J^\mu - ie[A_\mu, J^\mu] = 0 \quad , \tag{11.64a}$$

$$\tilde\partial_\mu J^\mu - ie[\tilde A_\mu, J^\mu] = -\frac{e}{2\pi}\epsilon_{\mu\nu}F^{\mu\nu} \quad , \tag{11.64b}$$

---

5) The second (topological) term in (11.61*b*) is often referred to as the "Wess-Zumino term", and denoted by $\Gamma[G]$. We shall also refer to it as the "Wess-Zumino action" and denote it by $S_{WZ}[G]$. The renormalization of the theory with action (11.61*b*) can be discussed[391] along the lines of Coleman's work on the sine-Gordon equation.

one finds for the solution of Eq. (11.3),

$$J_+ = \frac{i}{2\pi} U^{-1}\partial_+ U \quad , \tag{11.65a}$$

$$J_- = -\frac{i}{2\pi} U^{-1}\partial_- U \quad . \tag{11.65b}$$

The effective action $W[A]$, is now obtained by integrating

$$\delta W = \frac{e}{2}\int d^2x \mathrm{tr}(J_- \delta A_+) \quad . \tag{11.66}$$

The variation of A_+ with respect to U is given by

$$e\delta A_+ = iU^{-1}\partial_+\delta U - iU^{-1}\delta U U^{-1}\partial_+ U = i\nabla_+(U^{-1}\delta U) \quad , \tag{11.67}$$

where $\nabla_+ f = \partial_+ f + [U^{-1}\partial_+ U, f]$ (compare with Eq. (9.20)). Therefore

$$\delta W = -\frac{1}{4\pi}\mathrm{tr}\int d^2x U^{-1}\delta U \nabla_+(U^{-1}\partial_- U) = -\frac{1}{4\pi}\mathrm{tr}\int d^2x U^{-1}\delta U \partial_-(U^{-1}\partial_+ U) \quad , \tag{11.68}$$

where use has been made of the identity (9.22).

Equation (11.68) for $W[A]$ may now be integrated. To this end note the following two identities:

$$\delta\int d^2x \mathrm{tr} U^{-1}\partial_+ U U^{-1}\partial_- U = \\ -\int d^2x \mathrm{tr} U^{-1}\delta U[\partial_+(U^{-1}\partial_- U) + \partial_-(U^{-1}\partial_+ U)] \quad , \tag{11.69a}$$

$$\delta\int_0^1 dr \int d^2x \, \mathrm{tr}\, U_r^{-1}\dot{U}_r[U_r^{-1}\partial_+ U_r, U_r^{-1}\partial_- U_r] = \\ -\int d^2x \, \mathrm{tr}\, U^{-1}\delta U[\partial_+(U^{-1}\partial_- U) - \partial_-(U^{-1}\partial_+ U)] \quad . \tag{11.69b}$$

where we have used (11.48). Subtracting (11.69b) from (11.69a) and dividing by 8π, one obtains, the right hand side of (11.68). We thereby conclude that in the gauge $A_- = 0$,

$$W[A] = -\frac{1}{8\pi}\int d^2x \mathrm{tr}\partial^\mu U^{-1}\partial_\mu U \\ -\frac{1}{4\pi}\int_0^1 dr \int d^2x \epsilon^{\mu\nu}\mathrm{tr}[U_r^{-1}\dot{U}_r U_r^{-1}\partial_\mu U_r U_r^{-1}\partial_\nu U_r] \quad . \tag{11.70}$$

Using the (known) gauge invariance of the effective QCD_2 action, Polyakov and Wiegman conclude that in a general gauge it is given by (11.70), with U replaced by the gauge invariant product $G = UV^{-1}$. Continuing analytically to Euclidian space, using the formulae (C.20) to (C.25) we recover the result (11.61).

11.3.3 A further invariance of the effective action

Expression (11.61) for the effective QCD_2 action is not particularly useful in the context of the QCD_2 generating functional, since the relation between A_μ and U, V is non-linear, and also non-unique. This is evident from (11.44) which shows that the value of $A_\pm$ is left unchanged, if we perform the transformations

$$U \to g_- U \quad , \quad V \to g_+ V \quad , \tag{11.71}$$

where $g_+(g_-)$ only depends on $x_+(x_-)$. Note that this does not correspond to a gauge transformation (compare with (11.56c)). In fact, G in (11.57) is not invariant under the substitution (11.71), but rather transforms as

$$G \to g_- G g_+^{-1} \quad . \tag{11.72}$$

Hence, if we go over from the functional integration measure $[\mathcal{D}A^\mu]\triangle_{\mathcal{F}}[A]\delta[\mathcal{F}(A)]$ to $[\mathcal{D}G]$ an *additional* δ-function will have to be introduced in order to eliminate the above non-gauge ambiguity[401].

The effective action (11.61) should also be invariant under the transformation (11.71). This property becomes evident if we use the following important result due to Polyakov-Wiegman.

Theorem[50]

$$\Gamma^{(E)}[AB] = \Gamma^{(E)}[A] + \Gamma^{(E)}[B] + \frac{1}{4\pi}\int d^2x(\delta_{\mu\nu} + i\epsilon_{\mu\nu})\mathrm{tr}(A^{-1}\partial_\mu A)(B\partial_\nu B^{-1}). \tag{11.73}$$

Proof: It is convenient to introduce for (11.61*b*) the notation

$$\Gamma^{(E)}[G] = S^{(E)}_{P\sigma M}[G] + S^{(E)}_{WZ}[G] \quad , \tag{11.74a}$$

where $S^{(E)}_{P\sigma M}$ is the Euclidean version of the principal sigma model action, (9.16),

$$S^{(E)}_{P\sigma M}[G] = \frac{1}{8\pi}\int d^2x\, \mathrm{tr}[(\partial_\mu G^{-1})(\partial_\mu G)] \tag{11.74a}$$

and the functional $S^{(E)}_{WZ}[G]$ is the Euclidean Wess-Zumino action

$$S^{(E)}_{WZ}[G] = -\frac{i}{4\pi}\int_0^1 dr \int d^2x \epsilon_{\mu\nu}\mathrm{tr}[(G_r^{-1}\partial_r G_r)(G_r^{-1}\partial_\mu G_r)(G_r^{-1}\partial_\nu G_r)] \quad . \tag{11.74c}$$

Replacing $G_r(G)$ by $A_r B_r(AB)$ and using the cyclic properties of the trace one readily finds

$$S^{(E)}_{P\sigma M}[AB] = S^{(E)}_{P\sigma M}[A] + S^{(E)}_{P\sigma M}[B] + \frac{1}{4\pi}\int d^2x \mathrm{tr}[(A^{-1}\partial_\nu A)(B\partial_\mu B^{-1})]. \tag{11.75}$$

The corresponding calculation for the Wess-Zumino action (11.74c) requires some work. Using $(AB)^{-1}\partial(AB) = B^{-1}[A^{-1}\partial A + (\partial B)B^{-1}]B$, one finds after a lengthy calculation, that

$$S^{(E)}_{WZ}[AB] = S^{(E)}_{WZ}[A] + S^{(E)}_{WZ}[B] - \frac{i}{4\pi}\int_0^1 dr \int d^2x \epsilon_{\mu\nu}\mathcal{W}_{\mu\nu} \quad , \tag{11.76a}$$

where $\mathcal{W}_{\mu\nu}$ can be written in the form

$$\mathcal{W}_{\mu\nu} = \frac{d}{dr}\text{tr}[(A^{-1}\partial_\mu A)(B\partial_\nu B^{-1})] - \partial_\mu[(B\partial_\nu B^{-1})A^{-1}\dot{A}] - \partial_\nu[(A^{-1}\partial_\mu A)B\dot{B}^{-1}] \quad . \tag{11.76b}$$

The second and third term in (11.76b) contribute a surface term to $S^{(E)}_{WZ}[AB]$, which we drop. The first term gives a contribution which can be trivially integrated in r . Hence

$$S^{(E)}_{WZ}[AB] = S^{(E)}_{WZ}[A] + S^{(E)}_{WZ}[B] + \frac{i}{4\pi}\int d^2x \epsilon_{\mu\nu}\text{tr}[(A^{-1}\partial_\mu A)(B\partial_\mu B^{-1})]. \tag{11.77}$$

Combining (11.75) and (11.77) we obtain the announced result (11.73), which may also be written in the form

$$\Gamma[AB] = \Gamma[A] + \Gamma[B] + \frac{1}{4\pi}\int d^2x \text{tr}[(A^{-1}\partial_+ A)(B\partial_- B^{-1})]. \tag{11.78}$$

Here we have dropped the superscript "E", since in the above form the Polyakov-Wiegman result applies to the Minkowskian case as well.

We can now use this result in order to prove the invariance of $\Gamma[G]$ under the transformation (11.71). Since g_+ (g_-) only depends on x_+ (x_-), repeated application of (11.78) yields

$$\Gamma[g_- G g_+] = \Gamma[g_-] + \Gamma[G] + \Gamma[g_+^{-1}] = \Gamma[G] \quad .$$

The above considerations show that the result (11.61) for the effective action cannot be written in terms of A_μ itself. On the other hand, the transition $\mathcal{D}A_\mu \to \mathcal{D}U\mathcal{D}V$ in the functional measure cannot be performed without introducing δ-functions beyond the ones already required by the usual gauge-fixing procedure. We now show how this procedure can be avoided by the introduction of additional degrees of freedom. To achieve this we shall use the trick due to di Vecchia, Durhuus and Petersen[343]

11.4 Bosonization of the QCD_2 Fermionic Action

In two dimensions, the fermionic part of the Lagrangian (11.1a) is invariant not only under local gauge transformations $SU(N)$, but also under the local $SU(N)_L \times SU(N)_R$ transformations $\psi_{\substack{R\\L}} \to g_{\substack{R\\L}}\psi_{\substack{R\\L}}$, $A_\pm \to g_{\substack{R\\L}}(A_\pm + \frac{i}{e}\partial_\pm)g_{\substack{R\\L}}^{-1}$. For $g_R = g_L = g$ this corresponds to a pure vector gauge transformation, while for $g_R = g_L^{-1} = g$ it corresponds to a pure axial vector transformation,

$$\psi \to e^{i\gamma_5\theta}\psi \tag{11.79a}$$

$$A_+ \to gA_+g^{-1} + \frac{i}{e}g(\partial_+ g^{-1}) \tag{11.79b}$$

$$A_- \to g^{-1}A_-g + \frac{i}{e}g^{-1}(\partial_- g) \quad , \tag{11.79c}$$

with g taking values in $SU(N)$.

With the decomposition (11.44), the above transformation also reads

$$U \to Ug^{-1} \quad , \quad V \to Vg \quad . \tag{11.79d}$$

However, because of the axial anomaly, the determinant $\det i\,\not{D}[A]$, and hence $W[A]$, is *not* invariant under the chiral transformations (11.79). This non-invariance is expressed by the fact, that

$$S_F^{(E)}[A,\omega] \equiv \Gamma^{(E)}[U\omega V^{-1}] - \Gamma^{(E)}[UV^{-1}] \neq 0 \quad , \tag{11.80}$$

where $\omega \sim g^{-2}$.

It now follows from the definition (11.61) and the invariance of the Haar measure, that

$$\int \mathcal{D}\omega e^{-S_F[A,\omega]} = \text{const.}e^{-W^{(E)}[A]} \quad , \tag{11.81}$$

where the constant does not depend on A_μ. Hence $S_F[A,\omega]$ plays the role of the equivalent bosonic action for the fermions in a background field A_μ. Its explicit form is obtained[343,402] by repeated use of the Polyakov-Wiegman formula (11.73)

$$S_F[A,\omega] = \Gamma[\omega] + \frac{1}{4\pi}\int d^2x \operatorname{tr}\mathcal{W}(x) \quad , \tag{11.82a}$$

where

$$\mathcal{W} = e^2A_\mu A_\mu - e^2A_+\omega A_-\omega^{-1} - eA_+\omega i\partial_-\omega^{-1} - eA_-\omega^{-1}i\partial_+\omega. \tag{11.82b}$$

We have thus achieved two things:

Instead of representing det $i\not{D}$ as a functional integral over fermionic degrees of freedom, we have succeeded in representing it in terms of *bosonic* degrees of freedom (bosonization). In doing so, we have "freed" A_μ from its connection to U and V, ω being now treated as an *independent* degree of freedom.

Finally we observe that we may obtain from (11.81) the vacuum expectation value of products of the currents j^μ of the *free* fermion theory, by functionally differentiating with respect to A_μ, and setting $A_\mu = 0$. In this way one finds for the correlators of $j_\pm = j_0 \pm j_1$,

$$\langle j_+^{i_1j_1}(x_1)\cdots j_+^{i_nj_n}(x_n)\rangle_F = \left(\frac{-1}{2\pi}\right)^n \langle(\omega^{-1}(x_1)i\partial_+\omega(x_1))^{i_1j_1}\cdots(\omega^{-1}(x_n)i\partial_+\omega(x_n))^{i_nj_n}\rangle_B \quad (11.83a)$$

$$\langle j_-^{i_1j_1}(x_1)\cdots j_-^{i_nj_n}(x_n)\rangle_F = \left(\frac{-1}{2\pi}\right)^n \langle(\omega(x_1)i\partial_-\omega^{-1}(x_1))^{i_1j_1}\cdots(\omega(x_n)i\partial_-\omega^{-1}(x_n))^{i_nj_n}\rangle_B \quad ,(11.83b)$$

where $\langle\ \ \rangle_{F(B)}$ denotes the vacuum expectation value in the fermionic (bosonic) formulation. From here we deduce the operator equivalences

$$j_+ = -\frac{1}{2\pi}\omega^{-1}i\partial_+\omega \quad (11.84a)$$

$$j_- = -\frac{1}{2\pi}\omega i\partial_-\omega^{-1} \quad , \quad (11.84b)$$

in agreement with the non-abelian bosonization formulae obtained by Witten.

11.4.1 Recovering the $U(1)$ case

If the group is $U(1)$, the Wess-Zumino term in (11.61b) vanishes, and we are left with

$$W[A] = -\frac{1}{8\pi}\int d^2x(\partial_\mu G^{-1})(\partial_\mu G)] \quad (11.85)$$

Writing

$$U = e^{i\varphi+\phi} \quad , \quad V = e^{i\varphi-\phi} \quad (11.86)$$

we have

$$eA_\mu = (\partial_\mu\varphi + \tilde{\partial}_\mu\phi) \quad (11.87)$$

or using (B.4)

$$eF_{\mu\nu} = (\partial_\mu\tilde{\partial}_\nu - \partial_\nu\tilde{\partial}_\mu)\phi = \epsilon_{\mu\nu}\Delta\phi \quad (11.88)$$

Solving (11.88) for ϕ in terms of $F_{\mu\nu}$, one has

$$\phi = \frac{e}{2}\int d^2y D(x-y)\epsilon_{\mu\nu}F_{\mu\nu}(y) \quad . \tag{11.89}$$

Hence, making the identification $G = e^{2\phi}$, one obtains

$$W[A] = \frac{1}{2\pi}\int d^2x(\partial_\mu\phi)^2 = \frac{1}{8\pi}\int d^2x\epsilon_{\mu\nu}F_{\mu\nu}(x)D_E(x-y)\epsilon_{\lambda\rho}F_{\lambda\rho}(y) \quad , \tag{11.90}$$

where we have used $\Delta D_E(x) = -\delta^2(x)$. Expression (11.90) is seen to coincide with the result for $-\omega_{\not{D}/\not{\partial}}$ in (4.124).

11.4.2 QCD_2 bosonized currents

We may rewrite (11.82b) in the form

$$\begin{aligned} \mathcal{W} =& e^2 A_\mu A_\mu - e^2 A_\mu\omega\left(\delta_{\mu\nu} - i\epsilon_{\mu\nu}\right)A_\nu\omega^{-1} \\ & - ieA_\mu\left(\delta_{\mu\nu} - i\epsilon_{\mu\nu}\right)\omega\partial_\nu\omega^{-1} - ieA_\mu\left(\delta_{\mu\nu} + i\epsilon_{\mu\nu}\right)\omega^{-1}\partial_\nu\omega \quad . \end{aligned} \tag{11.91}$$

Note that the first two terms cancel if the group is abelian.

Taking the variational derivative of $S_F[A,\omega]$ with respect to A^a_μ, we obtain for the current, the expression

$$J_\mu = \frac{e}{2\pi}A_\mu - \frac{e}{4\pi}\left\{(\delta_{\mu\nu} - i\epsilon_{\mu\nu})\omega iD_\nu\omega^{-1} + (\delta_{\mu\nu} + i\epsilon_{\mu\nu})\omega^{-1}iD_\nu\omega\right\} \quad . \tag{11.92}$$

We may write this result also in the form

$$\begin{aligned} J_+ &= \frac{i}{2\pi}\left\{U^{-1}\partial_+U - (U\omega)^{-1}\partial_+(U\omega)\right\} \\ J_- &= \frac{i}{2\pi}\left\{V^{-1}\partial_-V - \left(V\omega^{-1}\right)^{-1}\partial_-\left(V\omega^{-1}\right)\right\} \quad . \end{aligned} \tag{11.93}$$

From here it is evident that under the local transformation

$$U \to Ug\ , \quad V \to Vg\ , \quad \omega \to g^{-1}\omega g\ , \tag{11.94}$$

the currents (11.93) transform covariantly

$$J_\pm \to g^{-1}J_\pm g \quad . \tag{11.95}$$

Moreover one readily checks, that under the gauge transformation (11.94),

$$S_F[A^g, g^{-1}\omega g] = S_F[A,\omega] \quad . \tag{11.96}$$

Hence (11.94) realizes the gauge transformation (11.56*c*) in the extended bosonic space of U , V and ω.

For the abelian case, the current (11.92) reduces to the form of the Witten currents (9.27),

$$J_+ = -\frac{i}{2\pi}\omega^{-1}\partial_+\omega \quad , \quad J_- = -\frac{i}{2\pi}\omega\partial_-\omega^{-1} \quad , \tag{11.97}$$

in agreement with what we have referred to as *adiabatic principle* in chapter 10.

In the non abelian case, the current however involves the gauge field itself. This leads to an unexpected form for $S_F[A,\omega]$, which in terms of the current (11.92) reads

$$S_F[A,\omega] = \Gamma[\omega] + \int d^2x \,\mathrm{tr}\left\{eJ_\mu A_\mu - \frac{1}{4\pi}\left(A_+A_- - \omega^{-1}A_+\omega A_-\right)\right\} \quad . \tag{11.98}$$

Only in the abelian case the second term in the integrand cancels, and $S_F[A,\omega]$ reduces to a conventional form. This shows that the "adiabatic principle", does not work in the non-abelian case.

11.5 Fermion Greens Function[395]

We have seen, that the Dirac operator of QCD_2 factorizes as shown in Eq. (11.47*b*). Hence we expect a corresponding factorization of the external field Dirac Greens function, defined by

$$(i\,\partial\!\!\!/ + e\,A\!\!\!/)G(x,y\mid A) = \delta^2(x-y) \quad . \tag{11.99}$$

Now, from (11.44) follows that U and V satisfy

$$\partial_\mp h_\pm(x\mid A) = ieA_\mp(x)h_\pm(x\mid A) \quad , \tag{11.100}$$

where we have streamlined our notation, by defining $h_+ = V^{-1}$, $h_- = U^{-1}$. In this notation, (11.47b) reads, ($\partial_\pm$ act here on everything to the right!)

$$(i\,\partial\!\!\!/ + e\,A\!\!\!/) = \begin{pmatrix} 0 & h_+\partial_-h_+^{-1} \\ -h_-\partial_+h_-^{-1} & 0 \end{pmatrix} \tag{11.101}$$

Hence it follows that

$$G(x,y|A) = \begin{pmatrix} 0 & h_-(x|A)D_-(x-y)h_-^{-1}(y|A) \\ -h_+(x|A)D_+(x-y)h_+^{-1}(y|A) & 0 \end{pmatrix} \tag{11.102a}$$

where $D_\pm$ has been defined in (11.10) We write this in the form[395,398]

$$G(x,y \mid A) = \frac{i}{2}\sum \gamma_\pm h_\mp(x \mid A) h_\mp^{-1}(y \mid A) D_\mp(x-y) \quad . \tag{11.102b}$$

Now, one readily verifies that Eq. (11.100) has the power series solution

$$h_\pm(x \mid A) = 1 + \sum_{n=1}^{\infty}(-ie)^n \int d^2z_1 \cdots d^2z_n \times$$
$$[D_\pm(x-z_1)D_\pm(z_1-z_2)\cdots D_\pm(z_{n-1}-z_n)]A_\mp(z_1)\cdots A_\mp(z_n) \quad , \tag{11.103}$$

with $h_\pm^{-1}(x \mid A)$ given by the substitution $e \to -e$. Noting that $h_\pm(x \mid A)$ is covariantly constant, $iD_\pm h_\mp(x) = 0$, one immediately verifies that $G(x,y \mid A)$ satisfies (11.99). Under a gauge transformation,

$$h_\pm(x \mid A) \to g(x) h_\pm(x \mid A) \quad . \tag{11.104}$$

Therefore, $h_\pm$ transforms like the path ordered exponential of the integral $-ie\int_{\Gamma_x} dz_\mu A_\mu$ with Γ_x a contour extending from x_μ to ∞.

11.5.1 Specializing to the $U(1)$ case

For an abelian group we may explicitly carry out the sum in (11.103). Indeed, since the $A_\pm$ commute with each other in this case, we may write this expansion as,

$$h_\pm(x \mid A) = 1 + \sum_{n=1}^{\infty}\frac{(-ie)^n}{n!}\int d^2z_1 \cdots d^2z_n \times$$
$$\Big[\sum_{P[z_i]} D_\pm(x-z_1)\cdots D_\pm(z_{n-1}-z_n)\Big]A_\mp(z_1)\cdots A_\mp(z_n) \tag{11.105}$$

where the second sum extends over all permutations of the coordinates z_i. Using the identity[6]), valid in $d = 2$,

$$\sum_{P[z_i]} D_\pm(x-z_{i_1})\cdots D_\pm(z_{i_{n-1}}-z_{i_n}) = \prod_{i=1}^{n} D_\pm(x-z_i) \quad , \tag{11.106}$$

the above expression reduces to

$$h_\pm(x \mid A) = 1 + \sum_{n=1}^{\infty}\frac{(-ie)^n}{n!}\left[\int d^2z D_\pm(x-z)A_\mp(z)\right]^2 \quad , \tag{11.107}$$

[6]) This identity will still play an important role in our later discussion in chapter 12.

so that

$$h_{\pm}(x \mid A) = e^{H_{\pm}(x|A)} \quad , \quad h_{\pm}^{-1}(x \mid A) = e^{-H_{\pm}(x|A)} \quad . \tag{11.108a}$$

with

$$H_{\pm}(x \mid A) = -ie \int d^2 z D(x-z)(\partial_\mu A_\mu(z) \mp \frac{i}{2}\epsilon_{\mu\nu} F_{\mu\nu}(z)) \quad . \tag{11.108b}$$

Correspondingly we have in the abelian case

$$G(x,y \mid A) = \frac{i}{2}\sum_{\pm} \gamma_{\pm} e^{(H_{\mp}(x|A) - H_{\mp}(y|A))} D_{\mp}(x-y) \tag{11.109a}$$

$$G(x,y \mid A) = e^{i(\phi(x)-\phi(y))} S_E(x-y) \quad , \tag{11.109b}$$

where ϕ is the now matrix-valued field,

$$\phi(x) = -e \int d^2 z D(x-z)\left(\partial_\mu A_\mu(z) + \frac{i}{2}\gamma_5 \epsilon_{\mu\nu} F_{\mu\nu}(z)\right) \quad . \tag{11.110}$$

We shall make use of this result in the following chapter, where we return to the discussion of the $U(1)$ massless Schwinger model. One could of course have obtained the result (11.109) much more easily, by working from the outset with the $U(1)$ group. This will in fact be done in the following chapter.

11.6 Conclusion

Our discussion of QCD_2 is by no means complete. We have not discussed the canonical quantization of QCD_2, nor the related problem of the Poisson algebra obeyed by the currents (11.93) in the bosonized formulation of the theory. The reason is that such questions will be extensively discussed in chapters 13 and 14 in the context of chiral gauge theories. Except for questions of this nature, our discussion nevertheless covered most of what can rigorously be said about the non perturbative structure of QCD_2. In particular we have been able to show that the external field problem for the effective action, gauge current and fermionic Greens function could be reduced to the calculation of the tree diagrams shown in figure 11.1. This reduction of loop diagrams to tree diagrams is also known as cutting rules, and is a specific property of two dimensional space-time. Further non perturbative insight into two dimensional gauge theories can be obtained by restricting to the maximal abelian subgroup (torus) of $SU(N)$. Since this group is the direct product, $[U(1)]^{N-1}$, of $U(1)$ components, this case parallels the one of the $U(1)$-Schwinger model, and we have not bothered to discuss it here. Details can be found in refs. [395] and [141]. QCD_2 being a superrenormalizable theory, we have also not touched on questions relating to its renormalization, and refer to the reader to the work of Gepner[391]. Similarly we have refrained from discussing the bound state spectrum, for reasons stated in the introduction.

It has been argued[374,375] that in QCD_2 the $SU(N)$-gauge group may be broken down to the maximal abelian subgroup - the $SU(N)$ torus. This would mean a spontaneous breakdown of the gauge symmetry, for which we see no sign in vector-like theories[7)].

Though there always exists a gauge in which $F^a_{\mu\nu}$ points in the torus direction of the group $SU(N)$, this is not true for the vector potential A^a_μ. This does not exclude the possibility that torus configurations of A^a_μ may play a dominant role in special situations[141]. This would however depend on the gauge chosen for the problem in question. Nevertheless we expect some features of the Schwinger model to be also characteristic of QCD_2. Thus taking the covariant *pseudo*-divergence of the Maxwell equations (11.1*b*) and using the anomaly equation (11.3) we arrive at[398]

$$\left(-\mathcal{D}^2 + \frac{e^2}{\pi}\right) F^b_{12} = 0 \quad ,$$

which generalizes Eq. (10.13) to the non-abelian case. The implications of this equation for the spectrum of QCD_2 remain unclear. It however strongly suggests that an intrinsic Higgs mechanism similar to the one found in QED_2 is also a feature of QCD_2. As we already remarked in the introduction, this disagrees with 't Hooft's analysis in the leading $1/N$ approximation, and has been the source of speculations for the existence of two phases (the 't Hooft phase, and the Higgs phase) in QCD_2.

We would also mention, that QCD_2 has served[403] as a laboratory for gaining an insight into some phenomenological aspects of "real world" QCD_4, such as the Brodsky-Farrar scaling law[404] for hadronic form factors, and the Drell-Yan-West relation[405] as well as the Bloom-Gilman duality[406] for deep inelastic lepton scattering.

Finally, we should emphasize that we have completely ignored the problem of zero modes. Although unlike the $U(1)$ case, the zero modes are not predicted by topological arguments ($\pi_1\left(SU(N)\right) \approx 0$) their existence can be explicitly verified for gauge field configurations lying in the torus of $SU(N)$[395,141]. In their presence the anomaly equation (11.3) must be replaced by (see (4.215)),

$$\tilde{\mathcal{D}}^{ab}_\mu J'^b_\mu(x \mid A) = -\frac{e}{2\pi}\epsilon_{\mu\nu}F^a_{\mu\nu} + 2\Omega_R(x)\sum_n u^{(0)\dagger}_n(x)\gamma^5\tau^a u^{(0)}_n(x) \quad , \tag{11.111}$$

---

7) In *chiral* gauge theories, the local gauge symmetry is dynamically broken by the fermion vacuum polarization, as shown in chapters 13 and 14.

where $J'^b_\mu(x \mid A)$ has been defined in chapter 4, and $u_n^{(0)}$ are the zero modes of the Dirac operator. The corresponding effective action W' satisfies (4.160*b*), with J'_μ replacing J_μ and is thus obtained by integrating the modified anomaly equation

$$\tilde{\mathcal{D}}^{ab}_\mu \frac{\delta W'[A]}{\delta A^b_\mu(x)} = -\frac{e^2}{2\pi}\epsilon_{\mu\nu}F^a_{\mu\nu}(x) + 2\Omega_R(x)e\sum_n u_n^{(0)\dagger}(x)\gamma^5\tau^a u_n^{(0)}(x) \quad . \tag{11.112}$$

This equation, together with $\mathcal{D}^{ab}_\mu \frac{\delta W'}{\delta A^b_\mu(x)} = 0$, expressing the covariant conservation of the current J'^a_μ, may be integrated for configurations lying in the torus of $SU(N)$. We leave the demonstration of this to the analogous $U(1)$ case in chapter 12.

Chapter XII

QUANTUM ELECTRODYNAMICS: FUNCTIONAL APPROACH

12.1 Introduction

In this chapter we reconsider the massless Schwinger model, this time however within the path integral approach. The building stones have already been layed in the preceding chapter. There we have seen that the non-abelian results for the effective action and external field Dirac Green's function simplify dramatically upon restricting the gauge group to be $U(1)$, as witnessed in the corresponding results (11.90) and (11.109).

As our discussion in chapter 10 leads one to expect, the path integral formulation of the Schwinger model exhibits a rather rich structure, far from the simplicity suggested by Schwinger's original solution. The key property responsible for this is the existence of Euclidean gauge field configurations with non vanishing "winding number" or Pontryagin index (instantons[407]) as a result of the topologically non trivial mapping of S^1 onto $U(1)$ ($\Pi_1(U(1)) = Z$).

As it turns out, the existence of an infinitely degenerate ground states in the Schwinger model witnessed in chapter 10 is directly linked to the existence of so called induced instantons[408], and the chirality of these degenerate ground states is found[355,409] to be related to the Pontryagin index of these Euclidean gauge field configurations via the Atiyah-Singer theorem (see chapter 4). The "vanishing theorem" (absence of negative(positive) chirality zero modes for positive(negative) winding number) is checked to be a generic property of QED_2[1)]. Massless QED_2 shares the above properties with QCD_4, which makes the Schwinger model an extremely interesting toy model in its own right. As we have seen in chapter 10, there however exists a further vacuum degeneracy labelled by fermion number, which turns out to be linked to the existence of "merons"[368] with Pontryagin index one half. Such meron configurations do not contribute in QCD_4, where they correspond to gauge field configurations with infinite (Euclidean) action.

In this "toy" model one furthermore explicitly checks, using clustering arguments, the 't Hooft rules for obtaining the path integral representation of fermionic correlation function connecting topologically inequivalent vacuum sectors (tunneling amplitudes). And last, but not least, Crewther's solution[265] to the so called

---

[1)] It is also a property of QCD_4 in the instanton gas approximation[175].

$U(1)$ problem[364] can be studied by embedding the chiral Gross-Neveu model[178] in a $U(1)$ gauge theory[382].

In the following sections we embark to explore these subtleties of massless QED_2 within the functional approach. Throughout this chapter we shall work with the Euclidean formulation.

12.2 Equivalent Bosonic Action

The equivalent bosonic fermion-action of QED_2 may be directly obtained from that of QCD_2, Eq. (11.82), by setting $\omega = e^{-2i\sqrt{\pi}\theta}$, with θ now an abelian field. The Wess-Zumino-Witten action (11.61b) and equivalent di Vecchia-Durhuus-Petersen action (11.82) then take the simple form

$$\Gamma[\omega] = \frac{1}{2}\int d^2x(\partial_\mu\theta)^2 \quad . \tag{12.1a}$$

and

$$S_F^{(E)}[A, e^{i\theta}] = \int d^2x\left\{\frac{1}{2}(\partial_\mu\theta)^2 - i\frac{e}{\sqrt{\pi}}A_\mu\tilde{\partial}_\mu\theta\right\} \quad . \tag{12.1b}$$

From (11.81) we thus have for the path integral representation of the QED_2 determinant,

$$\frac{\det i\not{D}}{\det i\not{\partial}} = \int \mathcal{D}\theta e^{\int d^2x[\frac{1}{2}\theta\Delta\theta + i\frac{e}{\sqrt{\pi}}\theta\tilde{\partial}_\mu A_\mu]} \quad , \tag{12.2a}$$

which may be explicitly integrated to yield

$$\frac{\det i\not{D}}{\det i\not{\partial}} = e^{-\frac{e^2}{\pi}\int\int \frac{1}{2}A_\mu\frac{\tilde{\partial}_\mu\tilde{\partial}_\nu}{\Delta}A_\nu} \quad . \tag{12.2b}$$

This result agrees with (4.125). Note that it is manifestly gauge invariant. In the Lorentz gauge it reduces to the result obtained by Schwinger. It must be emphasized, however, that for the time being we have assumed zero modes to be absent !

The exact solubility of QED_2 now rests on the following observations:

i)The pure gauge action

$$S_G^{(E)}[A] = \int d^2x\frac{1}{4}F_{\mu\nu}F_{\mu\nu} = -\int d^2x A_\mu\tilde{\partial}_\mu\tilde{\partial}_\nu A_\nu$$

is quadratic in A_μ.

ii) The QED_2 effective action is of the Gaussian form

$$W[A] \equiv -\ln\ \det(i\not{\partial} + e\not{A}) = \frac{e^2}{2\pi}\int\int A_\mu\frac{\tilde{\partial}_\mu\tilde{\partial}_\nu}{\Delta}A_\nu \quad . \tag{12.2c}$$

iii) The external field fermionic Greens function $G(x,y \mid A)$ is given in terms of an exponential linear in A_μ (see Eq. (11.109)),

$$G(x,y \mid A)_{\alpha\beta} = e^{i(\phi_\alpha(x)-\phi_\alpha(y))} S_E(x-y)_{\alpha\beta} \tag{12.3a}$$

where

$$\phi_\alpha(x) = \varphi(x) + i\gamma^5_{\alpha\alpha}\tilde{\varphi}(x) \quad , \tag{12.3b}$$

with

$$\varphi(x) = -e\int d^2z D(x-z)\partial_\mu A_\mu(z) \quad , \tag{12.3c}$$

$$\tilde{\varphi}(x) = e\int d^2z D(x-z)\tilde{\partial}_\mu A_\mu(z) \quad . \tag{12.3d}$$

Note that the decomposition (12.3*b*) corresponds to a separation of $\phi_\alpha(x)$ into its gauge-variant (φ) and gauge-invariant ($\tilde{\varphi}$) part. Properties i) - iii) imply that the calculation of correlation functions in QED_2 only involves the evaluation of Gaussian functional integrals, at least as far as the sector without zero modes is concerned.

12.3 Gauge Invariant Correlation Functions

The calculation of correlations functions in the functional framework involves two steps:

i) Calculation of the corresponding *external* field quantities, such as

$$\langle\overline{\psi}(x_1)\cdots\overline{\psi}(x_n)\psi(y_1)\cdots\psi(y_n)\rangle_A \ , \tag{12.4a}$$

$$\rho_\pm(x \mid A) = \langle :\overline{\psi}(x)\frac{1}{2}(1\pm\gamma_5)\psi(x):\rangle_A \ , \tag{12.4b}$$

$$G_{GI}(x,y \mid A) = \langle\psi(x)e^{-ie\int_y^x dz_\mu A_\mu(z)}\overline{\psi}(y)\rangle_A \quad , \tag{12.4c}$$

where the subscript GI stands for "gauge invariant".

ii) Integration of the product of such quantities (including $F_{\mu\nu}$) with respect to the measure (see Mathews-Salam formula (4.1))

$$d\mu[A]\det\, i\not{D}[A]\exp\left(-\int d^2x\frac{1}{4}F_{\mu\nu}F_{\mu\nu}\right) \quad ,$$

where $d\mu[A]$ includes the usual gauge fixing delta functional, and Faddeev-Popov determinant, $d\mu[A] = \mathcal{D}A_\mu\Delta_{\mathcal{F}}[A]\delta(\mathcal{F}(A))$; we suppose for the time being, that there exist no zero modes of the Dirac operator. In the following we calculate some external field quantities of interest. We generally omit the superscript "E" for "Euclidean".

12.3.1 The external field current, and chiral densities

The external field current has already been calculated in (11.35) for the case of QCD_2. In the $U(1)$ case, this expression reduces to the first order term. It is however instructive to repeat this calculation following Schwinger's method, that is, by computing the external field current in terms of the short-distance limit

$$\begin{aligned} J_\mu(x \mid A) &= -\lim_{\epsilon\to 0} \operatorname{tr}\gamma_\mu[G_{GI}(x+\epsilon, x \mid A) - G_{GI}(x+\epsilon, x \mid 0)] \\ &= -\lim_{\epsilon\to 0} \operatorname{tr}\{\gamma_\mu G(x+\epsilon, x \mid A) e^{-ie\int_x^{x+\epsilon} dz_\mu A_\mu(z)}\}\Big|_0^{A_\mu} \quad . \end{aligned} \tag{12.5}$$

Following Schwinger, we work with the gauge invariant Greens function, in order to insure a gauge-invariant result. As an extra bonus we thereby also insure the independence of the result of the direction chosen for the limit $\epsilon \to 0$. The definition (12.5) corresponds to defining the current in terms of vacuum expectation value of the Wilson short distance expansion (10.23)

Remembering that $S(x) = i\,\partial\!\!\!/ D(x) = -\frac{i}{2\pi}\frac{\gamma_\mu x_\mu}{x^2}$, we have, using (12.5)[2],

$$\operatorname{tr}[\gamma_\mu G_{GI}(x+\epsilon, x \mid A)] = -\frac{i}{2\pi}\operatorname{tr}[\gamma_\mu(1 + i\epsilon_\nu\partial_\nu\varphi - \gamma_5\epsilon_\nu\partial_\nu\tilde{\varphi} - ieA_\nu)\frac{\gamma_\lambda\epsilon_\lambda}{\epsilon^2}] + \mathcal{O}(\epsilon) \quad .$$

This expression can be written in the form

$$\operatorname{tr}\{\gamma_\mu[G_{GI}(x+\epsilon, x \mid A) - G_{GI}(x+\epsilon, x \mid 0)]\} =$$
$$\frac{1}{\pi}\Big\{\frac{\epsilon_\mu\epsilon_\nu}{\epsilon^2}\Big(\partial_\nu\varphi(x) - \tilde{\partial}_\nu\tilde{\varphi}(x) - eA_\nu(x)\Big) - \Big(\frac{\tilde{\epsilon}_\mu\epsilon_\nu - \epsilon_\mu\tilde{\epsilon}_\nu}{\epsilon^2}\Big)\partial_\nu\tilde{\varphi}(x)\Big\} + \mathcal{O}(\epsilon).$$

Recalling now (12.3) and property (B.4), we have the (direction independent!) result

$$\operatorname{tr}\{\gamma_\mu G_{GI}(x+\epsilon, x \mid A)\}\Big|_0^A = -\frac{i}{\pi}\tilde{\partial}_\mu\tilde{\varphi} = -\frac{1}{\pi}\int d^2z\,\tilde{\partial}_\mu\tilde{\partial}_\nu D(x-z)A_\nu(z) \quad . \tag{12.6}$$

Hence, in agreement with (11.24), we obtain the current

$$J_\mu(x \mid A) = \frac{1}{\pi}\int d^2z\Big(\delta_{\mu\nu} - \frac{\partial_\mu\partial_\nu}{\Delta}\Big)A_\nu(z) \quad . \tag{12.7}$$

The calculation of the chiral densities in (12.4b) is much simpler. They are obtained as the limit

$$\rho_\pm(x \mid A) = -\lim_{\epsilon\to 0}\operatorname{tr}\Big[\frac{1\pm\gamma_5}{2}G_{GI}(x+\epsilon, x \mid A)\Big] \quad .$$

Using (12.3a), and the properties (B.15) of the γ matrices, it follows that $\rho_\pm(x|A) = 0$. This result reflects the fact, that the *perturbative* vacuum is chirally neutral. It is no longer correct if zero modes are present (non-perturbative instanton effects) as we shall see later on.

---

[2] Note that up to $\mathcal{O}(\epsilon^2)$ the line integral in (12.5) is independent of the path, since $\oint dz_\mu A_\mu = \int d^2z\,\epsilon_{\mu\nu}F_{\mu\nu} \equiv \epsilon^2\epsilon_{\mu\nu}F_{\mu\nu}$.

12.4 Vacuum Structure[409]

From the operator solution of the (massless) Schwinger model (see chapter 10) we know, that the global $U(1) \times \tilde{U}(1)$ symmetry of the classical Lagrangian is spontaneously broken as put in evidence by the existence of an infinite number of (degenerate) ground states labelled by a chiral and fermionic (bare charge) quantum number. In the following we shall try to rediscover this degeneracy within the functional framework, by relating it to a violation of clustering of the correlation functions. In this way we shall be able to establish a direct connection between the quantum numbers labelling the vacua (see chapter 10) and the existence of "induced" instantons belonging to a non-trivial Pontryagin class. We shall restrict our discussion to the *chiral* aspects of the vacuum. In fact, the chiral properties of the QED_2 vacuum are shared by that of QCD_4. The starting point for our discussion will be the expectation value of chirally and charge neutral operators with respect to the (perturbative) Fock vacuum, which are known to have a conventional path-integral representation.

12.4.1 Chirality of the vacuum[355,409]

Consider the zero-chirality correlation function

$$W_n(\xi; x, y) = \langle 0 \mid \rho_-(\xi_1) \cdots \rho_-(\xi_n) \psi_1(x_1) \cdots \psi_1(x_n) \overline{\psi}_1(y_1) \cdots \overline{\psi}_1(y_n) \mid 0 \rangle \quad . \tag{12.8}$$

It has functional representation

$$W_n(\xi \mid x; y) = \int d\mu[A] \det i \not{D}[A] e^{-S_G[A]} \mathcal{W}_n(\xi; x, y \mid A) \quad , \tag{12.9a}$$

with

$$\mathcal{W}_n(\xi; x, y \mid A) = e^{i\Lambda(x,y|A)} e^{-\phi(\xi; x,y|A)} W_n^{(0)}(\xi; x, y) \quad , \tag{12.9b}$$

Here $\Lambda(x; y \mid A)$ and $\phi(\xi; x, y \mid A)$ are the gauge dependent and gauge invariant functionals

$$\Lambda(x; y \mid A) = -e \sum_i \int d^2 z \Big(D(x_i - z) - D(y_i - z) \Big) \partial_\mu A_\mu(z) \quad , \tag{12.10a}$$

$$\phi(\xi; x, y \mid A) = e \int d^2 z \sum_{i=1}^{n} (2D(\xi_i - z) - D(x_i - z) - D(y_i - z)) \epsilon_{\mu\nu} \partial_\mu A_\nu \tag{12.10b}$$

respectively; $W_n^{(0)}$ is the corresponding correlation function for the free fermion field:[25]

$$W_n^{(0)}(\xi; x, y) = \left(-\frac{1}{2\pi} \right)^{2n} \frac{\prod_{i<j} (\xi_i - \xi_j)^2 (x_i - x_j)_+ (y_i - y_j)_-}{\prod_{i,j} (\xi_i - x_j)_+ (\xi_i - y_j)_-} \quad . \tag{12.11}$$

From the operator solution (10.45) we know that the chiral densities $\rho_+(\rho_-)$ contain a spurion which acts as raising (lowering) operator for the chiral quantum number $n_{ch} = n_1 - n_2$ labelling the vacuum states $\mid n_1, n_2\rangle$. Since ρ_- is gauge invariant, it acts as a projector on the gauge invariant subspace. Thus we expect,[3)]

$$W_n(\xi; x, y) \underset{\xi_i \to \infty}{\to} \langle 0 \mid \prod_{i=1}^{n} \rho_-(\xi_i) \mid n\rangle\langle n \mid \psi_1(x_1) \cdots \overline{\psi}_1(y_n) \mid 0\rangle \quad , \tag{12.12}$$

where the limit is to be taken with $\xi_i - \xi_j$ held fixed.

We shall work in the *Lorentz* gauge. In that case the Faddeev Popov determinant is just a c-number and $\Lambda(x, y \mid A) = 0$. Hence

$$W_n(\xi; x, y) = \int \mathcal{D}A_\mu \delta(\partial_\mu A_\mu) e^{-S^{ind}(\xi; x, y \mid A)} W_n^{(0)}(\xi; x, y) \quad , \tag{12.13}$$

where S^{ind} is the "induced" action given by

$$S^{ind}(\xi; x, y \mid A) = S_G[A] + W[A] + \phi(\xi; x, y \mid A) \quad . \tag{12.14}$$

It is convenient to write (12.10*b*) in the form

$$\phi(\xi; x, y \mid A) = e \int J_\mu^{ind}(z \mid \xi; x, y) A_\mu(z) \quad , \tag{12.15a}$$

where

$$J_\mu^{ind}(z \mid \xi; x, y) = \tilde{\partial}_\mu^z \sum_i (2D(\xi_i - z) - D(x_i - z) - D(y_i - z)) \tag{12.15b}$$

is the induced current. It is also useful to define

$$S^{eff}[A] = S_G[A] + W[A] = \frac{1}{2} \int d^2 z A_\mu(z) (-\Delta + \frac{e^2}{\pi}) A_\mu(z) \quad . \tag{12.16}$$

where S_G is the action associated with the gauge field, $S_G[A] = \frac{1}{4} \int d^2x \, F_{\mu\nu} F_{\mu\nu}$.

The functional integral in (12.13) can be performed explicitly, since it is of the Gaussian type. We want to demonstrate however, the connection between clustering violation in this model and the existence of gauge field configurations with nontrivial Pontryagin number. We therefore expand S^{ind} around the classical

---

3) We only label the vacua by (1/2) their chirality $n_{ch} = 2n$, since the fermion number plays here a spectator role.

field configuration $A_\mu^{cl}(x)$ which minimizes S^{ind} : $(\delta S^{ind}/\delta A_\mu)_{A_\mu^{cl}} = 0$. Thus, A_μ^{cl} is a solution to the equation

$$\left(-\Delta + \frac{e^2}{\pi}\right) A_\mu^{cl} + eJ_\mu^{ind} = 0 \quad . \tag{12.17}$$

It is given by[408]

$$A_\mu^{cl}(z) = -\frac{\pi}{e}\tilde{\partial}_\mu(2\overline{v}(z) - v(z)) \quad , \tag{12.18a}$$

where

$$\overline{v}(z) \equiv \sum_{i=1}^{n} \mathcal{D}(\xi_i - z) \quad , \qquad v(z) \equiv \sum_{i}^{n}(\mathcal{D}(x_i - z) - \mathcal{D}(y_i - z)) \quad , \tag{12.18b}$$

$$\mathcal{D}(z) = D(z) - \Delta(z; \frac{e^2}{\pi}) \quad . \tag{12.18c}$$

We now make use of the property (B.20a), and

$$\left(-\Delta + \frac{e^2}{\pi}\right) \mathcal{D}(z) = \frac{e^2}{\pi} D(z) \quad , \tag{12.19b}$$

$$-\Delta\mathcal{D}(z) = \frac{e^2}{\pi}\Delta\left(z; \frac{e^2}{\pi}\right) \quad . \tag{12.19c}$$

Note that A_μ^{cl} satisfies the Lorentz condition $\partial_\mu A_\mu^{cl} = 0$. Separating A_μ into A_μ^{cl} and fluctuations around this classical field configuration, $A_\mu = A_\mu^{cl} + a_\mu$, we have, exactly,

$$S^{ind}(\xi; x, y \mid A) = S^{ind}\left[A_\mu^{cl}\right] + \frac{1}{2}\int a_\mu \left(-\Delta + \frac{e^2}{\pi}\right) a_\mu \quad . \tag{12.20}$$

The term linear in a_μ is absent, since A_μ^{cl} minimizes the action. With

$$\mathcal{D}A_\mu \delta(\partial_\mu A_\mu) = \mathcal{D}a_\mu \delta(\partial_\mu a_\mu) \quad ,$$

the a_μ integration in (12.13) just contributes a constant, and we have

$$W_n(\xi; x, y) = W_n^{(0)}(\xi; x, y) e^{-S^{ind}(\xi; x, y \mid A^{cl})} \quad . \tag{12.21}$$

We shall be interested in studying the $\xi \to \infty$ limit (clustering). This suggest making the separation (see Eq. (12.18))

$$A_\mu^{cl}(z) = -\frac{2\pi}{e}\tilde{\partial}_\mu^z \overline{v}(z) + A_\mu^{[n]^{cl}}(z) \quad , \tag{12.22}$$

with

$$A_\mu^{[n]^{cl}}(z) = \frac{\pi}{e}\tilde{\partial}_\mu^z v(z) \quad . \tag{12.23}$$

Here $A_\mu^{[n]^{cl}}$ describes a gauge-field configuration with winding number n. Indeed, from (12.23) and (12.18c) we have for the corresponding Pontryagin density

$$\begin{aligned}\frac{e}{4\pi}\epsilon_{\mu\nu}F_{\mu\nu}^{[n]^{cl}}(z) &= \frac{1}{2}\tilde{\partial}_\nu^z\tilde{\partial}_\nu^z\sum_i[\mathcal{D}(x_i - z) + \mathcal{D}(y_i - z)] \\ &= -\frac{e^2}{2\pi}\sum_i[\Delta(x_i - z) + \Delta(y_i - z)] \quad .\end{aligned} \tag{12.24}$$

Let us compute the integral over a volume V of this density. We split V into small spheres of radius ϵ_i around the singular points $z = x_i$, $z = y_i$, and the rest, which we denote by V'.

Since the singularities at $z = x_i, z = y_i$ are of the logarithmic type, the limit $\epsilon_i \to 0$ is unproblematic, and we can convert the integral over the Pontryagin density into a surface integral over the boundary S_R^1 of the (spherical) volume V :

$$\int \frac{d^2z}{4\pi}e\epsilon_{\mu\nu}F_{\mu\nu}^{[n]^{cl}}(z) = e\int_0^{2\pi}\frac{Rd\theta}{2\pi}\hat{n}_\mu\epsilon_{\mu\nu}A_\nu^{[n]^{cl}}(z) \quad . \tag{12.25}$$

Now, from (12.23) we have

$$A_\nu^{[n]^{cl}}(z) \underset{|z|\to\infty}{\longrightarrow} -\frac{n}{e}\frac{z_\mu}{z^2} \quad . \tag{12.26}$$

Therefore $A_\mu^{[n]^{cl}}$ behaves at infinity like the vortex (4.77) and we find

$$\int \frac{d^2z}{4\pi}e\epsilon_{\mu\nu}F_{\mu\nu}^{[n]^{cl}}(z) = n \quad . \tag{12.27}$$

The field configuration $A_\mu^{[n]^{cl}}$ thus describes an "*induced* instanton"[408] with winding number n. By the same token one finds that A_μ^{cl} carries winding number zero. Our separation (12.22) thus anticipates the already known result concerning the infinite degeneracy of the ground state (see chapter 10).

We may simplify our calculation, by noting that

$$S^{ind}(\xi; x, y \mid A^{cl}) = \frac{e}{2}\int d^2z J_\mu^{ind}A_\mu^{cl}(z) \tag{12.28}$$

where we have used the equation of motion (12.17) for A^{cl}_μ. Hence, recalling (12.18) we have

$$S^{ind}(\xi;x,y \mid A^{cl}) = \frac{e}{2}\int d^2z \tilde{\partial}^z_\mu (2\overline{v}-v)(-\frac{2\pi}{e}\tilde{\partial}^z_\mu \sum_j \mathcal{D}(\xi_j - z) + A^{[n]^{cl}}_\mu (z)) \quad . \tag{12.29}$$

The first factor falls off at infinity like z^{-2} so that we may integrate by parts, to yield

$$\begin{aligned} S^{ind}(\xi;x,y \mid A^{cl}) &= -\pi \sum_{i,j}(2\mathcal{D}(\xi_j - \xi_i) - \mathcal{D}(\xi_j - x_i) - \mathcal{D}(\xi_j - y_i)) \\ &-\frac{e}{2}\int d^2z \sum_{i=1}^{n}(2D(\xi_i - z) - D(x_i - z) - D(y_i - z))\tilde{\partial}_\mu A^{[n]cl}_\mu (z) \quad . \end{aligned} \tag{12.30}$$

In the limit $\xi \to \infty, \{x_i, y_i\}$ and $\xi_i - \xi_j$ fixed,

$$S^{ind}(\xi;x,y \mid A^{cl}) \to f(\xi) + e\int d^2z \sum_i (D(x_i - z) + D(y_i - z))\tilde{\partial}_\mu A^{[n]^{cl}}_\mu (z) \quad , \tag{12.31a}$$

with

$$\begin{aligned} f(\xi) &= -2\pi \sum_{i,j}\mathcal{D}(\xi_i - \xi_j) - \frac{n^2}{2} ln\mu^2\xi^2 - \frac{n}{2} ln\mu^2\xi^2 \int \frac{d^2z}{4\pi} e\epsilon_{\mu\nu} F^{[n]^{cl}}_{\mu\nu}(z) \quad , \\ &= -2\pi \sum_{i,j}\mathcal{D}(\xi_i - \xi_j) - n^2 ln\mu^2\xi^2 \quad . \end{aligned} \tag{12.32}$$

Now, $\sum_{i,j}\mathcal{D}(\xi_i - \xi_j) = 2\sum_{i<j}\mathcal{D}(\xi_i - \xi_j) + n\mathcal{D}(0)$; recalling definition (12.18c), and using the short distance property (B.24), one has

$$\mathcal{D}(0) = -\frac{1}{2\pi}\left[\ln\left(\frac{2\mu\sqrt{\pi}}{e}\right) - \gamma\right] \quad . \tag{12.33}$$

Hence expression (12.32) takes the form

$$f(\xi) = -4\pi \sum_{i<j}\Delta(\xi_i - \xi_j) - 2n\pi\mathcal{D}(0) + 4\pi \sum_{i<j} D(\xi_i - \xi_j) - n^2 ln\mu^2\xi^2 \quad . \tag{12.34}$$

From (12.11) one furthermore obtains in the above limit,

$$W^{(0)}_n(\xi;x,y) \to \frac{1}{(\mu^2\xi^2)^{n^2}}\left(\frac{\mu}{2\pi}\right)^{2n} \prod_{j<k} \mu^2(\xi_j - \xi_k)^2 \mu(x_j - x_k)_+ \mu(y_j - y_k)_- \quad . \tag{12.35}$$

Putting everything together one finds that the correlation function factorizes as expected in the form (12.12) with

$$\langle 0 \mid \prod_{i=1}^{n} J_{-}(\xi_i) \mid n\rangle = \left(\frac{\mu}{2\pi}\right)^n e^{2n\pi\mathcal{D}(0)} e^{-4\pi\sum_{i<j}\Delta(\xi_i-\xi_j)} \tag{12.36}$$

and

$$\langle n \mid \psi_1(x_1)\cdots\psi_1(x_n)\overline{\psi}_1(y_1)\cdots\overline{\psi}_1(y_n) \mid 0\rangle =$$

$$\left(\frac{\mu}{2\pi}\right)^n \prod_{j<k} \mu(x_j - x_k)\mu(y_j - y_k) e^{\{\frac{e}{2}\int d^2z \sum_i^n (D(x_i-z)+D(y_i-z))\epsilon_{\mu\nu}F_{\mu\nu}^{[n]^{cl}}(z)\}} . \tag{12.37}$$

We have therefore rediscovered the well known degeneracy of the QED_2 vacuum, and the connection $\Delta n_{ch} = 2n$ between chirality and winding number predicted by the Atiyah-Singer index theorem (see chapter 4). In a similar way one can establish a direct link between the additional (infinite) vacuum degeneracy labelled by fermion number, peculiar to QED_2, and the existence of induced instantons with half integral winding number (merons).

12.5 Why Study Gauge Invariant Correlation Functions?

In the preceding two sections we have concentrated on gauge-invariant operators in order to obtain "tunneling amplitudes" using the violation of clustering. In this way, spurious gauge effects are avoided. A simple example demonstrates the importance of this. Thus consider for instance the gauge variant (Euclidean) fermion two-point function $\langle\psi_\alpha(x)\overline{\psi}_\beta(y)\rangle$. The corresponding path-integral is now saturated by the classical field configuration

$$A_\mu^{cl} = \frac{\pi}{e}\gamma^5_{\alpha\alpha}\tilde{\partial}^z_\mu(\mathcal{D}(x-z) - \mathcal{D}(y-z)) \quad .$$

Proceeding as before, one finds

$$\langle\psi_\alpha(x)\overline{\psi}_\beta(y)\rangle = e^{\pi(\mathcal{D}(0)-\mathcal{D}(x-y))} S_{\alpha\beta}(x-y) \tag{12.38a}$$

or recalling (12.18c), and (12.33),

$$\langle 0 \mid \psi_\alpha(x)\overline{\psi}_\beta(y) \mid 0\rangle = (\tilde{m} \mid x - y \mid)^{1/2} e^{\frac{1}{2}K_0(m|x-y|)} S_{\alpha\beta}(x-y) \quad , \tag{12.38b}$$

where $\tilde{m} = \frac{e^\gamma}{2}m$, $m \equiv e/\sqrt{\pi}$.

It is evident from (12.38a), that the two-point function exhibits a canonical short-distance behaviour. For $| x - y | \to \infty$ it however behaves like

$$\langle 0 \mid \psi_\alpha(x)\overline{\psi}_\beta(y) \mid 0\rangle \to (\tilde{m} \mid x - y \mid)^{1/2} S_{\alpha\beta}(x - y)$$

and thus appears to cluster. This is however a result of the "would be" Goldstone boson, which does not decouple from the gauge variant correlation functions, and hence cannot be taken as a signal for the absence of a vacuum degeneracy! In fact, in the light-cone gauge one finds that this two-point functions grows linearly with $| x - y |$ for large separations[4]. This emphasizes that physically meaningful conclusions can only be drawn from the study of gauge-invariant quantities.

12.6 Screening versus Confinement

The disorder operator $\mu_\Sigma(x)$ introduced in Eq. (10.81) is closely related to the Wilson-loop operator

$$\mu[C_W] = e^{ie\int_{C_W} dz^\mu A_\mu(z)} \quad , \tag{12.39}$$

where C_W denotes a closed loop. Indeed, in operator language we have, recalling (10.6)

$$\mu[C_W] =: e^{-i\sqrt{\pi}\int_{C_W} dz^\mu \epsilon_{\mu\nu}\partial^\nu \Sigma}: \quad .$$

In the case where C_W is replaced by an spin contour extending from x^1 to ∞ along the spacial axis, we recover the soliton operator (10.81).

Let us compute the vacuum expectation value of the operator $\mu[C_W]$ in the Euclidean Feynman path-integral formulation. It is convenient to write the exponential in (12.39) in the form

$$\int_{C_W} dz_\mu A_\mu(x) = \int d^2 z I_\mu(z) A_\mu(z) \quad . \tag{12.40a}$$

Here

$$\begin{aligned} I_\mu(z) &= \oint_{C_W} d\xi_\mu \delta^2(z - \xi) \\ &= \int \frac{d\xi_\mu(\tau)}{d\tau} \delta^2(z - \xi(\tau)) \quad , \end{aligned} \tag{12.40b}$$

[4] For a discussion of the Schwinger model in this gauge, see refs. [359,388]. The authors of ref. [388] also show how the summation of the perturbation series leads to the exact solution for the fermion two-point function in this gauge.

where τ parametrizes the point along the contour C_W. Recalling (12.2c) we thus have in the Lorentz gauge,

$$\langle 0|\mu[C_W]|0\rangle = \int \mathcal{D}A_\mu \delta[\partial_\nu A_\nu] e^{\int d^2z\{\frac{1}{2}A_\mu(\Delta - \frac{e^2}{\pi})A_\mu + ieI_\mu A_\mu\}} \quad . \tag{12.41}$$

The functional integral is saturated by the classical field configuration A_μ^{cl} satisfying

$$\left(\Delta - \frac{e^2}{\pi}\right) A_\mu^{cl}(z) = -ieI_\mu(z) \quad . \tag{12.42}$$

The solution of this equation is given by

$$A_\mu^{cl}(z) = ie \oint_{C_W} d\xi_\mu \Delta(z - \xi; \frac{e^2}{\pi}) \quad . \tag{12.43}$$

A_μ^{cl} can be viewed as the vector potential induced by the purely imaginary "Wilson" current in a background plasma of $q\bar{q}$ pairs. Since $K_0(m \mid z \mid)$ decays exponentially as $\sim e^{-m|z|}$, this potential is concentrated in a ring of radius $e/\sqrt{\pi}$ along the contour C_W. Substituting (12.43) into (12.41), one obtains

$$\langle 0 \mid \mu[C_W] \mid 0\rangle = e^{-\varepsilon[C_W]} \quad , \tag{12.44a}$$

where

$$\varepsilon[C_W] = \frac{e^2}{2} \oint dz_\mu \oint d\xi_\mu \Delta(z - \xi; \frac{e^2}{\pi}) \quad . \tag{12.44b}$$

In the "quenched" case of non-dynamical fermions the dynamical Higgs mechanism characteristic of the Schwinger model is absent. In this case $\Delta(z; \frac{e^2}{\pi})$ is to be replaced by $D(z)$, and we obtain for circular loop in the limit of large radius R the "area" law, $\varepsilon[C_W] \sim \pi R^2$, characteristic of a confining potential. In the case of dynamical fermions this behavior turns into a "perimeter" law, reflecting the screening of the test charges by the induced vacuum polarization. In order to demonstrate this, it is convenient to choose for C_W a rectangular loop extending from $-\frac{L}{2}$ to $\frac{L}{2}$ in the spacial direction, and from $-\frac{T}{2}$ to $\frac{T}{2}$ in the Euclidean time direction. It is then convenient to write the Wilson-current I_μ as the pseudo-gradient of a scalar function Φ as follows:

$$I_\mu(z) = \tilde{\partial}_\mu \Phi(z)$$

where

$$\Phi(z) = \begin{cases} 1, & \text{within } C_W \\ 0, & \text{outside } C_W \end{cases} \quad .$$

$\varepsilon[C_W]$ may then be written in the form

$$\varepsilon[C_W] = \frac{e^2}{2}LT - \frac{e^4}{2\pi}\int d^2x \int d^2y \Phi(x)\Delta(x-y;\frac{e^2}{\pi})\Phi(y) \quad . \tag{12.44c}$$

Going over the Fourier-space, one finds for the integral,

$$-\frac{e^4}{2\pi}\int d^2x \int d^2y \Phi(x)\Delta(x-y;\frac{e^2}{\pi})\Phi(y) =$$
$$-\frac{1}{2}e^2LT + \frac{\sqrt{\pi}e}{2}(1-e^{-\frac{e}{\sqrt{\pi}}L})T + \frac{e^4}{\pi}\int \frac{dp_1}{\pi}\frac{\sin^2 p_1\frac{L}{2}}{p_1^2}\frac{1}{\omega_1^3}(1-e^{-\omega_1 T}) \quad ,$$

where $\omega_1 = \sqrt{\frac{e^2}{\pi} + p_1^2}$. In the limit $T \to \infty$, L fixed, we thus obtain

$$\langle 0 \mid \mu[C_W] \mid 0\rangle \underset{T\to\infty}{\to} Ce^{-TV(L)} \quad , \tag{12.45}$$

where $V(L)$ is the "screened" potential (10.182), already obtained in chapter 10, and C is the constant

$$C = e^{-\left(\frac{e^2}{\pi}\right)^2 \int_{-\infty}^{\infty} dp_1 \frac{\sin^2 \frac{p_1 L}{2}}{p_1^2}\frac{1}{\omega_1^3}} \quad . \tag{12.46}$$

It is rewarding that this constant has the expected interpretation, as $|\langle\Omega[0]|\Omega[\Phi]\rangle|^2$, where $|\Omega[\Phi]\rangle$ ($|\Omega[0]\rangle$) is the ground state in the presence (absence) of the external source. Indeed, using the representation (10.101) for the ground state wave-functional, one finds, using completeness and doing the gaussian functional integration

$$\langle\Omega[0] \mid \Omega[\Phi]\rangle = \int \mathcal{D}\Sigma \langle\Omega[0] \mid \Sigma\rangle\langle\Sigma \mid \Omega[\Phi]\rangle$$
$$= e^{-\frac{1}{4}\left(\frac{e^2}{\pi}\right)^2 \int\int \Phi \frac{1}{(\sqrt{D})^3}\Phi} \quad ,$$

where D has been defined in (10.100). The evaluation of the integral in the exponential is best done in Fourier space. One obtains,

$$\langle\Omega[0] \mid \Omega[\Phi]\rangle = e^{-\frac{1}{2}\left(\frac{e^2}{\pi}\right)^2 \int \frac{dp_1}{\omega_1}\frac{\sin^2\frac{p_1 L}{2}}{p_1^2}} \quad . \tag{12.47}$$

This proves our claim.

The massless Schwinger model thus provides again a simple, exactly treatable example for demonstrating general principles commonly applied to four-dimensional gauge theories.

12.7 Quasi Periodic Boundary Conditions and the θ Vacuum

According to our discussion in chapter 10, result (12.37) is to be replaced by

$$\langle \theta \mid \psi_1(x_1) \cdots \psi_1(x_n) \overline{\psi}_1(y_1) \cdots \overline{\psi}_1(y_n) \mid \theta \rangle = \left(\frac{\mu}{2\pi}\right)^n \prod_{j<k} \mu(x_j - x_k)\mu(y_j - y_k) \times$$
$$\sum_n e^{in\theta} e^{\frac{e}{2}\int d^2z \sum_i (D(x_i - z) + D(y_i - z))\epsilon_{\mu\nu} F^{[n]}_{\mu\nu}(z)} , \qquad (12.48)$$

if we work in the irreducible representation of the θ-vacuum ($n_{ch} = 2n$). It is instructive to pause for a moment, in order to understand this result from a general point of view[410,411].

The contribution of gauge fixed configurations with non-trivial winding numbers to fermionic correlation functions, shows that the A_μ-manifold over which the functional integration is to be performed is not simply connected. In this connection we have seen the natural appearance of an arbitrary parameter θ. We wish to show next, that this can be understood as the requirement of quasi periodic boundary conditions on a multiply connected manifold.

Let us first consider a quantum mechanical analog of this statement[411,412]. Suppose that we live on a one dimensional compactified space, a circle. We could choose our coordinate q to be an angular variable in the interval $q \in [0, 2\pi]$. The requirement of quasi-periodic boundary condition

$$\psi(q + 2\pi) = e^{-i\theta}\psi(q) \qquad (12.49)$$

for the Schrödinger wave functions is perfectly consistent with the principles of quantum mechanics[413] and reflects an ambiguity in the realization of a canonical quantization of the theory, if the manifold is not simply connected.

Indeed, in the Schrödinger representation of Quantum Mechanics, the operator

$$p_i = -i\frac{\partial}{\partial q_i} + A_i(q) \qquad (12.50a)$$

with

$$\partial_i A_j - \partial_j A_i = 0 \qquad (12.50b)$$

provides a realization of the commutation relation $[q_i, p_j] = i\delta_{ij}$. The condition (12.50$b$) allows one to write A_i as the partial derivative of a scalar "potential" $\varphi(q)$, $A_i = \partial_i\varphi(q)$. If the space is simply connected, then by means of a gauge transformation

$$\psi(q) \to e^{-i\varphi(q)}\psi(q) = \psi'(q) \qquad (12.51)$$

one comes back to the usual representation of the momentum operator, $p_i = -i\frac{\partial}{\partial q_i}$, with $\psi'(q)$ single valued functions. On the other hand, if space is not simply connected, such a gauge transformation may lead to non single valued wave functions.

In the case of only one degree of freedom, the choice $A = \frac{\theta}{2\pi}$, with θ a constant "angle" taking values between zero and 2π, the transformation (12.51) leads to a new wave function $\psi'(q) = e^{-i\frac{\theta}{2\pi}q}\psi(q)$ satisfying the quasi periodic boundary condition (12.49). If the manifold is that of a circle where the points q and $q + 2n\pi$ are identified, then $\psi'(q)$ is now a multivalued function, in the case where $\psi(q)$ is single valued. Thus the transition in this case from the representation (12.50) to (12.51) involves the transition from a description in terms of single valued to multivalued wave functions, and parallels the transition from the temporal to the Coulomb gauge.

In the temporal gauge we write for the momentum conjugate to A_μ,

$$\pi_\mu(x) = -i\frac{\delta}{\delta A_\mu(x)} + \frac{\theta}{2\pi}\frac{\delta q[A]}{\delta A_\mu(x)} \tag{12.52}$$

where $q[A]$ is the angular variable defined by[410]

$$q[A] = 2\pi \int_{A_\mu=0}^{A_\mu^{[n]}(x)} dx Q\left(A(x)\right)$$

where $Q\left(A\right)$ is the Pontryagin density, and where the integration is performed along a path parametrized in time, connecting the field configuration $A_\mu = 0$ with a field configuration of winding number n. The integral is the same for all paths which can be continuously deformed into each other.

The momentum (12.52) follows in the standard way from the Lagrangian

$$\mathcal{L}[A_\mu, \dot{A}_\mu] = \mathcal{L}[A_\mu, \dot{A}_\mu] + \frac{\theta}{2\pi}\frac{d}{dt}q[A_\mu]\,. \tag{12.53}$$

The corresponding action

$$S[A] = S_0[A] + \frac{\theta}{2\pi}\int dx Q\left(A(x)\right)$$

will be recognized as being just the QCD action with θ describing the θ-sectors of the theory. The parameter θ can be viewed as measuring the flux through the "hole" in configuration space.

In order to explore this analogy still further, we return to quantum mechanics, and discuss the propagator of a free particle in one spatial dimension[411].

The propagation kernel

$$G(q,t) \equiv \langle q,t \mid 0,0\rangle \tag{12.54}$$

consistent with the quasi periodicity property (12.49) is constructed as usual in terms of the normalized eigenfunctions $\psi_n(q)$ of $-\frac{1}{2}\frac{d^2}{dq^2}$ satisfying the boundary condition (12.49)

$$\psi_m(q) = \frac{1}{\sqrt{2\pi}}e^{i(m-\frac{\theta}{2\pi})q} \tag{12.55}$$

with the corresponding eigenvalues $E_m = \frac{1}{2}(m - \frac{\theta}{2\pi})^2$. It is evidently given by

$$G(q,t) = \sum_{m=-\infty}^{\infty} \psi_m \psi_m^*(0) e^{-\frac{i}{2}(m-\frac{\theta}{2\pi})^2 t} \tag{12.56}$$

This expression has also the following representation in terms of a functional integral

$$G(q,t) = \sum_{n=-\infty}^{\infty} e^{in\theta} \int_{q(0)=0}^{q(t)=q+2n\pi} \mathcal{D}q e^{i\int_0^t dt' \frac{1}{2}\dot{q}^2(t)} \tag{12.57}$$

Indeed, performing the $\mathcal{D}q$ integration, we have

$$G(q,t) = \frac{1}{\sqrt{2\pi i t}} \sum_{n=-\infty}^{\infty} e^{in\theta} e^{\frac{i}{2t}(q+2n\pi)^2} \tag{12.58}$$

We can rewrite this as an expansion in terms of the eigenfunctions $\psi_m(q)$ as follows

$$G(q,t) = \sum_{m=-\infty}^{\infty} \mathcal{A}_m(t) \psi_m(t) \qquad 12.59a)$$

with $\mathcal{A}_m(t)$ calculated from (12.56) to be

$$\mathcal{A}_m(t) = \frac{1}{\sqrt{2\pi i t}} \int_{-\infty}^{\infty} \frac{dq}{\sqrt{2\pi}} e^{-i(m-\frac{\theta}{2\pi})q} e^{i\frac{q^2}{2t}} = e^{-\frac{i}{2t}(m-\frac{\theta}{2\pi}^2 t^2)} \quad , \tag{12.59b}$$

where we have used the summation over n to replace the integral in q over the interval $[0, 2\pi]$ by an integral from $-\infty$ to $+\infty$.

Combining (12.59a) and (12.59b), one reproduces (12.56). The functional integral (12.57) represents the sum over the infinite number of possible paths connecting the points $q(0)$ and $q(t)$ on a circle, paths covering the circle n times being weighted by a factor $e^{in\theta}$. This guarantees that the Greens function $G(q,t)$ satisfies also the periodicity condition

$$\langle q + 2\pi, t \mid 0, 0\rangle = e^{-i\theta} \langle q, t \mid 0, 0\rangle \quad . \tag{12.60}$$

In order to exhibits the close analogy with QED_2 we rewrite (12.60) as

$$\langle q + 2\pi, t \mid 0, 0\rangle = e^{-i\frac{\theta}{2\pi}q} \int_{q(0)=0}^{q(t)=q} \mathcal{D}[q'] e^{i\int_0^t dt' \mathcal{L}_{eff}(q', \dot{q})} \quad , \tag{12.61a}$$

where

$$\mathcal{L}_{eff}(q, \dot{q}') = \frac{1}{2}\dot{q}^2 + \frac{\theta}{2\pi}\dot{q} \quad , \tag{12.61b}$$

and where the integration is to be performed over all topological inequivalent configurations in q-space.

The situation is now seen to be analogous to that of QED_2 in the temporal gauge with the following correspondences:
i) the configuration space of $q(t)$ is replaced by that of the A_μ vector potential,
ii) the "topological inequivalent configurations in q correspond to *gauge inequivalent* configurations in A_μ, and
iii) the "topological" density $\frac{\dot{q}}{2\pi}$ is replaced by the Pontryagin density. Thus for QED_2 in the $A_0 = 0$ gauge, the propagator (12.57) is replaced by

$$\left\langle \vec{A}; t \mid 0; 0 \right\rangle = \sum_{n=-\infty}^{\infty} e^{in\theta} \int_0^{\vec{A}^{g_n}} \mathcal{D}A_\mu \delta(A_0) e^{i\int_0^t dt' \mathcal{L}_\theta} \tag{12.62a}$$

where

$$\mathcal{L}_\theta = \mathcal{L}_{QED_2} + \frac{\theta}{2\pi}\frac{d}{dt}q[A_\mu] \tag{12.62b}$$

and $\vec{A}^{g_n}$ is the gauge transform of $\vec{A}$ with winding number n. The periodicity property (12.49) now becomes a periodicity of the wave functional with respect to topologically non-trivial gauge transformations[269]

$$\psi_\theta[g_1 A] = T[g_1]\psi_\theta[A] = e^{-i\theta}\psi_\theta[A] \tag{12.63}$$

where $T[g_1]$ is the unitary operator (10.62) implementing gauge transformations. Hence, although the θ-dependent term in (12.62*b*) does not affect the equations of motion, it evidently plays an important role with respect to the quantum mechanical boundary conditions. In particular, for $\theta \neq 0$, one has a violation of parity and time-reversal.

12.8 Functional Representation of Tunneling Amplitudes[409]

Our study of specific correlation functions in the preceeding sections has shown that

i) The QED_2 vacuum is infinitely degenerate.

ii) The vacuum states are labelled by a chiral quantum number n_{ch}.

iii) With these vacua are associated "induced instantons". Instantons with winding number n mediate between ground states differing in their chirality by $\Delta n_{ch} = 2n$.

Let us return now to consider again the correlation function (12.8). Taking again the conventional "zero-winding" Feynman path integral as starting point, and using the clustering property (12.12) we now propose to obtain a functional representation for the "tunneling amplitude" $\langle n \mid \psi_1(x_1)\cdots\psi_1(x_n)\overline{\psi}_1(y_1)\cdots\overline{\psi}_1(y_n) \mid 0\rangle$, in the Lorentz gauge. This will provide for the case of QED_2 an answer for the generally unsolved problem of functionally integrating over different "principal bundles", which leaves a priori the relative normalization of "tunneling amplitudes" unspecified.

Instead of separating the (topologically trivial) configurations A_μ into a winding number $\nu = 0$ configuration A_μ^{cl} and the fluctuations around it, we now separate A_μ into a classical configuration $A_\mu^{[-n]^{cl}}(z)$ with winding number $\nu = -n$ and the fluctuations $A_\mu^{[n]}$ around it, which then necessarily carry winding number $\nu = n$:

$$A_\mu(z) = -\frac{2\pi}{e}\tilde{\partial}_\mu^z \sum_i \mathcal{D}(\xi_i - z) + A_\mu^{[n]}(z) \quad . \tag{12.64}$$

We shall work in the Lorentz gauge. It is useful to consider separately the contributions

$$S_G = -\frac{1}{2}\int d^2 z A_\mu(z)\Delta A_\mu(z) \tag{12.65a}$$

$$W[A] = -ln\ \det i\,\not{D} = \frac{e^2}{2\pi}\int d^2 z A_\mu(z)A_\mu(z) \tag{12.65b}$$

and

$$\phi(\xi; x, y \mid A) = -e\int d^2 z \sum_{i=1}^{n}(2D(\xi_i - z) - D(x_i - z) - D(y_i - z))\tilde{\partial}_\mu A_\mu(z) \tag{12.65c}$$

to the induced action

$$S^{ind} = S_G + W[A] + \phi(\xi; x, y \mid A) \quad . \tag{12.66}$$

Substituting (12.64) into (12.66), a careful, but straightforward analysis shows that in the limit $\{\xi\} \to \infty$, $\xi_i - \xi_j$ held fixed,

$$S_G[A] \underset{\xi\to\infty}{\longrightarrow} 2\frac{\pi^2}{e^2}\sum_j \int d^2 z \mathcal{D}(\xi_i - z)\Delta^2\mathcal{D}(\xi_j - z) + S_G[A^{[n]}] \tag{12.67}$$

and

$$\phi(\xi; x, y \mid A) \to -2n^2 ln\xi^2 - 4\pi\sum_{i,j}\mathcal{D}(\xi_i - \xi_j) + \phi(x, y \mid A^{[n]}) \quad , \tag{12.68}$$

where

$$\phi(x, y \mid A) = e\int d^2 z \sum_i (D(x_i - z) + D(y_i - z))\tilde{\partial}_\mu A_\mu(z) \quad . \tag{12.69}$$

As one would expect on the basis of the Atiyah-Singer index theorem, the corresponding result for the determinant requires a more careful analysis because of the existence of zero modes, which implies that $W[A^{[n]}]$ must diverge ($\det i\,\not{D}[A^{[n]}] = 0$)[5]. This is indeed the case because of the bad asymptotic behaviour of A_μ configurations with non-zero winding (see chapter 4). For the calculation of W the separation (12.64) is thus not a convenient one, and we return to the back ground field decomposition with $A_\mu^{cl}(z)$ given by (12.22). We shall be interested in extracting the asymptotic behavior in ξ. Because of the slow asymptotic decay of $A_\mu^{[n]cl}(z)$, the asymptotic result cannot be of the form

$$W \to \omega(\xi) + \frac{e^2}{2\pi}\int A_\mu^{[n]}(z)A_\mu^{[n]}(z) \quad . \tag{12.70}$$

Indeed, writing A_μ in (12.65*b*) as $A_\mu^{cl} + a_\mu$ with A_μ^{cl} given by (12.22), and using

$$\left(\int \frac{d^2z}{4\pi} e\epsilon_{\mu\nu}F_{\mu\nu}^{[n]}\right)\sum_i \ln \mu^2\xi_i^2 = n^2 \ln \mu^2 \xi^2 \quad , \tag{12.71}$$

one obtains from (12.65*b*) in the limit $\{\xi_i \to \infty, \xi_i - \xi_j = \text{const.}\}$, the result

$$W[A] \to n^2 \ln \mu^2\xi^2 - 2\pi \sum_{i,j}\int d^2z\mathcal{D}(\xi_i - z)\Delta\mathcal{D}(\xi_j - z) + W'[A^{[n]}] \quad , \tag{12.72}$$

with

$$W'[A^{[n]}] = \frac{e^2}{2\pi}\int d^2z\left\{A_\mu^{[n]}(z)A_\mu^{[n]}(z) - \left(\frac{2n\pi}{e}\right)^2 \partial_\mu(\hat{D}\partial_\mu\hat{D})\right\} \quad , \tag{12.73}$$

where $\hat{D}(z) \equiv -\frac{1}{4\pi}\ln(\mu^2z^2 + 1)$, and where we have identified $A_\mu^{[n]^{cl}} + a_\mu$, with an arbitrary configuration $A_\mu^{[n]}$ of winding number n.

Combining all terms and using (12.19), we finally obtain for (12.9) in the "clustering limit"

$$\begin{aligned} W_n(\xi; x, y) \to& \left(\frac{\mu}{2\pi}\right)^{2n} e^{2n\pi\mathcal{D}(0)} e^{-4\pi\sum_{i<j}\Delta(\xi_i - \xi_j)} \\ &\times \prod_{j<k}\mu(x_j - x_k)\mu(y_j - y_k)\int[\mathcal{D}A_\mu]_n\delta(\partial_\mu A_\mu)e^{-(S_G[A] + W'[A] + \phi(x,y|A))} \quad , \end{aligned} \tag{12.74}$$

where the integration only extends over configurations with winding number n. Note that unlike $W[A^{[n]}]$, the new effective action $W'[A^{[n]}]$ is now finite. As the

---

[5] The existence and importance of zero modes has first been emphasized by G.'t Hooft[264].

"prime" already indicates, it is related to $-\ln\det'[i\not{D}]$ in the notation of chapter 4.

Comparing with (12.37) we have thus arrived at the following functional representation for the tunneling amplitude in the Lorentz gauge,

$$\langle n \mid \psi_1(x_1)\cdots\psi_1(x_n)\overline{\psi}_1(y_1)\cdots\overline{\psi}_1(y_n)|0\rangle =$$
$$\left(\frac{\mu}{2\pi}\right)^n \int [\mathcal{D}A_\mu]_n \delta(\partial_\mu A_\mu)\, e^{-S_G[A]-W'[A]-\phi(x,y|A)} \prod_{j<k} \mu(x_j-x_k)\mu(y_j-y_k)\,. \quad (12.75)$$

This result applies to the "minimal" case, where the number of fermion fields is equal to the chirality ($2n$) of the correlation function in question. It can be shown[409] to generalize as follows:

$$\langle n \mid \psi_1(x_1)\cdots\psi_1(x_{n+l})\psi_2(x'_1)\cdots\psi_2(x'_k)\overline{\psi}_1(y_1)\cdots\overline{\psi}_1(y_{n+k})\overline{\psi}_2(y'_1)\cdots\overline{\psi}_2(y'_l) \mid 0\rangle =$$
$$N\left(\frac{\mu}{2\pi}\right)^{n+l+k} \int [\mathcal{D}A_\mu]_n e^{-S_G[A]-W'[A]-\phi(x,y;x',y'|A)} \times$$
$$\frac{\prod_{i<j}\mu(x_i-x_j)_+ \prod_{i<j}\mu(y'_i-y'_j)_+ \prod_{i<j}\mu(y_i-y_j)_- \prod_{i<j}\mu(x'_i-x'_j)_-}{\prod_{i,j}\mu(y'_i-x_j)_+ \prod_{i,j}\mu(x'_i-y_j)_-}\,, \quad (12.76)$$

with

$$\phi(x,y;x',y' \mid A) = \phi(x,y \mid A) - \phi(x',y' \mid A) \quad . \quad (12.77)$$

The case of negative winding number is obtained by the replacement $1 \leftrightarrow 2$, and $z_+ \leftrightarrow z_-$.

It is instructive to specialize (12.76) to the case $n = \pm 1$. One then has

$$\langle \pm 1 \mid \psi_{\alpha_1}(x_1)\cdots\psi_{\alpha_l}(x_l)\overline{\psi}_{\beta_1}(y_1)\cdots\overline{\psi}_{\beta_l}(y_l) \mid 0\rangle = N\left(\frac{\mu}{2\pi}\right) \int [\mathcal{D}A_\mu]_{\pm 1}\delta(\partial_\mu A_\mu) \times$$
$$e^{-S_G[A]-W'[A]-\phi(x,y|A)} \sum_P (-1)^P \left(\frac{1\pm\gamma_5}{2}\right)_{\alpha_k\beta_{k'}} \prod_{\substack{i\neq k\\ j_i\neq k'}} iS(x_i-y_{j_i})_{\alpha_i\beta_{j_i}} \quad (12.78)$$

Hence the Feynman rules for zero winding configurations are modified by replacing one of the propagators by the projector $P_\pm$ on positive(negative) chirality states and summing over all possible permutations. This agrees with general expectations.

12.9 Interpretation of the Result[141,414]

We now rederive the above results following a more tractable procedure. Consider the generating functional for fermionic correlation functions:

$$Z^{[n]}[\eta,\overline{\eta}] = \int d\mu[A]_n \mathcal{D}\overline{\psi}\mathcal{D}\psi e^{-\int d^2z(\frac{1}{4}F^2_{\mu\nu}+\overline{\psi}i\not{D}[A]\psi)} e^{\int d^2z(\overline{\eta}\psi+\overline{\psi}\eta)} \quad , \tag{12.79}$$

where η and $\overline{\eta}$ are Grassmann valued c-numbers sources, and where $d\mu[A]_n$ is the Faddeev-Popov measure for the A_μ integration extending over configurations with winding number n. We expand ψ in the complete set of orthogonal eigenfunctions (4.67) of the *associated* Dirac operator, $\psi(x) = \sum b_i u_i(x)$. Defining the functionals (we now write $\overline{u}_i = u_i^\dagger$ for convenience),

$$u_i[\overline{\eta}] \equiv \int d^2z\overline{\eta}(z)u_i(z) \quad \text{and} \quad \overline{u}_i[\eta] \equiv \int d^2z\overline{u}_i(z)\eta(z) \quad , \tag{12.80}$$

one has

$$Z^{(n)}[\eta,\overline{\eta}] =$$

$$\int d\mu[A]_n \prod_i{}' d\overline{b}_i db_i e^{-S_G} e^{-\sum_i{}' \lambda_i\overline{b}_i b_i} e^{\sum_i(\overline{b}_i u_i[\overline{\eta}]+\overline{u}_i[\eta]b_i)} e^{\sum_k(\overline{b}^{(0)}_k u^{(0)}_k[\overline{\eta}]+\overline{u}^{(0)}_k[\eta]b^{(0)}_k)} \tag{12.81}$$

where we have made the change $\mathcal{D}\overline{\psi}\mathcal{D}\psi \to \prod_i d\overline{b}_i db_i$ in the integration measure. The Jacobian associated with this change of variables has been taken to be one, since the expansion has been done with respect to an orthonormal basis. In (12.81) the superscript "zero" refers to the zero modes and the "prime" generally indicates omission of the zero mode contributions. Indeed we know already from the Atiyah-Singer index (plus vanishing) theorem (see chapter 4) that there exist precisely $|n|$ such zero modes of positive ($n > 0$) or negative ($n < 0$) chirality.

Integration, using quadratic completion in the exponent, yields

$$Z^{(n)}[\eta,\overline{\eta}] = \int d\mu[A]_n e^{-S_G}\Big(\prod_i{}' \lambda_i[A]\Big) e^{-\sum_i{}' \frac{u_i[\overline{\eta}]\overline{u}_i[\eta]}{\lambda_i}} P_n[A;\eta,\overline{\eta}] \quad , \tag{12.82}$$

where

$$P_n[A;\eta,\overline{\eta}] = \int \prod_k d\overline{b}^{(0)}_k db^{(0)}_k \prod_i(1+\overline{b}^{(0)}_i u^{(0)}_i[\overline{\eta}]) \prod_j(1+\overline{u}^{(0)}_j[\eta]b^{(0)}_j) \quad . \tag{12.83}$$

Performing the integration in (12.82) over the zero-modes, using the Berezin rules (C.15,16) one obtains

$$P_n[A;\eta,\overline{\eta}] = \frac{1}{(n!)^2}\Big(\sum_i \epsilon_{i_1\cdots i_n} u^{(0)}_{i_1}[\overline{\eta}]\cdots u^{(0)}_{i_n}[\overline{\eta}]\Big)\Big(\sum_j \epsilon_{j_1\cdots j_n}\overline{u}^{(0)}_{j_1}[\eta]\cdots \overline{u}^{(0)}_{j_n}[\eta]\Big) \quad . \tag{12.84}$$

The exponential factor in (12.82) is just $\exp\{-\int d^2z d^2z'\overline{\eta}(z)G'(z,z' \mid A)\eta(z)\}$, where G' is the Greens function with the spectral representation

$$G'(x,y \mid A) = {\sum}' \frac{u_n(x)\overline{u}_n(y)}{\lambda_n} \quad . \tag{12.85}$$

It satisfies

$$i\not{D}G'(x,y \mid A) = \delta^2(x-y) - P^{(0)}(x,y) \tag{12.86}$$

with $P^{(0)}(x,y)$ the projector on the zero modes (see chapter 4).

As we seek an interpretation of our result (12.75) we shall restrict our attention to that correlation function, and consider

$$\frac{\delta}{\delta\overline{\eta}_{\alpha_1}(x_1)}\cdots\frac{\delta}{\delta\overline{\eta}_{\alpha_n}(x_n)}\frac{\delta}{\delta\eta_{\beta_1}(y_1)}\cdots\frac{\delta}{\delta\eta_{\beta_n}(y_n)}Z^{[n]}[\eta,\overline{\eta}]\Big|_{\eta,\overline{\eta}=0}$$
$$= \int d\mu[A]_n (Det' i\not{D}[A]) e^{-S_G[A]} P_n(x,y;A)_{\alpha_1\cdots\beta_n} \quad , \tag{12.87}$$

where

$$P_n(x,y;A)_{\alpha_1\cdots\beta_n} =$$
$$\sum_{\{i\}} \epsilon_{i_1\cdots i_n} u^{(0)}_{i_1}(x_1)_{\alpha_1}\cdots u^{(0)}_{i_n}(x_n)_{\alpha_n} \sum_{\{j\}}\epsilon_{j_1\cdots j_n}\overline{u}^{(0)}_{j_1}(y_1)_{\beta_1}\cdots\overline{u}^{(0)}_{j_n}(y_n)_{\beta_n} \tag{12.88}$$

is just the result of performing the functional differentiation of (12.84) with respect to the sources.

Notice that we have restricted ourselves to the simplest, non-vanishing tunneling amplitude $\mid 0\rangle \rightarrow \mid n\rangle$ involving "induced instanton" configurations of winding number n . For a more general amplitude, the Greens function G' will also contribute via the functional derivatives of the exponential in (12.82).

We shall now examine in more detail the structure of P_n above.

12.9.1 Zero modes[141,414]

The zero modes of the $d = 2$ Dirac equation can be classified according to their chirality. Their number is precisely $|n|$. They are given by (see chapter 4)

$$\psi_{l+}^{(0)}(x) = \left(\frac{1}{2\pi R}\right)^{\frac{1}{2}} \left(\frac{x_1 + ix_2}{R}\right)^{l-1} \begin{pmatrix} h_- \\ 0 \end{pmatrix} \quad , \quad n > 0 \tag{12.89}$$

$$\psi_{l-}^{(0)}(x) = \left(\frac{1}{2\pi R}\right)^{\frac{1}{2}} \left(\frac{x_1 - ix_2}{R}\right)^{l-1} \begin{pmatrix} 0 \\ h_+ \end{pmatrix} \quad , \quad n < 0 \quad , \tag{12.90}$$

where $h_\pm$ is a solution to (11.100) and has the expansion (11.105). Normalizability[6] with respect to the measure $d^2x\Omega_R$ restricts l to the range $1 < l \leq |n|$. For reasons of "economy" we have used the parameter R already appearing in (4.67) in order to give ψ the dimensions of $\frac{1}{2}$ (in units of mass). In the abelian case (see (11.108))

$$h = \begin{pmatrix} h_- & 0 \\ 0 & h_+ \end{pmatrix} = e^{-ie\int d^2z D(x-z)[\partial_\mu A_\mu(z) - i\gamma^5 \tilde{\partial}_\mu A_\mu(z)]} \quad . \tag{12.91}$$

More compactly

$$\psi_{l\pm}^{(0)}(x) = \left(\frac{1}{2\pi R}\right)^{\frac{1}{2}} \left(\frac{x_\pm}{R}\right)^{l-1} h(x)\chi_\pm \quad , \tag{12.92}$$

with

$$\chi_+ = \begin{pmatrix} 1 \\ 0 \end{pmatrix} \quad , \quad \chi_- = \begin{pmatrix} 0 \\ 1 \end{pmatrix} \quad . \tag{12.93}$$

For A_μ the vortex field of the form (4.76), we recover from (12.92) the expression (4.222).

For a configuration A_μ with winding $n \neq 0$ we have, in the Lorentz gauge, using (12.27)

$$\psi_{l\pm}^{(0)}(x) \to \left(\frac{1}{2\pi R}\right)^{\frac{1}{2}} \left(\frac{1}{\mu R}\right)^{l-1} e^{\pm i(l-1)\theta} (\mu|x|)^{l\mp n-1} \quad , \quad (n \neq 0) \tag{12.94}$$

with

$$\theta = \arctan\frac{x_2}{x_1} \quad . \tag{12.95}$$

Note that the asymptotic form of the zero modes depends on the Pontryagin class to which the configuration A_μ belongs.

The zero modes (12.92) are not normalized. In order to establish the connection with the $u_i^{(0)}$, we define

$$\langle \psi_{l'_\pm}, \psi_{l_\pm} \rangle = (N_\pm^{-1})_{l'l} \quad . \tag{12.96}$$

---

[6] Note that the norm of these zero modes with respect to the measure d^2x is actually logarithmically divergent. A similar situation prevails in QCD_4[144].

Note that zero modes of opposite chirality are orthogonal with respect to each other. Hence we define the matrix

$$N[A] = \begin{pmatrix} N_+[A] & 0 \\ 0 & N_-[A] \end{pmatrix} \quad . \tag{12.97}$$

This matrix is a functional of A ! Expand $u^{(0)}_{i\pm}$ in terms of the above zero modes:

$$u^{(0)}_{i\pm}(x) = \sum_l a^{\pm}_{il}\psi^{(0)}_{l\pm}(x) \quad . \tag{12.98}$$

Define

$$a = \begin{pmatrix} a^+ & 0 \\ 0 & a^- \end{pmatrix} \quad . \tag{12.99}$$

From $(u^{(0)}_i, u^{(0)}_j) = \delta_{ij}$ then follows $aN^{-1}a^\dagger = 1$, or

$$\det a \det a^\dagger = \det N[A] \quad . \tag{12.100}$$

Noting that $\sum_{\{i_l\}} \epsilon_{i_1\cdots i_n} a^{\pm}_{i_1 l_1} \cdots a^{\pm}_{i_n l_n} = \det a^{\pm}\epsilon_{l_1\cdots l_n}$, we see that P_n in (12.88) can be rewritten in the form

$$P_n[x,y;A] = \det N_+[A]\overline{P}_n[x,y;A] \quad , \quad n \neq 0 \quad , \tag{12.101}$$

with

$$\overline{P}_n(x,y;A) = \frac{1}{(n!)^2}\Big(\sum_l \epsilon_{l_1\cdots l_n}\psi^{(0)}_{l_{1+}}(x_1)\cdots\psi^{(0)}_{l_{n+}}(x_n)\Big)\Big(\sum_l \epsilon_{l_1\cdots l_n}\overline{\psi}^{(0)}_{l_{1+}}(y_1)\cdots\overline{\psi}^{(0)}_{l_{n+}}(y_n)\Big) . \tag{12.102}$$

Hence we have,

$$\overline{P}_n(x,y;A) = \Big(\frac{1}{2\pi R}\Big)^n \mathcal{L}_n(x,y)_+ e^{-\phi(x,y|A)} \quad , \tag{12.103}$$

where $\phi(x,y \mid A)$ has been defined in (12.69), and

$$\mathcal{L}_n(x,y)_+ = \sum_l \epsilon_{l_1\cdots l_n}\Big(\frac{x_{1+}}{R}\Big)^{l_1-1}\cdots\Big(\frac{x_{n+}}{R}\Big)^{l_n-1}\sum_l{}' \epsilon_{l'_1\cdots l'_n}\Big(\frac{y_{1-}}{R}\Big)^{l'_1-1}\cdots\Big(\frac{y_{n-}}{R}\Big)^{l'_n-1} . \tag{12.104}$$

Putting things together, expression (12.87) thus reduces to

$$\int d\mu[A]_n(\det i\not{D}[A])\det N_+[A]\mathcal{L}_n(x,y)_+ e^{-S_G[A]-\phi(x,y|A)} \quad . \tag{12.105}$$

Note that the parameter $\frac{1}{R}$ plays the role of IR regulator mass, with the integration measure $d^2x\Omega_R(x)$ tending to the flat-space measure $2d^2x$ in the IR limit $R \to \infty$. Finally, using

$$\sum_l \epsilon_{l_1 \cdots l_n} x_1^{l_1-1} \cdots x_n^{l_n-1} = \prod_{j<k} (x_k - x_j) \quad , \tag{12.106}$$

we have

$$\mathcal{L}_n(x,y)_+ = \prod_{j<k} \left(\frac{x_{k_+} - x_{j_+}}{R}\right)\left(\frac{y_{k_-} - y_{j_-}}{R}\right) \quad . \tag{12.107}$$

Notice that $\mathcal{L}_n(x,y)$ is the product of eigenfunctions of the angular momentum operator

$$l_{\mu\nu} = -i(x_\mu \partial_\nu - x_\nu \partial_\mu) \quad . \tag{12.108a}$$

The only non-vanishing element of $l_{\mu\nu}$ is $l_{12} = l_3$,

$$l_3 = -i(x_1\partial_2 - x_2\partial_1) = \frac{1}{2}(x_+\partial_- - x_-\partial_+) \quad , \tag{12.108b}$$

with the property $l_3 x_\pm^l = \pm l x_\pm^l$. The different zero modes thus differ with respect to their angular momentum.

With the identification of $\mathcal{L}_n(x,y)$ in (12.105) with the expression (12.107), we recognize that the results (12.105) and (12.75) agree, provided that

$$\ln(\det\, i\not{D}[A] \det N[A]) = -W'[A] \quad , \tag{12.109}$$

and the replacement $\frac{1}{R} \to \mu$ is made. We now show, that this indeed holds true.

12.9.2 Calculation of det $i\not{D}$ from the anomaly equation[414]

Consider the anomaly equation (4.215). In terms of $W'[A]$ it reads

$$\tilde{\partial}_\mu \frac{\delta W'}{\delta A_\mu(x)} = -\frac{e^2}{2\pi}\epsilon_{\mu\nu}F_{\mu\nu}(x) + 2\Omega_R \mathrm{tr}\gamma_5 P_0(x,x) \quad . \tag{12.110}$$

Now,

$$\mathrm{tr}\gamma_5 P_0(x,x) = \sum u_i^{(0)\dagger}(x)\gamma_5 u_i^{(0)}(x) = \mathrm{tr}(N\rho(x)) \quad , \tag{12.111}$$

where

$$\rho = \begin{bmatrix} \rho_+ & 0 \\ 0 & -\rho_- \end{bmatrix} \tag{12.112}$$

with

$$\rho(x)_\pm = \psi_l^{(0)}(x)_\pm \psi_{l'}^{(0)\dagger}(x)_\pm \quad , \tag{12.113}$$

Integration of (12.110) yields,

$$\frac{\delta W'}{\delta A_\mu(x)} = \frac{e^2}{2\pi}\int d^2y \tilde{\partial}_\mu^x D(x-y)\epsilon_{\mu\nu}F_{\mu\nu}(y)$$

$$-2e\int d^2y \Omega_R(y)\tilde{\partial}_\mu^x D(x-y)\mathrm{tr}(N\rho(y)) + \partial_\mu \mathcal{F}(x) \tag{12.114}$$

Current conservation requires $\mathcal{F}$ to be a constant . We have the following *Proposition:*

$$2e\int d^2y \Omega_R(y)\tilde{\partial}_\mu D(x-y)\,\mathrm{tr}(N\rho(y)) = -\frac{\delta}{\delta A_\mu(x)}\mathrm{tr}(lnN) \quad . \tag{12.115}$$

Proof:

From the definition (12.96) of N it follows,

$$\frac{\delta(N_\pm^{-1})}{\delta A_\mu(x)} = \pm 2i\int d^2y \Omega_R(y)\rho_\pm(y)\frac{\delta\tilde{\varphi}(y;A)}{\delta A_\mu(x)} \quad , \tag{12.116}$$

where $\tilde{\varphi}$ is given by (12.3d). Now

$$\frac{\delta\tilde{\varphi}(y;A)}{\delta A_\mu(x)} = \mp e\tilde{\partial}_\mu D(x-y) \quad , \tag{12.117}$$

or

$$\mathrm{tr}\Big(\frac{\delta N}{\delta A_\mu(x)}N^{-1}\Big) = -2e\int d^2y \Omega_R(y)\tilde{\partial}_\mu D(x-y)\mathrm{tr}[N\rho(y)] \quad . \tag{12.118}$$

Now, from the definition (12.96) we see that $N^\dagger = N$. Hence we may diagonalize N by a unitary transformation, so that we may write

$$\mathrm{tr}\Big(\frac{\delta N}{\delta A_\mu(x)}N^{-1}\Big) = \frac{\delta}{\delta A_\mu(x)}\mathrm{tr}\ln N \quad . \tag{12.119}$$

This, together with (12.118), concludes the proof of proposition (12.115).

Returning to (12.114) we thus have

$$\frac{\delta W'}{\delta A_\mu(x)} = \frac{e^2}{2\pi}\int d^2y \tilde{\partial}_\mu^x D(x-y)\epsilon_{\mu\nu}F_{\mu\nu}(y) + \frac{\delta}{\delta A_\mu(x)}\mathrm{tr}\ln N \quad . \tag{12.120}$$

Integration yields

$$W'[A] = \frac{e^2}{8\pi}\int d^2x \int d^2y \epsilon_{\mu\nu}F_{\mu\nu}(x)D(x-x')\epsilon_{\nu\rho}F_{\nu\rho}(y) + \mathrm{tr}\ln N[A] + c_n \quad , \tag{12.121}$$

Here c_n is an integration constant whose value depends at most of the winding number of the field configuration $A_\mu^{[n]}$. We may thus evaluate it by considering a specific field configuration, such as the vortex (4.76), for example. This demonstrates, that Eq. (12.109) is indeed satisfied!

It is remarkable that we have been able to calculate the QED_2 fermionic determinant for a *generic* field configuration A^μ, also in the case of a non-trivial topology! This appears even more remarkable if one tries to calculate this determinant for a specific field configuration, such as the Euclidean vortex configuration (4.76) (we have set the scale parameter $R = 1$), $V_\mu(x) = -\frac{n}{e}\epsilon_{\mu\lambda}\frac{x_\lambda}{1+x^2}$, by departing from the definition (4.14*b*) of the determinant in terms of the product of its eigenvalues. Let us examine at first, what the formula (12.121) reads in this case.

Replacing A_μ by (4.76) in the expression (12.91) for $h(x)$ one finds after a straightforward calculation

$$h_\pm(x) = e^{\mp\frac{n}{2}\ln[\mu^2(1+x^2)]} \quad , \tag{12.122}$$

which, when substituted into (12.92) yields the result already quoted in (4.222). For $l = 1, \cdots n$, these zero-modes are normalizable with respect to the scalar product (4.69*a*). In terms of them, $N_\pm^{-1}$ is calculated to be

$$(N_\pm^{-1})_{l'l} = \delta_{ll'}N_l^{-1} \quad , \tag{12.123}$$

with

$$N_l^{-1} = \mu^{2l-2|n|-1}\frac{\Gamma(l)\Gamma(|n|-l+1)}{\Gamma(n+1)} \quad . \tag{12.124}$$

Since $N_\pm$ is diagonal we have

$$\mathrm{tr}(\ln N_\pm) = -\sum_l^{|n|}\ln N_l^{-1} \quad , \tag{12.125}$$

where the sum only extends up to $l = |n|$, the number of normalizable zero modes. One finds after a straightforward calculation[414]

$$\mathrm{tr}(\ln N[V]) = -\sum_{l=1}(|n|-2l)\ln l - n^2 ln\,\mu \quad , \tag{12.126}$$

where we have made use of the fact, that $N_{\mp}[V] = 0 \quad for \quad n \gtrless 0$. The evaluation of the first term in (12.121) is also straightforward, and gives

$$W[V] = -n^2 \ln \mu - \frac{n^2}{2} \quad .$$

Combining all terms, we finally obtain,

$$W'[V] = -\frac{n^2}{2} - \sum_{l-1}^{|n|}(|n| - 2l) \ln l + c_n \quad . \tag{12.127}$$

The value of c_n can now be determined by calculating $W'[V]$ directly from its definition (4.14*b*) in terms of the eigenvalue spectrum of the associated Dirac operator, and comparing with (12.127).

12.10 Eigenvalue Spectrum of the Associated Dirac Operator[414]

The determinant of a differential operator is only defined on a compact manifold, where the spectrum of the operator is discrete. Ways of achieving this in the case of interest, is to work either with the associated Dirac equation (4.67), or in a finite "box" with periodic boundary conditions. In both cases one will take the appropriate limit in order to describe the continuum situation in R_2. In this section we shall follow the first prescription, while the latter procedure will be discussed in the following section.

The functional $W'[A]$ is the logarithm of the determinant of the Dirac operator $i \not{D}$ normalized with respect to $i \not{\partial}$. We therefore consider first the eigenvalue problem for the non-interacting case:

$$i \not{\partial} \psi(x) = \frac{2\lambda}{1 + x^2} \psi(x) \quad . \tag{12.128}$$

Define the Euclidean light-cone variables

$$u = x_1 - i x_2 \quad , \quad v = x_1 + i x_2 \quad , \tag{12.129a}$$

and

$$\partial_v =: \frac{1}{2}(\partial_1 - i\partial_2) \quad , \quad \partial_u =: \frac{1}{2}(\partial_1 + i\partial_2) \quad . \tag{12.129b}$$

Writing

$$\psi = \begin{pmatrix} \phi \\ \chi \end{pmatrix} \quad , \tag{12.130}$$

the eigenvalue equation (12.128) takes the form

$$\partial_v \chi = -\frac{\lambda}{1+uv}\phi \tag{12.131a}$$

$$\partial_u \phi = \frac{\lambda}{1+uv}\chi \quad . \tag{12.131b}$$

These equations imply

$$H\phi = \lambda^2 \phi \quad , \tag{12.132a}$$

with

$$H = -(1+uv)\partial_v(1+uv)\partial_u \quad . \tag{12.132b}$$

The solution of this eigenvalue problem can be reduced to the solution of a standard problem: define the operators

$$\begin{aligned} L_+^{(\kappa)} &= -i(v^2\partial_v + \partial_u + \kappa v) \\ L_-^{(\kappa)} &= -i(u^2\partial_u + \partial_v - \kappa u) \\ L_3^{(\kappa)} &= v\partial_v - u\partial_u + \kappa \quad . \end{aligned} \tag{12.133}$$

They satisfy the commutator algebra of angular momentum operators:

$$[L_+^{(\kappa)}, L_-^{(\kappa)}] = 2L_3^{(\kappa)} \tag{12.134a}$$

$$[L_3^{(\kappa)}, L_\pm^{(\kappa)}] = L_\pm^{(\kappa)} \quad . \tag{12.134b}$$

We have

$$(\vec{L}^{(\kappa)})^2 = -(1+uv)^2\partial_u\partial_v + \kappa(1+uv)L_3^{(\kappa)} \quad . \tag{12.135}$$

Of particular interest are the operators for $\kappa = 0$ and $\kappa = \frac{1}{2}$: Setting

$$L_i \equiv L_i^{(0)} \quad , \quad J_i \equiv L_i^{(\frac{1}{2})} \tag{12.136}$$

we have, according to Eq. (12.135)

$$\vec{L}^2 = -(1+uv)^2\partial_u\partial_v$$

$$\vec{J}^2 = -(1+uv)^2\partial_u\partial_v + \frac{1}{2}(1+uv)J_3 \quad . \tag{12.137}$$

We now make the substitution

$$\phi = (1+uv)^{-1/2}\phi' \quad \text{and} \quad H = (1+uv)^{-1/2}H' \tag{12.138}$$

in Eq. (12.132). This equation then takes the form

$$H'\phi' = \lambda^2\phi' \quad , \tag{12.139}$$

with

$$H' = (\vec{J}^2 + \frac{1}{4}) \tag{12.140}$$

It follows as usual from the commutation relations (12.134) that the eigenvalues λ^2 of H' are of the form

$$\lambda^2 = j(j+1) + \frac{1}{4} \quad , \tag{12.141}$$

with j an integer or half-integer, and that the degree of degeneracy is $d_j = (2j+1)$. The corresponding eigenfunctions can be chosen to be simultaneous eigenfunctions of $\vec{J}^2$ and J_3 with eigenvalues labelled by j and m. They are of the form

$$\phi_{j,m} = u^m f_{j,m}(uv) = r^m e^{-im\theta} f_{j,m}(r^2) \quad , \tag{12.142}$$

with $u = re^{-i\varphi}$, and satisfy

$$J_3\phi_{j,m} = (m + \frac{1}{2})\phi_{j,m} \quad \text{and} \quad \vec{J}^2\phi_{j,m} = j(j+1)\phi_{j,m} \quad , \tag{12.143}$$

where $m = -j, -j+1, \cdots, j$.

We now examine the eigenvalue spectrum of the associated Dirac operator in the vortex field V_μ. The eigenvalue problem (12.131) is then replaced by

$$\left(\partial_v - \frac{n}{2}\frac{u}{1+uv}\right)\chi = -\frac{\lambda}{1+uv}\phi \tag{12.144a}$$

$$\left(\partial_u + \frac{n}{2}\frac{v}{1+uv}\right)\phi = \frac{\lambda}{1+uv}\chi \tag{12.144b}$$

Making the change of variable

$$\begin{pmatrix}\phi \\ \chi\end{pmatrix} = \begin{pmatrix}(1+uv)^{-\frac{n}{2}}\tilde{\phi} \\ (1+uv)^{\frac{n}{2}}\tilde{\chi}\end{pmatrix} \quad , \tag{12.145}$$

and substituting (12.144b) into (12.144a), one obtains[414]

$$\left(\vec{L}^2 - (1-n)(1+uv)u\partial_u\right)\tilde{\phi} = \lambda^2\tilde{\phi} \quad , \tag{12.146}$$

where $\vec{L}^2 = \left(\vec{L}^{(0)}\right)^2$ defined in (12.136). Hence we have for the eigenvalue spectrum of the $n = 1$ vortex, the $2l + 1$-fold degenerated eigenvalues[7)]

$$\lambda_l^2 = l(l+1) \quad , \quad l = 0, 1, \cdots \tag{12.147}$$

with eigenvalues $\tilde{\phi}_{l,m}$, where m is the magnetic quantum number labelling as usual the degeneracy.

The eigenfunctions corresponding to the highest weight is

$$\tilde{\phi}_{l,l} \sim u^l (1 + uv)^{-l} \quad . \tag{12.148}$$

From here we obtain the other eigenfunctions by application of the lowering operator L_-:

$$\tilde{\phi}_{l,m} \sim L_-^{l-m} u^l (1 + uv)^{-l} \quad , \quad m = -l, -l+1, \cdots l \quad . \tag{12.149}$$

We observe that all of the eigenfunctions $\phi_{l,m}$ are normalizable with respect to the scalar product (4.69a). This applies in particular to the zero mode eigenfunction ($l = 0$), which is seen to be of the form (4.222), with $l_\pm = n = 1$. For the $n = 1$ vortex there exist precisely one zero mode, as expected.

The solution for the eigenvalue spectrum, as well as the calculation of $W'[V]$ in the general case where n is an arbitrary integer, can be found in ref. [414]. With the definition (4.22) of $-W'[A]$ one finds the result (12.127) with $c_n = 0$. As we have seen in chapter 4, the ambiguity involved in this calculation is of the form $\zeta(0) \ln M^2 = n \ln M^2$, and accounts for the arbitrary dimensional multiplicative factor $\left(\frac{\mu}{2\pi}\right)^n$ appearing in (12.75). We have thus demonstrated, that the formal - but quite general- manipulations in section 9 reproduce the result (12.75) obtained via "clustering", and hence can be expected to be applicable also in higher dimensions.

12.11 Zero Modes of the Dirac Operator as a Boundary-Valued Problem

The "compactification" of the Dirac operator $\not{D}[A]$ via stereographic projection in order to be able to define an eigenvalue problem with a discrete spectrum is only applicable if the Chern-Pontryagin number of the gauge field A_μ is an integer. In $d = 2$ dimensions the Euclidean gauge-field action is finite also for non-integral winding numbers, so that there exists no "finite-action" argument prohibiting the contribution of such "instanton" configurations to the functional integral. In fact, we know from section 4, that "merons" are responsible for the non-vanishing of the expectation value of correlations functions carrying fermionic quantum numbers in the θ-vacuum. It is thus natural to look for alternative "compactification schemes" consistent with the Atiyah-Singer Index theorem, which has been demonstrated in section 4, to be consistent with the results obtained using clustering arguments.

---

7) The angular momentum index l is not to be confused with the index l labelling the zero mode eigenfunctions $\psi_{l_\pm}^{(0)}$.

The free Dirac operator and non local boundary conditions[170,415]

The most natural alternative to the "stereographic compactification" is the formulation of the eigenvalue problem for the Dirac operator as a boundary value problem in a two-dimensional "box". Considering this box to be a disc D with radius R, $D = \{x; |x| \leq R\}$, one is tempted to simply impose Dirichlet conditions on the boundary ∂D of the disc:

$$\psi(x)\Big|_{\partial D} = 0 \quad . \tag{12.150}$$

In the case of the gauge-invariant Laplace operator this procedure indeed defines an eigenvalue problem of a self-adjoint operator with a discrete spectrum. In the case of the massless Dirac operator in $d = 2$ dimensions the condition (12.150) is however too restrictive, and is only compatible with the trivial solution $\psi = 0$. This is easily seen for the free Dirac operator, by introducing light-cone coordenates; it is also true in the presence of an external field, as our discussion in this section will show. Hence the usual box quantization cannot be applied in the case in question. This explains why the attempt in ref. [416], to reproduce Schwingers result for the determinant in this framework had to fail.

The way out of this dilemma is not as simple as one might think. The following procedures are not acceptable, since they violate fundamental symmetries which we wish to preserve:
i) Impose the Dirichlet boundary condition only on one of the components of the Dirac spinor:

$$\psi_1\Big|_{\partial D} = 0 \quad .$$

This however destroys charge-conjugation invariance C, which requires that

$$\psi^c(x; A_\mu) = C\psi^*(x; -A_\mu) \tag{12.151a}$$

with

$$C = \begin{pmatrix} 0 & -1 \\ 1 & 0 \end{pmatrix} \tag{12.151b}$$

be also a solution satisfying the same boundary conditions.
ii) Require local boundary conditions of the "transmission" type $\psi_2\big|_{\partial D} \sim \psi_1\big|_{\partial D}$. This condition however destroys not only charge-conjugation invariance, but γ^5 invariance as well!

Similar problems arise if we try to impose Neumann, or mixed (Neumann plus Dirichlet) boundary conditions. This leads us to abandon the idea of imposing *local* conditions.

Reconsider first the case of the free Dirac operator, $i\,\not{\partial}$, which we rewrite it in terms of the polar coordenates r and θ,

$$i\,\not{\partial} = -i\gamma_r(\partial_r + i\gamma^5 \frac{1}{r}\partial_\theta) \tag{12.152}$$

with

$$\gamma_r = \begin{pmatrix} 0 & -ie^{-i\theta} \\ ie^{i\theta} & 0 \end{pmatrix} = \gamma_r^\dagger \quad . \tag{12.153}$$

The solution of the eigenvalue problem then take the form

$$\psi = \begin{pmatrix} \sum\limits_m u_m(r)e^{im\theta} \\ \sum\limits_m \frac{1}{\lambda}(\frac{\partial}{\partial r} - \frac{m}{r})u_m(r)e^{i(m+1)\theta} \end{pmatrix} \tag{12.154}$$

In this formulation the Dirichlet boundary condition reads

$$u_m(R) = 0 \tag{12.155a}$$

$$(\frac{\partial}{\partial r} - \frac{m}{r})u_m(r)\Big|_{r=R} = 0 \quad , \quad \forall m \quad , \tag{12.155b}$$

implying again $u_m(r) = 0$. Hence we replace this condition by the requirement, that the upper (lower) components of ψ only involve negative (positive) "frequency components" $e^{im\theta}$ on the boundary. For (12.154) this implies

$$\begin{aligned} u_m(R) &= 0 \quad , \quad m \geq 0 \\ (\frac{\partial}{\partial r} - \frac{m}{r})u_m(r)\Big|_{r=R} &= 0 \quad , \quad m < 0 \quad . \end{aligned} \tag{12.156}$$

This restriction in "frequency" space amount to non-local (mixed) boundary conditions in R^2. Although, they define, together with the integrability condition at $r = 0$, a discrete eigenvalue problem, they violate again charge-conjugation invariance, since the corresponding charge-conjugated Dirac spinor on the boundary, $\psi^c|_{\partial D}$ does not respect this frequency decomposition.

It is convenient to reformulate these boundary conditions for

$$\tilde{\psi} = e^{i\frac{\theta}{2}\gamma^5}\psi \quad , \tag{12.157}$$

which now reads

$$\tilde{\psi}\Big|_{\partial D} = \begin{pmatrix} \sum_{m<0} u_m(R)e^{i(m+\frac{1}{2})\theta} \\ \sum_{m\geq 0} \frac{1}{\lambda}(\frac{d}{dR} - \frac{m}{R})u_m(R)e^{i(m+\frac{1}{2})\theta} \end{pmatrix} \quad . \tag{12.158}$$

The charge conjugate spinor is again given by a transformation of the form (12.151),

$$\tilde{\psi}^c(x; A_\mu) = C\tilde{\psi}^*(x; -A_\mu) \quad .$$

The exponentials $\varphi_n = e^{i(m+\frac{1}{2})\theta}$ are eigenfunctions of the operator $-i\partial_\theta$ satisfying antiperiodic boundary conditions:

$$\varphi_m(\theta + 2\pi) = -\varphi_m(\theta) \quad . \tag{12.159}$$

Hence our non-local boundary condition amounts to the statement, that

$$\begin{aligned} \tilde{\psi}_1\Big|_{\partial D} &\in \mathcal{H}_{<0}\left(\frac{1}{i}\frac{\partial}{\partial\theta}\right) \\ \tilde{\psi}_2\Big|_{\partial D} &\in \mathcal{H}_{>0}\left(\frac{1}{i}\frac{\partial}{\partial\theta}\right) \quad , \end{aligned} \tag{12.160}$$

where $\mathcal{H}_{>0}(\frac{1}{i}\frac{\partial}{\partial\theta})$ and $\mathcal{H}_{<0}(\frac{1}{i}\frac{\partial}{\partial\theta})$ are the L^2 Hilbert spaces spanned by the eigenfunctions φ_m of $-i\partial_\theta$ belonging to positive and negative eigenvalues respectively, and satisfying the antiperiodic boundary conditions (12.159). The reason for having to formulate the eigenvalue problem for double valued functions (sections) $\tilde{\psi}$ has its origin in the associated spin structure on a circle.

The little Dirac operator[170,415]

We now extend the above considerations to the Dirac operator in an external field A_μ: Decomposing $\vec{A}$ into its polar coordenates as follows

$$(A_1, A_2) = (A_r \cos\theta - A_\theta \sin\theta, A_r \sin\theta + A_\theta \cos\theta) \tag{12.161}$$

one may rewrite the Dirac operator as follows,

$$\not{D} = -\gamma_r(\partial_r - ieA_r - \frac{\gamma^5}{r}B) \tag{12.162a}$$

with

$$B = (-i\partial_\theta - reA_\theta) \quad . \tag{12.162b}$$

One refers to B as the "small" Dirac operator. It replaces $-i\partial_\theta$, if $A_\mu \neq 0$. Viewing it in this way, it will be a simple matter to generalize the above ideas on the spectral decomposition on ∂D to the interacting case.

To begin with let us consider again the question of hermiticity of the Dirac operator (12.162). We compute

$$\Delta = \int d^2x[\psi^\dagger i \not{D}\psi - (i\not{D}\psi)^\dagger\psi] \quad . \tag{12.163}$$

For a hermitian Dirac operator, we should have $\Delta = 0$. In terms of $\tilde{\psi}$ and A_r expression (12.163) reduces to

$$\begin{aligned}\Delta &= \int_0^\infty r dr \int_0^{2\pi} d\theta \{\tilde{\psi}^\dagger (\partial_r + \frac{1}{2r}) C\tilde{\psi} + (\partial_r + \frac{1}{2r})\tilde{\psi}^\dagger \cdot C\tilde{\psi}\} \\ &= R\int_0^{2\pi} d\theta(-\tilde{\psi}_1^* \psi_2 + \tilde{\psi}_2^* \psi_1)_{r=R}\end{aligned} \tag{12.164}$$

Hence $\tilde{\psi}_1|_{\partial D}$ and $\tilde{\psi}_2|_{\partial D}$ must lie in mutually orthogonal spaces, if Δ is to vanish. This can be achieved in many ways. We seek however boundary conditions which respect γ^5 and C invariance. In analogy to the free case we thus require that on ∂D the upper (lower) spin components of $\tilde{\psi}$ lie in $\mathcal{H}_{<0}(B)$ and $\mathcal{H}_{<0}(B)$ respectively, where $\mathcal{H}_{<0}(B)$ $(\mathcal{H}_{>0}(B))$ are now the L^2 Hilbert spaces spanned by the eigenfunctions $\phi_\lambda(\theta)$ of $B = -i\partial_\theta - rA_\theta$, belonging to negative (positive semi-definite) eigenvalues

$$\tilde{\psi}\Big|_{\partial D} = \begin{pmatrix} \sum_{\lambda<0} a_\lambda \phi_\lambda(\theta; A) \\ \sum_{\lambda\geq 0} b_\lambda \phi_\lambda(\theta; A)\end{pmatrix} \tag{12.165}$$

with

$$B\phi_\lambda(\theta; A) = \lambda\phi_\lambda(\theta; A) \tag{12.166a}$$

$$\phi_\lambda(\theta + 2\pi; A) = -\phi_\lambda(\theta; A) \quad . \tag{12.166b}$$

Note that we allow this time for "zero" modes of the small Dirac operator, although such modes violate charge conjugation invariance. Indeed under charge conjugation

$$\tilde{\psi}^c\Big|_{\partial D} = \begin{pmatrix} -\sum_{\lambda\geq 0} b_\lambda^* \phi_\lambda^*(\theta; -A) \\ \sum_{\lambda<0} a_\lambda^* \phi_\lambda^*(\theta; -A)\end{pmatrix} \tag{12.167}$$

Now, from the eigenvalue equation (12.166a) follows that

$$B^*\big|_{A\to -A} = -B \tag{12.168a}$$

Hence

$$\phi_\lambda^*(\theta; -A) = \phi_{-\lambda}(\theta; A) \tag{12.168b}$$

or

$$\tilde{\psi}^c\Big|_{\partial D} = \begin{pmatrix} -\sum_{\lambda\leq 0} b_{-\lambda}^* \phi_\lambda(\theta; A) \\ \sum_{\lambda<0} a_{-\lambda}^* \phi_\lambda(\theta; A)\end{pmatrix}$$

i.e.

$$\tilde{\psi}_1^c\Big|_{\partial D} \in \mathcal{H}_{\leq 0}(B) \tag{12.169a}$$

$$\tilde{\psi}_2^c\Big|_{\partial D} \in \mathcal{H}_{>0}(B) \tag{12.169b}$$

Hence, comparing with (12.160), we see that charge-conjugation invariance is violated by zero modes.

The Dirac operator in a vortex field[170,415]

To see that such zero modes may indeed occur, we consider again the example of the vortex V_μ, with the integer n replaced by a real number,

$$A_\mu = -\frac{\nu}{e}\frac{\epsilon_{\mu\lambda}x_\lambda}{1+x^2} \quad . \tag{12.170}$$

In order to facilitate the discussion, we choose ν to be positive. In terms of the decomposition (12.161) we have

$$\vec{A} = A_\theta \hat{e}_\theta \quad , \quad A_\theta = \frac{\nu}{e}\frac{r}{1+r^2} \quad . \tag{12.171}$$

Hence the eigenvalue problem (12.166) reads,

$$(-i\partial_\theta - \gamma)\phi_\lambda = \lambda\phi_\lambda \quad , \quad \phi_\lambda(\theta + 2\pi) = -\phi_\lambda(\theta) \tag{12.172a}$$

with γ defined by

$$\gamma = \nu\frac{R^2}{1+R^2} \quad . \tag{12.172b}$$

Equation (12.166a) has the solution

$$\phi_{\lambda_m} = e^{i(\lambda_m+\gamma)\theta} \quad . \tag{12.173}$$

with

$$\lambda_m = m + \frac{1}{2} - \gamma \tag{12.174}$$

Hence, if $\gamma = n + \frac{1}{2}$ we have a zero mode for $m = n$. For $R \to \infty$ this corresponds to a half integral winding number

$$\nu = n + \frac{1}{2} \quad . \tag{12.175}$$

Using (12.174), the spectral boundary condition for ψ, reads, in the interacting case

$$\psi\Big|_{\partial D} = \begin{pmatrix} \sum_{m<\gamma-1/2} u_m(R)e^{im\theta} \\ \sum_{m\geq\gamma-1/2} \frac{1}{\lambda}(\frac{\partial}{\partial R} - \frac{m}{R} + \gamma)u_m(R)e^{i(m+1)\theta} \end{pmatrix} \quad . \tag{12.176}$$

Suppose γ to lie in the interval ($n \geq 0$),

$$n \leq \gamma \leq n+1 \quad . \tag{12.177}$$

For reasons that will become clear later on, we subdivide this interval as follows:

$$\begin{aligned} &i)\; n \leq \gamma < n+1/2 \\ &ii)\; \gamma = n+1/2 \\ &iii)\; n+1/2 < \gamma < n+1 \end{aligned} \tag{12.178}$$

Now, the zero modes of the big Dirac operator belonging to positive and negative chirality are of the form (see (4.222))

$$\begin{aligned} \psi_{l+}^{(0)}(x) &= \begin{pmatrix} \frac{r^{l-1}}{(1+r^2)^{\frac{\nu}{2}}} e^{i(l-1)\theta} \\ 0 \end{pmatrix} \\ \psi_{l-}^{(0)}(x) &= \begin{pmatrix} 0 \\ \frac{r^{l-1}}{(1+r^2)^{\frac{-\nu}{2}}} e^{-i(l-1)\theta} \end{pmatrix} \end{aligned} \tag{12.179}$$

They also must satisfy the boundary condition (12.176) above. This means that

$$l < \gamma + 1/2 \qquad \text{for pos. chirality} \tag{12.180a}$$

$$l \leq -\gamma - 1/2 \qquad \text{for neg. chirality} \quad . \tag{12.180b}$$

Integrability at $r = 0$ requires $l \geq 1$. Hence the boundary condition (12.176) implies for γ in the intervals (12.178),

i) n zero modes of positive chirality; no zero modes of negative chirality.

iii) $n+1$ zero modes of positive chirality; no zero modes of negative chirality.

Hence the index of the Dirac operator must experience a "jump" for $\gamma = n + \frac{1}{2}$. In order to describe the jump we define the η-invariant,[417]

$$\eta(s) = \sum_m{}' sgn\, \lambda_m |\lambda_m|^{-s} \quad , \tag{12.181}$$

where λ_m are the eigenvalues of the little Dirac operator, and the "prime" indicates as usual the omission of the zero modes. In our case, with λ_m given by (12.174), we find

$$\eta(s) = \begin{cases} \zeta(s, \frac{1}{2} - \rho) - \zeta(s, \rho - \frac{1}{2}) + \left(\frac{1}{\rho - \frac{1}{2}}\right)^s \\ 0 \\ \zeta(s, \frac{1}{2} - \rho) - \zeta(s, \rho - \frac{1}{2}) - \left(\frac{1}{\frac{1}{2} - \rho}\right)^s \end{cases} \qquad (12.182)$$

with

$$\rho \equiv \gamma - n \quad , \qquad (12.183)$$

where the three values in (12.182) correspond to the intervals

$$i)\ \rho = [0, \frac{1}{2}) \quad , \quad ii)\ \rho = \frac{1}{2} \quad \text{and} \quad iii)\ \rho = (\frac{1}{2}, 1],$$

respectively.Noting that

$$\zeta(0, \alpha) = \frac{1}{2} - \alpha \quad ,$$

we obtain for the respective intervals,

$$i)\ \eta(0) = 2\rho \quad , \quad ii)\ \eta(0) = 0 \quad \text{and} \quad iii)\ \eta(0) = 2(\rho - 1) \quad .$$

Now, let h = number of zero-modes of $B|_{\partial D}$. In terms of ρ, the eigenvalues (12.174) of the little Dirac operator are given by

$$\lambda_m = (m - n + \frac{1}{2} - \rho) \quad . \qquad (12.184)$$

We therefore have

$$i)\ h = 0 \quad , \quad ii)\ h = 1 \quad \text{and} \quad iii)\ h = 0 \quad ,$$

zero modes in the respective intervals. The combination $\frac{\eta(0)+h}{2}$ thus has the graphical representation shown in figure 12.1

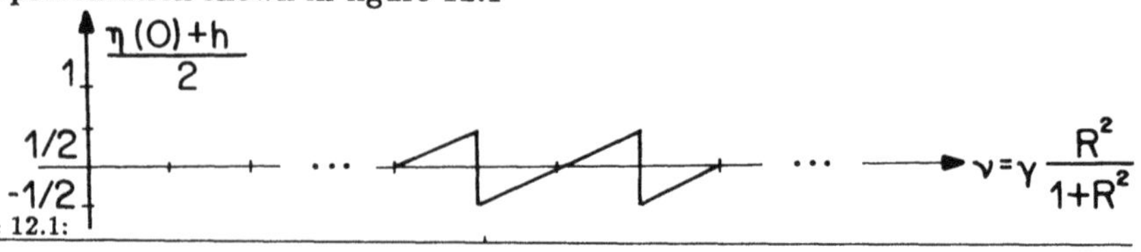

Figure 12.1:

We have thus for our example of a vortex (12.170), that (recall Eq. (12.183))

$$\text{index}\, i\, \not{D} = \gamma - \frac{\eta(0) + h}{2} \quad . \tag{12.185}$$

This is nothing but the Atiyah-Singer Index Theorem for the Dirac operator on a manifold in R^2 with boundary[417]. Indeed, noting that for our example

$$\int_D \frac{d^2x}{2\pi} \epsilon_{\mu\nu} \partial_\mu A_\nu = \nu \int \frac{d\theta}{2\pi} \frac{R^2}{1+R^2} = \gamma \tag{12.186}$$

we may write (12.185) also in the form

$$\text{index}\, i\, \not{D} = \dim \ker i\, \not{D}\Big|_{\gamma^5>0} - \dim \ker i\, \not{D}\Big|_{\gamma^5<0} = \int_D \frac{d^2x}{2\pi} \epsilon_{\mu\nu} F_{\mu\nu} - \frac{\eta(0)+h}{2} \quad . \tag{12.187}$$

For an integral winding number $\nu = n$, (see figure 12.1) we have

$$\frac{\eta(0)+h}{2} \underset{R\to\infty}{\longrightarrow} 0 \tag{12.188a}$$

and

$$\int_D \frac{d^2x}{2\pi} \epsilon_{\mu\nu} F_{\mu\nu} \underset{R\to\infty}{\longrightarrow} n \quad , \tag{12.188b}$$

so that one recovers in this case the form (4.211,212) of the Atiyah-Singer Theorem. The boundary-value formulation (12.185) of this theorem however allows also the discussion of non-integral winding numbers, such as in the case of "meron-configurations", which contribute to tunnelling amplitudes where changes in the fermionic quantum number are involved.

12.12 The $U(1)$ Problem Revisited[382]

It was stated in chapter 10 that instantons with fractional winding numbers $\nu = \pm 1/N$ are "responsible" for the spontaneous symmetry breakdown in the combined Schwinger chiral Gross-Neveu (S-GN) model, which thus serves as an illustration of the mechanism envisaged by Crewther[265] for the resolution of the so-called $U(1)$ problem in chiral QCD_4. We are now in the position of demonstrating the correctness of this statement.

The role of instantons with fractional winding number

Consider the θ-vacuum expectation value $\langle\theta|\rho_\pm(x)|\theta\rangle$ of the chiral densities (10.37) in the theory defined by (10.123). According to our discussion in section (1.10) of chapter 10 it is evident that this expectation value has the Euclidean functional representation

$$\langle\theta|\rho_\pm(x)|\theta\rangle = \langle\rho_\pm\rangle_{SU(N)} \int \mathcal{D}A^\mu \delta(\partial_\mu A^\mu) \cdot e^{-NW[A]} \times$$
$$e^{-\int d^2z \frac{1}{4} F_{\mu\nu}(z) F_{\mu\nu}(z) \mp e \int d^2z D(x-z)\epsilon_{\mu\nu} F_{\mu\nu}(z)} e^{i\theta e \int \frac{d^2z}{4\pi} \epsilon_{\mu\nu} F_{\mu\nu}(z)} \quad , \qquad (12.189a)$$

where $W[A]$ is given by (12.2c), and

$$\rho_\pm(x) \equiv \frac{1}{N} \sum_f \overline{\psi}_f(x) \frac{1 \pm \gamma_5}{2} \psi_f(x) \quad . \qquad (12.189b)$$

According to our discussion in section (1.10) of chapter 10, we expect A_μ-configuration with non-integral winding number to play a crucial role. The functional representation of $\langle\theta|\rho_\pm(x)|\theta\rangle$ will thus involve the determinant of the Dirac operator for a gauge-field with fractional winding number; its calculation therefore requires the formulation of the corresponding eigenvalue problem on a manifold with boundary as discussed in the previous sections. We can avoid these complications, by using clustering arguments for the (Euclidean) two-point function $\langle 0|\rho_+(x)\rho_-(y)|0\rangle$. Since $\rho_+\rho_-$ has zero chirality, only gauge-field configurations with winding number $\nu = 0$ will contribute.

From (10.123) we have

$$\langle 0|T\rho_-(x)\rho_+(x)|0\rangle = N^{-1} \int \mathcal{D}A_\mu \delta[\partial_\nu A^\nu] e^{-\int d^2z \frac{1}{4} F_{\mu\nu}(z) F_{\mu\nu}(z)} \mathcal{Z}[A] \quad , \qquad (12.190a)$$

where $\mathcal{Z}[A]$ is the external field correlation function of the GN model,

$$\mathcal{Z}[A] \equiv \langle T\rho_-(x)\rho_+(y) e^{-e\int d^2z j_\mu(z) A^\mu(z)} \rangle_{GN} \qquad (12.190b)$$

with

$$j_\mu \equiv \overline{\psi}\gamma_\mu\psi \quad .$$

Our discussion in section (1.10) of chapter 10 suggests that $\mathcal{Z}[A]$ factorizes into a $U(1)$ and a $SU(N)$ part. This factorization is easy to see in the bosonized representation (10.129-131) in chapter 10, where the (Euclidean) Lagrangian is seen to separates naturally into a $U(1)$ and a $SU(N)$ piece.

The calculation of the $U(1)$ part of Eq. (12.190) only involves a Gaussian integration and proceeds along the lines already witnessed in the Schwinger model: According to the discussion in sec. (10.1.10), it is given by

$$\langle 0|\rho_-(x)\rho_+(y)|0\rangle_{U(1)} = \langle 0|\rho_-(x)\rho_+(y)|0\rangle^{free}_{U(1)} \int \mathcal{D}A^\mu \delta[\partial_\nu A^\nu] e^{S_G[A]} e^{-NW[A]} \times$$
$$\exp\left\{-2e \int d^2 z A_\mu(z) \tilde{\partial}^z_\mu (D(x-z) - D(y-z))\right\} \quad . \quad (12.191)$$

The effective action is saturated by the Euclidean field configuration

$$A^{cl}_\mu(z) = \frac{2\pi}{Ne}\tilde{\partial}_\mu(\mathcal{D}(x-z) - \mathcal{D}(y-z)) \quad , \quad (12.192)$$

corresponding to an instanton/anti-instanton pair of fractional winding $\frac{1}{N}$ and $-\frac{1}{N}$ respectively. For the free piece one has

$$\langle 0|\rho_-(x)\rho_+(y)|0\rangle^{free}_{U(1)} = \left[\frac{1}{-\mu^2(x-y)^2 + i0}\right]^{1/N} \quad . \quad (12.193)$$

Doing the A^μ integration in (12.191) one obtains

$$\langle 0|\rho_-(x)\rho_+(y)|0\rangle_{U(1)} = e^{\frac{4\pi}{N}\Delta(x-y;\frac{Ne^2}{\pi})} \quad . \quad (12.194)$$

Hence the whole effect of the A^μ integration consisted in turning the "massless" $U(1)$ factor (12.193) into a massive one.

It is instructive to note, that the representation (12.191) can also be obtained from the following Ward identities (the remaining discussion in this section is in Minkowski space): We may compute the A^μ dependence of $\mathcal{Z}[A]$ in (12.190b) by observing that for a gauge invariant regularization the vector current is conserved, and the axial vector current in an external potential has the usual Adler-Bell-Jackiw anomaly for N flavors:

$$\partial_\mu j^\mu(x) = 0 \quad , \quad \partial_\mu j^\mu_5(x) = \frac{Ne}{2\pi}\epsilon_{\mu\nu}F^{\mu\nu}(x) \quad . \quad (12.195)$$

By functionally differentiating $\mathcal{Z}[A]$ in (12.190b) with respect to A_μ, and using (12.195) as well as the equal time commutation relations

$$[j^0(x), \rho_\pm(0)]_{ET} = 0 \quad (12.196a)$$

$$[j^0_5(x), \rho_\pm(0)]_{et} = \pm 2\rho_\pm(x)\delta(x^1) \quad , \quad (12.196b)$$

one obtains the Ward identities[382]

$$\partial_\mu \frac{\delta}{\delta A_\mu(z)} \mathcal{Z}[A] = 0 \tag{12.197a}$$

and

$$\tilde{\partial}_\mu \frac{\delta}{\delta A_\mu(z)} \mathcal{Z}[A] = 2ie\mathcal{Z}[A](\delta^2(z-x) - \delta^2(z-y)) - \frac{iNe^2}{\pi} \mathcal{Z}[A]\tilde{\partial}_\mu A^\mu(z) \quad , \tag{12.197b}$$

whose solution is given by

$$\mathcal{Z}[A] = \mathcal{Z}[0] e^{-NW[A]} \exp\left\{2ie \int d^2z (D(x-z) - D(y-z))\tilde{\partial}_\mu A^\mu(z)\right\} \quad , \tag{12.198a}$$

with $\mathcal{Z}[0]$ evidently given by

$$\mathcal{Z}[0] = \langle 0|T\rho_-(x)\rho_+(y)|0\rangle_{GN} = \langle 0|T\rho_-(x)\rho_+(y)|0\rangle_{U(1)}^{free} \langle 0|T\rho_-(x)\rho_+(y)|0\rangle_{SU(N)}^{GN} \tag{12.198b}$$

We thus reproduce the result (12.191).

The $SU(N)$-factor may be calculated in an $1/N$ expansion, as we now show.

$1/N$ expansion

As we have already seen in chapter 5, the $\tilde{U}(1) \times U(N)$ Gross-Neveu model admits an $1/N$ expansion only for the $SU(N)$ symmetric part of the fermionic correlation functions obtained after extracting the $\tilde{U}(1) \times U(1)$ factor associated with the infraparticle structure ("would-be" Goldstone boson) of the model. In the present case the $U(1)$ gauge–field serves to screen the $U(1) \times \tilde{U}(1)$ "charge" of the fermions, thus dynamically reducing them to $SU(N)$ multiplets.

In the following we calculate the (gauge–invariant) $SU(N)$ correlation function in (12.198b) to leading order in $1/N$. Introducing as in chapter 5 the auxiliary fields σ and π corresponding to $g\sum\bar{\psi}_f\psi_f$ and $g\sum\bar{\psi}_f i\gamma_5\psi_f$, respectively, we may replace Lagrangian (10.123) by (the discussion here is in Minkowski space)

$$\mathcal{L} = -\frac{1}{4}F^{\mu\nu}F_{\mu\nu} + \bar{\psi}(i\partial\!\!\!/ + e\, A\!\!\!/)\psi - \frac{1}{2g}(\sigma^2 + \pi^2) + \bar{\psi}(\sigma + i\pi\gamma^5)\psi \tag{12.199}$$

Doing the fermion integration in the corresponding partition function, leaves us with the effective action

$$S_{eff} = S_G[A] - \int d^2x \frac{1}{2g}(\sigma^2 + \pi^2) - iN\,\mathrm{tr}\ln(i\partial\!\!\!/ + eA\!\!\!/ + (\sigma + i\pi\gamma^5)) \quad . \tag{12.200}$$

Anticipating that $\langle\sigma\rangle \neq 0$, we redefine σ by $\sigma = \frac{\tilde{\sigma}}{\sqrt{N}} + \langle\sigma\rangle$, where now $\langle\sigma'\rangle = 0$. It is evident from (12.200) that $\langle\sigma\rangle \neq 0$ implies a spontaneous mass generation for the fermion. Hence, making the identification (5.23), and setting $\pi = \frac{\tilde{\pi}}{\sqrt{N}}$, we are left with

$$S_{eff} = S_G[A] - \frac{1}{2gN}\int d^2x(\tilde{\sigma}^2 + \tilde{\pi}^2) + \frac{1}{g}\frac{m}{\sqrt{N}}\int d^2x\tilde{\sigma}$$
$$-iN\,\mathrm{tr}\ln(i\,\partial\!\!\!/ - m + e\,A\!\!\!/ + \frac{1}{\sqrt{N}}(\tilde{\sigma} + i\tilde{\pi}\gamma^5)) \quad . \quad (12.201)$$

Holding gN and $e\sqrt{N}$ fixed, only the quadratic terms in the fields contribute to the effective action in leading order of $1/N$, leaving one with

$$iS_{eff} = \int\left\{\frac{1}{2}A_\mu\Gamma^{\mu\nu}_{AA}A_\nu + \frac{1}{2}\tilde{\sigma}\Gamma_{\sigma\sigma}\tilde{\sigma} + \frac{1}{2}\tilde{\pi}\Gamma_{\pi\pi}\tilde{\pi}\right\} + \int(\tilde{\sigma}\Gamma^\mu_{\sigma A}A_\mu + \tilde{\pi}\Gamma^\mu_{\pi A}A_\mu)$$
$$+\int d^2x\left(\frac{im}{g\sqrt{N}} + \sqrt{N}S_F(0)\right)\tilde{\sigma}(x) \quad (12.202)$$

with the Fourier transforms of the vertex functions given by the

$$\tilde{\Gamma}^{\mu\nu}_{AA}(p) = i\left(-p^2g^{\mu\nu} + p^\mu p^\nu\right) - i\Pi^{\mu\nu}_{AA}(p)$$
$$\tilde{\Gamma}_{\sigma\sigma}(p) = -\frac{i}{gN} - i\Pi_{\sigma\sigma}(p)$$
$$\tilde{\Gamma}_{\pi\pi}(p) = -\frac{i}{gN} + i\Pi_{\pi\pi}(p)$$
$$\tilde{\Gamma}^\mu_{\sigma A}(p) = -e\sqrt{N}\Pi^\mu_{\sigma A}(p)$$
$$\tilde{\Gamma}^\mu_{\pi A}(p) = -ie\sqrt{N}\Pi^\mu_{\pi A}(p) \quad . \quad (12.203)$$

Here $-i\Pi_{VV'}(p)$ represents the (single flavour) one loop diagram shown in figure 12.2, with the matrices Γ_V and $\Gamma_{V'}$ at the respective vertices, where $\Gamma_\sigma = 1$, $\Gamma_\pi = \gamma_5$, and $\Gamma_A = \gamma^\mu$.

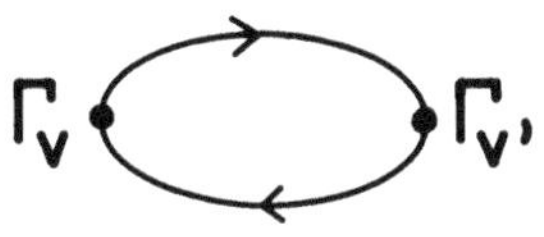

Figure 12.2: Fermion loop diagram representing $-i\Pi_{VV'}(p)$.

The requirement that the last term, linear in σ, be absent gives the relation (5.29) between the bare mass m and the cutoff Λ. The functions $\tilde{\Gamma}_{\sigma\sigma}(p)$ and $\tilde{\Gamma}_{\pi\pi}(p)$ have already been calculated in chapter 5. Using (5.30) they may be entirely expressed in terms of m (mass transmutation) and are given by (5.105,106). The function $\tilde{\Gamma}^{\mu}_{\sigma A}$ is found to vanish, while for $\tilde{\Gamma}^{\mu}_{\pi A}$ one finds

$$\begin{aligned}\tilde{\Gamma}^{\mu}_{\pi A}(p) &= -ie\sqrt{N}\int\frac{d^2k}{(2\pi)^2}\operatorname{tr}\frac{\gamma^{\mu}(\not{k}+\not{p}+m)\gamma^5(\not{k}+m)}{[(k+p)^2-m^2](k^2-m^2)}\\ &= -2ie\sqrt{N}m\epsilon_{\mu\nu}p^{\nu}A(p)\end{aligned} \tag{12.204}$$

where $A(p)$ is given by (5.37). Comparing (5.37) with (5.106) one verifies that

$$\tilde{\Gamma}^{\mu}_{\pi A}(p) = -2ie\sqrt{N}m\frac{\tilde{p}^{\mu}}{p^2}\tilde{\Gamma}_{\pi\pi}(p) \tag{12.205}$$

The result for $\tilde{\Gamma}^{\mu\nu}_{AA}(p)$ cannot be taken over from the corresponding result of the massless Schwinger model, since the fermion loop now involves massive fermions. Using (5.38) one obtains for a gauge-invariant Pauli-Villars regularization in the limit where the Pauli-Villars mass $M \to \infty$,

$$\tilde{\Gamma}^{\mu\nu}_{AA}(p) = \frac{iNe^2}{\pi}\left(g_{\mu\nu}-\frac{p_{\mu}p_{\nu}}{p^2}\right)\left(1-\frac{4\pi m^2}{p^2}\tilde{\Gamma}_{\pi}(p)\right) - i\left(p^2 g_{\mu\nu}-p_{\mu}p_{\nu}\right) . \tag{12.206}$$

Note that in the limit $m \to 0$, expression (12.206) reduces to the corresponding result in the Schwinger model. Making use of the above results, we see, that the effective action is diagonalized by the transformation $\tilde{\pi} = \pi' + 2\sqrt{N}em\frac{\tilde{\partial}_{\mu}}{\Box}A^{\mu}(x)$. This leaves one with the effective action,

$$S_{eff} = \frac{1}{2}\int\left\{A^{\mu}\left(\Box+\frac{Ne^2}{\pi}\right)\left(g_{\mu\nu}-\frac{\partial_{\mu}\partial_{\nu}}{\Box}\right)A^{\nu} - i\tilde{\sigma}\Gamma_{\sigma\sigma}\tilde{\sigma} - i\pi'\Gamma_{\pi\pi}\pi'\right\} \tag{12.207}$$

from where one directly reads off the Feynman rules:

$$\tilde{\Delta}_{\pi'}(p) = -i\tilde{\Gamma}^{-1}_{\pi\pi}(p) = 2\pi\frac{\operatorname{ctnh}\phi/2}{\phi} \tag{12.208a}$$

$$\tilde{\Delta}_{\tilde{\sigma}}(p) = -i\tilde{\Gamma}^{-1}_{\sigma\sigma}(p) = 2\pi\frac{\tanh\phi/2}{\phi} \tag{12.208b}$$

$$\tilde{D}_{\mu\nu}(p) = \frac{-1}{p^2-Ne^2/\pi}\left(g_{\mu\nu}-\frac{p_{\mu}p_{\nu}}{p^2}\right) . \tag{12.208c}$$

We are now in the position of calculating correlation functions of the scalar and pseudoscalar fermionic densities. When doing this we must keep in mind, that in terms of $\tilde{\sigma}$ and $\tilde{\pi}$, the relations are now given by

$$
\begin{aligned}
g\sum_f \bar{\psi}_f\psi_f &= \frac{\tilde{\sigma}}{\sqrt{N}} - m \\
g\sum_f \bar{\psi}_f i\gamma_5\psi_f &= \frac{\pi'}{\sqrt{N}} + 2em\frac{\tilde{\partial}_\mu}{\Box}A^\mu(x) \quad .
\end{aligned}
\tag{12.209}
$$

From $\langle\sigma\rangle = -m$ we have for the vacuum expectation value of the unrenormalized chiral density (12.189*b*), $\langle\rho^u_\pm\rangle = -\frac{1}{2}\frac{m}{gN}$, where the unrenormalized coupling constant g is related to the mass m and cutoff Λ via (5.29). Hence, defining the renormalized chiral density by $\rho_\pm = Z(\Lambda)\rho^u_\pm$ with $Z^{-1}(\Lambda) = \frac{1}{4\pi}\ln\left(\frac{\Lambda^2}{m^2}\right)$, we have

$$
\langle\rho_\pm\rangle = -\frac{m}{2} \quad . \tag{12.210}
$$

Since A_μ, $\tilde{\sigma}$ and π' are completely decoupled in (12.207), we may directly read off from the propagators (12.208) the result for the renormalized two-point function of $\rho_\pm$:

$$
\int d^2x e^{ip\cdot x}\langle T\rho_-(x)\rho_+(0)\rangle = \pi^2 m^2\delta^2(p) - \frac{i}{4N}\tilde{\Delta}_\sigma(p) - \frac{i}{4N}\tilde{\Delta}_{\pi'}(p) + \frac{im^2}{p^2N}\frac{Ne^2}{p^2 - Ne^2/\pi} . \tag{12.211}
$$

This expression approaches a finite limit as $p^2 \to 0$, showing that no trace of the "would be" Goldstone boson is left. In turn, the gauge field became massive as the last term in (12.211) explicitly shows. We have thus recovered in the functional language the results of section 2.10 of chapter 10.

12.13 Conclusion

We have shown that the results obtained in chapter 10 by the use of operator methods, can also be understood in the framework of Feynman path integrals. It is in the latter formulation, where the analogies between QED_2 and QCD_4 are most clearly revealed. In particular we have shown that massless QED_2 provides an exact realization of a number of properties commonly attributed to QCD_4 on he basis of perturbative and semi-classical arguments. It should however be kept in mind that, whereas in QCD_4 vacuum tunneling is attributed to generic instanton configurations, such tunneling is mediated in the Schwinger model by *induced* instantons, whose form depends on the particular correlation function in question.

This fact was first noted by B. Schroer and N. K. Nielsen[355]. These seemingly different properties however find their unifying element in the zero-modes of the Dirac operator, which ultimately determine the nature of the vacuum structure, and provide a linkage between the winding number of these instantons and the chirality of the vacua via the Atiyah-Singer theorem.

Our discussion here has been restricted to massless QED_2. The results of chapter 10 however show, that the basic features of the model concerning vacuum tunneling and confinement continue prevail in the massive case. In fact, the massive Schwinger model, as well as the methods discussed in section 4 of this chapter, have found an interesting application to the problem of monopole induced proton decay in $d = 3 + 1$ dimensions, known in this context as the "Rubakov effect"[418]. The interested reader should consult chapter 2 of ref. [419] for a general survey of this problem.

Chapter XIII

NON-ABELIAN CHIRAL GAUGE THEORIES

13.1 Introduction

In the chapters 10 through 12 we have studied in great detail gauge theories with pure vector-like coupling to a gauge field. In 3+1 dimensions such theories describe the electromagnetic ($U(1)$ gauge group) and strong ($SU(3)$ gauge group) interactions. However, unlike the case of 3+1 dimensions, where the photon and the gluons are described by massless fields, their corresponding analogues in d =1+1 dimensions acquire a mass via an intrinsic, dynamical Higgs mechanism, thus making the corresponding interactions short ranged. We have identified this phenomenon with the screening of the charge (color) of the fermions by the vacuum polarization induced by the respective interactions.

On the other hand, a ***dynamical*** Higgs mechanism turning the non-abelian gauge-bosons into massive vector bosons would be a welcome phenomenon in weak-interactions where the "intermediate" vector bosons $(\mathcal{W}^{\pm}, \mathcal{Z}^0)$ are known to have masses of the order of $m_{\mathcal{W}} \approx 80.8$GeV, and $m_{\mathcal{Z}} \approx 92.9$GeV.

In the Glashow-Salam-Weinberg model[19] of the weak interactions, the left- and right-handed fermions (quarks and leptons) are coupled asymmetrically to the gauge-fields ($V - A$ coupling), which in turn acquire a mass via spontaneous symmetry breaking induced by a Higgs potential[366] which is introduced explicitly in the theory.

It has long been known[201,210] that fermionic gauge theories with "chiral" coupling of the fermions to the gauge field exhibit an anomaly in the (covariant) divergence of the ***external*** field gauge-current $J^{\mu,a}_{ch}(x \mid A)$, which is being referred to as the "non-abelian" anomaly:

$$\mathcal{D}^{ab}_{\mu} J^{\mu,b}_{ch}(x, \mid A) = \mathcal{A}^a(x \mid A) \quad . \tag{13.1}$$

This anomaly has been considered as "dangerous" not only for being manifestly inconsistent with the equations of motion

$$\mathcal{D}^{ab}_{\mu} F^{\mu\nu,b}(x) + eJ^{\nu,a}_{ch}(x \mid A) = 0 \quad , \tag{13.2}$$

but also because it implies a breakdown of gauge invariance. However, contrary to earlier believes, this loss of gauge invariance on quantum level does not necessarily mean the loss of some other properties usually attributed to gauge theories, such as renormalizability, unitarity and Lorentz invariance.

Vector gauge theories

Chiral anomalies have played a central role in the development of quantum field theory in the last twenty years, but their deep mathematical significance was only realized recently[74]. They are an integral part of systems involving an infinite number of degrees of freedom. One speaks of an "anomaly", if a classical symmetry of the Lagrangian is spoiled when quantizing the theory. If one is dealing with a continuous classical symmetry, this means that the corresponding Noether current is no longer conserved.

In a purely *vectorial* (non-abelian) $SU(N)$ gauge theory, like chromodynamics ($N = 3$), a regularization prescription can be given, which preserves the gauge-invariance of the classical Lagrangian. Correspondingly, the (vector) gauge-current $J_V^{\mu,a}$ can be defined such as to be covariantly conserved:

$$\mathcal{D}_\mu^{ab} J_V^{\mu,b} = 0 \quad . \tag{13.3}$$

Moreover the corresponding external-field current $J_V^{\mu,a}(x \mid A)$ may also be chosen to be divergence free for *arbitrary* field configurations A_μ, if a gauge-invariant regularization is employed:

$$\mathcal{D}_\mu^{ab} J_V^{\mu,b}(x \mid A) = 0 \quad . \tag{13.4}$$

The same applies to the $U(1)$ vector currents $J_V^\mu$and $J_V^\mu(x \mid A)$ respectively, except that here the covariant divergence is replaced by the ordinary divergence

$$\partial_\mu J_V^\mu = 0 \tag{13.5a}$$

$$\partial_\mu J_V^\mu(x \mid A) = 0 \quad . \tag{13.5b}$$

Note that the conservation law (13.5*b*) is *consistent* with the gauge invariance of the functional determinant $\det(i \not\partial + e \not A)$, if $J_\nu^{\mu,a}$ is defined by

$$J_V^{\mu,a}(x \mid A) = \frac{1}{ie} \frac{\delta}{\delta A_\mu^a(x)} \ln \det(i \not\partial + e \not A) \quad . \tag{13.6}$$

In the vector $U(N)$ gauge theories, J_ν^μ and $J_\nu^{\mu,a}$ are the sources of $F_{\mu\nu}$ and $F_{\mu\nu}^a$. It is thus natural to adopt a regularization which ensures the conservation of these currents, as given by (13.4) and (13.5). This may only be achieved at the expense of an anomaly in the corresponding *axial* vector currents, having the form,

$$\partial_\mu J_A^\mu = \frac{e^2}{8\pi^2} \mathrm{tr}\, {}^*F_{\mu\nu} F^{\mu\nu} \tag{13.7a}$$

$$\mathcal{D}^{ab}_{\mu} J^{\mu,b}_{A} = \frac{e^2}{8\pi^2} \mathrm{tr}(\tau^{a\,*}F_{\mu\nu}F^{\mu\nu}) \quad , \tag{13.7b}$$

where ${}^*F_{\mu\nu}$ is the dual of $F_{\mu\nu}$. ${}^*F_{\mu\nu} = \frac{1}{2}\epsilon_{\mu\nu\lambda\rho}F^{\lambda\rho}$. The r.h.s. of Eq. (13.7) are the Adler-Bell-Jackiw (ABJ) and Bardeen anomalies[200,201].

Some remarks are in order: The (gauge invariant) $U(1)$-anomaly in (13.7a) is just twice the Pontryagin density, and can thus be written as the (ordinary) divergence of the Chern secondary characteristic class[1)],

$$\frac{1}{8\pi^2} \mathrm{tr}({}^*F_{\mu\nu}F^{\mu\nu}) = \partial_\mu C^\mu \quad . \tag{13.8a}$$

with

$$C^\mu = \frac{1}{4\pi^2} \epsilon^{\mu\nu\lambda\rho} \mathrm{tr}(A_\nu \partial_\lambda A_\rho + \frac{2}{3i} A_\nu A_\lambda A_\rho) \quad . \tag{13.8b}$$

This plays an important role in the so called $U(1)$-problem[364]; on the other hand, the anomaly in the $SU(N)$ axial vector current cannot be written as a total (covariant) divergence!

Since the anomalies (13.7a), (13.7b) are a result of the non-invariance of the fermionic measure[156] under abelian (non-abelian) local γ_5-transformations

$$\psi(x) \to e^{i\theta(x)\gamma^5}\psi(x) \quad , \tag{13.9}$$

($\theta \in$ Lie algebra of $U(1)$ resp. $SU(N)$), they are referred to as the abelian (non-abelian) anomaly.

Both anomalies, (13.7a) and (13.7b) are *gauge invariant*. They do not occur in the gauge current, and hence do not affect the equations of motion, which could put in question the existence of QCD as a bonafide quantum theory. In fact, from the phenomenological point of view such anomalies may be desirable. Indeed, in the absence of an anomaly, the hypothesis of partial conservation of the axial vector current (PCAC) implies a $\pi^0 \to 2\gamma$ decay rate which cannot be accounted for by the small mass of the pion[420]. Taking account of the anomaly of the form (13.7a) with $F^{\mu\nu}$ the electromagnetic field-strength tensor, one obtains excellent agreement with the experimentally measured decay rate for the case of three quark colors. This constitutes experimental evidence for $SU(3)$-color as the (gauge) group of the strong interactions.

[1)] When comparing with the literature it is to be kept in mind that we are working here with hermitian Lie-algebra valued fields, and with $SU(N)$ generators t^a normalized as in (C.1a).

Chiral gauge theories

The situation is quite different in so called "chiral" gauge theories, where only left or right handed fermions couple to the gauge field. Since the gauge-current $J^{\mu,a}_{ch}$ involves both a vector and axial vector part, it is impossible to regularize in this case the external field gauge current, such as to be covariantly conserved for arbitrary gauge-field configurations, and one generally has the anomalous divergence, Eq. (13.1). The result for $\mathcal{A}^a$ is however non-unique, reflecting the fact, that either the vector, or axial, current could still be chosen to be conserved or not[200].

If $J^{\mu,a}_{ch}(x \mid A)$ is defined in terms of the corresponding one-loop vacuum functional $W^{(ch)}[A]$, this now implies a non-trivial consistency condition on $\mathcal{A}^a$, which is then referred to as the "*consistent* (non-abelian) anomaly".

In a chiral gauge theory with only right (left) handed coupling of the fermions it is convenient to adopt a regularization procedure which treats the right (left) handed currents symmetrically. The non-abelian consistent anomaly reads in this case[2)]

$$\mathcal{A}^a_{\binom{R}{L}} = \partial_\mu C^{\mu,a}_{\binom{R}{L}} \quad , \tag{13.10a}$$

with

$$C^{\mu,a}_{\binom{R}{L}} = \mp\frac{fe^2}{24\pi^2}\epsilon^{\mu\nu\lambda\rho}\mathrm{tr}\{\tau^a(A_\nu\partial_\lambda A_\rho + \frac{e}{2i}A_\nu A_\lambda A_\rho)\} \tag{13.10b}$$

$$= \mp\frac{fe^2}{24\pi^2}d_{abc}\epsilon^{\mu\nu\lambda\rho}(A^b_\nu\partial_\lambda A^c_\rho + \frac{e}{4}f_{cde}A^b_\nu A^d_\lambda A^e_\rho) \quad , \tag{13.10c}$$

where $d_{abc} = \mathrm{tr}(t^a\frac{1}{2}\{t^b,t^c\})$ and where the proportionality constant f depends on the number of flavors. The fact that $C^{\mu,a}_L = -C^{\mu,a}_R$ shows that the anomaly (13.10) corresponds to a regularization for which the $SU(N)$ *vector* current is covariantly conserved. Note the factor 1/2 in (13.10*b*) as compared to 2/3 in (13.8*b*). This means that the consistent anomaly (13.10*a*) is not gauge covariant, in contrast to the axial anomaly (13.7*b*). It indicates that gauge-invariance is broken on quantum level in the chiral case, and thus raises the question as to the unitarity and renormalizability of the theory, properties usually linked to the Ward-Takahashi identities following from gauge invariance.

A further property, that makes the chiral anomaly (13.10) potentially "dangerous" is the fact that it occurs in the gauge current; Since (see C.8*b*)

$$D_\mu D_\nu F^{\mu\nu} = 0 \quad , \tag{13.11}$$

[2)] The fact that the anomaly can be written as a total divergence is not surprising. The conservation of the Noether current, following from the invariance under rigid chiral transformations, is equivalent to the Yang-Mills equations plus the covariant conservation of the source current, which in turn follows from the invariance under local chiral transformations. Both conservation laws are broken by the anomaly (13.1). If the anomaly is however expressible as a total divergence, then a new Noether-current may be defined, which is conserved. Thereby only the local, but not global chiral symmetry is broken.

this means that the *external field* equations of motion

$$\mathcal{D}_\mu^{ab} F_b^{\mu\nu}(x) + eJ^{\nu,a}_{\binom{R}{L}}(x \mid A) = 0 \quad , \tag{13.12}$$

are only consistent on the *subspace* of gauge field configurations A_μ on which the chiral anomaly $A^a_{\binom{R}{L}}$ vanishes. This puts in question the existence of a solution on full quantum level. Note that this question of consistency plays no role in the anomalous breakdown of conservation laws involving other Noether currents, which generally only implies a reduction of the global symmetries of the original Lagrangian.

A sufficient condition for the anomaly (13.10a) to vanish is that the fermions belong to a representation for which d_{abc} vanishes. Two cases may be distinguished:
i) $d_{abc} = 0$ for all representations of the group. Such groups are called "safe". Examples are $SU(2)$ (no other special unitary groups), all orthogonal groups, except $SO(6) \approx SU(4)$ and all simpletic groups.
ii) If the group is not safe, there may still exist particular representations for which $d_{abc} = 0$; these are then called safe representations.

The condition of anomaly cancellation when applied to the (non-safe) electroweak group $U(1) \times SU(2)_W$, requires the fermions to lie in safe representations of this group. This leads to the prediction that the number of quarks balances against the number of leptons. Although the discovery of charmed quarks and heavy leptons points toward an experimental confirmation of this prediction, the discovery of the partner of the bottom quark, the *top* quark, is still outstanding. It is thus of interest to reexamine the question of whether anomalies in the gauge current are admissible or not. We shall do this in the context of $d = 1 + 1$ dimensions.

It has been known for some time that a unified treatment of chiral anomalies can be given in a purely mathematical, differential-geometric context. Faddeev[72] has applied the methods of group theory cohomology to the analysis of such anomalies. The gauge-group cocycles introduced by him[412] can be obtained as integrals of the differential forms entering the Zumino "descent equations"[74]. We shall not dwell much on this elegant approach and rather discuss these anomalies in the framework of the canonical approach to the quantization of Feynman path integrals, with due reference, of course, to some basic mathematical concepts, such as cocycles.

The breakdown of gauge invariance, as twofold signalized by the anomalous divergence (13.1) in the gauge current, as well as the non gauge-covariance of the consistent anomaly itself, further manifests itself in the constraints being second class, with a corresponding anomalous extension in the (anomalous) Poisson algebra of the (would be) gauge-group generators in terms of an infinitesimal 2-cocycle. This algebra, as well as the ensuing problems with respect to the quantization of such anomalous gauge theories shall occupy most of our attention. The possibility of functionally bosonizing fermionic theories in two dimensions will help us to understand a number of subtleties, as for instance, the definition of *anomalous* Poisson brackets[422,423] referred to above. Of course, some subtleties, such as the appearance of 3-cocycles witnessed in d=3+1 dimensions[424,425] will be found to be absent in d=1+1 dimensions, where one is dealing with a superrenormalizable theory.

The "common wisdom" that gauge invariance of "gauge theories" on quantum level is intimately linked to their renormalizability and unitarity is put here in question, by embedding the anomalous theory into a bonafide gauge-theory on quantum level[72,426], at the expense of additional degrees of freedom - the Wess-Zumino fields. We shall refer to the two formulations as the *gauge-non-invariant* (GNI) and *gauge-invariant* (GI) formulations respectively, and shall study the quantization of the chiral theory in both formulations, side by side.

Since the non-abelian model is not explicitly soluble, we shall leave the explicit computation of correlation functions, as well as the discussion of the "physical" implications, to chapter 14. Since this and the following chapter emphasize the Hamiltonian approach to quantization, the discussion will refer to Minkowski space.

13.2 Anomalies and Cocycles

13.2.1 Consistent anomaly

In this chapter we shall study gauge theories in which massless fermions of only one chirality couple to a gauge field A_μ. Subsections (13.2.1) to (13.2.5) make no reference to the dimensionality of space-time.

The chirality of a fermion is defined by the corresponding eigenvalue of γ^5. Our convention will be

$$\gamma_5 \psi_{\substack{R\\L}} = \pm \psi_{\substack{R\\L}} \quad , \tag{13.13}$$

where $R(L)$ refers to a right (left) handed fermion. We denote by

$$P_\pm = \frac{1 \pm \gamma_5}{2} \quad , \quad P_\pm^2 = P_\pm \quad , \quad P_+ P_- = 0 \quad , \tag{13.14a}$$

the projectors in Dirac-space on the states of positive and negative chirality. We have

$$\gamma_5 P_\pm = \pm P_\pm \quad . \tag{13.14b}$$

To be specific we shall restrict ourselves to the case, where only left-handed fermions are coupled to A_μ. The corresponding Lagrangian density and action read respectively

$$\mathcal{L} = -\frac{1}{4}\mathrm{tr} F_{\mu\nu} F^{\mu\nu} + \bar{\psi}(i \not\partial + e \not A P_-)\psi \tag{13.15a}$$

$$S = S_G[A] + S_F^{(L)}[\psi, \bar{\psi}, A] \quad , \tag{13.15b}$$

where

$$S_G = -\frac{1}{4}\int d^dx \operatorname{tr} F_{\mu\nu}F^{\mu\nu} \tag{13.16a}$$

is the Yang-Mills action, and

$$S_F^{(L)} = \int d^dx \overline{\psi}(i \not\partial + e \not{A} P_-)\psi \tag{13.16b}$$

is the fermionic action. The corresponding Euler-Lagrange equations read

$$\mathcal{D}_\mu^{ab} F_b^{\mu\nu} + e\overline{\psi}\gamma^\nu P_- \tau^a \psi = 0 \tag{13.17}$$

$$(i \not\partial + e \not{A} P_-)\psi = 0 \quad , \tag{13.18}$$

with $\mathcal{D}_\mu^{ab}$ the covariant derivative (C.6). The fermions are taken to belong to the *fundamental* representation of the gauged group of $SU(N)$-color.

The action (13.16*a*) is invariant under the chiral gauge transformation

$$\psi \rightarrow {}^G\psi = G_-\psi \tag{13.19a}$$

$$\overline{\psi} \rightarrow \overline{{}^G\psi} = \overline{\psi} G_+^{-1} \tag{13.19b}$$

$$A_\mu \rightarrow A_\mu^{G^{-1}} = G A_\mu G^{-1} + \frac{i}{e} G \partial_\mu G^{-1} \quad , \tag{13.19c}$$

where

$$G_\pm(x) = e^{i\theta(x)P_\pm} \quad , \quad G(x) = e^{i\theta(x)} \in \quad SU(N) \tag{13.19d}$$

with $\theta(x)$ matrix valued fields lying in the Lie algebra of $SU(N)$.

In Minkowski space, transformation (13.19*b*) follows from (13.19*a*) and the definition $\overline{\psi} = \psi^\dagger \gamma^0$. In terms of the right and left handed components (13.13), the transformation laws (13.19*a*, *b*) reads

$$\psi_L \rightarrow G\psi_L \quad , \quad \psi_R \rightarrow \psi_R$$

$$\overline{\psi}_L \rightarrow \overline{\psi_L} G^{-1} \quad , \quad \overline{\psi}_R \rightarrow \overline{\psi}_R \quad . \tag{13.19e}$$

The Dirac operator in (13.18) transforms under (13.19*c*) as follows

$$G_+(i \not\partial + e \not{A} P_-)G_-^{-1} = (i \not\partial + e \not{A}^{G^{-1}} P_-) \quad . \tag{13.20}$$

This transformation law, already suggests that $Z[A^G] \neq Z[A]$, where $Z[A]$ is the "external" field partition function

$$Z[A] = \int \mathcal{D}\overline{\psi}\mathcal{D}\psi e^{iS_F^{(L)}[A,\psi,\overline{\psi}]} = e^{iW^{(L)}[A]} \quad . \tag{13.21}$$

Making the identification

$$iW^{(L)}[A] = \ln\det(i\,\not{\partial} + e\not{A}P_-) \quad , \tag{13.22}$$

we correspondingly expect

$$\alpha_1(A, G^{-1}) \equiv W^{(L)}[A^{G^{-1}}] - W^{(L)}[A] \quad , \tag{13.23}$$

not to vanish. The functional $\alpha_1(A, G^{-1})$ is seen to have the property of a one-cocycle:

$$\alpha_1(A^{G_2}, G_1^{-1}) - \alpha_1(A, G_2G_1^{-1}) + \alpha_1(A, G_2) = 0 \quad . \tag{13.24}$$

We shall refer to $\alpha_1(A, G^{-1})$ as the Wess-Zumino functional[52]. It may be calculated in different ways, to be discussed later.

The non-vanishing of $\alpha_1(A, G^{-1})$ shows that gauge-invariance is broken at the one-fermion-loop level. This implies an anomaly in the external field source-current of $F^{\mu\nu}$ as we now show. Define

$$J_L^{a,\mu}(x \mid A) = \frac{1}{e}\frac{\delta W^{(L)}[A]}{\delta A_\mu^a(x)} \quad . \tag{13.25}$$

Now, for an infinitesimal gauge-transformation

$$A_\mu^a \to (A_\mu^{G^{-1}})^a \approx A_\mu^a + \frac{1}{e}\mathcal{D}_\mu^{ab}\delta\theta^b \quad , \tag{13.26}$$

we have, using (13.25)

$$W^{(L)}[A^{G^{-1}}] = W^{(L)}[A] - \int dx\,\delta\theta^a(x)\mathcal{D}_\mu^{ab}J_L^{\mu,b}(x \mid A) \quad , \tag{13.27a}$$

or, in terms of (13.23),

$$\mathcal{D}_\mu^{ab}J_L^{\mu,b}(x \mid A) = -\frac{\delta\alpha_1[A, e^{-i\theta}]}{\delta\theta^a(x)}\bigg|_{\theta=0} = \frac{\delta\alpha_1[A, e^{i\theta}]}{\delta\theta^a(x)}\bigg|_{\theta=0} \equiv \mathcal{A}^a(x \mid A) \quad . \tag{13.28}$$

The functional $\mathcal{A}$ represents the anomaly. Eq. (13.27) may also be written as

$$\alpha_1(A, G) = \int dx\,\theta^a(x)\mathcal{A}^a(x) + O(\theta^2) \quad . \tag{13.27b}$$

Eq.(28) can be viewed as an external field Ward-Takahashi identity. In terms of fermions it states that

$$\mathcal{D}_\mu^{ab}\langle\overline{\psi}(x)\gamma^\mu\tau^b P_-\psi(x)\rangle_A = \mathcal{A}^a(x \mid A) \quad , \tag{13.29}$$

where the subscript indicates, that the expectation value is to be computed for a c-number field A_μ. In the following section we generalize this result to arbitrary external-field correlation functions.

The anomaly (13.28), defined in terms of the functional (13.23) satisfying the 1-cocycle condition (13.24), is referred to as the *consistent* anomaly. An elegant characterization of the "consistent" anomaly has been given by Faddeev and Shatashvili[421].

Consider the generator $T^a(x)$ of infinitesimal gauge-transformations, (13.19c) on the background field A^a_μ:

$$T^a(x) = -\mathcal{D}^{ab}_\mu \frac{\delta}{\delta A^b_\mu(x)} = -\left(\partial_\mu \frac{\delta}{\delta A^a_\mu(x)} + e f_{acb} A^c_\mu \frac{\delta}{\delta A^b_\mu(x)}\right) \quad . \tag{13.30}$$

These first-order homogeneous differential operators satisfy the commutation relations

$$[T^a(x), T^b(y)] = e f_{abc} T^c(y) \delta(x-y) \quad . \tag{13.31}$$

Now, under a gauge transformation, $Z[A] = e^{iW^{(L)}[A]}$ transforms as follows

$$Z[A] \to Z[A^G] = e^{i\alpha_1(A,G)} Z[A] \quad , \tag{13.32a}$$

or infinitesimally

$$T^a(x) Z[A] = -ie\mathcal{A}^a(x) Z[A] \quad . \tag{13.32b}$$

It is important to realize that this anomaly cannot be "renormalized a way" by adding suitable *local* counter terms to the Lagrangian, since there exists no local functional $\Omega[A]$ with the property $\mathcal{A}^a(x) = T^a(x)\Omega[A]$.

Equation (13.32) is commonly regarded as an indication for the breaking of gauge invariance, since it implies that different A_μ-configurations in the functional integral are gauge inequivalent.

Now, it follows from (13.31) and (13.32) that

$$T^a(x)\mathcal{A}^b(y) - T^b(y)\mathcal{A}^a(x) = e f_{acb} \mathcal{A}^c(x) \delta(x-y) \quad . \tag{13.33}$$

This relation is called the *Wess-Zumino (WZ) self consistency condition*[52], and an anomaly having this property is called *"consistent anomaly"*. Since the WZ consistency condition is a necessary condition for the integrability of (13.28), with $J^{\mu,a}_L$ given by (13.25), the terminology "consistent" is also synonymous with "integrable".

An anomaly satisfying (13.33) allows for an alternative point of view[421] of the "breaking of gauge invariance" as expressed by (13.32). Define the new "generators"

$$X^a(x) = T^a(x) + ie\mathcal{A}^a(x) \quad , \tag{13.34a}$$

where the second term just acts as an operator of multiplication. Using (13.33) one finds that $X^a(x)$ lies in the same Lie algebra as $T^a(x)$:

$$[X^a(x), X^b(y)] = ef_{abc}X^c(x)\delta^2(x-y) \quad . \tag{13.34b}$$

We have thus a different representation of the Lie algebra. In order to see the significance of this, we digress to make some general considerations.

Let G be a group with elements g, and let M be the space on which this group acts, with points m. Let $U(g)$ be a representation of G on the space of functions $\psi(m)$ on M,

$$\psi(m) \to U(g)\psi(m) = e^{i\alpha_1(m,g)}\psi(mg) \quad . \tag{13.35}$$

The condition that the representation satisfies the usual group property[3)]

$$U(g_1)U(g_2) = U(g_1g_2) \quad , \tag{13.36a}$$

then implies,

$$\alpha_1(mg,h) - \alpha_1(m,gh) + \alpha_1(m,g) = 0 \quad (mod\, 2\pi) \quad . \tag{13.36b}$$

The function $\alpha_1(m,g)$ is called a 1-cocycle of the group G. With the identification

$$m \to A_\mu \quad , \quad g \to G_2 \quad , \quad h \to G_1^{-1} \quad , \quad mg \to A^{G_2} \quad ,$$

the 1-cocycle condition (13.36b) just reduces to (13.24) and $Z[A]$ is seen to transform like $\psi(m)$ in (13.35).

The realization of the 1-cocycle in the form (13.23), or in the above notation, by

$$\alpha_1(m,g) = \alpha_0(mg) - \alpha_0(m) \quad , \tag{13.37}$$

is trivial in the mathematical sense, since we may remove in this case the exponential in (13.35) via a unitary transformation

$$\tilde{\psi}(m) = e^{i\alpha_0(m)}\psi(m) \quad . \tag{13.38a}$$

Indeed, the transformation (13.35) then takes the form

$$\widetilde{U}[g]\widetilde{\psi}(m) = \widetilde{\psi}(mg) \tag{13.38b}$$

where

$$\widetilde{U}[g] = e^{i\alpha_0(m)}U[g]e^{-i\alpha_0(m)} \quad . \tag{13.38c}$$

---

3) To be explicit, one should really write $U[g] \to U[m;g]$. Following Faddeev[421], we have suppressed the dependence on the manifold.

With the identification $\alpha_0(m) \to W^{(L)}[A]$, an infinitesimal transformation $U[g] \approx 1 + \int dx \delta\theta^a(x) T^a(x)$ implies

$$\tilde{U}[g] \approx 1 + \int dx \delta\theta^a(x) X^a(x) \quad . \tag{13.38d}$$

Now, in the notation of (13.38*a*), $Z[A]$ corresponds to $\tilde{\psi}(m)$, with $\psi(m) = 1$. Hence (13.38*b*) implies in this case,

$$X(x) Z[A] = 0 \quad . \tag{13.39}$$

which is just Eq. (13.32*b*).

This leads to an alternative interpretation of the "symmetry breakdown"; Gauge invariance is not broken, but is realized in a different way.

Let us finally remark, that though the cocycle (13.23) is trivial from the mathematical standpoint, we shall nevertheless consider it to be non-trivial, since it cannot be "renormalized away" by the introduction of a suitable set of local counter terms.

13.2.2 More about cocycles

Let g_i $(i = 1, 2)$ be elements of the group G, and suppose that the group property $g_1 g_2 = g_{12}$ is realized in the following way[427]:

$$U(g_1) U(g_2) = e^{i\alpha_2(m; g_1, g_2)} U(g_1 g_2) \quad . \tag{13.40}$$

Assuming associativity of the composition law,

$$(U(g_1) U(g_2)) U(g_3) = U(g_1)(U(g_2) U(g_3)) \quad , \tag{13.41}$$

one arrives at the condition

$$\delta\alpha_2(m; g_1, g_2, g_3) = 0 \quad , \tag{13.42}$$

where

$$\begin{aligned} \delta\alpha_2(m; g_1, g_2, g_3) \equiv \alpha_2(m g_1, g_2, g_3) - \alpha_2(m; g_1 g_2, g_3) + \alpha_2(m; g_1, g_2 g_3) \\ -\alpha_2(m; g_1, g_2) \quad , \quad (\text{mod } 2\pi) \quad ; \end{aligned} \tag{13.43}$$

δ is called the ***coboundary*** operator, and α_2 is a two-cochain on the gauge group.

The vanishing of $\delta\alpha_2$, Eq. (13.42), is the condition for $\alpha_2(m; g_1, g_2)$ to be a 2-cocycle. As Eq. (13.40) shows, such cocycles are relevant for the projective representations of the gauge group[413]. Denoting by G the generator of the transformation $U[g]$,

$$U = e^{iG} \tag{13.44}$$

the associativity, Eq. (13.41) implies that the Jacobi identity

$$[G_1,[G_2,G_3]]+c.p.=0 \tag{13.45}$$

is satisfied. It follows from (13.42) that $\alpha_2(A;1,g_2)=\alpha_2(A;g_1,1)=0$. Hence, for infinitesimal gauge transformations $g_{1,2}\approx 1+i\theta^a_{1,2}\tau^a$, α_2 is of second order in the parameters $\theta^a_{1,2}$:

$$\alpha_2(A;g_1,g_2)=\frac{1}{2}\int dx\int dy\theta^a_1(x)\theta^b_2(y)\mathcal{A}^{ab}(x,y)+\mathcal{O}(\theta^3) \quad . \tag{13.46}$$

Representations that involve 2-cocycles are called projective or ray representations. Such representations occur naturally in Quantum Mechanics if we try to represent Galilean transformations including spacial translations. They represent symmetry transformations for the free particle and correspond to simple translations in phase space, $\vec{r}\to\vec{r}+\vec{a}$,and $\vec{p}\to\vec{p}+\vec{b}$; these transformations are unitarily implemented by[427]

$$U[\vec{a},\vec{b}]=e^{i(\vec{a}\cdot\vec{p}-\vec{b}\cdot\vec{r})} \quad . \tag{13.47}$$

The operator (13.47) obeys the composition law (13.41) with

$$\alpha_2[\vec{r},\vec{p};\vec{a}_1,\vec{b}_1;\vec{a}_2,\vec{b}_2]=\frac{1}{2}(\vec{a}_1\cdot\vec{b}_2-\vec{a}_2\cdot\vec{b}_1) \quad , \tag{13.48}$$

where α_2 satisfies the consistency condition (13.42).

One may generalize the definition (13.42) to higher cocycles. As one might anticipate, the 3-cocycle involves abandoning associativity (13.41). This has dramatic consequences, since non-associating quantities cannot be represented by well-defined linear operators acting on a vector or Hilbert space.

Let us modify the associativity law (13.41) in the following way[427]

$$(U(g_1)U(g_2))U(g_3)=e^{i\alpha_3(m;g_1,g_2,g_3)}U(g_1)(U(g_2)U(g_3)) \quad . \tag{13.49}$$

By considering fourfold products, and associating them in different ways, one finds that consistency requires

$$\delta\alpha_3(m;g_1,g_2,g_3,g_4)=0 \pmod{2\pi} \quad , \tag{13.50}$$

with

$$\begin{aligned}\delta\alpha_3(m;g_1,g_2,g_3,g_4)\equiv{}&\alpha_3(mg_1;g_2,g_3,g_4)-\alpha_3(m;g_1g_2,g_3,g_4)+\\&\alpha_3(m;g_1,g_2g_3,g_4)-\alpha_3(m;g_1,g_2,g_3g_4)+\alpha_3(m;g_1,g_2,g_3,) \quad .\end{aligned} \tag{13.51}$$

Condition (13.50) states that α_3 is a 3-cocycle. There exist hints that 3-cocycles may play a role in chiral gauge theories[424]. They arise again naturally in the Quantum Mechanics of monopoles[427]. The implementation by linear operators then leads to the well known Dirac's quantization condition $2eg = n$, where e and g are the charges of the electron and monopole, respectively.

In general we define a n-cocycle by the condition

$$\delta\alpha_n(m; g_1, \cdots, g_{n+1}) = 0 \quad , \tag{13.52}$$

where $\alpha_0 = \alpha_0(m)$, $\alpha_1 = \alpha_1(m, g)$, etc, and

$$\begin{aligned} &\delta\alpha_n(m; g_1, \cdots, g_{n+1}) \equiv \\ &\alpha_n(mg_1; g_2, g_3, \cdots, g_{n+1}) - \alpha_n(m; g_1 g_2, g_3, \cdots, g_{n+1}) + \alpha_n(m; g_1, g_2 g_3, g_4, \cdots, g_{n+1}) \\ &\qquad + (-1)^n \alpha_n(m; g_1, \cdots, g_{n-1}, g_n g_{n+1}) + (-1)^{n+1} \alpha_n(m; g_1, \cdots, g_n) . \end{aligned} \tag{13.53}$$

$\alpha_n(m; g_1, \cdots, g_n)$ is called a cochain. Note that the operator δ acts in the *direction of the group.* (Think of δ as a gauge transformation). One verifies that $\delta^2\alpha_n = 0$.

The operation δ is the standard coboundary operator in homological algebra[428]. The function $\alpha = \delta\beta$ is called a *coboundary.* If $\delta\alpha = 0$, then α is called a cocycle. A cocycle is non-trivial if it is not a coboundary, i.e. $\delta\alpha = 0$, but α cannot be written in the form $\delta\beta$. In this sense the 1-cocycle defined in (13.23) is trivial, since in this case $\alpha_1(A, G) = \delta W^{(L)}[A]$.

13.2.3 Gauss anomaly

In the section 2.1 we have established a direct connection between the consistent (current) anomaly (13.28), and the 1-cocycle (13.23) for an infinitesimal gauge transformation, Eq. (13.27b). The existence of this anomaly manifests itself in a number of other ways, which we now highlight briefly. A more precise an detailed exposition of this question will be given further on in this chapter.

In a bonafide gauge theory, Gauss's law

$$G^a \equiv -\vec{D}^{ab} \cdot \vec{E}^b + eJ^a = 0 \tag{13.54}$$

is a first-class constraint. Here $E^{i,b}$ is the chromoelectric field; in the Hamiltonian formulation it is given by $\pi^{i,b}$, the momentum conjugate to A_i^b. On the quasi-classical level, G^a is also the generator of infinitesimal , time-independent gauge transformations with respect to the Poisson-bracket (PB) operation:

$$\{G^a(x), G^b(y)\} = ef_{abc}G^c(y)\delta(\vec{x} - \vec{y}) \quad , \tag{13.55a}$$

$$\{G[\theta], \vec{A}^a(x)\} = \vec{D}^{ab}\theta^b(x) \quad , \tag{13.55b}$$

$$\{G[\theta], \psi_\alpha(x)\} = ie\theta(x)\psi_\alpha(x) \quad . \tag{13.55c}$$

Here

$$G[\theta] \equiv \int d\vec{x}\theta^a(x)G^a(x) \quad . \tag{13.55d}$$

When going over to the quantum theory, two options are commonly considered:
i) The Gauss-law constraint is implemented *strongly* as an operator relation. This requires the complete fixation of a gauge. In the so called "non-relativistic gauges" (gauge condition does not involve the Lagrange multipliers A_0^a) this is achieved by Dirac procedure[338]. In the case of so-called "relativistic gauges" this is achieved via the Becchi-Rouet-Stora (BRS)[429] formalism.
ii) The Gauss-law constraint is implemented weakly as a condition on the states. This is preferably done by setting $A_0^a = 0$ (Weyl gauge) in the Lagrangian; the quantization of the resulting unconstrained system then involves only canonical commutators, obtained via the substitution $\{\,,\,\} \to -i[\,,\,]$, and the Gauss-law constraint is imposed as a condition on the physical states: $G^a(x) \mid \psi_{phys}\rangle = 0$.

Since in approach i) the gauge is fixed completely, the operator G^a now no longer generate gauge transformations, which finds its expression in the vanishing of the equal-time commutator $[G^a(x), G^b(y)]_{ET}$.

In the case of approach ii) the commutators $-i[\,G^a(x)\,,\,G^b(y)\,]_{ET}$, $-i[G[\theta], \vec{A}^a(x)]_{ET}$, and $-i[G[\theta], \psi_\alpha(x)]_{ET}$ continue to be given by the right-hand side of Eqs. (13.55), so that $G[\theta]$ generates the time-independent gauge transformations with respect to the canonical commutator algebra. This freedom still exists in Weyl gauge, since the conditions $A_0^a = 0$ still allow for arbitrary time-independent gauge transformations.

In the case of chiral gauge theories, the anomalous breakdown of gauge invariance no longer allows for a choice of gauge[4)]. As we shall learn in section 16, the (tree-graph) Poisson brackets in (13.55), are now to be replaced by anomalous Poisson brackets (APB) $\{\,,\,\}_{AP}$ incorporating the effects of the anomaly due to one-fermion loop graphs, and the transition to the quantum theory is achieved via the substitution $\{\,,\,\}_{AP} \to -i[\,,\,]$. In particular, the Poisson brackets (13.55*a*) are now replaced by

$$\{G^a(x), G^b(y)\}_{AP} = ef_{abc}G^c(x)\delta(\vec{x}-\vec{y}) + \mathcal{A}^{ab}(x,y) \tag{13.56}$$

where $\mathcal{A}^{ab}(x,y)$ is related to a two-cocycle via (13.46). This means that, despite this anomalous extension in the algebra, the Jacobi identity (13.45) continues to be satisfied by G^a. It also means that Gauss-law[5)] has turned into a 2nd class constraint, and thus can no longer be implemented consistently as a condition on the states, since in general $\mathcal{A}^{ab}(x,y) \mid \psi\rangle \neq 0$.

[4)] Nevertheless, the choice $A_0^a = 0$ has been generally used also in this case, to discuss the algebra of constraints and currents. See section 16 for the justification of such procedure.

[5)] This is however no longer the case for $\{E^{i,a}(x), E^{j,b}(y)\}_{AP}$. Evidence for this was first provided by the calculations of *Jo*[424] in the Weyl gauge.

Although the choice $A_0^a = 0$ is no longer allowed, it is remarkable that the results for the APB's $\{\,,\,\}_{AP}$ agree with those obtained by setting $A_0^a = 0$ in the Lagrangian, provided that the operators nowhere involve A_0^a, itself. The reason for this will be discussed in section 16, and provides an a posteriori justification for the results obtained in the literature.

13.2.4 Relation between consistent and covariant anomaly

Gauge anomalies are usually defined in terms of a gauge-variation of the connected one-loop vacuum functional in the presence of an external field, Eq. (13.27). As we have seen, the gauge current is no longer covariantly conserved, if this variation does not vanish[6)]

We have also seen, that the anomaly $\mathcal{A}^a$ defined by (13.28) satisfies the consistency condition (13.33), which restricts its functional form. Hence the name "consistent anomaly". In fact, the consistency conditions imply that the consistent non-abelian chiral anomaly *cannot* be given by a gauge-covariant expression. Correspondingly, the non-singlet, external field gauge currents defined by (13.25) will not transform covariantly under gauge transformations.

By adding of a suitable local functional of the gauge potential to the current defined by (13.25), one may however arrive at a new current $\tilde{J}^{\mu,a}$, whose covariant divergence is given by a gauge-*covariant* functional of A_μ. We refer to this functional as the *covariant* anomaly.

The consistent anomaly has an immediate significance, since it reflects the gauge-dependence of the vacuum functional, and refers to the source of the gauge-field. On the other hand, the "covariant" current $\tilde{J}^a_\mu$ has no interpretation in terms of an effective action, but may nevertheless be suitable for constructing gauge invariant couplings to other fields.

It is possible to construct covariant forms of the current and the anomaly from the knowledge of the consistent anomaly[430]. To this end we seek a *local* polynomial in the gauge potential, $\mathcal{P}^{\mu,a}[A]$, with an anomalous gauge transformation property opposite to that of the non-covariant current J_L^μ, and define[7)]

$$\tilde{J}_L^{\mu,a} = J_L^{\mu,a} + \mathcal{P}^{\mu,a} \quad . \tag{13.57}$$

We now show, how to construct $\mathcal{P}^{\mu,a}$

---

6) The standard point of view is that the effective chiral action cannot be defined in a gauge-covariant way. This is indeed the case from the point of view of perturbation theory, which generally leads to the consistent anomaly. Nevertheless, the covariant current itself may be shown to have a perturbative interpretation, provided Bose symmetry with respect to the external gauge field is abandoned.

7) Our discussion follows closely that of ref. [430]. In the following, $J_L^{\mu,a}(x)$ stands for the "consistent" *external* field current (13.25).

Under an infinitesimal gauge transformation

$$A_\mu^a \to A_\mu^a + T_\Lambda A_\mu^a \tag{13.58}$$

$$f[A] \to f[A] + T_\Lambda f[A]$$

where[8)]

$$T_\Lambda \equiv \int dy \Lambda^a(y) T^a(y) = -\int dy \Lambda^a(y) \mathcal{D}_\mu^{ab} \frac{\delta}{\delta A_\mu^b(y)} \tag{13.59}$$

We have

$$[T_{\Lambda'}, T_\Lambda] = \int dy f_{abc} \Lambda^a(y) \Lambda^b(y) T^c(y) \equiv T_{[\Lambda',\Lambda]} \tag{13.60}$$

Now

$$T_\Lambda W^{(L)}[A] = e\int dx J_L^{\mu,a}(x|A) \mathcal{D}_\mu^{ab} \Lambda^b(x) = -e\int dx \mathcal{A}^a(x|A) \Lambda^a(x) \tag{13.61}$$

The following trick proves useful[430]. Define the operator δ_B by

$$\delta_B f[A] \equiv \int dx \frac{\delta f[A]}{\delta A_\mu^a(x)} B_\mu^a(x) \quad . \tag{13.62}$$

If we were to choose $B_\mu^a(x) = \mathcal{D}_\mu^{ab} \Lambda^b(x)$, then $\delta_B = T_\Lambda$. Here B_μ^a is to be regarded however as some arbitrary field configuration, *independent* of A_μ. Thus δ_B induces some local transformation on A_μ which is not a gauge transformation! One immediately verifies that

$$[\delta_B, T_\Lambda] = \delta_{[B,\Lambda]} \tag{13.63}$$

where

$$\delta_{[B,\Lambda]} = e\int dx f_{abc} B_\mu^b(x) \Lambda^c(x) \frac{\delta}{\delta A_\mu^a(x)} \quad . \tag{13.64}$$

Applying (13.63) on the vacuum functional $W^{(L)}[A]$, one has

$$(T_\Lambda \delta_B + \delta_{[B,\Lambda]}) W^{(L)}[A] = \delta_B T_\Lambda W^{(L)}[A] \quad , \tag{13.65}$$

which upon using (13.61) and the definition (13.25) for the current, yields

$$\int dx \left[(T_\Lambda J_L^{\mu,a}(x)) + e f_{abc} \Lambda^b(x) J_L^{\mu,c}(x) \right] B_\mu^a(x) = -\int dx (\delta_B \mathcal{A}^a) \Lambda^a \quad . \tag{13.66}$$

---

8) All operations are understood to be performed at a fixed time.

If the current were to transform covariantly, the left hand side of (13.66) would vanish. Hence the r.h.s. expresses the precise nature of the non-covariance of the current.

For $\tilde{J}_L^{\mu,a}$, defined by (13.57) to transform covariantly, we must seek a local functional $\mathcal{P}^{\mu,a}$, with the transformation law

$$\int dx \left\{ (T_\Lambda \mathcal{P}^{\mu,a}) + e f_{abc} \Lambda^b \mathcal{P}^{\mu,c} \right\} B_\mu^a = \int dx (\delta_B \mathcal{A}^a) \Lambda^a \quad , \tag{13.67}$$

such as to cancel the anomalous contribution in the transformation law (13.66) of $J_L^{\mu,a}$. It is not a priori obvious, that such a functional can always be found. Nevertheless one can find such a polynomial $\mathcal{P}^{\mu,a}$ in $d = 2$ and $d = 4$ dimensions[430]. In $d = 2$, the consistent anomaly has the form

$$\mathcal{A}^a(x) = \frac{e}{4\pi}[(a-1)\partial^\mu - \tilde{\partial}^\mu] A_\mu^a(x) \quad ; \tag{13.68}$$

hence

$$\delta_B \mathcal{A}^a = \frac{e}{4\pi}[(a-1)\partial_\nu + \tilde{\partial}_\nu] B^{\nu,a} \quad . \tag{13.69}$$

The solution to (13.67) is given by

$$\mathcal{P}^{\mu,a} = -\frac{e}{4\pi}[(a-1)g^{\mu\nu} + \epsilon^{\mu\nu}] A_\nu^a \quad . \tag{13.70}$$

One explicitly verifies,

$$\mathcal{D}_\mu^{ab} \mathcal{P}^{\mu,b} = -\mathcal{A}^a - \frac{e}{4\pi}\epsilon^{\mu\nu} F_{\mu\nu}^a \quad , \tag{13.71}$$

which leads for $\tilde{J}^{\mu,b}$ to the *covariant* anomaly[9]:

$$\mathcal{D}_\mu^{ab} \tilde{J}_L^{\mu,b}(x) = -\frac{e}{4\pi}\epsilon^{\mu\nu} F_{\mu\nu}^a(x) \quad . \tag{13.72}$$

---

[9] The anomaly in d=4 dimensions is given by Eq. (13.10), with the choice[430]

$$\mathcal{P}^{\mu,a} = -\frac{e^2}{48\pi^2}\epsilon^{\mu\nu\rho\sigma} \mathrm{tr}\{\tau^a (A_\nu F_{\rho\sigma} + F_{\rho\sigma} A_\nu + ie A_\nu A_\rho A_\sigma)\}$$

one arrives at a covariant current with the covariant anomaly

$$\mathcal{D}_\mu^{ab} \tilde{J}^{\mu,b} = \tilde{\mathcal{A}}^a = -\frac{e^2}{32\pi^2}\epsilon^{\mu\nu\rho\sigma} \mathrm{tr}\,(\tau^a F_{\mu\nu} F_{\rho\sigma}) \quad .$$

Note that the r.h.s. of (13.72) is identical (except for a factor of $-1/2$) with the non-abelian axial anomaly in two dimensions, Eq. (11.3b). The same applies to four dimensions. This is not fortuitous, but can be shown to be the case in any even space-time dimensional[431], the reason being that $\tilde{J}_\mu$ corresponds to the definition

$$\tilde{J}^{\mu,a}(x) = -\frac{1}{2}\mathrm{tr}\tau^a\gamma^\mu(1-\gamma^5)G(x,x|eA) = \frac{1}{2}(J_V^{\mu,a}(x) - J_A^{\mu,a}(x)) \quad , \tag{13.73}$$

with $\mathcal{D}_\mu^{ab}J_V^{\mu,b} = 0$ and $\mathcal{D}_\mu^{ab}J_A^{\mu,b}$ the ABJ anomaly.

The anomaly (13.72) does not satisfy the consistency condition (13.33), and hence cannot be obtained as the gauge variation of the vacuum functional $W^{(L)}[A]$. This means in particular that it cannot be obtained by adding suitable (renormalization) counter-terms to $W^{(L)}[A]$. In fact: In $d = 2$ dimensions the most general ambiguity in $W^{(L)}$ is of the form[71] aA_μ^2. We see from (13.24) that, in the *non-abelian* case, there exists no value of the parameter a for which the r.h.s. in (13.68) becomes covariant.

It is interesting that in the Pauli definition (4.185) of the effective action we can equally well use for $J^{\mu,a}$ the *covariant* current. The result for $W^{(L)}$ will nevertheless be non-covariant as a result of the e' integration. In particular, the functional differentiation of $W^{(L)}$ with respect to $A^{\mu,a}$ will in this case yield the "minimal' form of the consistent current[10)].

13.3 Isomorphic Representations of Chiral QCD_2

Because of the non-invariance of the fermionic measure under the chiral transformations (13.19a,b), all dynamical variables are observable, and gauge fixing is neither required, nor allowed. We thus take the generating functional in configuration space to be given by the formula

$$Z[J,\eta,\bar{\eta}] \equiv \int \mathcal{D}A_\mu \int \mathcal{D}\bar{\psi}\mathcal{D}\psi e^{iS[A,\psi,\bar{\psi}]} e^{i\int(J^a.A^a+\bar{\eta}\psi+\bar{\psi}\eta)} \quad . \tag{13.74}$$

Invariance of the measure $\mathcal{D}A_\mu$ under translations $A_\mu(x) \to A_\mu(x) + \delta A_\mu(x)$ implies the equation of motion

$$\mathcal{D}_\mu^{ab}F_b^{\mu\nu} + e\bar{\psi}\tau^a\gamma^\nu P_-\psi = 0 \quad . \tag{13.75}$$

The validity of the representation (13.74) is by no means obvious, since as we shall see, one is dealing with a system with second class constraints, since gauge-invariance is broken on quantum level. Following Fujikawa[156], we may attribute

---

10) For the "minimal" current the anomaly only involves "old parity" terms. For $d = 2$, the minimal anomaly is given by (13.68) with $a = 1$.

this breakdown to the non-invariance of the fermionic measure under the chiral transformation (13.19a, b), as conveniently summarized by the following statement:

$$\mathcal{D}\bar{\psi}\mathcal{D}\psi e^{iS_F[A,\psi,\bar{\psi}]} = J\mathcal{D}\overline{^G\psi}\mathcal{D}^G\psi e^{iS_F[A,\psi,\bar{\psi}]} \quad , \tag{13.76}$$

with J the Jacobian of the transformation given by:

$$J = e^{-i\alpha_1[A,G^{-1}]} \quad . \tag{13.77}$$

Note that the Jacobian is given with reference to the fermionic action! For the infinitesimal gauge transformation (13.26) (see (13.27) and (13.28))

$$J \sim e^{i\int dx\delta\theta^a(x)A^a(x)} \quad . \tag{13.78}$$

It is immediately clear, that the generating functional does not allow one to define a "gauge-field" propagator, since the differential operator associated with S_G is not invertible, without choosing a gauge (which we are not allowed to do). The way out of this dilemma is to "embed" our chiral theory into a gauge-invariant one, which then allows us to choose a suitable gauge. The way to do this has been shown by Harada et al and Babelon et al[426], and relies on the Faddeev-Popov (gauge-fixing) procedure usually applied to true gauge theories. To this end we introduce in (13.74) the identity written in the usual form

$$\Delta_{\mathcal{F}}[A]\int \mathcal{D}g\delta[\mathcal{F}(A^g)] = 1 \quad , \tag{13.79}$$

where $\mathcal{F}$ is a suitable "section" and $\Delta_{\mathcal{F}}$ is a corresponding gauge invariant functional of A (Faddeev-Popov determinant) defined by (13.79). Doing the usual changes of variable, and using (13.76) we find, [11)]

$$Z[J,\eta,\bar{\eta}] = \int \mathcal{D}g\int d\mu[A]\int \mathcal{D}\psi\mathcal{D}\bar{\psi}e^{iI(A,g,\psi,\bar{\psi})}e^{i\int(J.{}^gA+\bar{\eta}^g\psi+{}^g\bar{\psi}\eta)} \quad , \tag{13.80}$$

where $d\mu$ is the Faddeev-Popov measure

$$d\mu[A] \equiv \mathcal{D}A_\mu\Delta_{\mathcal{F}}[A]\delta[\mathcal{F}(A)] \quad , \tag{13.81}$$

---

11) Note that we use interchangeably the notation

$${}^gA_\mu = A_\mu^{g^{-1}} = gA_\mu g^{-1} + \frac{i}{e}g\partial_\mu g^{-1}$$

and $I(A,g,\psi,\overline{\psi})$ is the new action

$$I(A,g,\psi,\overline{\psi}) = S[A,\psi,\overline{\psi}] + \alpha_1[A,g^{-1}] \tag{13.82}$$
$$= S_G[A] + I_F[A,g,\psi,\overline{\psi}] \quad , \tag{13.83}$$

with $\alpha_1(A,g^{-1})$ the 1-cocycle defined earlier, also referred to as the Wess-Zumino (WZ) term in the literature. The classical equation of motion associated with (13.82) reads

$$\mathcal{D}_\mu^{ab} F_b^{\mu\nu}(x) + e\overline{\psi}(x)\tau^a\gamma^\nu P_-\psi(x) + \frac{\delta\alpha_1[A,g^{-1}]}{\delta A_\nu^a(x)} = 0 \quad . \tag{13.84}$$

which replaces (13.75) in the new formulation[12)].

Recalling (13.76) it is easy to see that the (gauge variant) classical action (13.82) now defines at quantum level, a theory invariant under the *extended* gauge transformation

$$\begin{aligned} A_\mu \to A_\mu^{G^{-1}} \quad &, \quad g \to gG^{-1} \\ \psi \to G_-\psi \equiv {}^G\psi \quad &, \quad \overline{\psi} \to \overline{\psi}G_+^{-1} \equiv \overline{{}^G\psi} \quad . \end{aligned} \tag{13.85a}$$

For an infinitesimal transformation,

$$\begin{aligned} \delta A_\mu^a = \frac{1}{e}\mathcal{D}_\mu^{ab}\theta^b \quad &, \quad \delta g = -ig\theta \\ \delta\psi = i\theta P_-\psi \quad &, \quad \delta\overline{\psi} = -i\overline{\psi}P_+\theta \end{aligned} \tag{13.85b}$$

We have thus arrived at two representations for the same generating functional $Z[J,\eta,\overline{\eta}]$, which we refer to as the gauge non-invariant (GNI) and gauge invariant (GI) representation respectively. It follows immediately that correlation functions of A_μ, ψ and $\overline{\psi}$ in the GNI formulation are equal to the corresponding correlation functions of ${}^gA_\mu, {}^g\psi$ and $\overline{{}^g\psi}$ in the GI formulation[434]:

$$\langle 0|\mathcal{P}(A_\mu,\psi,\overline{\psi})|0\rangle_{GNI} = \langle 0|\mathcal{P}({}^gA_\mu,{}^g\psi,\overline{{}^g\psi})|0\rangle_{GI} \quad . \tag{13.86}$$

Since ${}^gA_\mu$ and ${}^g\psi$ are individually ***invariant*** under the extended gauge transformation (13.85), it follows that the ***rhs*** exhausts all possible "observables" of the "embedded" formulation, thus demonstrating that the two formulations are indeed isomorphic to each other[13)].

[12)] The modification of the fermionic action by addition of a Wess-Zumino term was first proposed by Faddeev[72]. As it was shown by D'Hoker and Farhi[433], the Wess-Zumino term also emerges from a chiral theory in which the fermion content has been adjusted such as to be anomaly free, in the limit of decoupling the heavy quark sector from the theory.

[13)] Other ways of embedding an anomalous chiral gauge theory into a bonafide gauge-theory have also been contemplated. Such alternative embeddings are not systematic and are generally limited to very special second class systems[435]. The $U(1)$ chiral Schwinger model for the JR parameter $a = 1$ provides an example.

It should be emphasized, that this isomorphism, that is, the identity of (13.80) and (13.74) only holds on full quantum level, where the integration over A_μ has also been performed. The result (13.80) is however independent of the choice of section, $\mathcal{F}(A) = 0$. Hence we are free to choose a gauge to define now the A_μ propagator. This identity becomes particularly transparent, if we generalize the result (13.80) to include also gauge-conditions which include g itself. This is easily done following once more the Faddeev-Popov procedure, by introducing in (13.80) the identity

$$\Delta_{\tilde{\mathcal{F}}}[A,g]\int \mathcal{D}\tilde{g}\delta[\tilde{\mathcal{F}}(A^{\tilde{g}},g\tilde{g})] = 1 \quad . \tag{13.87}$$

This again defines the functional $\Delta_{\tilde{\mathcal{F}}}[A,g]$, which is seen to be invariant under the extended gauge transformation (13.85):

$$\Delta_{\tilde{\mathcal{F}}}[A^{G^{-1}},gG^{-1}] = \Delta_{\tilde{\mathcal{F}}}[A,g] \quad . \tag{13.88}$$

Introducing the identity (13.87) in (13.80) and using (13.88), (13.79), we arrive at the generalization of (13.80), with $d\mu$ replaced by

$$d\mu[A,g] = \mathcal{D}A_\mu\Delta_{\tilde{\mathcal{F}}}[A,g]\delta[\tilde{\mathcal{F}}(A,g)] \quad . \tag{13.89}$$

We may now choose, in particular,

$$\tilde{\mathcal{F}}(A,g) = g - 1 \quad . \tag{13.90}$$

We refer to this as the *unitary* gauge. In this gauge $\alpha_1(A,g) = 0$, and representation (13.80) reduces identically to (13.74), thus making the equivalence of the two representations explicit. This fact has been referred to by stating that the correlation functions of the GNI formulation are just the correlation functions of the GI formulation in the "unitary" gauge. This statement is misleading in view of the isomorphism (13.86).

Let us emphasize that the identity of (13.74) and (13.80) does not exist on the level where A_μ is an external field. In fact, we shall see that the vacuum expectation value of the current of the GI formulation, in the presence of a c - number field A_μ, is covariantly conserved. This indicates that the inconsistency of the external field Maxwell equations is an artifact of the GNI formulation, and is expected to be absent on full quantum level[14)].

[14)] This conclusion is of course based on formal considerations and could turn out to be invalid, after proper renormalization. As we see, this expectation is nevertheless verified in the two-dimensional case.

13.4 External Field Ward Identities

The generating functional $Z[J,\eta,\overline{\eta}]$ in the GNI and GI formulation may be written in the form

$$Z[J,\eta,\overline{\eta}] = \int \mathcal{D}A^{\mu} e^{iS_G[A]} Z_F^{GNI}[A;\eta,\overline{\eta}] e^{i\int J^a \cdot A^a} \tag{13.91a}$$

$$= \int d\mu[A,g] e^{iS_G[A]} Z_F^{GI}[A;\eta,\overline{\eta}] e^{i\int J^a\cdot A^a} \tag{13.91b}$$

respectively, where

$$Z_F^{GNI}[A;\eta,\overline{\eta}] = \int \mathcal{D}\overline{\psi}\mathcal{D}\psi e^{iS_F[A,\psi,\overline{\psi}]+i\int(\overline{\eta}\psi+\overline{\psi}\eta)} \tag{13.92}$$

$$Z_F^{GI}[A;\eta,\overline{\eta}] = \int \mathcal{D}g \int \mathcal{D}\overline{\psi}\mathcal{D}\psi e^{iI_F[A,g,\psi,\overline{\psi}]+i\int(\overline{\eta}^g\psi+\overline{{}^g\psi}\eta)} \tag{13.93}$$

We consider separately the GNI and the GI case,

GNI formulation

Consider the generating functional $Z_F^{GNI}[A,\eta,\overline{\eta}]$ defined in (13.92). Using (13.77) as well as the gauge invariance of the classical fermionic action, we equivalently have, after suitable relabelling,

$$Z_F^{GNI}[A,\eta,\overline{\eta}] = \int \mathcal{D}\overline{\psi}\mathcal{D}\psi e^{iS_F^{(L)}[A^G,\psi,\overline{\psi}]-i\alpha_1(A,G)} e^{i\int(\overline{\eta}G_-\psi+\overline{\psi}G_+^{-1}\eta)} \quad . \tag{13.94}$$

Equating (13.92) and (13.94) for the case of an infinitesimal transformation,

$$\left(A_\mu^a\right)^G \sim A_\mu^a - \frac{1}{e}\mathcal{D}_\mu^{ab}\delta\theta^b \tag{13.95}$$

$$G_\pm \sim 1 + i\delta\theta^a\tau^a P_\pm \tag{13.96}$$

we arrive at the following infinite set of external field Ward identities,

$$0 = \left\langle\left\{\mathcal{D}_\mu^{ab}(\overline{\psi}(z)\tau^b\gamma_\mu\psi(z) - \mathcal{A}^a(z\mid A) - i\overline{\psi}(z)P_+\tau^a\eta(z) + i\overline{\eta}(z)\tau^a P_-\psi(z)\right\}\right\rangle_{A,\eta,\overline{\eta}}^{GNI} \tag{13.97}$$

This generalizes (13.29) to arbitrary external field correlation functions. Indeed, it immediately follows that

$$\langle T[\mathcal{D}_\mu^{ab}\overline{\psi}(z)\tau^b\gamma^\mu P_-\psi(z) - \mathcal{A}^a(z\mid A)]\mathcal{P}[\psi,\overline{\psi}]\rangle_A^{GNI} = 0 \quad ; \tag{13.98}$$

for any local polynomial $\mathcal{P}$ of ψ and $\overline{\psi}$ with arguments different from z. This generalizes Eq. (13.28).

GI formulation

Consider now the external field generating functional (13.93) of the GI formulation. Renaming the integration variables ψ, $\overline{\psi}$ and g by $\psi' = {}^G\psi$, $\overline{\psi}' = \overline{{}^G\psi}$, $g' = gG^{-1}$, and using the gauge invariance of $S_F^{(L)}$ and of the Haar measure $\mathcal{D}g$, we obtain

$$Z^{GI}[A;\eta,\overline{\eta}] = \int \mathcal{D}g \int \mathcal{D}\psi\mathcal{D}\overline{\psi}\, e^{iI_F[A^G,\psi,\overline{\psi}]} e^{i\int(\overline{\eta}\,{}^g\psi + \overline{{}^g\psi}\eta)} \quad , \tag{13.99}$$

where we have used the cocycle property (13.24).

Comparing again (13.99) with (13.93) for an infinitesimal gauge transformation we conclude this time, that

$$\left\langle T\mathcal{D}_\mu^{ab}\left\{\overline{\psi}(z)\tau^a\gamma^\mu P_-\psi(x) + \frac{1}{e}\frac{\delta\alpha_1(A,g^{-1})}{\delta A_\mu^b(x)}\right\}\mathcal{P}[{}^g\psi,\overline{{}^g\psi}]\right\rangle_A^{GI} = 0 \tag{13.100}$$

for any polynomial $\mathcal{P}$ of the (gauge invariant) fermion fields ${}^g\psi$ and $\overline{{}^g\psi}$ with arguments non-coincident with z.

We recognize the expression in curly brackets as being the source current of $F^{\mu\nu}$ in the GI formulation (see Eq. (13.84)). In particular, we have

$$\mathcal{D}_\mu^{ab}\mathcal{J}^{\mu,b}(x \mid A) = 0 \quad . \tag{13.101a}$$

where

$$\mathcal{J}^{\mu,a}(x \mid A) = \left\langle\overline{\psi}(x)\tau^a\gamma^\mu P_-\psi(x) + \frac{1}{e}\frac{\delta\alpha_1(A,g^{-1})}{\delta A_\mu^a(x)}\right\rangle_A^{GI} \tag{13.101b}$$

Hence the external field current in the GI formulation is covariantly conserved, and no inconsistency with the corresponding field equation (13.84) exists[15].

13.5 Ward-Takahashi Identities

We now consider the Ward-Takahashi (WT) identities satisfied on the full "operator level". We again examine separately the case of the GNI and GI formulation.

---

[15] An alternative approach for deriving these identities is to introduce a Pauli-Villars regulator field into the action, thereby rendering the action invariant under chiral transformations, and transferring the anomalous contribution to an explicit breaking of the chiral symmetry by the Pauli-Villars regulator mass term, which becomes the source of the chiral anomaly in the limit of infinite Pauli Villars mass. (See section 5.3 of chapter 4).

GNI formulation

We now show that the anomaly $\mathcal{A}^a$ actually vanishes on full quantum level, as it has to be the case, if the equations of motion (13.17) are to be consistent. Our starting point this time is the representation (13.74) of the generating functional.

The invariance of the measure $\mathcal{D}A_\mu$ under the infinitesimal gauge transformation (13.95) now directly leads to (no BRS analyses is needed, since the effective (quantum) action in this case is not gauge invariant!)

$$\langle \mathcal{D}_\mu^{ab}(\overline{\psi}(z)\tau^b\gamma^\mu P_-\psi(z)) + \mathcal{D}_\mu^{ab} J^{\mu,b}(z)\rangle_{J,\eta\overline{\eta}}^{GNI} = 0 \quad , \tag{13.102}$$

where $J^{\mu,b}$ is the external source current. Equation (13.102) implies

$$\langle \mathcal{D}_\mu^{ab}(\overline{\psi}(z)\tau^b\gamma^\mu P_-\psi(z))\mathcal{P}[A,\psi,\overline{\psi}]\rangle^{GNI} = 0 \quad , \tag{13.103}$$

as long as the arguments of $\mathcal{P}$ are not coincident with z. This shows that, on the operator level, we expect the anomaly $\mathcal{A}^a$ to vanish.

GI formulation

Since we are dealing this time with a true gauge theory, a BRST-type analysis is needed. Following common practise, we consider instead of the measure (13.81), the one obtained by performing the gaussian average

$$\int \mathcal{D}\kappa e^{-i\frac{\kappa^2}{2\alpha}}\mathcal{D}\mu_\kappa \quad , \tag{13.104a}$$

over the measure $\mathcal{D}\mu_\kappa$ defined by

$$d\mu_\kappa[A,g] = \mathcal{D}A_\mu \Delta_{\mathcal{F}}[A,g]\prod_a \delta[\mathcal{F}^a(A,g) - \kappa^a] \quad . \tag{13.104b}$$

This amounts to the substitution

$$d\mu[A,g] \to \mathcal{D}A_\mu \Delta_{\mathcal{F}}[A,g]e^{-i\frac{(\mathcal{F}^a(A,g))^2}{2\alpha}} \tag{13.105}$$

in the generating functional (13.91b). The simplest way to proceed now is to use the BRS symmetry of the effective Lagrangian obtained by exponentiating the Faddeev-Popov determinant $\Delta_{\mathcal{F}}$ involved in the measure (13.105). To this end we expand $\mathcal{F}^a$ in (13.87) around $\mathcal{F}^a(A,g) - \kappa^a = 0$. We suppose $\mathcal{F}^a$ to be a local function of A_μ^a and g. Then

$$\mathcal{F}^a(A^{\tilde{g}}(x), g\tilde{g}(x)) \sim \mathcal{F}^a(A,g) + \int dy \mathcal{M}^{ab}(x,y)\delta\tilde{\theta}^b(y) \quad , \tag{13.106}$$

where

$$\mathcal{M}_{ab}(x,y) \equiv \frac{\partial \mathcal{F}^a(A^{\tilde{g}}(x), g\tilde{g}(x))}{\partial \tilde{\theta}^b(y)}\Big|_{\tilde{\theta}=0} \quad . \tag{13.107}$$

Then

$$\Delta_{\mathcal{F}}[A,g] = (\det \mathcal{M}) = \int \mathcal{D}\bar{c}\mathcal{D}c e^{-\int \bar{c}^a \mathcal{M}_{ab} c^b} \quad . \tag{13.108}$$

Hence the partition function (13.80) now reads

$$Z[J,\eta,\bar{\eta}] = \int \mathcal{D}g \int \mathcal{D}A_\mu \int \mathcal{D}\psi\mathcal{D}\bar{\psi} e^{iI_{eff}[A,g,\psi,\bar{\psi}]} e^{i\int (J\cdot{}^gA + \bar{\eta}{}^g\psi + {}^g\bar{\psi})} \tag{13.109}$$

where $I_{eff} = I + S_{gh} + S_{gf}$, with I given by (13.82), and

$$S_{gh} = \int dx \bar{c} i \mathcal{M} c \tag{13.110a}$$

$$S_{gf} = -\int dx (\mathcal{F}^a(A,g))^2/2\alpha \quad . \tag{13.110b}$$

The BRS transformation, is chosen such as to leave $S = S_G + S_F$ and $S_{gh} + S_{gf}$ separately invariant (B stands for BRS)

$$\delta_B(S_G + S_F) = 0 \tag{13.111a}$$

$$\delta_B(S_{gh} + S_{gf}) = 0 \quad . \tag{13.111b}$$

For (13.111a) it is given by a gauge transformation (13.85b) with $\theta^a(x) = \delta\epsilon c^b(x)$ and $\delta\epsilon$ Grassmann-valued constant parameters ($c(x) = \tau^a c^a(x)$)[16],

$$\delta_B A^a_\mu(x) = \delta\epsilon \mathcal{D}^{ab}_\mu(x) c_b(x) \quad , \quad \delta_B g = -ie\delta\epsilon c(x) g(x)$$

$$\delta_B \psi(x) = ie\delta\epsilon c(x) P_- \psi(x) \quad , \quad \delta_B \bar{\psi}(x) = ie\delta\epsilon \bar{\psi}(x) P_+ c(x) \quad , \tag{13.112}$$

[16] The transformation laws can be streamlined[437] by introducing the operator s inducing the transformations

$$sA_\mu = D_\mu c \quad , \quad sg = -icg$$

$$sc = \frac{i}{2}[c,c] \quad , \quad s\bar{c} = -i\frac{\mathcal{F}(A,g)}{\alpha}$$

$$s\psi = iP_- c\psi \quad , \quad s\bar{\psi} = iP_+ \bar{\psi} c \quad ,$$

where $[\,,\,]$ stands for the ordinary Lie commutator $[X,Y]^a = if_{abc}X^bY^c$, and where s is supposed to satisfy the graded Leibnitz rule $s(XY) = (sX)Y + (-1)^{\epsilon(X)}X(sY)$, where $\epsilon(X)$ is the "Grassmann-signature" of X; s can be shown to be nilpotent ($s^2 = 0$) and thus acts as an exterior differential operator.

while Eq. (13.111*b*) further requires

$$\delta_B \bar{c}^a(x) = -\frac{i}{\alpha}\delta\epsilon \mathcal{F}^a(A(x), g(x))$$
$$\delta_B c^a(x) = -\frac{e}{2}\delta\epsilon f_{abc} c^b(x) c^c(x) \quad . \tag{13.113}$$

The only BRS non-invariant term is the 1-cocycle $\alpha_1[A, g^{-1}]$; using (13.24), is seen to transform as follows:

$$\delta_B I = \delta_B \alpha_1(A, g^{-1}) = -\int dx \delta\epsilon c^a(x) \mathcal{A}^a(x) \quad . \tag{13.114}$$

This change in I is compensated by the Jacobian (13.78), so that the partition function is BRS invariant. Hence the ST identities can be generated in the usual way.

13.6 The Effective Bosonic Action in the *GNI* Formulation

Our discussion so far has made no reference to the dimensionality of space-time. From now on we shall restrict ourselves to 1 + 1 dimensions.

In this case there are a number of techniques and results available to us. Thus the functional determinant appearing in the Mathews-Salam representation (4.1*b*) of the generating functional may be calculated explicitly in both the abelian and non-abelian case. The same applies to the external-field fermionic Greens function $G^{(L)}(x, y; A)$, if the gauge group is abelian (see chapter 14); in the non-abelian case this is not possible, but the answer is known to all orders of perturbation theory.

As we have witnessed already in the case of QCD_2, the fermionic functional determinant of chiral QCD_2 is also non-local and its dependence on A_μ is not explicit. As in the case of QCD_2, bosonization techniques can be used to overcome these difficulties. In fact such techniques will prove very important in order to arrive at a Hamiltonian formulation, which incorporates the anomalous behaviour of the fermionic measure under local chiral transformations. It will thus be our first objective, to arrive at a representation for the fermionic determinant analogous to that given in (11.81).

Consider first the QCD fermionic Lagrangian, which we may write in the form

$$\bar{\psi}(i\,\partial\!\!\!/ + e\,A\!\!\!/)\psi = \bar{\psi}(i\,\partial\!\!\!/ + e\,A\!\!\!/)P_+\psi + \bar{\psi}(i\,\partial\!\!\!/ + e\,A\!\!\!/)P_-\psi \tag{13.115}$$
$$= \bar{\psi}_R(i\partial_+ + eA_+)\psi_R + \bar{\psi}_L(i\partial_- + eA_-)\psi_L \quad . \tag{13.116}$$

Together with $\mathcal{D}\bar{\psi}\mathcal{D}\psi = \mathcal{D}\bar{\psi}_L\mathcal{D}\psi_L\mathcal{D}\bar{\psi}_R\mathcal{D}\psi_R$, this suggests the factorization of

$\det(i\not\partial + e\not A)$ into a left-and right-handed piece. This property can however not be preserved for a *gauge-invariant* regularization, as we have seen in chapter 11.

In a chiral theory, gauge-invariance is broken at the quantum level, which means, that there exists no regularization procedure preserving this local invariance. Hence we may obtain $W^{(L)}[A]$ directly from the effective action (11.61) by simply setting $A_+ = 0$. In terms of U and V, defined now in Minkowski space by (compare with (11.44))

$$eA_\mu = \frac{i}{2}(g_{\mu\nu} - \epsilon_{\mu\nu})U^{-1}\partial^\nu U + \frac{i}{2}(g_{\mu\nu} + \epsilon_{\mu\nu})V^{-1}\partial^\nu V \quad , \tag{13.117}$$

or

$$eA_+ = iU^{-1}\partial_+ U \quad , \quad eA_- = iV^{-1}\partial_- V \quad , \tag{13.118}$$

this corresponds to setting $U = 1$, i.e. $W^{(L)}[A] = -\Gamma[V^{-1}]$, where $\Gamma[G]$ is the WZW action

$$\Gamma[G] = S_{P\sigma M}[G] + S_{WZ}[G] \quad , \tag{13.119}$$

with $S_{P\sigma M}$ and S_{WZ} defined in (9.15) and (9.16). This is just one possible regularization procedure. The collection of all possible such regularizations can be resumed in the addition of a mass term of the form $\frac{ae^2}{8\pi}\mathrm{tr}A_\mu^2$ to the effective action (13.119), with a an arbitrary real parameter[71]. Hence the regularization ambiguity is taken into account by writing

$$\begin{aligned} W^{(L)}[A] = {} & \frac{ae^2}{8\pi}\mathrm{tr}A_\mu^2 - \frac{1}{8\pi}\int d^2x \mathrm{tr}(\partial_\mu V)(\partial^\mu V^{-1}) \\ & + \frac{1}{4\pi}\int_0^1 dr \int d^2x \epsilon^{\mu\nu}\mathrm{tr}[(V_r^{-1}\partial_r V_r)(V_r^{-1}\partial_\mu V_r)(V_r^{-1}\partial_\nu V_r) \quad . \end{aligned} \tag{13.120}$$

The e^2 dependence of the mass term reflects the fact that this "counterterm" results from one fermion-loop graphs in 2nd order perturbation theory.

In order to arrive at a bosonic functional representation for the fermionic determinant, we follow the procedure analogous to that in section 4 of chapter 11, and consider the identity

$$\int \mathcal{D}h e^{i(\Gamma[hV^{-1}] - \Gamma[V^{-1}])} = \mathrm{const.}\, e^{-i\Gamma[V^{-1}]} \tag{13.121}$$

which follows from the invariance of the Haar measure, $\mathcal{D}h$, under gauge transformations.

Notice that under a gauge transformation $A_\mu \to {}^hA_\mu$, we have according to (13.118), $V^{-1} \to hV^{-1}$, so that unlike the case of QCD_2, we are exploring this time the non-invariance of $\Gamma[V^{-1}]$ under ordinary gauge transformations. Now, the exponent in the integrand may be computed, using the Polyakov-Wiegman formula (11.73), which, in Minkowski space, reads

$$\Gamma[AB] = \Gamma[A] + \Gamma[B] + \frac{1}{4\pi}\int d^2x (g^{\mu\nu} + \epsilon^{\mu\nu}) \mathrm{tr}(A^{-1}\partial_\mu A)(B\partial_\nu B^{-1}) \tag{13.122}$$

$$= \Gamma[A] + \Gamma[B] + \frac{1}{4\pi}\int d^2x \mathrm{tr}(A^{-1}\partial_+ A)(B\partial_- B^{-1}) \quad . \tag{13.123}$$

Defining

$$\gamma_1[A,h] \equiv \Gamma[hV^{-1}] - \Gamma[V^{-1}] \quad , \tag{13.124}$$

one has

$$\gamma_1[A,h^{-1}] = \Gamma[h] - \frac{ie}{4\pi}\int d^2x \mathrm{tr}[A^\mu h^{-1}(\partial_\mu - \tilde{\partial}_\mu)h]$$

$$= S_{P\sigma M} + S_{WZ}[h] - \frac{ie}{4\pi}\int d^2x\, \mathrm{tr}[A^\mu h^{-1}(\partial_\mu - \tilde{\partial}_\mu)h] \quad , \tag{13.125}$$

From (13.119) and (13.121) we thus arrive at the following functional representation for the fermionic determinant:

$$e^{iW^{(L)}[A]} = e^{i\frac{ae^2}{4\pi}\mathrm{tr}A_\mu^2}\int \mathcal{D}h e^{i\gamma_1[A,h^{-1}]} \quad . \tag{13.126}$$

Note that $\gamma_1[A,h]$ constructed from $\Gamma[V^{-1}]$, also satisfies the 1-cocycle condition (13.24). It is important to keep in mind, that it differs from $\alpha_1[A,g]$ defined in (13.23), which is constructed from $W^{(L)}[A]$, containing the mass term, and which plays a role in the effective action (13.82) of the *GI* formulation. We have

$$\alpha_1[A,g^{-1}] = W^{(L)}[A^{g^{-1}}] - W^{(L)}[A] =$$
$$-(1-a)S_{P\sigma M}[g] - S_{WZ}[g] + \frac{ie}{4\pi}\int d^2x \mathrm{tr}[A^\mu g^{-1}((1-a)\partial_\mu - \tilde{\partial}_\mu)g] \quad . \tag{13.127}$$

Notice that for $a \to 0$, $\alpha_1[A,g^{-1}] \to -\gamma_1[A,g^{-1}]$[17].

From (13.127) we obtain for the consistent anomaly (see Eq. (13.28))

$$\mathcal{A}^a(x) = -\frac{\delta\alpha_1(A,g^{-1})}{\delta\theta^a(x)}\bigg|_{\theta^a=0} = \frac{e}{4\pi}[(a-1)\partial_\mu + \tilde{\partial}_\mu]A^{\mu,a}(x) \quad . \tag{13.128}$$

We now show how to construct $\alpha_1[A,g^{-1}]$ from the knowledge of the anomaly alone.

---

[17] This puts in evidence the cancellation of fermion loop contributions in formally gauge invariant quantities of the *GNI* formulation[438].

13.7 Construction of the one-Cocycle from the Anomaly

Let us consider the following general problem: Suppose we are given a functional $\alpha[A,G]$ in d-dimensions:

First check, whether it satisfies the 1-cocycle condition

$$\alpha[A^g,G] - \alpha[A,gG] + \alpha[A,g] = 0 \quad , \tag{13.129a}$$

which in particular implies

$$\alpha[A^g,g^{-1}] = -\alpha[A,g] \quad . \tag{13.129b}$$

Make the substitutions $A_\mu \to A_\mu^{g_r}$, $g \to \delta g_r$, and $G \to g_r g_{\delta r}$ in Eq. (13.129), where

$$g_r = e^{ir\theta} \quad , \quad g_{\delta r} = e^{i\delta r\theta} \quad . \tag{13.130}$$

This results in

$$\alpha[A^{g_r g_{\delta r}},(g_r g_{\delta r})^{-1}] - \alpha[A^{g_r},g_r^{-1}] = -\alpha[A^{g_r},g_{\delta r}] \tag{13.131}$$

Suppose now we are given the behavior of $\alpha[A,G]$ near the identity $G=1$:

$$\alpha[A,1+i\delta r\theta] \sim \delta r \int dx \mathrm{tr}\theta(x)\mathcal{A}(A(x)) \tag{13.132}$$

We then have from (13.130),

$$\frac{d}{dr}\alpha[A^{g_r},g_r^{-1}] = -\int dx \mathrm{tr}\theta(x)\mathcal{A}(A^{g_r}(x)) \quad . \tag{13.133}$$

Integration leads to the unique result

$$\alpha[A,g] = \int_0^1 dr \int dx \mathrm{tr}[\theta(x)\mathcal{A}(A^{g_r}(x))] \quad , \tag{13.134}$$

where we have once again used the cocycle property in the form of (13.129b).

The corresponding 1-cocycle in chiral QCD_4 can be obtained in an analogous way[439] by integrating the consistent anomaly (13.10a) over a one parameter family of gauge transformations.

The above procedure allows, in principle, for the construction of the cocycle $\alpha_1[A,g]$ from the anomaly in arbitrary dimensions![18)] To construct $\alpha_1[A,g]$ of chiral QCD_2 we choose for $\mathcal{A}^a$ the consistent anomaly (13.128); hence,

$$\alpha_1[A,g] = \frac{e}{4\pi}\int_0^1 dr \int d^2x \mathrm{tr}\{\theta[(a-1)\partial^\mu + \tilde{\partial}^\mu]A_\mu^{g_r}(x)\} \quad . \tag{13.135}$$

[18)] Note that this is not possible for the effective action.

Some algebra shows, that the r.h.s. can be rewritten in the form (recall that $\theta = -ig_r^{-1}\partial_r g_r$)

$$\frac{e}{4\pi}\int_0^1 dr\frac{\partial}{\partial r}\int d^2x\Big\{\frac{i}{4\pi}\mathrm{tr}A^\mu(x)g_r[(1-a)\partial_\mu-\tilde{\partial}_\mu]g_r^{-1}+\frac{a-1}{2e}(\partial_\mu g_r^{-1}\partial^\mu g_r)\Big\}$$

$$-\frac{1}{4\pi}\int_0^1 dr\int d^2x\epsilon^{\mu\nu}\mathrm{tr}[(g_r^{-1}\partial_r g_r)(g_r^{-1}\partial_\mu g_r)(g_r^{-1}\partial_\nu g_r)] \quad , \tag{13.136}$$

where we have dropped a surface term arising from an integration by parts. Integrating the last term once more by parts, we arrive at expression (13.127) with g replaced by g^{-1}. The corresponding expression for the 1-cocycle $-\gamma_1[A,h]$ in (13.125) is obtained by setting $g = h$ and $a = 0$ in (13.127). Note that the parameter r is essentially the parameter t in the algebraic construction of the Zumino descent equation[74].

13.8 Bosonic Action in the *GNI* and *GI* Formulation

Recalling (13.82) as well as (13.126), we arrive at the following equivalent bosonic actions for the GNI and GI case, respectively:

$$S_{bos} = \int d^2x\mathrm{tr}\Big\{-\frac{1}{4}F_{\mu\nu}F^{\mu\nu}+\frac{ae^2}{8\pi}A_\mu A^\mu\Big\}+\gamma_1[A,h^{-1}] =$$

$$\int d^2x\mathrm{tr}[-\frac{1}{4}F_{\mu\nu}F^{\mu\nu}+\frac{ae^2}{8\pi}A_\mu A^\mu-\frac{ie}{4\pi}(A^\mu h^{-1}(\partial_\mu-\tilde{\partial}_\mu)h)]+S_{P\sigma M}[h]+S_{WZ}[h] \tag{13.137}$$

and

$$I_{bos} = \int d^2x\mathrm{tr}\Big\{-\frac{1}{4}F_{\mu\nu}F^{\mu\nu}+\frac{ae^2}{8\pi}A_\mu A^\mu\Big\}+\gamma_1[A,h^{-1}]+\alpha_1[A,g^{-1}]$$

$$= S_{bos}+\frac{ie}{4\pi}\int\mathrm{tr}[A^\mu g^{-1}((1-a)\partial_\mu-\tilde{\partial}_\mu)g]-(1-a)S_{P\sigma M}[g]-S_{WZ}[g] \tag{13.138}$$

with $S_{P\sigma M}$ and S_{WZ} given by (9.15,16). We shall refer to $g(x)$ as the *Wess-Zumino field.*

The classical equations of motion in the GNI and GI formulation are given by

$$\frac{\delta S_{bos}}{\delta A_\mu^a(x)} = 0 \quad , \quad \frac{\delta S_{bos}}{\delta h_{ij}(x)} = 0 \tag{13.139}$$

and

$$\frac{\delta I_{bos}}{\delta A_\mu^a(x)} = 0 \quad , \quad \frac{\delta I_{bos}}{\delta h_{ij}(x)} = 0 \quad , \quad \frac{\delta I_{bos}}{\delta g_{ij}(x)} = 0 \tag{13.140}$$

respectively.

The computation of the variation of $S_{P\sigma M}[h]$ and $S_{WZ}[h]$ with respect to h proceeds as in chapter 9, the result being given by (9.18) and (9.23), with g replaced by h. Using these results, we easily obtain the equations motion. We discuss separately the case of the GNI and GI formulation.

GNI formulation

Using the results (9.18) and (9.23) one finds from (13.137)

$$\begin{aligned}\delta S_{bos} = &\int d^2x\delta A_{\nu,a}\Big(\mathcal{D}_\mu^{ab}F_b^{\mu\nu} + \frac{ae^2}{4\pi}A_a^\nu - \frac{ie}{4\pi}\mathrm{tr}(\tau^a h^{-1}(\partial^\nu - \tilde{\partial}^\nu)h)\Big)\\ &+\frac{1}{4\pi}\int d^2x \mathrm{tr}\Big((h^{-1}\delta h)\Big[D_\nu(h^{-1}(\partial^\nu - \tilde{\partial}^\nu)h + ie(\partial_\nu - \tilde{\partial}_\nu)A^\nu\Big]\Big) \quad , \qquad (13.141)\end{aligned}$$

where $\mathcal{D}_\mu$ is the usual covariant derivative, and where D_μ acts on elements in the Lie algebra as $D_\mu = \partial_\mu + [A_\mu, ...]$. The requirement $\delta S_{bos} = 0$ thus leads to the equations of motion,

$$D_\mu F^{\mu\nu} + \frac{ae^2}{4\pi}A^\nu - \frac{ie}{4\pi}h^{-1}(\partial^\nu - \tilde{\partial}^\nu)h = 0 \qquad (13.142a)$$

$$D_\mu(-ih^{-1}(\partial^\mu - \tilde{\partial}^\mu)h) + e(\partial_\mu - \tilde{\partial}_\mu)A^\mu = 0 \quad . \qquad (13.142b)$$

Equation (13.142) is of the form

$$D_\mu F^{\mu\nu} + eJ_L^\nu = 0 \qquad (13.143)$$

where

$$J_L^\nu = \frac{ae}{4\pi}A^\nu - \frac{i}{4\pi}h^{-1}(\partial^\nu - \tilde{\partial}^\nu)h \quad . \qquad (13.144)$$

We identify J_L^ν with the left-handed part of the fermionic current.

It follows from (13.142-144) that

$$D_\nu J_L^\nu = \frac{e}{4\pi}[(a-1)\partial_\nu + \tilde{\partial}_\nu]A^\nu \quad . \qquad (13.145)$$

The r.h.s. of this equation has exactly the form of the (consistent, non-abelian) anomaly (13.128), with A^ν now an operator valued field. Following the discussion of section 2.4, we arrive at the covariant anomaly, by considering the (covariant) divergence of a new current (13.57), where $\mathcal{P}^{\mu,a}$ is a determined by condition (13.67). As it was shown in section 2.4, we have

$$\tilde{J}_L^\mu = -\frac{i}{4\pi}h^{-1}(\partial^\nu - \tilde{\partial}^\nu)h + \frac{e}{4\pi}(g^{\mu\nu} - \epsilon^{\mu\nu})A_\nu \quad . \qquad (13.146)$$

Using the equations of motion (13.142) above, one verifies

$$\mathcal{D}_\mu^{ab}\tilde{J}_L^{\mu,b} = -\frac{e}{4\pi}(\partial_\mu - \tilde{\partial}_\mu)A^{\mu,a} + \frac{e}{4\pi}(g^{\mu\nu} - \epsilon^{\mu\nu})\mathcal{D}_\mu^{ab}A_\nu^b = -\frac{e}{4\pi}\epsilon^{\mu\nu}F_{\mu\nu}^a \quad . \tag{13.147}$$

Except for the term proportional to A^ν in (13.144), arising from the ambiguity in regularization, the left-handed current (13.144) has the same form as the corresponding current (9.27) for the case of free fermions, in agreement with the "adiabatic" principle of form invariance (see chapter 10). Invoking once again this principle, we make the following Ansatz for the right handed fermionic current

$$j_R^\mu = -\frac{i}{4\pi}h(\partial^\mu + \tilde{\partial}^\mu)h^{-1} \quad . \tag{13.148}$$

This current is however *not* conserved! Indeed, rewriting the variation $\delta_h S_{eff}$ in the alternative form (compare with (13.141)),

$$\delta_h S_{bos} = -\frac{1}{4\pi}\int d^2x \mathrm{tr}\left\{(\delta h h^{-1})[\partial_\mu(h(\partial^\mu + \tilde{\partial}^\mu)h^{-1}) - ie(g_{\mu\nu} + \epsilon_{\mu\nu})\partial^\mu(hA^\nu h^{-1})]\right\} \tag{13.149}$$

we see that

$$\partial_\mu j_R^\mu = \frac{e}{4\pi}(g^{\mu\nu} + \epsilon^{\mu\nu})\partial_\mu(hA_\nu h^{-1}) \quad . \tag{13.150}$$

Hence j_R^μ cannot represent the right handed current, which should be conserved, since the right-handed fermion does not interact with the gauge field[19]. We note however that the right hand-side of Eq. (13.150) is a divergence, so that one may define a new current[20]

$$J_R^\mu = j_R^\mu - \frac{e}{4\pi}(g^{\mu\nu} + \epsilon^{\mu\nu})(hA_\nu h^{-1}) = -\frac{1}{4\pi}(g^{\mu\nu} + \epsilon^{\mu\nu})hiD_\nu h^{-1} \quad , \tag{13.151a}$$

which is conserved:

$$\partial_\mu J_R^\mu = 0 \quad . \tag{13.151b}$$

Notice that J_R^μ is a gauge singlet, and is self-dual $J_R^0 = J_R^1$, as expected.

[19] The "J-R ambiguity" is attributed entirely to the left-handed part.
[20] D_ν applied on group-valued elements acts by simple matrix multiplication.

GI formulation

In order to obtain the equations of motion in the GI formulation, we need to compute δI_{bos}. We have this time

$$\delta I_{bos} = \int d^2x \delta A^a_\nu [\mathcal{D}^{ab}_\mu F^{\mu\nu}_b + e\,\mathrm{tr}(\tau^a \mathcal{J}^\nu)]$$
$$+\frac{1}{4\pi}\int d^2x \mathrm{tr}\Big\{h^{-1}\delta h[D_\mu(h^{-1}(\partial^\mu - \tilde{\partial}^\mu)h) + ie(\partial_\mu - \tilde{\partial}_\mu)A^\mu]\Big\}$$
$$-\frac{1}{4\pi}\int d^2x \mathrm{tr}\Big\{g^{-1}\delta g[D_\mu(g^{-1}[(1-a)\partial^\mu - \tilde{\partial}^\mu]g) + ie[(1-a)\partial_\mu - \tilde{\partial}_\mu]A^\mu]\Big\} \quad , \qquad (13.153)$$

where we have made already the identification,

$$\mathcal{J}^\mu = -\frac{i}{4\pi}h^{-1}(\partial^\mu - \tilde{\partial}^\mu)h + \frac{i}{4\pi}g^{-1}((1-a)\partial^\mu - \tilde{\partial}^\mu)g + \frac{ae}{4\pi}A^\mu \quad . \qquad (13.154)$$

Hence, from the condition $\delta I_{bos} = 0$ we obtain the equations of motion[440,441],

$$D_\mu F^{\mu\nu} + e\mathcal{J}^\mu = 0 \qquad (13.155a)$$
$$D_\mu[-ih^{-1}(\partial^\mu - \tilde{\partial}^\mu)h] = -e(\partial_\mu - \tilde{\partial}_\mu)A^\mu \qquad (13.155b)$$
$$D_\mu[-ig^{-1}((1-a)\partial^\mu - \tilde{\partial}^\mu)g] = -e[(1-a)\partial_\mu - \tilde{\partial}_\mu]A^\mu \quad . \qquad (13.155c)$$

It is easy to check that equations (13.155b,c) by themselves already imply the covariant conservation of $\mathcal{J}^\mu$,

$$D_\mu \mathcal{J}^\mu = 0 \quad . \qquad (13.156)$$

As a result, the current (13.154) is also covariantly conserved for A_μ an *external* field, which is nothing but the content of the external field Ward identity (13.101), if one notes that from (13.127), one has

$$\frac{\delta\alpha_1[A,g]}{\delta A^a_\mu(x)} = \frac{ie}{4\pi}\mathrm{tr}\tau^a g^{-1}[(1-a)\partial_\mu - \tilde{\partial}_\mu]g \quad . \qquad (13.157)$$

Thus, the seeming inconsistency of Maxwells equations has disappeared already on this intermediate level, where A_μ is treated as external field!

There exist two further currents which are conserved in the usual sense. Following the previous line, we rewrite $\delta_g I_{bos}$ in the form (compare with Eq. (13.153))

$$\delta_g I_{bos} = \frac{1}{4\pi}\int d^2x \mathrm{tr}\delta g g^{-1} \times$$
$$\Big\{[\partial_\mu(g[(1-a)\partial^\mu + \tilde{\partial}^\mu]g^{-1} + ie[(1-a)g_{\mu\nu} + \epsilon_{\mu\nu}]\partial^\mu(gA^\nu g^{-1})\Big\} \quad . \qquad (13.158)$$

This shows that in addition to [21)]

$$\mathcal{J}_R^\mu = -\frac{1}{4\pi}(g^{\mu\nu} + \epsilon^{\mu\nu})hiD_\nu h^{-1} \tag{13.159a}$$

with $\partial_\mu \mathcal{J}_R^\mu = 0$, there still exists another conserved current[442]

$$\mathcal{J}_{WZ}^\mu = -\frac{1}{4\pi}[(1-a)g^{\mu\nu} + \epsilon^{\mu\nu}]giD_\nu g^{-1} \quad , \tag{13.159b}$$

with $\partial_\mu \mathcal{J}_{WZ}^\mu = 0$. We shall refer to $\mathcal{J}_{WZ}^\mu$ as the *Wess-Zumino current.*

13.9 Symmetries of the Model

GNI formulation - Global symmetries

The fermionic action (13.16*b*) is invariant under *global* chiral transformations

$$\begin{aligned} \psi_L(x) &\to \Omega_L \psi_L(x) \\ \psi_R(x) &\to \Omega_R \psi_R(x) \\ A_\mu &\to \Omega_L A_\mu \Omega_L^{-1} \end{aligned} \tag{13.160a}$$

with $(\Omega_L, \Omega_R) \in U_L(N) \times U_R(N)$. Correspondingly the bosonic action (13.137) is invariant under the transformation

$$h(x) \to \Omega_R h \Omega_L^{-1} \quad , \quad A_\mu \to \Omega_L A_\mu \Omega_L^{-1} \tag{13.160b}$$

which in turn implies the transformation law

$$J_L^\mu \to \Omega_L J_L^\mu \Omega_L^{-1} \tag{13.161a}$$

$$J_R^\mu \to \Omega_R J_R^\mu \Omega_R^{-1} \tag{13.161b}$$

for the currents.

GNI formulation - Local symmetries

The local left-symmetry is broken at the quantum level. As a result we find that, when gauging $SU(N)_L$ $((\Omega_L \to G(x))$

$$h(x) \to \Omega_R h(x) G^{-1}(x) \tag{13.162}$$

$$A_\mu(x) \to G(x) A_\mu(x) G^{-1}(x) + \frac{i}{e} G(x) \partial_\mu G^{-1}(x) \tag{13.163}$$

---

21) The use of script $\mathcal{J}_R^\mu$ is to remind the reader, that $\mathcal{J}_R^\mu$, though of the same form as (13.151*a*), now satisfies a different dynamics in the *GI* formulation.

the left-handed GNI current transforms non-covariantly:

$$\begin{aligned} J_L^\mu \to & -\frac{i}{4\pi} G h^{-1}(\partial^\mu - \tilde{\partial}^\mu)(hG^{-1}) + \frac{ae}{4\pi} G A^\mu G^{-1} + \frac{ia}{4\pi} G \partial^\mu G^{-1} \\ & = G J_L^\mu G^{-1} + \frac{i}{4\pi} G[(a-1)\partial^\mu + \tilde{\partial}^\mu] G^{-1} \end{aligned} \quad , \qquad (13.164)$$

while the right-handed current transforms in a covariant way

$$J_R \to G J_R G^{-1} \quad . \qquad (13.165)$$

Eq. (13.164) shows, that the local transformation is not a symmetry of the equation of motion (13.142).

GI formulation – Global symmetries

In the GI formulation, we have found three conserved currents, $\mathcal{J}^\mu$, $\mathcal{J}_R^\mu$ and $\mathcal{J}_{WZ}^\mu$. Their conservation follows from the invariance of the cocycles $\gamma_1[A, h^{-1}]$ and $\alpha_1[A, g^{-1}]$ under the global transformations

$$g \to \Omega_{WZ}\, g\, \Omega_L^{-1} \quad , \quad h \to \Omega_R\, h\, \Omega_L^{-1} \qquad (13.166a)$$

$$A_\mu \to \Omega_L A_\mu \Omega_L^{-1} \quad . \qquad (13.166b)$$

Correspondingly we have from (13.159) the covariant transformation laws

$$\begin{aligned} \mathcal{J}^\mu &\to \Omega_L \mathcal{J}^\mu \Omega_L^{-1} \\ \mathcal{J}_R^\mu &\to \Omega_R \mathcal{J}_R^\mu \Omega_R^{-1} \\ \mathcal{J}_{WZ}^\mu &\to \Omega_{WZ} \mathcal{J}_{WZ}^\mu \Omega_{WZ}^{-1} \end{aligned} \qquad (13.167)$$

GI formulation – Local symmetries

The invariance of I_{bos} under local $SU(N)_L$ transformations, $(\Omega_L \to G(x) \in SU(N)_L)$,

$$\begin{aligned} A_\mu \to A_\mu^{G^{-1}} &= G A_\mu G^{-1} + \frac{i}{e} G \partial_\mu G^{-1} \\ h \to h G^{-1} \quad &, \quad g \to g G^{-1} \end{aligned} \qquad (13.168)$$

follows from the fact, that

$$\gamma_1[A, G^{-1}] + \alpha_1[A, G^{-1}] = -\frac{ae^2}{8\pi}[\mathrm{tr}(A_\mu^{G^{-1}})^2 - \mathrm{tr}(A_\mu)^2] \qquad (13.169)$$

and the cocycle property, which implies

$$\alpha_1[A^{G^{-1}}, Gg^{-1}] + \gamma_1[A^{G^{-1}}, Gh^{-1}] = \\ (\alpha_1[A, g^{-1}] + \gamma_1[A, h^{-1}]) - (\alpha_1[A, G^{-1}] + \gamma_1[A, G^{-1}]) \quad . \tag{13.170}$$

As a result, the global $SU(N)_L$ symmetry is "lifted" to a local $SU_L(N)$-symmetry via the gauge-interaction, the symmetry group of I_{bos} being in fact $U_{WZ}(N) \times U_R(N) \times SU_L(N)^{gauge}$. Since the current $\mathcal{J}^\mu$ is the source of $F^{\mu\nu}$, it should transform covariantly under a gauge transformation. Indeed, under the transformation (13.168) we have

$$\mathcal{J}^\mu \underset{G}{\rightarrow} G^{-1}\mathcal{J}^\mu G - \frac{i}{4\pi}G^{-1}(\partial^\mu - \tilde{\partial}^\mu)G + \frac{i}{4\pi}G^{-1}[(1-a)\partial^\mu - \tilde{\partial}^\mu]G + \frac{ia}{4\pi}G^{-1}\partial_\mu G \\ = G^{-1}\mathcal{J}^\mu G \quad . \tag{13.171}$$

On the other hand, the quantities $h(i\partial_\nu + eA_\nu)h^{-1}$ and $g(i\partial_\nu + eA_\nu)g^{-1}$ are left invariant under the transformations (13.168) so that $\mathcal{J}_R^\mu$ and $\mathcal{J}_{WZ}^\mu$ are gauge singlets:

$$\mathcal{J}_R^\mu \underset{G}{\rightarrow} \mathcal{J}_R^\mu \quad , \quad \mathcal{J}_{WZ}^\mu \underset{G}{\rightarrow} \mathcal{J}_{WZ}^\mu \quad . \tag{13.172}$$

13.10 Relation of Source Currents in *GNI* and *GI* Formulations

Eq. (13.171) shows, that the source-current (13.154) of the *GI* formulation is not gauge invariant - and hence, not an observable. The source-current (13.144) of the *GNI* formulation is, on the other hand, an observable. According to the isomorphism (13.86) it corresponds in the *GI* formulation to

$$\begin{aligned} {}^gJ_L^\mu &= -\frac{i}{4\pi}gh^{-1}(\partial^\mu - \tilde{\partial}^\mu)(hg^{-1}) + \frac{ae}{4\pi}{}^gA^\mu \\ &= gJ_L^\mu g^{-1} + \frac{i}{4\pi}g[(a-1)\partial^\mu + \tilde{\partial}^\mu]g^{-1} \end{aligned} \tag{13.173}$$

One readily checks, that ${}^gJ_\mu$ is invariant under the extended gauge transformation (13.85). It is easy to see, what the relation between $\mathcal{J}^\mu$ and J_L^μ should be. Indeed, according to the isomorphism (13.86), the equation of motion (13.143) of the *GNI* formulation ***should*** read

$$D_\mu({}^gA){}^gF^{\mu\nu} + e\,{}^gJ_L^\nu = 0 \tag{13.174}$$

in the *GI* formulation. Using

$$D_\mu({}^gA) = gD_\mu(A)g^{-1} \quad , \quad {}^gF^{\mu\nu} = gF^{\mu\nu}g^{-1} \tag{13.175}$$

we obtain from (13.174)

$$D_\mu(A)F^{\mu\nu} + eg^{-1}({}^gJ_L^\nu)g = 0 \quad . \tag{13.176}$$

One verifies immediately that

$$g^{-1}({}^gJ_L^\mu)g = \mathcal{J}^\mu \quad . \tag{13.177}$$

This establishes the desired relation between the two currents.

From the fermionic point of view, the result (13.177) is a consequence of the anomalous transformation properties of the local operator $\overline{\psi}\tau^a\gamma^\mu P_-\psi$ under a gauge transformation. Let us examine, how this result emerges in a functional approach.

From the GI representation (13.80) for the generating functional (13.74) we obtain, after performing the integration over the fermions

$$Z[J,\eta,\overline{\eta}] = \int \mathcal{D}g \int d\mu[A] e^{i(S_G[A]+W^{(L)}[A])+i\alpha_1[A,g^{-1}]}$$
$$\times e^{-i\int d^2x \int d^2y \overline{\eta}(x)G(x,y|{}^gA)\eta(y)} e^{i\int d^2x J_\mu {}^gA_\mu} \quad , \tag{13.178}$$

where $G(x,y \mid {}^gA)$ is the inverse of the Dirac operator $i\,\not{D}({}^gA) = i\,\not{\partial} + e{}^g\not{A}P_-$. From (13.178) we obtain

$$\langle 0 \mid T\overline{\psi}(x)\tau^a\gamma^\mu P_-\psi(x)\mathcal{P}[A] \mid 0\rangle^{GNI} = -\langle 0 \mid T\,\mathrm{tr}\{\tau^a\gamma^\mu P_-\not{D}^{-1}[{}^gA]\}\mathcal{P}[{}^gA] \mid 0\rangle^{GI} \tag{13.179}$$

where $\mathcal{P}[A]$ is some polynomial functional of A^μ, and where the trace is taken with respect to spin and internal symmetry indices. Using the identity

$$-e\mathrm{tr}(\tau^a\gamma^\mu P_-\not{D}^{-1}[A])_{xx} = -i\frac{\delta}{\delta A_\mu^a(x)}\mathrm{trln}\,\not{D}[A] = \frac{\delta}{\delta A_\mu^a(x)}W^{(L)}[A] \tag{13.180}$$

we may rewrite (13.179) in the form

$$\langle 0 \mid T\overline{\psi}(x)\tau^a\gamma^\mu P_-\psi(x)\mathcal{P}[A] \mid 0\rangle^{GNI} = \frac{1}{e}\left\langle 0\middle| T\frac{\delta}{\delta {}^gA_\mu^a(x)}W^{(L)}[{}^gA]\mathcal{P}[{}^gA]\middle|0\right\rangle^{GI}$$
$$= \frac{1}{e}\mathrm{tr}(\tau^b g^{-1}\tau^a g)\left\langle 0\middle|\frac{\delta}{\delta A_\mu^b(x)}W^{(L)}[{}^gA]\mathcal{P}[{}^gA]\middle|0\right\rangle^{GI} \tag{13.181}$$

or recalling definition (13.127) of $\alpha_1[A,g^{-1}]$,

$$\langle 0 \mid T\overline{\psi}(x)\tau^a\gamma^\mu P_-\psi(x)\mathcal{P}[A] \mid 0\rangle^{GNI} =$$
$$\mathrm{tr}(\tau^a g\tau^b g^{-1})\left\langle 0\middle| T\left\{\frac{1}{e}\frac{\delta W^{(L)}[A]}{\delta A_\mu^b(x)} + \frac{1}{e}\frac{\delta\alpha_1[A,g^{-1}]}{\delta A_\mu^b(x)}\right\}\mathcal{P}[{}^gA]\middle|0\right\rangle^{GI} \quad . \tag{13.182}$$

It is easy to see that this result is just a restatement of (13.177), the l.h.s. corresponding to ${}^gJ_\mu^{\mu,a}$, and the r.h.s. to ${}^gJ_L^\mu = g\mathcal{J}_\mu g^{-1}$.

13.11 Poisson Algebra of the Currents

The algebra of the currents plays a central role in the characterization of the symmetries involved in a dynamical theory. Two types of algebras are of interest:
i) the algebra defined in terms of ***Poisson brackets*** and
ii) the algebra defined in terms of ***Dirac brackets***[338]

In the case where no (first or second class) constraints are involved, the two types of brackets coincide.

If one wants to discuss the behaviour of a dynamical system under local symmetry transformations, it is the Poisson algebra which is of interest. If the classical symmetry is broken on quantum level, this Poisson algebra will have to take account of the quantum effects responsible for this symmetry breakdown. In two dimensional anomalous gauge theories this can be done by working with the effective bosonic action (S_{bos} or I_{bos}), which already incorporates the anomaly on classical level.

GNI formulation

The r-integration in the WZ-action (9.15) cannot be carried out explicitly. We recognize however, that $S_{WZ}[h]$ is a linear functional of $\partial_0 h$. Hence, performing the r-integration will lead to an expression of the form (9.34) in terms of $A(h)$, which is some unknown matrix-valued function of h. Fortunately we need not know $A(h)$ itself, but only the antisymmetric tensor $F_{ij;kl}$, given in (9.36c).

As we have seen in chapter 9, this tensor arises naturally when computing the variation of S_{WZ} with respect to h[337]:

$$\delta_h S_{WZ}[h] = \frac{1}{4\pi}\int d^2x \epsilon^{\mu\nu} F_{ij;kl}(x)\partial_0 h_{ji}\delta h_{lk} \quad . \tag{13.183}$$

Comparing with the explicit result for the variation one sees that $F_{ij;kl}(x)$ is given by (9.36c), with g replaced by h.

Making the replacement (9.34) in (13.137) we are lead to the following equivalent bosonic Lagrangian:

$$\mathcal{L}_{bos}^{GNI} = \mathcal{L}_G + \mathcal{L}_{P\sigma M} + \mathcal{L}_{WZ} + \mathcal{L}_I \quad , \tag{13.184}$$

with

$$\begin{aligned}
\mathcal{L}_G &= -\frac{1}{4}\mathrm{tr}F_{\mu\nu}F^{\mu\nu} + \frac{ae^2}{8\pi}\mathrm{tr}(A_\mu A^\mu) \\
\mathcal{L}_{P\sigma M} &= \frac{1}{8\pi}(\partial h^{-1})(\partial^\mu h) \\
\mathcal{L}_{WZ} &= \frac{1}{4\pi}\mathrm{tr}\mathcal{A}(h)\partial_o h \\
\mathcal{L}_I &= \frac{-ie}{4\pi}\mathrm{tr}[h^{-1}\partial^\mu h(g_{\mu\nu} + \epsilon_{\mu\nu})A^\nu] \quad .
\end{aligned} \tag{13.185}$$

From here we obtain for the momenta conjugate to $A^{\mu,a}$ and h_{ij}, respectively:

$$\pi^a_\mu = F^a_{\mu o} \tag{13.186a}$$

$$\Pi_{ij} = \frac{1}{4\pi}\partial_o h^{-1}_{ji} + \frac{1}{4\pi}A_{ji}(h) - \frac{ie}{4\pi}(A_- h^{-1})_{ji} \quad . \tag{13.186b}$$

They obey the usual (equal time) Poisson-bracket relations:

$$\{A^{\mu a}(x), \pi^b_\nu(y)\} = \delta^{ab} g^\mu_\nu \delta(x^1 - y^1) \tag{13.187a}$$

$$\{h_{ij}(x), \Pi_{kl}(y)\} = \delta_{ik}\delta_{jl}\delta(x^1 - y^1) \quad . \tag{13.187b}$$

Now, since $\mathcal{L}_{WZ}$ depends only linearly on the velocity $\partial_0 h$, its contribution to Π_{ij} $(A_{ji}/4\pi$) will cancel in the Hamiltonian H, which thus depends only on $A^\mu_a, h_{ij}, \pi^a_\mu$ and $\hat{\Pi}_{ij}$ defined by,

$$\hat{\Pi}_{ij} = \Pi_{ij} - \frac{1}{4\pi}A_{ji}(h) = \frac{1}{4\pi}\partial_0 h^{-1}_{ji} - \frac{ie}{4\pi}(A_- h^{-1})_{ji} \quad . \tag{13.188}$$

From, $\{\Pi_{ij}(x), \Pi_{kl}(y)\} = 0$ and (13.187) we deduce

$$\begin{aligned}
\{\hat{\Pi}_{ij}(x), \hat{\Pi}_{kl}(y)\} &= \frac{-1}{4\pi}F_{ji;lk}(h(x))\delta(x^1 - y^1) \\
\{h_{ij}(x), \hat{\Pi}_{kl}(y)\} &= \delta_{ik}\delta_{jl}\delta(x^1 - y^1) \\
\{A^{\mu a}(x), \hat{\Pi}_{kl}(y)\} &= 0 \quad .
\end{aligned} \tag{13.189}$$

In terms of $\hat{\Pi}$, the zero components of the currents (13.144), (13.148) and (13.151a) take the form

$$J^0_L = i\hat{\Pi}^T h - \frac{i}{4\pi}h^{-1}\partial_1 h + \frac{e}{4\pi}[(a-1)A_o + A_1] \tag{13.190}$$

$$J^1_L = -J^0_L + \frac{ae}{4\pi}A_- \tag{13.191}$$

$$j^0_R = -ih\hat{\Pi}^T + \frac{i}{4\pi}h\partial_1 h^{-1} + \frac{e}{4\pi}(hA_- h^{-1}) \tag{13.192}$$

$$J^0_R = J^1_R = -ih\hat{\Pi}^T + \frac{i}{4\pi}h\partial_1 h^{-1} \quad . \tag{13.193}$$

The Poisson brackets of these currents are readily calculated with the aid of the auxiliary formulae listed in appendix L:

$$\{J^{0,a}_L(x), J^{0,b}_L(y)\} = f_{abc}(J^{0,c}(x) - \frac{e}{4\pi}[aA^c_0 - A^c_-])\delta(x^1 - y^1) + \frac{1}{2\pi}\delta^{ab}\delta'(x^1 - y^1)$$

$$\{J_L^{1,a}(x), J_L^{1,b}(y)\} = f_{abc}(-J^{1,c}(x) - \frac{e}{4\pi}[aA_1^c - A_-^c])\delta(x^1 - y^1) + \frac{1}{2\pi}\delta^{ab}\delta'(x^1 - y^1)$$
$$\{J_R^{0,a}(x), J_R^{0,b}(y)\} = \frac{-1}{2\pi}\delta^{ab}\delta'(x^1 - y^1) + f_{abc}J_R^{0,c}(x)\delta(x^1 - y^1)$$
$$\{J_R^{0,a}(x), J_L^{0,b}(y)\} = 0 \quad .$$
$$\{J_L^{0,a}(x), J_L^{1,b}(y)\} = \frac{-1}{2\pi}\delta^{ab}\delta'(x^1 - y^1) + f_{abc}J_R^{1,c}(x)\delta(x^1 - y^1) + \frac{e}{4\pi}[aA_1^c(x) - A_-^c(x)]\delta(x^1 - y^1)$$
$$\{J_R^{0,a}(x), J_L^{0,b}(y)\} = 0 \quad . \tag{13.194}$$

Comparing the Poisson brackets (13.194) with the corresponding expression for the commutators obtained[424] by Jo in the BJL limit, we see that they agree for $A_0 = 0$ and J_0 the (unphysical) choice $a = 0$ for the JR parameter. The Poisson bracket of the right handed current corresponds to that of a free right-handed fermions, as given by the Poisson brackets (9.38*b*) of the Witten current j_-^a.

GI formulation

The bosonic Lagrangian now reads

$$\mathcal{L}_{(bos)}^{GI} = \mathcal{L}_{(bos)}^{GNI}[A, h] - \frac{1-a}{8\pi}\mathrm{tr}(\partial_\mu g \partial^\mu g^{-1}) - \mathrm{tr}A(g)\partial_0 g + \frac{ie}{4\pi}\mathrm{tr}A^\mu g^{-1}[(1-a)\partial_\mu - \tilde{\partial}_\mu]g \quad . \tag{13.195}$$

In addition to (13.186) we have now also the momenta P_{ij} conjugate to g_{ij}.

$$P_{ij} = \frac{a-1}{4\pi}\partial_0 g_{ji}^{-1} - \frac{1}{4\pi}A_{ji}(g) - \frac{ie}{4\pi}([(a-1)A_o + A_1]g^{-1})_{ji} \quad , \tag{13.196}$$

with

$$\{g_{ij}(x), P_{kl}(y)\} = \delta_{ik}\delta_{jl}\delta(x^1 - y^1) \quad . \tag{13.197}$$

As before, only

$$\hat{P}_{ij} = \frac{a-1}{4\pi}\partial_0 g_{ji}^{-1} - \frac{ie}{4\pi}([(a-1)A_o + A_1]g^{-1})_{ji} \tag{13.198}$$

will enter into the Hamiltonian. It has again a non-trivial Poisson bracket with respect to itself (notice the change in sign in (13.199) relative to (13.189)):

$$\{\hat{P}_{ij}(x), \hat{P}_{kl}(y)\} = \frac{1}{4\pi}F_{ji;lk}(g(x))\delta(x^1 - y^1) \quad . \tag{13.199}$$

The other Poisson brackets reads as in (13.189):

$$\{g_{ij}(x), \hat{P}_{kl}(y)\} = \delta_{ik}\delta_{jl}\delta(x^1 - y^1)$$
$$\{A^{\mu,a}(x), \hat{P}_{kl}(y)\} = 0 \quad .$$

In terms of phase-space variables, the covariantly conserved source-current now has the form

$$\mathcal{J}^0 = i\hat{\Pi}^T h - \frac{i}{4\pi}h^{-1}\partial_1 h + i\hat{P}^T g + \frac{i}{4\pi}g^{-1}\partial_1 g \tag{13.200a, b}$$
$$\mathcal{J}^1 = -i\hat{\Pi}^T h + \frac{i}{4\pi}h^{-1}\partial_1 h + \frac{i}{a-1}\hat{P}^T g + \frac{i(a-1)}{4\pi}g^{-1}\partial_1 g - \frac{ea^2 A_1}{4\pi(a-1)}$$

Their Poisson brackets are easily computed with the aid of the formulae of appendix L:

$$i\{\mathcal{J}_0^a(x), \mathcal{J}_0^b(y)\} = if_{abc}\mathcal{J}_0^c(x)\delta(x^1 - y^1)$$
$$i\{\mathcal{J}_0^a(x), \mathcal{J}_1^b(y)\} = if_{abc}\mathcal{J}_1^c(x)\delta(x^1 - y^1) + i\frac{a^2}{4\pi(a-1)}\mathcal{D}_1^{ab}\delta(x^1 - y^1)$$
$$i\{\mathcal{J}_1^a(x), \mathcal{J}_1^b(y)\} = if_{abc}\mathcal{J}_0^c(x)\delta(x^1 - y^1) + \alpha f_{abc}[\mathrm{tr}(\hat{P}^T g\tau^c)$$
$$- \frac{1}{4\pi}(g^{-1}\partial_1 g\tau^c)]\delta(x^1 - y^1) \tag{13.201}$$

where

$$\alpha = \frac{1}{(a-1)^2} - 1 \quad . \tag{13.202}$$

Notice the occurrence of a Schwinger term. The commutator relations (13.202) simplify considerably, if we choose $a = 2$. We shall return to this point later in section (13.12.2).

The other conserved currents $\mathcal{J}_R^\mu$ and $\mathcal{J}_{WZ}^\mu$ (Eqs. (13.159)) have the following phase-space representation:

$$\mathcal{J}_R^0 = \mathcal{J}_R^1 = -ih\hat{\Pi}^T + \frac{i}{4\pi}h\partial_1 h^{-1}$$
$$\mathcal{J}_{WZ}^0 = ig\hat{P}^T + \frac{i}{4\pi}g\partial_1 g^{-1} \quad . \tag{13.203}$$
$$\mathcal{J}_{WZ}^1 = -\frac{i}{a-1}g\hat{P}^T - \frac{i}{4\pi}(a-1)g\partial_1 g^{-1} + \frac{e}{4\pi}\Big(\frac{1}{a-1} - 1\Big)gA_1 g^{-1} \quad .$$

Using again the results of appendix L, we obtain for the corresponding Poisson brackets:

$$i\{\mathcal{J}_{WZ}^{0,a}(x), \mathcal{J}_{WZ}^{0,b}(y)\} = if_{abc}\mathcal{J}_{WZ}^{0,c}(x)\delta(x^1 - y^1) + \frac{i}{2\pi}\delta^{ab}\delta'(x^1 - y^1)$$
$$i\{\mathcal{J}_R^{0,a}(x), \mathcal{J}_R^{0,b}(y)\} = if_{abc}\mathcal{J}_R^{0,c}(x)\delta(x^1 - y^1) - \frac{i}{2\pi}\delta^{ab}\delta'(x^1 - y^1) \tag{13.204}$$
$$i\{\mathcal{J}_{WZ}^{0,a}(x), \mathcal{J}_R^{0,b}(y)\} = 0 \quad .$$

One readily checks that $\mathcal{J}^0$ "commutes" with $\mathcal{J}^0_R$ and $\mathcal{J}^0_{WZ}$:

$$i\{\mathcal{J}^{0,a}(x), \mathcal{J}_R^{0,b}(y)\} = i\{\mathcal{J}_R^{0,a}(x), \mathcal{J}_{WZ}^{0,b}(y)\} = 0 \tag{13.205}$$

This is not true for $\mathcal{J}^{1,a}$ (unless $a = 2$!)

13.12 Hamiltonian Quantization

GNI formulation[337,443]

From (13.184) we obtain the ***primary*** constraints

$$\Omega_1^a \approx 0 \quad , \quad \Omega_1^a = \pi_0^a \quad . \tag{13.206}$$

where the sign $\approx$ of weak equality is to be understood in the sense of Dirac[338]. The condition $\Omega_1^a = 0$ defines the subspace Γ_p of the primary constraints. Following Dirac's terminology, the ***total*** Hamiltonian reads

$$H_T = H_c + \int dx^1 v^a(x)\pi_0^a(x) \quad , \tag{13.207a}$$

where $v^a(x)$ are Lagrange multipliers, to be determined below, and H_c is the "canonical" Hamiltonian given by

$$\begin{aligned} H_c^{GNI} = \int dx^1 \{ & \frac{1}{2}\pi_1^a\pi_1^a + A_0^a\mathcal{D}_1^{ab}\pi_1^b - 2\pi \mathrm{tr}\hat{\Pi}^T h\hat{\Pi}^T h + \frac{1}{8\pi}\mathrm{tr}(\partial_1 h \partial_1 h^{-1}) \\ & - ie\mathrm{tr}\hat{\Pi}^T hA_- + \frac{ie}{4\pi}\mathrm{tr}(h^{-1}\partial_1 hA_-) + \frac{e^2}{8\pi}\mathrm{tr}A_-^2 - \frac{ae^2}{8\pi}A_+A_-\} \quad . \end{aligned} \tag{13.207b}$$

Now

$$\partial_0\Omega_1^a(x) \approx \{\Omega_1^a(x), H_T\}_{\Gamma_p} = -\mathcal{D}_1^{ab}\pi_1^b + eJ_L^{0,a} \quad ,$$

with $J_L^{0,a}$ given by (13.190). Hence the requirement that the primary constraint be preserved in time leads to Gauss law as a secondary constraint,

$$\Omega_2^a = -\mathcal{D}_1^{ab}\pi_1^b + eJ_L^{0,a} \approx 0 \quad , \tag{13.208}$$

We have so far

$$\{\Omega_1^a(x), \Omega_1^b(y)\} = 0 \tag{13.209a}$$

$$\{\Omega_1^a(x), \Omega_2^b(y)\} = \frac{e^2}{4\pi}(1-a)\delta^{ab}\delta(x^1 - y^1) \quad . \tag{13.209b}$$

The cases $a \neq 1$ and $a = 1$ must be considered separately.

The case $a \neq 1$

In the case $a \neq 1$ there are no further constraints, since the condition $\partial_0 \Omega_2 \approx 0$ just serves to fix the Lagrange multipliers in (13.207a). One computes[337]

$$\{\Omega_2^a(x), \Omega_2^b(y)\} = e f_{abc} \Omega_2^c(x) \delta(x^1 - y^1) + \mathcal{A}^{ab}(x,y) \quad . \tag{13.209c}$$

where $\mathcal{A}^{ab}(x,y)$ is the "Gauss anomaly"

$$\mathcal{A}^{ab}(x,y) = \frac{e^3}{4\pi} f_{abc}[A_1^c(x) - (a-1)A_0^c(x)]\delta(x^1 - y^1) \quad . \tag{13.209d}$$

It is interesting to note that the form of the Gauss anomaly calculated by Jo[424] using the BJL technique corresponds to (13.209d) with $a = 0$[22)].

Different expressions for the Gauss anomaly are occasionally quoted in the literature. They can generally be related by a suitable redefinition of the Gauss operator in terms of local counter terms. Thus consider for instance the Gauss operator

$$G^a = \Omega_2^a + \alpha A_0^a + \beta A_1^a \quad .$$

Using (13.209c), one obtains

$$\begin{aligned} i\{G^a(x), G^b(y)\} = &ie f_{abc}\{G^c(x) - (\alpha + e^3 \frac{a-1}{4\pi})A_0^c(x) + \\ &(\beta + \frac{e^2}{4\pi})A_1^c(x)\}\delta(x^1 - y^1) - 2i\beta\delta^{ab}\delta'(x^1 - y^1) \quad . \end{aligned}$$

The choice $\alpha = -\beta = e^2(1-a)/4\pi$ leads to the Gauss anomaly quoted by Shatashvili[72],

$$\tilde{\mathcal{A}}^{ab}(x,y) = e^2 \frac{1-a}{2\pi} \delta^{ab}\delta'(x^1 - y^1) + \frac{ae^3}{4\pi} f_{abc} A_1^c(x)\delta(x^1 - y^1) \tag{13.210}$$

and amounts to a redefinition of the current in terms of a local counterterm:

$$\tilde{J}_L^{\mu,a} = J_L^{\mu,a} - e\frac{a-1}{4\pi}(g^{\mu\nu} + \epsilon^{\mu\nu})A_\nu^a \quad . \tag{13.211}$$

One computes

$$\mathcal{D}_\mu^{ab} \tilde{J}^{\mu,b} = -\frac{ea}{4\pi}\epsilon^{\mu\nu}\partial_\mu A_\nu^a - e^2 \frac{a-1}{4\pi} f_{acb}\epsilon^{\mu\nu} A_\mu^c A_\nu^b \quad . \tag{13.212}$$

It is interesting that for $a = 2$, the rhs just reduces to the covariant anomaly (13.72).

---

22) That A_+^a appears rather than A_-^a, as in the case of ref. [424] results from the fact, that Jo considers a right-handed coupling.

The Poisson brackets (13.209) show that we are dealing with second class constraints. This shows that gauge invariance is broken by the anomaly. Hence, the Gauss operators $\{\Omega^a\}$ are no longer generators of (time independent) gauge transformations. The Poisson bracket (13.209) together with the condition that $\dot{\Omega}_2 \approx \{\Omega_2, H_T\} \approx 0$ fix the Lagrange multipliers v^a in (13.207a) to be

$$\begin{aligned} v^a(x) &= \frac{4\pi}{e^2}\frac{1}{a-1}\{H_c^{GNI}, \Omega_2^a(x)\}_{\Gamma_\Omega} \\ &= \partial_1 A_1^a + \frac{1}{a-1}(\pi_1^a(x) - ef_{abc}A^{0,b}(x)A^{1,c}(x)) \\ &= \partial_1 A_1^a + \frac{1}{a-1}(\partial_1 A_0^a - \partial_0 A_1^a) \end{aligned} \tag{13.213}$$

where Γ_Ω is the hypersurface defined by the constraints $\{\Omega_\alpha^a = 0\}$. Recalling that on the other hand $v^a = \partial_0 A_0^a$, we have from (13.213)

$$[(a-1)g^{\mu\nu} - \epsilon^{\mu\nu}]\partial_\mu A_\nu = 0 \quad . \tag{13.214}$$

This just states, that consistency of the equations of motion requires the current anomaly (13.145) to vanish.

An alternative way of viewing the above result, is to recognize that on Γ_p,

$$\partial_0 \Omega_2^a = \{\Omega_2^a, H_T\}_{\Gamma_p} = \mathcal{A}^a - ef_{abc}A_0^b\Omega_2^c + \frac{e^2(a-1)}{4\pi}(v^a - \partial_0 A_0^a) \tag{13.215a}$$

Making the identification $v^a = \partial_0 A_0^a$ this may be written in the form

$$\mathcal{D}_0^{ab}\Omega_2^a = \{\Omega_2^a(x), H_T\}_{\Gamma_p} = \mathcal{A}^a(x) \quad , \tag{13.215b}$$

which is Fujikawa's result[444], relating the current anomaly to Gauss's law[23)]. Let us remark, that this result holds whenever the Hamilton's equations for A_1^a and π_1^a are equivalent to Maxwell's equations. One then obtains, after some algebra, and making use of the Jacobi identity, Fujikawa's relation in the form

$$\{\Omega_2^a(x), H_T\}_{\Gamma_p} = e\mathcal{D}_\mu^{ab} J^{\mu,b}(x) \quad . \tag{13.216}$$

The current- and Gauss- anomaly are different manifestations of the same anomaly. They may be obtained by following a purely algebraic procedure by

---

23) From Eq. (13.215b) and (13.209c) one derives the following condition[461]

$$\begin{aligned} &\partial_0 \mathcal{A}^{ab}(x,y) + f_{aec}A_0^e(x)\mathcal{A}^{cb}(x,y) + f_{bec}A_0^e(y)\mathcal{A}^{ac}(x,y) = \\ &\qquad \{\Omega^a(x), \mathcal{A}^b(y)\} + \{\mathcal{A}^a(x), \Omega^b(y)\} + f_{abc}\mathcal{A}^c(x)\delta(x-y) \quad . \end{aligned}$$

It is easily checked to be satisfied in $d = 2$. It can be shown to hold also in $d = 4$[461].

"descending" from the Pontryagin density in $d+2$ dimensions[421]. In the framework of Poisson brackets considered here, one is led to consider de Jacobi identity

$$\{\pi_0^2(x), \{\Omega_2^b(y), H_T\}\}_{\Gamma_\Omega} + \text{c.p.} = 0 \quad ,$$

which leads, in our case to

$$e\{\pi_0^a(x), \mathcal{A}^b(y)\}_{\Gamma_\Omega} = \mathcal{A}^{ab}(x,y) \\ -\frac{e^3(1-a)}{4\pi} f_{abc} A_0^c(x)\delta(x^1-y^1) + \frac{e^2(1-a)}{4\pi}\{\pi_0^a(x), v^b(y)\} \quad .$$

Identifying v^b with $\partial_0 A_0^b$, $\mathcal{A}^{ab}(x,y)$ with (13.209), and using the canonical Poisson brackets (13.187a), this just reduces to an identity. For $a=1$ this identity takes the form

$$\mathcal{A}^{ab}(x,y) = e\{\pi_0^a(x), \mathcal{A}^b(y)\}_{\Gamma_\Omega} \quad , \quad (a=1) \quad . \tag{13.217}$$

An analogous relation is also obtained in QCD_4, which actually resembles QCD_2 for $a=1$. It is also obtained if we set $A_0^a = 0$, as Fujikawa does.

Since we are dealing with a second class system, the constraints cannot be realized consistently as conditions on the states. The commutators which realize the constraints strongly may be computed following the algorithm of Dirac. We first construct the Dirac brackets in the standard way

$$\{A,B\}_D = \{A,B\} - \int \{A, \Omega_\alpha^a\} Q^{-1}_{\alpha a;\beta b}\{\Omega_\beta^b, B\} \tag{13.218a}$$

with $Q^{-1}_{\alpha a;\beta b}$ the inverse of

$$Q_{\alpha a;\beta b}(x,y) = \{\Omega_\alpha^a(x), \Omega_\beta^b(y)\} \quad . \tag{13.218b}$$

Using (13.209) we find

$$Q^{-1}(x,y) = \left(\frac{4\pi}{(1-a)e^2}\right)^2 \begin{pmatrix} -e(\hat{\Omega}_2 + \frac{e^2}{4\pi}[\hat{A}_1 - (a-1)\hat{A}_0]) & -\frac{1-a}{4\pi}e^2 \\ \frac{1-a}{4\pi}e^2 & 0 \end{pmatrix} \delta(x^1-y^1) \tag{13.219}$$

where $\hat{\Omega}_2, \hat{A}_\mu$ are anti-hermitian, Lie algebra valued matrices in the *adjoint* representation: $\hat{\Omega}_2 = \Omega_2^c T^c$ *etc.*, $T^c_{ab} = f_{acb}$. Making the usual transition $i\{A,B\}_D \to [A,B]$

from Dirac-brackets to (equal time) commutators one finds[337], using the auxiliary formulae in appendix L:

$$
\begin{aligned}
[A^1_a(x), \pi^b_1(y)]_{ET} &= i\delta^{ab}\delta(x^1 - y^1) \\
[h_{ij}(x), \hat{\Pi}_{kl}(y)]_{ET} &= i\delta_{ik}\delta_{jl}\delta(x^1 - y^1) \\
[A^0_a(x), A^1_b(y)]_{ET} &= \frac{4\pi i}{e^2(1-a)}\mathcal{D}^{ab}_1(x)\delta(x^1 - y^1) \\
[A^0_a(x), h_{ij}(y)]_{ET} &= \frac{4\pi}{e(1-a)}(h\tau^a)_{ij}\delta(x^1 - y^1) \\
[A^0_a(x), \pi^b_1(y)]_{ET} &= -\frac{i}{1-a}[\delta^{ab} - \frac{4\pi}{e}f_{abc}\pi^c_1]\delta(x^1 - y^1) \\
[A^0_a(x), \hat{\Pi}_{ij}(y)]_{ET} &= \frac{1}{(1-a)e}\{[\tau^a g^{-1}(y)]_{ji}\delta'(x^1 - y^1) - 4\pi(\tau^a\hat{\Pi}^T)_{ji}\delta(x^1 - y^1)\} \\
[\hat{\Pi}_{ij}(x), \hat{\Pi}_{kl}(y)]_{ET} &= \frac{i}{4\pi}(g^{-1}_{jk}\partial_1 g^{-1}_{li} - g^{-1}_{li}\partial_1 g^{-1}_{jk})\delta(x^1 - y^1) \\
[A^a_0(x), A^b_0(y)]_{ET} &= \frac{i(4\pi)^2}{(1-a)^2 e^3} f_{abc}\{[A^c_1 - (a-1)A^c_0]\frac{e^2}{4\pi} + \Omega^c_2\}\delta(x^1 - y^1) \quad .
\end{aligned}
\tag{13.220}
$$

In what follows we shall make formal use of these commutation relations without dwelling on properly defining the operator products appearing on the right hand side.

The commutator algebra of the currents $J^{0,a}_{\substack{L\\R}}$ follows from the commutation relations (13.220). One finds

$$[J^{0,a}_L(x), J^{0,b}_L(y)]_{ET} = -if_{abc}J^{0,c}_L(x)\delta(x^1 - y^1) \tag{13.221a}$$

$$[J^{0,a}_R(x), J^{0,b}_R(y)]_{ET} = -\frac{i}{2\pi}\delta^{ab}\delta'(x^1 - y^1) + if_{abc}J^{0,c}_R(x)\delta(x^1 - y^1) \tag{13.221b}$$

$$[J^{0,a}_R(x), J^{0,b}_L(y)]_{ET} = 0 \quad . \tag{13.221c}$$

Here we have used the fact that the constraint (13.208) is realized strongly, and have set $\Omega_2 = 0$. The fact that $J^{0,a}_R$ commutes with $J^{0,b}_L$ follows from the property

$$[A^{0,a}(x), \mathrm{tr}(\hat{\Pi}^T g(y)\tau^b) - \frac{1}{4\pi}(g\partial_1 g^{-1}(y)\tau^b)]_{ET} = 0 \quad . \tag{13.222}$$

Notice that according to (13.221) the currents $J^{0,a}_L$ obey an ordinary current algebra (without Schwinger term), although with the wrong sign! This change in sign is a result of the specific non-canonical form of the commutators involving A^a_0. This means, that in the *GNI* formulation, the Lie algebra structure of the commutation relations (13.221*a*) must be regarded as a coincidence. Indeed, there exists no obvious principle by which such a structure is to be expected: although the anomaly (13.145) must vanish on full quantum level (recall (13.103)) in order for the operator equations of the motion (13.142a) to be consistent, this does not mean that gauge-invariance is restored on full quantum level! The Hamiltonian system is second-class, and hence there is no local symmetry left.

It is interesting to examine the constraint-algebra obtained by replacing the "consistent" current J_L^μ in (13.208) by the "covariant" current $\tilde{J}_L^\mu$ as given by Eq. (13.146). In terms of phase-space variables

$$\tilde{J}_L^0 = i\hat{\Pi}^T h - \frac{i}{4\pi} h^{-1}\partial_1 h + \frac{e}{2\pi} A_1 \tag{13.223a}$$

with the corresponding definition

$$\tilde{\Omega}_2^a = -\mathcal{D}_1^{ab}\pi_1^b + e\tilde{J}_L^{0,a} \tag{13.223b}$$

one finds

$$\{\tilde{J}_L^{0,a}(x), \tilde{J}_L^{0,b}(y)\} = f_{abc}\tilde{J}_L^{0,c}(x)\delta(x^1 - y^1) + \frac{1}{2\pi}\delta^{ab}\delta'(x^1 - y^1) \tag{13.224a}$$

and

$$\{\tilde{\Omega}_2^a(x), \tilde{\Omega}_2^b(y)\} = e f_{abc}\Omega_2^c(x)\delta(x^1 - y^1) - \frac{e}{\pi}\mathcal{D}_1^{ab}(x)\delta(x^1 - y^1) \quad . \tag{13.224b}$$

Hence the covariant Gauss anomaly reads in this case

$$\tilde{\mathcal{A}}^{ab}(x, y) = -\frac{e}{\pi}\mathcal{D}_1^{ab}(x)\delta(x^1 - y^1) \quad . \tag{13.225}$$

The anomalous Schwinger term in (13.224b) has been identified by Faddeev as a 2-cocycle on the space-gauge-group[72].

The case $a = 1$

The r.h.s. of (13.209b) now vanishes, and we must look for further constraints. The total Hamiltonian in this case reads as in (13.207a), with H_c now replaced by[337]

$$H_c^{GI} = \int dx^1 \left\{ \frac{1}{2}\pi_1^a\pi_1^a - \right.$$
$$\left. 2\pi \mathrm{tr}(\hat{\Pi}^T h \hat{\Pi}^T h) + \frac{1}{8\pi}\mathrm{tr}(\partial_1 h \partial_1 h^{-1}) + A_1^a \mathcal{D}_1^{ab}\pi_1^b - A_-^b \Omega_2^b \right\} \quad , \tag{13.226a}$$

where Ω_2^a is the "Gauss" operator (13.208) for $a = 1$,

$$\Omega_2^a = -\mathcal{D}_1^{ab}\pi_1^b + ie\,\mathrm{tr}(\hat{\Pi}^T h\tau^a) - \frac{ie}{4\pi}\mathrm{tr}(h^{-1}\partial_1 h\tau^a) + \frac{e^2}{4\pi}A_1^a \quad . \tag{13.226b}$$

Since $\{\Omega_1(x), \Omega_2(x)\} = 0$, the condition $\partial_0\Omega_2 \approx 0$ no longer fixes the Lagrange multipliers in H_T. One finds[458]

$$\{\Omega_2^a(x), H_T\} = -\Omega_3^a(x) \tag{13.227}$$

with

$$\Omega_3^a = \frac{e^2}{4\pi}\pi_1^a + \frac{e^3}{4\pi}f_{abc}A_0^b A_1^c + ef_{abc}A_0^b\Omega_2^c \quad . \tag{13.228}$$

Hence the requirement that $\Omega_2^a \approx 0$ be preserved in time leads to the new constraint $\Omega_3^a \approx 0$. Although

$$\{\Omega_1^a(x), \Omega_3^b(y)\} = -ef_{abc}(\frac{e^2}{4\pi}A_1^c + \Omega_2^c)\delta(x^1 - y^1) \quad , \tag{13.229}$$

the requirement $\partial_0\Omega_3^a \approx 0$ does not yet determine the Lagrange multipliers, since the determinant of the corresponding Q-matrix (13.218b) is found to vanish. Hence we need to carry the algorithm one step further. One calculates

$$\{\Omega_3^a(x), H_T\} = -B^a + ef_{abc}v^b(\Omega_2^c + \frac{e^2}{4\pi}A_1^c) \tag{13.230}$$

with

$$\begin{aligned} B^a = &-\frac{e^2}{4\pi}(\mathcal{D}_1\pi_1)^a + \left(\frac{e^2}{4\pi}\right)^2 A_-^a \\ &-\frac{e^3}{4\pi}f_{abc}A_0^b\mathcal{D}_1^{cd}A_0^d + \frac{2e^3}{4\pi}f_{abc}A_0^b\pi_1^c - \frac{e^2}{4\pi}\Omega_2^a + ef_{abc}A_0^b\Omega_3^c \quad . \end{aligned} \tag{13.231}$$

Recalling (13.191), we may write (13.231) also in the form

$$^*B^a = \frac{e^3}{4\pi}\left(J_L^{1,a} + 2f_{abc}A_0^b\pi_1^c - f_{abc}A_0^b\mathcal{D}_1^{cd}A_0^d\right) \quad , \tag{13.232}$$

where the "star" indicates, that $^*B^a$ is B^a on the constraint surface defined by $\Omega_1^a \approx \Omega_2^a \approx \Omega_3^a \approx 0$. The requirement $\partial_0\Omega_3^a \approx 0$ thus leads to the following equations for the Lagrange multipliers:

$$^*B^a = \frac{e^3}{4\pi}f_{abc}v^b A_1^c \quad . \tag{13.233}$$

This equation implies $N-1$ constraints corresponding to the dimension of the Cartan subalgebra. One of these constraints is given by $B^a A_1^a \approx 0$, which upon using $\Omega_2^a \approx 0$ and $\Omega_3^a \approx 0$, can be written in the form

$$\Omega_4 \equiv e A_\mu^a J_L^{\mu,a} + 2\pi_1^a \pi_1^a \approx 0 \quad . \tag{13.234}$$

In the case of $SU(2)$, the most general solution of (13.233) is given by (the arrow refers to isospin)

$$\vec{v} = -\frac{4\pi}{e^3}\frac{\vec{B}\times\vec{A}_1}{|\vec{A}_1|^2} + \lambda\vec{A}_1 \quad ,$$

where λ is an arbitrary parameter which is fixed by requiring $\Omega_4 \approx 0$ for all times, since

$$\{\vec{A}_1(x)\cdot\vec{\Omega}_1(x), \Omega_4(y)\} \approx -e\vec{A}_1\cdot\vec{J}_{0,L} + \frac{e^2}{4\pi}\vec{A}_1^2 \not\approx 0$$

In the $U(1)$ case, $\Omega_3 \approx \Omega_4 \approx 0$ imply $\pi_1 \approx J_L^1 \approx 0$ which, upon using the Gauss constraint $\Omega_2 \approx 0$ and (13.191) also imply $J_L^0 = 0$ and $A_- = 0$. We shall rediscover this result in chapter 14.

We do not dwell on repeating the construction of the Dirac brackets for $a = 1$ and leave it to the following chapter, where this analysis is carried out for the simpler $U(1)$ case. Our analysis has nevertheless shown that for the special value of $a = 1$ the number of constraints doubles! That this value of a is special indeed was already to be expected from the singular nature of the commutators (13.220) for $a = 1$.

GI formulation

We again have a primary constraint $\omega_1^a \approx 0$, with

$$\omega_1^a = \pi_0^a \quad . \tag{13.235}$$

The total Hamiltonian is again of the form (13.207), with the canonical Hamiltonian H_c given by[454],

$$\begin{aligned} H_c^{GI} &= \int dx^1 \left\{\frac{1}{2}(\pi_1^a)^2 + A_0^a(\mathcal{D}_1\pi_1)^a\right\} \\ &+ \int dx^1 \{ \frac{1}{8\pi}\mathrm{tr}(\partial_1 h)(\partial_1 h^{-1}) - 2\pi\mathrm{tr}(\hat{\Pi}^T h\hat{\Pi}^T h) - ie\mathrm{tr}(\hat{\Pi}^T h A_-) + \frac{ie}{4\pi}\mathrm{tr}(h^{-1}\partial_1 h A_-) \} \\ &+ \int dx^1 \{\frac{a-1}{8\pi}\mathrm{tr}(\partial_1 g)(\partial_1 g^{-1}) - \frac{2\pi}{a-1}\mathrm{tr}(\hat{P}^T g\hat{P}^T g) - \frac{ie}{a-1}\mathrm{tr}(\hat{P}^T g[(a-1)A_0 + A_1]) \\ &\qquad - \frac{ie}{4\pi}\mathrm{tr} g^{-1}\partial_1 g([(a-1)A_1 + A_0])\} + \int dx^1 \frac{a^2 e^2}{8\pi(a-1)}(A_1^a)^2 \quad . \end{aligned} \tag{13.236}$$

We note again, that for $a = 2$, H_c exhibits a remarkable symmetry with respect to A_+ and A_- (except for the last term!)

The case $a \neq 1$

For this case we have

$$\partial_0 \omega_1^a(x) = \{\omega_1^a, H_T\}_{\Gamma_p} = -\mathcal{D}_1^{ab}\pi_1^b + e\mathcal{J}^{0,a} \tag{13.237a}$$

with $\mathcal{J}^{0,a}$ given by (13.200a). The requirement of persistency in time of the constraint $\omega_1 \approx 0$ leads to the secondary constraint

$$\omega_2^a \equiv -\mathcal{D}_1^{ab}\pi_1^b + e\mathcal{J}^{0,a} \approx 0 \quad . \tag{13.237b}$$

One finds this time

$$\{\omega_1^a(x), \omega_2^b(y)\} = 0 \tag{13.238a}$$

$$\{\omega_2^a(x), \omega_2^b(y)\} = ef_{abc}\omega_2^c(x)\delta(x^1 - y^1) \quad . \tag{13.238b}$$

Now

$$\{\omega_2^a(x), H_T\} = -ef_{abc}A_0^b(x)\omega_2^c(x) \quad . \tag{13.238c}$$

Hence no further constraints arise. The constraints are now first class, as one expects in a theory with gauge invariance. Indeed, one easily checks that ω_2^a generates the correct infinitesimal transformations on A_1^a, h and g,

$$\{\omega_2^a(x), A_1^b(y)\} = -\mathcal{D}_1^{ab}(x)\delta(x^1 - y^1) \tag{13.239a}$$

$$\{\omega_2^a(x), h(y)\} = -ie(h(x)\tau^a)\delta(x^1 - y^1) \tag{13.239b}$$

$$\{\omega_2^a(x), g(y)\} = -ie(g(x)\tau^a)\delta(x^1 - y^1) \quad , \tag{13.239c}$$

and on the currents

$$\{\omega_2^a(x), \mathcal{J}_\mu^b(y)\} = ef_{abc}\mathcal{J}_\mu^c(x)\delta(x^1 - y^1) \quad . \tag{13.240}$$

In order to implement the constraints strongly one would need to fix the gauge. If the gauge condition $\chi^a(A) = 0$ does not involve the Lagrange multipliers $A^{0,a}$, this can be done by introducing the Dirac brackets constructed this time in terms of the inverse of the matrix (we suppress indices)

$$Q(x,y) = \begin{pmatrix} \{\omega(x), \omega(y)\} & \{\omega(x), \chi(y)\} \\ \{\chi(x), \omega(y)\} & \{\chi(x), \chi(y)\} \end{pmatrix} \quad .$$

If, on the other hand we wish to quantize the theory in the so called "covariant" gauges, one will have to do this in terms of the BRS Hamiltonian. The constraints being now first class, one can also chose to impose these constraints on the states. Since $\omega_{1,2}^a$ are generators of the gauge transformations this means, as usual, that only gauge invariant states and "observables" are to be considered. In this case the commutator algebra of the currents is just that given in (13.201). Because of (13.240), this algebra transforms "covariantly" under transformations generated by $\omega_{1,2}^a$. We shall return to the quantization in a covariant gauge in section 14.

The case $a = 2$

Eqs. (13.201) show, that the Poisson brackets of the gauge current simplify drastically. It is convenient to define

$$\mathcal{J}_+ \equiv \mathcal{J}_0 + \mathcal{J}_1 = 2i\hat{\Pi}^T h - \frac{i}{2\pi} h^{-1}\partial_1 h + \frac{e}{\pi} A_1 \tag{13.241a}$$

$$\mathcal{J}_- \equiv \mathcal{J}_0 - \mathcal{J}_1 = 2i\hat{P}^T g + \frac{i}{2\pi} g^{-1}\partial_1 g - \frac{e}{\pi} A_1 \quad . \tag{13.241b}$$

In terms of these currents the Poisson brackets (13.201) are replaced by

$$\{\mathcal{J}_+^a(x), \mathcal{J}_-^b(y)\} = 2f_{abc}\mathcal{J}_+^c(x)\delta(x^1 - y^1) + \frac{2}{\pi}\mathcal{D}_1^{ab}\delta(x^1 - y^1) \tag{13.242a}$$

$$\{\mathcal{J}_+^a(x), \mathcal{J}_-^b(x)\} = 0 \quad . \tag{13.242b}$$

We similarly define

$$\mathcal{J}_R^- \equiv \mathcal{J}_R^0 - \mathcal{J}_R^1 = 0$$

$$\mathcal{J}_R^+ \equiv \mathcal{J}_R^0 + \mathcal{J}_R^1 = -2ih\hat{\Pi}^T + \frac{i}{2\pi} h\partial_1 h^{-1} \tag{13.243a}$$

and

$$\mathcal{J}_{WZ}^- \equiv \mathcal{J}_{WZ}^0 - \mathcal{J}_{WZ}^1 = 2ig\hat{P}^T + \frac{i}{2\pi} g\partial_1 g^{-1}$$

$$\mathcal{J}_{WZ}^+ = 0 \quad . \tag{13.243b}$$

These currents satisfy the equal time Kac-Moody algebra (compare with (9.38)),

$$\{\mathcal{J}_R^{+a}(x), \mathcal{J}_R^{+b}(y)\} = 2f_{abc}\mathcal{J}_R^{+c}(x)\delta(x^1 - y^1) - \frac{2}{\pi}\delta^{ab}\delta'(x^1 - y^1) \tag{13.244a}$$

$$\{\mathcal{J}_{WZ}^{-a}(x), \mathcal{J}_{WZ}^{-b}(y)\} = 2f_{abc}\mathcal{J}_{WZ}^{-c}(x)\delta(x^1 - y^1) + \frac{2}{\pi}\delta^{ab}\delta'(x^1 - y^1) \tag{13.244b}$$

The Hamiltonian can now be written in the Sugawara[340] form

$$H_c^{GI} = H_c^{(1)} + H_c^{(2)} \quad , \tag{13.245a}$$

with

$$H_c^{(1)} = \int dx^1 \mathrm{tr}\left\{\frac{1}{2}\pi_1^2 + A_0\omega_2 + \frac{\pi}{4}(\mathcal{J}^+\mathcal{J}^+ + \mathcal{J}^-\mathcal{J}^-)\right\} \tag{13.245b}$$

and

$$H_c^{(2)} = \frac{\pi}{4}\int dx^1 \mathrm{tr}\left\{(\mathcal{J}_R^+)^2 + (\mathcal{J}_{WZ}^-)^2\right\} \quad . \tag{13.245c}$$

The currents $\mathcal{J}_R, \mathcal{J}_{WZ}$ have the remarkable property of "commuting" with the gauge currents $\mathcal{J}^{\pm}$ (see Eq. (13.205)). Their Poisson bracket with π_1, Ω_2, A_0 and H_c vanishes. Hence $H_c^{(2)}$ which evidently describes a conformally invariant sector of free massless fermions decouples from the theory. $H^{(1)}$ has exactly the form of the Hamiltonian for a pure vector coupling with π_1, ω_2 and $\mathcal{J}^+$ satisfying the same algebra under the Poisson bracket operation as the one obtained above for the $a = 2$ case. We therefore conclude that for $a = 2$, chiral QCD_2 is equivalent to a pure vector theory plus free massless Dirac fermions associated with the currents $\mathcal{J}_R^\mu$ and $\mathcal{J}_{WZ}^\mu$. We shall next rediscover this equivalence from a slightly different point of view.

13.13 Fermionization of $\alpha_1[A, g^{-1}]$

We have seen in section 6, that the 1-cocycle $\gamma_1[A, h^{-1}]$, defined by (13.125) was related to the fermionic determinant via (13.126) (bosonization of fermions). Looked at from the inverse point of view, the integral $\int \mathcal{D}h \exp(i\gamma_1[A, h^{-1}])$ admits a functional representation in terms of fermions[442] (inverse problem: fermionization of bosons). Now, from (13.127) we see that

$$\alpha_1[A, g^{-1}]\Big|_{a=2} = \Gamma[g^{-1}] - \frac{ie}{4\pi}\int d^2x \operatorname{tr}[A^\mu g^{-1}(\partial_\mu + \tilde{\partial}_\mu)g] \qquad (13.246)$$

where we have used the property $S_{WZ}[g^{-1}] = -S_{WZ}[g]$. Recalling the Polyakov-Wiegman identity (13.122) we see that

$$\alpha_1[A, g^{-1}]\Big|_{a=2} = \Gamma[Ug^{-1}] - \Gamma[U] \qquad (13.247)$$

or we have for $a = 2$,

$$\int \mathcal{D}e^{i\alpha_1[A,g^{-1}]} = \text{const}e^{-i\Gamma[U]} \quad , \qquad (13.248)$$

where $-\Gamma[U]$ is the effective action of *right*-handed fermions for a regularization with *vanishing* JR-parameter, with U the variable defined in (13.118). We therefore conclude that for $a = 2$ there exists an equivalence between the fermionic part I_F of the effective bosonic action (13.138) and the fermionic action

$$\tilde{I}_F = \int d^2x\{\overline{\psi}(i\,\partial\!\!\!/ + e\,A\!\!\!/P_-)\psi + \overline{\chi}(i\,\partial\!\!\!/ + e\,A\!\!\!/P_+)\chi\} \qquad (13.249a)$$

as given by

$$\int \mathcal{D}h \int \mathcal{D}g e^{iI_F[A,h,g]} = \int \mathcal{D}\psi\mathcal{D}\overline{\psi}\int \mathcal{D}\chi\mathcal{D}\overline{\chi}e^{i\tilde{I}_F(A,\psi,\overline{\psi},\chi,\overline{\chi})} \quad . \qquad (13.249b)$$

The fields ψ and χ represent fermionic degrees of freedom with left and right-handed coupling to the vector potential A_μ, respectively.

In (13.249b) the ψ-integration is to be regularized with the JR-parameter $a = 2$, whereas the χ integration is to be regularized with a JR-parameter equal to zero! Correspondingly we have for the anomalies

$$\mathcal{D}_\mu^{ab}(\overline{\psi}\tau^a\gamma^\mu P_-\psi)\Big|_{a=2} = \frac{e}{8\pi}(\partial_\mu + \tilde{\partial}_\mu)A^{\mu,a} \qquad (13.250a)$$

$$\mathcal{D}_\mu^{ab}(\overline{\chi}\tau^a\gamma^\mu P_+\chi)\Big|_{a=0} = -\frac{e}{8\pi}(\partial_\mu + \tilde{\partial}_\mu)A^{\mu,a} \quad , \qquad (13.250b)$$

thus demonstrating the cancellation of the anomaly in the total current, in agreement with the conservation law (13.156).

Let us regroup the fermionic fields in (13.249) as follows:

$$\psi' = \begin{pmatrix} \chi_R \\ \psi_L \end{pmatrix} \quad , \quad \chi' = \begin{pmatrix} \psi_R \\ \chi_L \end{pmatrix} \quad . \qquad (13.251)$$

We may then rewrite I_F as follows[442]

$$I_F = \int \left\{ -\frac{1}{4}\mathrm{tr}(F_{\mu\nu}F^{\mu\nu}) + \overline{\psi}'(i\,\partial\!\!\!/ + e\,A\!\!\!/)\psi' + \overline{\chi}' i\,\partial\!\!\!/\chi' \right\} \quad , \qquad (13.252)$$

which puts in evidence the equivalence with a non-abelian vector theory plus decoupled massless fermions, as witnessed already in the previous section. The currents $\mathcal{J}_R^+$ and $\mathcal{J}_{WZ}^-$ in (13.245c) are just the currents associated with the fermions $(\frac{1\pm\gamma_5}{2})\chi'$ of negative and positive chirality which are seen to decouple from the theory, in accordance with the remarks made in the previous section.

13.14 BRS Quantization of *GI* Formulation

The *GI* formulation represents a true gauge theory on quantum level. One is generally interested in quantizing gauge theories either in the (ghost free[24)]) temporal gauge $A^0 = 0$, or in a class of covariant gauges (such as the α-gauges, for example). A essential characteristic of such gauges consists in the appearance of the "Lagrange multiplier" A_0 in the gauge condition. There exist several approaches to the quantization in such gauges. The most general one is that proposed by Fradkin, Vilkovisky and Batalin[445], who have developed a canonical approach to BRS quantization. For the case of interest their approach is equivalent to a simpler one based on the construction of the BRS Hamiltonian associated with the effective

[24)] Actually the question of whether the non-abelian temporal gauge is really ghost free is still an open question.

ghost-Lagrangian obtained after fixing the gauge in the configuration-space generating functional, using the Faddeev-Popov trick. The reason is, that the structure functions $U^{c,\gamma}_{\alpha a;\beta b}$ and $V^{\beta,b}_{\alpha,a}$ associated with the first class constraints (13.235) and (13.237b), as defined by the Poisson brackets,

$$\{\omega^a_\alpha(x), \omega^b_\beta(y)\} = \omega^c_\gamma U^{\gamma,c}_{\alpha a,\beta b} \tag{13.253a}$$

$$\{H_c, \omega^a_\alpha\} = \omega^b_\beta V^{\beta,b}_{\alpha a} \quad , \tag{13.253b}$$

are actually constants. One says that the algebra is "closed"[446]. In this case, the BRS transformations, is of rank 1. The same applies to the BRS Hamiltonian. We shall therefore follow the less general BRS quantization procedure to be described next.

Consider the class of covariant "α-gauges" obtained by performing the gaussian average (13.104) over the Lorentz-type gauge conditions $\mathcal{F}^a(A) - \kappa^a = 0$, with $\mathcal{F}^a(A) \equiv \partial_\mu A^{\mu,a}$. This amounts to a gauge-fixing and ghost-Lagrangian of the form

$$\mathcal{L}_{gf} = -\frac{1}{2\alpha}\mathrm{tr}(\partial_\mu A^\mu)^2 \quad . \tag{13.254}$$

and

$$\mathcal{L}_{gh} = -i(\partial^\mu \bar{c}^a)(\mathcal{D}^{ab}_\mu c^b) \quad , \tag{13.255}$$

where the factor i is required in order that $\mathcal{L}_{gh}$ be hermitian. After gauge-fixing the effective Lagrangian of the GI formulation thus is,

$$\mathcal{L}_B = \mathcal{L}^{GI} + \mathcal{L}_{gf} + \mathcal{L}_{gh} \tag{13.256}$$

and the corresponding Euler-Lagrange equations read (compare with (13.155)),

$$\mathcal{D}^{ab}_\mu F^{\mu\nu}_b + \frac{1}{\alpha}\partial^\nu(\partial \cdot A^a) + i f_{abc}(\partial^\nu \bar{c}^b)c^c + e\mathcal{J}^{\nu,a} = 0 \tag{13.257a}$$

$$D_\mu[-ih^{-1}(\partial^\mu - \tilde{\partial}^\mu)h] = -e(\partial_\mu - \tilde{\partial}_\mu)A^\mu \tag{13.257b}$$

$$D_\mu[-ig^{-1}((1-a)\partial^\mu - \tilde{\partial}^\mu)g] = -e((1-a)\partial_\mu - \tilde{\partial}_\mu)A^\mu \tag{13.257c}$$

$$\partial_\mu(\mathcal{D}^\mu_{ab} c^b) = 0 \tag{13.257d}$$

$$\mathcal{D}^{ab}_\mu(\partial^\mu \bar{c}^b) = 0 \quad . \tag{13.257e}$$

One may also adopt an alternative formulation, by introducing an auxiliary field B via the identity

$$e^{-i\int d^2x \frac{(\partial \cdot A^a)^2}{2\alpha}} = \mathrm{const}\int \mathcal{D}B e^{i\int d^2x \mathcal{L}_{gf}(B,A)} \quad , \tag{13.258}$$

where

$$\mathcal{L}_{gf} = B^a \partial_\mu A^{\mu,a} + \frac{\alpha}{2}(B^a)^2 \quad . \tag{13.259}$$

We shall first discuss the formulation with $\mathcal{L}_{gf}$ given by (13.254), and then comment on a second formulation.

With $\mathcal{L}_{gf}$ given by (13.254), we now have

$$\pi_0^a = -\frac{1}{\alpha}(\partial_\mu A^{\mu,a}) \quad . \tag{13.260}$$

The absence of a primary constraint implies that we are now dealing with an unconstrained system. Hence quantization proceeds via the generalized Poisson brackets[445,446]

$$\{A, B\} = \sum (-1)^{gh(q)} \left(\frac{\partial^{(r)} A}{\partial q} \frac{\partial^{(l)} B}{\partial p} - (-1)^{\epsilon(A)\epsilon(B)} \frac{\partial^{(r)} B}{\partial q} \frac{\partial^{(l)} A}{\partial p} \right) \quad , \tag{13.261a}$$

where $\epsilon(Q)$ represents the "Grassmann signature" of Q

$$\epsilon(Q) = \begin{cases} 1, & Q \quad \text{Grassmann-valued} \\ 0, & Q \quad \text{Bosonic} \end{cases} \tag{13.261b}$$

and $gh(q)$ represents the "ghost-number" of the coordinate q. $\partial^{(l)}$ and $\partial^{(r)}$ stands for "left" and "right" derivative, respectively; they are conveniently defined by

$$df = dq \frac{\partial^{(l)} f}{\partial q} = \frac{\partial^{(r)} f}{\partial q} dq \quad . \tag{13.262}$$

The extra factor $(-1)^{gh(q)}$ reflects the quantization with negative metric of ghost-degrees of freedom. The factor $(-1)^{\epsilon(A)\epsilon(B)}$ assumes that the quantization of fermions proceeds via anti-commutators. Collecting all fields $A^{\mu,a}, h_{ij}, g_{ij}, c^a, \bar{c}^a$ into a vector q^A, and denoting collectively by p_A the corresponding conjugate momenta, the canonical Poisson-brackets can compactly be written in the form (note the ordering of the factors!)

$$\{p_A, q^B\} = -\delta_A^B \quad , \tag{13.263}$$

where

$$p_A \equiv \frac{\partial^{(l)} \mathcal{L}_B}{\partial q^A} \quad . \tag{13.264}$$

The BRS Hamiltonian obtained from $\mathcal{L}_B$ via Legendre transformation is given by

$$H_B = H_c^{GI} + H_{gf} + H_{gh} \tag{13.265a}$$

$$H_{gf} = \int dx^1 \left\{ -\frac{\alpha}{2}(\pi_0^a)^2 + \pi_0^a(\partial_1 A_1^a) \right\} \tag{13.265b}$$

$$H_{gh} = \int dx^1 \left\{ i\pi_{c^a}\pi_{\bar{c}^a} + e f_{acb} \pi_c^a A_0^c c^b - i\partial_1 \bar{c}^a \mathcal{D}_1^{ab} c^b \right\} \quad , \tag{13.265c}$$

where π_0^a is given by (13.260), and

$$\begin{aligned} \pi_c^a &= i\partial_0 \bar{c}^a \\ \pi_{\bar{c}}^a &= -i\mathcal{D}_0^{ab} c^b \quad . \end{aligned} \tag{13.265d}$$

Note that $gh(\pi_c) = -gh(\pi_{\bar{c}}) = -1$. The momenta conjugate to $A^{\mu,a}, h_{ij}$ and g_{ij} continue to be given by (13.186) and (13.196) respectively. The BRS Lagrangian (13.256) is invariant under the infinitesimal BRS transformations,

$$\begin{aligned} \delta_B A_\mu^a &= \delta\epsilon \mathcal{D}_\mu^{ab} c^b \\ \delta_B \bar{c}^a &= -i\delta\epsilon\left(\frac{\partial \cdot A^a}{\alpha}\right) \quad , \quad \delta_B h = -ie\delta\epsilon h\tau^a c^a \\ \delta_B c^a &= -\delta\epsilon\frac{e}{2} f_{abc} c^b c^c \quad , \quad \delta_B g = -ie\delta\epsilon g\tau^a c^a \quad , \end{aligned} \tag{13.266a}$$

with $\delta\epsilon$ a Grassmann valued constant parameter, as well as the rigid transformation

$$c \rightarrow e^\theta c \quad , \quad \bar{c} \rightarrow e^{-\theta}\bar{c} \quad , \tag{13.266b}$$

associated with the conservation of the ghost-number, with θ a constant , Lie-algebra valued "bosonic" matrix.

One verifies that δ_B defined by the transformation laws (13.266) has the nilpotency property

$$\delta_B^2 = 0 \quad . \tag{13.267a}$$

It is convenient to define a (nilpotent) operator s by[437]

$$\delta_B = \delta\epsilon s \quad . \tag{13.267b}$$

With this ordering of the factors, s obeys the graded Leibnitz rule

$$s \cdot AB = (sA)B + (-1)^{\epsilon(A)} A(sB) \quad . \tag{13.268}$$

Denoting by $c(\bar{c})$ the $SU(N)$ Lie-algebra valued fields $c = \sum c^a\tau^a \quad (\bar{c} = \sum \bar{c}^a\tau^a)$, the transformation laws (13.266a) may be conveniently written in the form:

$$\begin{aligned} sA_\mu &= D_\mu c \\ sh &= -iehc \quad , \quad s\bar{c} = -i\frac{(\partial \cdot A)}{\alpha} \quad , \\ sg &= -iegc \quad , sc = \frac{i}{2}[c, c] \end{aligned} \tag{13.269}$$

where D_μ acts on elements of the Lie algebra (group) as in (C.11) and [,] denotes the ordinary Lie commutator[16]. The invariance of $\mathcal{L}_B$ under the transformation is easily verified once one realizes that

$$sD_\mu c = 0 \tag{13.270}$$

on account of the generalized Jacobi identity,

$$[[A_\mu, c], d] - [[d, A_\mu], c] + [[c, d], A_\mu] = 0 \quad , \tag{13.271}$$

where c and d are Grassmann-valued fields. One thereby easily verifies the nilpotency property $s^2 = 0$ [25)]

The conserved Noether current implied by the symmetry transformation Eq. (13.269) is defined as usual by (notice the ordering of the factors!)

$$e\mathcal{J}^\mu_B = \frac{\partial \mathcal{L}_B}{\partial(\partial_\mu A^a_\nu)} sA^a_\nu + \\ \frac{\partial \mathcal{L}_B}{\partial(\partial_\mu h_{ij})} sh_{ij} + \frac{\partial \mathcal{L}_B}{\partial(\partial_\mu g_{ij})} sg_{ij} + sc^a \frac{\partial \mathcal{L}_B}{\partial(\partial_\mu c^a)} + s\bar{c}^a \frac{\partial \mathcal{L}_B}{\partial(\partial_\mu \bar{c}^a)} \tag{13.272}$$

and is given by

$$e\mathcal{J}^\mu_B = -\mathrm{tr}(F^{\mu\nu} D_\nu c) - \frac{ie}{4\pi}\mathrm{tr}\left\{[g^{-1}((1-a)\partial^\mu - \tilde{\partial}^\mu)g]c\right\} - \mathrm{tr}\left\{\frac{(\partial \cdot A^a)}{\alpha} D^\mu c\right\} - \\ \frac{e^2 a}{4\pi}\mathrm{tr}(A^\mu c) + \frac{ie}{4\pi}\mathrm{tr}\left\{[h^{-1}(\partial^\mu - \tilde{\partial}^\mu)hc]\right\} - \frac{ie}{2} f_{abc}(\partial^\mu \bar{c}^a)c^b c^c \quad . \tag{13.273}$$

The corresponding BRS charge is given by

$$Q_B = e\int dx^1 \mathcal{J}^0_B = -\int dx^1 [\mathcal{D}^{ab}_1 F^{10}_b + \frac{1}{\alpha}\partial^0(\partial \cdot A^0) + ief_{abc}(\partial^0 \bar{c}^b)c^c + e\mathcal{J}^{0,a}]c^a \\ + \int dx^1 [\frac{1}{\alpha}\partial^0(\partial \cdot A^a)c^a - \frac{(\partial \cdot A^a)}{\alpha}\mathcal{D}^\mu_{ab} c^b + \frac{ie}{2} f_{abc}(\partial_0 \bar{c}^a)c^b c^c] \quad . \tag{13.274}$$

The first term on the r.h.s. vanishes on account of the equation of motion (13.257) for $F^{\mu 0}$ (Gauss law). The remaining contribution is conveniently expressed in terms of the canonical momenta:

$$Q_B = \int dx^1 \left\{ -(\partial_0 \pi^a_0)c^a + \frac{e}{2} f_{abc}\pi^a_c c^b c^c + i\pi^a_0 \pi^a_{\bar{c}} \right\} \quad . \tag{13.275}$$

---

25) Actually $s^2\bar{c} = 0$ only on the space of solutions of the equations of motion, which imply $\partial_\mu D_\mu c = 0$.

From the equation of motion (13.257a) one has,

$$\partial_0\pi_0^a - \omega_2^a - ef_{abc}\pi_c^b c^c = 0 \quad . \tag{13.276}$$

We may thus rewrite (13.275) as

$$Q_B = \int dx^1 \left\{ -\omega_2^a c^a - \frac{e}{2} f_{abc}\pi_c^a c^b c^c + i\pi_0^a \pi_{\bar{c}}^a \right\} \quad . \tag{13.277}$$

Using the Poisson-brackets (13.238) for $\omega_{1,2}^a$, we easily verify the nilpotency of Q_B

$$\{Q_B, Q_B\} = 0 \quad . \tag{13.278}$$

It is interesting to note, that the last term in (13.277) does not play any role in verifying (13.278), since, when written in terms of phase-space variables, ω_2^a neither depends on A_0^a nor on $\bar{c}$. It does however play a crucial role for generating the transformation laws (13.266a)($\{\ ,\ \}$) stands for the generalized Poisson bracket (13.261))

$$\begin{aligned}
&\{Q_B, A_\mu^a\} = -\mathcal{D}_\mu^{ab} c^b \\
&\{Q_B, h\} = iehc \quad , \quad \{Q_B, c^c\} = \frac{e}{2} f_{abc} c^b c^c \\
&\{Q_B, g\} = iehc \quad , \quad \{Q_B, \bar{c}\} = -i\pi_0^a = i\frac{\partial \cdot A}{\alpha} \quad .
\end{aligned} \tag{13.279}$$

The Faddeev-Popov ghost charge Q_{gh} is similarly constructed, by considering the Noether current associated with the rigid symmetry transformation (13.266b):

$$Q_{gh} = -\int dy^1 (c^b \pi_c^b - \bar{c}^b \pi_{\bar{c}}^b) \quad . \tag{13.280}$$

It generates the transformation laws,

$$\{Q_{gh}, c^a\} = c^a \tag{13.281a}$$
$$\{Q_{gh}, \bar{c}^a\} = -\bar{c}^a \tag{13.281b}$$

We have

$$\{H_B, Q_B\} = \{H_B, Q_{gh}\} = 0 \tag{13.282}$$

and

$$\{Q_B, Q_B\} = 0 \tag{13.283a}$$
$$\{Q_{gh}, Q_B\} = Q_B \tag{13.283b}$$
$$\{Q_{gh}, Q_{gh}\} = 0 \quad . \tag{13.283c}$$

These commutation relations define a BRS supersymmetry[161]. Eq. (13.283a) expresses the nilpotency property already referred to, while relations (13.283b, c) state that Q_B and Q_{gh} carry ghost numbers one and zero respectively. Geometrically they represent the translation and dilatation in the real element of the Grassmann algebra c. The indefinite metric states are excluded from the space of physical states by requiring that

$$Q_B|phys\rangle = 0 \quad , \tag{13.284a}$$

whereas the absence of ghost states in the physical spectrum is imposed by the requirement

$$Q_{gh}|phys\rangle = 0 \quad . \tag{13.284b}$$

The physical space evidently contains zero norm states. Let V_0 be the space of these zero-norm states. We have

$$V_0 = \{|\Psi^{(0)}\rangle = Q_B|\Psi\rangle\} \quad . \tag{13.285}$$

This property is nowadays referred to as BRS cohomology. The zero norm states have no effect on physical processes, i.e. $|phys\rangle$ and $|phys\rangle + Q_B|\Psi\rangle$ are not distinguishable. Hence we can construct a positive Hilbert space[26)] as the factor space of V_{phys} with respect to V_0:

$$\mathcal{H}_{phys} = \frac{V_{phys}}{V_0} \quad . \tag{13.286}$$

As is well known[447], the physical S-matrix S_{phys} can be defined consistently in this factor space, and the unitarity of S_{phys} can be proven[27)]:

$$S^\dagger_{phys}S_{phys} = S_{phys}S^\dagger_{phys} = 1 \tag{13.287}$$

We now turn to the second formulation of the BRS quantization.

As we already have remarked, we could have chosen to quantize the Lagrangian with $\mathcal{L}_{gf}$ given by (13.259). Although this Lagrangian is equivalent to the previous one, it implies this time ***two primary*** constraints:

$$\phi_1^a = \pi_0^a - B^a \approx 0 \tag{13.288a}$$

$$\phi_2^a = \pi_B^a \approx 0 \quad . \tag{13.288b}$$

Since they are ***second*** class,

$$\{\phi_1^a, \phi_1^b\} = \{\phi_2^a, \phi_2^b\} = 0 \tag{13.289a}$$

$$\{\phi_1^a, \phi_2^b\} = -\delta^{ab}\delta(x-y) \quad , \tag{13.289b}$$

---

26) Actually this can be achieved only for $a > 1$. More on this in chapter 14.

27) The proof of unitarity is based on the property that the BRS invariant "ghost" state $|\,n\rangle$ is generated by the BRS charge Q_B itself ($Q_B\,|\,n\rangle = 0\,,\,|\,n\rangle = Q_B\,|\,n'\rangle$). This property is referred to as the BRS cohomology, and plays an essential role in string theory.

there are no further (secondary) constraints.

The quantization is done following the algorithm of Dirac[338]. Define:

$$Q^{ab}_{\alpha\beta}(x,y) = \{\phi^a_\alpha(x), \phi^b_\beta(y)\} \quad . \tag{13.290a}$$

Then

$$Q = \begin{pmatrix} 0 & -1 \\ 1 & 0 \end{pmatrix} \quad , \quad Q^{-1} = \begin{pmatrix} 0 & 1 \\ -1 & 0 \end{pmatrix} \quad . \tag{13.290b}$$

Hence we obtain

$$\{A^{0,a}(x), B^b(y)\}_D = \delta^{ab}\delta(x-y) \tag{13.291a}$$

$$\{B^a(x), \pi^b(y)\}_D = 0 \quad , \tag{13.291b}$$

where $\{\ \ ,\ \ \}_D$ stands for the Dirac bracket defined in (13.218). All remaining Dirac brackets of the canonical fields in question are equal to their Poisson brackets.

The canonical Hamiltonian, evaluated on the constrained surface now reads,

$$\hat{H}_B = H^{GI}_c + H_{gh} + \hat{H}_{gh} \quad , \tag{13.292a}$$

with H^{GI}_c and H_{gh} again given by equations (13.236) and (13.265c) respectively, and $\hat{H}_{gf}$ the new gauge-fixing Hamiltonian

$$\hat{H}_{gf} = \int dx^1 \left\{ B^a \partial_1 A^a_1 - \frac{\alpha}{2}(B^a)^2 \right\} \quad , \tag{13.292b}$$

where we have used the constraints (13.288). The BRS Lagrangian is now invariant under

$$\begin{aligned} sA_\mu &= D_\mu c \quad , \quad & sc &= -\frac{e}{2}[c,c] \\ s\bar{c} &= iB \quad , \quad & sh &= -iehc \\ sB &= 0 \quad , \quad & sg &= -iegc \quad . \end{aligned} \tag{13.293}$$

Note that $\delta B = 0$ implies $s^2\bar{c} = 0$. Correspondingly the BRS current now reads

$$\begin{aligned} \hat{\mathcal{J}}^{\mu,a}_B &= -F^{\mu\nu}_a \mathcal{D}^{ab}_\nu c^b - \frac{e^2 a}{4\pi}\mathrm{tr}(A^\mu c) + \frac{ie}{4\pi}\mathrm{tr}\left\{[h^{-1}(\partial^\mu - \tilde{\partial}^\mu)hc\right\} \\ &\quad - \frac{ie}{4\pi}\mathrm{tr}\left\{[g^{-1}((1-a)\partial^\mu - \tilde{\partial}^\mu)g]c\right\} + B^a \mathcal{D}^\mu_{ab} c^b - \frac{ie}{2} f_{abc}(\partial^\mu \bar{c}^a)c^b c^c \quad . \end{aligned} \tag{13.294}$$

Comparing this expression with that of the previous formulation, Eq. (13.273), we see that they agree, provided we can make the identification

$$B^a = -\frac{\partial \cdot A^a}{\alpha} \quad . \tag{13.295}$$

This relation is just one of Hamilton equations in the present formulation:

$$\partial_0 A_0^a = \{A_0^a, H_B\}_{\mathcal{D}} = \partial_1 A_1^a - \alpha B^a \quad . \tag{13.296}$$

The BRS charge is now given by

$$Q_B = \int dx^1 \left\{ -\dot{B}^a c^a + B^a \mathcal{D}_0^{ab} c^b - \frac{ie}{2} f_{abc} (\partial_0 \bar{c}^a) c^b c^c \right\} \quad , \tag{13.297}$$

or, written in phase space

$$Q_B = \int dx^1 \left\{ -(\partial_0 \pi_0^a) c^a + i \pi_0^a \pi_{\bar{c}^a} - \frac{e}{2} f_{abc} \pi_c^a c^b c^c \right\} \quad , \tag{13.298}$$

where we have used the constraint (13.288a) as well as (13.265d). Comparing with our earlier expression (13.275) we see that they agree. This establishes the equivalence of the two formulations.

13.15 Chiral QCD_2 in Terms of Chiral Bosons

As we have seen, the construction of equivalent bosonic Lagrangians for theories involving Dirac fermions has been well understood for some time. This is not the case for theories involving chiral (Weyl) fermions. The equivalent bosonic theory in this case must be described in terms of "self-dual" scalar fields satisfying the self-duality condition $\partial_- \varphi = 0$ (right movers) or $\partial_+ \varphi = 0$ (left movers).

Siegel[448] has proposed a manifestly Lorentz covariant action,which provides a natural starting point for coupling these chiral bosons to other fields. The classical reparametrization invariance of the Siegel action (Siegel symmetry) is however broken on quantum level by the gravitational anomaly. This circumstance has led to modifications of this action resulting in the absence of such an anomaly.

An alternative Lagrangian has been proposed by Floreanini and Jackiw[448]; its lack of manifest gauge invariance makes it however an inappropriate candidate for coupling the scalars to a gauge field.

There exists by now a very extensive literature on chiral bosons. For a brief introductory discussion, the reader is referred to appendix M. In this chapter we are particularly interested in the role chiral bosons play in two-dimensional chiral gauge theories. Various schemes have been proposed. We shall follow here the work of Harada[449], which naturally fits into the quantization procedure followed so far in this chapter.

We have based our analysis on the bosonized action of chiral gauge theories of left-handed *Dirac* fermions coupled to a gauge field. As we have seen, the right-handed fermions propagate freely, and thus merely play the role of "spectators". It would thus seem natural to exclude from the discussion. We now show how this may be achieved.

Instead of (13.15) we consider the Lagrangian

$$\mathcal{L}^{ch} = -\frac{1}{4}\mathrm{tr}F_{\mu\nu}F^{\mu\nu} + \overline{\psi}_L(i\,\partial\!\!\!/ + e\,A\!\!\!/P_-)\psi_L \tag{13.299}$$

with

$$\psi_L = P_-\psi \quad . \tag{13.300}$$

This Lagrangian differs from (13.15) by the absence of the kinetic part for the right-handed fermions.

For $d = 2$, the free left (right) handed fermions is a left (right) mover. For a vanishing gauge field interaction, the absence of right-handed fermions can thus be viewed as the vanishing of the right-handed current (13.193):

$$\Omega^a \equiv \mathrm{tr}(t^a h\hat{\Pi}^T) - \frac{1}{4\pi}\mathrm{tr}(t^a h\partial_1 h^{-1}) \approx 0 \quad . \tag{13.301}$$

We shall impose (13.301) as an *external* constraint. On the subspace defined by this constraint, the Hamiltonian (13.207) reduces to

$$\begin{aligned} H_c^{ch} = \int dx^1 \mathrm{tr}\,\{\, &\frac{1}{2}\pi_1^2 + A^0\partial_1\pi_1 - \frac{1}{4\pi}h^{-1}\partial_1 h h^{-1}\partial_1 h \\ &-\frac{ie}{2\pi}h^{-1}\partial_1 h A_- + \frac{e^2}{8\pi}A_-^2 - \frac{ae^2}{8\pi}A_\mu A^\mu \,\} \end{aligned} \tag{13.302}$$

It is easily checked that this Hamiltonian is obtained from the following action:

$$S^{ch}[h,A] = \int d^2x \left\{-\frac{1}{4}\mathrm{tr}F_{\mu\nu}F^{\mu\nu} + \frac{e^2 a}{8\pi}\mathrm{tr}A_\mu^2\right\} + S_F^{ch}[A,h] \tag{13.303a}$$

where

$$\begin{aligned} S_F^{ch}[A,h] = S_{WZ}[h] + & \\ \int d^2x\{-\frac{1}{4\pi}\mathrm{tr}(h^{-1}\partial_- h)(h^{-1}\partial_1 h) &- \frac{ie}{2\pi}\mathrm{tr}h^{-1}\partial_1 h A_- - \frac{e^2}{8\pi}\mathrm{tr}A_-^2\} \end{aligned} \quad .(13.303b)$$

This expression replaces $\gamma_1[A,h]$, defined by (13.125), in the action (13.137) of the GNI formulation.

We note that $S_F^{ch}[A,h]$ does not vanish for $h = 1$, so that, unlike $\gamma_1[A,h]$, it cannot be a 1-cocycle. Hence, in contrast to the case of the (usual) bosonization discussed so far in this chapter, $S_F^{ch}[A,h]$ cannot be understood in terms of the gauge transform of the vacuum functional, and its relation to Feynman graphs is thus a priori unclear.

In order to demonstrate that the action (13.303) and (13.137) lead to the same effective action, we use a variant of the "di Vecchia et al trick" by explicitly showing that one nevertheless continues to have a property of the form (13.123):

$$\int \mathcal{D}h e^{iS_F^{ch}[A,h]} = \text{const}\, e^{-i\Gamma[V^{-1}]} \quad . \tag{13.304}$$

To demonstrate this, we compute[450]

$$\begin{aligned}\Delta \equiv S_{ch}[A,h] + \Gamma[V^{-1}] &= (\Gamma[V^{-1}] + \Gamma[h]) - \\ &\frac{1}{4\pi}\int d^2x \mathrm{tr}(h^{-1}\partial_- h h^{-1}\partial_1 h - \frac{1}{2}h^{-1}\partial_\mu h h^{-1}\partial^\mu h) + \\ \frac{1}{2\pi}\int d^2x \mathrm{tr}(h^{-1}\partial_1 h)&(V^{-1}\partial_- V) + \frac{1}{8\pi}\int d^2x \mathrm{tr}(V^{-1}\partial_- V)(V^{-1}\partial_- V) \quad .\end{aligned}$$

Using the Polyakov-Wiegman formula (13.122), this expression may be written in the form

$$\begin{aligned}\Delta &= \Gamma[hV^{-1}] + \frac{1}{8\pi}\int d^2x \mathrm{tr}(h^{-1}\partial_- h - V^{-1}\partial_- V)^2 \\ &= \Gamma[hV^{-1}] + \frac{1}{8\pi}\int d^2x \mathrm{tr}\{(hV^{-1})^{-1}\partial_-(hV^{-1})\} \\ &= f(hV^{-1}) \quad .\end{aligned}$$

Using the invariance of the Haar measure under gauge transformations, we arrive at (13.304).

An alternative demonstration is obtained as follows[451]: Define the chiral Wess-Zumino-Witten action for right/left handed fermions by

$$\Gamma_{ch}^{(R/L)} = \pm\frac{1}{4\pi}\int d^2x \mathrm{tr}(h^{-1}\partial_\pm h)(h^{-1}\partial_1 h) + S_{WZ}[h] \quad , \tag{13.305}$$

where $S_{WZ}[h]$ is the Wess-Zumino action. In terms of $\Gamma_{ch}^{(L)}[h]$, Eq. (13.303b) reads,

$$S_{ch}^{(L)}[A,h] = \Gamma_{ch}^{(L)}[h] - \frac{ie}{2\pi}\int d^2x \mathrm{tr}(h^{-1}\partial_1 h A_-) - \frac{e^2}{8\pi}\int d^2x \mathrm{tr} A_-^2 \quad . \tag{13.306}$$

$\Gamma_{ch}^{(R/L)}[h]$ are related to the WZW action by

$$\Gamma_{ch}^{(R/L)}[h] = \Gamma[h] + \frac{1}{8\pi}\int d^2x \mathrm{tr}(h^{-1}\partial_\pm h)^2 \quad . \tag{13.307}$$

Using the Polyakov-Wiegman formula (13.122) one obtains an analogous formula for $\Gamma_{ch}^{(R/L)}[h]$.

$$\Gamma_{ch}^{(R)}[AB] = \Gamma_{ch}^{(R)}[A] + \Gamma_{ch}^{(R)}[B] - \frac{1}{2\pi}\int d^2x \mathrm{tr}(A^{-1}\partial_+ A)(B\partial_1 B^{-1}) \quad (13.308a)$$

and

$$\Gamma_{ch}^{L}[AB] = \Gamma_{ch}^{L}[A] + \Gamma_{ch}^{L}[B] + \frac{1}{2\pi}\int d^2x \mathrm{tr}(A^{-1}\partial_1 A)(B\partial_- B^{-1}) \quad .(13.308b)$$

In particular, choosing $A = h$ and $B = V^{-1}$ with V defined by (13.118), we obtain from here

$$\Gamma_{ch}^{(L)}[hV^{-1}] = \Gamma_{ch}^{(L)}[h] + \Gamma_{ch}^{(L)}[V^{-1}] - \frac{ie}{2\pi}\int d^2x \mathrm{tr}(h^{-1}\partial_1 h)A_- \quad . \quad (13.309)$$

Hence (13.306) can also be written as

$$\begin{aligned} S_{ch}^{(L)}[A,h] &= \Gamma_{ch}^{(L)}[hV^{-1}] - \Gamma_{ch}^{(L)}[V^{-1}] - \frac{e^2}{8\pi}\int d^2x \mathrm{tr} A_-^2 \\ &= \Gamma_{ch}^{(L)}[hV^{-1}] - \Gamma[V^{-1}] \quad . \end{aligned} \quad (13.310)$$

From here Eq. (13.304) again follows upon using the gauge invariance of the Haar measure. This demonstrates the equivalence of actions (13.303) and (13.137). We thus conclude that there exist two equivalent formulations. However, only in the first case one has the 1-cocycle property. In particular, only the action (13.137) can be obtained by integration of the anomaly equation (see section 7).

From (13.303) one obtains the equations of motion[450]

$$\mathcal{D}_\mu^{ab} F_b^{\mu\nu} + eJ_{ch}^{\nu,a} = 0 \quad (13.311a)$$

$$D_-(h^{-1}\partial_1 h) = \partial_1 A_- \quad , \quad (13.311b)$$

with

$$J_{ch}^{a,\mu} = (-1)^\mu \left\{\frac{1}{2\pi}\mathrm{tr}(\tau^a h^{-1}\partial_1 h) - \frac{1}{8\pi}A_-^a\right\} - \frac{a}{8\pi}A^{\mu,a} \quad . \quad (13.311c)$$

Notice the absence of a time derivative! Taking the covariant divergence of $J^{a,\mu}$ we have

$$\mathcal{D}_\mu^{ab} J^{b,\mu} = \frac{1}{4\pi}[(a-1)\partial_\mu + \tilde{\partial}_\mu]A^{\mu,a} \quad . \quad (13.312)$$

The r.h.s. is just the chiral anomaly.

The procedure so far has been "heuristic" since we have imposed the chiral constraint (13.301) from "outside" in order to arrive at the bosonized action (13.303), which was proven to be equivalent to (13.137). We now demonstrate the self-consistency of our results by proceeding to quantize canonically this theory.

Following the Dirac algorithm[338], we obtain from $\mathcal{L}^{ch}$ two sets of primary constraints:

$$\Omega_1^a \equiv \pi_0^a \approx 0 \tag{13.313a}$$

$$\Omega_{ij} \equiv (h\hat{\Pi}^T)_{ij} + \frac{1}{4\pi}(\partial_1 h h^{-1})_{ij} \approx 0 \quad . \tag{13.313b}$$

Note that condition (13.313b) is just the chiral constraint (13.301), which now emerges as an "internal" constraint of the new Lagrangian $\mathcal{L}^{ch}$. The total Hamiltonian thus reads

$$H_T = H^{ch} + \int d^2x(v^a\Omega_1^a + u_{ij}\Omega_{ij}) \quad . \tag{13.314}$$

Now

$$Q_{ij;kl} = \{\Omega_{ij}(x), \Omega_{kl}(y)\} \approx \frac{1}{2\pi}\delta_{il}\delta_{jk}\delta'(x^1 - y^1) \tag{13.315}$$

and

$$\{\Omega_1^a(x), \Omega_{ij}(y)\} = 0 \quad . \tag{13.316}$$

Hence the condition $\partial_0\Omega_{ij} \approx 0$ only serves to fix the Lagrange multiplier u_{ij}, and generates no new constraint. The condition $\partial_0\Omega_1^a \approx 0$, on the other hand leads to a secondary ("Gauss Law") constraint

$$\Omega_2^a \equiv -(\mathcal{D}_1\pi_1)^a - \frac{ie}{2\pi}\mathrm{tr}\tau^a h^{-1}\partial_1 h + \frac{e^2}{4\pi}\mathrm{tr}\tau^a\{(a-1)A_0 + A_1\} \approx 0 \tag{13.317}$$

which coincides with (13.208) on the surface defined by the "chiral" constraint (13.313b). Since

$$\{\Omega_1^a(x), \Omega_2^b(y)\} = -\frac{e^2(a-1)}{4\pi}\delta^{ab}\delta(x^1 - y^1) \quad , \tag{13.318}$$

the condition $\partial_0\Omega_2^a \approx 0$, merely serves to fix the remaining multipliers u^a and there are no further constraints for $a \neq 1$.

In order to compare with the algebra of constraints of the non-chiral Lagrangian we define a Poisson bracket $\{\quad\}^*$ in the subspace defined by the constraint (13.313). Although the inverse of the matrix Q is a priori not unique, we have for "suitable" boundary conditions

$$(Q^{-1})_{ij;kl}(x,y) = \pi\delta_{ij}\delta_{jk}\epsilon(x^1 - y^1) \quad , \tag{13.319}$$

where $\epsilon(z)$ is the sign function with the property $\epsilon(z) = 1$ for $z > 0$ and $\epsilon(z) = -1$ for $z < 0$. We define

$$\{A(x), B(y)\}^* \equiv$$
$$\{A(x), B(y)\} - \int dz^1 d\omega^1 \{A(x), \Omega_{ij}(z)\} Q^{-1}_{ij;kl}(z,\omega)\{\Omega_{kl}(\omega), B(y)\} \quad . \tag{13.320}$$

One then finds

$$\begin{aligned} \{A^{\mu,a}(x), \pi^b_\nu(y)\}^* &= \delta^\mu_\nu \delta^{ab} \delta(x^1 - y^1) \\ \{h_{ij}(x), h_{kl}(y)\}^* &= \pi h_{kj}(x)\epsilon(x^1 - y^1) h_{il}(y) \quad . \end{aligned} \tag{13.321}$$

We need not calculate the new Poisson brackets involving $\hat{\Pi}_{ij}$, since we may eliminate these degrees of freedom by invoking the constraints (13.313*b*), which have now been implemented strongly. The remaining modified Poisson brackets are canonical. As is to be expected, one recovers the constraint algebra (13.209):

$$\{\Omega^a_1(x), \Omega^b_2(y)\}^* = (1-a)\frac{e^2}{4\pi}\delta^{ab}\delta(x^1 - y^1) \tag{13.322}$$
$$\{\Omega^a_2(x), \Omega^b_2(y)\}^* = ef_{abc}\{\Omega^c_2(x) + \frac{e^2}{4\pi}[A^c_1(x) - (a-1)A^c_0(x)]\}\delta(x^1 - y^1) \quad .$$

Introducing Dirac brackets with respect to the Poisson brackets $\{\ \ \}^*$ and going over to the quantum theory it is thus not surprising that the chiral current $J^{0,a}_{ch}$ satisfies the commutator relations already encountered in (13.221);

$$[J^{0,a}_{ch}(x), J^{0,b}_{ch}(y)]_{ET} = -if_{abc}J^{0,c}_{ch}(x)\delta(x^1 - y^1) \tag{13.323}$$

To conclude this section let us remark, that the determinant of QCD_2 can also be written in the form

$$\det(i \not\partial + e \not A) = e^{-i\Gamma[UV^{-1}]} = e^{\frac{ie^2}{2\pi}\int d^2x \mathrm{tr} A_1^2} e^{-i\Gamma^{(R)}_{ch}[U]} e^{-i\Gamma^{(L)}_{ch}[V^{-1}]} \quad . \tag{13.324}$$

This follows from the identity

$$\begin{aligned} \Gamma[UV^{-1}] &= \Gamma[U] + \Gamma[V^{-1}] + \frac{1}{4\pi}\int d^2x \mathrm{tr}(U^{-1}\partial_+ U)(V^{-1}\partial_- V) \\ &= \Gamma^{(R)}_{ch}[U] + \Gamma^{(L)}_{ch}[V^{-1}] - \frac{1}{8\pi}\int d^2x \mathrm{tr}[(U^{-1}\partial_- U)^2 + (V^{-1}\partial_- V)^2 \\ &\quad - 2(U^{-1}\partial_+ U)(V^{-1}\partial_- V)] \\ &= \Gamma^{(R)}_{ch}[U] + \Gamma^{(L)}_{ch}[V^{-1}] - \frac{e^2}{2\pi}\int d^2x \mathrm{tr} A_1^2 \end{aligned} \tag{13.325}$$

where we have used Eq. (13.307). Hence the Dirac fermions of QCD_2 cannot be described in terms of the direct product of two independent algebras. This reflects the fact that the invariance under separate transformations for the left and right movers represents an obstruction to the gauge invariance of QCD_2. This is of course well known since the work of Witten[51] and Polyakov and Wiegman[50].

13.16 Constraint Structure from Fermionic Hamiltonian

On the level of the fermionic Lagrangian (13.15*a*), there is a priori no indication for the existence of second-class constraints. These, we have seen, are a result of anomalies present on quantum level. By going over to the bosonized form of the theory, we have been able to incorporate explicitly these anomalies into a Dirac algorithm operating on a classical level.

In higher dimensions, such bosonization techniques are in general not available to us, and we must seek a method for obtaining the constraint structure and equations of motion on *quantum* level by working with the *fermionic* Hamiltonian. Chiral QCD_2 presents in this respect an excellent laboratory for testing such methods, since, as we have seen, the constraint structure one should obtain, is exactly known.

In the past, such methods have been based on the calculation of the "commutators" in the $A^0 = 0$ gauge, using the BJL limiting procedure[452]. Actually, this procedure does not yield the true commutators of a chiral gauge theory, but rather the "anomalous Poisson brackets" of such a theory, where "anomalous" means, that the classical Poisson brackets of the fermionic formulation are modified by the presence of the anomaly on quantum level. Although this fact was implicitly understood in the literature, the validity for performing the calculations in the $A_0 = 0$ gauge, in a theory where gauge invariance is broken on quantum level, though questioned[453], was tacitly accepted.

The gauge $A_0 = 0$ was first used in electrodynamics by H. Weyl, and hence is also referred to as the Weyl gauge. In this gauge, Gauss law has to be imposed separately. If there are no anomalous contributions due to fermion loops, the quantum commutators are canonical and the Gauss law operator G^a generates time-independent gauge transformations corresponding to the residual symmetry in the Weyl gauge:

$$[G^a(x), \vec{A}^b(y)]_{ET} = -i\vec{\mathcal{D}}^{ab}(x)\delta(\vec{x} - \vec{y}) \tag{13.326a}$$

$$[G^a(x), \vec{E}^b(y)]_{ET} = ief_{abc}\vec{E}^c(x)\delta(\vec{x} - \vec{y}) \quad . \tag{13.326b}$$

Since in the anomaly-free case, one has the commutation relations

$$[G^a(x), G^b(y)]_{ET} = ief_{abc}G^c(x)\delta(\vec{x} - \vec{y}) \quad , \tag{13.327}$$

we have here the option of consistently imposing Gauss' law as a condition on the physical states $G^a(x)|Phys\rangle = 0$.

For the abelian case this means that the physical wave functional can only depend on the transverse components of $\vec{A}$; this means that such a wave functional is not normalizable, which may lead to contradictions if proper care is not taken. Of course, one also has the option of realizing Gauss' law as a strong operator relation, by fixing the gauge. In the case of the Weyl gauge, which represents a restriction on the Lagrange multiplier, this can only be done at the expense of introducing additional Grassmann degrees of freedom (ghosts) and finds its most elegant formulation in the BRS formalism. For the axial gauges, such as the $A_3 = 0$ gauge, this is achieved[455] within the framework of the "canonical" Dirac algorithm, by fixing the gauge completely, using for instance Halpern's subsidiary conditions[456]. In a chiral gauge theory, no such gauge fixing is required nor allowed, and Gauss law has to be imposed strongly.

In what follows we present a method for generating the constraints and their anomalous Poisson algebra from the *fermionic* Hamiltonian, following a Dirac type algorithm.

The basic idea[422,423] of the method is to add to the (fermionic) Lagrangian (13.15) a term involving $(\partial_0 A_0)^2$. As a result there no longer exists the primary constraint $\pi_0 \approx 0$, and Gauss law becomes a dynamical equation for the theory described by the new Lagrangian $\mathcal{L}_{(\alpha)}$. For reasons of Lorentz invariance it is convenient to choose $\mathcal{L}_{(\alpha)}$ as

$$\mathcal{L}_{(\alpha)} = \mathcal{L} - \frac{1}{2\alpha}\mathrm{tr}(\partial^\mu A_\mu)^2 \quad . \tag{13.328}$$

The new Lagrangian $\mathcal{L}$ describes an unconstrained, though still anomalous, theory. The fundamental phase–space variables obeying the dynamics described by $\mathcal{L}_{(\alpha)}$ would satisfy canonical commutation relations, if there was no anomaly. The contribution from the one–fermion loops however modifies these commutation relations in a way that is exactly calculable in perturbation theory. This leads us to define anomalous Poisson brackets (APB) by[422]

$$\{A, B\}_{AP}^{(\alpha)} \equiv -i[A, B]^{(\alpha)} \quad , \tag{13.329a}$$

where the superscript (α) reminds us, that the commutator $[A, B]$ is calculated in the theory defined by $\mathcal{L}_{(\alpha)}$.

The correlation functions of the theory of interest, as described by $\mathcal{L}$, are obtained from those associated with $\mathcal{L}_{(\alpha)}$, by taking the limit $\alpha \to \infty$. Note that this is just the opposite of what one would do, if one wanted to arrive at the Lorentz gauge $\partial_\mu A^\mu = 0$. It is clear that if $\mathcal{L}$ were to describe a true gauge theory, this limit $\alpha \to \infty$ would in general not exist for the correlation functions, because of the existence of an infinite number of gauge–equivalent A^μ–configurations. In the case in question this gauge symmetry is however broken on quantum level. Hence the limit $\alpha \to \infty$ exists. This observation can be used to argue[457], that the commutator (13.329) is actually independent of α. The reason lies in the structure of the A^μ-propagator associated with $\mathcal{L}_{(\alpha)}$, which, in momentum space, reads,

$$D_F^{\mu\nu}(k) = -\frac{1}{k^2}\left(g^{\mu\nu} - (1-\alpha)\frac{k^\mu k^\nu}{k^2}\right) \quad . \tag{13.330}$$

Thus perturbative graphs will show a polynomial dependence on the parameter α. Since only a few of such graphs generally contribute in a BJL calculation of the commutator (13.329), its dependence on α will be polynomial–like. Since the limit $\alpha \to \infty$ however exists, this polynomial in α must be of zeroth degree. Hence we come to the important conclusion, that we can drop the superscript (α) in (13.329). This defines for us the anomalous Poisson brackets of the theory described by the Lagrangian $\mathcal{L}$ of interest, as

$$\{A, B\}_{AP} = -i[A,B]^{(\alpha)} = -i[A,B]^{(\infty)} \quad . \tag{13.329b}$$

In chapter 14 we shall examine how this independence of (α) is realized on the operator level for the chiral Schwinger model. We shall discover, that extra degrees of freedom associated with the dynamics described by $\mathcal{L}_{(\alpha)}$, while decoupling from the correlation functions for $\alpha \to \infty$, do not decouple in the commutation relations thereby ensuring that the quasi–canonical structure of the APB does not turn into that of anomalous Dirac brackets (true quantum commutators of the chiral gauge theory) in this limit.

We now apply the method described above to the theory described by the Lagrangian (13.15) by calculating the constraints and their "algebra" via a Dirac algorithm on quantum level.

The difference between the algorithm to be described now[458], and the usual Dirac algorithm, as applied to non-anomalous theories, is that the anomalous Poisson brackets (APB) replace the usual canonical Poisson brackets.

We begin by computing the anomalous Poisson bracket of π_0^a with the total Hamiltonian of chiral QCD$_2$,

$$H_T = H_c + \int dx^1 v^a \pi_0^a \tag{13.331a}$$

where,

$$H_c = \int dx^1 \left[\frac{1}{2}(\pi_1^a)^2 + \bar{\psi} i\gamma_1 \partial_1 \psi + e\bar{\psi}\tau^a\gamma_1 A_1^a P_- \psi - A_0^a G^a\right] \quad , \tag{13.331b}$$

with G^a the Gauss operator $G^a = -D_1^{ab}\pi_1^b + e\bar{\psi}\tau^a\gamma^0 P_-\psi$.

Now, in the auxiliary theory described by $\mathcal{L}_{(\alpha)}$, the momentum conjugate to A_0^a is given by

$$\pi_0^a = -\frac{1}{\alpha}(\partial_\mu A^{\mu,a}) \quad . \tag{13.332}$$

Hence the first step in the algorithm consists in calculating

$$i\{\pi_0^a(x), H_T\}_{AP} = \left[-\frac{1}{\alpha}\partial_\mu A^{\mu,a}(x), H_T\right]^{(\alpha)} \quad . \tag{13.333}$$

The computation of the r.h.s. is done using the BJL limit[452], which allows one to calculate the $\alpha - \beta$ matrix element of the equal time commutator of two operators A and B from the corresponding two–point function via the limit

$$\lim_{q_0 \to \infty} q_0 T(q) = i \int d\vec{x} e^{-i\vec{q}\cdot\vec{x}} \langle\alpha|[A(0,\vec{x}), B(0)]|\beta\rangle \tag{13.334a}$$

where,

$$T(q) := \int dx e^{iqx} \langle\alpha|TA(x)B(0)|\beta\rangle \quad , \tag{13.334b}$$

with d the dimension of space–time. In words: For $q_0 \to \infty$, the matrix element $\langle\alpha|[A,B]_{ET}|\beta\rangle$ is determined by the coefficient of $1/q_0$ in the asymptotic expansion of $T(q)$.

In a normal gauge theory, only the last term in (13.331*b*) contributes to $\{\pi_0^a, H_T\}$, yielding G^a. In the anomalous case one discovers[458] the following unusual situation: The α–commutator of the fermionic kinetic term

$$T_F = \int dx^1 \mathcal{T}_F = \int dx^1 \bar{\psi} i\gamma_1\partial_1\psi \quad , \tag{13.335}$$

with π_0^a and π_1^a is anomalous! The relevant graphs are shown in figure 13.1. Their regularization is ambiguous, because of the lack of gauge invariance. It is convenient to parametrize this ambiguity by adding to $\mathcal{L}_{(\alpha)}$ a term of the form

$$\Delta\mathcal{L} = \frac{1}{2}\bar{\phi}\left(\frac{\Box + M^2}{M}\right)\phi + e\sqrt{\frac{a}{2}}\bar{\phi}\,\not{A}\phi \quad , \tag{13.336}$$

where a is an arbitrary parameter, ϕ is a bosonic spinor auxiliary field, and M is the Pauli–Villars mass. Since $\Delta\mathcal{L}$ *naively* gives a vanishing contribution to the one-loop correlation functions in the limit $M \to \infty$, it can be considered as a regularization ambiguity. Note that the effective action associated with $\Delta\mathcal{L}$, as obtained by performing the Gaussian integration over ϕ in the corresponding partition function, tends to the one–loop contribution $\frac{ae^2}{8\pi}\mathrm{tr}A_\mu^2$ in the limit $M \to \infty$. The parameter a in (13.336) thus plays the role of the JR parameter.

The term $\Delta\mathcal{L}$ generates an additional contribution to the current and kinetic term:

$$
\begin{aligned}
J^{\mu,a}(x) =& [\bar{\psi}(x)\tau^a\gamma^\mu P_-\psi(x)]_{P.V.} \\
&+ \frac{iea}{2}\lim_{M\to\infty}\int d^2y\langle 0|T\bar{\phi}(x)\tau^a\gamma^\mu\phi(x)\bar{\phi}(y)\tau^b\gamma^\nu\phi(y)|0\rangle A^b_\nu(y) \\
=& [\bar{\psi}(x)\tau^a\gamma^\mu P_-\psi(x)]_{P.V.} + \frac{ae}{4\pi}A^{\mu,a} \qquad (13.337) \\
\mathcal{T}_F(x) =& [\bar{\psi}(x)\gamma_1\partial_1\psi(x)]_{P.V.} \\
&- \frac{ae^2}{8}\lim_{M\to\infty}\int d^2y\int d^2z \\
&\langle 0|\bar{\phi}(x)\left(\frac{\Delta - M^2}{M}\right)\phi(x)\bar{\phi}(y)\tau^a\gamma^\mu\phi(y)\bar{\phi}(z)\tau^b\gamma^\nu\phi(z)|0\rangle\cdot A^a_\mu(y)A^b_\nu(z) \\
=& [\bar{\psi}(x)\gamma_1\partial_1\psi(x)]_{P.V.} + \frac{e^2a}{8\pi}\mathrm{tr}A^2_\mu \qquad (13.338)
\end{aligned}
$$

where the subscript $P.V.$ stands for "Pauli–Villars regularization". Since this regularization preserves the chiral $(1-\gamma_5)$ structure of the current, we have from (13.337),

$$
J^{0,a}_L = -J^{1,a}_L + \frac{ae}{4\pi}(A^a_0 - A^a_1) \quad . \qquad (13.339)
$$

Let us first consider the α–commutators involving A_μ, π_μ and T_F. The relevant graphs are shown in figure 13.1. The commutators $[A^{\mu,a}, T_F]^{(\alpha)}$ and $[\pi^a_\mu, T_F]^{(\alpha)}$ are obtained from the diagrams depicted in figure 13.1, which imply a linear dependence of these commutators on the vector-potential.

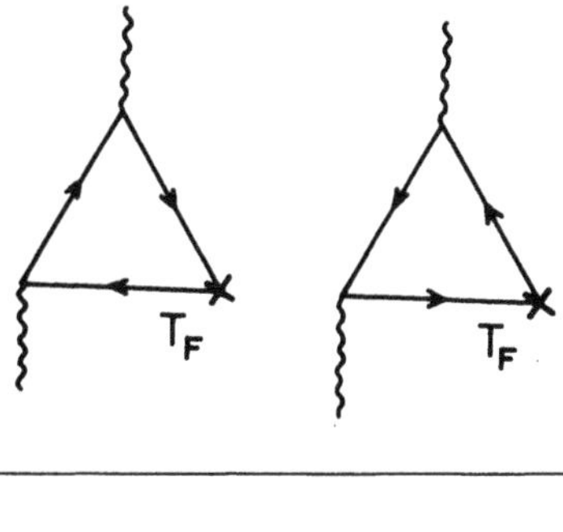

Figure 13.1: Diagrams contributing to $\{\pi^a_\mu, T_F\}_{AP}$.

Using Pauli–Villars regularization and Feynman parametrization, one obtains for the vertex function associated with figure 13.1a,

$$\begin{aligned}\Gamma^{\mu a,\nu b}_{P.V.}(k,q) =& -ie^2 \lim_{M\to\infty} \int \frac{d^2p}{(2\pi)^2} \int_{M^2}^{0} dm^2 \frac{\partial}{\partial m^2} \\ &\times \frac{\mathrm{tr}[(\not k + \not p + \not q + m)(\gamma^1 p^1 + m)(\not p + m)\tau^b\gamma^\nu P_-(\not k + \not q + m)\tau^a\gamma^\mu P_-]}{[p^2 - m^2][(p+q)^2 - m^2][(k+p+q)^2 - m^2]} \\ =& -\frac{e^2}{16\pi}\delta^{ab}\mathrm{tr}[\gamma^\nu P_-\gamma^1\gamma^\mu\gamma^1] \\ &+ \frac{\delta^{ab}}{8\pi}\int_0^1 d\alpha \int_0^1 d\beta \frac{\alpha}{[\alpha(1-\alpha)q^2 + \alpha\beta(1-\alpha\beta)k^2 + 2\alpha\beta(1-\alpha)k\cdot q]} \\ &\times \Big[\mathrm{tr}[\gamma^\nu P_-((1-\alpha)\not q - \alpha\beta\not k)\gamma^\mu((1-2\alpha)\not q + (1-2\alpha\beta)\not k)] \\ &+ \mathrm{tr}[\gamma^\nu P_-\gamma^1\gamma^\mu((1-\alpha)\not q + (1-\alpha\beta)\not k)\gamma^1(\alpha\not q + \alpha\beta\not p)]\Big] \quad .\end{aligned}$$

In the limit $q_0 \to \infty$, the second term just reproduces the first term, so that

$$\Gamma^{\mu a,\nu b}_{P.V.}(k,q) \underset{q^0\to\infty}{\longrightarrow} -\frac{2e^2}{16\pi}\delta^{ab}\mathrm{tr}[\gamma^\nu P_-\gamma^1\gamma^\mu\gamma^1] \quad . \tag{13.340}$$

Figure 13.1b gives the same contribution in the BJL limit.

Introducing the dependence on the JR parameter via (13.338), and contracting with the A_μ-propagator (13.330), one obtains from (13.340) in the standard way,

$$i\{A^{\lambda,a}(x), T_F\}_{AP} = 0 \tag{13.341a}$$

$$i\{\pi_0^a(x), T_F\}_{AP} = \frac{ie^2}{4\pi}((1-a)A_0^a(x) - A_1^a(x)) \tag{13.341b}$$

$$i\{\pi_1^a(x), T_F\}_{AP} = \frac{ie^2}{4\pi}(A_0^a(x) - (1+a)A_1^a(x)) \quad . \tag{13.341c}$$

The commutator $[A^{\mu,a}, \pi_\nu^b]^{(\alpha)}$ is determined by the tree graph ($D^{\mu\nu}$ propagator), alone. Hence we have in this case the canonical results

$$\begin{aligned} i\{\pi^{\mu,a}(x), \pi_\nu^b(y)\}_{AP} &= 0 \\ i\{A^{\mu,a}(x), \pi_\nu^b(y)\}_{AP} &= ig_\nu^\mu\delta^{ab}\delta(x^1 - y^1) \quad . \end{aligned} \tag{13.342}$$

The non-canonical structure (13.341c) at first suggests that a Dirac algorithm based on the APB will no longer reproduce the Gauss constraint. Surprisingly, this expectation turns out to be wrong. Indeed, using (13.342), we obtain,

$$\{\pi_0^a(x), H_T\}_{AP} = G^a(x) + \Delta^a(x) \tag{13.343a}$$

where,

$$\Delta^a(x) = \{\pi_0^a(x), T_F\}_{AP} - e\int dy^1 \{\pi_0^a(x), J_L^{\mu,b}(y)\}_{AP} A_\mu^b(y) \quad . \tag{13.343b}$$

We are thus left to compute $\{\pi_0^a(x), J^{\mu,b}(y)\}_{AP}$. The relevant graphs are shown in figure 13.2. Using again a Pauli–Villars regularization and introducing the dependence on the JR parameter via (13.338), one obtains

$$\begin{aligned} i\{\pi_0^a(x), J_L^{0,b}(y)\}_{AP} &= -\frac{ie}{4\pi}(a-1)\delta^{ab}\delta(x^1 - y^1) \\ i\{\pi_0^a(x), J_L^{1,b}(y)\}_{AP} &= -\frac{ie}{4\pi}\delta^{ab}\delta(x^1 - y^1) \quad . \end{aligned} \tag{13.344}$$

Putting all these results together one finds that a remarkable cancellation of the anomalous contributions takes place, leading to

$$\Delta^a = 0 \quad .$$

We thus recover from $\dot{\pi}_0 \approx 0$ the cherished Gauss constraint $G^a \approx 0$ on *quantum* level!

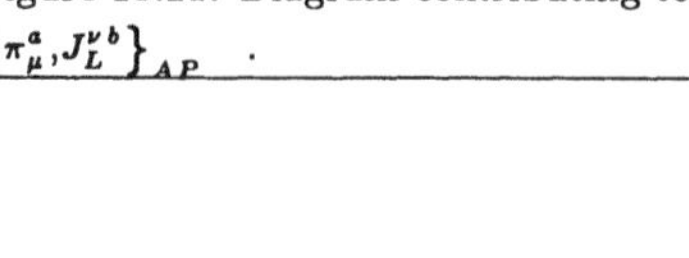

(a)

Figure 13.2a: Diagram contributing to $\{\pi_\mu^a, J_L^{\nu b}\}_{AP}$.

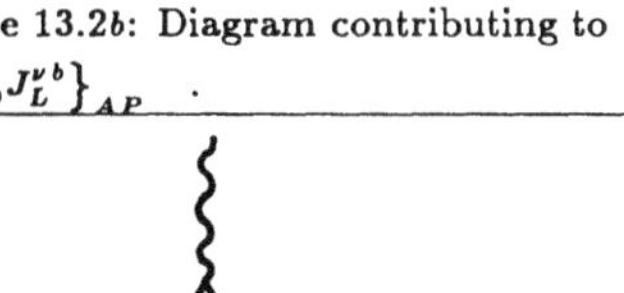

(b)

Figure 13.2b: Diagram contributing to $\{J_L^{\mu a}, J_L^{\nu b}\}_{AP}$.

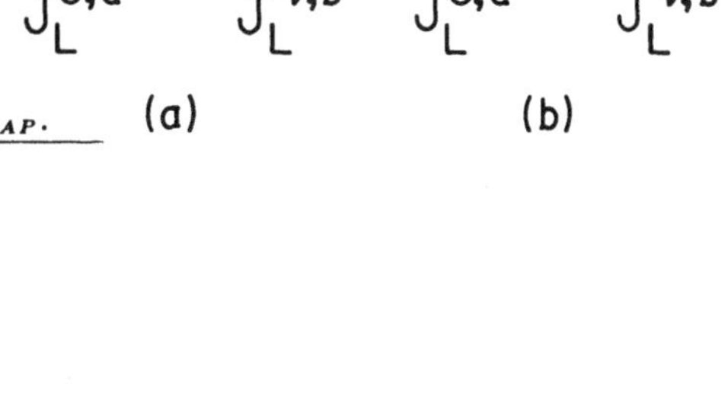

(a) (b)

Figure 13.3: Diagrams contributing to $\{J_L^{\mu,a}, J_L^{\nu,b}\}_{AP}$.

In order to see, whether there exist further constraints, we must compute some further α–commutators. From the diagrams of figures 13.2 and 13.3 one finds[28)]

$$
\begin{aligned}
i\{\pi_1^a(x), J_L^{0,b}(y)\}_{AP} =& \frac{ie}{4\pi}\delta^{ab}\delta(x^1-y^1)\\
i\{\pi_1^a(x), J_L^{1,b}(y)\}_{AP} =& -\frac{ie}{4\pi}\delta^{ab}(a+1)\delta(x^1-y^1)\\
i\{J_L^{0,a}(x), J_L^{0,b}(y)\}_{AP} =& if_{abc}\, J_L^{0,c}(x)\delta(x^1-y^1)+\frac{i}{2\pi}\delta^{ab}\partial_1\delta(x^1-y^1)\\
&+\frac{ie}{4\pi}f_{abc}[(a+1)A_1^c(x)-A_0^c(x)]\delta(x^1-y^1)\\
i\{J_L^{0,a}(x), J_L^{1,b}(y)\}_{AP} =& if_{abc}\, J_L^{1,c}\delta(x^1-y^1)-\frac{i}{2\pi}\delta^{ab}\partial_1\delta(x^1-y^1)\\
&+\frac{ie}{4\pi}f_{abc}[A_0^c(x)+(a-1)A_1^c(x)]\delta(x^1-y^1) \quad .
\end{aligned}
\tag{13.345}
$$

Using these results, one finds

$$
\{\pi_0^a(x), G^b(y)\}_{AP} = \frac{e^2}{4\pi}(1-a)\delta^{ab}\delta(x^1-y^1) \tag{13.346}
$$

so that the algorithm ends at this point if $a \neq 1$, since the constraints are second class in this case.

The results obtained here coincide with those obtained previously using the bosonic formulation of the theory. Hence it is clear that Maxwell's equations are also satisfied on quantum level, when written in terms of the APB's. To complete the analysis, let us check, whether the non-canonical structure of the APB's (13.341*b*), involving the kinetic term, can also be understood from the bosonic point of view.

Comparing the Hamiltonian (13.331*b*) and current (13.337) with their respective bosonized versions (13.207*b*) and (13.190), one is led to the identification

$$
\bar{\psi}i\gamma_1\partial_1\psi = -2\pi\mathrm{tr}(\hat{\Pi}^T h\hat{\Pi}^T h)+\frac{1}{8\pi}\mathrm{tr}(\partial_1 h\partial_1 h^{-1})-\frac{e^2}{8\pi}\mathrm{tr}(A_1^2-aA_+A_-) \quad . \tag{13.347}
$$

Using the Poisson brackets (13.189) we recover from here the anomalous Poisson brackets (13.341).

We have thus been able to check the method proposed in ref. [422], for calculating APB in a chiral gauge theory, in terms of a non–trivial example. At the same time, we have thereby given support to the identification of the "algebra of constraints" with the algebra of ABP as defined via (13.329). This opens the way for developing a Dirac algorithm for chiral gauge theories in higher dimensions,

---

28) The results involving only the currents have also been obtained in ref. [453], for the case $a=0$. The $a \neq 0$ results follow from here upon using (13.337).

where bosonization techniques are not available. In fact it has been shown that cancellations similar to the ones witnessed above also occur in QCD_4[459].

The possibility of calculating the exact APB's perturbatively allows one to construct the exact commutators of an anomalous chiral gauge theory using the Dirac construction. In the fermionic formulation this amounts to the replacement of (13.218) by

$$\{A,B\}_{AD} = \{A,B\}_{AP} - \int \{A,\Omega_\alpha\}_{AP} Q^{-1}_{\alpha\beta}\{\Omega_\beta, B\}_{AP} \quad , \tag{13.348}$$

where "AD" stands for "anomalous Dirac bracket". The commutators are then obtained via the substitution $i\{A,B\}_{AD} \to [A,B]$. This circumvents two difficulties:
i) Unlike the case of the APB, the computation of the commutators require the knowledge of the exact Greens-functions, and thus involve the summation over an infinite set of diagrams. This summation is here contained in the inversion of $Q_{\alpha\beta}(x,y)$.
ii) In the GNI formulation we are not allowed to choose a gauge. As a result, a perturbative scheme can only be formulated by explicitly breaking the gauge invariance of the classical Lagrangian, summing the resulting perturbation series to all orders, and then "undoing" the gauge fixation, by taking the appropriate limit in the final result. Though this difficulty can be overcome by going over to the GI formulation, the $a = 1$ case continues to pose a problem, because of the absence of the kinetic term $S_{P\sigma M}$ for the WZ field g in the action (13.138). We leave the discussion of this problem to chapter 14, where we consider the abelian case.

13.17 Conclusion

We have discussed in this chapter general aspects of chiral QCD_2, as they have been investigated up to date. A very interesting structure has emerged, much of which could also be a feature of chiral QCD_4. Unlike the case of chiral QCD_4, where the observed violation[424,460] of the Jacobi identity (3-cocycle) in the double commutator of the electric field strength raises the question as to the algebraic consistency of chiral gauge theories in $d = 3+1$ dimensions, no such violation occurs in $d = 1+1$ dimensions. Also, since QCD_2 is a superrenormalizable theory, we expect it to respect unitarity and Lorentz-covariance, properties which can be explicitly verified in the $U(1)$ case to be discussed in the following chapter.

At present no exact solutions of QCD_2 are available, so that a calculation of the corresponding correlation functions will have to be based on some perturbative scheme. Since it was our principal aim to gain a structural insight into the theory, we have not dwelled on any dynamical calculation. Although the main body of our discussion was based on the equivalent bosonic formulation, we have seen in the last section, how the same results are obtained in the framework of the fermionic formulation. This is very important, since it opens the possibility of gaining a similar insight into the structural properties of chiral QCD in $d = 3+1$ dimensions. Already in QCD_2, a superrenormalizable theory, we have witnessed on the fermionic level a delicate interplay of anomalous contributions with the result that the form of the constraints and equations of motion remained unchanged. The corresponding situation in QCD_4 remains uncertain. Recent calculations[459] based on the method described in the last section nevertheless show, that also in the case of QCD_4, cancellations of anomalous contributions similar to the ones witnessed in the case of QCD_2, occurs. Calculations performed in the $A^0 = 0$ gauge lead to similar conclusions[436].

In the case of an abelian group, the analysis of this chapter can be carried further, and the theory can be solved completely. We shall therefore turn in the following chapter to the study of the exact solutions of the corresponding $U(1)$ theory, chiral QED_2.

Chapter XIV

CHIRAL QUANTUM ELECTRODYNAMICS

14.1 Introduction

Chiral QCD_2 cannot be solved exactly at present. The discussion of the preceding chapter therefore concentrated on general aspects, with special emphasis on fermion-boson equivalences, Hamiltonian quantization, and current conservation laws associated with the global and local symmetries. This has been done in the framework of two isomorphic representations. By going over to the equivalent bosonic theory one could discuss on a semiclassical level the breaking of gauge invariance by the fermion loops, as well as the implications of this breakdown. Finally we showed how this analysis could also be carried out on the fermionic level, without recourse to bosonization.

Actually, the renewed interest in chiral gauge theories with anomalous breakdown of gauge invariance was the result of work due to Jackiw and Rajaraman[71], who demonstrated that chiral QED_2 – the so-called "chiral Schwinger model" – though exhibiting an anomalous breakdown of the local U(1) symmetry of the classical Lagrangian, admitted exact solutions in a positive metric Hilbert space, respecting unitarity, provided that the Jackiw-Rajaraman parameter a, also introduced in the context of chiral QCD_2 in chapter 13, was restricted to the range $a \geq 1$.

Our analysis in this chapter will be mainly concerned with the construction and "physical" interpretation of the exact solutions in Fock space of the equations of motion for the vector potential and fermion field of chiral QED_2. This will be done for both, the gauge-non-invariant (GNI) and gauge-invariant (GI) formulations. We shall further show how these two formulations are related in phase space. Having at our disposal the exact solutions, we shall also be able to understand, on operator level, the "mechanism" behind the procedure discussed in the last section of chapter 13 for obtaining the constraint algebra in the fermionic formulation.

The lack of gauge invariance on quantum level invalidates the use of specific gauges. This, as we shall see, makes a perturbative discussion in the GNI formulation difficult. When comparing the exact results with perturbation theory, we are thus naturally led to consider the GI formulation, where Feynman rules can be formulated after choosing a gauge.

The general results for chiral QCD_2, as obtained in the previous chapter, also apply to (massless) chiral QED_2 if we replace the SU(N) generators τ^a by the identity. In order to simplify the expressions, we follow the literature and redefine the charge by[1])

$$e/2\sqrt{\pi} \to e \quad .$$

Making the further replacement $\tau^a \to 1$, Eqs. (13.17,18) read

$$\partial_\mu F^{\mu\nu} + 2\sqrt{\pi}e\bar{\psi}\gamma^\nu P_-\psi = 0 \qquad (14.1a)$$

$$(i\partial\!\!\!/ + 2\sqrt{\pi}eA\!\!\!/P_-)\psi = 0 \quad , \qquad (14.1b)$$

with A_μ now the (abelian) U(1) vector potential, and

$$F^{\mu\nu} = \partial^\mu A^\nu - \partial^\nu A^\mu \qquad (14.1c)$$

the field–strength tensor. For later reference we also give here the corresponding classical Hamiltonian:

$$H_c^{cl} = \int dy^1 \left[\frac{1}{2}\pi_1^2 - \bar{\psi}\gamma^1\partial_1\psi - \Omega_2 A^0 + 2\sqrt{\pi}e\bar{\psi}\gamma^1 P_-\psi A^1\right] + 2\sqrt{\pi}e\bar{\psi}\gamma^1 P_-\psi A^1] \quad , \qquad (14.2)$$

with π_1 the momentum conjugate to A^1,

$$\pi_1 = F^{01} \quad , \qquad (14.3)$$

and Ω_2 the Gauss operator,

$$\Omega_2 \equiv -\partial_1\pi_1 + 2\sqrt{\pi}e\bar{\psi}\gamma^0 P_-\psi \quad . \qquad (14.4)$$

We now proceed to construct the exact solutions of the corresponding quantum theory. The discussion in this chapter is again done in Minkowski space.

14.2. The JR Model

In the abelian case the "topological" (Wess-Zumino) term in expression (13.120) for $W^{(L)}[A]$ vanishes because of the antisymmetry of $\epsilon_{\mu\nu}$. Recalling the parametrization (13.117), we may write the effective action in the form[2])

$$W^{(L)}[A] = \frac{ae^2}{2}\int d^2x A^\mu A_\mu - \frac{e^2}{2}\int\int (g^{\mu\lambda} + \epsilon^{\mu\lambda})A_\lambda \frac{\tilde{\partial}_\mu\tilde{\partial}_\nu}{\Box}(g^{\nu\rho} + \epsilon^{\nu\rho})A_\rho \quad . \qquad (14.5)$$

---

[1]) The currents in this chapter are thus scaled by a factor $2\sqrt{\pi}$ with respect to those in chapter 13.

[2]) Note that this result could also be obtained directly from the Minkowski version of (12.4c) via the substitution $A_\mu \to \frac{1}{2}(g_{\mu\lambda} + \epsilon_{\mu\lambda})A^\lambda$ suggested by the identity $\gamma^\mu(1-\gamma_5)A_\mu = \gamma^\mu(g_{\mu\nu} + \epsilon_{\mu\nu})A^\nu$.

As we already remarked in the context of chiral QCD_2, the additional mass term proportional to A_μ^2 takes full account of the ambiguity in the regularization procedure resulting from the lack of gauge invariance as a guiding principle, and amounts to including a local counterterm in the Lagrangian. The inclusion of such a term is absolutely crucial since, as we shall demonstrate, unitarity will only be respected provided[71] $a \geq 1$. Note that there exists no value of the parameter a, for which expression (14.5) can be made gauge invariant.

With the parametrization

$$h = e^{i2\sqrt{\pi}\phi}, \quad g = e^{i2\sqrt{\pi}\theta} \tag{14.6}$$

of the (now $U(1)$-valued) fields h and g introduced in chapter 13, the 1–cocycles (13.125,127) take the form

$$\gamma_1[A, h^{-1}] = \int d^2x \frac{1}{2}\partial_\mu\phi\partial^\mu\phi + e\int d^2x A^\mu(\partial_\mu - \tilde{\partial}_\mu)\phi \tag{14.7}$$

$$\alpha_1[A, g^{-1}] = -(1-a)\int d^2x \frac{1}{2}\partial_\mu\theta\partial^\mu\theta + e\int d^2x A^\mu[(a-1)\partial_\mu + \tilde{\partial}_\mu]\theta\,. \tag{14.8}$$

In this abelian case, representation (13.126) can thus be trivially checked by explicitly performing the Gaussian integration over ϕ, without recourse to the trick of di Vecchia et al[343].

We summarize the results in the GNI and GI formulations.

GNI formulation

The bosonic Lagrangian in the GNI formulation, as obtained from (13.137), reads

$$\mathcal{L}_{bos}^{GNI} = -\frac{1}{4}F_{\mu\nu}F^{\mu\nu} + \frac{a}{2}e^2 A_\mu A^\mu + \frac{1}{2}(\partial_\mu\phi)^2 + eA^\mu(\partial_\mu - \tilde{\partial}_\mu)\phi \tag{14.9}$$

and equations (13.142) are now replaced by

$$\partial_\mu F^{\mu\nu} + (\partial^\nu - \tilde{\partial}^\nu)\phi + aeA^\nu = 0 \tag{14.10a}$$

$$\Box\phi + e(\partial_\mu - \tilde{\partial}_\mu)A^\mu = 0 \quad . \tag{14.10b}$$

The currents and their corresponding divergencies now read (compare with (13.145) and (13.151))

$$J_L^\mu = (\partial^\mu - \tilde{\partial}^\mu)\phi + aeA^\mu\,, \qquad \partial_\mu J^\mu = \mathcal{A} \quad , \tag{14.11a}$$

$$J_R^\mu = -(g^{\mu\nu} + \epsilon^{\mu\nu})(\partial_\nu\phi - eA_\nu) \quad , \quad \partial_\mu J_R^\mu = 0 \quad , \tag{14.11b}$$

where $\mathcal{A}$ is the anomaly

$$\mathcal{A} = e[(a-1)\partial_\mu + \tilde{\partial}_\mu]A^\mu \quad . \tag{14.12}$$

Under a gauge transformation $G = e^{-i2\sqrt{\pi}\Lambda}$,

$$A_\mu \rightarrow A_\mu + \frac{1}{e}\partial_\mu \Lambda \quad , \qquad \phi \rightarrow \phi + \Lambda \tag{14.13a}$$

$$J_L^\mu \underset{G}{\rightarrow} J_L^\mu + [(a-1)\partial^\mu + \tilde{\partial}^\mu]\Lambda \quad , \quad J_R^\mu \underset{G}{\rightarrow} J_R^\mu \,. \tag{14.13b}$$

which now replaces the transformation laws (13.162) to (13.165). The choice $a = 1$ in (14.12) corresponds to what has been referred to in the literature as the *minimal* anomaly.

GI formulation

The bosonic Lagrangian, as obtained from (13.138), now reads [3)]

$$\mathcal{L}_{bos}^{GI} = \mathcal{L}_{bos}^{GNI} + \frac{a-1}{2}(\partial_\mu\theta)^2 + eA^\mu[(a-1)\partial_\mu + \tilde{\partial}_\mu]\theta \tag{14.14}$$

and equations (13.155) are replaced by

$$\partial_\mu F^{\mu\nu} + e\mathcal{J}^\nu = 0 \tag{14.15a}$$

$$\Box\phi + e(\partial_\mu - \tilde{\partial}_\mu)A^\mu = 0 \tag{14.15b}$$

$$(1-a)\Box\theta - e[(a-1)\partial_\mu + \tilde{\partial}_\mu]A^\mu = 0 \quad . \tag{14.15c}$$

The currents (13.154), (13.159) now read

$$\mathcal{J}^\mu = (\partial^\mu - \tilde{\partial}^\mu)\phi + aeA^\mu + [(a-1)\partial^\mu + \tilde{\partial}^\mu]\theta \tag{14.16a}$$

$$\mathcal{J}_R^\mu = -(g^{\mu\nu} + \epsilon^{\mu\nu})(\partial_\nu\phi - eA_\nu) \tag{14.16b}$$

$$\mathcal{J}_{WZ}^\mu = -[(1-a)g^{\mu\nu} + \epsilon^{\mu\nu}](\partial_\nu\theta + eA_\nu) \quad . \tag{14.16c}$$

All currents are now conserved and gauge singlets.

The above linear equations evidently admit an exact solution. These are to be constructed subject to 'initial conditions', which are provided as usual by the equal time commutators. Since in the abelian case it is natural to choose ϕ and θ rather than h and g as 'coordinates' to be quantized, it is more practical to develop here separately the canonical formalism without reference to our earlier non-abelian results. The results may be checked against those of chapter 13 by noting that

$$\Pi_h = \frac{-i}{2\sqrt{\pi}}h^{-1}\pi_\phi \quad , \quad \pi_\phi = \frac{\partial\mathcal{L}}{\partial\partial_0\phi} \tag{14.17}$$

with similar expressions involving the "Wess-Zumino" field θ.

[3)] The ad hoc inclusion of a Wess–Zumino term to obtain a gauge invariant formulation of chiral QED_2 was first discussed in ref.[462].

14.3 Quantization in the *GNI* Formulation

14.3.1 Hamiltonian and constraints[463,73]

From Eq. (14.9) we obtain for the momenta π_μ and π_ϕ conjugate to A^μ and ϕ:

$$\begin{aligned} \pi_0 &= 0 \qquad \pi_1 = F^{01} \\ \pi_\phi &= \partial^0\phi + e(A_0 - A_1) \quad . \end{aligned} \tag{14.18}$$

We recognize in $\pi_0 = 0$ a primary constraint to be imposed weakly:

$$\Omega_1 \equiv \pi_0 \approx 0 \quad . \tag{14.19}$$

Following the terminology of Dirac, the *total* Hamiltonian obtained from the *GNI* Lagrangian (14.9) is

$$H_T = H_c^{GNI} + \int dy^1 v(y)\pi_0(y) \tag{14.20a}$$

with v a Lagrange multiplier, and H_c^{GNI} the *canonical* Hamiltonian defined in terms of the Legendre transform of $\mathcal{L}_{bos}^{GNI}$:

$$H_c^{GNI} = \int dy^1 \left[\frac{1}{2}\pi_1^2 + \frac{1}{2}\pi_\phi^2 + \frac{1}{2}(\partial_1\phi)^2 + A^0\partial_1\pi_1 - \frac{ae^2}{2}A_\mu^2 \right.$$
$$\left. + \frac{e^2}{2}A_-^2 - e\pi_\phi A_- - eA_-(\partial_1\phi)\right] \, , \tag{14.20b}$$

which is easily recognized to be the abelian version of (13.207).

Following the algorithm of Dirac, we find that the requirement of persistence in time of the primary constraint (14.19) leads to the secondary constraint

$$\Omega_2 \equiv -\partial_1\pi_1 + eJ_L^0 \approx 0 \quad , \tag{14.21a}$$

with

$$J_L^0 = \pi_\phi + \partial_1\phi + e[(a-1)A_0 + A_1] \tag{14.21b}$$

the phase-space version of the time-component of the current defined in (14.11a). We recognize in (14.21a) the Gauss constraint. One calculates[464]

$$\begin{aligned} \{\Omega_1(x),\, \Omega_1(y)\} &= 0 \quad , \qquad \{\Omega_2(x),\, \Omega_2(y)\} = 0 \, , \\ \{\Omega_1(x),\, \Omega_2(y)\} &= -\, e^2(a-1)\delta(x^1 - y^1) \quad , \end{aligned} \tag{14.22}$$

where it is understood that the Poisson brackets are computed for equal times.

For $a \neq 1$, the constraints (14.19,21) are evidently second class, and the requirement that they be satisfied at all times,

$$\partial_0 \Omega_\alpha(x) \approx \{\Omega_\alpha(x), H_T\} \approx \{\Omega_\alpha(x), H_c^{GNI}\} + \int dy^1 \{\Omega_\alpha(x), \Omega_1(y)\} v(y) \approx 0 .$$

thus serves to fix the so far undetermined Lagrange multiplier v :

$$v \approx \partial_1 A_1 + \frac{1}{a-1}\pi_1 \quad . \tag{14.23}$$

Noting that $v \approx \partial_0 A_0$, and recalling (14.18), we see that (14.23) implies the vanishing of the anomaly on the space of solutions, as consistency of the equations of motion (14.10) requires.

For $a = 1$ we need to carry the algorithm further. H_c and Ω_2 are now given by (14.20*b*) and (14.21*b*) with $a = 1$. The condition $\partial_0 \Omega_2 \approx 0$ requires $\{\Omega_2, H_T\} \approx 0$ which leads to the further constraint

$$\Omega_3 \equiv \pi_1 \approx 0 \tag{14.24a}$$

expressing the vanishing of the "electric field". Since $\{\Omega_1, \Omega_\alpha\} = 0$, $(\alpha = 1,2,3)$, the condition $\partial_0 \Omega_3 \approx 0$ still does not fix the Lagrange multiplier v, but rather leads to a further constraint

$$\Omega_4 = e(A_0 - A_1) - J_L^0 \approx 0 \tag{14.24b}$$

where J_L^0 is now the current in (14.21) for $a = 1$. We recognize in Ω_4 the constraint (13.234) of the non-abelian case. The constraints (14.19,21*a*) and (14.24) now imply

$$J_L^0 = 0 \quad , \quad A_- = 0 \quad . \tag{14.25}$$

Since $\{\Omega_1, (x), \Omega_4(y)\} \not\approx 0$, the requirement $\partial_0 \Omega_4 \approx 0$ now determines the Lagrange multiplier v, and no further constraints are generated. One easily verifies that the constraints $\Omega_\alpha, \alpha = 1,...4$ are again of *second* class, so that gauge invariance continues to be broken for $a = 1$. Because of this drastic change in the constraint structure for $a = 1$, the cases $a \neq 1$ and $a = 1$ require a separate discussion, as it was also the case for chiral QCD_2.

For later reference we state here also some Poisson bracket relations involving the source current (14.21*b*):

$$\{\pi_0(x), J_L^0(y)\} = -\{\pi_1(x), J_L^1(y)\} = -e(a-1)\delta(x^1 - y^1) \tag{14.26a}$$

$$\{\pi_1(x), J_L^0(y)\} = -\{\pi_0(x), J_L^1(y)\} = e\delta(x^1 - y^1) \tag{14.26b}$$

$$\{J_L^0(x), J_L^0(y)\} = -\{J_L^0(x), J_L^1(y)\} = \{J_L^1(x), J_L^1(y)\} = 2\delta'(x^1 - y^1) \tag{14.26c}$$

In section 7 we shall see that, in the fermionic formulation, these Poisson brackets correspond to the "anomalous Poisson brackets" defined in (13.329).

14.3.2 Commutation relations[73]

We first consider the case $a \neq 1$. Since the constraints are second-class, we cannot impose these constraints consistently on the states, i.e. we *must* implement the constraints strongly. One thus needs to compute the Dirac brackets $\{,\}_{\mathcal{D}}$, defined by (13.218). Recalling (14.17), we may read off the results directly from (13.220). The non-vanishing equal time commutators read

$$
\begin{aligned}
&[A^1(x),\pi_1(y)]_{ET} = i\delta(x^1-y^1) && [A^0(x),\pi_\phi(y)]_{ET} = \frac{-i\delta'(x^1-y^1)}{e(a-1)}\\
&[A^0(x),A^1(y)]_{ET} = \frac{-i\delta'(x^1-y^1)}{e^2(a-1)} && [\phi(x),\pi_\phi(y)]_{ET} = i\delta(x^1-y^1)\\
&[A^0(x),\pi_1(y)]_{ET} = \frac{i\delta(x^1-y^1)}{(a-1)} && [A^0(x),\phi(y)]_{ET} = \frac{i\delta(x^1-y^1)}{e(a-1)}
\end{aligned}
\tag{14.27a}
$$

where the 'prime' on δ indicates differentiation of δ with respect to its argument. In particular, the equal-time commutators of the constraints vanish, by construction. Note that the commutators are singular for $a = 1$, which corresponds to the case of the minimal anomaly.

We now seek the solution of the equations of motion (14.10) subject to the "initial conditions" (14.27a), which are preferably rewritten in terms of the Lagrangian variables[5]:

$$
\begin{aligned}
&[A^0(x),A^1(y)]_{ET} = \frac{-i\delta'(x^1-y^1)}{e^2(a-1)}, && [A^0(x),\phi(y)]_{ET} = \frac{i\delta(x^1-y^1)}{e(a-1)},\\
&[A^0(x),\partial^0\phi(y)]_{ET} = 0, && [A^1(x),\partial^0\phi(y)]_{ET} = \frac{i}{e(a-1)}\delta'(x^1-y^1),\\
&[\phi(x),\partial^0 A^1(y)]_{ET} = \frac{-i\delta'(x^1-y^1)}{e(a-1)}, && [\phi(x),\partial^0\phi(y)]_{ET} = i\frac{a\delta(x^1-y^1)}{(a-1)},\\
&[\partial^0\phi(x),\partial^0\phi(y)]_{ET} = 0, && [A^0(x),\partial^0 A^1(y)]_{ET} = \frac{i\delta(x^1-y^1)}{(a-1)},
\end{aligned}
$$

$$
[A^1(x),\partial^0 A^1(y)]_{ET} = i\Big[1-\frac{1}{e^2(a-1)}\partial_x^1\partial_x^1\Big]\delta(x^1-y^1) \quad . \tag{14.27b}
$$

The singularity at $a = 1$ in the above commutation relations is a reflection of change in the constraint structure at this value of a. The most general solution of the equations of motion (10) is found to be given by[71,73]

$$
\begin{aligned}
A^\mu &= -\frac{1}{ae}[\partial^\mu\phi + (a-1)\tilde{\partial}^\mu\phi + a\tilde{\partial}^\mu h]\\
&= -\frac{1}{ae}[a\tilde{\partial}^\mu\sigma + (\partial^\mu - \tilde{\partial}^\mu)(\sigma - h)]\\
\phi &= \sigma - h \quad ,
\end{aligned}
\tag{14.28a}
$$

where σ and h are free fields satisfying

$$(\Box + m^2)\sigma = 0 \quad , \qquad \Box h = 0 \tag{14.28b}$$

with

$$m^2 = \frac{a^2 e^2}{a-1} \quad . \tag{14.28c}$$

Note that

$$F^{\mu\nu} = -ea\epsilon^{\mu\nu}\sigma \quad . \tag{14.28d}$$

One again checks that the anomaly (14.12) vanishes on the space of solutions.

From (14.28c) it is evident that for $a < 1$ the model does not define a bonafide theory. Hence we restrict our attention to $a \geq 1$. Imposing the "initial conditions" (14.27), one finds that σ and h are to be quantized with the commutation relations

$$\begin{aligned} &[h(x), h(y)]_{ET} = 0, \quad [h(x), \partial_0 h(y)]_{ET} = i\delta(x^1 - y^1) \\ &[\sigma(x), \sigma(y)]_{ET} = 0, \quad [\sigma(x), \partial_0 \sigma(y)]_{ET} = \frac{i}{a-1}\delta(x^1 - y^1) \quad . \end{aligned} \tag{14.29}$$

The above commutation relations are singular for $a = 1$, reflecting the change in the constraint structure for this value of the JR parameter. Proceeding as before, one finds that for $a = 1$ the commutation relations (14.26a) are replaced by[73]

$$\begin{aligned} &[A^0(x), A^1(y)]_{ET} = \tfrac{-2i}{e^2}\delta'(x^1 - y^1) \qquad [A^0(x), \phi(y)]_{ET} = \frac{i}{e}\delta(x^1 - y^1) \\ &[A^0(x), \pi_\phi(y)]_{ET} = \tfrac{-i}{e}\delta'(x^1 - y^1) \qquad [A^1(x), \phi(y)]_{ET} = \frac{-i}{e}\delta(x^1 - y^1) \\ &[A^1(x), \pi_\phi(y)]_{ET} = \tfrac{i}{e}\delta'(x^1 - y^1) \qquad [\phi(x), \pi_\phi(y)]_{ET} = i\delta(x^1 - y^1). \end{aligned} \tag{14.30}$$

All other commutators vanish. The canonical Hamiltonian may now conveniently be written in the form

$$H_c^{GNI} = \frac{1}{2}\int dx^1[\pi_\phi^2 + (\partial_1\phi)^2] + \int dx^1[\frac{1}{2}\Omega_3^2 + A^1\partial^1\Omega_3 - (A_0 - A_1)\Omega_2] \quad . \tag{14.31}$$

The second term is seen to vanish on the constraint surface, so that the $a = 1$ theory corresponds to a free zero mass field ϕ. Recalling (14.25), one obtains from (14.31) and (14.30) the equations of motion

$$A^0 = -A^1 = -\frac{2}{e}\partial_+\phi \quad . \tag{14.32}$$

The field A^μ thus corresponds to a regressive wave propagating with the velocity of light. Given the triviality of this case, we shall limit ourselves to the case $a \neq 1$ in the remaining part of this section[4])

---

[4]) Actually, from the point of view of the constraint structure, chiral QED and QCD in four dimensions resemble the case $a = 1$[459].

14.3.3 Current–potential and bosonic representation of the fermion field

Before we can proceed to calculate the correlation functions of the fermionic theory in question, we must establish the role of the harmonic field h in relation to ψ. Since σ describes a free, massive field, we can treat it as a field that commutes with the massless fermion $\psi^{(0)}$ at all space-time points. On the other hand, since h is a harmonic (massless) field, it may, and indeed does, play the role of the "potential" associated with the free fermionic current:

$$:\bar{\psi}^{(0)}\gamma^{\mu}\psi^{(0)}:= \frac{1}{\sqrt{\pi}}\epsilon^{\mu\nu}\partial_{\nu}h \quad . \tag{14.33}$$

In order to establish this, we note that from (14.28a) follows

$$\partial_{\mu}F^{\mu\nu} = ea\tilde{\partial}^{\nu}\sigma \quad . \tag{14.34}$$

Comparing with (14.1a), we conclude that there must exist a regularization prescription for which

$$2\sqrt{\pi}\mathcal{N}[\bar{\psi}\gamma^{\mu}P_{-}\psi](x) = -a\tilde{\partial}^{\mu}\sigma(x) \quad , \tag{14.35}$$

where $\mathcal{N}$ denotes a generalized "normal product".

Now, replacing A_{μ} in the Dirac equation (14.1b) by (14.28) and integrating, one finds for the renormalized fermion field[464]

$$\psi(x) =: e^{-2i\sqrt{\pi}\sigma(x)P_{-}} : \psi^{(0)}(x) \quad , \tag{14.36}$$

where $\psi^{(0)}(x)$ stands for the free Dirac field[5)]. We see that the right-handed component of ψ remains free, as it should, since it does not couple to the gauge field. The left-handed component only involves the σ-field; the term involving $\sigma - h$ in (14.28a) drops out due to the kinematical relation, $\gamma^{\mu}P_{-}(\partial_{\mu} - \tilde{\partial}_{\mu}) = 0$.

Following Mandelstam[47] we represent the (canonical) fermion field in terms of bosons as follows:

$$\psi_{\alpha}^{(0)}(x) = \left(\frac{\mu}{2\pi}\right)^{1/2} e^{-\frac{i\pi}{4}\gamma_5} : e^{i\sqrt{\pi}\gamma^{5}_{\alpha\alpha}\varphi(x) - i\sqrt{\pi}\int_{x}^{\infty} dz^{\mu}\tilde{\partial}_{\mu}\varphi(z)} : \tag{14.37a}$$

with the corresponding left–handed current

$$:\bar{\psi}^{(0)}(x)\gamma_{\mu}P_{-}\psi(x) := -\frac{1}{2\sqrt{\pi}}(\tilde{\partial}_{\mu} - \partial_{\mu})\varphi \quad . \tag{14.37b}$$

---

5) Eq. (14.36) implies a non–trivial wave–function renormalization constant for the unrenormalized fermion field, $\psi_u = Z_{\psi}^{1/2}\psi$, with $Z_{\psi}(\epsilon) = \exp(-4\pi\Delta^{(+)}(\epsilon; m^2))$ and ϵ an ultraviolet cutoff. There is no coupling constant renormalization.

Here φ is a zero mass field. We define the normal product (14.35) in terms of the short-distance expansion

$$Z^{-1}(\epsilon)[\bar{\psi}(x+\epsilon)\gamma^\mu P_- e^{i\sqrt{\pi}eP_- \int_x^{x+\epsilon} dz^\mu(\kappa_1 A_\mu + \kappa_2 \tilde{A}_\mu)}\psi(x) - V.E.V.]$$
$$= \mathcal{N}[\bar{\psi}\gamma^\mu P_- \psi](x) - \frac{be}{2\sqrt{\pi}} A^\mu(x) + O(\epsilon) \tag{14.38}$$

where b is a constant which is to be determined, and where — in analogy to the case of QED_2 —we have introduced a Schwinger-line integral with κ_1 and κ_2 free parameters reflecting the freedom one has in regularizing the operator product. The term involving A^μ takes account of the fact that, as a result of the dimensionality of e, such a term should be included among the leading terms since gauge invariance is broken. For a suitable choice of the renormalization constant $Z(\epsilon)$, one finds for the $\mu = 0$ component of the l.h.s. of Eq.(14.38),

$$(l.h.s.)_{\mu=0} = \frac{\epsilon^\nu}{2\sqrt{\pi}\epsilon^+}\left[(2-\kappa_1-\kappa_2)\partial_\nu\sigma + \left(\kappa_1 - \frac{\kappa_1-\kappa_2}{a}\right)(\partial_\nu - \tilde{\partial}_\nu)\sigma + (\partial_\nu - \tilde{\partial}_\nu)\varphi\right] \tag{14.39}$$

where we have used (A.27a).

The choice $\kappa_1 = \kappa_2 = 1$ corresponds to a gauge invariant regularization. Making this choice, one sees that (14.39) has the direction independent limit $(l.h.s.)_{\mu=0} = \frac{1}{2\sqrt{\pi}}[\partial_+(\sigma+\varphi)]$. Because of the identity $\gamma^0 P_- = -\gamma^1 P_-$, we thus obtain for the l.h.s. of Eq.(14.38),

$$(l.h.s.) = \frac{1}{2\sqrt{\pi}}[(\partial^\mu - \tilde{\partial}^\mu)(\sigma+\varphi)] \quad . \tag{14.40}$$

Finally choosing $b = a$ on the r.h.s. of Eq. (14.38), and making the identification

$$\varphi = -h \quad , \tag{14.41}$$

one obtains the desired result (14.35). For a corresponding calculation of the external field current using Schwinger's point splitting technique see ref. [465]. We have thus shown that the harmonic field h is to be identified according to (14.41) with the "potential" of the free fermionic current[6)].

To conclude, we remark that, unlike the case of the Schwinger model, the chiral charge, constructed from (14.35), is conserved and commutes with the Hamiltonian, as well as ψ. Hence, ψ is chirally neutral, much as in the Schwinger model, where the physical fermion operator is electrically neutral. ψ, nevertheless, carries the selection rules of the fermion[464].

---

6) In our analysis, Maxwell's equations are satisfied as operator equations. Our analysis differs in this respect with that of Boyanowsky[466], which relies on the introduction of redundant degrees of freedom.

14.3.4 Energy-momentum tensor

The energy momentum tensor is defined by

$$\begin{aligned}\tilde{\Theta}^{\mu\nu} =& \frac{\partial \mathcal{L}^{GNI}_{bos}}{\partial\partial_\mu\phi}\partial^\nu\phi + \frac{\partial \mathcal{L}^{GNI}_{bos}}{\partial\partial_\mu A_\lambda}\partial^\nu A_\lambda - g^{\mu\nu}\mathcal{L}^{GNI}_{bos} \\ =& -F^{\mu\lambda}\partial^\nu A_\lambda + \partial^\mu\phi\partial^\nu\phi + e(A^\mu + \tilde{A}^\mu)\partial^\nu\phi - g^{\mu\nu}\mathcal{L}^{GNI}_{bos}\ . \end{aligned} \tag{14.42}$$

Making use of Maxwell's equation (14.10a) and of (14.28) one has, respectively,

$$F^{\mu\lambda}\partial^\nu A_\lambda = F^{\mu\lambda}F^\nu_\lambda - eJ^\mu_L A^\nu + \partial_\lambda(F^{\mu\lambda}A^\nu) \tag{14.43}$$

$$(\tilde{\partial}^\mu\phi)A^\nu - \tilde{A}^\mu(\partial^\nu\phi) - g^{\mu\nu}A^\lambda\tilde{\partial}_\lambda\phi = 0\quad . \tag{14.44}$$

Making use of these relations, one finds upon dropping in (14.43) the term involving a total divergence, we arrive at the symmetric energy momentum tensor

$$\Theta^{\mu\nu} = [-F^{\mu\nu}F^\nu_\lambda + \frac{1}{4}g^{\mu\nu}F^{\lambda\rho}F_{\lambda\rho}] + (t^{\mu\nu} - \frac{1}{2}g^{\mu\nu}t^\lambda_\lambda)\quad , \tag{14.45a}$$

where

$$t^{\mu\nu} = \partial^\mu\phi\partial^\nu\phi + e(\partial^\mu\phi A^\nu + A^\mu\partial^\nu\phi) + ae^2A^\mu A^\nu\quad . \tag{14.45b}$$

One checks that Θ^{00} coincides with the Hamiltonian density in (14.20b), on the surface Γ_Ω defined by the constraints (14.19,21).

The tensor $t^{\mu\nu}$ decomposes into a sum of two terms involving only the σ and h field, respectively, which we identify with the energy-momentum tensor of a (non-canonically normalized) zero mass σ and a canonical harmonic field h. Remembering that $F^{\mu\nu}$ and σ are related by (14.34), one sees that the first term in (14.45) contributes a (mass) term of the form $\frac{a^2e^2}{2}g^{\mu\nu}\sigma^2$ to $\Theta^{\mu\nu}$. One thus obtains the expected result[71]

$$\Theta^{\mu\nu} = \Theta^{\mu\nu}_{(\sigma)} + \Theta^{\mu\nu}_{(h)} \tag{14.46}$$

where $\Theta^{\mu\nu}_{(\sigma)}(\Theta^{\mu\nu}_{(h)})$ are the energy momentum tensor of the (free) σ and h fields with the normalization implied by the commutation relations (14.29):

$$\begin{aligned}\Theta^{\mu\nu}_{(\sigma)} =& (a-1)[\partial^\mu\sigma\partial^\nu\sigma - \frac{1}{2}g^{\mu\nu}(\partial^\lambda\sigma\partial_\lambda\sigma - m^2\sigma^2)] \\ \Theta^{\mu\nu}_{(h)} =& \partial^\mu h\partial^\nu h - \frac{1}{2}g^{\mu\nu}\partial^\lambda h\partial_\lambda h\quad . \end{aligned} \tag{14.47}$$

Having shown that — up to a total divergence — expression (14.42) can be written in the form (14.46) on the space of solutions, it is immediately clear that the Lorentz generators constructed in the standard way from $\Theta^{\mu\nu}$ will generate the correct transformation laws corresponding to scalar fields σ and h, and hence of the vector field (14.28a). This is a priori not entirely obvious because of the non-canonical form (14.27) of the commutation relations, but can be demonstrated to hold also by working with the original field variables A_μ and ϕ[467].

14.3.5 A_μ two-point function

It is convenient to write (14.28*a*) in the form

$$A^\mu(x) = \frac{1}{ae}[\Gamma^{\mu\alpha}\partial_\alpha\sigma(x) - \Lambda^{\mu\alpha}\partial_\alpha h(x)] \tag{14.48a}$$

where

$$\begin{aligned}\Gamma^{\mu\nu} &= g^{\mu\nu} + (1-a)\epsilon^{\mu\nu} \quad ,\\ \Lambda^{\mu\nu} &= g^{\mu\nu} + \epsilon^{\mu\nu} \quad .\end{aligned} \tag{14.48b}$$

One then finds[73]

$$\begin{aligned}&< 0|TA^\mu(x)A^\nu(y)|0 >= \text{C.T.}+\\ &= \frac{-i}{a^2e^2}\left[\frac{1}{a-1}\Gamma^{\mu\alpha}\Gamma^{\nu\beta}\partial^x_\alpha\partial^y_\beta\Delta_F(x-y;m) + \Lambda^{\mu\alpha}\Lambda^{\nu\beta}\partial^x_\alpha\partial^y_\beta D_F(x-y)\right] ,\end{aligned} \tag{14.49a}$$

where we have used

$$i < 0|Th(x)h(y)|0 > = D_F(x-y) \tag{14.49b}$$

$$i < 0|T\sigma(x)\sigma(y)|0 > = \frac{1}{a-1}\Delta_F(x-y;m) \tag{14.49c}$$

and where "*C.T.*" denotes "contact terms" arising from the non-commutativity of the time-derivative with the time-ordering operation. They are given by

$$\begin{aligned}C.T. &= \frac{1}{a^2e^2}\left(\frac{1}{a-1}\Gamma^{\mu 0}\Gamma^{\nu 0} + \Lambda^{\mu 0}\Lambda^{\nu 0}\right)\delta^2(x-y)\\ &= \frac{1}{e^2}\left(\frac{-1}{a}g^{\mu\nu} + \frac{1}{a-1}g^{\mu 0}g^{\nu 0}\right)\delta^2(x-y) \quad .\end{aligned} \tag{14.49d}$$

A little algebra shows that

$$-i < 0|TA^\mu(x)A^\nu(0)|0 >= K^{\mu\nu}(x) + \frac{1}{e^2(a-1)}g^{\mu 0}g^{\nu 0}\delta^2(x) \tag{14.50a}$$

where $K^{\mu\nu}$ is the JR propagator[71]

$$K^{\mu\nu}(x) = \left[g^{\mu\nu} + \frac{1}{a-1}\left(\frac{1}{e^2} + \frac{2}{\Box}\right)\partial^\mu\partial^\nu - \frac{1}{a-1}\left(\frac{\partial^\mu\tilde{\partial}^\nu + \partial^\nu\tilde{\partial}^\mu}{\Box}\right)\right]\Delta_F(x;m) \quad . \tag{14.50b}$$

Thus the JR propagator is obtained from (14.49) by dropping the non-covariant part of the contact term (14.49*d*) in the time-ordered product. For later reference we already note at this point that $K^{\mu\nu}(x)$ satisfies

$$\Omega_{\mu\lambda}K^{\lambda\nu}(x) = g^\nu_\mu\delta^2(x-y) \tag{14.51a}$$

with

$$\Omega_{\mu\nu} = g_{\mu\nu}\Box - \partial_\mu\partial_\nu + e^2[ag_{\mu\nu} - (\partial_\mu - \tilde{\partial}_\mu)\frac{1}{\Box}(\partial_\nu - \tilde{\partial}_\nu)] \quad . \tag{14.51b}$$

From (14.49*a*) and (14.28*c*) it is manifest that for $a < 1$, $K^{\mu\nu}$ contains a tachyonic ghost. For $a > 1$, $K^{\mu\nu}$ contains a massless excitation, as well as a massive one with the mass (14.28*c*).Unlike the case of the vector Schwinger model, the massless excitation lives in a positive metric Hilbert space, and is thus physical. This is a result of the underlying broken gauge invariance.

14.3.6 Fermionic two-point function

From (14.36) one similarly calculates[464] for the renormalized fermion two-point function

$$\langle 0|T\psi(x)\bar{\psi}(y)|0\rangle = e^{-\frac{4i\pi}{a-1}\Delta_F(x-y;m)P_-} iS_F(x-y) \tag{14.52}$$

with S_F the free fermion propagator. In contrast to the (vector) Schwinger model, we see that the theory contains ***asymptotically*** zero-mass fermions (no confinement),

$$\langle 0 \mid T\psi(x)\bar{\psi}(y) \mid 0\rangle \longrightarrow iS_F(x-y) \quad , \tag{14.53}$$

and scales with an anomalous dimension which can be read off from the short-distance behaviour

$$\langle 0 \mid T\psi(x)\bar{\psi}(y) \mid 0\rangle \longrightarrow [-\tilde{m}^2(x-y)^2 + i0]^{-\frac{1}{a-1}} iS_F(x-y) \quad , \tag{14.54a}$$

where $\tilde{m} = \frac{m}{2}e^\gamma$, with γ the Euler constant. Since gauge-invariance is broken, these are physically meaningful statements. Furthermore, the correlation functions of fermion fields and chiral densities are found to cluster, so that, unlike in the Schwinger model, the vacuum is non-degenerate. The chiral version of the Schwinger model thus exhibits features which are quite distinct from those of its vector-counterpart.

14.4 Quantization in the *GI* Formulation

14.4.1 Hamiltonian and constraints

We distinguish again between $a \neq 1$ and $a = 1$.

For $a \neq 1$ we obtain from (14.14) for the momenta conjugate to A_μ, ϕ and θ expressions (14.18) as well as

$$\pi_\theta = (a-1)\partial_0\theta + e[(a-1)A_0 + A_1] \quad . \tag{14.55}$$

We have again one primary constraint

$$\omega_1 \equiv \pi_0 \approx 0 \quad . \tag{14.56a}$$

The canonical Hamiltonian now reads

$$H_c^{GI} = H_c^{GNI} + \int dx^1 \Big[\frac{1}{2}\frac{1}{a-1}[\pi_\theta - e((a-1)A_0 + A_1)]^2 + \frac{a-1}{2}(\partial_1\theta)^2 + e[A_0 + (a-1)A_1]\partial_1\theta\Big] \,. \tag{14.57}$$

Thus the dynamics is governed by the total Hamiltonian

$$H_T = H_c^{GI} + \int dy^1 v(y)\pi_0(y) \tag{14.58}$$

with $v(y)$ a Lagrange multiplier. Persistency in time of (14.56a) leads to the secondary constraint

$$\omega_2 \equiv -\partial_1\pi_1 + e\mathcal{J}^0 \approx 0 \tag{14.56b}$$

with $\mathcal{J}^0$ the phase-space version of the time-component of the (gauge) current defined in (14.16a):

$$\mathcal{J}^0 = \pi_\phi + \partial_1\phi + \pi_\theta - \partial_1\theta \quad . \tag{14.59}$$

We recognize in (14.56b) the Gauss constraint. In contrast to H_c^{GNI}, the Hamiltonian H_c^{GI} only depends linearly onA_0. This becomes evident once we rewrite (14.57) in the form,

$$H_c^{GI} = \int dz^1 [\frac{1}{2}\pi_1^2 + \frac{1}{2}\pi_\phi^2 + \frac{1}{2}\frac{1}{a-1}\pi_\theta^2 - e\pi_\phi A_1 - \frac{e}{a-1}\pi_\theta A_1 + \frac{1}{2}(\partial_1\phi)^2 + \frac{1}{2}(a-1)(\partial_1\theta)^2 + e(\partial^1\phi)A_1 - e(a-1)(\partial^1\theta)A_1 + \frac{e^2a^2}{2(a-1)}A_1^2 - A^0\omega_2] \tag{14.60}$$

Thus, A^0 plays the role of a Lagrange multiplier. One checks that

$$\{\mathcal{J}^0(x), \mathcal{J}^0(y)\} = 0 \quad . \tag{14.61}$$

Hence it immediately follows that the constraints $\omega_1 \approx 0, \omega_2 \approx 0$ are *first class*. Since

$$\{\omega_2, H_T\} = 0 \tag{14.62}$$

there are no further constraints.

For $a = 1$ the Hamiltonian (14.60) is singular, which signalizes the change in the constrained structure already encountered in section 3. We thus need to return in this case to the Lagrangian (14.14), and set $a = 1$ from the beginning:

$$\mathcal{L}_{bos}^{GI} = -\frac{1}{4}F_{\mu\nu}F^{\mu\nu} + \frac{e^2}{2}A_\mu^2 + \frac{1}{2}(\partial_\mu\phi)^2 + eA^\mu(\partial_\mu - \tilde{\partial}_\mu)\phi + eA^\mu\tilde{\partial}_\mu\theta \quad . \tag{14.63}$$

The canonical momenta are this time

$$\begin{aligned} \pi_0 = 0, \quad \pi_1 = F^{01} \quad , \\ \pi_\theta = eA_1, \quad \pi_\phi = \partial_0\phi + e(A_0 - A_1) \quad , \end{aligned} \tag{14.64}$$

implying the existence of *two* primary constraints:

$$\omega_1 \equiv \pi_0 \approx 0 \quad , \quad \omega_3 \equiv \pi_\theta - eA_1 \approx 0 \quad ; \tag{14.65}$$

the canonical Hamiltonian now reads

$$\begin{aligned} H_c^{GI} = \int dx^1 (\frac{1}{2}\pi_1^2 + A^0\partial_1\pi_1 + \frac{1}{2}\pi_\phi^2 - e\pi_\phi A_- - e^2 A_1 A_- \\ + \frac{1}{2}(\partial_1\phi)^2 - e\partial_1\phi A_- - e\partial_1\theta A^0) \quad . \end{aligned} \tag{14.66}$$

Persistence in time of the primary constraints (14.64) leads to the secondary constraints

$$\omega_2 \equiv -\partial_1\pi_1 + e\mathcal{J}^{\,0} \approx 0 \quad , \quad \omega_4 \equiv \pi_1 \approx 0 \tag{14.67}$$

where $\mathcal{J}^0$ is given by (14.59). The Hamiltonian (14.66) may thus be written in the form

$$H_c^{GI} = \int dx^1 [\frac{1}{2}\pi_1^2 + \frac{1}{2}\pi_\phi^2 + eA_1(\pi_\phi + eA_1 + \partial_1\phi) + A^0\omega_2] + \frac{1}{2}(\partial_1\phi)^2 \tag{14.68}$$

which shows that A^0 now plays the role of a Lagrange multiplier.

14.4.2 Implementation of gauge conditions

In the case $a > 1$, we have only first-class constraints. For such Hamiltonian systems, Dirac's conjecture holds as a theorem[468], so that we may define the *extended* Hamiltonian (see appendix O)

$$H_E = H_c^{*GI} + \sum_{\alpha}^{2} \int dy^1 \xi^\alpha(y)\omega_\alpha(y) \tag{14.69}$$

where $\xi^\alpha, \alpha = 1, 2$ are Lagrange multipliers, and where H_c^{*GI} is H_c^{GI} evaluated on the constrained surface $\omega_1 = \omega_2 = 0$; thus it no longer depends on A_0. The generator of infinitesimal gauge transformations is given by

$$G = \sum_{\alpha=1}^{2} \int dx^1 \epsilon^\alpha(x^1)\omega_\alpha(x^1) \tag{14.70}$$

where ϵ^1 and ϵ^2 are infinitesimal independent parameters. G thus generates infinitesimal gauge transformations according to $\delta F = \{F, G\}$, that is[434],

$$\begin{aligned} &\delta A^0 = \epsilon^1,\ \delta A^1 = \partial_1 \epsilon^2,\ \delta\phi = e\epsilon^2,\ \delta\theta = e\epsilon^2, \\ &\delta\pi_0 = 0,\ \ \delta\pi_1 = 0,\ \ \delta\pi_\phi = e\partial_1\epsilon^2,\ \ \delta\pi_\theta = -e\partial_1\epsilon^2 \quad . \end{aligned} \tag{14.71}$$

Note, however, that the Lagrangian and the equations of motion deriving from it are invariant only under a restricted class of transformations obtained from (14.71) by imposing the condition $\epsilon^1 = -\dot{\epsilon}^2$.

In order to quantize the theory in the GI formulation with the constraints imposed strongly, we must choose a gauge. For the class of non-covariant gauges not involving the Lagrange multiplier A_0, this is achieved by imposing two subsidiary conditions

$$\chi_\alpha \approx 0, \quad \alpha = 1, 2 \tag{14.72}$$

subject to the restriction

$$\det\{\omega_\alpha(x), \chi_\beta(y)\} \not\approx 0 \quad , \tag{14.73}$$

where the sign of weak equality now refers to the constraints plus the subsidiary conditions, which we conveniently collect into a single vector (see appendix O)

$$(T_r) = \begin{pmatrix} \omega \\ \chi \end{pmatrix} \approx 0, \quad r = 1, 2, 3, 4. \tag{14.74}$$

The Dirac bracket of two functions of the phase-space variables is then defined by

$$\{f_1, f_2\}_D = \{f_1, f_2\} - \sum_{r=1}^{4} \int dx^1 \int dy^1\, \{f_1, T_r(x)\}\, R_{rs}(x, y)\, \{T_s(y), f_2\} \ , \tag{14.75}$$

where "equal time" is understood and $R = Q^{-1}$, with

$$Q_{rs}(x, y) \equiv \{T_r(x), T_s(y)\} \quad . \tag{14.76}$$

Note that this procedure only works for the so-called "non-covariant" gauges and excludes in particular the Lorentz-type "α-gauges" and the temporal gauge[469].

As a simple example consider the $\theta = 0$ gauge. Setting $\theta = 0$ in the GI Lagrangian (14.14), we recover the Lagrangian (14.9) of the GNI formulation. Hence we also expect to recover the commutation relations (14.27). This is indeed so. To see this, we define a field B^0 conjugate to π_0, $\{B^0(x), \pi_0(y)\} = \delta(x^1 - y^1)$, and impose the subsidiary conditions (14.72) with

$$\chi_1 = T_3 = B^0 \quad , \quad \chi_2 = T_4 = \theta \quad . \tag{14.77}$$

We then have for the matrix (14.76)

$$Q = \begin{pmatrix} 0 & 0 & -1 & 0 \\ 0 & 0 & 0 & -e \\ 1 & 0 & 0 & 0 \\ 0 & e & 0 & 0 \end{pmatrix} \quad . \tag{14.78}$$

It is then a matter of simple algebra to show that one indeed recovers the equal-time commutators (14.27) of the *GNI* formulation. Here it is to be kept in mind that A_0 appearing in the commutation relations (14.27) is the negative of the Lagrange multiplier ξ_2 appearing in (14.69), as may be read off from the Hamiltonian (14.60). This Lagrange multiplier is to be computed as a function of the phase space variables from Eq. (O.55). The result is $\xi_2 = -\frac{1}{a-1}(\frac{1}{e}\pi_\theta - A_1)$.

14.4.3 Isomorphism between *GI* and *GNI* formulation: phase space view[434]

It is interesting to examine the isomorphism between the *GI* and *GNI* formulation, witnessed in chapter 13, in the phase-space formulation. As the identity (13.86) already implies, only the Dirac brackets of the "observables" of the *GI* formulation are of interest. These must be invariant with respect to the infinitesimal transformation (14.71). We distinguish between the cases $a \neq 1$ and $a = 1$.

The case $a \neq 1$

As we already remarked, A^0 plays the role of a Lagrange multiplier in the *GI* formulation. We thus need to look for a candidate isomorphic to the A^0 of the *GNI* formulation. The hint comes from the identity (13.86) itself, which suggests the choice

$$A_0^\theta \equiv A_0 + \frac{1}{e}\partial_0\theta \quad . \tag{14.79a}$$

From (14.55) we then have

$$A_0^\theta = \frac{1}{a-1}[\frac{1}{e}\pi_\theta - A_1] \quad . \tag{14.79b}$$

Recalling the transformation laws (14.71), we see that (14.79*b*) is indeed one of the "observables" we are looking for.

The other gauge invariants are evidently

$$\begin{aligned} A_1^\theta &= A_1 + \frac{1}{e}\partial_1\theta \quad , \\ \pi_\phi^\theta &= \pi_\phi - \partial_1\theta \quad , \quad \phi^\theta = \phi - \theta \\ \pi_0^\theta &= \pi_0 \qquad\qquad , \quad \pi_1^\theta = \pi_1 \end{aligned} \tag{14.79c}$$

and correspond to those of the GNI formulation. The Dirac brackets of these quantities do not depend on the choice of gauge, and should in fact coincide with the corresponding Dirac brackets in the GNI formulation.

The first statement is contained in the following

Theorem: For a constrained system possessing only first-class constraints, $\{\omega_\alpha = 0\}$, $\alpha = 1, ...n$, the Dirac brackets of any two first-class (i.e. gauge invariant) quantities is also first class and coincides on Γ_ω with the Poisson brackets of the same quantities, where $\Gamma_\omega$is the hypersurface defined by the constraints $\omega_\alpha = 0$.

Proof: The proof of this theorem is trivial: Let f_1, f_2 be two such first-class quantities, i.e.

$$\{f_1, \omega_\alpha\} \approx 0, \quad \{f_2, \omega_\alpha\} \approx 0, \quad \alpha = 1, ..n$$

Their Dirac bracket (14.75) then reduces to

$$\{f_1, f_2\}_D \approx \{f_1, f_2\} - \sum_{r=n+1}^{2n} \int dx^1 \int dy^1 \, \{f_1, T_r(x)\} \, R^{rs}(x,y) \, \{T_s(y), f_2\} \quad .$$

Now, since ω_α are first class, one has from (14.76)

$$Q_{rs}\Big|_{\Gamma_\omega} = 0 \quad , \quad r, s = 1, ...n,$$

and, as a consequence

$$R_{rs}\Big|_{\Gamma_\omega} = 0 \quad , \quad r, s = n+1, ...2n \quad .$$

This completes the proof.

We can thus evaluate the Dirac brackets of the fields (14.79) without reference to a choice of gauge! One finds

$$\begin{aligned}
i\{A_0^\theta,(x)A_1^\theta(y)\}_D &= \tfrac{i\delta'(x^1-y^1)}{e^2(a-1)}, & i\{A_0^\theta(x),\phi^\theta(y)\}_D &= \frac{i\delta(x^1-y^1)}{e(a-1)}, \\
i\left\{A_0^\theta(x),\pi_\phi^\theta(y)\right\}_D &= \tfrac{-i\delta'(x^1-y^1)}{a-1}, & i\{A_1^\theta(x),\pi_1^0(y)\}_D &= -i\delta(x^1-y^1), \\
i\{A_0^\theta(x),\pi_1^\theta(y)\}_D &= \tfrac{i\delta(x^1-y^1)}{a-1}, & i\{\phi^\theta(x),\pi_\phi^\theta(y)\}_D &= i\delta(x^1-y^1), \quad (14.80)
\end{aligned}$$

while all other Dirac brackets vanish. The results (14.80) are seen to coincide with those obtained in the GNI formulation, Eqs. (14.27a).

The above isomorphism can be carried still further. We recognize that the first class constraints $\omega_\alpha \approx 0$, when written in terms of the new gauge invariant variables (14.79), take the form of the second class constraints (14.19,21) of the GNI formulation: $\omega_1 = \Omega_1$, and

$$\omega_2 = -\partial_1\pi_1 + e\pi_\phi^\theta + e\partial_1\phi^\theta + e[(a-1)A_0^\theta + A_1^\theta] = \Omega_2(A_\mu^\theta, \pi_\mu^\theta, \phi^\theta, \pi_\phi^\theta) \quad . \qquad (14.81)$$

Thus, if one thinks of (14.79) not as composite objects, but rather as the basic coordinates of a "reduced" phase space, and defines new Poisson brackets $\{\ ,\ \}_R$ with respect to these variables, then one obviously recovers the Poisson algebra (14.22) of the constraints in the GNI formulation. We furthermore have from (14.20*b*) and (14.66)

$$H_C^{GNI}[A^\theta, \pi^\theta, \phi^\theta, \pi_\phi^\theta]\Big|_{\Gamma_\Omega} = H_C^{GI}[A, \pi, \phi, \pi_\phi, \theta, \pi_\theta]\Big|_{\Gamma_\omega} \quad , \qquad (14.82a)$$

where $\Gamma_\Omega : \Omega = 0$, $\Gamma_\omega : \omega = 0$, respectively. Correspondingly we expect that the currents J_L^μ and $\mathcal{J}^{\,\mu}$ in the respective formulations are just "gauge transforms" of each other. Indeed, we see from (14.11*a*) and (14.16*a*) that

$$\mathcal{J}^{\,\mu}[A, \phi, \theta] = J_L^\mu[A^\theta, \phi^\theta] \quad , \qquad (14.82b)$$

which is just the statement (13.86) and (13.177) for the abelian case. This establishes the isomorphism in the Hamiltonian formulation.

The case $a = 1$

We have now four constraints, as given by (14.65) and (14.67). The constraints $\omega_{1,2} \approx 0$ are first class, whereas the other two constraints are second class. Hence the above theorem no longer applies, as it stands. We may, however, use the second-class constraints to eliminate explicitly the conjugate pair (A^1, π_1) by working with Poisson brackets $\{,\}^*$ defined in the subspace Γ_ω^* of the constraints $\omega_r \approx 0$, $r = 3, 4$. In the case in question they read

$$\{A, B\}^* = \{A, B\} - e \sum_{r,s=2,3} \{A, \omega_r\}\, \epsilon_{rs}\, \{\omega_r, B\} \quad . \qquad (14.83)$$

This leaves us with only the first-class constraints

$$\omega_1 \equiv \pi_0 \approx 0\ , \quad \omega_2 = -\partial_1\pi_1 + e(\pi_\phi + \pi_\theta + \partial_1\phi - \partial_1\theta) \approx 0 \quad . \qquad (14.84)$$

The theorem above thus again applies, provided we make the replacement $\{,\} \to \{,\}^*$. The new canonical Hamiltonian now reads

$$H_c^* = \int dx^1 [\frac{1}{2}\pi_\phi^2 + \frac{1}{2}(\partial_1\phi)^2 - \pi_\theta(\pi_\phi + \partial_1\phi + \pi_\theta) - A^0\omega_2] \quad . \qquad (14.85)$$

One checks that H_c^* is first class.

We may now proceed as before. In analogy to (14.79), we again have for the gauge-invariant fields, expressions (14.79c), while expression (14.79b) for A_0^θ is now replaced by[7)]

$$A_0^\theta = \frac{1}{e}(\partial_1\theta + \pi_\theta) \quad . \tag{14.86}$$

Their equal-time commutator algebra can again be abstracted from the respective Poisson-bracket algebra, and one finds

$$\begin{aligned} &[A_0^\theta(x), A_0^\theta(y)]_{ET} = \tfrac{2i}{e^2}\delta'(x^1 - y^1)\,, \qquad &&[\phi^\theta(x), \pi_\phi^\theta(y)]_{ET} = i\delta(x^1 - y^1)\,, \\ &[A_0^\theta(x), \phi^\theta(y)]_{ET} = \tfrac{i}{e}\delta(x^1 - y^1)\,, \qquad &&[A_0^\theta(x), \pi_\phi^\theta(y)]_{ET} = \frac{-i}{e}\delta'(x^1 - y^1) \end{aligned} \tag{14.87}$$

The commutators involving A_1^θ are obtained via the identification $A_0^\theta = A_1^\theta$, while all other commutators vanish. They are seen to coincide with those of the *GNI* formulation for $a = 1$, Eqs.(14.30). One again has a relation of the form (14.81) between the Gauss constraint in the *GI* and *GNI* formulation.

14.4.4 Alternative approach to quantization

One can avoid the first-class constraints of the *GI* formulation by adding from the outset a gauge-fixing term to the Lagrangian : $\mathcal{L} = \mathcal{L}^{GI} + \mathcal{L}_{gf}$. We shall examine two situations:

i) $\mathcal{L}_{gf} = B\theta \quad (\theta = 0 \text{ gauge})$

ii) $\mathcal{L}_{gf} = B\partial_\mu A^\mu + \frac{\alpha}{2}B^2$ (Lorentz-type gauges).

Here B is an auxiliary field. We consider separately the two types of gauges.

Operator solution in the $\theta = 0$ gauge

The momenta conjugate to A^0 and B, π_0 and π_B, vanish identically. We therefore have two primary constraints

$$\omega_1 \equiv \pi_0 \approx 0\,, \quad \omega_2 \equiv \pi_B \approx 0 \quad . \tag{14.88a}$$

The total Hamiltonian thus reads

$$H_T = H_c^{GI} - \int dx^1 B\theta + \int dx^1(u\pi_0 + v\pi_B) \quad ,$$

7) This Ansatz is suggested by the fact that in the *GNI* formulation for $a = 1$, we have $A_- = 0$ (see Eq. (2.5)). This should imply in the *GI* formulation $A_0^\theta = A_1^\theta = A_1 + \frac{1}{e}\partial_1\theta$, which becomes (14.86) upon using the constraint $\omega_3 \approx 0$ in Eq. (14.65). The Ansatz can also be obtained from the corresponding phase space generating functional.[434]

where H_c^{GI} is given by (14.57). The consistency of the primary (and secondary) constraints under time evolution leads to further constraints[470]:

$$\omega_3 \equiv \dot{\omega}_1 \approx -\partial_1\pi_1 + e[\pi_\phi + \pi_\theta + \partial_1\phi - \partial_1\theta] \approx 0\,, \qquad \omega_4 \equiv \dot{\omega}_2 \approx -\theta \approx 0 \qquad (14.88b)$$

$$\omega_5 \equiv \dot{\omega}_3 \approx -eB \approx 0\,, \qquad \omega_6 \equiv \dot{\omega}_4 \approx \frac{1}{a-1}[\pi_\theta - e[(a-1)A_0 + A_1]] \approx 0\,. \qquad (14.88c)$$

The requirement $\dot{\omega}_5 \approx \dot{\omega}_6 \approx 0$ now determines the Lagrange multipliers u and v uniquely. Since these constraints are all second-class, we can define the corresponding Dirac brackets, which then allow us to eliminate θ, B, π_θ and π_B using (14.88). The remaining variables A_μ, π_1, ϕ and π_ϕ are found to satisfy the commutation relations of the GNI formulation, as expected, and the Hamilton equations of motion turn out to be equivalent to the Lagrange equations of motion (14.10) of the GNI formulation, whose solutions have already been discussed in detail.

Operator solution in Lorentz-type gauges

In this case one has two primary constraints

$$\omega_1 \equiv \pi^0 - B \approx 0 \quad , \qquad \omega_2 \equiv \pi_B \approx 0 \quad . \qquad (14.89)$$

Since these primary constraints are already second class, there are no further constraints. The total Hamiltonian reads, this time,

$$H_T = H_c^{GI} + \int dx^1 (B\partial_1 A_1 - \frac{\alpha}{2}B^2) + \int dx^1 (u\omega_1 + v\omega_2) \quad . \qquad (14.90)$$

Standard Dirac quantization results in the equal-time commutation relations

$$[A_0(x), B(y)] = [A^1(x), \pi_1(y)] = [\phi(x), \pi_\phi(y)] = [\theta(x), \pi_\theta(y)] = i\delta(x^1 - y^1) \quad . \qquad (14.91)$$

All other "fundamental" commutators vanish. The equations of motion now read

$$\partial_\mu A^\mu + \alpha B = 0\,, \qquad \Box\phi = -e(\partial_\mu - \tilde{\partial}_\mu)A^\mu\,,$$
$$\partial_\mu F^{\mu\nu} - \partial^\nu B + e\mathcal{J}^\nu = 0, \qquad \Box\theta = -\frac{e}{a-1}((a-1)\partial_\mu + \tilde{\partial}_\mu)A^\mu \qquad (14.92)$$

where $\mathcal{J}^\nu$ is the current (16a). Note the presence of the B field and of an additional equation as compared to (14.15). Eqs. (14.92) admit an exact solution[470] given by [8)]

$$\begin{aligned} \phi &= -\frac{e}{m^2}F + \frac{a-1}{a}h + \chi \\ \theta &= \frac{1}{ea^2}F - \frac{1}{a}h + \chi \\ A_\mu &= \frac{1}{e^2 a}[\frac{a-1}{a}\tilde{\partial}_\mu F + \partial_\mu B + e\tilde{\partial}_\mu h - ea\partial_\mu\chi] \end{aligned} \qquad (14.93)$$

[8)] Actually Eq.(14.93) represents a particular solution of a two-parameter family of solutions[471].

where

$$(\Box + m^2)F = 0, \quad \Box B = 0, \quad \Box \chi = \alpha e B, \quad \Box h = 0 \quad . \tag{14.94}$$

Here χ is seen to represent a zero-mass 'dipole field' (for $\alpha \neq 0$), playing the role of a gauge-degree of freedom which may eliminated by an operator gauge transformation.

In terms of the canonical variables, the Fock-space fields read

$$\begin{aligned} B &= \pi^0 \quad , & F &= \pi^1 \\ h &= \phi - \theta - \tfrac{1}{ae}\pi_1 \quad , & \chi &= \frac{1}{a}\phi + \frac{a-1}{a}\theta \quad . \end{aligned} \tag{14.95}$$

Hence F is just the electric field. Recalling (14.91), one has for the non-vanishing ETC,

$$\begin{aligned} [F(x), \dot{F}(y)]_{ET} &= im^2\delta(x^1 - y^1)\,, & [h(x), \dot{h}(y)]_{ET} &= i\delta(x^1 - y^1)\,, \\ [B(x), \dot{\chi}(y)]_{ET} &= ie\delta(x^1 - y^1)\,, & [\chi(x), \dot{\chi}(y)]_{ET} &= \frac{1}{a}\delta(x^1 - y^1)\,. \end{aligned} \tag{14.96}$$

From (14.92) we see that Maxwell's equations are satisfied on the subspace $|\Psi\rangle$ defined by

$$\partial_\mu B^{(+)}|\psi\rangle = 0, \quad |\psi\rangle \in H_{phys} \quad . \tag{14.97}$$

Note that the gradient of the zero mass field B is a well defined operator. For $\alpha \to 0$ we arrive at the Lorentz gauge, and χ becomes an ordinary zero-mass field. This takes us back to a formulation analogous to that of chapter 10, with $\partial_\mu B$ now playing the role of the "longitudinal current" L_μ in the Schwinger model.

Since the zero mass free field B commutes with F and h for all space-time points, but not with χ, we conclude that the observables of the theory can only involve the massive and massless free bosonic fields F and h, respectively. This is in line with our observation that χ represents a gauge-degree of freedom. Since F and h are quantized with "positive metric", we conclude that in the physical sector the theory is unitary and positive definite.

One can now proceed as in the vector Schwinger model, and construct the corresponding operator solution for the fermion of the GI formulation in the Landau gauge[472]. Of course only the gauge-invariant fermionic operator ψ^θ is of real interest. In view of the isomorphism discussed above we shall not dwell on the construction of ψ^θ, though the precise relation between the renormalized field (14.36) and ψ^θ has been the matter of some controversy[473].

14.5 Path-Integral Formulation

GNI formulation

The usual procedure for arriving at a path integral representation consists in the following steps:

$$\mathcal{L} \to \mathcal{H} \to Z_{\text{phase}} \to Z_{\text{config.}} = \int e^{i\mathcal{L}'} \quad ,$$

where Z_{phase} ($Z_{\text{config.}}$) denote the partition function in phase (configuration) space.

In most cases the "circle"closes and one finds $\mathcal{L} = \mathcal{L}'$, but this need not be so, as we shall see later on.

The first step involves just the transition from (14.9) to the Hamiltonian (14.20*b*) via a Legendre transformation. Following Fradkin and Vilkovisky[445] one has for the phase-space partition function

$$Z = \int \mathcal{D}A^\mu \mathcal{D}\pi_\mu \mathcal{D}\phi \mathcal{D}\pi_\phi \prod_i \delta[\Omega_i] \prod_{x^0} [Det[\Omega_i(x^0, x^1)]]^{1/2} \times e^{i\int d^2x[\pi_\mu \partial_0 A^\mu + \pi_\phi \partial_0 \phi - \mathcal{H}_c(A^\mu, \pi_\mu, \phi, \pi_\phi)]} \tag{14.98}$$

where the δ-functions implement the constraints. We must again treat the case $a > 1$ and $a = 1$ separately[464].

In the case $a > 1$ the Hamiltonian and associated constraints are given by (14.20) and (14.19,21), respectively. The above Faddeev-Popov determinant is just a constant which we absorb in the normalization constant N. The π_0 integration is trivially performed. Using the Fourier representation

$$\delta[\Omega_2] = \int \mathcal{D}\xi e^{i\int d^2x \xi(x)\Omega_2(x)} \tag{14.99}$$

for the Gauss constraint, and performing the remaining π_1, π_ϕ and A^0 integration, one finds for the configuration space partition function

$$Z_{a\neq 1} = \int \mathcal{D}A^\mu \mathcal{D}\phi exp[i \int d^2x \mathcal{L}^{GNI}_{bos}] \tag{14.100}$$

where we have reintroduced the *Lagrangian* variable A^0 by making the identification $\xi = A_0$. Hence, $\mathcal{L}' = \mathcal{L}$ in this case. The (unrenormalized) correlation functions involving A_μ, ψ, and $\overline{\psi}$ are obtained by differentiating the generating functional

$$\begin{aligned} Z[J,\eta,\overline{\eta}] &= Z^{-1} \int \mathcal{D}A^\mu \mathcal{D}\psi \mathcal{D}\bar{\psi} e^{iS_G[A] + \int \bar{\psi} i \not{D}[A] \psi} e^{i\int \overline{\eta}\psi + i\int \bar{\psi}\eta} e^{i\int J^\mu A_\mu} \\ &= Z^{-1} \int \mathcal{D}A^\mu \mathcal{D}\phi e^{i\int \mathcal{L}^{GNI}_{bos}} e^{i\int J^\mu A_\mu} e^{i\int \overline{\eta} G^{(L)}(\ , \ |A)\eta} \end{aligned} \tag{14.101}$$

with respect to the external sources J, η, and $\overline{\eta}$, where $G^{(L)}(x,y|A)$ is the external field fermionic Green's function satisfying

$$(i\partial\!\!\!/ + eA\!\!\!/P_-)G^{(L)}(x,y|A) = \delta^2(x-y) \quad . \tag{14.102}$$

Its explicit form may be deduced directly from (12.5) via the identification $A_\mu \to (g_{\mu\nu} + \epsilon_{\mu\nu})A^\nu$,

$$G^{(L)}(x,y|A) = exp[-ieP_- \int d^2zA^\mu(z)S_\mu(z;x,y)]S_F(x-y) \quad , \tag{14.103a}$$

where $S_\mu(z;x,y)$ is the induced current

$$S_\mu(z;x,y) = (\partial_\mu - \tilde{\partial}_\mu)[D_F(z-x) - D_F(z-y)] \quad . \tag{14.103b}$$

Performing the ϕ integration, the generating functional (14.101) reduces to the form

$$Z[J,\eta,\overline{\eta}] = Z^{-1}\int \mathcal{D}A^\mu$$
$$e^{\frac{i}{2}\int d^2z \int d^2z' A^\mu(z)\Omega_{\mu\nu}(z-z')A^\nu(z')} e^{i\int J^\mu A_\mu} e^{i\int \overline{\eta}(x)G^{(L)}(x,y|A)\eta(y)} \quad , \tag{14.104}$$

with $\Omega_{\mu\nu}$ given by (14.51b).

Let us turn to the computation of the two-point functions. The result (14.50) follows from (14.103) by setting $\eta = \overline{\eta} = 0$ and performing the A_μ integration:

$$Z[J,0,0] = e^{-\frac{i}{2}\int d^2yJ^\mu(x)K_{\mu\nu}(x-y)J^\nu(y)} .$$

For the fermion two-point functions, one has

$$\langle 0|T\psi_u(x)\overline{\psi}_u(y)|0\rangle = \int \mathcal{D}A^\mu e^{\frac{i}{2}\int d^2z \int d^2zA^\mu(z)\Omega_{\mu\nu}(z-z')A^\nu(z')} iG_{(L)}(x,y|A).$$

Using the representation (14.103) for the Dirac Greens function, we may perform the (Gaussian) A_μ-integration to yield

$$\langle 0|T\psi_u(x)\overline{\psi}_u(y)|0\rangle = Z_\psi e^{-\frac{4\pi i}{a-1}\Delta_F(x-y,m^2)} iS_F(x-y) \quad , \tag{14.105a}$$

where

$$Z_\psi = \exp\left[\frac{4\pi i}{a-1}\Delta_F(0;m^2)\right] \tag{14.105b}$$

is a renormalization constant, and the subscript "u" stands for "unrenormalized". This confirms our earlier result (14.52).

It is instructive to see how the (quantum) equation of motion (14.1a) emerges within the functional approach. In order to illustrate this, we consider the expectation value

$$E^{\mu}_{\alpha\beta}(z;x,y) =< 0|T(\partial_\mu F^{\mu\nu}(z) + e\overline{\psi}(z)\gamma^\nu P_-\psi(z))\psi_\alpha(x)\overline{\psi}_\beta(y)|0 > \qquad (14.106)$$

for non-coincident arguments. From (14.101) one has

$$E^{\mu}_{\alpha\beta}(z;x,y) = \frac{1}{Z[0]}\frac{1}{i}\frac{\vec{\delta}}{\delta\overline{\eta}_\alpha(x)}\Bigg[\int \mathcal{D}A^\mu \frac{1}{i}\frac{\delta}{\delta A_\mu(z)}$$
$$\times e^{[iS_0[A]+iW^{(L)}[A]-i\int d^2x'\int d^2y'\overline{\eta}(x')G^{(L)}(x',y'|A)\eta(y')]}\Bigg]\frac{1}{i}\frac{\overleftarrow{\delta}}{\delta\eta_\beta(y)}\Bigg|_{\eta=\overline{\eta}=0}.$$

Performing all differentiations and the A^μ integration in presence of an external source term, one obtains[474]

$$E^{\mu}_{\alpha\beta}(x,y,z) = \frac{1}{z[0]}\Bigg\{\Bigg[-ieP_-S^\mu(z;x,y) + i\int d^2\zeta\Omega^{\mu\nu}(z,\zeta)\frac{1}{i}\frac{\delta}{\delta J^\nu(\zeta)}\Bigg]$$
$$\times e^{-\frac{i}{2}\int d^2z'\int d^2z''(J^\lambda(z')-eP_-S^\lambda(z';x,y))K_{\lambda\sigma}(z'-z'')(J^\sigma(z'')-eP_-S^\sigma(z'';x,y))}\Bigg\}_{J=0},$$

where S^μ and $K^{\mu\nu}$ are given by (14.103*b*) and (14.50*b*), respectively. Doing the differentiation with respect to J^μ, one finds that $E^\mu_{\alpha\beta}$ vanishes. This confirms our previous findings, that on full quantum level, Maxwell's equations are consistent, despite the anomaly.

Consider the vacuum expectation value of the *Euclidian* Wilson-loop functional

$$W_C[A] = exp[ie\oint_C dz_\mu A_\mu(z)] \qquad (14.107a)$$

where C is a rectangular contour in $\Re^2$ with "spacial" and "temporal" extensions R and T, respectively. One has[475] (see appendix C, part e)

$$\left\langle W_C[\hat{A}]\right\rangle = Z^{-1}\int \mathcal{D}A_\mu\ e^{-\frac{1}{2}\int\int A_\mu\Omega^{(E)}_{\mu\nu}A_\nu}W_C[A] \qquad (14.107b)$$

where $\Omega^{(E)}_{\mu\nu}$ is the analytic continuation of $\Omega_{\mu\nu}$ defined in (14.51*b*) to Euclidian space-time. Its inverse is given by

$$\Omega^{(E)-1}_{\mu\nu} = \left[-\delta_{\mu\nu} + \frac{1}{a-1}\left[\frac{1}{e^2}-\frac{2}{\Delta}\right]\partial_\mu\partial_\nu + \frac{i}{a-1}\frac{\partial_\mu\tilde{\partial}_\nu + \tilde{\partial}_\mu\partial_\nu}{\Delta}\right]\frac{1}{m^2-\Delta}. \qquad (14.108)$$

It is convenient to write W_C in the form

$$W_C[A] = exp[ie \int d^2 z j_\mu(z) A_\mu(z)] \tag{14.109}$$

where $j_\mu(z) = \epsilon_{\mu\nu}\partial_\nu \Phi$, with $\Phi = 0$ ($\Phi = 1$) outside (inside) the area enclosed by the contour C. Since $\partial_\mu j_\mu = 0$, only the first term of the propagator (14.108) contributes to the functional integral. Hence we obtain a result analogous to that found in the Schwinger model:

$$< \hat{W}_C >= exp[-\frac{e^2}{2} \oint dz_\mu \oint dz'_\mu \Delta_E(z - z'; m^2)] \quad . \tag{14.110a}$$

Here

$$\Delta_E(x; m) = \frac{1}{2\pi} K_0(m\sqrt{x^2}) \tag{14.110b}$$

with m the JR mass, Eq. (14.28c).

Expression (14.110) coincides with Eq.(12.44) and hence exhibits for $t \to \infty$ the "perimeter" law characteristic of screened quarks in the Schwinger model. This has led some authors to conclude that in the chiral Schwinger model the fermions are screened as well as confined, just as in the vector Schwinger model. This is not so, as the asymptotic behavior (14.53) demonstrates. Since ψ_α is an observable of the GNI formulation, the conclusions drawn in section 3.6 are physically meaningful. This is unlike the case of the Schwinger model, where the physical fermion is represented by an operator of the form (10.45c). As we now show, the analogue of this operator,

$$\hat{\psi}(x) = \psi(x) e^{ieP_- \int_x^\infty dz^\mu A_\mu(z)} \quad , \tag{14.111}$$

does *not* create asymptotic states.

Consider the (*formally* gauge-invariant) Minkowski-space two-point fermionic Green's function

$$iG_{inv}(x,y)_{\alpha\beta} =< 0|T\left[e^{ieP_- \int_x^y dz^\mu A_\mu(z)} \psi_\alpha(x)\overline{\psi}_\beta(y)\right]|0> \quad . \tag{14.112a}$$

The right-hand part propagates freely, whereas the left-hand part has the functional representation

$$G_{inv}(x,y)_{21} = \frac{\int \mathcal{D}A^\mu G_{21}^{(L)}(x,y)|A) exp[ie \int_x^y dz^\mu A_\mu(z) + \frac{i}{2} \int\int A^\mu \Omega_{\mu\nu} A^\nu]}{\int \mathcal{D}A^\mu exp[\frac{i}{2} \int\int A^\mu \Omega_{\mu\nu} A^\nu]} \quad . \tag{14.112b}$$

It is convenient to rewrite the line integral as follows (compare with (12.40))

$$\int_x^y dz^\mu A_\mu(z) = \int d^2z J^\mu(z;x,y)A_\mu(z) \quad , \tag{14.113}$$

where

$$J^\mu(z;x,y) = (y-x)^\mu \int_0^1 dt\delta^2(\xi(t;x,y)-z) \quad . \tag{14.114}$$

Here $\xi_\mu(t;x,y) = (y-x)_\mu t + x_\mu$ parametrizes the straight line which connects the points x and y. The result of the A^μ integration can be written in the form[475]

$$G_{inv}(x,y)_{21} = exp\left[-\frac{ie^2}{2}\int\int R^\mu K_{\mu\nu}R^\nu\right] S_F(x-y)_{21} \quad , \tag{14.115}$$

where $K_{\mu\nu}$ is again the JR propagator (14.50), and

$$R^\mu = S^\mu - J^\mu \tag{14.116}$$

with S^μ and J^μ defined by (14.103*b*) and (14.114), respectively. Noting that

$$\partial_\mu J^\mu = \partial_\mu S^\mu = \delta^2(x-z) - \delta^2(y-z) \tag{14.117}$$

i.e. $\partial_\mu R^\mu = 0$, we see that again only the $g_{\mu\nu}$ part of $K_{\mu\nu}$ contributes in (14.115), thus leading to

$$G_{inv}(x;y)_{21} = exp\left[-\frac{ie^2}{2}\int\int R^\mu \frac{g_{\mu\nu}}{\Box + m^2} R^\nu\right] S_F(x-y)_{21} \quad . \tag{14.118a}$$

Since $\partial_\mu - \tilde{\partial}_\mu$ is a vector of zero length, it follows that the term diagonal in S_μ does not contribute in (118a). We thus find[475]

$$G_{inv}(x,y)_{21} = e^{-\frac{ie^2}{2}(I_1+I_2)} S_F(x-y)_{21} \tag{14.118b}$$

with

$$I_1 = -2\int d^2z \int d^2z' S^\mu(z)\Delta_F(z-z';m^2)J_\mu(z') \, , \tag{14.118c}$$

$$I_2 = \int d^2z \int d^2z' J^\mu(z)\Delta_F(z-z';m^2)J_\mu(z') \, . \tag{14.118d}$$

Recalling (14.103*b*) and (14.114) these expressions reduce to the parametric integral

$$I_1 = \frac{4}{m^2}(y-x)^\mu \int_0^1 \frac{dt}{t}(\partial_\mu + \tilde{\partial}_\mu)[\Delta_F((x-y)t;m^2) - D_F((x-y)t)]$$

$$I_2 = (y-x)^2 \int_0^1 dt \int_0^1 dt' \Delta_F((y-x)(t'-t);m^2) \quad . \tag{14.119a, b}$$

Going over to momentum space, one finds

$$\frac{1}{2}(I_1(z)+I_2(z)) = \int \frac{d^2q}{(2\pi)^2}\left(\frac{2}{q^2}+\frac{z^2}{(q\cdot z)^2}\right)\frac{(1-e^{-iq\cdot z})}{m^2-q^2}$$

which is to be substituted into (14.118*b*). The resulting expression for $G_{inv.}(x,y)_{21}$ coincides[476] with the result one obtains for the Schwinger model, with the substitution $\frac{e}{\sqrt{\pi}} \to m$. In particular, this means, that the operator (14.111) does not create asymptotic states. This has mistakingly been interpreted as confinement of quarks in the chiral Schwinger model.

We turn now to the case $a = 1$. It turns out that from the point of view of the constraint structure, chiral QED in $d = 4$ dimensions resembles the case $a = 1$ of chiral QED_2.

For $a = 1$, the inverse of $\Omega_{\mu\nu}$ no longer exists, as is evident from (14.50), and the analysis has to be repeated, starting with $\mathcal{L}_{bos}^{GNI}$ now given by (14.63). Following the procedure of the preceding section, we find for the partition function[464], after integration over the phase-space momenta,

$$Z = \int \mathcal{D}A^\mu \mathcal{D}\phi\, \delta[A_-] e^{i\int \mathcal{L}'^{GNI}} \tag{14.120a}$$

with

$$\mathcal{L}'^{GNI} = -\frac{1}{2}(\partial_+\phi + eA_1)^2 + \frac{1}{2}\partial_\mu\phi\partial^\mu\phi \quad . \tag{14.120b}$$

Hence $\mathcal{L}'^{GNI}(A,\phi) \neq \mathcal{L}^{GNI}(A,\phi)$, also on the surface $A_- = 0$, showing that in this case the "circle" referred to in the beginning of this section does not close! One finds for the A_μ-two point function

$$\langle 0|TA^\mu(x)A^\nu(y)|0\rangle = \frac{i}{e^2}(\partial^\mu - \tilde{\partial}^\mu)(\partial^\nu - \tilde{\partial}^\nu)D_F(x-y) \;+\; C.T. \tag{14.121a}$$

where

$$C.T. = -\frac{i}{e^2}(g^{\mu 0} - \epsilon^{\mu 0})(g^{\nu 0} - \epsilon^{\nu 0})\delta^{(2)}(x-y) \tag{14.121b}$$

are contact terms, which we drop. This propagator respects the constraint $\pi_1 = F^{01} = 0$. One furthermore verifies that it agrees with the result obtained from the corresponding operator solution.

It is remarkable that the above propagator, though derived from the bosonic Lagrangian (14.63), does not possess an inverse, and is thus not derivable from an effective Lagrangian depending on A_μ alone. It is nevertheless obtained as the $a \to 1$ limit of the propagator (14.50), after dropping a covariant contact term.

GI Formulation

We shall limit ourselves to just a few comments concerning the $a \neq 1$ case. The correlation functions of $A^{\theta}_{\mu}, \psi^{\theta}$ and $\overline{\psi}^{\theta}$ are independent of the choice of section $\mathcal{F}[A,\theta] = 0$ (see section 13.3) and are expected to coincide with the corresponding ones of the GNI formulation. Hence the *observable* ψ^{θ} of the GI formulation represents the unconfined fermion of the GNI formulation. This is particularly evident in the $\theta = 0$ gauge where ψ of the GNI and GI formulations coincide.

Aside from the $\theta = 0$ gauge, there exists another 'preferred' gauge singled out by the requirement that the external-field anomaly (14.12) of the GNI formulation should vanish:

$$H(A) \equiv h^{\mu} A_{\mu} = [(a-1)\partial_{\mu} + \tilde{\partial}_{\mu}]A^{\mu} = 0 \quad . \tag{14.122}$$

In fact, the operator solution (14.28) for A_{μ} has this property.

The possibility of making the choice (14.122) of gauge does not by itself mean that the current anomaly of the GI formulation vanishes on full quantum level, since this requires that $[(a-1)\partial^{\mu} + \tilde{\partial}^{\mu}]A^{\theta}_{\mu} = 0$. This property is nevertheless satisfied, as Eq.(14.15c) shows.

In the gauge (14.122) the θ-field decouples as seen from (14.14). This gauge has therefore been referred to as the JR gauge. In the corresponding generating functional θ thus only couples to the external sources. Setting $\eta = \bar{\eta} = 0$, and integrating over θ, one obtains

$$Z[J,0,0] = Z^{-1} \int \mathcal{D}A^{\mu} \delta[H(A)] e^{i\int d^2x \frac{1}{2} A_{\mu}\Omega^{\mu\nu}A_{\nu} + i\int d^2x J^{\mu}A_{\mu} - \frac{i}{2}\frac{4\pi}{e^2}\frac{1}{a-1}\iint J^{\mu}\frac{\partial_{\mu}\partial_{\nu}}{\Box}J^{\nu}} \quad . \tag{14.123}$$

Exponentiating the δ–function and integrating over A_{μ} one recovers from here the JR propagator (14.50b).

To demonstrate the independence of the choice of gauge, it is instructive to consider the generalization of (14.122) to the one-parameter set of gauge conditions

$$\mathcal{F}(A) = [(a-1)\partial_{\mu} + (b-1)\tilde{\partial}_{\mu}]A^{\mu} = 0 \quad . \tag{14.124}$$

For $b \neq 0$ the θ-field no longer decouples from A^{μ}. A long and tedious calculation[474] shows that the two-point function-$i\langle TA^{\theta}_{\mu}(x)A^{\theta}_{\nu}(y)\rangle_{GI}$ is indeed independent of the parameter b and again coincides with the JR-propagator.

Correlation functions of A^{μ}, ψ and $\overline{\psi}$ in the GI formulation depend, on the other hand, on the choice of gauge. In particular one finds for the class of gauges (14.124)

$$-i\langle TA^{\mu}(x)A^{\nu}(y)\rangle_{GI} =$$

$$\left[g_{\mu\nu} - \frac{(a-1)^2 + (b-1)^2}{(a-1)^2}\frac{\partial_{\mu}\partial_{\nu}}{\Box} + \frac{b-1}{a-1}\frac{\partial_{\mu}\tilde{\partial}_{\nu} + \tilde{\partial}_{\mu}\partial_{\nu}}{\Box}\right]\Delta_F(x-y;m) \quad . \tag{14.125}$$

Note, however, that this two-point function exhibits the correct pole at the JR mass (14.28c), independent of the gauge parameter b. We shall comment on this again in the following section.

14.6 Perturbative Analysis in the Fermionic Formulation

In this section we attempt a perturbative analysis of chiral QED_2. This will reveal some difficulties which one would have to deal with in $d = 4$ dimensions, where only perturbative methods are accessible.

In the fermionic GNI formulation, the specification of the gauge–field propagator requires a breaking of the gauge invariance of the classical Lagrangian, and its subsequent restoration at the end of the calculation, once the relevant perturbative contributions have been summed to all orders. This procedure leads to the correct result; it fails however in finite order of perturbation theory.

The difficulty is avoided in the GI formulation, which allows one to choose a gauge in which the Feynman propagator of the vector-potential may be defined. This constitutes the main advantage of the GI formulation.

The choice $a = 1$ in the chiral Schwinger model nevertheless continues to pose a problem, since the kinetic term in the Wess–Zumino field θ is absent, in this case, as Eq.(14.14) shows. This difficulty happens not to be purely academic, since chiral QED_4 and QCD_4 turn out to exhibit this same problem.

A serious disadvantage of the GI formulation is that here the physical fermion and gluon field, ψ^θ, A^θ_μ, are composite operators. This problem is particularly problematic in the non–abelian case, where it is still enhanced by the complicated structure of the Wess–Zumino action. We now proceed to a perturbative analysis of chiral QED_2.

14.6.1 Perturbative analysis in the GNI formulation

In the following we limit ourselves to a perturbative study of the two–point function $-i\langle 0|TA_\mu(x)A_\nu(y)|0\rangle_{GNI} = K_{\mu\nu}(x-y)$ for $a > 1$. It is known to be given by Eq.(14.50). In Fourier space, it has a Laurent expansion in powers of e^2 of the form

$$K_{\mu\nu}(k) = \frac{1}{a-1}\frac{1}{e^2}\frac{k_\mu k_\nu}{k^2} + \frac{1}{k^2}\left\{-g_{\mu\nu} + \right.$$
$$\left.[1+\left(\frac{1}{a-1}\right)^2]\frac{k_\mu k_\nu}{k^2} + \frac{1}{a-1}\frac{k_\mu \tilde{k}_\nu + \tilde{k}_\mu k_\nu}{k^2}\right\}\sum_{n=0}^{\infty}\left(\frac{m^2}{k^2}\right)^n \quad . \quad (14.126)$$

The first term is singular in e^2 and seems to defy a perturbative interpretation.

Figure 14.1: Geometric series representing A_μ two-point function.

Figure 14.2: Polarization tensor $-iW^{\mu\nu}$ of vector-potential.

From the point of view of perturbation theory, the full propagator should be obtained by summing the geometric series depicted in Fig. 14.1. Ingredients of this series are the Feynman propagator and polarization tensor of the vector potential.

It has been frequent practise to choose for $D_F^{\mu\nu}(k)$ the propagator in the Weyl gauge $A_0 = 0$, although the broken gauge invariance on quantum level prohibits such a choice. In order to exhibit the disastrous implications of choosing a gauge, it is instructive to consider the family of gauges implied by the gauge fixing term[9)]

$$\mathcal{L}_{gf} = -\frac{\lambda}{2}\left((p\partial_\mu + q\tilde{\partial}_\mu)A^\mu\right)^2 \quad . \tag{14.127}$$

This gauge fixing term includes in particular the Lorentz gauge $(p = 1, q = 0)$ as well as the so-called JR gauge $(p = a - 1, q = 1)$. The Feynman propagator associated with $\mathcal{L}^{GI} + \mathcal{L}_{gf}$ is given by

$$D^{(0)}_{\mu\nu}(k) = -\frac{1}{k^2}\left\{g_{\mu\nu} + c_1\frac{k_\mu k_\nu}{k^2} + c_2\frac{k_\mu\tilde{k}_\nu + \tilde{k}_\mu k_\nu}{k^2}\right\} \tag{14.128a}$$

with

$$c_1 := \frac{1 - \lambda(p^2 + q^2)}{\lambda p^2}, \quad c_2 := \frac{q}{p} \quad . \tag{14.128b}$$

In the JR gauge (14.122), this propagator takes the form

$$D^{(0)}_{\mu\nu}(k) = -\frac{1}{k^2}\left\{g_{\mu\nu} - [1 + \left(\frac{1}{a-1}\right)^2]\frac{k_\mu k_\nu}{k^2} + \frac{1}{a-1}\frac{k_\mu\tilde{k}_\nu + \tilde{k}_\mu k_\nu}{k^2}\right\} \quad . \tag{14.128c}$$

We next compute the contribution coming from the fermion loop-graph depicted in figure 14.2, which we denote by $-iW_{\mu\nu}(x - y)$.

9) Our discussion in this section follows that of ref. [477]. See also [441].

If the regularization procedure respects the identity $\gamma^\mu\gamma^5 = \epsilon^{\mu\nu}\gamma_\nu$, then

$$W^{\mu\nu}(x) = \frac{1}{4}(g^{\mu\lambda} + \epsilon^{\mu\lambda})(g^{\nu\rho} + \epsilon^{\nu\rho})\pi_{\lambda\rho}(x),$$

where $\pi_{\lambda\rho}$ is the polarization tensor for a pure vector coupling. An ambiguity in $\pi_{\lambda\rho}$ of the form $\delta\pi_{\lambda\rho} = ag_{\lambda\rho}$ will therefore not manifest itself in $W^{\mu\nu}$. Das[478] has used this observation to argue that there actually does not exist any ambiguity of the JR type. This would correspond to the JR parameter a being set equal to zero. Unfortunately this leads to a non-unitary theory, as we have learned.

Actually there does exist an ambiguity in the calculation of $W^{\mu\nu}$, arising from the need for regularizing the logarithmically divergent diagram of figure 14.2 It is introduced by using the Pauli Villars regularization

$$\begin{aligned}
-iW^{\mu\nu}(k)_{reg} = \lim_{M\to\infty} -(ie)^2 \int \frac{d^2p}{(2\pi)^2} \mathrm{tr}[\gamma^\mu P_- \frac{i}{\not p}\gamma^\nu P_- \frac{i}{\not p} \quad - \\
- \gamma^\mu(P_- + \alpha P_+)\frac{i}{\not p - M}\gamma^\nu(P_- + \beta P_+)\frac{i}{\not p - \not k - M} \qquad (14.129a) \\
+ \alpha\beta\gamma^\mu P_+ \frac{i}{\not p - M}\gamma^\nu P_+ \frac{i}{\not p - \not k - M}] \quad .
\end{aligned}$$

Using $P_+P_- = 0$, one checks that this integral converges. This expression may also be written in the form

$$\begin{aligned}
-iW^{\mu\nu}(k)_{reg} = \lim_{M\to\infty}\Big\{\int \frac{d^2p}{(2\pi)^2}\Big[\mathrm{tr}[\gamma^\mu P_- \frac{i}{\not p}\gamma^\nu P_- \frac{i}{\not p - \not k} \\
-\gamma^\mu P_- \frac{i}{\not p - M}\gamma^\nu P_- \frac{i}{\not p - \not k - M}] - \frac{(\alpha+\beta)g^{\mu\nu} + (\alpha - \beta)\epsilon^{\mu\nu}}{(p^2 - M^2 + i0)((p-k)^2 - M^2 + i0)}\Big]\Big\} (14.129b)
\end{aligned}$$

For $\alpha \neq \beta$, there will thus occur a term proportional to $\epsilon^{\mu\nu}$. This term will however not contribute to the effective action $W^{(L)}[A] = \int A_\mu W^{\mu\nu} A_\nu$, and hence may be ignored. Alternatively, one may choose $\alpha = \beta$. With the identification $\alpha = \beta = a/2$, a straightforward calculation leads to

$$W^{\mu\nu}(k) = -e^2\left[ag^{\mu\nu} - \frac{(k^\mu - \tilde{k}^\mu)(k^\nu - \tilde{k}^\nu)}{k^2}\right] \quad . \qquad (14.130)$$

This result has also been obtained using dimensional regularization[441]. In fact, it can be shown to incorporate the most general ambiguity in the effective action. In particular this shows that there exists no regularization which respects the gauge invariance of the classical Lagrangian, since $k^\mu W_{\mu\nu} \neq 0$ for any value of the JR parameter a.

We now compute the A_μ two–point function for the gauge fixing term (14.127). The geometric series of figure 14.1 with the propagator (14.128) and polarization tensor (14.130) amounts to summing

$$D_{\mu\nu}(k) = D^{(0)}_{\mu\nu}(k) + D^{(0)}_{\mu\lambda}(k)W^{\lambda\rho}(k)D^{(0)}_{\lambda\nu}(k) + ...$$

and leads to

$$D_{\mu\nu}(k) = \frac{1}{N(k)}\Big\{[e^2(a-1) - \lambda p^2 k^2]g_{\mu\nu} + [\lambda(p^2+q^2) - 1 + \frac{2e^2}{k^2}]k_\mu k_\nu - [\lambda pq + \frac{e^2}{k^2}](k_\mu \tilde{k}_\nu + \tilde{k}_\mu k_\nu)\Big\} \quad , \qquad (14.131a)$$

where

$$N(k) = \lambda p^2 k^4 - [\lambda(p-q)^2 + a\lambda(p^2-q^2) + (a-1)e^2]e^2k^2 + a^2(e^2)^2 \quad . \qquad (14.131b)$$

This result makes clear that for the computation of correlation functions the restriction to a special gauge has catastrophic consequences: The pole in the A_μ two–point function is gauge dependent! This would not be the case in a true gauge theory, and thus reflects the presence of the anomaly. In fact, taking the limit $\lambda \to \infty$ $((p\partial_\mu + q\tilde{\partial}_\mu)A^\mu = 0$ gauge) yields

$$D_{\mu\nu}(k) = \frac{-g_{\mu\nu} + (1+\eta^2)\frac{k_\mu k_\nu}{k^2} - \eta\frac{(k_\mu \tilde{k}_\nu + \tilde{k}_\mu k_\nu)}{k^2}}{\{k^2 - e^2[(1-\eta)^2 + a(1-\eta^2)]\}} \qquad (14.132)$$

where $\eta = q/p$. In the Lorentz gauge $(\eta = 0)$

$$D_{\mu\nu}(k) = \frac{-g_{\mu\nu} + \frac{k_\mu k_\nu}{k^2}}{k^2 - e^2(1+a)} \quad . \qquad (14.133)$$

The pole is seen not to correspond to the JR mass (14.28c).

Similar problems have been noted by Halliday et al., who showed that the restriction to non–covariant gauges leads to a non–relativistically invariant[462] spectrum.

Nevertheless, expression (14.131) has an interesting property: in the limit $\lambda \to 0$ (*no* gauge fixing) it reduces to the JR two–point function (14.125). This observation will play an important role in the discussion of section 14.7. Nevertheless, the limit $\lambda \to 0$ does not exist in finite order perturbation theory, as one verifies by expanding (14.131) in powers of e^2. Hence, a procedure, consisting in first adding the gauge fixing term (14.127) and then removing it again by letting $\lambda \to 0$ will not work in finite order perturbation theory.

In the case of chiral QED_2 there nevertheless exists a possibility of giving a diagrammatic interpretation to the Laurent expansion (14.126). The idea is to[477,479] start from a Proca theory, defined by the Lagrangian

$$\mathcal{L}' = -\frac{1}{4} F_{\mu\nu} F^{\mu\nu} + \frac{\mu^2}{2} A_\mu^2 + \bar{\psi} i \not{D} \psi \quad . \tag{14.134}$$

The propagator of the vector potential is now given by

$$\Delta_{\mu\nu}^{(0)}(k;\mu^2) = -\frac{1}{k^2-\mu^2}(g_{\mu\nu} - \frac{k_\mu k_\nu}{\mu^2}) \quad . \tag{14.135}$$

Using this propagator in the geometric series depicted in figure 14.1 now leads to

$$\Delta_{\mu\nu}(k) = \frac{1}{k^2 - \tilde{m}^2} \left\{ -g_{\mu\nu} + \frac{1}{\tilde{a}-1} \left(\frac{1}{e^2} - \frac{2}{k^2} \right) k_\mu k_\nu + \frac{1}{\tilde{a}-1} \frac{k_\mu \tilde{k}_\nu + \tilde{k}_\mu k_\nu}{k^2} \right\} \tag{14.136a}$$

with

$$\tilde{m}^2 = \frac{e^2 \tilde{a}^2}{\tilde{a}-1} \quad , \quad \tilde{a} = a + \frac{\mu^2}{e^2} \quad . \tag{14.136b}$$

Comparing with the Fourier transform of (14.50*b*) we see that we recover the JR propagator in the limit $\mu \to 0$.

The limit $\mu \to 0$ does not exist, however, in each order of perturbation theory, as the $1/\mu^2$ singularity in the propagator (14.135) already demonstrates.

We nevertheless obtain a diagrammatic interpretation of the Laurent expansion (14.126) by recognizing, that with the identification $\mu^2 = ae^2$, the mass term in the Proca–Lagrangian (14.134) just corresponds to the regularization ambiguity of the one–loop diagram shown in figure 14.2. Hence we proceed by using the Feynman rules corresponding to the Lagrangian (14.134), choosing a regularization for the one–loop diagram which respects the chiral structure of the Lagrangian,

$$W_0^{\mu\nu}(k) = e^2 \frac{(k^\mu - \tilde{k}^\mu)(k^\nu - \tilde{k}^\nu)}{k^2} \quad , \tag{14.137}$$

and setting $\mu^2 = ae^2$ in the results. This results in the geometric series

$$K_{\mu\nu}(k) = \Delta_{\mu\nu}^{(0)}(k) + \Delta_{\mu\lambda}^{(0)}(k) W_0^{\lambda\rho}(k) \Delta_{\rho\nu}^{(0)}(k) + \ldots \tag{14.138}$$

where

$$\Delta_{\mu\nu}^{(0)}(k) = \Delta_{\mu\nu}^{(0)}(k; ae^2) = -\frac{1}{k^2 - ae^2} \left(g_{\mu\nu} - \frac{k_\mu k_\nu}{ae^2} \right) \quad . \tag{14.139}$$

We have denoted the result of summing the series by $K_{\mu\nu}(k)$, in anticipation of the fact, that the geometric series (14.138) just corresponds to a reordering of the Laurent expansion (14.126).

Indeed, each term in (14.138) can be represented by a Laurent expansion in the coupling constant, whose leading term is of order $O(e^{-2})$. The result of summing the contributions of the leading terms in (14.138) is easily computed. To this end we write (14.136) in the form

$$K_{\mu\nu}(k) = \sum_{n=0}^{\infty} \Delta_{\mu\nu}^{(n)}(k) \tag{14.140}$$

where $\Delta_{\mu\nu}^{(n)}(k)$ is iteratively computed from

$$\Delta_{\mu\nu}^{(n)}(k) = \Delta_{\mu\lambda}^{(0)}(k) W_0^{\lambda\rho}(k) \Delta_{\rho\nu}^{(n-1)}(k) \quad . \tag{14.141}$$

Noting that

$$\Delta_{\mu\nu}^{(0)}(k) = \frac{1}{ae^2}\frac{k_\mu k_\nu}{k^2} + O(e^0)$$

$$\Delta_{\mu\nu}^{(0)}(k) W_{0\rho}^{\lambda}(k) = \frac{1}{a}(k_\mu k_\rho - k_\mu \tilde{k}_\rho) + O(e^2)$$

we have

$$\Delta_{\mu\nu}^{(n)}(k) = \frac{1}{e^2}\frac{1}{a^{n+1}}\frac{k_\mu k_\nu}{k^2} + O(e^0) \quad . \tag{14.142}$$

Substituting (14.142) into (14.140), one obtains

$$K_{\mu\nu}(k) = \frac{1}{e^2}\frac{1}{a-1}\frac{k_\mu k_\nu}{k^2} + O(e^0),$$

in agreement with the leading term in the Laurent expansion (14.126).

We have thus succeeded in reproducing the $1/e^2$ singularity of the JR propagator within a perturbative framework based on the Proca–Lagrangian (14.134). This has been motivation for dismissing the claim made in ref. [73], that the $\frac{1}{e}$ singularity in the commutation relations (14.27) is of non–perturbative origin.

The above procedure, based on the identification of μ^2 with ae^2, is only applicable to two–dimensional (abelian and non–abelian) chiral gauge theories. Moreover, it is well–known that the introduction of a mass term for the vector potential spoils renormalizability of non–abelian gauge theories in $d = 3 + 1$ dimensions. Hence we must conclude, that in the framework of GNI formulation, we have no satisfactory perturbative scheme available.

As we now show, this difficulty can be resolved by embedding the chiral gauge theory into a bonafide gauge theory following the procedure discussed in chapter 13.

14.6.2 Perturbative analysis in the *GI* formulation[477]

We restrict our attention to the two–point function of the (gauge invariant) vector potential A^θ_μ, which corresponds to A_μ in the *GNI* formulation.

In terms of fermions, the Lagrangian of the *GI* formulation reads:

$$\mathcal{L}^{GI} = -\frac{1}{4}F_{\mu\nu}F^{\mu\nu} + \bar{\psi}(i\partial\!\!\!/ + eA\!\!\!/)\psi + \frac{a-1}{2}(\partial_\mu\theta)^2 + eA^\mu[(a-1)\partial_\mu + \tilde{\partial}_\mu]\theta \quad . \quad (14.143)$$

The building blocks of the relevant Feynman diagrams are given by:

A_μ–propagator:	$iD^{(0)}_{\mu\nu}(k)$,	(14.144*a*)
θ–propagator:	$\frac{1}{a-1}\frac{i}{k^2}$,	(14.144*b*)
$A_\mu - \theta$ vertex:	$ie\left((a-1)k_\mu + \tilde{k}_\mu\right)$,	(14.144*c*)
Vacuum polarization tensor:	$W^{\mu\nu}(k)$.	(14.144*c*)

We shall work with the gauge fixing term (14.127), where $D^{(0)}_{\mu\nu}(k)$ is given by (14.128). The result should however be independent of the parameters p and q. Anticipating the isomorphism (13.86), we use the notation

$$-i\langle 0|TA^\theta_\mu(x)A^\theta_\nu(y)|0\rangle = K_{\mu\nu}(x-y).$$

The diagrams contributing up to order $O(e^0)$ are shown in figure 14.3

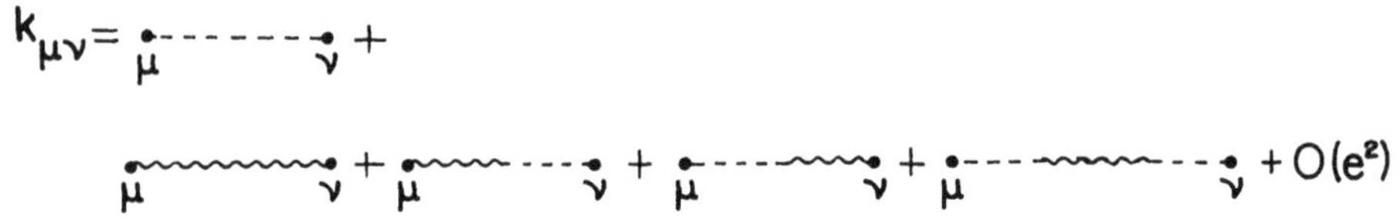

Figure 14.3: Diagrams contributing to $K_{\mu\nu}(k)$ up to order $O(e^0)$.

An index μ on an *external* θ–line indicates that an extra factor $-\frac{i}{e}k_\mu(+\frac{i}{e}k_\mu)$ must be included for a momentum flowing to (away from) the point labelled by μ. Hence the first diagram is seen to contribute the term $\frac{1}{e^2}\frac{1}{a-1}\frac{k_\mu k_\nu}{k^2}$, which is seen to coincide with the $O(e^{-2})$ term of the Laurent expansion (14.126) of the JR propagator.

The remaining four $O(e^0)$ diagrams sum up to

$$-\frac{1}{k^2}\left\{g_{\mu\nu}-[1+\left(\frac{1}{a-1}\right)^2]\frac{k_\mu k_\nu}{k^2}+\frac{1}{a-1}\frac{k_\mu\tilde{k}_\nu+\tilde{k}_\mu k_\nu}{k^2}\right\}$$

which again agrees with the $O(e^0)$ term of the Laurent expansion (14.126), and also coincides with expression (14.128c).

Note that these results are independent of the gauge parameters p, q and λ. This shows that the $\frac{1}{e^2}$ singularity in the JR propagator is *not* an artifact of the $\theta = 0$ gauge.

In higher order, fermion-loop diagrams now contribute. As an *internal* insertion, they always occur in the combination shown in fig. 14.4, which we denote by $-i\Pi_{\mu\nu}(k)$, which has the transversality property characteristic of a gauge theory:

$$\Pi_{\mu\nu}(k) = m^2(-g_{\mu\nu} + \frac{k_\mu k_\nu}{k^2}) \quad . \tag{14.145}$$

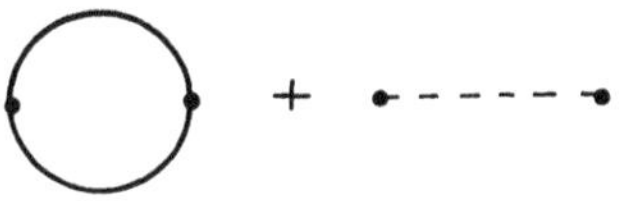

Figure 14.4: Diagrammatic representation of $\Pi_{\mu\nu}(k)$.

The full propagator $K_{\mu\nu}(k)$ is given by the sum of diagrams given in figure 14.5.

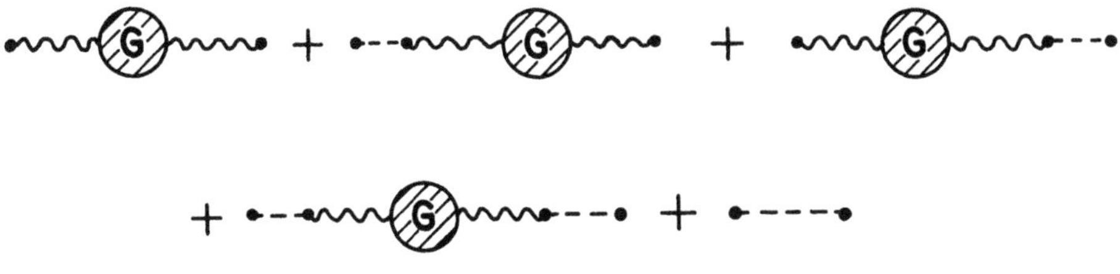

Figure 14.5: Diagrammatic contributions to the full propagator $K_{\mu\nu}(k)$.

Here the first diagram represents the gauge-*dependent* propagator $-i\langle 0|TA_\mu(x)A_\nu(y)|0\rangle_{GI} \equiv G_{\mu\nu}$. It is obtained by summing the geometric series

$$G_{\mu\nu} = D^{(0)}_{\mu\nu} + D^{(0)}_{\mu\lambda}\Pi^{\lambda\rho}D^{(0)}_{\rho\nu} + D^{(0)}_{\mu\lambda}\Pi^{\lambda\rho}D^{(0)}_{\rho\sigma}\Pi^{\sigma\tau}D^{(0)}_{\tau\nu} + \cdots$$

The result is

$$G_{\mu\nu}(k) = \frac{1}{m^2 - k^2}\left\{g_{\mu\nu} + \left(c_1 - \frac{m^2}{\lambda p^2}\frac{1}{k^2}\right)\frac{k_\mu k_\nu}{k^2} + c_2\frac{k_\mu \tilde{k}_\nu + \tilde{k}_\mu k_\nu}{k^2}\right\} \quad . \tag{14.146}$$

As expected, this result differs from the corresponding result (14.131) obtained in the GNI formulation. Adding the contribution of the remaining three diagrams above, the dependence on the gauge parameters q, p and λ cancels, and one recovers[477] the JR propagator. This confirms the isomorphism (13.86). Such a restoration of the correct relativistic spectrum by the addition of a WZ term has also been noted in ref.[462].

In view of this (expected) gauge independence of the result, we could have simplified the calculation considerably, by choosing the gauge parameters to have the values $p = a - 1, q = 1$, and letting λ tend to infinity. This corresponds to choosing the JR gauge. As we have already remarked previously, the θ-field decouples completely in this gauge, and propagates freely. Following our previous notation, this means

$$K_{\mu\nu}(k) = G_{\mu\nu}(k) + \frac{1}{e^2(a-1)}\frac{k_\mu k_\nu}{k^2} \quad . \tag{14.147}$$

The gauge-*variant* Green function $G_{\mu\nu}(k)$ is now given by the geometric series depicted in figure 14.1, with the A_μ propagator and polarization tensor given by (14.128c) and (14.130), respectively. This series can be shown[477] to sum up to the result (14.125) with $b - 1 = 1$, after having taken the limit $\lambda \to \infty$. Upon adding to this the second term in (14.147), one recovers the JR propagator.

In view of the isomorphism (13.86), this solves the problem of obtaining a diagrammatic and perturbative representation of $< 0|TA^\mu(x)A^\nu(y)|0 >_{GNI}$ for $a \neq 1$. For $a = 1$, expression (14.147) is singular, reflecting the fact that for this value of the JR parameter the kinetic term of the θ-field is absent in (14.143). Since chiral gauge theories in $d = 3+1$ dimensions actually resemble their two-dimensional analogues for $a = 1$, this poses a serious problem, whose solution is still outstanding.

14.7 Anomalous Poisson Brackets Revisited[422]

In section 13.16 we have discussed a method for implementing the Dirac algorithm in the GNI, *fermionic* formulation of chiral QCD_2. We learned that the Dirac algorithm had to be understood as referring to "anomalous Poisson brackets" (APB) defined in terms of an auxiliary Lagrangian $\mathcal{L}^{(\alpha)}$ parametrized by α, which could be chosen as in Eq. (13.328). We shall now illustrate how the method works on the *operator* level, leaving a brief discussion from the perturbative point of view to the end of this section.

14.7.1 Operator view of anomalous Poisson brackets

The auxiliary Lagrangian reads in the present case,

$$\mathcal{L}^{(\alpha)} = -\frac{1}{4}F_{\mu\nu}F^{\mu\nu} + \bar{\psi}(i\partial\!\!\!/ + e\,A\!\!\!/P_-)\psi - \frac{1}{2\alpha}(\partial_\mu A^\mu)^2 \quad . \tag{14.148a}$$

On the bosonic level, $\mathcal{L}^{(\alpha)}$ is evidently given by

$$\mathcal{L}^{(\alpha)}_{bos} = \mathcal{L}^{GNI}_{bos} - \frac{1}{2\alpha}(\partial_\mu A^\mu)^2 \quad , \tag{14.148b}$$

where $\mathcal{L}^{GNI}_{bos}$ is the JR Lagrangian (14.9). In the limit $\alpha \to \infty$ we formally recover the JR model. The momenta conjugate to A^0 A^1 and ϕ are given by

$$\pi_0 = -\frac{1}{\alpha}\partial_\mu A^\mu \quad , \qquad \pi_1 = F^{01} \quad , \tag{14.149a, b}$$

$$\pi_\phi = \partial_0\phi + eA_- \quad . \tag{14.149c}$$

Hence there is no primary, and thus also no secondary constraint.

In order to obtain the solution to the quantum theory, we first solve the classical equations of motion derived from (14.148b),

$$\partial_\mu F^{\mu\nu} + \frac{1}{\alpha}\partial^\nu(\partial_\mu A^\mu) + eJ_L^0 = 0, \tag{14.150a}$$

$$\Box\phi + e(\partial_\mu - \tilde{\partial}_\mu)A^\mu = 0 \quad , \tag{14.150b}$$

where J_L^ν is given by (14.11a). Note that

$$\Omega_1 = \pi_0 \tag{14.151a}$$

$$\Omega_2 = -\partial_1\pi_1 + eJ_L^0 \tag{14.151b}$$

are no longer constrained to vanish!

One finds after a lengthy calculation[422] that the solutions for A_μ and ϕ can be expressed in terms of two massive free fields $\sigma_\pm$ and a harmonic field h as follows,

$$\begin{aligned} A^\mu &= -\frac{1}{ae}\left[(\partial^\mu\sigma_+ + \kappa_+\tilde{\partial}^\mu\sigma_+) + (\partial^\mu\sigma_- + \kappa_-\tilde{\partial}^\mu\sigma_-) - (\partial^\mu - \tilde{\partial}^\mu)h\right] \quad , \\ \phi &= \frac{1}{a}(1+\kappa_+)\sigma_+ + \frac{1}{a}(1+\kappa_-)\sigma_- - h \quad , \end{aligned} \tag{14.152a}$$

where

$$\begin{aligned} (\Box + m_\pm^2)\sigma_\pm &= 0 \\ \Box h &= 0 \end{aligned} \tag{14.152b}$$

and

$$\begin{aligned} m_\pm^2 &= \frac{e^2}{2}\left[[(a+1)+(a-1)\alpha] \pm \sqrt{[(a+1)+(a-1)\alpha]^2 - 4\alpha a^2}\right] \\ \kappa_\pm &= a - 1 - \frac{m_\pm^2}{\alpha e^2} \quad . \end{aligned} \tag{14.152c}$$

We note that in the limit $\alpha \to \infty$,

$$m_+^2 \to \infty \quad , \quad m_-^2 \to m^2 = \frac{e^2 a^2}{a-1} \quad , \tag{14.153}$$

so that the massive field σ_+ decouples from $\mathcal{H}$ in this limit, while σ_- tends to the σ–field (14.28b) of the JR model with mass (14.28c). Quantization of the solution (14.152) now requires the specification of the "initial value data" for σ_+, σ_- and h. This is most easily obtained by solving Eqs.(14.152a) for σ_+, σ_- and h in terms of A^0, A^1 and ϕ, and their canonically conjugate momenta, and imposing the *canonical* commutation relations

$$[A^0(x), \pi_0(y)]_{ET} = [A^1(x), \pi_1(y)]_{ET} = [\phi(x), \pi_\phi(y)]_{ET} = i\delta(x^1 - y^1) \tag{14.154}$$

corresponding to the unconstrained theory defined by the bosonic Lagrangian given by (14.148b). One finds after some calculation

$$\begin{aligned} [h(x), \partial_0 h(y)]_{ET} &= i\delta(x^1 - y^1) \\ [\sigma_+(x), \partial_0\sigma_+(y)]_{ET} &= im^2\left(\frac{\alpha}{\Delta m^2}\right)^2\left(\frac{1}{\alpha} - \kappa_-^2\right)\delta(x^1 - y^1) \\ [\sigma_-(x), \partial_0\sigma_-(y)]_{ET} &= im^2\left(\frac{\alpha}{\Delta m^2}\right)^2\left(\frac{1}{\alpha} - \kappa_+^2\right)\delta(x^1 - y^1) \end{aligned} \tag{14.155a}$$

with

$$\Delta m^2 = m_+^2 - m_-^2 \quad . \tag{14.155b}$$

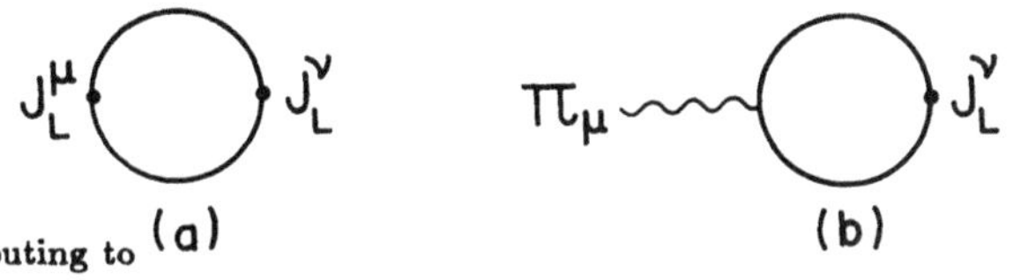

Figure 14.6: Only diagrams contributing to a) $[J^\mu_L(x), J^\nu_L(y)]_{ET}$; b) $[\pi_\mu(x), J^\nu_L(y)]_{ET}$.

All other canonical commutators vanish. Notice again that, as expected,

$$[\sigma_-(x), \partial_0\sigma_-(y)]_{ET} \underset{\alpha\to\infty}{\longrightarrow} \frac{i}{a-1}\delta(x^1 - y^1) \tag{14.156a}$$

in agreement with (14.29), while

$$[\sigma_+(x), \partial_0\sigma_+(y)]_{ET} \underset{\alpha\to\infty}{\longrightarrow} -i\frac{a^2}{a-1}\delta(x^1 - y^1) \quad . \tag{14.156b}$$

Since σ_+ decouples from $\mathcal{H}$ in this limit, its only role in this limit thus is to maintain the canonical structure of the commutation relations (14.154); that is, the "Poisson algebra character" of these commutation relations is maintained for $\alpha \to \infty$, while the σ_+ excitation decouples from the correlation functions in this limit, as it must, if we are to recover the JR model.

Using the canonical commutation relations (14.154), we may proceed to calculate, in the auxiliary theory, various commutators involving J^0_L and Ω_α, $\alpha = 1, 2$. It is obvious that one obtains for the commutators $-i[\,\Omega_\alpha(x), \Omega_\beta(y)\,]_{ET}$, $-i[\pi_\mu(x), J^\nu_L(y)]_{ET}$ and $-i[J^\mu_L(x), J^\nu_L(y)]_{ET}$ the Poisson bracket results (14.22) and (14.26), respectively.

14.7.2 Bjorken-Johnson-Low view of anomalous Poisson brackets

As was shown in chapter 13, the "canonical" commutators of the auxiliary theory should correspond, in the *fermionic* formulation of chiral QED_2, to anomalous Poisson brackets (APB) defined by (13.329), which may be calculated using the BJL prescription (13.334). The A_μ-propagator in the theory defined by (14.143) is given by

$$-i\langle 0|TA^\mu(x)A^\nu(y)|0\rangle = D^{\mu\nu}_F(k) = -\frac{1}{k^2}[g^{\mu\nu} - (1-\alpha)\frac{k^\mu k^\nu}{k^2}] \quad . \tag{14.157}$$

As an example, we calculate, in the fermionic formulation, $i\{\pi_0(x), J^\nu_L(y)\}_{AP}$, $i\{\pi_1(x), J^\nu_L(y)\}_{AP}$ and $i\{J^0_L(x), J^0_L(y)\}_{AP}$, by studying the BJL limit of the vacuum expectation values $\langle T(-\frac{1}{\alpha}\partial_\mu A^\mu(x)\, J^\nu_L(y))\rangle$, $\langle T(F^{01}(x)\, J^\nu_L(y))\rangle$ and $\langle T(J^0_L(x)\, J^0_L(y))\rangle$, respectively. The relevant diagrams exhibiting a $1/q^0$ contribution for $q^0 \to \infty$ are shown in figures (14.6a, b). The fermion loop is given by the vacuum polarization tensor (14.130).

A simple BJL-type calculation analogous to that of chapter 13 reproduces[422] the results (14.26) obtained earlier via bosonization. They are independent of α, in agreement with our general expectations (see (13.329*b*)). This shows that their perturbative calculation poses no problem. On the other hand, a perturbative calculation of correlation functions based on the auxiliary Lagrangian (14.148) leads to divergent results in the limit $\alpha \to \infty$, and thus suffers from similar problems as the formulation in terms of the Proca-Lagrangian (14.134), in the limit $\mu \to 0$.

14.7.3 Reconstruction of commutators of the *GNI* formulation

As was pointed out in chapter 13, we may use the APB to calculate the commutators of the *GNI* formulation, using Dirac's construction (13.348). This is important, since a BJL calculation of such *commutators* requires in general the exact knowledge of the leading q_0^{-1} contribution to the corresponding two-point functions.

In the *GNI* formulation, the computation of the APB was based on the auxiliary Lagrangian (14.148*a*), describing a dynamics different from the one of actual interest.

In the *GI* formulation, on the other hand, we are free to choose a gauge, and the APB correspond to the *commutators* of the actual Lagrangian of interest if we work in the so-called "covariant α- gauges", i.e.

$$\mathcal{L} = \mathcal{L}^{GI} - \frac{1}{2\alpha}(\partial_\mu A^\mu)^2 \quad , \tag{14.158}$$

with $\mathcal{L}^{GI}$ given by (14.143). This streamlines the procedure in the *GNI* and *GI* formulation, the APB being calculated by the BJL limiting procedure based on formally similar Lagrangians. In the latter formulation, the reconstruction of the commutators of the *GNI* formulation, however proceeds along the lines discussed in section 4.2. These will be independent of the choice of gauge, since they refer to gauge-invariant quantities of the *GI* formulation. From a diagrammatic point of view this is a consequence of the transversal character of the effective polarization tensor $\Pi^{\mu\nu}(k)$ as given by (14.145).

The reconstruction of the commutators of the *GNI* formulation for the JR parameter $a = 1$, poses a problem in the *GI* formulation: The absence of the kinetic term of the θ-field in (14.143) makes a perturbative calculation impossible, as expression (14.144*b*) for the θ-propagator shows. We shall thus have to resort to a hybrid procedure to be described next.

For $a \neq 1$ the equal time commutators (ETC) of the *GI* formulations in the "α-gauge" can be exactly calculated in perturbation theory, using the Feynman rules of section (6.2) with $D^{(0)}_{\mu\nu}(k)$ given by (14.157), as well as the BJL definition

(13.334). The ETC of the A_μ and θ fields are found to have a "Poisson-bracket" structure. In particular,

$$[A^\mu(x), \pi_\nu(y)]_{ET} = ig^\mu_\nu \delta(x^1 - y^1)$$
$$[\theta(x), \pi_\theta(y)]_{ET} = i\delta(x^1 - y^1) \quad . \tag{14.159a}$$

For the ETC involving the source current

$$\mathcal{J}^\mu = \bar{\psi}\gamma^\mu P_- \psi + [(a-1)\partial^\mu + \tilde{\partial}^\mu]\theta$$

a BJL calculation yields

$$[\theta(x), \mathcal{J}^0(y)]_{ET} = i\delta(x^1 - y^1), \qquad [\pi_\theta(x), \mathcal{J}^0(y)]_{ET} = -i\delta'(x^1 - y^1),$$
$$[A^\mu(x), \mathcal{J}^0(y)]_{ET} = [\pi_\mu(x), \mathcal{J}^0(y)]_{ET} = 0, \quad [\mathcal{J}^0(x), \mathcal{J}^0(y)]_{ET} = 0, \tag{14.159b}$$

as well as

$$[\mathcal{J}^0(x), \mathcal{J}^1(y)]_{ET} = -i\frac{a^2\delta'(x^1-y^1)}{a-1}, \qquad [\pi_1(x), \mathcal{J}^1(y)]_{ET} = \frac{-ia^2e}{a-1}\delta(x^1 - y^1),$$
$$[\pi_\theta(x), \mathcal{J}^1(y)]_{ET} = -i(a-1)\delta'(x^1 - y^1), \quad [\theta(x), \mathcal{J}^1(y)]_{ET} = \frac{i}{a-1}\delta(x^1 - y^1)$$
$$[A^\mu(x), \mathcal{J}^1(y)]_{ET} = [\pi_0(x), \mathcal{J}^1(y)]_{ET} = 0, \quad [\mathcal{J}^1(x), \mathcal{J}^1(y)]_{ET} = 0 \tag{14.159c}$$

A BJL calculation shows that

$$[A^\mu(x), J^\nu_F(y)]_{ET} = [\pi_\theta(x), J^\nu_F(y)]_{ET} = 0 \quad , \tag{14.159d}$$

where J^ν_F denotes the fermionic part of the source current. Using the relations (14.79c) as well as (13.82b), implying $[\mathcal{J}^\mu]_{GI} = [J^\mu_L]_{GNI}$, one recovers from here the commutators (14.80), as well as those involving the current J^μ_L of the GNI formulation.

From the point of view of QED_4, the case $a = 1$ is the more interesting one[459]. The singularity of the θ-propagator (14.144b) makes a direct calculation of the ETC along the above lines impossible. This is evident from the $a = 1$ singularity of the commutators (14.159c) containing $\mathcal{J}^1(x)$. Indeed, $\mathcal{J}^1(x)$ involves $\partial_0\theta(x)$, which for $a = 1$ can no longer be expressed in terms of canonical variables via Eq. (14.55). This circumstance is intimately related to the existence of a primary constraint, in this case:

$$\omega_1 \equiv \pi_\theta - eA_1 \approx 0 \quad . \tag{14.160}$$

We must thus compute $\partial_0\theta$ from the Hamiltonian equations

$$\partial_0\theta(x) = -i[\theta(x), H_T] \quad , \tag{14.161}$$

where H_T is the total Hamiltonian

$$H_T = H_c + \int dx^1 v\omega_1 \tag{14.162a}$$

with

$$H_c = \int dx^1 \left\{ \bar{\psi} i\gamma_1 \partial_1 \psi + \frac{1}{2}\pi_1^2 - \frac{\alpha}{2a}\pi_0^2 - A^0(\partial_1\pi_1 + e\mathcal{J}^0) + A^1 J_F^1 + \pi^0 \partial_1 A_1 \right\} \quad , \tag{14.162b}$$

$$\mathcal{J}^\mu = J_F^\mu + \tilde{\partial}^\mu \theta \quad . \tag{14.163}$$

In order that the constraint (14.160) be conserved in time, we demand $\partial_0\omega_1 \approx -i[\omega_1, H_T] \approx 0$. The computation of this commutator only requires knowledge of the "finite ETC" obtained from (14.159*a*, *b*, *d*). One is thus led to the secondary constraint

$$\omega_2 = \pi_1 \approx 0 \quad . \tag{14.164}$$

Note that the constraints (14.160) and (14.164) coincide with $\omega_3 \approx 0$ in (14.65*b*) and $\omega_4 \approx 0$ in (14.67*b*), respectively. They are second class,

$$-i[\omega_r(x), \omega_s(y)]_{ET} = e\epsilon_{rs}\delta(x^1 - y^1) \quad , \tag{14.165}$$

and therefore serve to fix the Lagrange multiplier v. One finds

$$\partial_0\theta = v = -\frac{1}{e}(eJ_F^1 + \partial_1\pi^0) \quad . \tag{14.166}$$

Hence we conclude from (14.163) that

$$\mathcal{J}^1 = -\frac{1}{e}\partial_1\pi_0 \quad . \tag{14.167}$$

We are now in the position of computing also the 'singular' ETC, from our knowledge of the "non-singular" commutators. They are all found to vanish:

$$[\mathcal{J}^0(x), \mathcal{J}^1(y)]_{ET} = [\pi_1(x), \mathcal{J}^1(y)]_{ET} = [\theta(x), \mathcal{J}^1(y)]_{ET} = 0 \quad . \tag{14.168}$$

The ETC calculated in this way agree with (i times) the corresponding Poisson brackets in the bosonic formulation.

Following our discussion in section 4.2, we must next construct the corresponding commutators on the subspace Γ^* defined by the second-class constraints (14.160) and (14.164). Because of (14.165), they are obtained via Eq. (14.83), with $\{,\}$ replaced by $-i[,]_{ET}$. Since these two brackets have been found to coincide, it immediately follows from the theorem quoted in section 4.2 that we recover in this way the commutation relations (14.87), upon using the isomorphic relations (14.86) and (14.79c).

Though the GNI formulation provides a simple framework for computing the $a = 1$ commutators (14.87), our treatment here also offers an interesting insight into the quantization of chiral QED_4, where the r.h.s. of these ETC are found to exhibit a non-polynomial dependence on the vector potential. It is gratifying that the procedure just described also leads to the correct results for chiral QED_4 in the GI formulation[459].

14.8 Chiral QED_2 in terms of Chiral Bosons

The meaning of "chiral bosons", as well as their role in chiral gauge theories and string models has already been discussed in the preceding chapter, where we have shown the chiral action (13.306) to provide a representation of chiral–QCD$_2$ which is equivalent to that given by the usual bosonized action (13.137). The corresponding result for the $U(1)$ Lagrangian obtained by omitting the trace, and by making the replacement (14.6), with ϕ now an abelian field[480]:

$$\begin{aligned} \mathcal{L} &= -\frac{1}{4}F_{\mu\nu}F^{\mu\nu} + \mathcal{L}_{ch} \\ \mathcal{L}_{ch} &= \partial_-\phi\partial_1\phi + 2eA_-\partial_1\phi - \frac{e^2}{2}A_-^2 + \frac{ae^2}{2}A_\mu^2 \quad . \end{aligned} \tag{14.169}$$

This Lagrangian is a gauged version of the chiral Lagrangian of Floreanini and Jackiw[448]. Unlike in the non–abelian case, the equivalence of the action $S_{ch}[A,\phi]$, associated with $\mathcal{L}_{ch}$, to the effective action $W^{(L)}[A]$ as given by (14.5), can be explicitly verified by performing the Gaussian integration. One has

$$\int D\phi e^{i\int d^2x \mathcal{L}_{ch}(A,\phi)} = \text{const } e^{iW^{(L)}[A]} \tag{14.170a}$$

where

$$\text{const} = \int D\phi e^{i\int d^2x(\partial_1\phi\partial_-\phi)} \quad . \tag{14.170b}$$

The quantization proceeds as usual. The primary constraints (13.313) and canonical Hamiltonian (13.302) are now replaced by

$$\begin{aligned} \Omega_1 &\equiv \pi_0 \approx 0 \\ \Omega_3 &\equiv \pi_\phi - \partial_1\phi \approx 0 \end{aligned} \tag{14.171}$$

and

$$H_{ch} = \int d^2x \left[\frac{1}{2}\pi^2 + A_0 \partial_1 \pi_1 + (\partial_1 \phi)^2 - 2e\partial_1 \phi A_- + \frac{1}{2}e^2 A_-^2 - \frac{1}{2}ae^2 A_\mu^2 \right] \quad . \tag{14.172a}$$

Hence the total Hamiltonian reads

$$H_T = H_{ch} + \int dx^1 (u_1 \Omega_1 + u_3 \Omega_3) \quad . \tag{14.172b}$$

The requirement $\dot{\Omega}_1 \approx 0$ leads to Gauss's law as a secondary constraint

$$\Omega_2 \equiv -\partial_1 \pi_1 + eJ^0_{ch} \approx 0 \tag{14.173}$$

where

$$J^0_{ch} = 2\partial_1 \phi + e[(a-1)A_0 + A_1] \quad . \tag{14.174}$$

The set of constraints $\Omega_i \approx 0$, $i = 1, 2, 3$, are second class, since the matrix

$$Q(x,y) = \{\Omega_i(x), \Omega_j(y)\} = \begin{pmatrix} 0 & -e^2(a-1) & 0 \\ e^2(a-1) & 0 & 0 \\ 0 & 0 & -2e\partial_1 \end{pmatrix} \delta^2(x-y) \tag{14.175}$$

has a non-vanishing determinant. The conditions $\dot{\Omega}_2 \approx 0$ and $\dot{\Omega}_3 \approx 0$ thus merely serve to fix the Lagrange multipliers in (172b), and there are no further constraints. With the aid of[10])

$$Q^{-1}(x,y) = \begin{pmatrix} 0 & \frac{\delta(x^1-y^1)}{e^2(a-1)} & 0 \\ -\frac{\delta(x^1-y^1)}{e^2(a-1)} & 0 & 0 \\ 0 & 0 & -\frac{1}{4e}\epsilon(x^1-y^1) \end{pmatrix} \delta(x^0 - y^0) \tag{14.176}$$

we construct the Dirac brackets in the usual way.

Choosing ϕ, A_1 and π_1 as independent variables, one has for the non-vanishing commutators

$$\begin{aligned} [\phi(x), \phi(y)]_{ET} &= -\frac{i}{4}\epsilon(x^1 - y^1) \\ [A^1(x), \pi_1(y)]_{ET} &= i\delta(x^1 - y^1) \\ [\phi(x), \pi_\phi(y)] &= \frac{i}{2}\delta(x^1 - y^1) \quad . \end{aligned} \tag{14.177}$$

The commutators involving A^0 may be computed from here by using the constraints.

10) The inverse of ∂_1 is not unique. The choice $\partial_1^{-1} \to \frac{1}{2}\epsilon$ with $\langle x|\epsilon|y\rangle = \epsilon(x^1 - y^1)\delta(x^0 - y^0)$ amounts to a principal value prescription.

The equations of motion following from (14.169) read

$$\begin{aligned}\partial_\mu F^{\mu 0} + eJ^0_{ch} &= 0 \\ \partial_\mu F^{\mu 1} + eJ^1_{ch} &= 0\end{aligned} \tag{14.178a}$$

$$\partial_- \partial_1 \phi + e\partial_1 A_- = 0 \quad , \tag{14.178b}$$

where J^0_{ch} has already been defined in (14.174), and

$$J^1_{ch} = -2e\partial_1\phi + eA_- - aeA_1 \quad . \tag{14.179}$$

Despite the non-covariant look of these equations, they admit a covariant[480] solution of the same form as (14.28),

$$A_\mu = -\frac{1}{a\tilde{e}}[\partial_\mu\phi + (a-1)\tilde{\partial}_\mu\phi - a\tilde{\partial}_\mu h] \quad , \tag{14.180a}$$

with

$$\begin{aligned}&\phi = \sigma - h, \\ &(\Box + m^2)\sigma = 0, \quad m^2 = \frac{a^2e^2}{a-1} \quad ,\end{aligned} \tag{14.180b}$$

the only difference being, that this time h is a *self-dual* field, satisfying

$$\partial_- h = 0 \quad . \tag{14.181}$$

This is in agreement with the fact that in the present formulation the free right-handed fermions (left movers) have been excluded, which amounts to removing one-half of the original number of degrees of freedom in h. The field σ has the same commutation relations as in the usual (non-chiral) formulation

$$\begin{aligned}[\sigma(x), \sigma(y)]_{ET} &= 0 \\ [\sigma(x), \partial_0\sigma(y)]_{ET} &= \frac{i}{a-1}\delta(x^1 - y^1)\end{aligned} \tag{14.182a}$$

whereas $h(x)$ now satisfies

$$\begin{aligned}[h(x), h(y)]_{ET} &= -\frac{i}{4}\epsilon(x^1 - y^1) \\ [h(x), \partial_0 h(y)]_{ET} &= \frac{i}{2}\delta(x^1 - y^1) \quad .\end{aligned} \tag{14.182b}$$

These commutation relations follow from (14.177), if one makes use of the solution for A_μ, Eq.(14.180).

Eq. (14.181) evidently holds in an arbitrary Lorentz frame. This is a property of two-dimensional space-time, where the left and right light cones are disconnected. Hence A_μ given by (14.180) transforms like a Lorentz vector. This can be made more precise by explicitly constructing the generators of Lorentz transformations in terms of the energy momentum tensor for this model. Using the constraints to express all quantities in terms of the independent variables A_1, π_1 and ϕ, one finds for the generators of the Poincaré group[471]

$$\begin{aligned} P^0 &= \int dx^1 \mathcal{H}^*_{ch} \\ P^1 &= \int dx^1 [(\partial_1\phi)^2 - \pi_1 \partial_1 A_1] \\ M &= x^0 P^1 + \int dx^1 [x^1 \mathcal{H}^*_{ch} + \frac{1}{e^2(a-1)}\pi_1(-\partial_1\pi_1 + 2e\partial_1\phi - e^2 A_1)] \quad , \end{aligned} \tag{14.183}$$

where $\mathcal{H}^*$ is the Hamiltonian density corresponding to (14.172a), evaluated on the constraint surface Γ^*:

$$\begin{aligned} \mathcal{H}^*_{ch} =& \frac{1}{2}(\pi_1)^2 - \frac{1}{a-1}\pi_1\partial_1 A_1 + \frac{1}{2e^2(a-1)}(\partial_1\pi_1)^2 \\ &+ \frac{a+1}{a-1}(\partial_1\phi)^2 - \frac{2}{e(a-1)}\partial_1\pi_1\partial_1\phi + \frac{2ea}{a-1}A_1\partial_1\phi + \frac{m^2}{2}(A_1)^2 \\ &- \frac{1}{e^2(a-1)}\partial_1(\pi_1\partial_1\pi_1) - \frac{2}{e(a-1)}\partial_1(\pi_1\partial_1\phi) \quad . \end{aligned} \tag{14.184}$$

A little algebra shows that the operators (14.183) satisfy the correct Poincaré algebra,

$$\begin{aligned} [P^0, P^1]_{ET} &= 0 \\ [M, P^1]_{ET} &= -iP^0 \\ [M, P^0]_{ET} &= -iP^1 \quad . \end{aligned} \tag{14.185}$$

This explicitly confirms the Lorentz covariance of the theory.

14.9 Conclusion

We have seen that the chiral Schwinger model is completely soluble in its GNI and GI formulation. The use of a chirally asymmetric regularization (JR parameter $a \neq 0$) was crucial in order to obtain a meaningful quantum field theoretical model. In fact, the requirement of unitarity and absence of tachyonic ghosts, restricted the JR parameter to satisfy $a \geq 1$.

We have seen that in the GNI formulation no satisfactory perturbation theory can be formulated. Although the temporary introduction of an explicit mass μ for the vector potential allows for a formulation of Feynman rules, the individual terms in the resulting perturbation series diverge as we let μ tend to zero. Although the result obtained by *summing* this series tends to the correct result as $\mu \to 0$, the failure of perturbation theory in this limit makes this method impracticable in $d = 3+1$ dimensions, where the perturbation series cannot be summed. Besides this, the non-abelian Proca theory is known to be non-renormalizable. From the perturbative point of view one is thus literally forced to *embed* an anomalous chiral gauge theory into a bonafide gauge theory on quantum level. This does not mean, however, that in the resulting GI formulation the theory is non-anomalous. This is seen from the, in general, non-canonical form of the APB in this formulation, and is also implied by the isomorphism existing between the (gauge-invariant) observables of the GI formulation and the fields of the GNI formulation, as represented by Eq. (13.86). Although the manipulations establishing this isomorphism were admittedly formal, we have verified it explicitly for the case of the two-point function of the vector potential.

The GI formulation allowed for a perturbative treatment, with exception of the case $a = 1$, where the kinetic term of the Wess-Zumino field is absent in the Lagrangian (14.14). As we have seen, the number of constraints doubled at this particular value of the JR parameter. A similar situation is seen to prevail also in the non-abelian case (see chapter 13). In the GNI formulation one has four types of second-class constraints, while in the GI formulation two of the four constraints are first class. Although the case $a = 1$ appears to be only of academic interest, this is not so: As was shown by F. Otto[459], chiral QED_4 and chiral QCD_4 resemble in this respect their two-dimensional analogues with $a = 1$. Henceforth, the perturbative treatment in these cases remains an open problem.

Nonetheless, this circumstance does not pose a problem for discussing the constraint algebra of these theories, which in all cases turn out to be second class in the GNI formulation. We have learned that in the fermionic formulation this algebra has to be understood in terms of *anomalous* Poisson brackets. These can be exactly calculated in perturbation theory defined in terms of an auxiliary Lagrangian $\mathcal{L}^\alpha$, which is obtained from the Lagrangian of interest by addition of —preferably — a term $-\frac{1}{\alpha}(\partial_\mu A^\mu)^2$. This term not only breaks explicitly the classical gauge-invariance of $\mathcal{L}$, but also turns the theory of interest into an unconstrained though still anomalous theory. Both properties are important! In this way we were able to develop the *complete* Dirac algorithm on fermionic level, which was shown to agree completely with results first obtained in the equivalent bosonic formulation (see chapter 13). This method has recently been also successfully applied by F. Otto [459] to QED_4 and QCD_4, which closes an important gap between two-dimensional and four-dimensional chiral gauge theories.

Chapter XV

CONFORMALLY INVARIANT FIELD THEORY IN TWO DIMENSIONS

15.1 Introduction

Conformally invariant quantum field theories were first considered in the sixties[482], and used to understand the small distance behavior of field operator products[31]; their consideration was also motivated by the fact that in the high energy limit, renormalizable theories are approximately conformally invariant[483], thus allowing for the perturbative computation of several high energy processes[21] involving QCD, Considerable effort has been dedicated to gaining a deeper understanding of conformally invariant theories in any space-time dimension[484]. However they have not found extensive applications. There is a conceptual difficulty, concerning the fact that conformal transformations can turn space-like into time-like distances, which leads to a contradiction with Einstein's causality[485]. Later on, when the connection between Euclidean quantum field theory and statistical mechanics was established, the above clash could however be ignored. Although the causality paradox has by now been solved[81,98,486], the main application of conformal quantum theory has still been in the area of statistical mechanics. Indeed, classical scale invariance is a quite general feature shared by many field theories, having no dimensional parameter, a property generally met by statistical systems at criticality[85]. Polyakov and Migdal were the first to use extensively these ideas[487]. The theoretical framework of the renormalization group provided a satisfactory description of the theory at criticality[85]. On the other hand, the computational power of such methods is rather limited, and there is no general framework based on these methods, to study general results, except for some properties of specific models[488].

The fact that a number of two dimensional models exhibit not only scale, but also conformal invariance, is of great importance, since the conformal group in two dimensions is infinite dimensional. The recognition that in general, scale invariant theories are also conformally invariant[489], is a major development in recent years. Polyakov thus proposed to use conformal invariance as an essential ingredient in the study of criticality[487].

15.2 Conformal Transformations in Field Theory

15.2.1 The conformal algebra

Consider the general coordinate transformations $x \to x'(x)$, with the property that the metric tensor $g_{\mu\nu}(x)$ is left invariant except for a local multiplicative factor (Weyl transformation[490]):

$$g_{\mu\nu}\partial_\lambda x'^\mu \partial_\rho x'^\nu = \lambda(x) g_{\lambda\rho} \quad , \tag{15.1}$$

with $\lambda(x)$ a continuous, non vanishing function of x.

In order to determine the corresponding infinitesimal transformation we look for solutions of the Killing Cartan equation, obtained from (15.1) by the substitutions $x'_\mu = x_\mu + \xi_\mu$, and $\lambda = 1 + \delta\lambda$, such that

$$\partial_\mu \xi_\nu + \partial_\nu \xi_\mu = (\delta\lambda) g_{\mu\nu} \quad . \tag{15.2}$$

Expressing $\delta\lambda$ in terms of ξ by contracting (15.2) with $g^{\mu\nu}$, we obtain

$$\partial_\mu \xi_\nu + \partial_\nu \xi_\mu = \frac{2}{D} \partial^\rho \xi_\rho g_{\mu\nu} \quad . \tag{15.3}$$

In $(D \neq 2)$ we find the following Killing vectors[490], that is, solutions of Eq. (15.3)[1)]:
for translations, we have the Killing vectors

$$t^\alpha_\mu = -\delta^\alpha_\mu \quad ; \tag{15.4a}$$

with the transformation parameter $\xi_\mu = -\delta^\alpha_\mu \epsilon_\alpha = -\epsilon_\mu$;
for Lorentz transformations, we have

$$\xi^\alpha_{\mu\nu} = x_\nu \delta^\alpha_\mu - x_\mu \delta^\alpha_\nu \quad , \tag{15.4b}$$

with $\xi_\mu = \xi^\alpha_{\mu\nu} \epsilon^\nu_\alpha$; for dilatations, we get

$$d^\alpha = -x^\alpha \quad , \tag{15.4c}$$

implying $\xi_\mu = -\epsilon x_\mu$; and, for special conformal transformations, we obtain

$$\xi^\alpha_\mu = 2x_\mu x^\alpha - x^2 \delta_{\mu\alpha} \quad , \tag{15.4d}$$

with the parameter ξ_μ given by $\xi_\mu = \xi^\alpha_\mu \epsilon_\alpha$. This completes the set of Killing vectors, in D space-time dimensions.

---

1) The infinitesimal ξ_μ in Eq. (15.3) are obtained multiplying the Killing vectors in Eqs. (15.4) by appropriate infinitesimal parameters.

Holomorphic realization

When restricting the dimension of space time to D=2, Eq. (15.3) may be rewritten, in the Euclidean case[2)], as:

$$(\partial_1 + i\partial_2)(\xi_1 + i\xi_2) = 0 \tag{15.5a}$$
$$(\partial_1 - i\partial_2)(\xi_1 - i\xi_2) = 0 \quad . \tag{15.5b}$$

Therefore, any analytic (antianalytic) transformation $z = x_1 + ix_2 \rightarrow f(z)$, $(\overline{z} = x_1 - ix_2 \rightarrow \overline{f}(\overline{z}))$ is allowed. Expanding $f(z)$ in a Laurent series in z as

$$f(z) = z + \sum_{-\infty}^{\infty} \epsilon_n z^{n+1} \tag{15.6a}$$

$$\overline{f}(\overline{z}) = \overline{z} + \sum_{-\infty}^{\infty} \overline{\epsilon}_n \overline{z}^{n+1} \quad , \tag{15.6b}$$

with ϵ_n small parameters, we can write

$$f(z) = z + \sum \epsilon_n (L_n^c z) \quad . \tag{15.6c}$$

The classical generators of the transformations (15.6) are thus,

$$L_n^c = z^{n+1} \frac{d}{dz} \tag{15.7a}$$

$$\bar{L}_n^c = \overline{z}^{n+1} \frac{d}{d\overline{z}} \quad . \tag{15.7b}$$

They obey the classical loop algebra

$$[L_n^c, L_m^c] = -(n-m) L_{n+m}^c \quad , \tag{15.8a}$$
$$[\bar{L}_n^c, \bar{L}_m^c] = -(n-m) \bar{L}_{n+m}^c \quad , \tag{15.8b}$$

as one easily obtains from Eq. (15.7).

Notice that $L_0^c, L_{+1}^c, L_{-1}^c$ (together with $\bar{L}_0^c, \bar{L}_{+1}^c, \bar{L}_{-1}^c$) correspond to the generators of the special conformal group. In two dimensions, this group is described by $SO(2,2) \approx SL(2,R)_+ \otimes SL(2,R)_-$, each connected piece being generated by 3 operators, conveniently labeled as L_0, L_{+1}, L_{-1} (and of course $\bar{L}_0, \bar{L}_{+1}, \bar{L}_{-1}$ for the antiholomorphic component).

[2)] In the Euclidean formulation $x_2^E = ix_0$, $x_1^E = x^1$, $\partial_1^E = \partial_1$; we drop the index E in this chapter.

Consider the change of variables

$$z = \frac{1 + \frac{i}{2}x^-}{1 - \frac{i}{2}x^-} \quad , \quad \bar{z} = \frac{1 - \frac{i}{2}x^+}{1 + \frac{i}{2}x^+} \quad . \tag{15.9}$$

Under this transformation the left moving part of the energy momentum tensor $\theta(\xi)$ turns into

$$T(z) = 2\pi \left(\frac{2}{1+z}\right)^4 \theta\left(2i\frac{1-z}{1+z}\right) \quad . \tag{15.10a}$$

The definition of the generators of translations, dilatations and special conformal transformations are, respectively

$$P_- = \int \theta(x^-) dx^- \tag{15.10b}$$

$$\mathcal{D}_- = \int x^- \theta(x^-) dx^- \tag{15.10c}$$

$$\mathcal{K}_- = \int (x^-)^2 \theta(x^-) dx^- \quad , \tag{15.10d}$$

with corresponding expressions for x^+; they satisfy

$$[\mathcal{D}, P] = -iP \tag{15.11a}$$

$$[\mathcal{D}, \mathcal{K}] = i\mathcal{K} \tag{15.11b}$$

$$[P, \mathcal{K}] = 2i\mathcal{D} \quad . \tag{15.11c}$$

We have

$$L_0 = P + \frac{1}{4}\mathcal{K} \tag{15.11d}$$

$$L_{\pm 1} = P \pm i\mathcal{D} - \frac{1}{4}\mathcal{K} \quad . \tag{15.11e}$$

The operators $L_{0,\pm 1}$, satisfy the $SL(2,R)$ Virasoro algebra with, P_μ generating the infinitesimal translations, D the infinitesimal dilatations, and K_μ the infinitesimal holomorphic special conformal transformations.

A representation of this algebra is given by the restriction of (15.7) to $n = 0, \pm 1$, with analogous expressions for the antiholomorphic fields. In general, conformal invariance is broken due to renormalization, since one has to introduce some mass parameter, which spoils scale invariance. However, if conformal invariance happens to survive quantization, it imposes severe restrictions to the theory.

Conformal dimensions

In a conformally invariant field theory, the transformation of the elementary field is defined by its conformal dimension Δ, which naively corresponds to the scale dimension. The two and three point functions in a conformally invariant theory are completely determined up to constants[484,490] given in terms of the conformal dimensions.

As we already mentioned in two dimensional space time we have a simplification due to the fact that the conformal group factorizes in terms of the $SL(2,R)$ group of Möbius transformations, $G = SL(2,R) \bigotimes SL(2,R)$. Correspondingly there exists a special class of factorizable fields, that transforms in the following simple way[79]

$$\phi(z,\bar{z}) = \left(\frac{d\zeta}{dz}\right)^{\Delta} \left(\frac{d\bar{\zeta}}{d\bar{z}}\right)^{\bar{\Delta}} \phi(\zeta(z),\bar{\zeta}(\bar{z})) \quad . \tag{15.12a}$$

For Möbius transformations this reads

$$\phi(z,\bar{z}) = (cz+d)^{-2\Delta}(\overline{cz}+\bar{d})^{-2\bar{\Delta}}\phi(\zeta(z),\bar{\zeta}(\bar{z})) \quad . \tag{15.12b}$$

These are called primary fields[79], and are the building blocks of the construction of two dimensional conformally invariant quantum field theory.[3)]

At this point it is convenient to choose new complex coordinates for the problem, mapping the usual (Euclidean) space-time into a space such that constant "Euclidean" time corresponds to circles. Thus, if we take the map[129] (15.9), with inverse $x^- = 2i\frac{1-z}{1+z}$, we are mapping the real axis into the circle $|z| = 1$, $-\pi \leq \arg z \leq \pi$. Such transformations, in which lines of constant Euclidean time are mapped into circles around the origin are very useful in the study of conserved generators. The best example is

$$z = e^{(x_2 - ix_1)} \quad , \tag{15.13}$$

where $x_2 = -\infty$ corresponds to the origin, $x_2 = +\infty$ to the circle at infinity, and a line of constant time is still a circle.

If we consider a conserved current, we know from general arguments of conformal quantum field theory, that its scale dimension is given by the naive value[491]. Indeed, the current obeys a current algebra with a Schwinger term[11]

$$[J_0^a(t,x^1), J_1^b(t,y^1)] = if^{abc}J_1^c\delta(x^1-y^1) + k\frac{i}{\pi}\delta^{ab}\delta'(x^1-y^1) \quad . \tag{15.14}$$

If we make a rescaling, we readily see that the Schwinger term matches the remaining terms in the equation only if $\dim J = 1$; therefore, the dimension of J

3) For the time being, we are only considering Möbius transformations. See section 15.2.2.

is fixed. In general, if the current is a primary field as in Eq. (15.12), we have the transformation property

$$J(z,\bar{z}) = \left(\frac{d\zeta}{dz}\right)^{\Delta_J} \left(\frac{d\bar{\zeta}}{d\bar{z}}\right)^{\bar{\Delta}_J} J(\zeta,\bar{\zeta}) \quad , \tag{15.12c}$$

For dimension one, we have for the left (holomorphic) and right (antiholomorphic) movers (fields),

$$J(z) = \frac{d\zeta}{dz} J(\zeta) \tag{15.15a}$$

$$J(\bar{z}) = \frac{d\bar{\zeta}}{d\bar{z}} J(\bar{\zeta}) \quad . \tag{15.15b}$$

Thus we obtain for the charge

$$Q = \int J(x^0, x^1) dx^1 \to \frac{1}{2\pi i} \oint_{|z|=const} J(z) dz + \frac{1}{2\pi i} \oint_{|z|=const} J(\bar{z}) d\bar{z} \quad , \tag{15.16}$$

where the integration over constant radius (z) corresponds to an integration at constant time, as dictated by the new time interpretation following Eq. (15.12).

Two and four point functions

One can use conformal invariance to compute the two and three point function of a scalar theory, up to some dimensionless parameters. These functions are given by simple expressions; for the two point function, Poincaré and scale invariance imply

$$\langle 0|\phi_i(x^+)\phi_j(y^+)|0\rangle = \lambda[(x^+ - y^+)]^{-2\Delta_+} \tag{15.17a}$$

$$\langle 0|\phi_i(x^-)\phi_j(y^-)|0\rangle = \lambda[(x^- - y^-)]^{-2\Delta_-} \quad , \tag{15.17b}$$

where we separated right and left movers (λ is an arbitrary constant)[4)].

Indeed, Poincaré invariance implies that it is a function of the difference of the coordinates; scale invariance implies that it is a homogeneous function; the scale transformation fixes the power behavior (notice that the generator L_0 of scale transformations allows only for powers of the argument). Finally, special conformal transformations generated by L_1 force the two scaling parameters to be equal ($\Delta_i = \Delta_j = \Delta$). For the three-point function we have[484,490]:

$$\langle 0|\phi_1(x_1)\phi_2(x_2)\phi_3(x_3)|0\rangle =$$

[4)] In what follows we use the notation $x^\pm$ also for holomorphic and antiholomorphic, unless otherwise explicitly stated. However, in general, the light cone notation implies that we are working in Minkowski space.

$$\lambda[(x_{12})^2]^{\frac{1}{2}(-\Delta_3+\Delta_1+\Delta_2)}[(x_{13})^2]^{\frac{1}{2}(-\Delta_2+\Delta_1+\Delta_3)}[(x_{23})^2]^{\frac{1}{2}(-\Delta_1+\Delta_2+\Delta_3)} \quad , \tag{15.18}$$

where $x_{ij} = x_i - x_j$, and λ is an arbitrary constant.

Indeed, Poincaré invariance again implies dependence only on the difference of coordinates, and invariance under scale and special conformal transformations implies (15.18).

When considering a four point function, there is a further invariant, namely the Möbius invariant

$$x = \frac{x_{12}x_{34}}{x_{13}x_{24}} \quad . \tag{15.19}$$

Indeed, under the transformation

$$\delta_f \phi = \left\{\Delta_i \frac{\partial f(x_i)}{\partial x_i} + f(x_i)\frac{\partial}{\partial x_i}\right\}\phi \quad , \tag{15.20}$$

with $f(x) = 1$ for Poincaré transformations, $f(x) = x$ for scale transformations and $f(x) = x^2$ for special conformal transformations, we have for the 4-point function the following identity

$$\sum_{i=1}^{4}\left(\Delta_i \frac{\partial f_i}{\partial x_i} + f_i \frac{\partial}{\partial x_i}\right)\langle 0 \mid \phi_1(x_1)\phi_2(x_2)\phi_3(x_3)\phi_4(x_4) \mid 0\rangle = 0 \quad . \tag{15.21}$$

One easily verifies that $\sum_i f_i \frac{\partial}{\partial x_i} x = 0$, for $f_i = 1, x_i, x_i^2$, i.e., functions of x are Möbius invariant. Therefore the most general conformally invariant 4-point function is given by

$$\langle 0|\phi_1(x_1)\phi_2(x_2)\phi_3(x_3)\phi_4(x_4)|0\rangle = \prod_{i<j} x_{ij}^{-(\Delta_i+\Delta_j-\frac{1}{3}\sum_k \Delta_k)}\mathcal{F}\left(\frac{x_{12}x_{34}}{x_{13}x_{24}}\right) \quad , \tag{15.22}$$

where $\mathcal{F}$ is an arbitrary function depending on the (Möbius invariant) harmonic ratio x.

15.2.2 The BPZ construction

We next verify that the field algebra is tremendously enlarged in a 2 dimensional field theory. First notice that Poincaré invariance implies the conservation of the energy momentum tensor,

$$\partial^\mu T_{\mu\nu} = 0 \quad , \tag{15.23}$$

while scale invariance implies its tracelessness

$$T^\mu_\mu = 0 \tag{15.24}$$

Moreover, the following current is conserved

$$\mathcal{J}_\mu = T_{\mu\nu}\epsilon^\nu \quad , \tag{15.25}$$

whenever $\xi_\mu = \epsilon_\mu$ satisfies Eq. (15.3), and $T_{\mu\nu}$ is traceless.

Conservation and tracelessness of the energy momentum tensor in (Euclidean) 2-dimensional space-time imply that the combination

$$\overline{T} = T_{\overline{z}\overline{z}} = T_{11} + iT_{12} \quad , \tag{15.26}$$

satisfies

$$(\partial_1 - i\partial_2)\overline{T} = 0 \quad . \tag{15.27}$$

Therefore $\overline{T}$ is a function of $\overline{z} = x_1 - ix_2$ only. Analogously:

$$T = T_{zz} = T_{11} - iT_{12} \tag{15.28}$$

is a function of $z = x_1 + ix_2$. However, $T_{\mu\nu}$ generates a space-time symmetry, since each of the above components can be expanded in a Laurent series as

$$T(z) = \sum_{n=-\infty}^{\infty} z^{-n-2} L_n \tag{15.29a}$$

and

$$\overline{T}(\overline{z}) = \sum_{n=-\infty}^{\infty} \overline{z}^{-n-2} \overline{L}_n \quad . \tag{15.29b}$$

It will be very useful to work with complex variables. In this case, quantization has to be effected in a way already indicated previously. This is useful, since the definition of a *time* direction would break the independence of z and $\overline{z}$ variables. We map space-time into circles, through the change of variables (15.12). This means that constant *time* is associated with a constant radius: a circle. *Time* flows from the origin outwards, the origin being the far past, as previously explained. In this case, the conserved charge, analogous to $\int \mathcal{J}(x)dx^1$, is now given by the angular integration of the radial component of the current. This prescription is known as radial quantization[492]. We obtain, for the infinitesimal generator of diffeomorphic transformations (15.16),

$$Q_\epsilon = \frac{1}{2\pi i}\int dx^1 T_{0\nu}\epsilon^\nu \quad , \tag{15.30a}$$

or, when expressed in terms of the holomorphic variables z and $\overline{z}$,

$$Q_\epsilon = \frac{1}{2i\pi}\oint T(\xi)\epsilon(\xi)d\xi + \frac{1}{2i\pi}\oint \overline{T}(\overline{\xi})\overline{\epsilon}(\overline{\xi})d\overline{\xi} \quad , \tag{15.30b}$$

where the integration contour encircles the origin. The commutator of Q_ϵ with some field $\phi(z)$, is then obtained by using the following rule: the integration contour around the origin of the complex ξ-plane is chosen such that z lies inside (outside) of the contour if $\phi(z)$ appears on the right (left) of Q_ϵ. This corresponds to the usual time ordering prescription in quantum field theory, as we mentioned before (see also discussion after Eq. (15.12)). Indeed, when we continue a given Greens function from Minkowski to Euclidean space[493], and use $\langle \phi(x_1)e^{-H(\tau_1-\tau_2)}\phi(x_2)e^{-H(\tau_2-\tau_3)}\ldots\phi(x_n)\rangle$ with the Hamiltonian H bounded from below, the above expression converges only for $\tau_i > \tau_{i+1}$. An explicit computation of the function $\langle :e^{i\alpha\varphi(z)}::e^{-i\alpha\varphi(\omega)}:\rangle$, with $\varphi(z) = \sum \frac{a_n}{n} z^{-n}$, $a_{-n} = a_n^\dagger$, and the Wick product defined as in chapter 2, shows that the sum is well defined only for $|z| > |\omega|$. Indeed we find $e^{-\alpha^2 \sum \frac{1}{n}(\frac{\omega}{z})^n} = e^{-\alpha^2 \ln(1-\frac{\omega}{z})}$. This is clearly valid only for $|z| > |\omega|$.

Using the above prescription one obtains for an infinitesimal transformation of $\phi(z)$, generated by Q_ϵ,

$$\delta\phi(z) = [Q_\epsilon, \phi(z)] = \frac{1}{2i\pi}\oint T(\xi)\phi(z)\epsilon(\xi)d\xi \quad , \tag{15.31}$$

where ξ is integrated around z.

The energy momentum tensor has the Laurent expansion (15.29) which may be inverted to give

$$L_n = \frac{1}{2i\pi}\oint T(z)z^{n+1}dz \quad . \tag{15.32}$$

We now look for the commutator algebra satisfied by the operators L_n. This algebra is obtained by considering the operator product expansion of the energy momentum tensor. In order to obtain this expansion, we first discuss the transformation properties of $T(z)$ under the conformal group. For an object of dimension γ, one usually has (see Eq. (15.20))

$$\delta_f J = f\frac{dJ}{dz} + \gamma f' J \quad . \tag{15.33}$$

Since T is conserved, there is no room for an anomalous dimension[491]. Hence it has dimension $\gamma = 2$. Allowing for an anomaly compatible with dimension two (the well known Schwinger term[11]), we write

$$\delta_f T(z) = \frac{c}{12}f'''(z) + f(z)\frac{dT(z)}{dz} + 2\frac{df(z)}{dz}T(z) \quad . \tag{15.34}$$

This implies that the transformation law for the energy momentum tensor for a finite transformation is given by the expression

$$T(z) = \left(\frac{d\xi}{dz}\right)^2 T(\xi(z)) + \frac{c}{12}\{\xi, z\} \quad , \tag{15.35}$$

where[5])

$$\{\xi, z\} = \frac{d^3\xi/dz^3}{d\xi/dz} - \frac{3}{2}\left(\frac{d^2\xi/dz^2}{d\xi/dz}\right)^2 \quad . \tag{15.36}$$

is the Schwarz derivative of ξ with respect to z.

In particular, for an inversion, $\xi = \frac{1}{z}$, we obtain the behaviour of the energy momentum tensor for large argument, $T(z) \sim \frac{1}{z^4}$. Comparing with the expression (15.31) with $\phi(z) = T(z)$, we obtain

$$T(z)T(w) = \frac{\frac{c}{2}}{(z-w)^4} + \frac{T(z)+T(w)}{(z-w)^2} \quad . \tag{15.37}$$

In particular, the transformation law (15.35) implies that T is not a primary field in the sense of (15.12). However, it is *quasiprimary* in the sense that (15.12) is obeyed by $T(z)$ if we restrict ourselves to Möbius transformations, for which the inhomogeneous term in (15.35) can be shown to vanish.

The constant c is the so called central charge. It plays a fundamental role in the Virasoro algebra implied by (15.37),

$$[L_n, L_m] = (n-m)L_{n+m} + \frac{c}{12}n(n^2-1)\delta_{n,-m} \quad , \tag{15.38}$$

where use has been made of the representation (15.32).

Realization of the conformal group

Our aim is to construct a theory which is invariant under the conformal transformations. It will be based upon scalar fields $\phi_{\Delta,\overline{\Delta}}(z,\overline{z})$, defined by the anomalous dimensions Δ and $\overline{\Delta}$. At this point some remarks are in order. Local fields transforming under the conformal group have an anomalous dimension γ, defined via dilatations by

$$A(x) \to A(\lambda x) = \lambda^{-\gamma}A(x) \quad , \tag{15.39}$$

[5]) This is obtained from the fact that invariance under finite transformations is a consequence of the closure of the algebra; thus one considers a transformation $T \to T^{(1)} \to T^{(2)}$, where $z \to \xi \to \omega$, which satisfies

$$T^{(2)}(z) = \left(\frac{d\xi}{dz}\right)^2 T^{(1)}(\xi) + \frac{c}{12}\{\xi,z\} = \left(\frac{d\xi}{dz}\right)^2\left\{(\frac{d\omega}{d\xi})^2 T(\omega) + \frac{c}{12}\{\omega,\xi\}\right\} + \frac{c}{12}\{\xi,z\}$$

$$= \left(\frac{d\omega}{dz}\right)^2 T(\omega) + \frac{c}{12}\{\omega,z\} \quad ,$$

from which one obtains[494] $\{\omega,z\} = (\frac{d\xi}{dz})^2\{\omega,\xi\} + \{\xi,z\}$, which defines (15.36).

with $\gamma \geq 0$. In terms of Δ and $\overline{\Delta}$ the dimension γ and the spin s are given by $\gamma = \Delta + \overline{\Delta}$, and $s = \Delta - \overline{\Delta}$. [6)]

We suppose there is a complete set of fields A_i spanning the representation of the conformal algebra. By "complete" we mean that any state may be obtained by application of a suitable linear combination of those fields on the vacuum. This implies also that a product of two of them may be written as a linear combination with finite, single valued c-number coefficients $C_{ij}^k(\xi)$ as[79]

$$A_i(z)A_j(0) = \sum C_{ij}^k(z)A_k(0) \tag{15.40}$$

(all fields actually depend on z and $\overline{z}$, but because of conformal invariance, it is sufficient to consider holomorphic fields; full results may be obtained at the end of the calculation by forming products of holomorphic and antiholomorphic contributions). Conformal invariance imposes severe constraints on the coefficients $C_{ij}^k(z)$, as we see later. We shall now restrict ourselves to an important subset of fields to which we have already alluded, having particularly simple transformation laws, (see Eq. (15.12)).

We will consider the transformation of conformal fields under the substitution

$$z \to \zeta(z) = z + \epsilon(z)$$

For purely holomorphic fields we have

$$\phi(z) = \left(\frac{d\zeta}{dz}\right)^{\Delta} \phi(\zeta) \quad , \tag{15.41}$$

where Δ is a non-negative parameter. Such a field will be called primary, and will play a central role in the sequel.

Let us discuss the Ward identities following from conformal invariance. For a general quantum field theory we have the transformation law (15.31) for $A(w)$. For a local field A, its variation must be a linear combination of a finite number of derivatives of ϵ, so that we write

$$\delta_\epsilon A_j(z) = \sum_{k=0}^{\nu_j} B_j^{(k-1)}(z)\left(\frac{d}{dz}\right)^k \epsilon(z) \quad , \tag{15.42}$$

where $B_j^{(k-1)}(z)$ are local fields in the set $\{A_j\}$, and ν_j are integers. The first two coefficients are fixed by the translation and dilatation properties of the fields, such that

$$B_j^{-1}(z) = \frac{\partial}{\partial z}A_j(z) \quad , \tag{15.43a}$$

6) γ is the mass dimension of the theory. The above value for the spin arises from the Lorentz transformations properties of a spin-s field, that is $\phi(e^{\lambda}z, e^{-\lambda}\overline{z}) = e^{\lambda s}\phi(z,\overline{z})$.

and

$$B_j^0(z) = \Delta_j A_j(z) \quad . \tag{15.43b}$$

The dimension of the field B_j^{k-1} can be read from expression (15.42), to be

$$\Delta_j^{(k-1)} = \Delta_j + 1 - k, \quad k = 0, 1, ..., \nu_j \quad . \tag{15.44}$$

Thus, the set of fields B_j^k have dimensions which are equally spaced by integers, $\Delta_n^k = \Delta_n - k$. It is also clear that for a primary field [7]

$$\delta_\epsilon \phi(z) = \epsilon(z) \frac{\partial \phi(z)}{\partial z} + \Delta \epsilon'(z) \phi(z) \quad . \tag{15.45}$$

As a result, the operator product $T(z)\phi(w)$ is given by the following series involving only non-negative values of k

$$T(z)\phi(w) = \sum_{k=0}^{\infty} (z-w)^{-2+k} \phi^{(-k)}(w) \quad . \tag{15.46}$$

Comparison of (15.46) with (15.45) shows, using (15.31), that

$$\phi^{(0)} = \Delta \phi \quad , \quad \phi^{(-1)} = \frac{\partial \phi}{\partial z} \quad . \tag{15.47}$$

The remaining fields $\phi^{(-k)}, k = 2, 3, ...$ are local fields generated by the application of L_{-n} on ϕ, and are called the "descendants" of the "ancestor" ϕ, or simply secondary fields

$$L_{-n}\phi_a(z) = \phi_a^{(-n)}(z) \quad . \tag{15.48}$$

We also define

$$\phi_a^{(-n_1, ..., -n_A)}(z) = L_{-n_1}(z) ... L_{-n_A}(z) \phi_a(z) \quad , \tag{15.49}$$

where

$$L_{-n}(z) = \oint d\xi \frac{T(\xi)}{(\xi - z)^{n+1}} \tag{15.50}$$

with the inverse transformation $T(z) = \sum \frac{L_n(w)}{(z-w)^{n+2}}$.

---

[7] The above equation implies that there is no operator $B^{(k)}$ with dimension smaller than Δ, (in analogy with the lower bound of the magnetic quantum number). Thus Δ plays the role of the highest root in representation theory. Quasi primary field are defined in terms of the representative of the highest root, which is the primary field.

It is clear that the fields (15.49) are in general not primary since the application of L_n, with $n > 0$, does not annihilate them; they are eigenstates of L_0 with eigenvalue

$$\Delta_a^{n_1 \cdots n_A} = \Delta_a + n_1 + \cdots + n_A \quad . \tag{15.51}$$

We call $\Delta^{n_1 \cdots n_A} - \Delta_a$ the *level* of a given state.

We are able to learn more about the operator product expansion of a local field with the energy momentum tensor. In view of (15.31) and (15.42) we have

$$\langle 0|T(z)A_{j_1}(z_1)...A_{j_n}(z_n)|0\rangle = \sum_{l-1}^{n}\sum_{k=0}^{\nu_l} k!(z-z_l)^{k-1}\times$$

$$\langle 0|A_{j_1}(z_1)...A_{j_{l-1}}(z_{l-1})B_{j_l}^{(k-1)}(z_l)A_{j_{l+1}}(z_{l+1})...A_{j_n}(z_n)|0\rangle \quad . \tag{15.52}$$

Going back to the primary fields, we shall use them in order to build a representation of the conformal group. This representation is built in terms of the ancestor together with its descendants, a construction similar to the one well known in group theory, where the representation is constructed by starting with a state with the highest weight[495]. Consider the state vector[8)]

$$|a\rangle = \phi_a(0)|0\rangle \quad , \tag{15.53}$$

where the vacuum $|0\rangle$ is defined by

$$L_n|0\rangle = 0 \qquad n \geq -1 \quad , \tag{15.54}$$

which means that the vacuum is conformally invariant. From (15.31), (15.45) and (15.46), we easily obtain

$$[L_n, \phi_a(z)] = z^{n+1}\frac{\partial}{\partial z}\phi_a(z) + \Delta_a(n+1)z^n\phi_a(z) \quad . \tag{15.55}$$

Thus we have, for the primary states $|a\rangle$,

$$L_n|a\rangle = 0 \quad \text{for} \quad n > 0 \tag{15.56a}$$

$$L_0|a\rangle = \Delta_a|a\rangle \quad . \tag{15.56b}$$

The above set of states define a huge Hilbert space. Our next aim is to restrict this space to a "physical" one, a task which will be achieved by imposing constraints, as we shall now see.

---

8) Recall that $z = 0$ is related to the fact that $|a >$ is an *in* state, at $t = -\infty$ – see comment after Eq. (15.29).

The Kac determinant and physical Hilbert space

The classification of the conformal representations is done by imposing constraints on the theory in the form of relations which some of the operators in (15.48) must obey.

The constraints involve only states of a given level. At level one, there is only one operator, namely L_{-1}, and imposing the condition $L_{-1}\phi = 0$ gives as a trivial result a constant field. At level 2 there are two operators, L_{-2} and L_{-1}^2, and we require

$$(L_{-2} + aL_{-1}^2)\phi = 0 \quad . \tag{15.57}$$

This condition must be invariant under applications of $L_n, n > 0$; this determines a and the dimension of ϕ. Let us work out this case in detail; we have

$$L_1(L_{-2} + aL_{-1}^2)\phi = (3 + 2a + 4a\Delta)L_{-1}\phi = 0 \tag{15.58}$$

and

$$L_2(L_{-2} + aL_{-1}^2)\phi = (4\Delta + \frac{c}{2} + 6a\Delta)\phi = 0 \quad , \tag{15.59}$$

where equation (15.38) has been used; for $n \geq 3$, L_n annihilates the state on the left hand side of (15.57) identically. Equation (15.58) implies

$$a = \frac{-3}{2(2\Delta + 1)} \tag{15.60a}$$

which can be used to calculate c from (15.59):

$$c = \frac{2\Delta(5 - 8\Delta)}{(2\Delta + 1)} \, , \quad \text{or} \quad \Delta = -\frac{c-5}{16} \pm \frac{1}{16}\sqrt{(c-1)(c-25)} \quad . \tag{15.60b}$$

This is a special example out of a series of very profound results obtained by Kac[495]. States such as ϕ above constitute the so called *Verma module*[496]. The condition for obtaining a well defined finite representation fixes the conformal dimension Δ to take the following values, labeled by two integers, p and q,

$$\Delta_{p,q} = \frac{c-1}{24} + \left(\frac{1}{2}p\alpha_+ + \frac{1}{2}q\alpha_-\right)^2 \quad , \tag{15.61}$$

where

$$\alpha_\pm = \frac{\sqrt{-c+1} \pm \sqrt{(-c+25)}}{\sqrt{24}} \quad . \tag{15.62}$$

The equations above show the first steps towards defining the so called degenerate representation of the conformal family. In general there is degeneracy at a given level, and one looks for the ***singular vectors***, which are linear combinations of states of a given level, perpendicular to any given state.

The argument leading to the above formulae is sketched below (for details one should refer to the mathematical literature[495]).

Physical Hilbert space

A family of states may be constructed acting with a tower of L_{-n}'s on a given state labelled by Δ, the eigenvalue of L_0[79]:

$$|\Delta; n_1, \cdots, n_l\rangle = L_{-1}^{n_1} \cdots L_{-l}^{n_l} |\Delta\rangle \tag{15.63}$$

$$L_0 |\Delta\rangle = \Delta |\Delta\rangle \quad . \tag{15.64}$$

If $\sum_{i=1}^{l} i n_i = N$ we have, using (15.38)

$$L_0 |\Delta; n_1, \cdots n_l\rangle = (\Delta + N) |\Delta; n_1, \cdots, n_l\rangle \quad . \tag{15.65}$$

We call (15.63) a state of level N. There are $P(N)$ = (number of partitions of N into a sum of positive integers) of such states. The states defined in (15.63) are not always linearly independent, for general values of Δ and c. In fact, the cases where there exist linear relations among them are the ones of physical interest, the reason being that states with c and Δ arbitrary contain states of negative norm. Moreover, we are interested in states belonging to realizations containing a finite number of fields.

We also require that physical states be annihilated by L_n , $n > 0$

$$L_n |\text{phys}\rangle = 0 \quad , \quad n > 0 \quad . \tag{15.66}$$

The procedure to be followed, depends now on the type of theory we are going to describe, and is twofold. In the case of string theories, the states are obtained by applying the creation operators from the mode expansion of the position operator $X_\mu(\tau, \sigma)$ (see (15.244) later) on the vacuum, and (15.66) must be obeyed[18,91]. The full Hilbert space will not be positive definite; negative norm states are discarded imposing a transversality condition (if the central charge assumes the critical value). We shall not use this procedure, but rather build new states as in (15.63), starting from a given highest weight state. The procedure now is to obtain linear combinations of states of the same level N. We define a null vector $|\chi\rangle$ as being a linear combination of states of the same level. As any other state, it must be stable under the operations of L_n and L_0. Since it is of a given order N, it satisfies

$$L_n |\chi\rangle = 0 \, , \, n > 0 \tag{15.67a}$$

$$L_0 |\chi\rangle = (\Delta + N) |\chi\rangle \quad . \tag{15.67b}$$

The first non trivial example is given in (15.57) (in that case $N = 2$). Moreover, as in that case, we must have definite values for the central charge c and anomalous dimension Δ. The procedure in the general case, is to compute the Kac determinant[495], formed by matrix elements which are scalar products of the states. In the case $N = 2$ we have

$$D_k^{(2)} = \begin{vmatrix} \langle\Delta|L_2L_{-2}|\Delta\rangle & \langle\Delta|L_1^2L_{-2}|\Delta\rangle \\ \langle\Delta|L_2L_{-1}^2|\Delta\rangle & \langle\Delta|L_1^2L_{-1}^2|\Delta\rangle \end{vmatrix} = \begin{vmatrix} 4\Delta + \frac{c}{2} & 6\Delta \\ 6\Delta & 4\Delta(1+2\Delta) \end{vmatrix} \quad . \tag{15.68}$$

This determinant vanishes when $\Delta = 0$, and when (15.60a,b) hold. For a general level, the Kac determinant is given by [495,87]

$$\det D_\alpha^{(N)} = \prod_{j=1}^{N} [\psi_j(c,\Delta)]^{P(N-j)} \quad , \tag{15.69}$$

where

$$\psi_j(c,\Delta) = \prod_{pq=j} (\Delta - \Delta_{p,q}^{(c)}) \quad , \tag{15.70}$$

with $\Delta_{p,q}^{(c)}$ given by (15.61).

The central charge in unitary representations

The determinant has no zeros for $c > 1$, and $\Delta > 0$. If the determinant vanishes, there exist linear combinations of states at a given level which are zero, or a null state. We shall see that we obtain linear differential equations for the correlators as a consequence of this property.

In order to find out the degeneracies, we look at the Kac determinant (15.69), built out of all matrix elements obtained from Eq. (15.63). The zeros of this determinant determine the values of the conformal dimensions in finite dimensional representations of the conformal group, as given in Eq. (15.61).

A very important property of quantum field theory is unitarity, since it allows the usual probabilistic interpretation of quantum mechanics. Although it is not mandatory that statistical mechanical systems obey such a requirement, there is an analogous property, which is reflexion positivity[497]. Such unitary theories represent a rather large class of models, as for example those with a hermitian transfer matrix. A detailed study of the Kac determinant (15.69) done by[497,498] Friedan, Qiu and Shenker reveals that unitarity imposes stringent restrictions. By looking for the zeros of the Kac determinant in the plane Δ vs. c, they studied the positions of the zero norm states, which separated regions of positive and negative norm states, and proved that in order to have unitarity, for $c < 1$, we should have

$$c = 1 - \frac{6}{n(n+1)} \quad , \tag{15.71}$$

with $n = 3, 4, 5\ldots$

In this case, we have for the conformal dimension the expression

$$\Delta_{p,q} = \frac{(p(n+1)-qn)^2 - 1}{4n(n+1)} \quad , \tag{15.72}$$

where $1 \le q \le p \le n-1$. The primary field having dimension $\Delta_{p,q}$ is denoted by $\psi_{(p,q)}$. In ref. [79] the general Kac's result (15.72) was studied in great detail.

The central charge, in general

In the general case one obtains, setting the null fields equal to zero, the corresponding (infinite dimensional) degenerate representation. Inspection of (15.61) shows that for a suitable choice of $\alpha_\pm$ one arrives at a finite dimensional degenerate representation of the conformal group. This is achieved for the choice

$$\frac{\alpha_-}{\alpha_+} = -\frac{m}{n} \quad , \tag{15.73a}$$

with the corresponding solutions $\frac{m-1}{n+1} = -\frac{p}{q}$ or $\frac{m+1}{n-1} = -\frac{p}{q}$. For this case, we solve (15.62) and (15.73*a*), obtaining for c

$$c = 1 - 6\frac{(p-q)^2}{pq} \quad ; \tag{15.73b}$$

this result reproduces (15.71) if $q = p+1$, which is the unitary case. The non unitary case is however also very important in statistical mechanics[80], and in two dimensional gravity[499], which need not be unitary.

The primary fields $\psi_{p,q}$ are finitely many in this case, providing a realization of the conformal algebra in terms of a finite number of fields. This is the key ingredient to identify, in this case, statistical models of physical interest at criticality, using these ideas.

Constraints and correlators

When computing correlators containing null-fields, such as defined in (15.57), one is led to differential equations, which must be obeyed by the correlators. Consider the expression

$$\langle L_{-2}\phi(z)\phi_1(z_1)\cdots\phi_n(z_n)\rangle = \frac{1}{2\pi i}\oint \frac{dw}{w-z}\langle T(w)\phi_1(z_1)\cdots\phi_n(z_n)\rangle \quad . \tag{15.74}$$

where the contour encircles only the point z, and where ϕ_i are primary fields. We can deform it in such a way as to encircle the points $z_1 \cdots z_n$ instead. Using the operator product expansion for the product of ϕ with the energy momentum tensor we obtain, after using (15.7)

$$\langle L_{-2}\phi(z)\phi_1(z_1)\dots\phi_n(z_n)\rangle = \sum_a \{\frac{\Delta_a}{(z_a-z)^2} + \frac{1}{z-z_a}\partial_a\}\langle\phi(z)\phi_1(z_1)\dots\phi_n(z_n)\rangle \quad . \tag{15.75}$$

For the level 2 constraint (15.57) we thus find

$$\frac{3}{2(2\Delta+1)}\frac{\partial^2}{\partial z^2}\langle\phi(z)\phi_1(z_1)\dots\phi_n(z_n)\rangle = \sum_a \{\frac{\Delta_a}{(z_a-z)^2} + \frac{1}{z-z_a}\partial_a\}\langle\phi(z)\phi_1(z_1)\dots\phi_n(z_n)\rangle \quad , \tag{15.76}$$

where we specialized to $\phi(z) = \psi_{(1,2)}(z)$ (see definition following (15.72)); it has dimension $\Delta_{(1,2)}$ (see (15.61)), and obeys (15.57).

The above second order differential equations can be used to determine completely the four point functions of the theory, in terms of the scale parameters Δ_a. Consider the most general conformally invariant four point function, which is given by

$$\langle\phi_0(z_0)\phi_1(z_1)\phi_2(z_2)\phi_3(z_3)\rangle = \prod(z_0-z_i)^{-(\Delta+\Delta_i-\frac{1}{3}\sum_{j=0}^{3}\Delta_j)}(z_i-z_j)^{-(\Delta_i+\Delta_j-\frac{1}{3}\sum_{k=0}^{3}\Delta_k)}\mathcal{F}(z) \quad , \tag{15.77}$$

where $z = \frac{(z_0-z_1)(z_2-z_3)}{(z_0-z_3)(z_2-z_1)}$. The function $\mathcal{F}(z)$ obeys a differential equation of the hypergeometric type. If we set $z_1 = 0$, $z_2 = 1$, $z_3 = \infty$, $z = z_0$, by using appropriate Möbius transformations, we obtain the following differential equation for $\hat{\mathcal{F}}(z) = z^\alpha(1-z)^{\alpha'}\mathcal{F}(z)$:

$$\hat{\mathcal{F}}''(z) + \left\{\frac{A}{z} + \frac{B}{z-1}\right\}\hat{\mathcal{F}}'(z) + \left\{\frac{C}{z^2} + \frac{D}{(z-1)^2} + \frac{E}{z(z-1)}\right\}\hat{\mathcal{F}}(z) = 0 \quad , \tag{15.78}$$

where

$$A = \frac{2}{3}(2\Delta+1) - 2\tilde{\Delta}_{0,1} - 2\alpha \tag{15.79a}$$

$$B = \frac{2}{3}(2\Delta+1) - 2\tilde{\Delta}_{0,2} - 2\alpha' \tag{15.79b}$$

$$C = \alpha^2 + \alpha + 2\alpha\tilde{\Delta}_{0,1} + \tilde{\Delta}_{0,1}\left(\tilde{\Delta}_{0,1}+1\right) - \frac{2}{3}(\alpha+\Delta_1)(2\Delta+1) - \frac{2}{3}\Delta_{0,1}(2\Delta+1) \tag{15.79c}$$

$$D = \alpha'^2 + \alpha' + 2\alpha'\tilde{\Delta}_{0,2} + \tilde{\Delta}_{0,2}(\tilde{\Delta}_{0,2}+1) - \frac{2}{3}(\alpha' + \Delta_2)(2\Delta+1)$$
$$- \frac{2}{3}\Delta_{0,2}(2\Delta+1) \tag{15.79d}$$

$$E = 2\alpha\alpha' + 2\alpha\tilde{\Delta}_{0,2} + 2\alpha'\tilde{\Delta}_{0,1} + 2\tilde{\Delta}_{0,2}\tilde{\Delta}_{0,1} - \frac{2}{3}(\alpha+\alpha')(2\Delta+1) \tag{15.79e}$$

$$\tilde{\Delta}_{0,2} = \frac{2}{3}(\Delta_0 + \Delta_2) - \frac{1}{3}(\Delta_1 + \Delta_3) \tag{15.79f}$$

$$\tilde{\Delta}_{0,1} = \frac{2}{3}(\Delta_0 + \Delta_1) - \frac{1}{3}(\Delta_2 + \Delta_3) \quad . \tag{15.79g}$$

If we choose α, and α' in such way that $C = D = 0$ we have a hypergeometric wave equation; in the case of fields of level 2, with $\Delta_1 = \Delta_2 = \Delta_3 = \Delta$ and $\tilde{\Delta}_{0,2} = \tilde{\Delta}_{0,1} = \frac{2}{3}\Delta$ one has $\alpha = \alpha' = \frac{4}{3}\Delta$ and

$$\hat{\mathcal{F}}(z) = F(z;\alpha,\beta,\gamma) \quad . \tag{15.80}$$

At level 3, we obtain third order differential equations. In the BPZ's original paper[79] the differential equation obeyed by the four point function involving $\psi_{1,3}$ was derived (see comment after (15.72). From the completeness of the OPE[79] (15.40) we obtain for $z_1 \approx z$,

$$\phi(z)\phi_1(z_1) = C(z-z_1)^{\lambda}\phi_{\Delta'}(z)\cdots \quad , \tag{15.81}$$

where $\lambda = \Delta' - \Delta_1 - \Delta$ and Δ' is the scale dimension of the primary field $\phi_{\Delta'}(z)$. Consider now the most divergent term as $z_1 \to z$, in (15.76); requiring its coefficient to vanish, we obtain

$$\frac{3}{2(2\Delta+1)}\lambda(\lambda-1) - \Delta + \lambda = 0 \quad . \tag{15.82}$$

The above equation determines the critical behavior of the new field obtained by fusing ϕ and ϕ_1 . The so called fusion rules are fundamental in the study of conformal field theory. Their knowledge permits to obtain all relevant information of the theory. Ref. [79] contains tables for the fusion of some simple representations. The procedure permits the identification of the critical behavior of several statistical mechanics models. Procedures to obtain fusion rules are discussed in the literature[500].

15.3 Operator Realization of the Conformal Algebra

Let us discuss now the issue of a realization of a conformal algebra in terms of field operators[99]. We discuss the role of exponential fields in conformal field theories, and show that they are natural objects to be considered in this case. They have already appeared in the case of the massless Thirring model, which is conformally invariant. Conformal invariance in Minkowski space requires special care[485,81].

We consider the exponential fields (see 2.13,60)

$$V_{\Delta\overline{\Delta}}(z,\overline{z}) = C : e^{\left(i\alpha\varphi(z)+i\overline{\alpha}\varphi(\overline{z})\right)} : \quad . \tag{15.83}$$

We can separate the right and left movers as

$$V_\alpha(z,\overline{z}) = V_\alpha(z)V_{\overline{\alpha}}(\overline{z}) \quad . \tag{15.84}$$

The fields φ and $\overline{\varphi}$ are the (Euclidean) right and left moving components of a free massless bosonic field

$$\phi(z,\overline{z}) = \varphi(z) + \overline{\varphi}(\overline{z}) \tag{15.85}$$

with an action

$$S = \frac{1}{32\pi}\int dz d\overline{z} \partial_z\phi\partial_{\overline{z}}\phi = \frac{1}{16\pi}\int d^2x(\partial_\mu\phi)^2 \quad . \tag{15.86}$$

Because of the selection rules discussed in chapter 2, we have

$$\langle V_\alpha(z)V_\beta(z')\rangle = (z-z')^{-2\alpha^2}\delta_{\alpha,-\beta} \quad . \tag{15.87}$$

This is a conformally invariant two point correlator. Further correlators can be computed analogously, and we obtain, using (15.83,84,85), and (2.14,18)

$$\langle V_{\alpha_1}(z_1)V_{\alpha_2}(z_2)V_{\alpha_3}(z_3)V_{\alpha_4}(z_4)\rangle = \prod_{i>j}(z_i - z_j)^{2\alpha_i\alpha_j} \quad , \tag{15.88}$$

with $\sum \alpha_i = 0$, which is nothing but the selection rule (2.19).

The conformal dimension of $V_\alpha(z)$ follows immediately from (15.87) above,

$$\Delta_\alpha = \alpha^2 \quad . \tag{15.89}$$

The operator product expansion of the (holomorphic part of the) energy momentum tensor $T(z) = \frac{-1}{4}(\partial\phi)^2$ with the exponential field is given by the expression

$$T(z)V_\alpha(\omega) = \frac{\Delta_\alpha}{(z-w)^2}V_\alpha(\omega) + \frac{1}{z-w}\partial_w V_\alpha(\omega) + \text{reg. terms} \quad , \tag{15.90}$$

where the first term corresponds to the double contraction of $(\partial\phi)^2$ with $e^{i\alpha\phi}$, the coefficient being α^2. The second term corresponds to a single contraction, the uncontracted field giving rise to the derivative. The central charge has the special value $c = 1$, as is seen by computing the two point function of the energy momentum tensor,

$$\langle T(z)T(w)\rangle = \frac{1}{2(z-w)^4} \quad , \tag{15.91}$$

and comparing with Eq. (15.37).

The above example is particularly simple. It is however possible to modify the usual prescription and define a theory with $c \neq 1$[99]. Although the procedure is rather formal, it is useful for arriving at the results.

The idea is to introduce background charges at infinity[99]. Consider as an example the correlator defined by

$$\langle V_{\alpha_1}(z_1)\dots V_{\alpha_n}(z_n)\rangle_{\alpha_o} = \lim_{R\to\infty} R^{4\alpha_o{}^2}\langle V_{\alpha_1}(z_1)\dots V_{\alpha_n}(z_n)V_{-2\alpha_o}(R)\rangle_o \tag{15.92}$$

where the right hand side is computed in the usual way. The "zero charge" selection rule $\sum\alpha_i = 0$ in (15.88) is now replaced by

$$\sum \alpha_i = 2\alpha_o \quad . \tag{15.93}$$

In particular, for $n = 2$ we have

$$\langle V_\alpha(z)V_{2\alpha_o-\alpha}(z')\rangle_{\alpha_o} = (z-z')^{-2\alpha(\alpha-2\alpha_o)} \quad . \tag{15.94}$$

But we know that the 2-point function of a conformal theory is non vanishing only if the conformal dimensions of the fields are equal[484,485,490]. Thus we should have

$$\Delta_\alpha = \Delta_{2\alpha_o-\alpha} = \alpha(\alpha-2\alpha_o) \quad . \tag{15.95}$$

We see that expression (15.94) indeed reflects this property, being symmetric under the exchange $\alpha \leftrightarrow 2\alpha_o - \alpha$.

In order to obtain such an anomalous dimension, we have to modify the energy momentum tensor in an appropriate way. One notices that the boundary conditions for the fields at infinity are quite different now. They may be incorporated in the theory, by allowing for total derivatives in the energy momentum tensor. We thus make the Ansatz

$$T(z) = \frac{-1}{4}(\partial_z\phi)^2 + A\partial_z{}^2\phi \quad , \tag{15.96}$$

where the constant A is determined below. Notice that the term we added is a total derivative. Therefore we are allowing for nontrivial boundary conditions, if the new term is to be relevant at all. Consider the operator product expansion for $T(z)V_\alpha(\omega)$:

$$\begin{aligned} T(z)V_\alpha(\omega) =&: \left(\frac{-1}{4}(\partial_z\phi)^2 + A\partial_z{}^2\phi\right) : :e^{i\alpha\phi(w)}: = \\ =& \frac{\alpha^2 + 2i\alpha A}{(z-w)^2} :e^{i\alpha\phi(w)}: + \frac{i\alpha}{z-w} :\partial_z\phi(z)e^{i\alpha\phi(w)}: + \\ &:\left(-\frac{1}{4}(\partial_z\phi)^2 + A\partial_z{}^2\phi\right) e^{i\alpha\phi(w)}: \quad , \end{aligned} \tag{15.97}$$

which written in the form of equation (15.90) is equal to

$$T(z)V_\alpha(\omega) = \frac{\alpha^2 + 2i\alpha A}{(z-w)^2} V_\alpha(\omega) + \frac{1}{z-w}\partial_w V_\alpha(\omega) \quad . \tag{15.98}$$

Comparing with (15.45-47) one obtains from above the dimension of $V_\alpha(\omega)$,

$$\Delta_\alpha = \alpha^2 + 2i\alpha A \quad . \tag{15.99}$$

Thus, to match (15.99) with (15.95) we need

$$A = i\alpha_o \quad . \tag{15.100}$$

Correspondingly the transformation properties of $\phi(z)$ under a conformal transformation

$$z \to f(z) = z + \epsilon(z) \quad . \tag{15.101}$$

also change. The field $V_\alpha(z)$ has conformal dimension given in (15.95), that is, under the transformation (15.101), V_α transforms as

$$V_\alpha(z) \to (f'(z))^{\alpha^2 - 2\alpha\alpha_o} V_\alpha(f(z)) = (f')^{\alpha^2} :e^{\{i\alpha(\phi(f(z)) + i\alpha_o \ln f'(z))\}}: \quad . \tag{15.102}$$

The quantity inside the parentheses in Eq. (15.102) gives the transformation law of $\phi(z)$:

$$\phi(z) \to \phi(f(z)) + i\alpha_o \ln f'(z) \quad . \tag{15.103}$$

We can also calculate the central charge of the Virasoro algebra by calculating the energy momentum two point function

$$\langle T(z)T(w)\rangle = \frac{\frac{c}{2}}{(z-w)^2} \quad . \tag{15.104}$$

Using (15.96), with $A = i\alpha_o$ we find

$$\left\langle \left(\frac{-1}{4} :(\partial_z\phi)^2: + i\alpha_o \partial_z{}^2\phi(z) \right) \left(\frac{-1}{4} :(\partial_w\phi)^2: + i\alpha_o \partial_w{}^2\phi(w) \right) \right\rangle =$$

$$= \frac{1}{8}\left(\frac{2}{(z-w)^2} \right)^2 + 2\alpha_o{}^2 \partial_z^2 \partial_w^2 \ln(z-w) = \frac{\frac{1}{2}}{(z-w)^4}\{1 - 24\alpha_o{}^2\} \quad , \qquad (15.105)$$

where the first term corresponds to the double contraction of the terms $(\partial\phi)^2$, and the second, to the contraction of the terms $\partial^2\phi$.

Thus we get, for the central charge, the value

$$c = 1 - 24\alpha_o{}^2 \quad . \qquad (15.106)$$

Four point correlators for $c \neq 1$

With the above results, we are able to obtain integral representation for the various correlators. We analyse now the 4-point correlators $\langle V_{\alpha_1}(z_1)V_{\alpha_2}(z_2)V_{\alpha_3}(z_3)V_{\alpha_4}(z_4)\rangle$, subject to the condition

$$\sum_{i=1}^{4} \alpha_i = 2\alpha_o \quad . \qquad (15.107)$$

The conformal dimension of a multivalued function in a conformal field theory, cannot be fixed at will, for all fields. In the case of the four point correlator (15.88) we must have equality (at least two by two) of the conformal dimensions. In the case where the conformal dimensions are all equal, we have $\alpha_i = \alpha$, or $\alpha_i = 2\alpha_o - \alpha$. Thus we have the following possibilities

$$i)\ \alpha_1 = \alpha_2 = \alpha_3 = \alpha_4 = \alpha \qquad (15.108a)$$

$$ii)\ \alpha_1 = \alpha_2 = \alpha_3 = \alpha, \quad \alpha_4 = 2\alpha_o - \alpha \qquad (15.108b)$$

$$iii)\ \alpha_1 = \alpha_2 = \alpha, \quad \alpha_3 = \alpha_4 = 2\alpha_o - \alpha \quad . \qquad (15.108c)$$

None of them satisfies (15.107). The alternative found in [99] is to use conformally invariant operators not depending on space-time, but with non zero charge, in order that (15.107) be met. Such an operator is provided by

$$Q = \oint dz J(z) \quad , \qquad (15.109)$$

where, in order for Q to be conformally invariant $(\Delta_Q = 0)$, $J(z)$ must have conformal dimension one. We take

$$J_\pm(z) = V_{\alpha_\pm}(z) \quad , \qquad (15.110a)$$

where

$$\Delta_{\alpha_\pm} = {\alpha_\pm}^2 - 2\alpha_\pm\alpha_o = 1 \quad . \tag{15.110b}$$

Thus

$$\alpha_\pm = \alpha_o \pm \sqrt{1+{\alpha_o}^2} \quad . \tag{15.110c}$$

If we express α_o^2 in terms of c using equation (15.106), we immediately recognize in (15.110c) Kac's formula. The introduction of these charges can resolve the above inconsistency with (15.108) only if $\sum_{i=1}^{4} \alpha_i - 2\alpha_o > 0$. This eliminates possibility (15.108c). For (15.108a) we need to cancel

$$\sum \alpha_i - 2\alpha_o = 4\alpha - 2\alpha_o \quad . \tag{15.111}$$

Adding $(n_+ - 1)\,Q_+{}'s$ and $(n_- - 1)\,Q_-{}'s$ we have a quantization of $\alpha(n_\pm \geq 1)$:

$$\alpha = \frac{1}{4}(1-n_+)\alpha_+ + \frac{1}{4}(1-n_-)\alpha_- \quad . \tag{15.112}$$

For (15.108b) we have

$$\sum_{i=1}^{4} \alpha_i - 2\alpha_o = 2\alpha \quad , \tag{15.113}$$

and the quantization of α reads instead

$$\alpha = \frac{1}{2}(1-n_+)\alpha_+ + \frac{1}{2}(1-n_-)\alpha_- \quad . \tag{15.114}$$

Let us deal with this last case, following [99]. The conformal dimension obeys the Kac's formula (15.61), that is

$$\Delta_{mn} = {\alpha_{mn}}^2 - 2\alpha_o\alpha_{mn} = \frac{1}{4}[(n\alpha_+ - m\alpha_-)^2 - (\alpha_+ + \alpha_-)^2] \quad . \tag{15.115}$$

We now finally obtain an integral representation for the four point function, given by

$$< V_{m_1n_1}(z_1)V_{m_2n_2}(z_2)V_{m_3n_3}(z_3)V_{2\alpha_o-\alpha_{m_4n_4}}(z_4)\rangle = \oint dw_1 \cdots dw_M dv_1 \cdots dv_N \times$$

$$< J_+(w_1)\cdots J_+(w_M)J_-(v_1)\cdots J_-(v_N)V_{m_1n_1}(z_1)\cdots V_{m_4n_4}(z_4) >_{2\alpha_o} \quad , \tag{15.116a}$$

where

$$M = \frac{m_1+m_2+m_3-m_4}{2} - 1 \tag{15.116b}$$

$$N = \frac{n_1+n_2+n_3-n_4}{2} - 1 \quad . \tag{15.116c}$$

The correlators inside the integral are immediately computed, taking into account the factorization of the integrand in terms of two-point functions (see equation (2.67)), and using (15.87) and (15.94). Before performing the computation in full generality, we make a specific choice for three of the arguments in the correlator, using the projective invariance. We know that the four point function is given by the scaling factors (powers of $z_i - z_j$) times a function of $z = \frac{z_{12}z_{43}}{z_{13}z_{42}}$. Extracting these scaling factors, and choosing $z_1 = 0, z_2 = z, z_3 = 1$ and $z_4 = \infty$, we consider a simple example,

$$
\begin{aligned}
G_{12}^{mn}\ {}_{12}^{\bar{m}\bar{n}}(z) &= < V_{12}(z)V_{12}(1)V_{mn}(0)V_{2\alpha_o-\alpha_{mn}}(\infty) >^{n.s.} = \\
&\oint dw < V_{12}(z)V_{12}(1)V_{mn}(0)V_{2\alpha_o-\alpha_{mn}}(\infty)V_{\alpha_+}(w) >^{n.s.} \quad , \qquad (15.117)
\end{aligned}
$$

where the superscript $n.s.$ means omission of the scaling factors. Thus we obtain

$$
G(z) = \oint dw z^{\alpha_{12}\alpha_{mn}}(1-z)^{\alpha_{12}}w^{\alpha+\alpha_{mn}}(w-1)^{\alpha+\alpha_{12}}(w-z)^{\alpha+\alpha_{12}} \quad . \qquad (15.118)
$$

The integral

$$
I_C(a,b,c;z) = \oint_C dw w^a (w-1)^b (w-z)^c \qquad (15.119)
$$

is a hypergeometric function[501]. There exists an explicit dependence on the path of integration, and I_C vanishes if no singularity is included. The various possible integration paths correspond to different solutions of the hypergeometric equation. Two independent solutions are

$$
\begin{aligned}
I_1(a,b,c;z) &= \int_1 dw w^a (w-1)^b (w-z)^c = \\
&= \frac{\Gamma(-a-b-c-1)\Gamma(b+1)}{\Gamma(-a-c)}\mathcal{F}(-c,-a-b-c-1,-a-c;z) \qquad (15.120a)
\end{aligned}
$$

and

$$
\begin{aligned}
I_2(a,b,c;z) &= \int_0^z dw w^a(-w+1)^b(-w+z)^c = z^{1+a+c}\int_0^1 dw w^a (1-w)^c (1-zw)^b \\
&= z^{1+a+c}\frac{\Gamma(a+1)\Gamma(c+1)}{\Gamma(a+c+2)}\mathcal{F}(-b,a+1,a+c+2;z) \quad , \qquad (15.120b)
\end{aligned}
$$

where $\mathcal{F}(\alpha,\beta,\gamma;z)$ is the usual hypergeometric function, defined by

$$
\mathcal{F}(\alpha,\beta,\gamma;z) = 1 + \frac{\alpha\beta}{1!\gamma}z + \frac{\alpha(\alpha+1)\beta(\beta+1)}{2!\gamma(\gamma+1)}z^2 + \cdots \quad . \qquad (15.120c)
$$

Our problem has thus been reduced to the solution of the hypergeometric equation, the solution being given by (15.118).

The above procedure can be seen to be equivalent to the introduction of a gravitational interaction of the bosonic field ϕ, described by

$$S = \int d^2x \left\{ \frac{1}{16\pi} (\partial_\mu \phi)^2 + \frac{A}{4\pi} R\phi \right\} \quad . \tag{15.121}$$

Indeed, the energy momentum tensor obtained from the above action in the conformal gauge is simply given by (15.96), up to an overall constant.

15.4 Non Abelian Conformal Theory

15.4.1 WZNW theory revisited

The non abelian generalization of a conformally invariant field theory in two dimensions is realized by the principal σ model with a Wess-Zumino term[52], the so called Wess-Zumino-Novikov-Witten action[51,336,50]

$$S = \frac{1}{2\lambda^2} tr \int d^2\xi \partial^\mu g^{-1} \partial_\mu g + k S_{WZ}(g) \quad , \tag{15.122}$$

where

$$S_{WZ}(g) = \frac{1}{4\pi} \epsilon^{\mu\nu} \int_0^1 dr \int d^2\xi g^{-1} \dot{g} g^{-1} \partial_\mu g g^{-1} \partial_\nu g \quad , \tag{15.123}$$

which has already been discussed in chapter 9. We know also that the corresponding β function has a zero at the point $\lambda^2 = \frac{4\pi}{k}$.[51] We suppose that λ assumes the critical value, from now on.

15.4.2 Kac Moody algebra

The action (15.122) at the critical point is invariant under the transformation[502,503]

$$g(\xi) \to \Omega(z) g(\xi) \overline{\Omega}^{-1}(\bar{z}) \quad , \tag{15.124}$$

where $z = \xi^1 + i\xi^2$ and $\bar{z} = \xi^1 - i\xi^2$, and Ω is independent of $\overline{\Omega}$.

The Noether currents associated with these symmetries are

$$J = J^a \tau^a = -\frac{1}{2} k \partial_z g g^{-1} \tag{15.125a}$$

and

$$\overline{J} = \overline{J}^a \tau^a = -\frac{1}{2} k g^{-1} \partial_{\overline{z}} g \quad , \tag{15.125b}$$

where τ^a are the generators of the group. These currents satisfy (compare with (9.25))

$$\partial_{\overline{z}} J = 0 = \partial_z \overline{J} \quad , \tag{15.126}$$

as can be deduced from the equation of motion obtained from (15.122) with $\lambda^2 = \frac{4\pi}{k}$. Thus their dependence on the space time variable is that of a holomorphic, or antiholomorphic function, respectively. We can generate a loop algebra for the currents by developing $J(z)$ and $\overline{J}(\overline{z})$ in a Laurent series

$$J^a(z) = \sum_{n=-\infty}^{\infty} J_n^a z^{-n-1} \tag{15.127a}$$

$$\overline{J}^a(\overline{z}) = \sum_{n=-\infty}^{\infty} \overline{J}_n^a \overline{z}^{-n-1} \quad . \tag{15.127b}$$

The quantum theory associated with (15.122) at the critical point, may be analysed by various methods. Witten discussed its equivalence to a free theory, as well as the algebra obeyed by the currents (15.127). A detailed discussion of this point has been given in chapter 9 with the result (9.38,41), which in terms of the components J_n^a and $\overline{J}_n^a$, takes the form of two mutually commuting Kac-Moody algebras:

$$\left[J_n^a, J_m^b\right] = i f^{abc} J_{m+n}^c + \frac{1}{2} k n \delta^{ab} \delta_{n+m,0} \tag{15.128a}$$

$$\left[\overline{J}_n^a, \overline{J}_m^b\right] = i f^{abc} \overline{J}_{m+n}^c + \frac{1}{2} k n \delta^{ab} \delta_{n+m,0} \quad . \tag{15.128b}$$

Notice that under an infinitesimal transformation of the form $\Omega(z) = 1 + w^a(z)\tau^a$, the current $J(z)$ transforms like

$$\delta_\omega J(z) = \left[\omega(z), J(z)\right] + \frac{1}{2} k \omega'(z) \quad . \tag{15.129}$$

which in turn implies the commutation relations (15.128), for consistency. The energy momentum tensor is found to be of the Sugawara form (see Eq. (9.51)).

$$T(z) = \frac{1}{2\chi} : J^a(z) J^a(z) := \frac{1}{2\chi} \sum_{n,m=-\infty}^{\infty} J_{n-m}^a J_m^a z^{-n-2} = \sum_{n=-\infty}^{\infty} L_n z^{-n-2} \quad , \tag{15.130}$$

with $\chi = c_V + k$, and c_V the Casimir, defined by $f^{abc} f^{dbc} = c_V \delta^{ad}$.

The normalization of the energy momentum tensor above, has been chosen such that the L_n satisfy the usual Kac-Moody-Virasoro algebra (15.38). Proceeding as before, one finds

$$[L_n, J_m^a] = -mJ_{n+m}^a \quad . \tag{15.131}$$

as well as

$$[L_n, L_m] = (n-m)L_{n+m} + \frac{c}{12}n(n^2-1)\delta_{n+m,0} \quad . \tag{15.132}$$

The first term in (15.132) follows from the classical algebra. For the central term we have

$$c = \frac{k\dim G}{c_V + k} \tag{15.133}$$

We now construct a representation of the (non abelian) conformal algebra, by seeking primary fields ϕ_j transforming as[502,503]

$$\delta_\epsilon\phi_j(z) = \frac{1}{2\pi i}\oint T(\xi)\epsilon(\xi)\phi_j(z)d\xi \tag{15.134}$$

$$\delta_\omega\phi_j(z) = \frac{1}{2\pi i}\oint J^a(\xi)\omega^a(\xi)\phi_j(z)d\xi \quad , \tag{15.135}$$

under reparametrizations and local group transformation, respectively, with $\delta_\epsilon\phi_j$ given by (15.45), and $\delta_\omega\phi_j$ by

$$\delta_\omega\phi_j(z) = \tau^a\omega^a\phi_j(z) \tag{15.136a}$$

This implies the following operator product expansion[502,503]

$$J^a(\omega)\phi_j(z) = \frac{\tau^a}{\omega - z}\phi_j(z) + \text{regular terms} \tag{15.136b}$$

which one obtains by comparing (15.135) and (15.136a). Since ϕ_j is a primary field, we have also

$$T(\omega)\phi_j(z) = \frac{\Delta_j}{(\omega-z)^2}\phi_j(z) + \frac{1}{\omega - z}\frac{\partial}{\partial z}\phi_j(z) + \text{regular terms} \tag{15.137}$$

where Δ_j is the scale dimension of ϕ_j.

15.4.3 Ward identities and equations of motion

Ward identities may again be derived by applying the same techniques as already used in connection with the pure Virasoro algebra. As previously discussed, we have that correlators of the energy momentum tensor and primary fields may be written as a sum of terms, each corresponding to the expansion (15.137), that is

$$< T(z)\phi_1(z_1)\ldots\phi_n(z_n) >= \sum_{j=1}^{n}\Big\{\frac{\Delta_j}{(z-z_j)^2}+\frac{1}{z-z_j}\partial_j\Big\} < \phi_1(z_1)\ldots\phi_n(z_n) > \quad , \tag{15.138}$$

while for the current we obtain

$$< J^a(z)\phi_1(z_1)\ldots\phi_n(z_n) >= \sum_{j=1}^{n}\frac{\tau_j^a}{z-z_j} < \phi_1(z_1)\ldots\phi_n(z_n) > \quad . \tag{15.139}$$

Notice that regular terms do not contribute to the contour integral.

In order to obtain the behavior of $T(z)$ and $J^a(z)$ for large arguments, we consider the mapping $z \to 1/z$, using the relations

$$T\left(\frac{1}{z}\right) = \left[-\frac{1}{z^2}\right]^{-2} T(z) \tag{15.140a}$$

$$J^a\left(\frac{1}{z}\right) = \left[\frac{1}{z}\right]^{-2} J^a(z) \quad , \tag{15.141a}$$

which follow from the transformation properties of the conformal fields.[9] Thus we learn that for $z \to \infty$ the energy momentum tensor and current behave as $T(z) \sim 1/z^4$ and $J^a(z) \sim 1/z^2$ respectively. Therefore, multiplying (15.138,139) by z^{m+1} and integrating (15.138) for $m = -1, 0, 1$ and (15.139) for $m = -1$, we obtain, using the theorem of residues,

$$\sum_{j=1}^{\infty}\Big\{z_j^{m+1}\partial_j + (m+1)\Delta_j z_j^m\Big\} < \phi_1(z_1)\ldots\phi_n(z_n) > = 0 \quad , \tag{15.140b}$$

with $m = -1, 0, 1$, and

$$\sum_{j=1}^{\infty}\tau_j^a < \phi_1(z_1)\ldots\phi_n(z_n) > = 0 \quad , \tag{15.141b}$$

which express the invariance under the restricted conformal group (Möbius transformations) and the symmetry group G with generators τ_j^a, respectively.

---

9) Since the inversion is a particular Möbius transformation, no contribution of the central term arises.

Besides primary fields, we define further local fields descendents belonging to the same family, by applying L_{-n}'s and J^a_{-n}'s on the primary field

$$\phi^{\left(-n_1 \ldots -n_A,(-m_1,a_1)\ldots(-m_B,a_B)\right)} = L_{-n_1}\ldots L_{-n_A} J^{a_1}_{-m_1}\ldots J^{a_B}_{-m_B}\phi \quad . \tag{15.142}$$

It is not difficult to see that the conformal dimension of these fields is given by

$$\Delta_{(m,n)} = \Delta + \sum_{i=1}^{A} n_i + \sum_{j=1}^{B} m_j \quad . \tag{15.143}$$

The field equations of the WZW theory have the form

$$\kappa\frac{\partial}{\partial z}g(z,\overline{z}) = \; : J^a(z)\tau^a g(z,\overline{z}) : \tag{15.144a}$$

$$\kappa\frac{\partial}{\partial \overline{z}}g(z,\overline{z}) = \; : \overline{J}^a(\overline{z})\tau^a g(z,\overline{z}) : \quad . \tag{15.144b}$$

The parameter κ will be defined below. We need to first define the r.h.s. in the quantum theory. Let us consider the operator product expansion of the current with the field $g(z,\overline{z})$, which we suppose to be a primary field in the conformal theory. We have

$$J^a(\omega)\tau^a g(z,\overline{z}) = \frac{c_g}{\omega - z}g(z,\overline{z}) + \sum_{n=1}^{\infty}(\omega - z)^{n-1}\tau^a J^a_{-n}g(z,\overline{z}) \quad , \tag{15.145}$$

where c_g is defined by

$$\tau^a\tau^a = c_g \quad . \tag{15.146}$$

The first term in the sum ($n = 1$) in (15.145) must be the derivative, as given by (15.144a)

$$: J^a(\omega)\tau^a g(\overline{z},z) : \; = \lim_{\omega\to z}\left\{J^a(\omega) - \frac{\tau^a}{\omega - z}\right\}\tau^a g(\overline{z},z) = \kappa\frac{\partial}{\partial z}g(z,\overline{z}) \quad . \tag{15.147}$$

In terms of the OPE, we have

$$J^a(\omega)\tau^a g(\overline{z},z) = \frac{c_g}{\omega - z}g(\overline{z},z) + \kappa\frac{\partial}{\partial z}g(\overline{z},z) + \ldots \quad ; \tag{15.148}$$

taking into account the fact that $L_{-1} = \frac{\partial}{\partial z}$ we obtain from (15.144) the constraint

$$\chi \equiv \left(J^a_{-1}\tau^a - \kappa L_{-1}\right)g(\overline{z},z) = 0 \quad , \tag{15.149}$$

which in view of subsection (15.2.2) means that we have a null field, and the family derived from the primary field $g(\bar{z}, z)$ is degenerate. Notice that the WZW field $g(z, \bar{z})$ is the simplest possible nontrivial primary field and obeys a first order differential equation, in contrast with the abelian case, where the simplest non trivial condition on a degenerate family implies a second order differential equation. In the non abelian case, the existence of the current component J^a_{-1} makes the difference. In order that condition (15.149) be stable, it must be zero under application of L_n, $J^a_n, n > 0$, and must satisfy

$$L_0\chi = (\Delta + 1)\chi \tag{15.150a}$$
$$J^a_0\chi = \tau^a\chi \quad , \tag{15.150b}$$

since χ is of level 1. Thus we obtain

$$L_1\chi = (\tau^a\tau^a - 2\kappa\Delta)g = 0 \tag{15.151a}$$
$$J^b_1\chi = \left(if^{bac}J^c_0\tau^a + \frac{k}{2}\tau^b - \kappa J^b_0\right) g = 0 \quad . \tag{15.151b}$$

Therefore, defining the invariants $c_g = \tau^a\tau^a$, and $f^{abc}f^{dbc} = c_V\delta^{ad}$ we have

$$c_g - 2\kappa\Delta = 0 \tag{15.152a}$$
$$-c_V - k + 2\kappa = 0 \quad . \tag{15.152b}$$

We solve these equations for Δ and κ to obtain

$$\Delta = \frac{c_g}{c_V + k} \tag{15.153a}$$
$$\kappa = \frac{1}{2}(c_V + k) \quad . \tag{15.153b}$$

Relations (15.153) must be satisfied in order that the equations (15.149) be consistent. As in the case of Eqs. (15.57-59), the application of L_n and J^a_n with $n > 1$ yields no further conditions.

15.4.4 Correlation functions

The above methods permit also the exact computation of some correlation functions. Consider as an example

$$G(z_i, \bar{z}_i) = \langle g^{\beta_1}_{\alpha_1}(z_1, \bar{z}_1) g^{-1\alpha_2}_{\beta_2}(z_2, \bar{z}_2) g^{-1\alpha_3}_{\beta_3}(z_3, \bar{z}_3) g^{\beta_4}_{\alpha_4}(z_4, \bar{z}_4)\rangle \quad , \tag{15.154}$$

where we now fix the symmetry groups to be $G = SU(N)$. Conformal invariance restricts the Green function to be a conformal factor times a function of the harmonic quotients

$$x = \frac{(z_1 - z_2)(z_3 - z_4)}{(z_1 - z_4)(z_3 - z_2)} \tag{15.155a}$$

$$\overline{x} = \frac{(\overline{z}_1 - \overline{z}_2)(\overline{z}_3 - \overline{z}_4)}{(\overline{z}_1 - \overline{z}_4)(\overline{z}_3 - \overline{z}_2)} \quad . \tag{15.155b}$$

such that

$$G(z_i, \overline{z}_i) = \left\{(z_1 - z_4)(z_2 - z_3)(\overline{z}_1 - \overline{z}_4)(\overline{z}_2 - \overline{z}_3)\right\}^{-2\Delta} G(x, \overline{x}) \quad . \tag{15.156}$$

The $SU(N)_R \otimes SU(N)_L$ invariant decomposition of $G(x, \overline{x})$ is given by

$$G(x, \overline{x}) = \sum_{A,B=1}^{2} I_A \overline{I}_B G_{AB}(x, \overline{x}) \quad , \tag{15.157}$$

where

$$I_1 = \delta^{\alpha_2}_{\alpha_1} \delta^{\alpha_4}_{\alpha_3} \qquad \overline{I}_1 = \delta^{\beta_1}_{\beta_2} \delta^{\beta_3}_{\beta_4} \tag{15.158a, b}$$

$$I_2 = \delta^{\alpha_3}_{\alpha_1} \delta^{\alpha_4}_{\alpha_2} \qquad \overline{I}_2 = \delta^{\beta_3 \beta_1} \delta_{\beta_4 \beta_2} \quad . \tag{15.158c, d}$$

The analog of (15.139) now reads

$$\tau^a_i \langle J^a(z) g(z_1, \overline{z}_1) \ldots g(z_n, \overline{z}_n) \rangle = \left\{ \frac{c_g}{z - z_i} + \sum_{j \neq i} \frac{\tau^a_i \tau^a_j}{z - z_j} \right\} \langle g(z_1, \overline{z}_1) \ldots g(z_n, \overline{z}_n) \rangle \tag{15.159}$$

Using the O.P.E. (15.148) on the l.h.s. we obtain the following differential equation upon taking the limit $z \to z_i$:

$$\left\{ \kappa \frac{\partial}{\partial z_i} - \sum_{j \neq i} \frac{\tau^a_i \tau^a_j}{z_i - z_j} \right\} \langle g(z_1, \overline{z}_1) \ldots g(z_n, \overline{z}_n) \rangle = 0 \quad . \tag{15.160}$$

Let us choose $n = 4$. We write this equation in terms of $G(x, \overline{x})$, Eq. (15.156). This will require a change in sign for the second term in Eq. (15.160) above, since we are now dealing with g^{-1}. Choosing $i = 1$, we obtain for the first term

$$\frac{\partial G}{\partial z_1}(z_i, \overline{z}_i) = \left[z_{14} z_{23} \overline{z}_{14} \overline{z}_{23}\right]^{-2\Delta} \left\{ \frac{-2\Delta}{z_1 - z_4} G(x, \overline{x}) + x\left(\frac{1}{z_1 - z_2} - \frac{1}{z_1 - z_4}\right) \frac{\partial G}{\partial x}(x, \overline{x}) \right\} \tag{15.161}$$

For the computation of the second term it is convenient to parametrize the dependence of $G(x,\bar{x})$ on the holomorphic coordinates by (we suppress the dependence on $\bar{x}$ and the indices β)

$$G_{\alpha_1\alpha_2\alpha_3\alpha_4}(x) = G_1(x)\delta_{\alpha_1\alpha_2}\delta_{\alpha_3\alpha_4} + G_2(x)\delta_{\alpha_1\alpha_3}\delta_{\alpha_2\alpha_4} \tag{15.162}$$

Choosing again $i = 1$, we find

$$\begin{aligned}
\sum \frac{\tau_1^a\tau_j^a}{z_1 - z_j}\langle g(z_1)\cdots g(z_4)\rangle = &-\frac{1}{z_1 - z_2}\tau^{a(1)}_{\alpha_1\alpha_1'}\tau^{a(2)}_{\alpha_2'\alpha_2}G_{\alpha_1'\alpha_2'\alpha_3\alpha_4}(x) \\
&-\frac{1}{z_1 - z_3}\tau^{a(1)}_{\alpha_1\alpha_1'}\tau^{a(3)}_{\alpha_3'\alpha_3}G_{\alpha_1'\alpha_2\alpha_3'\alpha_4}(x) + \frac{1}{z_1 - z_4}\tau^{a(1)}_{\alpha_1\alpha_1'}\tau^{a(2)}_{\alpha_4\alpha_4'}G_{\alpha_1'\alpha_2\alpha_3\alpha_4'}(x) = \\
&-\frac{1}{z_1 - z_2}\left\{\frac{N^2-1}{2N}\delta_{\alpha_1\alpha_2}\delta_{\alpha_3\alpha_4}G_1 + \frac{1}{2}\left(\delta_{\alpha_1\alpha_2}\delta_{\alpha_3\alpha_4} - \frac{1}{N}\delta_{\alpha_1\alpha_3}\delta_{\alpha_2\alpha_4}\right)G_2\right\} \\
&-\frac{1}{z_1 - z_3}\left\{\frac{1}{2}\left(\delta_{\alpha_1\alpha_3}\delta_{\alpha_2\alpha_4} - \frac{1}{N}\delta_{\alpha_1\alpha_2}\delta_{\alpha_3\alpha_4}\right)G_1 + \frac{N^2-1}{2N}\delta_{\alpha_1\alpha_3}\delta_{\alpha_2\alpha_4}G_2\right\} \\
+\frac{1}{z_1 - z_4}&\left\{\frac{1}{2}\left(\delta_{\alpha_1\alpha_3}\delta_{\alpha_2\alpha_4} - \frac{1}{N}\delta_{\alpha_1\alpha_2}\delta_{\alpha_3\alpha_4}\right)G_1 + \frac{1}{2}\left(\delta_{\alpha_1\alpha_2}\delta_{\alpha_3\alpha_4} - \frac{1}{N}\delta_{\alpha_1\alpha_3}\delta_{\alpha_2\alpha_4}\right)G_2\right\}
\end{aligned} \tag{15.163}$$

where we used the formulae

$$c_g = \frac{N^2-1}{2N} \tag{15.164}$$

and

$$\tau^a_{\alpha_1\alpha_2}\tau^a_{\alpha_3\alpha_4} = \frac{1}{2}\left(\delta_{\alpha_1\alpha_4}\delta_{\alpha_2\alpha_3} - \frac{1}{N}\delta_{\alpha_1\alpha_2}\delta_{\alpha_3\alpha_4}\right) \quad . \tag{15.165}$$

valid for $SU(N)$. Collecting all terms we get, for $\kappa = \frac{1}{2}(N+k)$

$$\kappa\frac{\partial G_1}{\partial x} = -\frac{N^2-1}{2N}\frac{1}{x}G_1 + \frac{1}{2N}\frac{1}{x-1}G_1 - \frac{1}{2x}G_2 \quad , \tag{15.166a}$$

$$\kappa\frac{\partial G_2}{\partial x} = -\frac{N^2-1}{2N}\frac{1}{x-1}G_2 + \frac{1}{2N}\frac{1}{x}G_2 - \frac{1}{2(x-1)}G_1 \quad , \tag{15.166b}$$

which may also be written in 2×2 matrix form

$$\frac{\partial G}{\partial x} = \left\{\frac{1}{x}P + \frac{1}{x-1}Q\right\}G \tag{15.167a}$$

$$\frac{\partial G}{\partial \bar{x}} = G\left\{\frac{1}{\bar{x}}P^t + \frac{1}{\bar{x}-1}Q^t\right\} \quad , \tag{15.167b}$$

where $G = (G)_{A\overline{B}}$, and

$$P = -\frac{1}{2N\kappa}\begin{pmatrix} N^2-1 & N \\ 0 & -1 \end{pmatrix} \tag{15.168a}$$

$$Q = -\frac{1}{2N\kappa}\begin{pmatrix} -1 & 0 \\ N & N^2-1 \end{pmatrix} \quad , \tag{15.168b}$$

with t denoting transposition. The general solution of the above system can be expressed in terms of hypergeometric functions. We write

$$G_{AB}(x,\bar{x}) = \sum_{p,q=0,1} U_{pq}\mathcal{F}_A^{(p)}(x)\mathcal{F}_B^{(q)}(\bar{x}) \quad . \tag{15.169}$$

The functions $\mathcal{F}_A^{(p)}(x)$ and $\mathcal{F}_B^{(q)}(\bar{x})$ obey hypergeometric equations, and are given by

$$\mathcal{F}_1^{(0)}(x) = x^{-2\Delta}(1-x)^{\Delta}F\left(-\frac{1}{N+k},\frac{1}{N+k},1+\frac{1}{N+k};x\right) \tag{15.170a}$$

$$\mathcal{F}_2^{(0)}(x) = -(2N+k)^{-1}x^{1-2\Delta}(1-x)^{\Delta_1-2\Delta}\times$$
$$F\left(1-\frac{1}{N+k},1+\frac{1}{N+k},2+\frac{1}{N+k};x\right) \tag{15.170b}$$

$$\mathcal{F}_1^{(1)}(x) = x^{\Delta_1-2\Delta}(1-x)^{\Delta_1-2\Delta}$$
$$F\left(-\frac{N-1}{N+k},-\frac{N+1}{N+k},1-\frac{N}{N+k};x\right) \tag{15.171a}$$

$$\mathcal{F}_2^{(1)}(x) = -Nx^{\Delta_1-2\Delta}(1-x)^{\Delta_1-2\Delta}$$
$$F\left(-\frac{N-1}{N+k},-\frac{N+1}{N+k},-\frac{N}{N+k};x\right) \quad , \tag{15.171b}$$

where the function $F(a,b,c;x)$ is the hypergeometric function defined by the series expansion (15.120c).

The final step is to fix the constants U_{pq}. First, we know that in the usual Euclidean domain where x is the complex conjugate to $\bar{x}$, the Green function must be single valued. Now for x in the vicinity of the origin (where $F(a,b,c;x) \approx 1$) we see that the terms involving U_{01} and U_{10} do no satisfy this requirement. Hence we must have

$$U_{01} = U_{10} = 0 \quad . \tag{15.172}$$

A second restriction comes from crossing symmetry. By this we mean the following: we have, in the conformal theories, the "fusion property", where a product of two conformal fields can be expanded in terms of a third conformal field. For the 4 point function, this expansion can be done for 2 blocks of 2 fields. And this can be done in two different ways, which must both give the same result. When written in one way or another, this amounts to substituting x by $1-x$, and $\bar{x}$ by $1-\bar{x}$, with a corresponding interchange of the legs. Thus we have

$$G_{AB}(x,\bar{x}) = \sum_{A'B'=1,2} E_{AA'}E_{B'B}G_{A'B'}(1-x,1-\bar{x}) \quad , \tag{15.173}$$

with $E_{12}=E_{21}=1$; $\quad E_{11}=E_{22}=0$.

We now use the following property of the hypergeometric functions

$$\mathcal{F}_A^{(p)}(x) = \sum_{q,A'} C^{(p,q)} E_{A,A'}\mathcal{F}_{A'}^{(q)}(1-x) \quad , \tag{15.174}$$

where

$$C^{00} = -C^{11} = N\frac{\Gamma(-\frac{N}{N+k})\Gamma(\frac{N}{N+k})}{\Gamma(-\frac{1}{N+k})\Gamma(\frac{1}{N+k})} \tag{15.175a}$$

$$C^{10} = -N\frac{\Gamma(-\frac{N}{N+k})^2}{\Gamma(-\frac{N-1}{N+k})\Gamma(-\frac{N+1}{N+k})} \tag{15.175b}$$

and

$$C^{01}C^{10} - C^{00}C^{11} = 1 \quad . \tag{15.175c}$$

Substituting (15.169) and (15.174) into (15.173) one obtains

$$\begin{aligned} U_{00}\mathcal{F}_1^{(0)}(x)\mathcal{F}_1^{(0)}(\bar{x}) + U_{11}\mathcal{F}_1^{(1)}(x)\mathcal{F}_1^{(1)}(\bar{x}) &= \\ = U_{00}\mathcal{F}_2^{(0)}(1-x)\mathcal{F}_2^{(0)}(1-\bar{x}) &+ U_{11}\mathcal{F}_2^{(1)}(1-x)\mathcal{F}_2^{(1)}(1-\bar{x}) \\ = U_{00}C^{0p}\mathcal{F}_2^{(p)}(1-x)C^{0q}\mathcal{F}_2^{(q)}(1-\bar{x}) &+ U_{11}C^{1p}\mathcal{F}_2^{(p)}(1-x)C^{1q}\mathcal{F}_2^{(q)}(1-\bar{x}) \quad . \end{aligned} \tag{15.176}$$

Therefore we have

$$U_{11} = U_{00}{C^{01}}^2 + U_{11}{C^{11}}^2 \tag{15.177a}$$

$$U_{00} = U_{00}{C^{00}}^2 + U_{11}{C^{10}}^2 \tag{15.177b}$$

$$0 = U_{00}C^{01}C^{00} + U_{11}C^{11}C^{10} \quad . \tag{15.177c}$$

One gets for $h = \frac{U_{11}}{U_{00}}$ the result

$$h = \frac{1-{C^{00}}^2}{{C^{10}}^2} = \frac{1}{N^2}\frac{\Gamma(\frac{N-1}{N+k})\Gamma(\frac{N+1}{N+k})}{\Gamma(\frac{N}{N+k})^2}\frac{1-N^2\left\{\frac{\Gamma(\frac{N}{N+k})\Gamma(-\frac{N}{N+k})}{\Gamma(\frac{1}{N+k})\Gamma(-\frac{1}{N+k})}\right\}^2}{\frac{\Gamma(\frac{N}{N+k})^2}{\Gamma(\frac{N-1}{N+k})\Gamma(\frac{N+1}{N+k})}} \quad . \tag{15.178}$$

Inspection of the zeros and poles of the last term in (15.178), implies that we may write the result in a simpler way. One notices that there is a zero at $\frac{k+1}{N+k}$ a non positive integer, and for $\frac{k-1}{N+k}$ a non positive integer, and double poles for $\frac{k}{N+k}$. Then, looking at the residuum for $k=1$, one obtains

$$h = \frac{1}{N^2}\frac{\Gamma(\frac{N-1}{N+k})\Gamma(\frac{N+1}{N+k})\{\Gamma(\frac{k}{N+k})\}^2}{\Gamma(\frac{k+1}{N+k})\Gamma(\frac{k-1}{N+k})\{\Gamma(\frac{N}{N+k})\}^2} \quad . \tag{15.179}$$

Normalizing the Greens function in the same way as the two point function one arrives at

$$G_{AB}(x,\overline{x}) = M^{-8\Delta}\left\{\mathcal{F}_A^{(0)}(x)\mathcal{F}_B^{(0)}(\overline{x}) + h\mathcal{F}_A^{(1)}(x)\mathcal{F}_B^{(1)}(\overline{x})\right\} \quad . \tag{15.180}$$

An analysis of the singularity structure reveals the contributions of the various composite fields. One has, as an example, contributions from the symmetric and antisymmetric[502] fields

$$\phi_S = g_{\{\alpha\beta}g_{\gamma\delta\}}(z,\overline{z}) \tag{15.181}$$
$$\phi_A = g_{[\alpha\beta}g_{\gamma\delta]}(z,\overline{z}) \quad , \tag{15.182}$$

which have the correct dimensions. For $k=1$, we have $h=0$, and the function (15.180) simplifies considerably:

$$G(x,\overline{x}) = \left[x\overline{x}(1-x)(1-\overline{x})\right]^{\frac{1}{N}}\left\{\frac{1}{x}\delta_{\alpha_1\alpha_2}\delta_{\alpha_3\alpha_4} + \frac{1}{1-x}\delta_{\alpha_1\alpha_3}\delta_{\alpha_2\alpha_4}\right\}\times$$
$$\left\{\frac{1}{\overline{x}}\delta_{\beta_1\beta_2}\delta_{\beta_3\beta_4} + \frac{1}{1-\overline{x}}\delta_{\beta_1\beta_3}\delta_{\beta_2\beta_4}\right\} \quad , \tag{15.183}$$

which is the way Witten's bosonization works in this language if we consider the bound state

$$g_{\alpha\beta} = i\mu^{-1} : \psi^\dagger_{1\alpha}(z)\psi_{2\beta}(\overline{z}) : \quad . \tag{15.184}$$

We shall see how these ideas carry over to the case $k \neq 1$. Some general correlation functions have already been computed in the literature[504].

15.5 The Non Abelian Thirring Model at the Critical Point, and the WZW Theory

15.5.1 Definition of the conformally invariant theory

In this section we shall study the properties of the $SU(N) \times U(1)$ Thirring model, showing that at a special value of the coupling constant, it has a conformally invariant solution. As we shall show this solution is related to that of the WZW theory in the sense, that the bosonic fields of the latter correspond to bound states of the fermionic theory in question.

Consider the $SU(N) \times U(1)$ invariant Thirring model, defined by the Lagrangian density[505]

$$L = \overline{\psi} i \not{\partial} \psi - \frac{1}{2} g \overline{\psi} \gamma^\mu \tau^a \psi \overline{\psi} \gamma_\mu \tau^a \psi - \frac{1}{2} g' \overline{\psi} \gamma^\mu \psi \overline{\psi} \gamma_\mu \psi \quad ; \tag{15.185}$$

by a Fierz transformation, this is exactly the Gross-Neveu model (see chapter 5)

The field equation is

$$i \not{\partial} \psi = g \not{j}^a \tau^a \psi + g' \not{j} \psi \quad , \tag{15.186}$$

where the currents are defined by

$$j^a_\mu = \overline{\psi} \gamma_\mu \tau^a \psi \quad , \tag{15.187a}$$

$$j_\mu = \overline{\psi} \gamma_\mu \psi \quad . \tag{15.187b}$$

These are the Noether currents corresponding to the $SU(N)$ and $U(1) \otimes U(1)$ symmetry, respectively, while $j^5_\mu = \epsilon_{\mu\nu} j^\nu$ is associated with the pseudo charge. All three currents are conserved

$$\partial^\mu j^a_\mu = 0 \tag{15.188a}$$

$$\partial^\mu j_\mu = 0 \tag{15.188b}$$

$$\partial^\mu j^5_\mu = 0 \quad . \tag{15.188c}$$

However, the current $j^{5a}_\mu = \epsilon_{\mu\nu} j^{\nu a}$ is not conserved. Using the equations of motion we find

$$\partial_\mu j^{5\mu a} = g f_{abc} j^{\lambda b} j_\lambda{}^{5c} \quad . \tag{15.189}$$

The $SU(N)$ axial current j^{5a}_μ is a priori not conserved, corresponding to the fact that chiral $SU(N)$ transformations are not a symmetry of (15.185). Nonetheless, it is possible to show that there exists a particular value of the coupling constant were quantum fluctuations cancel the right hand side of (15.189). In the following we study the conditions under which the quantum model defined by (15.185) exhibits conformal invariance[505]. Our starting point is the current algebra

$$[j_0^a(t,x), j_0^b(t,y)] = if^{abc} j_0^c(t,x)\delta(x-y) \tag{15.190a}$$

$$[j_0^a(t,x), j_1^b(t,y)] = if^{abc} j_1^c(t,x)\delta(x-y) + i\frac{k}{2\pi}\delta^{ab}\delta'(x-y) \tag{15.190b}$$

$$[j_1^a(t,x), j_1^b(t,y)] = if^{abc} j_0^c(t,x)\delta(x-y) \quad . \tag{15.190c}$$

This is also the algebra one obtains in the case of free fermions, when $k = 1$.

If the theory is scale invariant, it follows from above that the current $j^{a\mu}$ has scale dimension -1. We can prove now that j_μ^a is divergenceless and curl free. In order to see that, consider the two point function of the combination $j_+^a(x) = (j_0^a(x) + j_1^a(x))$. Under Lorentz transformations it transforms like $\frac{1}{x^+}$. Thus its two point function is fixed up to a multiplicative constant to be

$$\langle 0|j_+^a(x) j_+^b(y)|0\rangle = \frac{c\delta^{ab}}{(x^+ - y^+ + i\epsilon)^2} \quad . \tag{15.191}$$

We may analogously consider functions of j_-. From these expressions it immediately follows that

$$\partial_\mu j^{a\mu} = 0 \quad , \tag{15.192a}$$

as well as

$$\epsilon^{\mu\nu}\partial_\mu j_\nu^a = 0 \quad . \tag{15.192b}$$

Thus, the current algebra may be rewritten in terms of light-cone components as

$$[j_\pm^a(x^\pm), j_\pm^b(y^\pm)] = 2if^{abc} j_\pm^c(x^\pm)\delta(x^\pm - y^\pm) + i\frac{k}{\pi}\delta^{ab}\delta'(x^\pm - y^\pm) \tag{15.193a}$$

$$[j_\pm(x^\pm), j_\pm(y^\pm)] = iC_0\delta'(x^\pm - y^\pm) \quad , \tag{15.193b}$$

where C_0 is an arbitrary constant.

The action of the currents on the fundamental fields is assumed to be given by

$$[j_\pm(x), \psi(y)] = -(a \pm \overline{a}\gamma^5)\psi(y)\delta(x^\pm - y^\pm) \tag{15.194a}$$

$$[j_\pm^a(x), \psi(y)] = -\sigma\tau^a(1 \pm \delta\gamma^5)\psi(y)\delta(x^\pm - y^\pm) \quad . \tag{15.194b}$$

with a, $\overline{a}$ and δ some constants. Due to Jacobi identity one must have $\sigma = 1$ and $\delta^2 = 1$. The energy momentum tensor can also be computed, and is found to be of the Sugawara form[340]

$$T(x^\pm) = \frac{1}{2C_0}{:}j_\pm(x^\pm)^2{:} + \frac{\pi}{n+k}{:}j_\pm^a(x^\pm)^2{:} \quad . \tag{15.195}$$

Indeed, this has been checked in several situations[506]. The correctness of (15.195) can also be checked by computing the commutator of the energy-momentum tensor with the fermionic field, which turns out to be given by the derivative of the field (see (15.202) below). Thus, $T(x)$ is the generator of translations, a property that characterizes it completely.

Equation (15.195) implies that the Fourier coefficients of the energy momentum tensor satisfy a Virasoro-Kac-Moody algebra. Together with the Fourier coefficients of the currents,we arrive at

$$[L_n, L_m] = (n-m)L_{n+m} + \frac{c}{12}n(n^2-1)\delta_{n,-m} \tag{15.196a}$$

$$[L_n, J^a_m] = -mJ_{n+m} \tag{15.196b}$$

$$[J^a_m, J^b_n] = if^{abc}J^c_{m+n} + nk\delta^{ab}\delta_{n,-m} \quad . \tag{15.196c}$$

The central charge c in Eq. (15.196a) may be computed using the Sugawara form of the energy momentum tensor. It is given by the expression

$$c = \frac{k\dim G}{c_V + k} \quad , \tag{15.197}$$

which for $SU(N)$ reads

$$c_{SU(N)} = \frac{k(N^2-1)}{N+k} \quad . \tag{15.198}$$

The (conformally invariant) two point function is given in terms of the spin-s and anomalous dimension-γ, just as in the case of a general conformally invariant theory, by

$$\langle 0|\psi_2^\dagger(x)\psi_2(y)|0\rangle = f_0\{i(x^+-y^+)+\epsilon\}^{-2s}\{-(x-y)^2+i\epsilon(x^++x^--y^+-y^-)\}^{(\gamma-s)} , \tag{15.199}$$

where f_0 is an arbitrary constant fixed by the normalization of the fields.

15.5.2 The solution at the conformally invariant fixed point

The four point function has constraints arising from the fermionic field equations. These can be set up by means of the following construction. Since the energy-momentum tensor is of the Sugawara form (15.195), we may use (15.194) in order to find the commutator of the energy momentum tensor with the fermionic field,

$$[T_\pm, \psi] = -\frac{1}{C_0}j_\pm(a \pm \bar{a}\gamma_5)\psi + \frac{2\pi}{N+k}(-\tau^a)\, j^a_\pm(1 \pm \delta\gamma_5)\psi \quad . \tag{15.200}$$

Thus for the first component we have

$$[T_\pm, \psi_1] = -\frac{(a \pm \bar{a})}{C_0} j_\pm \psi_1 - 2\pi \frac{1 \pm \delta}{N + k} \tau^a j_\pm^a \psi_1 \quad . \tag{15.201}$$

with analogous expressions for ψ_2. Since $T_\pm$ should generate translations, we tentatively identify the r.h.s. of (15.200) with $\partial_\pm \psi$, thus obtaining the equations

$$i\partial_+ \psi_1 = \left\{ 2\pi \frac{1+\delta}{N+k} \tau^a : j_+^a \psi_1 : + \frac{(a+\bar{a})}{C_0} : j_+ \psi_1 : \right\} \tag{15.202a}$$

$$i\partial_+ \psi_2 = \left\{ 2\pi \frac{1-\delta}{N+k} \tau^a : j_+^a \psi_2 : + \frac{(a-\bar{a})}{C_0} : j_+ \psi_2 : \right\} \tag{15.202b}$$

$$i\partial_- \psi_1 = \left\{ 2\pi \frac{1-\delta}{N+k} \tau^a : j_-^a \psi_1 : + \frac{(a-\bar{a})}{C_0} : j_- \psi_1 : \right\} \tag{15.202c}$$

$$i\partial_- \psi_2 = \left\{ 2\pi \frac{1+\delta}{N+k} \tau^a : j_-^a \psi_2 : + \frac{(a+\bar{a})}{C_0} : j_- \psi_2 : \right\} \quad . \tag{15.202d}$$

In order that (15.202) agrees with the equations of motion (15.186) we must require (the abelian part is a trivial Thirring coupling, and we shall not dwell on it)

$$g = 2\pi \frac{1+\delta}{N+k} \quad . \tag{15.203}$$

The value $\delta = -1$ corresponds to a free theory (abelian Thirring model), while $\delta = +1$ gives rise to a non trivial fix point. We now proceed as in section 1. The four point function can be written in terms of the Möbius invariant variables u and v, given by

$$u = \frac{[x^+ - x'^+ - i\epsilon][y^+ - y'^+ - i\epsilon]}{[x^+ - y'^+ - i\epsilon][y^+ - x'^+ - i\epsilon]} \tag{15.204a}$$

$$v = \frac{[x^- - x'^- - i\epsilon][y^- - y'^- - i\epsilon]}{[x^- - y'^- - i\epsilon][y^- - x'^- - i\epsilon]} \quad . \tag{15.204b}$$

Thus we write

$$\begin{aligned} \langle 0 | \psi_1^{a\dagger}(x) \psi_1'^{a}(y) \psi_1^{b\dagger}(x') \psi_1'^{b}(y') | 0 \rangle = & \\ & f_0^2 u^{\gamma+s} v^{\gamma-s} \{[i(x^+ - y^+) + \epsilon][i(x'^+ - y'^+) + \epsilon]\}^{-(\gamma+s)} \\ & \times \{[i(x^- - y^-) + \epsilon][i(x'^- - y'^-) + \epsilon]\}^{-(\gamma-s)} G_{aa'bb'}(u,v) \quad . \end{aligned} \tag{15.205}$$

For the symmetry group $SU(N)$,the function G can be written as

$$G_{aa',bb'}(u,v) = \delta_{aa'} \delta_{bb'} \tilde{H}_1(u,v) + \delta_{ab'} \delta_{ba'} \tilde{H}_2(u,v) \quad . \tag{15.206}$$

Differential equations are obeyed by $\tilde{H}_1$ and $\tilde{H}_2$ as a consequence of Eqs. (15.200) and (15.201). For the sake of comparison with the bosonic theory of section 3, it is convenient, at this point to rewrite the four point function in Euclidean space. We then have

$$\langle 0|\psi_1^{\dagger a}(z_1)\psi_1^{\dagger b}(z_2)\psi_1^c(z_3)\psi_1^d(z_4)|0\rangle =$$
$$[(z_1 - z_4)(z_2 - z_3)]^{-2\Delta}\{\delta^{ad}\delta^{cb}A_1(x) + \delta^{ac}\delta^{bd}A_2(x)\} \quad , \qquad (15.207)$$

where $z = x_1 + ix_2$, and x is the Möbius invariant combination

$$x = \frac{(z_1 - z_2)(z_3 - z_4)}{(z_1 - z_4)(z_3 - z_2)} \quad . \qquad (15.208)$$

The functions $A_1(x)$ and $A_2(x)$ obey hypergeometric differential equations. This is due to the fact that the fermionic fields equations of motion also define a representation of the non abelian conformal group. We have

$$\frac{\partial}{\partial z}\psi_1(z) = \frac{1}{N+\kappa}\tau^a{:}j^a(z)\psi_1(z){:} \quad . \qquad (15.209)$$

The last piece of information comes from the short distance expansion

$$j^a(z)\psi_1(\omega) = \frac{\tau^a}{z-\omega}\psi_1(\omega) + {:}j^a(z)\psi_1(z){:} \quad , \qquad (15.210)$$

which permits defining

$$:j^a(z)\psi_1(\omega):= \oint_\omega dz j^a(z)(z-\omega)^{-1}\psi_1(\omega) \quad , \qquad (15.211)$$

which should be compared with Eq. (15.147), with ψ_1 replacing g. Following the same steps as in (15.44-49), one finds

$$:j^a(z)\psi_1(z):= J^a_{-1}\psi_1(z) \quad , \qquad (15.212)$$

which, upon using the equation of motion (15.186) is seen to be equivalent to the definition of the Verma module, requiring certain linear combinations of states to be a null vector (see Eq. (15.57)). In the present example we have

$$\chi = \left(J^a_{-1}\tau^a - \frac{1}{2}(N+\kappa)L_{-1}\right)\psi = 0 \quad . \qquad (15.213)$$

This means that we obtain a fermionic representation of non abelian conformal theory. In terms of the Minkowski variables, we use the equivalent equation

$$j_+^{a(-)}(x^+)\psi(y) = \frac{1+\delta}{2\pi}\frac{1}{i(x^+ - y^+) + \varepsilon}\tau^a\psi(y) \qquad (15.214)$$

and analogous expressions coming from (15.194b).

Using (15.214), (15.211) and (15.209) one obtains, in Euclidean space

$$\frac{\partial}{\partial z_1}\langle 0|\psi_1^i(z_1)\psi_1^{j\dagger}(z_2)\psi_1^{k\dagger}(z_3)\psi_1^l(z_4)|0\rangle = \langle 0|\psi_1^{i'}(z_1)\psi_1^{j'\dagger}(z_2)\psi_1^{k'\dagger}(z_3)\psi_1^{l'}(z_4)|0\rangle \times$$

$$\frac{2}{N+k}\left\{\frac{\tau^a_{ii'}\tau^a_{j'j}\delta_{kk'}\delta_{ll'}}{z_2-z_1}+\frac{\tau^a_{ii'}\tau^a_{k'k}\delta_{jj'}\delta_{ll'}}{z_3-z_1}+\frac{\tau^a_{ii'}\tau^a_{l'l}\delta_{kk'}\delta_{jj'}}{z_1-z_4}\right\} \tag{15.215}$$

From above, one arrives at differential equations which should be fulfilled by the A's in (15.207):

$$x(x-1)\frac{\partial A_1}{\partial x}(x) = \left\{(x-1)\frac{\frac{1}{N}-N}{2(N+k)}+\frac{x}{2N(N+k)}\right\}A_1(x)-\frac{x-1}{2(N+k)}A_2(x) \tag{15.216a}$$

$$x(x-1)\frac{\partial A_2}{\partial x}(x) = \left\{x\frac{\frac{1}{N}-N}{2(N+k)}+\frac{x-1}{2N(N+k)}\right\}A_2(x)-\frac{x}{2(N+k)}A_1(x) \quad . \tag{15.216b}$$

The solutions of this set of equations are given in terms of hypergeometric functions as

$$A_1(x) = \mathcal{F}_1^{(0)}(x)+h\mathcal{F}_1^{(1)}(x) \tag{15.217a}$$

$$A_2(x) = \mathcal{F}_2^{(0)}(x)+h\mathcal{F}_2^{(1)}(x) \quad , \tag{15.217b}$$

where the functions $\mathcal{F}_A^{(a)}(x)$ have been defined in subsection (15.4.4) and h is determined by crossing to be the value given in Eq. (15.180). These results may be compared with those of section (15.4). Actually, both theories present solutions derived from equations defining a representation of the conformal group, given by the condition

$$\left(J^a_{-1}\tau^a-\kappa L_{-1}\right)|\psi\rangle = 0 \quad . \tag{15.218}$$

We verify that the two and four point functions of the operator product[507,508]

$$g_{ij}(z,\bar z) = i\mu^{-1}\psi^\dagger_{1i}(z)\psi_{2j}(\bar z) \tag{15.219}$$

are analogous to those of chapter 9, Eqs. (9.14).[10)] is correct, with the crossing symmetry being obeyed now by the composite field. The fact that the Gross-Neveu model has a nontrivial fix point has been discussed in the literature[509]. Different results may be obtained as compared to (15.203).

[10)] This is not surprising, since the solution of (15.218), and (15.149) as well as the analogous expressions in chapter 9 are realizations of the conformal algebra[507,508].

Path integral interpretation

One notices that the model may be written in terms of the functional integral

$$Z = \int \mathcal{D}\overline{\psi}\mathcal{D}\psi\mathcal{D}A_\mu e^{i\int d^2x\left[\overline{\psi}(\partial\!\!\!/ - A\!\!\!/)\psi + \frac{1}{2g}A^{a\mu}A^a_\mu\right]} \quad . \tag{15.220}$$

We suppose that the path integral is performed respecting chiral transformations. we can use the result for the fermion determinant in terms of the WZW action (9.14), and the change of variables[510] as

$$\begin{aligned}\mathcal{D}A_\pm &= \det(\delta^{ab}\partial + f^{acb}A^b_\pm)\mathcal{D}d_\pm \\ &= e^{-ic_V S_{WZW}(g)a}\mathcal{D}g_\pm\end{aligned} \quad , \tag{15.221}$$

valid for A_+ and A_- separately.

We also use the identity[50]

$$S(u) + S(v) = S(uv) - \frac{1}{4\pi}\int d^2x u^{-1}\partial_+ uv\partial_- v^{-1} \quad . \tag{15.222}$$

Therefore we verify that at a certain value of the coupling constant the non conformally invariant term $A^\mu A_\mu$ has vanishing coefficient.

15.6 More on Kac Moody Algebras

Although our aim is not to provide a general review on conformal invariance, it is worth mentioning how the previous construction concerning the WZW theory and conformally invariant non abelian Thirring model generalize and have a widespread application to statistical systems at criticality[511], and also to string theories[503,512,91].

We shall discuss the current algebra realization of the conformal algebra, analyzing it using mathematically more elaborated concepts. The idea is based on the construction of vertex operators of string theory. Let us suppose that we have a set of free scalar fields

$$\varphi^\mu(z) = \varphi^\mu_0 - ip^i \ln z + \sum_{n\neq 0}\frac{a^\mu_n}{n}z^{-n} \quad , \tag{15.223}$$

where z is the complex variable which is defined, in string theory, in terms of the proper time τ and position in the string σ, by $z = e^{i(\tau-\sigma)} = e^{i(x^0-x^1)}$, or in the Euclideanized version, by $z = e^{x_2 - ix_1}$. We have the commutation relations $[a^\mu_m, a^\nu_n] = mg^{\mu\nu}\delta_{m,-n}$.

The energy-momentum tensor is given by

$$T(z) = \frac{1}{2} z^2 \left(\frac{d\varphi}{dz}\right)^2 \quad . \tag{15.224}$$

In the following construction, we consider a lattice Λ spanned by the discrete infinite series of $\vec{e}_i$, that is $\Lambda = \{\sum_{i=1}^N n_i e_i\, ,\, n_i \in Z\}$. Via the vertex operator construction we obtain a Kac-Moody algebra[513,514] associated with this lattice. A full mathematical account is given in ref. [515], while the necessary facts, sufficient to understand the construction and its applications to physics are concisely gathered in refs. [516,517]. We shall try to provide a simple account of relevant facts connected with the main subject of this book.

The vertex operator is constructed in terms of the exponential of φ^μ. We define first the vertex operators[518]

$$V(\vec{\alpha}, z) =: e^{i\alpha^\mu \varphi_\mu} := e^{i\alpha^\mu \varphi_\mu^<(z)}\, e^{i\alpha^\mu \varphi_{0\mu}(z)}\, e^{i\alpha^\mu \varphi_\mu^>(z)} \quad , \tag{15.225}$$

where $\varphi^>$ ($\varphi^{<0}$ contains only $n > 0\,(n < 0)$ in (15.223), and φ_0 the first two terms. We now define[516,517]

$$A_{\vec{\alpha}} = \frac{1}{2\pi i} \oint \frac{dz}{z} V(\vec{\alpha}, z) \quad , \tag{15.226}$$

where the integration encircles the origin. As we shall see, the above operators satisfy simple algebraic relations. We can use the relations (2.14,68) to obtain

$$A_{\vec{\alpha}} A_{\vec{\beta}} - (-1)^{\vec{\alpha}\cdot\vec{\beta}} A_{\vec{\beta}} A_{\vec{\alpha}} = \frac{1}{(2\pi i)^2} \left\{ \oint \frac{dz}{z} \oint_{|z|>|\omega|} \frac{d\omega}{\omega} - \oint \frac{dz}{z} \oint_{|\omega|>|z|} \frac{d\omega}{\omega} \right\} z^{\frac{\vec{\alpha}^2}{2}} \omega^{\frac{\vec{\beta}^2}{2}}$$

$$\times : e^{(i\alpha^\mu \varphi_\mu(z) + i\beta^\mu \varphi_\mu(\omega))} : (z - \omega)^{\vec{\alpha}\cdot\vec{\beta}}$$

$$= \frac{1}{(2\pi i)^2} \oint \frac{d\omega}{\omega} \oint_{around\, \omega} \frac{dz}{z} (z - \omega)^{\alpha\cdot\beta} z^{\frac{\alpha^2}{2}} \omega^{\frac{\beta^2}{2}} e^{i(\alpha\cdot\varphi(z) + \beta\cdot\varphi(\omega))} \quad . \tag{15.227}$$

The integration over z may be performed, depending on the value of $\alpha \cdot \beta$. In order to treat the simplest operators, it is enough to consider the sublattice Λ_2, generated by vectors of length squared 2[516,517,519]. In this case, we have three possibilities for the above; if $\vec{\alpha}\cdot\vec{\beta} \geq 0$, there is no singularity, and the r.h.s. vanishes. If $\alpha\cdot\beta = -1$, the right hand side is simply obtained by setting $z = \omega$ in the integrand. Thus we obtain

$$\frac{1}{2\pi i} \oint \frac{d\omega}{\omega} \omega^{\frac{(\alpha+\beta)^2}{2}} e^{i(\alpha+\beta)\varphi(\omega)} = A_{\alpha+\beta} \quad . \tag{15.228}$$

If $\alpha \cdot \beta = -2$, we have $(\alpha + \beta)^2 = 0$; therefore $\alpha + \beta = 0$ (in the Euclidean case). Using the theorem of residues we compute the integral, taking the derivative of the integrand of the r.h.s. of (15.227) at $z = \omega$. As a result we obtain

$$r.h.s. = \frac{1}{2\pi i} \oint \frac{d\omega}{\omega} i\alpha\omega \frac{d\varphi}{d\omega} = \alpha \cdot p \quad . \tag{15.229}$$

Moreover using the same techniques, we can compute also the following relation

$$[p^\mu, A_\alpha] = \alpha^\mu A_\alpha \quad . \tag{15.230}$$

The above relations are neither commutation nor anticommutation relations. Nevertheless, we can transform the above relations into commutation relations by means of the non-commuting quantities c_α, $\alpha \in \Lambda$, whose commutator algebra reads

$$c_\alpha c_\beta = (-1)^{\alpha\cdot\beta} c_\beta c_\alpha \tag{15.231a}$$

$$c_\alpha c_\beta = \epsilon(\alpha \cdot \beta) c_{\alpha+\beta} \quad . \tag{15.231b}$$

We define the new operator

$$\chi(\alpha) = c_\alpha A_\alpha \quad , \tag{15.232}$$

which satisfies

$$[\chi(\alpha), \chi(\beta)] = 0 \quad , \quad \alpha \cdot \beta \geq 0 \quad , \tag{15.233a}$$

$$[\chi(\alpha), \chi(\beta)] = \epsilon(\alpha \cdot \beta)\chi(\alpha + \beta) \quad , \quad \alpha \cdot \beta = -1 \quad , \tag{15.233b}$$

$$[\chi(\alpha), \chi(\beta)] = \alpha \cdot p \quad , \quad \alpha = -\beta \quad . \tag{15.233c}$$

Therefore we have a Lie algebra g_Λ generated by the ladder operators $\chi(\alpha)$ and the Cartan subalgebra p_μ. The roots of g_Λ are the points of Λ with length squared 2; for lattices with non positive definite metric (non Euclidean), there are complications whose discussion lies outside the scope of this text, and we refer to [516,517] for a comprehensible and full account.

After defining the Lie algebra, we can also extend it to an affine Lie algebra, generalizing (15.226) to

$$A^n_{\vec{\alpha}} = \frac{1}{2\pi i} \oint \frac{dz}{z} z^n V(\vec{\alpha}, z) \quad . \tag{15.234}$$

All calculations in (15.228) through (15.233c) follow the previous pattern, except for the equation (15.229) where an extra (normal order) term appears when taking the derivative with respect to z^n. Nevertheless, the coefficient of that term is unity at $z = \omega$, and we obtain

$$[\chi_n(\alpha), \chi_m(\beta)] = 0 \qquad \alpha \cdot \beta \geq 0 \tag{15.235a}$$

$$[\chi_n(\alpha), \chi_m(\beta)] = \epsilon(\alpha, \beta)\chi_{m+n}(\alpha + \beta) \qquad \alpha \cdot \beta = -1 \tag{15.235b}$$

$$[\chi_n(\alpha), \chi_m(\beta)] = \alpha \cdot p_{m+n} + n\delta_{n,-m} \qquad \alpha = -\beta \tag{15.235c}$$

The above is a strongly simplified version of the Frenkel Kac construction[515].

There exist rather interesting relations of the previous construction with important theories. If the lattice is not Euclidean, one has also to compute $[A_\alpha, A_\beta]$ when $(\alpha + \beta)^2 = 0$ but $\alpha \neq -\beta$; One obtains

$$[A_\alpha, A_\beta] = \frac{1}{2\pi i} \oint \frac{dz}{z} \frac{\alpha - \beta}{2} iz \frac{d\varphi}{dz} e^{i(\alpha+\beta)\varphi(z)} \quad . \tag{15.236}$$

The r.h.s. of (15.236) plays an important role in string theories; it is a so called DDF operator, and describes "emission of elementary particles" from the string[91,518,521].

Moreover, we can also repeat the previous computations when the square length of the lattice vectors is unity, rather than 2. One has to be careful about the cuts introduced by the factor $z^{\alpha^2/2}$ in the integrals. In fact we have

$$e^{i\alpha\cdot\varphi(z)} = z^{\frac{\alpha^2}{2}+\alpha\cdot p} e^{i\alpha\cdot\varphi_0} e^{i\alpha\cdot\varphi_<(z)} e^{i\alpha\cdot\varphi_>(z)} \quad . \tag{15.237}$$

If $\alpha^2 = 2$, p must be contained in the dual Λ_R^* in order not to generate singularities other than isolated poles[11)]. If however $\alpha^2 = 1$ we need $\alpha \cdot p = Z + 1/2$. It is not clear how to deal with this situation in general; in the case of $O(2N)$ symmetry the situation may be completely mastered. In this case, we can simply define p to be in the sum of the lattice Λ_R with the spinor (or anti spinor) representation

$$p \in (\Lambda_R + \lambda_s) \cup (\Lambda_R + \lambda_{\bar{s}}) \quad . \tag{15.238}$$

For the $O(2N)$ case, these representations are related with the two extremes of the Dynkin diagram as seen in figure 15.1

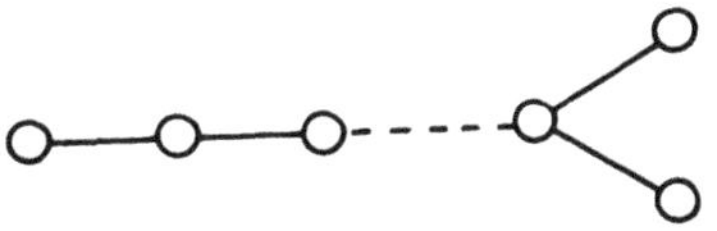

Figure 15.1: Dynkin diagram for the $O(2N)$ case.

11) Given an inner product for the lattice vectors, we define the dual lattice Λ^* as the set $\Lambda^* = \{y \,|y \cdot v \in Z\,, \forall v \in \Lambda\}$.

In this case, the operators $A_{\vec{\alpha}}$ obey commutation as well as anticommutation relations. The previous calculations (15.227-233) may again be performed without encountering new difficulties.

When computing A_α^2 we obtain no singularity in the integral (15.227). Therefore

$$A_\alpha^2 = 0 \quad , \quad \alpha^2 = 1 \quad . \tag{15.239}$$

Likewise

$$A_\alpha A_\beta - A_\beta A_\alpha = 0 \quad , \quad if \quad \alpha \cdot \beta = 0 \quad , \quad \alpha^2 = \beta^2 = 1 \quad . \tag{15.240}$$

In the case $\alpha = -\beta$, $\alpha^2 = 1$, we obtain $A_0 = 1$; hence, summarizing we have

$$A_\alpha A_{-\alpha} + A_{-\alpha} A_\alpha = 1 \quad , \quad \alpha^2 = 1 \quad . \tag{15.241}$$

Finally, we can compute also mixed commutators, involving A_{α_2}, $\alpha_2^2 = 2$ and A_{α_1}, $\alpha_1^2 = 1$. It is by now simple to obtain

$$A_{\alpha_1} A_{\alpha_2} - (-1)^{\alpha_1 \alpha_2} A_{\alpha_2} A_{\alpha_1} = 0 \quad , \quad \alpha_1 \cdot \alpha_2 = 0, 1 \tag{15.242a}$$

$$A_{\alpha_1} A_{\alpha_2} + A_{\alpha_2} A_{\alpha_1} = A_{\alpha_1 + \alpha_2} \quad , \quad \alpha_1 \cdot \alpha_2 = -1 \quad . \tag{15.242b}$$

If we define new quantities c_u, such that

$$c_\alpha c_\beta = (-1)^{(\alpha \cdot \beta)^2 - \alpha^2 \beta^2} c_\beta c_\alpha \tag{15.243a}$$

$$c_\alpha c_\beta = \epsilon(\alpha, \beta) c_{\alpha+\beta} \tag{15.243b}$$

$$\chi(\alpha) = c_\alpha A_\alpha \quad . \tag{15.243c}$$

we obtain only anticommutation relations for $\chi(\alpha)$. This is again, a form of the (non abelian) bosonization discussed in chapter 9, and represented by Eq. (15.184). Generalizing the Frenkel-Kac construction to this case, one obtains[516,517] the Neveu-Schwarz[521] and Ramond fields[522]. The above factors c_α are also known, in field theory, as Klein factors.

15.7 Further Applications to String Theories

One important issue of string theory, is the compactification process, where the theory, at the beginning defined on a d-dimensional manifold, has $d - d_f$ of its dimensions curved to form a compact manifold, leaving d_f-dimensional Minkowskian space-time components[91,523]. The physically interesting case corresponds to $d_f = 4$. This Kaluza-Klein type compactification[524] is crucial in order for the theory to provide experimental predictions. Now, the σ model Lagrangian (6.69) and (7.164) may be seen as a model for strings, if the fundamental field is taken as the string coordinate. Hence, it is necessary to understand the sigma model defined on the compactified manifold. Since conformal invariance is a necessary ingredient of string theory, it turns out to be important to investigate sigma models having this property. The best understood is the previously discussed WZW model.

The main problems one has to tackle concern the definition of physical states, and the vertex operators[518,520].

The bosonic string theory is basically described by the position operator

$$X^\mu(\tau,\sigma) = x^\mu + 2p^\mu\tau - i\sum_{n\neq 0}\frac{1}{n}\alpha_n^\mu z^{-n} - i\sum_{n\neq 0}\frac{1}{n}\tilde{\alpha}_n^\mu \bar{z}^{-n} \quad , \tag{15.244}$$

where the $z, \bar{z}$ variables are

$$z = e^{ix^+} \quad , \quad \bar{z} = e^{ix^-} \quad , \tag{15.245}$$

and are taken as being independent. Moreover α_n^μ satisfies the algebra

$$[\alpha_m^\mu, \alpha_n^\nu] = -m\delta_{m,-n}g^{\mu\nu} \tag{15.246a}$$

$$[\tilde{\alpha}_m^\mu, \alpha_n^\nu] = 0 \tag{15.246b}$$

$$[\tilde{\alpha}_m^\mu, \tilde{\alpha}_n^\nu] = -m\delta_{m,-n}g^{\mu\nu} \quad . \tag{15.246c}$$

The $z, \bar{z}$ derivatives of X^μ

$$\partial_z X^\mu = i\sum_{-\infty}^{\infty}\alpha^\mu(n)z^{-n-1} \tag{15.247}$$

$$\partial_{\bar{z}} X^\mu = i\sum_{-\infty}^{\infty}\tilde{\alpha}^\mu(n)\bar{z}^{-n-1} \quad , \tag{15.248}$$

are nothing but the currents $J^\mu(z;a)$ for a group $U(1)^d$, where a is a parameter, $a = 1,\cdots,d$ labeling each $U(1)$ factor. An energy momentum tensor presenting both contributions (of a group manifold G and $U(1)^d$) is

$$T(z) = -\frac{1}{2}\partial_z X^\mu \partial_z X^\mu + \frac{1}{c_V+\kappa}J^a(z)^2 \quad . \tag{15.249}$$

The central charge of the Virasoro algebra is given by[503]

$$c = d_f + \frac{\kappa \dim G}{c_V + \kappa} \quad . \tag{15.250}$$

and includes a contribution from each of the d_f (free) coordinates, and from the compactified non abelian group manifold Eq. (15.133). Covariant quantization of strings requires, in the bosonic case, that we deal with the central charge having the value 26:

$$d_f + \frac{\kappa\, \mathrm{dim} G}{c_V + \kappa} = 26 \quad . \tag{15.251}$$

The ground state is characterized by the fact that it is annihilated by J^a_n, $n > 0$, and such that

$$J^a_0 |i\bar{i}\rangle = (\tau^a_R)_{ij} |j\bar{i}\rangle \tag{15.252a}$$

$$\overline{J}^a_0 |i\bar{i}\rangle = (\tau^a_{\overline{R}})_{\bar{i}\bar{j}} |j\bar{i}\rangle \quad , \tag{15.252b}$$

where τ^a_R are Gell-Mann matrices belonging to representation R. New states are obtained by the action of J^a_{-n}'s. The spectrum follows from the application of L_0 to such states

$$L_0 \prod J^{a_i}_{-n_i} |R\rangle = \left(\frac{c_g}{c_V + \kappa} + \sum n_i \right) |R\rangle \quad . \tag{15.253}$$

Thus, the on shell condition is written as

$$m^2 = -2 + \frac{c_R + c_{\overline{R}}}{c_V + \kappa} + N + \overline{N} + M + \overline{M} \quad , \tag{15.254}$$

where $N, \overline{N}, M, \overline{M}$ are the number operators for $J, \overline{J}, \alpha$ and $\overline{\alpha}$. We also have to satisfy the (closed string) constraint $L_0 = \overline{L}_0$, that is,

$$M + N + \frac{c_g}{c_V + \kappa} = \overline{N} + \overline{M} + \frac{\overline{c}_g}{c_V + \kappa} \quad , \tag{15.255}$$

which has a solution, only if $\frac{c_g - \overline{c}_g}{c_V + \kappa}$ is an integer.

We have seen previously that some constraints may be imposed on the primary states, such as to define a finite representation of the current algebra; we call these representations integrable. In principle, almost all of the fields of the theory are non integrable. There are selection rules arising from the current algebra that are very restrictive, and it has been shown by Gepner and Witten[503] that correlators involving any non integrable field vanishes. Hence, the integrable fields provide a complete representation of the theory. We shall not give here the detailed proof. The idea is to consider an arbitrary correlator with an extra (integrable) field ϕ_λ of highest weight in the Kac-Moody algebra, and to act on it with a suitable power of J^θ_{-1}, such as to obtain zero:

$$\langle (J^\theta_{-1})^{M+1} \phi_\lambda(z) \phi_1(z_1) \cdots \phi_n(z_n) \rangle = 0 \quad , \tag{15.256a}$$

with

$$(J^\theta_{-1})^M \phi_\lambda(z) = \chi_n(z) \quad . \tag{15.256b}$$

After applying Ward identity (15.256) several times and performing some suitable manipulations one obtains[503]

$$(\tau_r)^{m+1}\langle\phi_1(z_1)\cdots\phi_n(z_n)\rangle = 0 \quad . \tag{15.257}$$

If the field ϕ_r were not integrable, Eq. (15.257) would show that the correlator vanishes identically.

There are also consequences for the correlators involving integrable fields. If one considers the fusion of two fields in a given representation of the Kac-Moody algebra, one can work out the resulting field, analysing the behavior of the three point function, which must obey

$$(\tau_1^{\theta})^{l_1}(\tau_2^{\theta})^{l_2}\langle\phi_\lambda(z)\phi_1(z_1)\phi_2(z_2)\rangle = 0 \quad , \tag{15.258}$$

for $l_1 + l_2 \geq M + 1$.

Analysing modular invariance of the partition function, it is possible to obtain further information about the theory, and in particular, about the spectrum[503]. It is interesting to point out that besides the usual fundamental field g and its powers g^l, there may appear also the field ψ, the soliton of this theory (remember that the soliton of the sine-Gordon theory is the elementary Thirring field). In such case we are referred back to the discussion of the $SU(N)$ conformal Thirring field, section 15.5.

15.8 Superconformal Symmetry

Supersymmetry seems to be in the realm of almost all attempts to obtain a unified theory of all interactions[525,526]. In particular, many problems of string theory are solved once supersymmetry is introduced[91,527]. The interplay of supersymmetry and conformal invariance is an interesting application of the rules presented up to now.

The simplest way to introduce the superconformal group and its realization in terms of a doublet (φ, χ) of a bosonic (φ) and a fermionic (χ) fields, is to represent the energy momentum tensor by the superfield[528]

$$T(z,\theta) = T_{z\theta}(z) + T_{zz}(z)\theta \quad , \tag{15.259}$$

where $T_{zz}(z)$ is the zz component of the energy momentum tensor. The matter superfield is

$$\phi(z,\theta) = \varphi(z) + \theta\chi(z) \quad , \tag{15.260}$$

with a supersymmetric derivative defined by

$$D = \frac{\partial}{\partial\theta} + \theta\frac{\partial}{\partial z} \quad . \tag{15.261}$$

In terms of components we write

$$T_{z\theta}(z) = \frac{1}{2} \sum_{n=-\infty}^{\infty} z^{-n-\frac{3}{2}} G_n \quad , \tag{15.262a}$$

$$T_{zz}(z) = \sum_{n=-\infty}^{\infty} z^{-n-2} L_n \quad , \tag{15.262b}$$

$$\varphi(z) = z^{-h} \sum_{n=-\infty}^{\infty} z^{-n} \varphi_n \quad , \tag{15.262c}$$

$$\chi(z) = z^{-h-1/2} \sum_{n=-\infty}^{\infty} z^{-n} \chi_n \quad , \tag{15.262d}$$

where h is the anomalous dimension of the matter supermultiplet.

The operator product expansion of the energy momentum tensor and the matter field is given by[528]

$$T(z,\theta)\phi(z,\theta') = \Delta \frac{\theta-\theta'}{(z-z')^2}\phi(z',\theta') + \frac{1}{z-z'}\frac{1}{2}D\phi(z',\theta') + \frac{\theta-\theta'}{z-z'}\partial\phi \quad , \tag{15.263}$$

where the first term corresponds to the usual term describing the dimension of φ and χ, the second term contains the supersymmetry transformation and where the third term, completes the conformal transformation, as one readily verifies. For $\phi(z,\theta)$ replaced by $T(z,\theta)$ itself, we also have a central term

$$T(z,\theta)T(z',\theta') = \frac{\frac{1}{4}\hat{c}}{(z-z')^3} + \frac{\frac{3}{2}(\theta-\theta')}{(z-z')^2}T(z',\theta') + \frac{\frac{1}{2}}{z-z'}d'T + \frac{\theta-\theta'}{z-z'}\partial T \quad . \tag{15.264}$$

In terms of the components (15.262*a*, *b*) we obtain from (15.264),

$$[L_n, L_m] = (n-m)L_{n+m} + \frac{1}{8}\hat{c}n(n^2-1)\delta_{n,-m} \tag{15.265a}$$

$$\{G_n, G_m\} = 2L_{m+n} + \frac{1}{2}c\left(n^2 - \frac{1}{4}\right)\delta_{n,-m} \tag{15.265b}$$

$$[L_n, G_m] = \left(\frac{1}{2}n - m\right)G_{m+n} \quad , \tag{15.265c}$$

and using (15.262*c*, *d*), we have

$$[L_n, \varphi] = z^{n+2}\partial\varphi + \Delta(n+2)z^n\varphi \tag{15.266a}$$

$$[G_n, \varphi] = z^{n+\frac{1}{2}}\chi \tag{15.266b}$$

$$[L_n, \chi] = z^{n+2}\partial\varphi + \left(\Delta + \frac{1}{2}\right)(n+1)z^n\chi \tag{15.266c}$$

$$[G_n, \chi] = z^{n+\frac{1}{2}}\partial\varphi + 2\left(n+\frac{1}{2}\right)\Delta z^{n-\frac{1}{2}}\varphi \quad . \tag{15.266d}$$

Notice that (15.265) corresponds to the Neveu-Schwarz-Ramond algebra[521,522]. Actually, the theory has two subspaces. In the Ramond subspace[522] we have for the fermionic component

$$\chi(e^{2\pi i}z) = \chi(z) \quad , \tag{15.267}$$

while in the Neveu-Schwarz (NS) sector[521], the fermions satisfy antiperiodic boundary conditions:

$$\chi(e^{2\pi i}z) = -\chi(z) \quad ; \tag{15.268}$$

Thus, in the NS sector the index n in G_n is half integer, while in the Ramond sector it is integer valued.

The theory is realized by the Lagrangian density[529]

$$\mathcal{L} = \sqrt{-g}\left[\frac{1}{2}g^{\alpha\beta}\partial_\alpha\varphi^\mu\partial_\beta\varphi_\mu + \frac{i}{2}\overline{\psi}^\mu\gamma^\alpha D_\alpha\psi_\mu + \frac{i}{2}\overline{\chi}_\alpha\gamma^\beta\gamma^\alpha\psi^\mu(\partial_\beta\varphi_\mu - \frac{i}{4}\chi_\beta\psi_\mu)\right] \quad , \tag{15.269}$$

which is invariant under the local supersymmetry transformations

$$\delta e^\alpha_a = 2i\overline{\chi}^\alpha\gamma_a\epsilon \tag{15.270a}$$

$$\delta\chi^a_\alpha = -D_\alpha\epsilon^a = -\left(\partial_\alpha\epsilon^a + \frac{1}{2}\omega^{ab}_\alpha\epsilon^b\right) \tag{15.270b}$$

$$\delta\varphi^\mu = \overline{\epsilon}\psi^\mu \tag{15.270c}$$

$$\delta\psi^\mu = -i\left[\partial_\alpha\varphi^\mu + \overline{\chi}_\alpha\psi^\mu\right]\gamma^\alpha\epsilon + \frac{1}{2}\varphi^\mu\overline{\psi}^n u\psi_\nu \quad , \tag{15.270d}$$

where the zweibein e^α_a was introduced in such a way that $g^{\alpha\beta} = e^\alpha_a e^\beta_b\eta^{ab}$, with η^{ab} the Minkowskian metric. Moreover, if γ_a is the Minkowskian Dirac matrix in flat space, $\gamma^\alpha = e^\alpha_a\gamma^a$. The spin connection $\omega^{ab}_\alpha = \gamma_5^{ab}\omega_\alpha$ does not appear in the action since in 2 dimensions the Majorana field satisfies $\overline{\psi}\gamma^\alpha\gamma_5\psi = 0$.

In the super conformal gauge, defined by[529]

$$g_{\alpha\beta} = \rho\delta_{\alpha\beta} \tag{15.271a}$$

$$\chi_\alpha = \gamma_\alpha\xi \quad , \tag{15.271b}$$

the graviton and gravitino fields ρ and χ decouple, and the Lagrangian density is given by

$$\mathcal{L} = \frac{1}{2}(\partial_\alpha\varphi^\mu)^2 - \frac{i}{2}\overline{\psi}^\mu\,\partial\!\!\!/\psi_\mu \quad . \tag{15.272}$$

The fact that the above Lagrangian with two dimensional supersymmetry describes a D-dimensional superstring with space time supersymmetry, that is with supersymmetric generators acting on the states rather than on the fields φ^μ, ψ^μ, is a feature that may be understood from the bosonization techniques[91].

Indeed, the fermion bilinear

$$j^{\mu\nu}(z) = \overline{\psi}^{\mu}(z)\psi^{\nu}(z) \quad , \tag{15.273}$$

obeys a current algebra

$$[j_n^{\mu\nu}, j_m^{\kappa\delta}] = f_{\lambda\sigma}^{\mu\nu\kappa\rho} j_{m+n}^{\lambda\sigma} + kn(\delta^{\mu\kappa}\delta^{\nu\delta} - \delta^{\mu\delta}\delta^{\nu\kappa})\delta_{n,-m} \quad , \tag{15.274}$$

where

$$j_n^{\mu\nu} = \frac{1}{2\pi i}\oint dz z^n j^{\mu\nu}(z) \quad . \tag{15.275}$$

The Sugawara representation of L_{-1} derived from[528] $T = -\frac{1}{4(d-1)}:j^{\mu\nu}j_{\mu\nu}:$ suggests the constraint

$$\left(\frac{1}{d-1}j_{-1}^{\mu\nu}M_{\mu\nu} - L_{-1}\right)\phi_i = 0 \quad , \tag{15.276}$$

where $j_0^{\mu\nu} = M^{\mu\nu}$ are the Lorentz generators. This implies differential equations which should be obeyed by the correlation functions, analogous to Eq. (15.76) in the abelian case, or (15.160), in the non abelian case.

In the light-cone gauge, the little group $SO(8)$ is associated with j^{ij}. The three representations $v, s, \bar{s}$ of $SO(8)$ are related by the triality of the Dynkin diagram, and the current (15.272) may be written in terms of the true spinor field $S_\alpha(z)$; (Indeed, $S_\alpha(z)$ is true in the sense that it behaves as a spinor under Lorentz transformations, while ψ^μ transforms as a vector). In this case the ψ^μ plays the role of generalized γ matrices, with the operator algebra

$$\psi^\mu(z)S_\alpha(\omega) = (z-\omega)^{-\frac{1}{2}}\gamma^\mu_{\alpha\beta}S^\beta(\omega) + \cdots \tag{15.277a}$$

$$S^\alpha(z)S_\beta(\omega) = (z-\omega)^{-\frac{5}{4}}\delta^\alpha_\beta + (z-\omega)^{-\frac{1}{4}}\left(\frac{1}{2}\gamma_\mu\gamma_\nu\right)^\alpha_\beta \psi^\mu(z)\psi^\nu(\omega) \tag{15.277b}$$

$$S_\alpha(z)S_\beta(\omega) = (z-\omega)^{-\frac{3}{4}}\gamma^\mu_{\alpha\beta}\psi_\mu(\omega) + \cdots \quad . \tag{15.277c}$$

The relation between single valuedness of ψ and double valuedness of S is obtained from the bosonic representation[91].

A construction similar to the Coulomb gas presented in section (15.3) may be built once again. As it turns out[530], the value of the central charge can be modified by means of the introduction of a charge $2\alpha_0$ at infinity, in the computation of expectation values, such that $c = 1 - 16\alpha_0^2$. We have to introduce supersymmetric operators of zero dimension given by

$$R_\pm = \int d\theta dz V_{\alpha_\pm}(z) \quad , \tag{15.278}$$

which have zero dimension and are supercovariant if

$$\alpha_\pm = \alpha_0 \pm \sqrt{\alpha_0^2 + 1/2} \quad , \tag{15.279}$$

or, in terms of the central charge

$$\alpha_\pm = \frac{1}{4}\left[\sqrt{1-c} \pm \sqrt{9-c}\right] \quad . \tag{15.280}$$

If we impose that α_+ and α_- be related by integer numbers[79,87]

$$\frac{\alpha_+}{\alpha_-} = -\frac{p}{q} \quad , \tag{15.281}$$

we have , for the central charge, the value

$$c = 1 - 2\frac{(p-q)^2}{pq} \quad . \tag{15.282}$$

For $q = p + 2$ we obtain the unitary series[497,498]

$$c = 1 - \frac{8}{p(p+2)} \quad . \tag{15.283}$$

15.9 Conformally Invariant Two Dimensional Gravity

15.9.1 The general theory of gravity in two dimensions

A very important application of the ideas of conformal invariance is the solution of induced two dimensional gravity. This theory is rather peculiar in two dimensional space-time. The equation of motion of Einstein's theory of gravitation coupled to matter fields is

$$R_{\mu\nu} - \frac{1}{2}g_{\mu\nu}R = \kappa T_{\mu\nu} \quad , \tag{15.284}$$

where $T_{\mu\nu}$ is the energy momentum tensor of matter, and κ is the Newton constant. In two dimensional space-time this equation does not imply a dynamics for the graviton, since one has the relation

$$R_{\mu\nu} = \frac{1}{2}g_{\mu\nu}R \quad , \tag{15.285}$$

valid identically in two dimensions. This is a consequence of the Gauss-Bonnet theorem[531], which states that Einstein's Lagrangian is a total derivative, and the action is a topological invariant, which measures the genus of the manifold on which one integrates[91,532,533]. If the manifold is compact, we have the following expression for the Einstein action

$$S = \int_m d^2x\sqrt{-g}R = 2\pi n = 2\pi(1-g) \quad , \tag{15.286}$$

where for a compact M, g is the number of handles (for a sphere $g = 0$, for a 1-torus, $g = 1$, etc.). In this case, the equation of motion of the gravitational field turns out to be an equation of constraint, rather than a dynamical equation. This is the case in string and superstring theories at the critical dimension: the graviton and gravitino equations of motion ensure reparametrization and super-reparametrization invariance, by means of the Virasoro (and super Virasoro) conditions[18].

Moreover, Eqs. (15.284,285) imply that

$$g^{\mu\nu}T_{\mu\nu} = 0 \quad , \tag{15.287}$$

which is true if the theory is scale invariant. In general, this is not true at the quantum level, due to the trace anomaly[156,157,158].

15.9.2 Induced gravity; contribution from the matter fields to the determinant

Although Eqs. (15.284-285) imply at the classical level equations of constraint for the matter fields, at the quantum level, integration over the latter may give rise to non trivial self interactions for the graviton field $g_{\mu\nu}$. In the case of scalar matter, we must compute the following determinant

$$\det\Box = \det\frac{1}{\sqrt{-g}}\partial^\mu\sqrt{-g}g_{\mu\nu}\partial^\nu \quad , \tag{15.288}$$

One would expect, at first, this determinant to be trivial, since in the conformal gauge

$$\Box = e^\sigma\partial^2 \quad , \tag{15.289a}$$

where

$$g^{\mu\nu} = e^\sigma\eta^{\mu\nu} \quad , \tag{15.289b}$$

and

$$\partial^2 = \partial^\mu\partial^\nu\eta_{\mu\nu} \quad . \tag{15.289c}$$

Therefore, the σ contribution would decouple and have no propagation. However, due to the conformal anomaly[156,157,158], the above conclusion is incorrect, and we have a non trivial contribution for this determinant[82].

Teichmüller parameters

In fact, there are three types of contributions, due to the bosonic fields, fermionic fields, and ghosts. All contributions turn out to provide the same sort of action for the σ field, with different coefficients, which we shall compute at the end. The effective action itself may be computed by the heat kernel technique[138,139]. First, let us separate the pure gauge piece for the integration over the metric. An integral over the metric may be performed by integrating over the conformal factor and the traceless symmetric part h of $g_{\alpha\beta}$, that is[154,534],

$$dg_{\alpha\beta} = 2g_{\alpha\beta}d\sigma + dh_{\alpha\beta} \quad , \tag{15.290}$$

with $g^{\alpha\beta}dh_{\alpha\beta} = 0$. Hence the scalar product

$$||dg||^2 = \int d^2x\sqrt{-g}(g^{\alpha\beta\gamma\delta} - g^{\alpha\beta}g^{\gamma\delta})dg_{\alpha\beta}dg_{\gamma\delta} \quad , \tag{15.291}$$

where $g^{\gamma\delta}_{\alpha\beta}$ is given by

$$g^{\gamma\delta}_{\alpha\beta} = \frac{1}{2}(\delta^{\gamma}_{\alpha}\delta^{\gamma}_{\beta} + \delta^{\gamma}_{\alpha}\delta^{\gamma}_{\beta} - g_{\alpha\beta}g^{\gamma\delta}) \quad , \tag{15.292}$$

decomposes as follows

$$||dg||^2 = \int d^2x\sqrt{-g}g^{\alpha\beta\gamma\delta}dh_{\alpha\beta}dh_{\gamma\delta} + 8u\int d^2x\sqrt{-g}(d\sigma)^2 \quad , \tag{15.293}$$

where we introduced an arbitrary real parameter u. The functional integral over $g_{\alpha\beta}$ may be performed by integrating over $d\sigma$ and $dh_{\alpha\beta}$. At this point we go to the Euclidian space.

A gauge transformation on g acts as

$$dg_{\alpha\beta} = D_\alpha dV_\beta + D_\beta dV_\alpha \quad , \tag{15.294}$$

where dV_α is the infinitesimal generator of local transformations[534]. One thus obtains for a gauge transformation on $dh_{\alpha\beta}$

$$dh_{\alpha\beta} = 2g^{\gamma\delta}_{\alpha\beta}D_\gamma dV_\delta \equiv (P_1 dV)_{\alpha\beta} \quad . \tag{15.295}$$

The transformation of the conformal factor is trivial. In order to separate the gauge volume, using the Faddeev-Popov method, we need to compute the following determinant[138,534]

$$\frac{\partial dh_{\alpha\beta}}{\partial dV} = \det|P_1| = (\det P_1^+ P_1)^{1/2} \quad . \tag{15.296}$$

In the above, P_1 is a special case of the operator P_n, corresponding in holomorphic variables, to the derivative operator mapping real symmetric traceless vectors of order n into real symmetric traceless vectors of order $n+1$. Two remarks are now in order. The first concerns zero modes which must be extracted from (15.296). The second concerns gauge transformations which have no infinitesimal realization. They are given in terms of a finite number of parameters (see appendix N), called Teichmüller parameters[154]. We will restrict ourselves to a space without handles or holes, where these gauge transformations do not show up.

Heat kernel computation

We now compute the determinant using the heat kernel defined by[138,139]

$$G(x,x';\tau) = \langle x|e^{-\tau\Delta}|x'\rangle \quad , \tag{15.297}$$

with $\lim_{\tau\to 0} G(x,x';\tau) = \delta(x-x')$. From (15.297) we find that G obeys the equation of motion

$$\left(\frac{\partial}{\partial\tau} + \Delta\right) G = 0 \quad . \tag{15.298}$$

For the free theory, the above equation may be solved. We choose a perturbative approach, which is sufficient for our purposes. Actually, to calculate the induced Lagrangian, the procedure allows for the exact result. Thus, we write:

$$\Delta = \partial^2 + V \quad , \tag{15.299}$$

and obtain the formal solution

$$G = G_0 + G_0 V G \quad , \tag{15.300}$$

where G_0 obeys the free equation ($V = 0$). The determinant is given by the expression (4.127). Since we need the expansion of (15.297) in τ only up to first order to compute the determinant (see (4.132)), it is sufficient to consider

$$G = G_0 + G_0 V G_0 + \mathcal{O}(\sqrt{\tau}) \quad . \tag{15.301}$$

We have

$$G_0(x,0;\tau) = \frac{1}{4\pi\tau} e^{-\frac{x^2}{4\tau}} \quad , \tag{15.302}$$

and from (15.289*a*),

$$V(x) = (e^{-2\sigma} - 1)\Delta \quad . \tag{15.303}$$

Since we only need the trace of the diffusion operator, it suffices to know G for almost coinciding points. We choose $\sigma(0) = 0$, $\partial\sigma(0) = 0$, by performing a suitable conformal transformation. Thus,

$$V = \frac{1}{2} x^\alpha x^\beta \partial_\alpha \partial_\beta \sigma(0) \partial^2 + \mathcal{O}(x^2) \tag{15.304}$$

and

$$\begin{aligned} G(0,0;\tau) &= G_0(0,0;\tau) \\ &\quad + \int d^2x \int_0^\tau d\tau' G_0(0,x;\tau-\tau')[-x^\alpha x^\beta \partial_\alpha\partial_\beta\sigma(0)]\partial^2 G_0(x,0;\tau') + \mathcal{O}(\sqrt{\tau}) \\ &= \frac{1}{4\pi\tau} + \frac{1}{12\pi}\partial^2\sigma(0) + \mathcal{O}\left(\sqrt{\tau}\right) \quad . \end{aligned} \tag{15.305}$$

We use now Eq. (4.132) and integrate the above result (15.301), to obtain first the variation of the determinant, as given by

$$\delta \ln \det \Delta = -\frac{1}{4\pi\epsilon}\int d^2x e^{-\sigma}\delta\sigma - \frac{1}{24\pi}\int d^2x\sqrt{g}R\delta\sigma \tag{15.306}$$

Since we are in the conformal gauge (15.289*b*), we immediately integrate the above equation to obtain

$$\begin{aligned}\ln \det\Box &= -\int d^2x\sqrt{g}\int_\epsilon^\infty \frac{d\tau}{\tau}G(x,x;\tau) = \\ &= \frac{1}{4\pi\epsilon}\int d^2x\sqrt{g} + \frac{1}{12\pi}\int d^2x\partial^\mu\sigma\partial_\mu\sigma \\ &= \frac{1}{12\pi}\int d^2x\{(\partial^\mu\sigma\partial_\mu\sigma) + \mu^2 e^{-\sigma}\} \quad ,\end{aligned} \tag{15.307}$$

up to terms of the order ϵ, where μ is a regularization dependent constant. For the contribution of the ghost we have (for a long discussion see [62])

$$-\frac{26}{48\pi}\int d^2x\{\partial^\mu\sigma\partial_\mu\sigma + \mu^2 e^{-\sigma}\} \quad . \tag{15.308}$$

With D bosonic fields and for the ghosts we have the famous combination $D-26$. An alternative to the above tricky calculation is to use Feynman diagrams, which give the ghost result (-26) in a simpler way[534].

The effective action for the gravitational field is readily obtained, once we integrate out the matter fields. It is given by the functional integral

$$G(g) = \int \mathcal{D}x^\alpha e^{i\int \sqrt{-g}g^{\alpha\beta}\partial_\alpha x^a \partial_\beta x^a d^2\xi} \quad , \tag{15.309}$$

where we now returned to Minkowskian formulation.

Perturbative computation

In the weak field approximation, we decompose the gravitational field as

$$g_{\alpha\beta} = \eta_{\alpha\beta} + \delta g_{\alpha\beta} \quad . \tag{15.310}$$

Up to first order in δg we have for the action

$$S = S_0 + \delta S \quad , \tag{15.311a}$$

$$\delta S = \int \sqrt{-g}\delta g^{\mu\nu}T_{\mu\nu}d^2x \quad , \tag{15.311b}$$

and

$$T_{\alpha\beta} = \partial_\alpha x \partial_\beta x - \frac{1}{2} g_{\alpha\beta}(g\partial x \partial x) \quad (15.311c)$$

is the energy momentum tensor; g is the determinant of $g_{\mu\nu}$. The action S_0 is the flat space contribution. The effective action obtained by performing the x integration in (15.309) can be calculated, at one loop order from the diagram 15.2, with the result

$$i\delta S = -\frac{D}{8}\int \frac{d^2k}{(2\pi)^2}\times \frac{[k_\alpha(p+k)_\beta + k_\beta(p+k)_\alpha - \eta_{\alpha\beta}k\cdot(p+k)][(\alpha\beta)\leftrightarrow(\gamma\delta)]}{k^2(p+k)^2}\delta g_{\alpha\beta}(p)\delta g_{\gamma\delta}(-p) \quad (15.312)$$

The diagram 15.3 vanishes due to the tracelessness of δg.

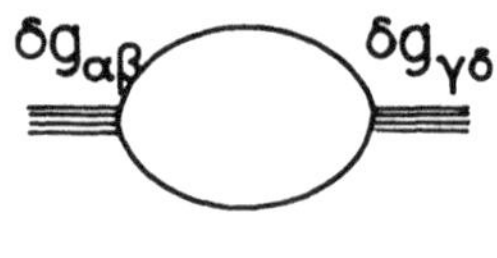

Figure 15.2: One loop contribution to the graviton effective action

Figure 15.3: Vanishing contribution to the one loop effective action.

The integral (15.312) is strongly singular, and requires a careful treatment. Using light cone variables we have

$$\delta S = -\frac{D}{2}\delta g_{++}\delta g_{++}\int \frac{d^2k}{(2\pi)^2}\frac{k_-^2(p_-+k_-)^2}{k^2(p+k)^2} - \frac{D}{2}\delta g_{--}\delta g_{--}\int \frac{d^2k}{(2\pi)^2}\frac{k_+^2(p_++k_+)^2}{k^2(p+k)^2}$$
$$-D\delta g_{++}\delta g_{--}\int \frac{d^2k}{(2\pi)^2}\frac{k_-^2(p_-+k_-)^2}{k^2(p+k)^2} \quad (15.313)$$

where $d^2k = dk_+dk_-$.

The self energy is given by the expression

$$\Pi_{++,++} = -\frac{D}{8}\int \frac{d^2k}{(2\pi)^2}\frac{k_+(p_++k_+)}{(k_-+i\epsilon\mathrm{sign}\,k_+)(p_-+k_-+i\epsilon\mathrm{sign}(p_++k_+))}$$
$$= -\frac{D}{16\pi^2}\int dk_+(p_++k_+)\int \frac{dk_-}{(k_-+i\epsilon\mathrm{sign}k_+)(p_-+k_-+i\epsilon\mathrm{sign}\,k_+)} \quad (15.314a)$$

The above integral is performed by elementary methods, resulting in[534]

$$\Pi_{++,++} = -\frac{D}{48\pi}\frac{p_+^3}{p_-} \quad . \tag{15.314b}$$

Thus we obtain for the quadratic part of the action, a sum of terms consisting of (15.314), the corresponding term with + replaced by −, and further regularization dependent terms containing g_{+-} as well[534]:

$$i\delta S = -\frac{D}{48\pi}\int\frac{d^2p}{(2\pi)^2}\Big(\delta g_{--}\frac{p_+^3}{p_-}\delta g_{--} + \delta g_{++}\frac{p_-^3}{p_+}\delta g_{++} + \Lambda\delta g_{++}p^2\delta g_{--} + \Lambda'\delta g_{+-}p^2\delta g_{+-} + \Lambda''\delta g_{+-}p_-^2\delta g_{++} + \delta g_{+-}p_+^2\delta g_{--}\Big) \ , \tag{15.315}$$

where $\Lambda, \Lambda', \Lambda''$ are quadratically divergent counterterms[531]. For the gauge invariant piece of the above contribution, we have a finite result, which is part of the gauge invariant action

$$\delta S = -\frac{D}{48\pi}\int\frac{d^2p}{(2\pi)^2}R(p)\frac{1}{p^2}R(p) \quad . \tag{15.316}$$

In the conformal gauge it is the free part of the Liouville action

$$\delta S = -\frac{D}{48}\int\frac{d^2x}{(2\pi)^2}(\partial_\mu\varphi)^2 \quad . \tag{15.317}$$

In the light cone gauge $g_{--} = 0 \, , g_{+-} = \frac{1}{2}$, it is a non local action, but with the advantage of not requiring a cosmological term, since $\sqrt{g} = 1$, while in the conformal gauge $\sqrt{g} = e^{-\varphi}$, and we need to include a counterterm to (15.317) in the the Liouville action.

We still need to calculate the ghost contribution to the determinant (15.296) of the operator

$$(P_1^\dagger P_1)_\alpha^\beta = D_\alpha^\beta + D^2\delta_\alpha^\beta - D^\gamma g_{\alpha\gamma}D^\beta \quad . \tag{15.318}$$

In terms of a functional integral this determinant is given by $\left(\text{we take } g = \begin{pmatrix} g_{++} & \frac{1}{2} \\ \frac{1}{2} & 0\end{pmatrix}\right)$

$$(\det P^+P)^{-\frac{1}{2}} = \int \mathcal{D}\omega_+\mathcal{D}\omega_- e^{i\int d^2\xi\{\partial_+\omega_+\partial_-\omega_- + \frac{1}{2}\partial_-g_{++}\omega_+\partial_-\omega_- - \frac{1}{2}g_{++}\partial_-\omega_+\partial_-\omega_-\}} \quad . \tag{15.319}$$

where $\omega_\pm$ are (Grassmann-valued) ghost fields. The functional integration amounts to a one loop diagram. To second order in δg one finds (see figure 15.3)

$$i\delta S_{gh} = \frac{1}{2}\int\frac{d^2k}{(2\pi)^2}\Big(\frac{k_-^2(k_-+p_-)^2}{k^2(k+p)^2} + 2p_-\frac{k_-^2(k_-+p_-)^2}{k^2(k+p)^2} - p_-^2\frac{(k_-+p_-)^2}{k^2(k+p)^2}\Big)\delta g_{++}\delta g_{++} \tag{15.320}$$

which, upon integration yields,

$$i\delta_{gh}S = \frac{i}{2}\frac{26}{48\pi}\frac{2p_-^3}{p_+}\delta g_{++}\delta g_{++} = i\frac{26}{48\pi}g_{++}\frac{p_-^3}{p_+}g_{++} \quad . \tag{15.321}$$

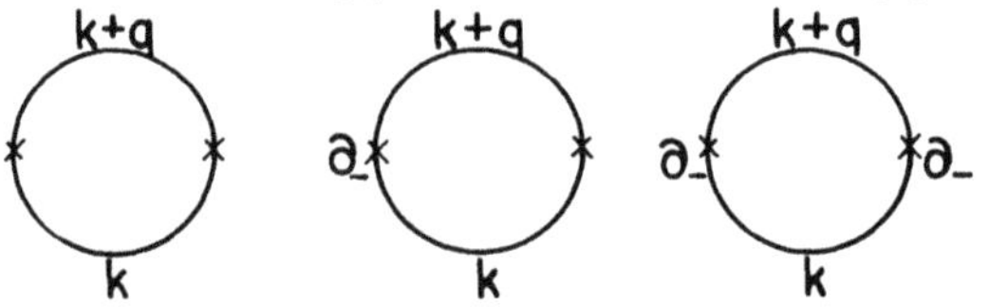

Figure 15.4: One loop ghost contribution to the graviton effective action.

Therefore we finally have for the effective action the expression

$$S_{eff} = \frac{-D+26}{48\pi}\int d^2x R\frac{1}{\Box}R \quad . \tag{15.322}$$

Calculation of central charges

The last and simplest way to arrive at the coefficient of the effective action is by means of the computation of the central charge, which measures the departure from the reparametrization invariance (for a modern and rigorous statement one has to resort to a BRST analysis[446]). In that case, we write the total energy momentum tensor (we shall take only the ++ component):

$$T^B_{++} = \partial_+X\partial_+X \tag{15.323}$$

$$T^{gh}_{++} = \frac{1}{2}c^+\partial_+b_{++} + \partial_+c^+b_{++} \quad , \tag{15.324}$$

where the ghost action was taken as[12)]

$$S_g = \int d^2x\sqrt{g}g^{\alpha\beta}c^\gamma D_\alpha b_{\beta\gamma} \quad . \tag{15.325}$$

We use the two point functions (see appendix A)

$$\underbrace{X^\alpha(x)X^\beta(0)} = -\ln(x^2)\delta^{\alpha\beta} \tag{15.326}$$

$$\underbrace{c^+(x)b_{++}(0)} = \frac{1}{x_-} \quad , \tag{15.327}$$

to obtain

[12)] From $\delta g = 2D\xi$ we obtain for the F.P. determinant

$$\det D_+ \det D_- = \int \mathcal{D}c^+\mathcal{D}b_{++}\mathcal{D}c^-\mathcal{D}b_{--}e^{-\int d^2x(c^+D_-b_{++}+c^-D_+b_{++})} \quad ,$$

whose covariant form is (15.325).

$$T_{++}(x^+)T_{++}(0) = \frac{c/2}{(x^+)^4} + \frac{T_{++}(x) + T_{++}(0)}{x_+^2} \quad , \tag{15.328}$$

with $c = \frac{(D-26)}{4}$.

If we use the well known identities

$$\partial_- \frac{1}{x^+} = i\pi\delta^2(x) \quad , \tag{15.329}$$

and

$$\partial_-(x^+)^{-4} = -\frac{i\pi}{6}\partial_+^3\delta^2(x) \quad , \tag{15.330}$$

and take the Fourier transform of (15.328), we obtain

$$\langle 0|T_{++}(p)T_{++}(-p)|0\rangle = -\pi c\frac{p_+^3}{p_-} \quad , \tag{15.331}$$

which should be compared with (15.322).

At last, we wish to comment on the contribution to the conformal anomaly coming from ghost fields of general conformal spin $(0, j)$, with the Laplacian[13)]

$$\mathcal{L}_j = \rho^{j-1}\partial_-\rho^{-j}\partial_+ \quad , \tag{15.332}$$

where ρ is the conformal factor of the metric.

The energy momentum tensor for the ghosts is obtained by considering variations of $[S^{(gh)}]$ with respect to reparametrizations, $\delta\xi_-$; using

$$\delta b_{++} = j\frac{\partial\delta\xi_-}{\partial\xi_-}b_{++}(\xi_-, \xi_+) + \partial_+ b_{++}\delta\xi_- \tag{15.333}$$

$$\delta c^+ = (1-j)\frac{\partial\delta\xi_-}{\partial\xi_-}c^+(\xi_-, \xi_+) + \partial_+ c^+\delta\xi_- \tag{15.334}$$

one obtains

$$\delta S^{(gh)} = \int (c^+\partial_+ b_{++} - j\partial_+(c^+ b_{++}))\partial_-\delta\xi_- \quad . \tag{15.335}$$

Thus, the $++$ component of the energy momentum tensor reads

$$T_{++} = \partial_+ c^+ b_{++} - j\partial_+(c^+ b_{++}) \quad . \tag{15.336}$$

For $j = -1$ we recover (15.324).

13) This corresponds to more general operators $P_n^+ P_n$; see [62].

For the usual fermionic field, $j = 1/2$. The corresponding central charge is computed to be

$$c_j = \pm\frac{1-3(1-2j)^2}{2} = \mp[1+6j(j-1)] \quad , \tag{15.337}$$

where the lower value is valid for particles obeying anticommutation relations.

In the supersymmetric case, we need ghosts for the supersymmetric partner of the graviton, the gravitino; these ghosts must have conformal dimension $j = -1/2$, since e^a_+ has conformal dimension 1, and the gravitino has conformal dimension 1/2. Thus we obtain

$$c^B_{j=0} = D \quad , \quad c^{gh}_{j=-1} = -13 \quad (\text{or } -26 \text{ for complex fields}) \tag{15.338a}$$

$$c^F_{j=1/2} = \frac{D}{2} \quad , \quad c^{gh}_{j=-1/2} = \frac{11}{2} \quad (\text{or } 11 \text{ for complex fields}) \tag{15.338b}$$

$$c^{tot} = \frac{3(D-10)}{2} \quad , \tag{15.338c}$$

and the critical dimension is 10.

Continuing (15.322), back to Minkowski space, we obtain the following effective action

$$S_g = \kappa \int_M d^2x\sqrt{-g}R\frac{1}{\Box}R \quad , \tag{15.339}$$

where κ is the central charge. It plays the role of a coupling constant.

Scale invariance of the above action will be used to obtain information about the operator product expansions and consequently about the Green functions of the theory defined by (15.339).

In the following we canonically quantize theory, exploring as far as we can, the symmetries of the theory. Working in the light cone gauge, we shall discover a hidden $SL(2,R)$ invariance, which will give rise to a Kac-Moody algebra[82,84]. Afterwards, using the Ward identities of the conformally invariant theory, we then obtain the OPE's, and constraints enough to fix the Greens functions. It will be possible to obtain the short distance (critical) behavior of the theory, in terms of the anomalous dimensions which turn out to be computable in terms of the scale dimensions[82].

15.9.3 Canonical quantization and $SL(2,R)$ symmetry

In order to study the symmetries of (15.339) it is convenient to work with a local Lagrangian. We thus introduce an auxiliary field φ defined by[83,535]

$$\varphi = -\frac{\alpha}{2}\Box^{-1}R \quad , \tag{15.340}$$

where $\alpha^2 = 8\kappa$.

We have thus classically

$$\Box\varphi + \frac{1}{2}\alpha R = 0 \tag{15.341}$$

and

$$\kappa R\frac{1}{\Box}R = -\frac{1}{2}(\varphi\Box\varphi + \alpha R\varphi) \tag{15.342}$$

identically. Equations (15.341) and (15.342) imply that the action described by (15.339) may be replaced by

$$S[\varphi, g_{\mu\nu}] = \int d^2x\mathcal{L} = \int d^2x\sqrt{-g}\left(-\frac{1}{2}\varphi\Box\varphi - \frac{\alpha}{2}R\varphi\right) \quad , \tag{15.343}$$

supplemented by an external constraint $\hat{\Omega} \sim 0$, where

$$\hat{\Omega} = \varphi + \frac{1}{2}\alpha\Box^{-1}R \sim 0 \quad . \tag{15.344}$$

As long as we are dealing with the classical properties of the model, it is legitimate to replace the above constraint by

$$\Omega = \Box\tilde{\Omega} = \Box\varphi + \frac{1}{2}\alpha R \sim 0 \quad . \tag{15.345}$$

For the quantum theory, we have, identically[14)]

$$\int \mathcal{D}g_{\mu\nu}e^{i\int d^2x\sqrt{-g}\kappa R\Box^{-1}R} = \int \mathcal{D}g_{\mu\nu}\mathcal{D}\varphi\delta(\varphi + \frac{1}{2}\alpha\Box^{-1}R)e^{i\int d^2x\sqrt{-g}(-\frac{1}{2}\varphi\Box\varphi - \frac{1}{2}\alpha R\varphi)} \tag{15.346}$$

---

14) We can as well eliminate the function due to the fact that the integration over the auxiliary field is Gaussian. One should in that case take into account the determinant of the Laplace operator, which would amount to a redefinition of the central charge by one, that is[535]: $\kappa = \frac{c}{24\pi} \to \frac{c-1}{24\pi}$.

In the light cone gauge[15] one has

$$S = \int d^2x \mathcal{L} = \int d^2x \left(\partial_+\varphi\partial_-\varphi + g_{++}(\partial_-\varphi)^2 - \alpha\partial_- g_{++}\partial_-\varphi\right) \quad , \tag{15.347}$$

and the external constraint (15.345) takes the form

$$\Omega = 2\partial_- \left(\partial_+\varphi + g_{++}\partial_-\varphi - \frac{1}{2}\alpha\partial_- g_{++}\right) \sim 0 \quad . \tag{15.348}$$

Note that the φ-field is defined only up to harmonic functions, i.e. we have the symmetry

$$\varphi \to \varphi + \varepsilon \quad , \quad \Box\varepsilon = 0 \quad . \tag{15.349}$$

Since both the $\Box$ operator and the curvature R transform like scalars under coordinate transformations, we conclude from (15.340) that φ is also a scalar under such transformations. We also have the Weyl symmetry, defined by

$$\delta g_{\mu\nu} = -\Lambda g^{\mu\nu} \quad . \tag{15.350}$$

The transformations on $\Box$, R and φ induced by (15.350) are, using the definitions of the Laplacian $\Box$, the curvature R, and (15.350),

$$\delta\Box = -\Lambda\Box + \frac{d-2}{2} g^{kl}\partial_l\Lambda\partial_k \tag{15.351a}$$

$$\delta R = -\Lambda R + (1-d)\Box\Lambda \tag{15.351b}$$

$$\delta\varphi = \frac{1}{2}\alpha(d-1)\Lambda + \frac{2-d}{2}\Box^{-1}\partial_\mu\Lambda\partial^\mu\varphi \quad , \tag{15.351c}$$

where d is the space-time dimension (we shall take $d=2$). However Weyl transformations are not a symmetry of the theory due to the last term in (15.343) which is not invariant, unless the parameter Λ obeys

$$\sqrt{-g}R\Lambda = \partial_\mu(\sqrt{-g}\xi^\mu) \Leftrightarrow R\Lambda = D_\mu\xi^\mu \tag{15.352}$$

[15] Since other results do not depend on those of these sections we shall use in the remaining of this section, and in section 15.10 the following convention:

$$x^\pm = \frac{x^0 \pm x^1}{\sqrt{2}} \qquad \partial_\pm = \frac{\partial_0 \pm \partial_1}{\sqrt{2}}$$

$$g^{\mu\nu} = \begin{pmatrix} g^{++} & g^{+-} \\ g^{-+} & g^{--} \end{pmatrix} = \begin{pmatrix} 0 & 1 \\ 1 & 2g_{++} \end{pmatrix} \quad , \quad \sqrt{-g} = 1 \quad .$$

where ξ^μ is any vector. Although this is a very unusual type of symmetry (where the parameter obeys the above relation) it will be useful in the light cone gauge where the theory describes conformal fields depending only on one light cone variable, which is equivalent to a differential equation. We will see more examples of that type while studying chiral bosonic fields.

The Lagrangian defined in (15.347) is explicitly covariant, and under a general coordinate transformation

$$x^\mu \to x^\mu + \epsilon^\mu(x) \tag{15.353}$$

the fields φ and $g^{\mu\nu}$ transforms as

$$\delta\varphi = \epsilon^\mu \partial_\mu \varphi \tag{15.354a}$$

$$\delta g^{\mu\nu} = \epsilon^\alpha \partial_\alpha g^{\mu\nu} - g^{\alpha\mu}\partial_\alpha \epsilon^\nu - g^{\alpha\nu}\partial_\alpha \epsilon^\mu \quad . \tag{15.354b}$$

We summarize the whole set of symmetry transformations by the following

$$\delta\varphi = \varepsilon + \epsilon^\mu \partial_\mu \varphi + \frac{1}{2}\alpha\Lambda \tag{15.355a}$$

$$\delta g^{\mu\nu} = \epsilon^\alpha \partial_\alpha g^{\mu\nu} - g^{\mu\alpha}\partial_\alpha \epsilon^\nu - g^{\alpha\nu}\partial_\alpha \epsilon^\mu - \Lambda g^{\mu\nu} \quad , \tag{15.355b}$$

where $\Box\varepsilon = 0$ and

$$g g^{\mu\nu}\partial_\mu \left(\frac{1}{\sqrt{-1}}\partial_\nu \Lambda\right) = \partial_\alpha \xi^\alpha \quad . \tag{15.355c}$$

In the light cone gauge we have

$$\delta\varphi = \epsilon^+\partial_+\varphi + \epsilon^-\partial_-\varphi - \frac{1}{2}\alpha(\partial_+\epsilon^+ + \partial_-\epsilon^-) \tag{15.356a}$$

$$\delta g_{++} = -\partial_+\epsilon^- + \epsilon^- \overleftrightarrow{\partial_-} g_{++} + \partial_+(\epsilon^+ g_{++}) \quad , \tag{15.356b}$$

where we disregarded the ε symmetry (15.349), which is trivial.

In order to maintain the light cone gauge we imposed

$$\delta g^{+-} = -(\partial_-\epsilon^- + \partial_+\epsilon^+ + \Lambda) = 0 \quad , \tag{15.357}$$

which fixes Λ, and

$$\delta g^{++} = -2g^{+-}\partial_-\epsilon^+ = -2\partial_-\epsilon^+ = 0 \quad . \tag{15.358}$$

Using now (15.352) with an appropriate choice of ξ^μ, we have

$$\partial_-^3 \epsilon^- = 0 \quad . \tag{15.359}$$

The Euler-Lagrange equations following from (15.347) are

$$L_\varphi = \frac{\partial\mathcal{L}}{\partial\varphi} - \partial_\mu \frac{\partial\mathcal{L}}{\partial\partial_\mu\varphi} = -2\partial_-[\partial_+\varphi_+g_{++}\partial_-\varphi - \frac{1}{2}\alpha\partial_-g_{++}] = -\Omega = 0 \quad (15.360a)$$

and

$$L_{g_{++}} = T_{--} = (\partial_-\varphi)^2 + \alpha\partial_-^2\varphi = 0 \quad , \quad (15.360b)$$

where one recognizes in the r.h.s. the constraint (15.348) and T_{--} respectively.

The momenta conjugate to φ and g_{++} as computed from (15.343) are

$$\pi_\varphi = \frac{\partial\mathcal{L}}{\partial\dot{\varphi}} = (1+g_{++})(\dot{\varphi}-\varphi') - \varphi' - \frac{1}{2}\alpha(\dot{g}_{++} - g'_{++}) \quad (15.361a)$$

$$\pi_{g_{++}} = \frac{\partial\mathcal{L}}{\partial\dot{g}} = -\frac{1}{2}\alpha(\dot{\varphi}-\varphi') \quad , \quad (15.361b)$$

with

$$\{\varphi(t,y), \pi_\varphi(t,y)\} = \{g(t,x), \pi_g(t,y)\} = \delta(x-y) \quad . \quad (15.361c)$$

The corresponding Hamiltonian then reads

$$\mathcal{H} = \varphi'\pi_\varphi + g'_{++}\pi_{g_{++}} - \frac{2}{\alpha}\left[\pi_\varphi - \varphi' + \frac{1}{\alpha}(1+g_{++})\pi_{g_{++}}\right]\pi_{g_{++}} \quad . \quad (15.362)$$

The symmetries (15.356) are related to the generators of the $SL(2,R)$ algebra. From the Noether's theorem, it is not difficult to obtain the generator of (15.356), after some algebra, we find

$$\mathcal{J}^o_{\epsilon^-} = \frac{-\epsilon^-}{\sqrt{2}}[(1+g_{++})L_{g_{++}} - \frac{1}{2}\alpha^2\partial_-^2 g_{++}] + \partial_-\epsilon^-[\frac{-\alpha^2}{2\sqrt{2}}\partial_- g_{++}] + \partial_-^2\epsilon^-[\frac{\alpha^2}{2\sqrt{2}}g_{++}] \quad . \quad (15.363)$$

Using the constraint (15.359) we make the decomposition

$$\epsilon^- = \zeta^-(x^+) - x^-\zeta^o(x^+) + (x^-)^2\zeta^+(x^+) \quad , \quad (15.364)$$

such that the current takes the form

$$\mathcal{J}^o_{\epsilon^-} = \zeta^-(x^+)J^+ + \zeta^o(x^+)J^o + \zeta^+(x^+)J^- \quad , \quad (15.365)$$

where

$$J^+ = -\frac{1}{\sqrt{2}}\left[(1+g_{++})L_{g_{++}} - \frac{1}{2}\alpha^2\partial_-^2 g_{++}\right] \quad (15.366a)$$

$$J^o = \frac{1}{2\sqrt{2}}\alpha^2\partial_- g_{++} - x^-J^+ \quad (15.366b)$$

$$J^- = \frac{1}{\sqrt{2}}\alpha^2 g_{++} - 2x^-J^o - (x^-)^2J^+ \quad . \quad (15.366c)$$

We compute the algebra of the $J^{a'}s$ using the Poisson brackets (15.361c), and arrive after some calculation at

$$\{J^a(x), J^b(y)\} = -2\varepsilon^{abc}\eta_{cd}J^d(x)\delta(x-y) + \frac{1}{2}\alpha^2\eta^{ab}\delta'(x-y) \quad , \tag{15.367}$$

where

$$\begin{aligned} \eta^{ab} &= \begin{pmatrix} 0 & 0 & 2 \\ 0 & -1 & 0 \\ 2 & 0 & 0 \end{pmatrix} \\ \eta_{ab} &= \begin{pmatrix} 0 & 0 & \frac{1}{2} \\ 0 & -1 & 0 \\ \frac{1}{2} & 0 & 0 \end{pmatrix} \end{aligned} \tag{15.368}$$

and $\varepsilon^{-0+} = 1$ define the $SL(2,R)$ structure constants. The quantum theory is defined by a Kac-Moody algebra following from (15.367). Therefore, the currents are $SL(2,R)$ generators. We now turn to the second generator associated with the symmetry parametrized by ϵ^+. It is given by the expression:

$$\mathcal{J}^o_{\epsilon^+} = \frac{1}{\sqrt{2}}\epsilon^+\left\{g_{++}\left[(1+g_{++})L_h - \frac{1}{2}\alpha^2\partial_-^2 g_{++}\right] + \frac{1}{\alpha^2}\left[\frac{1}{2}\alpha^2\partial_- g_{++}\right]^2\right\} \quad , \tag{15.369}$$

which can be written in terms of $J^a(x)$, Eqs. (15.366) as

$$\mathcal{J}^o_{\epsilon^+} = -\epsilon^+\frac{\sqrt{2}}{\alpha^2}\{J^-J^+ - J^{o2}\} = -\epsilon^+\frac{\sqrt{2}}{\alpha^2}\eta_{ab}J^aJ^b = -\epsilon^+\frac{\sqrt{2}}{\alpha^2}T_S \quad , \tag{15.370}$$

where T_S is the Sugawara tensor built from the $SL(2,R)$ currents obeying a Virasoro algebra.

We can get a further insight in the symmetry transformation, and set $\Lambda = 0$ in Eq. (15.357), since (15.350) is not a "bonafide" symmetry; due to (15.352)

$$\epsilon^+ = v(x^+) \tag{15.371a}$$

$$\epsilon^- = w(x^+) - x^-\partial_+ v(x^+) \quad . \tag{15.371b}$$

Equation (15.359) is still valid, but since we now have only two independent parameters, we do not have as many independent operators as shown in (15.365) and (15.370).

As it turns out

$$\mathcal{J}^o = w(x^+)J^+ + \partial_+ v(x^+)J^o - \frac{\sqrt{2}}{\alpha^2}v(x^+)T_S \quad . \tag{15.372}$$

We thus have two independent symmetries. One associated with $w(x^+)$,

$$\delta_w F = \int dx w(x^+)\{J^+(x), F\} \quad , \tag{15.373}$$

with generator $J^+(x^+)$, and the other associated with $v(x)$,

$$\delta_v F = \int dx v(x)\left\{-\frac{\sqrt{2}}{\alpha^2}T_S(x) - \sqrt{2}\partial_1 J^0(x), F\right\} \quad . \tag{15.374}$$

with generator

$$\hat{J}^0 = -v\frac{\sqrt{2}}{\alpha}(T_S + \alpha^2\partial_1 J^0) = -v\frac{\sqrt{2}}{\alpha}T_{--} \quad . \tag{15.375}$$

Returning to the constraint (15.348) and differentiating it with respect to x^-, we obtain,

$$\frac{1}{2}\alpha^2\partial_-^3 g_{++} = (\partial_+ + g_{++}\partial_- + 2\partial_- g_{++})L_g = \nabla_+ T_{--} \quad , \tag{15.376}$$

where $L_{g_{++}}$ is the equation of motion of g_{++}, Eq. (15.360). Using $L_{g_{++}} = 0$, we have

$$\partial_-^3 g_{++} = 0 \quad . \tag{15.377}$$

Therefore, on the space of solutions we find

$$\frac{\alpha^2}{\sqrt{2}}g_{++} = J^- + 2x^- J^o + (x^-)^2 J^+ \quad , \tag{15.378}$$

with $J^a = J^a(x)$, and

$$T_{++} = T_S + \frac{1}{\sqrt{2}}\partial_+ J^o \quad . \tag{15.379}$$

The canonical quantization could proceed along these lines, by quantizing the Kac Moody algebra obtained[535] above. We shall follow another line, which makes transparent the correlators of the quantum theory.

The fact that we have a linear term in the energy momentum tensor can be checked by an explicit computation. We have to compute

$$T_{++} = \left.\frac{\delta S}{\delta g_{--}}\right|_{g_{--}=0} \quad . \tag{15.380}$$

In order to evaluate the linear term, we write the full action, before going to the light cone gauge, and expand in g_{--}, around zero) picking up only the linear term. The linear term comes from the variation of R in $R\frac{1}{\Box}R$. The relevant term is

$$R_{+-+-} = -\frac{1}{2}\partial_+\partial_- g_{--} + \text{etc} \quad , \tag{15.381}$$

which gives

$$\frac{\delta S}{\delta g_{--}} = -2\partial_+^2 \frac{R}{\Box} + \text{etc} \quad , \tag{15.382}$$

implying

$$T_{++}^{\text{linear}} = \frac{\alpha^2}{8}\partial_+\partial_- g_{--} \quad . \tag{15.383}$$

A very important information may be obtained from the addition of a cosmological term to the Lagrangian, which turns into

$$\begin{aligned} \mathcal{L} =& \kappa\sqrt{-g}\left(R\frac{1}{\Box}R + \mu\right) \\ \approx& \sqrt{-g}\left(-\frac{1}{2}\varphi\Box\varphi - \frac{\alpha}{2}R\varphi + \frac{\alpha^2}{2}\mu\right) \end{aligned} \tag{15.384}$$

The equation of motion for g^{+-} may be obtained from the above; we have, in general,

$$T_{\mu\nu} = \theta_{\mu\nu} - \frac{1}{2}g_{\mu\nu}\left[\theta_\rho^\rho + \frac{\alpha^2}{8}\left(\mu + \frac{4}{\alpha}\Box\varphi\right)\right] \tag{15.385}$$

with

$$\theta_{\mu\nu} = \frac{1}{2}\partial_\mu\varphi\partial_\nu\varphi + \frac{\alpha}{2}\partial_\mu\partial_\nu\varphi \quad . \tag{15.386}$$

From the above, we obtain

$$T_{+-} = -h_{++}T_{--} - \frac{\kappa}{2}(\mu + 4\partial_-^2 h_{++} + \text{constraint} = 0 \quad . \tag{15.387}$$

Using the g^{--} equation of motion, we arrive at

$$\partial_-^2 h_{--} = -\frac{4\sqrt{2}}{\alpha^2}J^+ = -\frac{\mu}{4} \quad . \tag{15.388}$$

An analogous equation has been obtained by Bershadsky and Ooguri[510], who proved that $2-D$ gravitational is equivalent to an $SL(2,R)$ WZNW theory, with a constraint $\overline{J}^{(+)} \approx$ constant. However, the constraint was first discovered by A. A. Belavin.

15.9.4 Operator product expansions and quantum solution

The use of the symmetries discussed so far allow one to obtain the Green's functions of the quantum theory. The symmetry (15.356b) implies a useful Ward identity: under the transformation (15.356b) the gravity effective action changes as follows[82]

$$\delta S = \int T_{--}\delta g_{++} d^2x = -\int \nabla_+ T_{--}\epsilon^- d^2x \quad , \tag{15.389}$$

Using (15.376) we have

$$\delta S = -\frac{\alpha^2}{2}\int \partial_-^3 g_{++}\epsilon^- d^2x \quad . \tag{15.390}$$

For the n-point correlator of $h_\pm$, the infinitesimal variation (15.356b) implies

$$\int \mathcal{D}g\{\sum\langle g_{++}(x_1)\cdots\nabla_+\epsilon^-(x_i)\cdots g_{++}(x_n)e^{iS(g)}\rangle$$

$$+\langle g_{++}(x_1)\cdots g_{++}(x_n)(-i\frac{\alpha^2}{2})\int d^2x\partial_-^3 g_{++}\epsilon^- e^{iS(g)}\rangle\} = 0 \quad . \tag{15.391}$$

Therefore, for (15.391) to hold independent of ϵ, the following Ward identity must be satisfied:

$$\begin{aligned} i\frac{\alpha^2}{2}\partial_-^3\langle g_{++}(x)g_{++}(x_1)\cdots g_{++}(x_n)\rangle &= \\ &\sum\langle g_{++}(x_1)\cdots \not{g}_{++}(x_i)\cdots g_{++}(x_n)\rangle\partial_+\delta(x-x_i) \\ &+\sum\left[\delta(x-x_i)\partial_i\cdots\partial_-\delta(x-x_i)\right]\langle g_{++}(x_1)\cdots g_{++}(x_n)\rangle \end{aligned} \tag{15.392}$$

In order to obtain short distance expansions, we must get rid of the delta functions. We use[16)]

$$\partial_-\frac{1}{x^+} = 2i\pi\delta^{(2)}(x) \quad , \tag{15.393a}$$

or, as a consequence of this

$$\frac{1}{4\pi i}\partial_-^3\frac{(x^-)^2}{x^+} = \delta^{(2)}(x) \quad , \tag{15.393b}$$

16) In fact, one has to use the prescription $\frac{1}{x^+}\to\frac{1}{x^+-i\epsilon\,\mathrm{sign}(x^-)}$.

which is a more useful identity, because of the operator ∂_-^3 which also appears on the r.h.s. of Eq. (15.392); after rescaling $g \to \frac{1}{\alpha^2} g$,

$$\langle g_{++}(x) g_{++}(x_1) \cdots g_{++}(x_n) \rangle = \frac{c}{6} \sum_{j=1}^{n} \frac{(x^- - x_j^-)^2}{(x^+ - x_j^+)^2} \langle g_{++}(x_1) \cdots \not{g}_{++}(x_j) \cdots g_{++}(x_n) \rangle$$
$$- \sum_{j=1}^{n} \left[\frac{(x^- - x_j^-)^2}{x^+ - x_j^+} \partial_-^j + 2 \frac{x^- - x_j^-}{x^+ - x_j^+} \right] \langle g_{++}(x_1) \cdots g_{++}(x_n) \rangle \quad , \qquad (15.394)$$

where the slash in the jth field in the first term of (15.394) means that it has been deleted. This implies the operator product expansion

$$g(x)g(y) = \frac{c}{6} \left(\frac{x^- - y^-}{x^+ - y^+} \right)^2 - \frac{(x^- - y^-)^2}{x^+ - y^+} \partial_-^y g_{++}(y) - 2 \frac{x^- - y^-}{x^+ - y^+} g_{++}(y) \quad . \qquad (15.395)$$

If a matter field transforms as

$$\delta\chi = \epsilon^- \partial_- \chi + \lambda \partial_- \epsilon^- \chi \qquad (15.396)$$

we have

$$g_{++}(x)\chi(y) = - \frac{(x^- - y^-)^2}{x^+ - y^+} \partial_- \chi(y) - 2\lambda \frac{x^- - y^-}{x^+ - y^+} \chi(y) \quad . \qquad (15.397)$$

Moreover, if we use the decomposition (15.378), we arrive at the expression

$$J^a(x)\chi(y) = \frac{l^a(x)}{x^+ - y^+} \chi(y) \qquad (15.398)$$

where

$$l^{(-1)}(x) = \partial_- \qquad (15.399a)$$
$$l^{(0)}(x) = x^- \partial_- + \lambda \qquad (15.399b)$$
$$l^{(1)}(x) = (x^-)^2 \partial_- + 2\lambda x^- \quad , \qquad (15.399c)$$

are generators of $SL(2,R)$ and

$$J^a(x) J^b(y) = - \frac{c}{12} \eta^{ab} \frac{1}{(x^+ - y^+)^2} + \frac{\varepsilon^{abc} \eta_{cd}}{x^+ - y^+} J^d(y) \quad , \qquad (15.400)$$

which should be compared to the usual Kac-Moody algebra; defining

$$J^a(x) = \sum J_n^a e^{inx} \qquad (15.401a)$$

we have from (15.400),

$$[J_n^a, J_m^b] = f^{abc} J_{n+m}^c + nk\eta^{ab} \delta_{n,-m} \quad , \qquad (15.401b)$$

which corresponds to the loop algebra of the quantized version of Eq. (15.367).

15.9.5 Interaction of matter fields with gravity in two dimensions

We consider in this subsection a general matter field interacting with gravity. In 2-dimensional space time, it is sufficiently general to consider a set of Majorana fermions, interacting with gravity via the usual coupling of the zweibein $\hat{e}^{a\mu}$, with the Lagrangian[536]

$$i\overline{\chi}\hat{e}^{a\mu}\gamma_a\partial_\mu\chi = i\psi_-(\partial_+ - g_{++}\partial_-)\psi_- + \psi_+(\partial_- - g_{--}\partial_+)\psi_+ \quad , \tag{15.402a}$$

where

$$\psi_- = (\sqrt{2}\hat{e}^{++})^{1/2}\chi_- \tag{15.402b}$$
$$\psi_+ = (\sqrt{2}\hat{e}_{++})^{1/2}\chi_+ \tag{15.402c}$$
$$g_{--} = \hat{e}_{+-}/\hat{e}_{++} \tag{15.402d}$$
$$g_{++} = \hat{g}_{-+}/\hat{e}_{--} \tag{15.402e}$$

We can use reparametrization invariance to get rid of g_{--} by fixing it to be zero. In the gauge $g_{--} = 0$, $g_{+-} = 1$, the effective action (15.339) takes the form

$$S_{eff}(g_{++}) = \ln\det(\partial_+ - g_{++}\partial_-) \quad , \tag{15.403}$$

For this gauge to be implemented, we need a principle of gauge invariance: the generating functional must be independent of the values of g_{--} and g_{+-} at $g_{--} = 0\,,\, g_{+-} = 0$: that is

$$\left.\frac{\delta Z}{\delta g_{--}}\right|_{g_{--}=0} = \left.\frac{\delta Z}{\delta g_{+-}}\right|_{g_{+-}=0} = 0 \quad , \tag{15.404}$$

which are equivalent to saying that the total energy momentum tensor is zero (or else, that the equation of motion of $g_{\mu\nu}$ holds).

In order to fix the gauge we need a set of ghost fields, which we introduce as follows. The metric is described by

$$ds^2 = g_{--}(dx^-)^2 + 2g_{+-}dx^+dx^- + g_{++}(dx^+)^2 \quad , \tag{15.405}$$

with $g_{--} = 0\,,\, g_{+-} = 1$. The following transformations are symmetries of the action (reparametrization invariance)

$$\delta g_{--} = 2\nabla_-\epsilon_- \tag{15.406a}$$
$$\delta g_{+-} = \nabla_+\epsilon_- + \nabla_-\epsilon_+ \quad , \tag{15.406b}$$

requiring a ghost pair (-1,2) (that is, ghosts with Lorentz spin -1 and 2, respectively) for g_{--}, with Lagrangian

$$\mathcal{L}^{(1)}_{gh} = \eta_{++}\nabla_-\epsilon_- \quad , \tag{15.407}$$

and a ghost (1,0) for g_{+-}, with Lagrangian

$$\mathcal{L}_{gh}^{(2)} = \zeta(\nabla_+ \epsilon_- + \nabla_- \epsilon_+) \quad . \tag{15.408}$$

Going back to (15.404) we have for the first equation

$$T_{++}^{tot}(g, \text{ghosts}) = 0 \tag{15.409}$$

and for the second,

$$\theta(g, \text{ghosts}) = T_{+-}^{tot} = 0 \quad . \tag{15.410}$$

where $T_{\pm}^{tot}$ is the total energy momentum tensor of matter, gravity and ghosts. We have thus

$$T_M + T_{gr} + T_{gh}^{(1)} + T_{gh}^{(2)} = 0 \quad . \tag{15.411}$$

and as a consequence, also the sum of the central charges vanishes,

$$c_M + c_{gr} + c_{gh}^{(1)} + c_{gh}^{(2)} = 0 \quad . \tag{15.412}$$

The above central charges are ($c_0 = \#\text{of bosons} + \frac{1}{2}\#\text{of fermions}$),

$$c_{gh}^{(1)} = -26 \quad (\text{reparametrization ghost}) \tag{15.413}$$

and

$$c_{gh}^{(2)} = -2 \quad (\text{dilatation ghost}) \quad . \tag{15.414}$$

For a ghost field of "spin" $(1-j, j)$, the central charge is given by Eq. (15.337) and (15.338). In order to complete the computations, we need c_{gr}, which depends on the explicit form of the gravity energy momentum tensor of pure gravity, T_{gr}.

Going back to Eq. (15.379), we obtain it in terms of the currents:

$$T_{gr} = \frac{1}{\kappa + 2} g_{ab} : J^a J^b : + \frac{1}{\sqrt{2}} \partial_+ J_0 \quad , \tag{15.415}$$

where κ is the renormalized coupling constant, and the constant $\frac{1}{\kappa+2}$ arises from the Kac-Moody algebra (15.128) obeyed by J^a's, and the Sugawara construction. The Sugawara term contributes to the central charge with the quantity (see Eq. (15.197), with $\dim SL(2,R) = 3$, $c_V|_{G=SL(2,R)} = 2$)

$$c_S = \frac{3\kappa}{\kappa + 2} \quad , \tag{15.416}$$

as usual. For the derivative contribution we have

$$J^{0'}(x) J^{0'}(y) = -\frac{\kappa \eta^{00} 2 \cdot 3}{(x-y)^4} \quad . \tag{15.417}$$

Thus

$$c' = -6\kappa \tag{15.418a}$$

or

$$c = c_S + c' = \frac{3\kappa}{\kappa+2} - 6\kappa \quad . \tag{15.418b}$$

Equation (15.412), together with (15.413), (15.414) and (15.418*b*) implies:

$$c_0 - 28 + \frac{3\kappa}{\kappa+2} - 6\kappa = 0 \quad , \tag{15.419}$$

which means

$$\kappa = -2 + \frac{1}{12}\left(c_0 - 13 \pm \sqrt{(c_0-1)(c_0-25)}\right) \quad . \tag{15.420a}$$

The plus sign reproduces the semi-classical result for $c_0 \to \infty$[536]. Thus

$$\kappa + 2 = \frac{1}{12}\left(c_0 - 13 + \sqrt{(c_0-1)(c_0-25)}\right) \quad . \tag{15.420b}$$

We can define "eigenstates" of the operators by the conformal equations

$$J(x^-,x^+)\phi^{(j)}(y^-,y^+) = \left(2j\frac{x^- - y^-}{x^+ - y^+} + \frac{(x^- - y^-)^2}{x^+ - y^+}\frac{\partial}{\partial y^+}\right)\phi^{(j)}(y^-,y^+) \quad . \tag{15.421}$$

From (15.415), we have the Virasoro operator

$$L_0^G = \frac{1}{\kappa+2}\sum : J^a_{-n} J^a_n : + J^0_{10} \quad . \tag{15.422}$$

The physical states are defined by

$$J_n|Phys\rangle = 0 \quad , \quad n \geq 0 \tag{15.423a}$$

and

$$L_n|Phys\rangle = 0 \quad , \quad n \geq 0 \quad . \tag{15.423b}$$

The eigenvalues of L_0^G corresponding to physical (zero ghost) states are given by

$$\Delta = \frac{j(j+1)}{k+2} + j \quad , \tag{15.424}$$

where $k = -\kappa - 4$. We thus have the relation

$$\Delta_0 = \Delta + \frac{\Delta(1-\Delta)}{k+2} \tag{15.425}$$

where $\Delta = -j$, thus completing the relation between the free matter theory defined by Δ_0 and c_0, and the gravity interacting theory defined by Δ and κ.

We examine now the relation between the conformal dimension of the matter fields with and without gravity. The conformal dimension will be obtained from Eq. (15.425), which is solved by

$$\Delta = -\frac{k+1}{2} \pm \frac{1}{2}\sqrt{(k+1)^2 + 4(k+2)\Delta^0} \quad . \tag{15.426}$$

Now we use Kac's formula (15.61), with

$$\Delta_0 = \frac{c-1}{24} = -\frac{(k+1)^2}{4(k+2)} \quad , \tag{15.427}$$

obtaining

$$\Delta = -\frac{k+1}{2} \pm [n(k+2) - m] \quad , \tag{15.428}$$

which acquires discrete values.

Correlation functions may be computed in closed form, following the work of [70] where the correlators for the Kac-Moody symmetry $SL(2)$ have been computed.

15.10 Two Dimensional Supergravity

15.10.1 General theory of supergravity in two dimensions

The issue of induced 2-D gravity has been addressed in the previous section, where conformal invariance was used to obtain information about the operator products and hence about the Green 's functions of the theory. As an example, 2-D gravity could be described in terms of a SL(2,R) Kac-Moody algebra, and the critical behavior of a given theory in the presence of gravity Eq. (15.426) could be related to the behavior in flat space Eq. (15.427).

In this section we generalize this construction[84], by including a graviton/-gravitino pair, described in the light cone gauge as[537]

$$g^{\mu\nu} = \begin{pmatrix} 0 & 1 \\ 1 & 2g_{++} \end{pmatrix}$$

$$\Psi_- = 0 \qquad \Psi_+ = \begin{pmatrix} 0 \\ \psi_+ \end{pmatrix} \quad , \tag{15.429}$$

with the notation (see footnote 15).

$$\gamma_+ = -i\sqrt{2}\begin{pmatrix} 0 & 1 \\ 0 & 0 \end{pmatrix}, \qquad \gamma_- = i\sqrt{2}\begin{pmatrix} 0 & 0 \\ 1 & 0 \end{pmatrix} \quad . \tag{15.430}$$

The root of the discussion is the issue of the diffeomorphism anomaly, which must vanish for consistency reasons, as will become clear below.

We suppose that the theory contains two main symmetries. The first one is a gauge symmetry associated with diffeomorphic transformations surviving the gauge fixing in (15.429). It is given by the transformations

$$\delta\Psi_+ = \epsilon_+\partial_-\Psi_+ - \frac{1}{2}\partial_-\epsilon_+\Psi_+ \tag{15.431a}$$

$$\delta g_{++} = -\partial_+\epsilon_+ + \epsilon_+\partial_- g_{++} - \partial_-\epsilon_+ g_{++} \quad , \tag{15.431b}$$

where ϵ_+ is a parameter labelling the diffeomorphisms. The second one is local supersymmetry, which after fixing the gauge as in (15.429), reads

$$\delta\Psi_+ = -\partial_+\epsilon + \frac{1}{2}\epsilon\partial_- g_{++} - \partial_-\epsilon g_{++} \tag{15.432a}$$

$$\delta g_{++} = 2i\overline{\Psi}_+\gamma_+\epsilon \quad , \tag{15.432b}$$

with $\gamma_-\epsilon = 0 \qquad \left(\epsilon = \begin{pmatrix} 0 \\ \epsilon \end{pmatrix}\right)$.

An action with matter fields may be defined as

$$S = \int d^2x\left\{\partial_+\phi\partial_-\phi + \frac{i}{2}\overline{\chi}\gamma_-\partial_+\chi + 2\overline{\chi}\Psi_+\partial_-\phi + g_{++}\left((\partial_-\phi)^2 + \frac{i}{2}\overline{\chi}\gamma_-\partial_-\chi\right)\right\} \quad , \tag{15.433a}$$

with the symmetries (15.431) and (15.432) supplemented by

$$\delta\phi = \epsilon_+\partial_-\phi \tag{15.433b}$$

$$\delta\chi = \epsilon_+\partial_-\chi + \frac{1}{2}\partial_-\epsilon_+\chi \tag{15.433c}$$

$$\delta\phi = \overline{\epsilon}\chi \tag{15.433d}$$

$$\delta\chi = -i\partial_-\phi\gamma_+\epsilon \tag{15.433e}$$

and $\chi = \begin{pmatrix} \chi \\ 0 \end{pmatrix}$.

The action (15.433a) may be obtained truncating two dimensional supergravity with a matter multiplet[537,538].

15.10.2 OPE and quantum solution

There is a well known anomaly in the quantum theory of 2-D supergravity, generalizing the reparametrization anomaly. It is given by the expression[537]:

$$\delta S = -\frac{c}{24\pi}\int d^2x\epsilon_+\partial_-^3 g_{++} + \frac{ic}{6\pi}\int d^2x\bar{\epsilon}\gamma_+\partial_-^2\Psi_+ \quad . \tag{15.434}$$

where δS is the anomalous response of (15.433) to the variations (15.431) and (15.431). Notice that we can eliminate the anomaly by requiring

$$g_{++} = \mathcal{J}^{(+)}(x^+) - 2\mathcal{J}^{(0)}(x^+)x^- + \mathcal{J}(x^+)(x^-)^2 \tag{15.435}$$

$$\psi_+ = \psi^{-1/2}(x^+) + \psi^{1/2}(x^+)x^- \quad . \tag{15.436}$$

We will come back to this later on.

The anomaly may be used to derive the short distance expansion of the fields. Starting from the correlator

$$\langle g_{++}(x_1)g_{++}(x_2)\cdots g_{++}(x_n)\psi_+(y_1)\cdots\psi_+(y_m)\rangle =$$
$$\int \mathcal{D}g_{++}\mathcal{D}\psi_+ g_{++}(x_1)g_{++}(x_2)\cdots g_{++}(x_n)\psi_+(y_1)\cdots\psi_+(y_m)e^{-\kappa W(g,\psi)} \tag{15.437}$$

we make variations of g_{++} and ψ_+ under (15.431) and (15.432); as a consequence, we have a variation in the action given by the anomaly, Eq. (15.434). Thus, we obtain the Ward identities

$$\frac{ic}{24\pi}\frac{\partial^3}{\partial z^{-3}}\langle g_{++}(z)\underline{X}\rangle = \sum_{i=1}^{n}\frac{\partial}{\partial z^+}\delta(z-x_i)\langle\underline{X}_{\hat{x}_i}\rangle$$
$$+\sum_{i=1}^{n}\left(\frac{\partial}{\partial z^-}\delta(z-x_i)+\delta(z-x_i)\frac{\partial}{\partial x_i^-}\right)\langle\underline{X}\rangle$$
$$+\sum_{j=1}^{n}\left(\frac{1}{2}\frac{\partial}{\partial z^-}\delta(z-y_j)+\delta(z-y_j)\frac{\partial}{\partial x_i^-}\right)\langle\underline{X}\rangle \tag{15.438}$$

and

$$\frac{c}{6\pi}\frac{\partial^2}{\partial z^{-2}}\langle\gamma_+\psi_+(z)\underline{X}\rangle =$$
$$-2i\sum_{i=1}^{n}\delta(z-x_i)\langle\gamma_+\psi_+(x_i)\underline{X}_{\hat{x}_i}\rangle + i\sum_{j=1}^{m}(-1)^j\partial_+\delta(z-y_j)\langle\underline{X}_{\hat{y}_i}\rangle$$
$$-i\sum_{j=1}^{m}(-1)^{j+1}\left\{\frac{1}{2}\delta(z-y_j)\frac{\partial}{\partial y_j^-}+\frac{\partial}{\partial z^-}\delta(z-y_j)\right\}\langle\underline{X}_{\hat{y}_i}g_{++}(y_j)\rangle \tag{15.439}$$

where $\underline{X} = g_{++}(x_1)g_{++}(x_2)\cdots g_{++}(x_n)\psi_+(y_1)\cdots\psi_+(y_m)$, and $\underline{X}_{\hat{x}_i}$ is $\underline{X}$ with the field corresponding to x_i being deleted.

Using the manipulations already used in (15.392) and (15.395), we obtain the short distance expansions

$$\frac{c}{6}g_{++}(z)g_{++}(x) = \left(\frac{z^- - x^-}{z^+ - x^+}\right)^2 - \left\{2\left(\frac{z^- - x^-}{z^+ - x^+}\right) + \frac{(z^- - x^-)^2}{z^+ - x^+}\frac{\partial}{\partial x^-}\right\}g_{++} \quad ,$$

$$\frac{c}{6}g_{++}(z)\psi_+(y) = -\left\{\frac{z^- - y^-}{z^+ - y^+} + \frac{(z^- - y^-)^2}{z^+ - y^+}\frac{\partial}{\partial y^-}\right\}\psi_+(y) \quad , \qquad (15.440a,b,c)$$

$$\frac{c}{6}\psi_+(z)\gamma_+\psi_+(y) = \frac{1}{2}\frac{z^- - y^-}{(z^+ - y^+)^2} - \frac{1}{2}\left\{\frac{1}{z^+ - y^+} + \frac{1}{2}\frac{z^- - y^-}{z^+ - y^+}\frac{\partial}{\partial y^-}\right\}g_{++}(y) \quad .$$

The above expressions imply a super Kac-Moody algebra generalizing the Neveu-Schwarz algebra of superstring theory; we define the expansions

$$\mathcal{J}^a(x^-) = \sum_{n=-\infty}^{\infty} \mathcal{J}^a_n x^{-n} \quad , \qquad (15.441)$$

$$\psi^r_+(x^-) = \sum_{n=-\infty}^{\infty} \psi^r_n x^{-n} \quad . \qquad (15.442)$$

In terms of the components J^a_n and ψ^r_n, the short distance expansions are given (after a suitable rescaling) by

$$[\mathcal{J}^a_n, \mathcal{J}^b_m] = f^{abc}\mathcal{J}^c_{n+m} + kn\eta^{ab}\delta_{n,-m} \quad , \qquad (15.443a)$$

$$[\mathcal{J}^a_n, \psi^r_m] = h^{ars}\psi^s_{n+m} \quad , \qquad (15.443b)$$

$$\{\psi^r_n, \psi^s_m\} = h^{rsc}\mathcal{J}^c_{n+m} + kn\eta^{rs}\delta_{n,-m} \quad , \qquad (15.443c)$$

where $f^{+-0} = 2$, (f antisymmetric), $\eta^{00} = \frac{1}{2}$, $\eta^{(1/2)(-1/2)} = 1$, $\eta^{+-} = 1$, $h^{+,-1/2,1/2} = 1$, $h^{-,1/2,1/2} = -1$, and $h^{0,\pm 1/2,1/2} = \pm 1$.

The finite algebra is realized by the following set of operators

$$\mathcal{J}^+ = \partial_- \qquad (15.444a)$$

$$\mathcal{J}^0 = y^-\partial_- + \frac{1}{2}\theta\partial_\theta - j \qquad (15.444b)$$

$$\mathcal{J}^- = (y^-)^2\partial_- + y^-\theta\partial_\theta - 2jy^- \qquad (15.444c)$$

$$\mathcal{P}^{1/2} = i\sqrt{\frac{1}{2}}(\partial_\theta + \theta\partial_-) \qquad (15.444d)$$

$$\mathcal{P}^{-1/2} = i\sqrt{\frac{1}{2}}(y^-\partial_\theta + y^-\theta\partial\_2j\theta) \qquad (15.444e)$$

with the Casimir c_g given by

$$c_g = \eta_{ab}\mathcal{J}^a\mathcal{J}^b + \eta_{rs}\mathcal{P}^r\mathcal{P}^s = j(j+\frac{1}{2}) \quad . \tag{15.445}$$

The energy momentum tensor generating the general coordinate transformations may be readily obtained. It has the Sugawara form, plus a linear term in $\mathcal{J}^0$

$$\mathcal{T}(x) = \frac{1}{k+\frac{3}{2}}\{:\mathcal{J}^a(x^+)\mathcal{J}^b(x^+):\eta_{ab}+:\psi^r(x^+)\psi^s(x^+):\eta_{rs}\} + \mathcal{J}^{0\prime}(x^+) \quad . \tag{15.446}$$

The coefficient $\frac{1}{k+\frac{3}{2}}$ has been obtained by requiring a super Kac-Moody Virasoro algebra

$$[L_n, \mathcal{J}^a_m] = m\mathcal{J}^a_{n+m} \quad , \tag{15.447}$$

where we have, for the Sugawara piece of the energy momentum tensor,

$$\mathcal{T}_{Sug}(x^+) = \sum_{n=-\infty}^{\infty} L_{-n}(x^+)^n \quad . \tag{15.448}$$

15.10.3 Interaction of matter fields and supergravity

The full Virasoro algebra may now be obtained by computing the expectation value of the product of the energy momentum tensors at nearby points. After a lengthy, although straightforward calculation one obtains the following result for the central charge:

$$c_{Sug} = \frac{k}{k+\frac{3}{2}} \quad . \tag{15.449}$$

In order to balance the central charges, we have to consider the following set of fields

i) matter fields, with central charge c_0;
ii) diffeomorphism ghosts, with central charge -15;
iii) superconformal ghosts, with central charge $-\frac{5}{2}$;
iv) the Sugawara piece of the supergravity multiplet contributing $\frac{k}{k+\frac{3}{2}}$;
and
v) a linear term contributing $-6k$ to the central charge.

With the above balance, we have (recalling in the supersymmetric case, we use as usual $c_0 = \frac{3}{2}d_0$)

$$k + \frac{3}{2} = \frac{d_0 - 5 \pm \sqrt{(d_0-1)(d_0-9)}}{8} \quad , \tag{15.450}$$

(recall the bosonic result $k+2=\frac{13-c_0\mp\sqrt{(c_0-1)(c_0-25)}}{12}$).

With the realization (15.444), the Casimir c_g given in (146), and $j=-\Delta$, the anomalous dimensions of the free theory Δ_0 has a relation with the gravitationally interacting theory given by

$$\Delta_0=\frac{\Delta(\frac{1}{2}-\Delta)}{k+\frac{3}{2}}+\Delta \quad . \tag{15.451}$$

Notice that $\Delta=\frac{1}{2}$ is a fixed point (in the same way as $\Delta=1$ was a fixed point in the bosonic theory). Conclusions similar to the bosonic case may be drawn. A new series of critical indices emerges, and may be compared with the usual results.

15.11 Non Perturbative 2D Gravity, Surfaces, and Critical Behavior

Our previous analysis of 2-D gravity is valid for genus zero Riemann surfaces, that is, the basis space is the ordinary two dimensional space-time.

In string theory, or when analysing random surfaces, we are faced with the problem of 2-D quantum gravity in an arbitrary genus surface. The sum over all the topologies is an essential ingredient in string theories; since the sum over genus is badly behaved, one expects that each term is not a good representative of the full theory, and that non perturbative effects play a major role. Moreover, as we have already seen, Eq. (15.420) is not real for $1<d<25$, and one expects a phase transition at $d=1$. Therefore, it is possible that string theory has a phase transition at this point, where the string *degenerates to a branched polymer or is in a crumpled phase*[539]. Therefore we pass to the study of random surfaces, and related critical phenomena[540].

The propagation of strings along random surfaces is a rather general problem in physics. It describes low energy QCD[315] in the large N limit. In fact it is not difficult to relate the large N limit to 2-D gravity (Liouville theory)[541]. Some statistical models have been studied in this context, such as the Ising model on a random lattice[542,543], and the $Q=0,1$ Potts models[544]; polymers on a random surface[545], the $O(N)$ model[546], the Yang Lee[547] edge singularity[548], as well as D=1 strings[549] including nonperturbative effects[550] and relations to the KdV hierarchy[551]; W gravity[552], non critical strings[553], and topological field theory[554].

The starting point is the consideration of 2-D gravity with a cosmological term

$$S=\frac{1}{2}\int d^2\xi\sqrt{-g}g^{ab}\partial_a\varphi^\mu\partial_b\varphi_\mu+\beta\int d^2\xi\sqrt{-g} \quad . \tag{15.452}$$

For a genus-χ surface, we have the Gauss-Bonnet identity

$$\frac{1}{8\pi}\int d^2\xi\sqrt{-g}R=1-\chi \quad , \tag{15.453}$$

such that the Einstein gravity term in two dimensions is trivial.

In order to take into account of Weyl and reparametrization transformations, we need to introduce ghosts, (the previously discussed b, c systems), and integrate over them. However, in general, the conformal mode only cancels if we tune the central charge of the matter field φ_μ, which is D, to cancel against the ghost contribution, -26, as is well known. Otherwise, we are forced to consider the Liouville mode, with action (15.452). We can supplement (15.453) by the term

$$S_Q = -\frac{Q}{2}\int d^2\xi \hat{R}\varphi \quad , \tag{15.454}$$

where $\hat{R}$ is computed from $\hat{g}_{ab} = e^{-\alpha\varphi} g_{ab}$. If we compute the energy momentum tensor, we find

$$T_{zz} = -\frac{1}{2}[(\partial\varphi)^2 + Q\partial^2\varphi] \quad ; \tag{15.455}$$

we immediately recognize the construction (15.96), from where we have the central charge $c_\varphi = 1 + 3Q^2$; since $c_{tot} = c_\varphi + c - 26 = 0$, we have

$$Q = \sqrt{\frac{25-c}{3}} \quad . \tag{15.456}$$

Critical exponents of the theory may be computed as follows. Consider the partition functions as a function of the area[555] A

$$Z(A) = \int \mathcal{D}\varphi \mathcal{D}X e^{-S}\delta\left(\int e^{\alpha\varphi}\sqrt{\hat{g}}d^2\xi - A\right) \sim K A^{\Gamma-3} \quad , \tag{15.457}$$

defining the string susceptibility[556] Γ. Consider now the shift

$$\varphi \to \varphi + \frac{\rho}{\alpha} \quad , \tag{15.458}$$

with ρ a constant[555]. We have a corresponding shift in the action given by

$$S \to S - Q(1-\chi)\frac{\rho}{\alpha} \quad ; \tag{15.459}$$

the δ function in (15.457) transforms as $\delta() \to e^{-\rho}\delta()$, if we shift the area A as $A \to e^{-\rho}A$. Therefore

$$Z(A) = e^{[\frac{Q(1-\chi)}{\alpha}-1]\rho} Z(e^{-\rho}A) \quad , \tag{15.460}$$

from which we obtain

$$Z(A) = K A^{(1-\chi)\frac{Q}{\alpha}-1} \quad . \tag{15.461}$$

For the susceptibility Γ as a function of the genus χ we find the expression[17)]

$$\Gamma(\chi) = \frac{(1-\chi)}{12}[c - 25 - \sqrt{(25-c)(1-c)}] + 2 \quad ; \tag{15.462}$$

compare with (15.420-426) for genus $\chi = 0$. We have agreement with the semiclassical result[557]

$$\Gamma(\chi) = (1-\chi)\frac{(c-19)}{6} + 2 \quad . \tag{15.463}$$

Further information may be obtained from the discretized version of the theory, from which we shall obtain an effective description. With this aim, we triangulate the basis manifold[544,553], characterizing its points by $\underline{i}$, links $\langle ij\rangle$ and triangle $\langle ijk\rangle$, such that to each $\langle ij\rangle$ we have a matrix

$$g_{ij} = \begin{cases} 1 & i,j \text{ nearest neighbours} \\ 0 & \text{otherwise} \end{cases} \quad .$$

The distance between points is given by

$$h_{\langle ij\rangle} = \sqrt{(\varphi_i - \varphi_j)^2} \quad , \tag{15.464}$$

which is the main ingredient to discretize the first term in action (15.452). The second term in Eq. (15.452) is discretized defining the volume σ_i of the dual image (to be defined below) as

$$\sigma_i = \frac{1}{3}(\#\text{of nearest neighbours}) = \frac{1}{3}n_i \quad . \tag{15.465}$$

The curvature is

$$R_i = \pi\frac{6-n_i}{n_i} \tag{15.466}$$

(for $n = 6$ we have the planar result, as we should). Thus we have for the discretized action, the expression

$$S_{\text{discr.}} = \frac{1}{2}\sum(\varphi_i - \varphi_j)^2 + \beta\sum\sigma_i \quad . \tag{15.467}$$

We define the dual lattice G^* as the one obtained from G, associating respectively faces, edges and vertices of G^* to vertices, edges and faces of G, as in figure 15.5.

---

17) In fact this expression is only correct if there are no negative metric states, and the behaviour is dominated by the above simple scaling of the area. Otherwise a correction arises, and we have $\gamma^{\chi=0} = \frac{c-1-24\Delta-\sqrt{(1-c)(25-c-24\Delta)}}{12(1-\Delta)}$, and $\gamma^{\chi} - 2 = (1-\chi)(\gamma^{0} - 2)$[555].

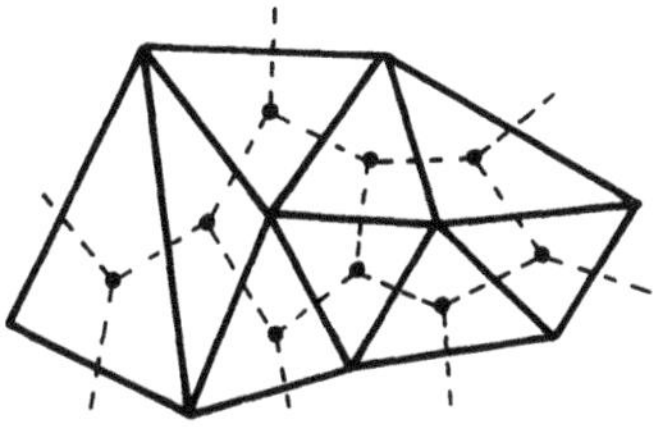

Figure 15.5: The lattice G and the dual lattice G^*.

We can rewrite the partition functions in terms of the dual diagram as follows. The velocity

$$V_a^\mu = \partial_a X^\mu \tag{15.468a}$$

c satisfies the constraint

$$\epsilon^{ab}\partial_a V_b^\mu = 0 \quad , \tag{15.468b}$$

which can be imposed by means of a Lagrange multiplier, resulting in

$$\int \mathcal{D}V_a^\mu \mathcal{D}Y^\mu e^{-\sigma\int d^2\xi\sqrt{g}g^{ab}V_a^\mu V_{b\mu}-\beta\int d^2\xi\sqrt{g}-i\int d^2\xi\sqrt{g}Y^\mu\epsilon^{ab}\frac{1}{\sqrt{g}}\partial_b V_{a\mu}} \tag{15.469}$$

In the discretized version, we can rewrite (15.469) as

$$V^\mu_{\langle ij\rangle} = X_i^\mu - X_j^\mu \tag{15.470a}$$

$$V^\mu_{\langle ij\rangle} + V^\mu_{\langle jk\rangle} + V^\mu_{\langle ki\rangle} = 0 \quad . \tag{15.470b}$$

We denote the vertices in the dual diagram G^* by $u^* = 1, \cdots, 2N$, and lines by $\langle u^* v^*\rangle$, associating to $\langle u^* v^*\rangle$ the variable $V^\mu_{u^*v^*} = V^\mu_{\langle ij\rangle}$ where $\langle ij\rangle$ is the link in G dual to $\langle u^* v^*\rangle$ in G^*. The Lagrange multipliers corresponding to (15.470b) act as

$$\delta\left(\sum_{v^*} V^\mu_{u^*v^*}\right) = \frac{1}{(2\pi)^D}\int d^D Y_{u^*} e^{Y^\mu_{u^*}\sum_{v^*} V^\mu_{u^*v^*}} \quad . \tag{15.471}$$

The sum over v^* corresponds to a sum over the triangles in G. Notice that one of the conditions is superfluous, and we omit it, obtaining

$$(2\pi)^{\frac{N-1}{2}} \int \prod_{v^*} d^D Y_{v^*} e^{-2\beta N - \sum \frac{1}{2}(Y_{u^*}-Y_{v^*})^2} \quad . \tag{15.472}$$

This case can be interpreted as a planar ϕ^3 theory with an exponential propagator $e^{-\frac{1}{2}(Y-Y')^2}$, and a coupling constant $g = (2\pi)^{D/4}e^{-\beta}$. Thus the planar expansion is a $1/N$ expansion, with

$$S[\phi] = N \,\mathrm{tr}\left\{\frac{1}{2}\int d^D Y d^D Y' \phi(Y) e^{\frac{1}{2}(Y-Y')^2}\phi(Y') - \frac{1}{3}g\int d^D Y \phi^3(Y)\right\} \tag{15.473}$$

This procedure can however be generalized to an arbitrary matrix model, describing a random surface. In order to simplify the matter, we shall consider from now on the problem in zero dimensions, that is $D = 0$. In any case, there is, as we have already mentioned, a potential singularity beyond $D = 1$ (see Eq. (15.420)), which must be properly considered, and we are obliged to deal with $D \leq 1$. For $D = 0$, we are faced with the problem of considering the partition function of the matrix model,

$$Z_N(\beta) = \int d\phi e^{-\beta \operatorname{tr} U(\phi)} \quad . \tag{15.474}$$

The large N expansion of the above theory, being given by the planar expansion, corresponds to a discrete version of the sum over two dimensional surfaces of given genus[393]. Indeed, suppose that

$$U(\phi) = \frac{1}{2}\phi^2 + \sum_{n \geq 3} c_n \phi^n = \sum U_l(\phi) \quad . \tag{15.475}$$

We rescale $\phi \to \phi\sqrt{\frac{N}{\beta}}$, such that for each link we have a propagator $1/N$, for the vertices $\left(\frac{N}{\beta}\right)^{\frac{l-2}{2}} N$, and for each face a factor N. We thus obtain the expansion

$$\ln Z_N(\beta) = \text{regular terms} + \sum N^{2(1-\chi)} \left(\frac{N}{\beta}\right)^A F_s(U) \quad , \tag{15.476}$$

with $A = \frac{1}{2}\sum_{\text{vertices}}(l-2)$.

Our aim is the study of the singularity properties of the above expression. In the continuum limit we should obtain a divergence due to the dominance of terms with an infinite number of triangles. We can conveniently choose the potential so that the singularity is adjusted to be at the point $\beta = N$. Thus we should have

$$\ln Z_N(\beta) \approx \beta^{2(1-\chi)} \left(\frac{\beta}{N} - 1\right)^{2-\gamma_\chi(U)} \quad . \tag{15.477}$$

Moreover, if we fix the genus, we have the previously discussed behavior

$$\sum_s F_s(U) \sim A^{\gamma_\chi(U)-2} \quad , \tag{15.478}$$

with[555,559]

$$\gamma_\chi(U) - 2 = (1-\chi)(\gamma_0(U) - 2) \quad . \tag{15.479}$$

This corresponds to string perturbation theory, with[560]

$$g^2_{\text{string}} = \left[\beta^2\left(\frac{\beta}{N}-1\right)^{2-\gamma_0}\right]^{-1} \quad . \tag{15.480}$$

Kazakov obtained $\gamma_0 = -1/k$, which corresponds to a central charge of the minimal models given by $c = 1 - \frac{6}{k(k+1)}$. However, several authors pointed out that there are reasons to believe that non unitary conformal models play a definite role, and a possible fitting of pure $2D$ gravity and of the Lie Yang edge singularity is given by the $(p,q) = (2, 2k-1)$ conformal models, which corresponds to[548]

$$c = 1 - \frac{3(2k-3)^2}{2k-1} \quad . \tag{15.481}$$

However, the above choice is not unique.

In order to study the theory on a definite ground, we shall adjust the parameters as

$$\ln Z_N(\beta) = F(t) \tag{15.482a}$$

$$t = (\beta - N)\beta^{-\frac{1}{2k+1}} \quad . \tag{15.482b}$$

The *specific heat* $f(t)$ is minus the second derivative of $F(t)$:

$$f = -\ddot{F} \quad . \tag{15.483}$$

In order to solve the large N limit of the above theory, we employ the very powerful method of orthogonal polynomials, which allow one to obtain several non-perturbative results.

Since $U(\phi)$ is supposed to be a trace of powers of ϕ, we factorize the fundamental field as

$$\phi = \Omega^\dagger \Lambda \Omega \quad , \tag{15.484}$$

where Λ is a diagonal matrix, $\Lambda_{ij} = \varphi_i \delta_{ij}$, and

$$\mathcal{D}\phi = \mathcal{D}\Omega \prod d\varphi_i \Delta_N(\varphi_i)^2 \quad , \tag{15.485}$$

where $\Delta_N(\varphi_i)$ is the Vandermond determinant

$$\Delta_N(\varphi_i) = \prod_{1 \le i < j \le N} (\varphi_j - \varphi_i) \quad . \tag{15.486}$$

By taking linear combinations, we can rewrite this determinant in terms of convenient polynomials

$$\Delta(\varphi) = \begin{vmatrix} 1 & \cdots & 1 \\ \varphi_1 & \cdots & \varphi_N \\ \varphi_1^2 & \cdots & \varphi_N^2 \\ \vdots & \cdots & \vdots \\ \varphi_1^{N-1} & \cdots & \varphi_N^{N-1} \end{vmatrix} = \begin{vmatrix} 1 & \cdots & 1 \\ P_1(\varphi_1) & \cdots & P_1(\varphi_1) \\ \vdots & \cdots & \vdots \\ P_{N-1}(\varphi_1) & \cdots & P_{N-1}(\varphi_N) \end{vmatrix} \tag{15.487}$$

where $P_n(\lambda) = \lambda^n + \mathcal{O}(\lambda^{n-1})$. We choose them such that

$$\int P_n(\lambda)P_m(\lambda)e^{-\beta U(\lambda)} = h_n\delta_{n,m} \quad . \tag{15.488}$$

With these preliminaries, we obtain for the partition function

$$Z_N(\beta) = N!h_0h_1\cdots h_{N-1} \quad . \tag{15.489}$$

We can use some recursion formulae to obtain more information about the theory. We shall suppose, in particular, that $U(\phi)$ is an even function. We have

$$P_n' = \sum_{j=0}^{n-1} A_{nj}P_j \tag{15.490a}$$

$$\lambda P_n = \sum_{j=0}^{n+1} R_{nj}P_j \quad . \tag{15.490b}$$

Therefore,

$$\int d\lambda e^{-U}\lambda P_m P_n = R_{nm}h_m = R_{mn}h_n \tag{15.491a}$$

$$\int d\lambda \frac{d}{d\lambda}\left(P_nP_me^{-\beta U}\right) = A_{nm}h_m + A_{mn}h_n - \beta\frac{dU_{mn}}{d\lambda} = 0 \quad . \tag{15.491b}$$

Thus R_{nm} is non zero iff $|n-m| \le 1$. For even U we have

$$\lambda P_n = P_{n+1} + R_nP_{n-1} \quad . \tag{15.492}$$

From (15.491a) we get

$$R_n = \frac{h_n}{h_{n-1}} \quad . \tag{15.493}$$

Therefore

$$F = \ln Z = \ln h_0^N N! + \sum_1^{N-1} (N-n) \ln R_n \quad . \tag{15.494}$$

Moreover, from (15.491) we find

$$\beta^{-1} = \frac{1}{n} U'_{n-1,n} \sqrt{R_n} \quad . \tag{15.495}$$

We can write $U'_{n,m}$ entirely in terms of R_n, obtaining a recursion relation for R_n; therefore we are able to write (15.494) entirely in terms of the potential. Moreover (15.495) will give us differential equations which lead to important informations about the problem.

We define the operator $\hat{l}$ as

$$\hat{l}|n\rangle = \frac{n}{\beta}|n\rangle \quad , \quad \hat{l} = -\frac{i}{\beta}\frac{d}{d\theta} \quad , \quad \theta = \frac{i}{\beta}\frac{d}{dl} \quad , \tag{15.496}$$

and change the labelling of indices by $x = n/\beta \quad , \quad R_n \to R(x)$.

Define now

$$\phi = e^{-i\theta} + e^{i\theta} R(\hat{l}) \quad . \tag{15.497}$$

We have for the matrix elements of $U'(\phi)$

$$\langle n - \sigma | U'(\phi) | n \rangle = \frac{n}{\beta} \delta_{\sigma,1} \quad . \tag{15.498}$$

We write $|n - \sigma\rangle = e^{i\sigma\theta}|n\rangle$, and the eigenvalue of $\hat{l}$; taking the factors $e^{i\theta}$ into account, one obtains

$$\int_{-\pi}^{\pi} \frac{d\theta}{2\pi} e^{-i\sigma\theta} U' \left[e^{-i\theta} + e^{i\theta} R \left(\frac{n}{\beta} + \frac{i}{\beta}\frac{\partial}{\partial\theta} \right) \right] .1 = \frac{n}{\beta}\delta_{\sigma,1} = x\delta_{\sigma,1} \tag{15.499}$$

We now consider the limit $\beta \cdot N \to \infty$, $\frac{\beta}{N} > 1$ a fixed number. Introduce the function

$$W(R) = \oint \frac{dz}{2\pi i} U'(z + R/z) = x \quad , \tag{15.500}$$

such that

$$U(\phi) = \int_0^1 \frac{du}{u} W(u(1-u)\phi^2) \quad . \tag{15.501}$$

The power series of both are related as

$$U(\phi) = \sum U_{2k} \phi^{2k} \tag{15.502a}$$

$$W(R) = \sum \frac{(2k)!}{k!(k-1)!} U_{1k} R^k \quad . \tag{15.502b}$$

The partition function may be seen to diverge as a function of β, as we shall see; taking Eq. (15.494) and writing

$$Z_N(\beta) = N! h_0^N e^{\sum (N-n)\ln R_n} \sim e^{\beta^2 \int_0^X (X-x)\ln R(x)dx} \tag{15.503}$$

At $x = 1$, we must have $R = 1$, and $W = 1$.

Now we make the Ansatz

$$W_k(R) = 1 - (1-R)^k \quad , \tag{15.504}$$

which corresponds to

$$U_2(\phi) = \phi^2 - \frac{1}{12}\phi^4 \quad , \quad (k=2) \tag{15.505}$$

or in general

$$U_k = \sum_1^k (-1)^{j-1} \phi^{2j} \frac{k!(j-1)!}{(k-j)!(2j)!} \tag{15.506}$$

and $R = 1 - (1-x)^{1/k}$.

Thus we have for the partition function the behavior

$$\ln Z_N(\beta) \sim \beta^2 (1-x)^{2+1/k} \quad . \tag{15.507}$$

The string anomalous dimension is given by

$$\gamma_0 = -1/k \quad . \tag{15.508}$$

The above analysis turns out to be even more powerful; indeed, if we consider the specific heat $f(t)$, an analysis of $f(t)$ shows that it obeys a Painlevé equation. The pure gravity case has been argued to correspond to case $k = 2$.

15.12 Conclusions

Conformally invariant field theory in two dimensions has turned into an extremely powerful tool to handle with a rich and important class of problems. It has encountered applications to fundamental interactions (string theory), critical theory in two dimensional statistical models, random surfaces, polymer physics and provides a method for obtaining solutions to important models, such as the WZNW theory, two dimensional gravity and supergravity. In the framework of mathematics, several new and elegant developments have been achieved. Also the case of Minkowskian conformal invariance in two dimensions has found applications recently, an important issue which we could not tackle in this book.

This chapter contains an immense plethora of non perturbative results of wide application and profound consequences, which has no parallel in physics of recent years.

To start with, general conformally invariant field theory in two dimensions could solve, and classify an extended class of statistical models, starting with the Ising model. They serve to classify general models showing critical behavior, supersymmetric models, generalizations to models invariant under non-abelian symmetry groups, as well as new identifications. A new mathematical branch of physics emerged from the fusion rules dictated by associativity in conformally invariant models, giving rise to the so called quantum groups, which contains a new multiplication rule obeying certain properties, such as quasi triangularity, meaning that there exists an *universal R matrix* interpolating two elements of the algebra, and obeying the Yang Baxter relation.

As it turns out, the quantum group is a symmetry group of rational conformal field theories; the conformal blocks are the equivalent of invariant tensors of the quantum group, and the fusion algebra is given by the decomposition rules of tensors product of irreducible representations. A common language between conformally invariant systems and integrable systems emerges quite naturally.

Furthermore, the other face of the problem, from the side of string theory, provides representations of conformal algebra with central charge $c \geq 1$, but in a different context. In this turn we have a mathematical framework for the study of extended objects represented by a quantum field .

The fact that both structures are underlined by the same theory has only very recently been better understood. As we have seen in the study of $2D$ gravity, the value $c = 1$ is a "borderline case for the applicability of the matrix model approach". in fact, the models with $c < 1$ have a finite number of degrees of freedom, a number which diverges as c approaches 1. Therefore, the theory with $c > 1$ must be a very difficult problem. The bordercase, $c = 1$ has been still treated in the literature. In fact, there are singularities when trying to study the string problem ($c > 1$) using matrix models techniques[79,555,559]. There is a possibility that beyond the value $a = 1$, a strong coupling emerges, where the string is in another phase. Usual string (and superstring) theory in the critical dimension (26,10,etc) has been studied extensively in the course of the years, and represents the other side of the present study. Further insight can also be gained introducing alternative formulations where the string action also contains terms depending on the intrinsic curvature. From the side of string theories however there are still many problems to be solved; especially the fact that the string perturbation theory is not Borel summable, requires nonperturbative treatments.

In this chapter we reviewed the construction of a general conformally invariant theory, realizing it in terms of the exponential operator introduced in chapter 2. Afterwards, we generalized the theory to include the non abelian case, realizing the above construction both for the bosonic WZNW theory and for a non abelian fermionic theory, thereby obtaining as a bonus, a generalization of the non abelian bosonization prescription to the interacting case. We then turned to more formal developments, which find application in mathematics and in string theory, especially the Kac Frenkel construction, and similarly, constructions in the heterotic string content.

In the last part of the chapter, we discussed induced gravity in two dimensions. This problem has a full quantum solution. Since it solves, in particular, the Liouville theory, we have a candidate to understand string theory away from the critical dimensions.

Chapter XVI

FINAL REMARKS

Two dimensional quantum field theory provides a very powerful laboratory for gaining a non perturbative understanding of quantum field theory. The kinematical simplifications resulting from two dimensional space time have allowed for the complete solution of a variety of models involving interacting fields. The non trivial nature of these solutions provide a deeper insight into the structure of quantum field theory, and has found useful applications in several areas of research, such as string theories and systems in statistical mechanics at criticality.

In all cases of completely soluble models, free bosonic fields, the exponential of such fields, as well as the boson-fermion equivalence for free fields play a key role in the explicit construction of the correlation functions. Hence we devoted chapter 2 to a detailed discussion of these topics, including a simple application to a model of scalar fields with derivative coupling to an axial vector fermionic current. As a further application of the exponential of a free field we have studied in chapter 3 the massless Thirring model, which provides the simplest example of a QFT model realizing all the Wightman axioms. The extension of this model to the case of massive fermions takes us, strictly speaking, outside the realm of free fields. Nevertheless, by doing an expansion in the mass parameter, the equivalence of the massive Thirring model to the sine Gordon equation could be proven on the basis of zero mass bosonic fields alone. Mandelstam's representation of fermions in terms of bosons, in turn showed, that the fermion of the massive Thirring model is to be identified with the soliton of the sine Gordon theory. The existence of such soliton sectors was seen to provide a link with the order disorder algebra in statistical mechanics. We established the existence of an infinite number of conservation laws on classical and quantum level, which eventually allowed for a construction of the exact S matrix in all sectors of the theory, that is, soliton-soliton, and soliton-bound state. These ideas prove useful in more algebraic approaches to QFT.

The study of two dimensional gauge theories in chapters 10 to 12 provided a deeper insight into features believed also to be characteristic of four dimensional chromodynamics, such as the θ-vacuum, the screening of "color", and confinement of quarks, as well as the evasion of the Nambu-Goldstone realization of the η'–the so called $U(1)$ problem– and the topology of the configuration space over which one functionally integrates.

The above model studies have provided an extraordinary amount of information about quantum field theory. Their success has motivated further developments leading to a direct application of two dimensional models to physical phenomena.

In chapter 4, we prepared the ground of subsequent parts of the book, making use of non perturbative functional techniques, by discussing methods for computing functional determinants, which constitute a basic building block of functional integrals. In particular, these techniques allow for a computation of the "Wess Zumino term" (albeit not of the functional determinant!) in any dimensions, and thus represent powerful tools in the construction of the anomalous chiral gauge theories, embedded into a gauge invariant theory at quantum level.

Conservation laws, $1/N$ expansion, S-matrix factorization and operator techniques, were essential elements of the non perturbative treatment of the $O(N)$ and $SU(N)$ chiral Gross-Neveu models, as well as of non linear sigma models, and $\mathbb{C}P^{N-1}$ models, and have been amply discussed in chapters 5 and 7. These models turned out to be much more complex than a $U(1)$ gauge theory of fermions interacting with gauge fields via minimal coupling. In the case of sigma models, a highly non linear interaction is present, and a very rich structure emerged. Geometry was here the main guideline since, following the ideas of gravitation, one wished to describe a theory containing the largest amount of symmetry as possible. This was discussed within a differential geometric framework in chapter 6 at the classical level, and the quantization of those ideas was presented in chapter 7. The outcome was a very elegant description of integrable theories, allowing for a Lax pair at the classical level, thus generalizing the ideas developed for the sine Gordon theory. At the quantum level, one obtained field theories with exactly computable S-matrices, as described in chapter 8. Integrability properties of these sigma models are found to hold also for a classical supersymmetric Yang-Mills theory and supergravity in 10 dimensions, as well as the respective theories obtained by dimensional reduction in 4 dimensions. This opens the possibility for non-trivial models in higher dimensions, which can be treated exactly.

Interestingly enough, one of the potentially most realistic applications of sigma models, is in the realm of string theories and quantum gravity, using the background field method, as discussed in chapter 7, which can be applied to the calculation of the quantum corrections to the Einstein's gravitational equations of motion. The requirement of conformal invariance in the framework of string theory, enforces the vanishing of the sigma model β function, which turns out to be the "string corrected" gravity equations of motion.

The construction of non-trivial S-matrices on the basis of factorization, and its relation to an infinite number of conservation laws, as discussed in chapter 8, should also find useful applications in algebraic quantum field theory. In fact, conformally invariant theories are known to possess a rich algebraic structure derived from the fusion rules of the elementary fields. These fusion rules being associative, and at the same time non-trivial, imply certain algebraic relations known from the mathematical literature as Artin Braid relations. These relations are similar to the factorization relations obtained for factorizable S-matrices. A general structure commonly known as "quantum groups", or to be more precise, "quantum algebras", emerge naturally from these constructions, and are today an active area of research. The construction of new integrable S-matrices from perturbations of conformally invariant theories, by conformal operators, is also of interest today.

The anomalous two dimensional chiral gauge theories considered in chapters 13 and 14, provide a very useful laboratory for understanding the role of gauge anomalies in chiral gauge theories, in general. Indeed, as was emphasized in chapter 13, chiral QCD_4 shares in this respect a variety of properties with its two dimensional analogue, QCD_2. In fact, any serious attempt to understand string theories away from criticality, or to deal with the Weinberg-Salam model of the weak interactions in the case that the "top" quark is not found, must deal with the problem of chiral anomalies. In the final sections of chapter 13 and 14 we presented a technique for obtaining, also in 4 dimensions, the constraint structure associated with these anomalies.

One of the most promising applications of two dimensional models to the description of fundamental interactions is that of string theory. This theory has a history which we could well call bizarre[561]. It was in the beginning conceived as an approach alternative to that of quantum field theory based on the concept of duality; hence also dual theory. The discovery by Veneziano of an amplitude describing most of the desired features of such a theory was the cornerstone of a complex story to follow. It was realized, that this theory could be understood as the movement of an extended one dimensional object in space time; hence the name "string theory". Later fermions have been incorporated into the theory, leading to the discovery of supersymmetry.

The old string model, as well as the Neveu-Schwarz-Ramond model, could be described in terms of a two dimensional QFT, with a local Lagrangian. The discovery of space-time supersymmetry by means of the[562] Gliozzi-Scherk-Olive construction, permitted one to envisage possible applications to grand unified theories, since supersymmetry was a requirement to solve the hierarchy problem in unified theories; this use of string theories had been proposed earlier, from the interpretation of the spin 2 (lowest) state of closed string theory, as the graviton, and strings would in such a case be a theory of quantum gravity[520]. This unification dream was later implemented at the theoretical level after a historical sequence of works on the cancellation of anomalies[563], which opened the possibility of "grand unification" in terms of a new superstring theory treating left and right movers differently, the "heterotic string"[564].

The strings describe a two dimensional world sheet embedded in the target manifold, on which the string is moving. Their dynamics is thus governed by two dimensional quantum field theory, which has thereby acquired the status of "physical relevance". Because of reparametrization invariance of string theory, our discussion in chapter 15 can be extensively used to understand the string. In fact, its interactions are based on the construction of vertex operators, which are a generalization of the exponential fields of chapter 2.

Of course, our ultimate objective is to understand the dynamics of reality. Many successes of quantum field theory merely rely on a "kinematical understanding", such as the (very important) idea of dynamical symmetry breaking, and the construction of representations of the gauge group in the matter sector. The actual non-perturbative dynamics of quantum field theory in four dimensions remains largely unknown. Hence, an application of the experience gained from the study of two dimensional QFT, to higher dimensions is highly welcome, and scores success. It is rewarding that the machinery of two dimensional QFT provides many of the expected results, sometimes just "taken for granted" in higher dimensions, due to the lack of methods for proving or disproving their validity. We have however also witnessed some surprising features which are non-perturbative in nature, and cannot be (or are difficult to be) understood in a perturbative context.

Some important results concerning generalizations to higher dimensions have been obtained. Bosonization of fermions is also possible in three dimensional space-time if we have a gauge field with a Chern-Simons density in the Lagrangian. Three dimensional models have been studied in this context; the gauge theory with a Chern-Simons topological action is related to the ideas presented in chapter 15. The ideas developed in the context of two dimensional quantum field theory thus appear to represent a step in the right direction. Moreover, recent progress made in the study of random surfaces, as presented at the end of chapter 15, shows that the ideas of string theory have a wider range of validity; indeed, there are hints showing a phase transition in string theory as a function of the space time dimensionality, such that the theory can accommodate the grand unification ideas, statistical models, and strong interactions, in a common framework.

Appendix A

Notation (Minkowski Space)

In this Appendix we summarize our notation and conventions in *Minkowski* space.

Kinematics

We use the metric

$$g^{00} = -g^{11} = 1, \quad g^{\mu\nu} = g_{\mu\nu} \quad . \tag{A.1}$$

The Levi-Civita tensor $\epsilon^{\mu\nu}$ is taken to be

$$\epsilon_{01} = -\epsilon^{01} = 1, \quad \epsilon^{\mu\nu} = -\epsilon^{\nu\mu} \quad , \tag{A.2}$$

and satisfies the relations $\epsilon^{\mu\nu}\epsilon^{\lambda\rho} = -g^{\mu\lambda}g^{\nu\rho} + g^{\mu\rho}g^{\nu\lambda}$, and $\epsilon_{\mu\lambda}\epsilon^{\lambda\nu} = g^{\nu}_{\mu}$. We use the compact notation

$$\tilde{a}^{\mu} = \epsilon^{\mu\nu}a_{\nu} \quad , \quad a^{\mu} = \epsilon^{\mu\nu}\tilde{a}_{\nu} \quad . \tag{A.3}$$

In particular, $\tilde{\partial}_{\mu} = \epsilon_{\mu\nu}\partial^{\nu}$, and $\partial^{\mu} = \epsilon^{\mu\nu}\tilde{\partial}_{\nu}$, with $\partial_{\mu} = \frac{\partial}{\partial x^{\mu}}$. One has the useful relations

$$\begin{aligned} a_{\mu}a_{\nu} - \tilde{a}_{\mu}\tilde{a}_{\nu} &= a^2 g_{\mu\nu} \\ \tilde{a}_{\mu}a_{\nu} - \tilde{a}_{\nu}a_{\mu} &= a^2 \epsilon_{\mu\nu} \quad . \end{aligned} \tag{A.4}$$

The d'Alembert operator $\Box$ is given by

$$\Box = \partial^{\mu}\partial_{\mu} = \partial_0^2 - \partial_1^2 \quad . \tag{A.5}$$

The combinations $\partial_{\mu} \pm \tilde{\partial}_{\mu}$ are lightlike 2-vectors, satisfying

$$(\partial_{\mu} \pm \tilde{\partial}_{\mu})(\partial^{\mu} \pm \tilde{\partial}^{\mu}) = 0 \quad . \tag{A.6}$$

The tensor

$$P^{(\pm)}_{\mu\nu} \equiv \frac{1}{2}(g_{\mu\nu} \pm \epsilon_{\mu\nu}) \tag{A.7}$$

is a projector:

$$P^{\pm)}_{\mu\lambda}P^{\lambda\nu}_{(\pm)} = P^{(\pm)\nu}_{\mu} \quad , \quad P^{(\pm)}_{\mu\lambda}P^{\lambda\nu}_{(\mp)} = 0 \quad . \tag{A.8}$$

Except for section 15.10, and part of 15.9, we use the definitions

$$\begin{aligned} a^{\pm} &= (a^0 \pm a^1)\,, \\ \tilde{a}^{\pm} &= (\tilde{a}^0 \pm \tilde{a}^1)\,; \end{aligned} \tag{A.9}$$

one has

$$\tilde{a}^{\pm} = \pm a^{\pm}$$
$$\tilde{a}_{\pm} = \mp a_{\pm} \quad . \tag{A.10a}$$

We also have

$$\begin{aligned} a_{+}b_{-} &= a_{\mu}(g^{\mu\nu} + \epsilon^{\mu\nu})b_{\nu} \ , \\ a_{-}b_{+} &= a_{\mu}(g^{\mu\nu} - \epsilon^{\mu\nu})b_{\nu} \ , \end{aligned} \tag{A.10b}$$

as well as the inverse relations,

$$\begin{aligned} a_{\mu}b^{\mu} &= \frac{1}{2}(a_{+}b_{-} + a_{-}b_{+}) \ , \\ \epsilon_{\mu\nu}a^{\mu}b^{\nu} &= \frac{1}{2}(a_{+}b_{-} - a_{-}b_{+}) \quad . \end{aligned} \tag{A.11}$$

In most of this book we work in the Weyl representation, where

$$\gamma^5 = \gamma_5 = \gamma^0\gamma^1 = \sigma_z = \begin{pmatrix} 1 & 0 \\ 0 & -1 \end{pmatrix} \ , \tag{A.12a}$$

and

$$\gamma^0 = \sigma_x = \begin{pmatrix} 0 & 1 \\ 1 & 0 \end{pmatrix} , \quad \gamma^1 = -i\sigma_y = \begin{pmatrix} 0 & -1 \\ 1 & 0 \end{pmatrix} \ . \tag{A.12b}$$

Here $\sigma_x, \sigma_y, \sigma_z$ are the usual Pauli matrices. In particular, we have

$$\gamma_{+} = \gamma^{-} = \begin{pmatrix} 0 & 2 \\ 0 & 0 \end{pmatrix} , \gamma_{-} = \gamma^{+} = \begin{pmatrix} 0 & 0 \\ 2 & 0 \end{pmatrix} , \gamma_{+}^2 = \gamma_{-}^2 = 0 \quad . \tag{A.13}$$

The γ matrices have the useful properties:

$$\begin{aligned} \gamma^{\mu}\gamma^5 &= \epsilon^{\mu\nu}\gamma_{\nu} \\ [\gamma^{\mu}, \gamma^{\nu}] &= 2g^{\mu\nu} \\ [\gamma^{\mu}, \gamma^{\nu}] &= -2\epsilon^{\mu\nu}\gamma_5 \ . \end{aligned} \tag{A.14}$$

One further has

$$\begin{aligned} \gamma^{\mu}\gamma^{\nu} &= g^{\mu\nu} - \epsilon^{\mu\nu}\gamma^5 \\ \operatorname{tr}\gamma^{\mu}\gamma^{\nu} &= 2g^{\mu\nu} \\ \operatorname{tr}\gamma^{\mu}\gamma^{\nu}\gamma^{\rho}\gamma^{\sigma} &= 2[g^{\mu\nu}g^{\rho\sigma} - g^{\mu\rho}g^{\nu\sigma} + g^{\mu\sigma}g^{\nu\rho}] \\ \operatorname{tr}\gamma^{\mu}\gamma^{\nu}\gamma^5 &= -2\epsilon^{\mu\nu} \\ \operatorname{tr}\gamma^{\mu}\gamma^{\nu}\gamma^{\rho}\gamma^{\sigma}\gamma^5 &= -2[g^{\mu\nu}\epsilon^{\rho\sigma} + g^{\rho\sigma}\epsilon^{\mu\nu}] \quad . \end{aligned} \tag{A.15}$$

With the definition $\tilde{j}^\mu = \bar{\psi}\gamma^\mu\gamma^5\psi$ for the axial vector current, the relation (A.3) is preserved, $\tilde{j}^\mu = \epsilon^{\mu\nu}j_\nu$. The projector on right (left) handed fermions is given by $P_+(P_-)$, where

$$P_\pm = \frac{1}{2}(1 \pm \gamma^5) \quad . \tag{A.16}$$

$P_\pm$ satisfies the relations $\gamma^\mu P_\pm = (g^{\mu\nu} \pm \epsilon^{\mu\nu})\gamma_\nu$.

It is sometimes convenient to introduce the Majorana representation, as we did in chapters 5, 6, 7, and 9, where

$$\gamma_0 = \sigma_2 = \begin{pmatrix} 0 & -i \\ i & 0 \end{pmatrix} \quad , \qquad \gamma_1 = i\sigma_1 = \begin{pmatrix} 0 & i \\ i & 0 \end{pmatrix} \tag{A.17a}$$

$$\gamma_5 = \sigma_3 = \begin{pmatrix} 1 & 0 \\ 0 & -1 \end{pmatrix} \tag{A17b}$$

$$\gamma_\mu\gamma_\nu = g_{\mu\nu} + \epsilon_{\mu\nu}\gamma_5 \quad , \qquad \gamma^\mu\gamma^5 = -\epsilon^{\mu\nu}\gamma_\nu \quad . \tag{A.17c}$$

Two-point functions

The free scalar and Dirac field satisfies the equations

$$(\Box + m^2)\phi(x) = 0 \ , \tag{A.18a}$$

$$(i\partial\!\!\!/ - m)\psi(x) = 0 \ , \tag{A.18b}$$

respectively. The corresponding Feynman propagators are defined by

$$-i\Delta_F(x-y) \equiv < 0|T\phi(x)\phi(y)|0 > \ , \tag{A.19a}$$

$$iS_F(x-y)_{\alpha\beta} \equiv < 0|T\psi_\alpha(x)\bar{\psi}_\beta(y)|0 > \quad . \tag{A.19b}$$

The canonical commutation relations then imply

$$(\Box + m^2)\Delta_F(x-y) = \delta^2(x-y) \ , \tag{A.20a}$$

$$(i\partial\!\!\!/ - m)S_F(x-y) = \delta^2(x-y) \quad . \tag{A.20b}$$

From here we obtain the Fourier representations

$$\Delta_F(x-y) = -\int \frac{d^2q}{(2\pi)^2}\frac{e^{-iq(x-y)}}{q^2 - m^2 + i0} \quad , \tag{A.21a}$$

$$S_F(x-y) = -(i\partial\!\!\!/ + m)\Delta_F(x) = \int \frac{d^2p}{(2\pi)^2}\frac{e^{-ip(x-y)}}{p\!\!\!/ - m + i0} \quad . \tag{A.21b}$$

We further define

$$\Delta^{(+)}(x-y) \equiv [\phi^{(+)}(x), \phi^{(-)}(y)] = < 0|\phi(x)\phi(y)|0 > \quad , \tag{A.22a}$$

$$i\Delta(x-y) \equiv [\phi(x), \phi(y)] = \Delta^{(+)}(x-y) - \Delta^{(+)}(y-x) \quad . \tag{A.22b}$$

One has the following integral representation

$$\Delta^{(+)}(x-y) = \int \frac{d^2q}{(2\pi)}\theta(q^0)\delta(q^2-m^2)e^{-iqx} \quad , \qquad (A.23a)$$

$$i\Delta(x-y) = \int \frac{d^2q}{(2\pi)}\epsilon(q^0)\delta(q^2-m^2)e^{-iqx} \quad . \qquad (A.23b)$$

Explicitly

$$\Delta^{(+)}(x) = \frac{1}{2\pi}K_0(m\sqrt{-x^2+ix^0\epsilon}) \qquad (A.24a)$$

$$\Delta(x) = -\frac{1}{2}\theta(x^2)\epsilon(x^0)J_0(m\sqrt{x^2}) \qquad (A.24b)$$

$$\Delta_F(x) = \frac{i}{2\pi}K_0(m\sqrt{-x^2+i\epsilon}) \quad . \qquad (A.24c)$$

For small argument

$$\Delta^{(+)}(x) \simeq -\frac{1}{4\pi}\ln\tilde{m}^2(-x^2+ix^0\epsilon) \quad , \qquad (A.25a)$$

$$\Delta_F(x) \simeq -\frac{i}{4\pi}\ln\tilde{m}^2(-x^2+i\epsilon) \quad , \qquad (A.25b)$$

where $\tilde{m} = \frac{m}{2}e^{\gamma}$, $\gamma = -\Gamma'(1)$. The corresponding zero mass two-point functions are denoted by $D^{(+)}, D$ and D_F, respectively. They are given by

$$D^{(+)}(x) = \frac{1}{2\pi}\int_{-\infty}^{\infty}\frac{dp^1}{2\mid p^1\mid}\left[e^{-ipx} - \theta(\lambda - \mid p^1\mid)\right] \qquad (A.26a)$$

$$= -\frac{1}{4\pi}\ln(-\mu^2x^2+ix^0\epsilon) = -\frac{1}{4\pi}\ln(i\mu x^+ + \epsilon)(i\mu x^- + \epsilon)) \qquad (A.26b)$$

$$D(x) = -\frac{1}{2}\theta(x^2)\epsilon(x^0) \qquad (A.26c)$$

$$D_F(x) = -\frac{i}{4\pi}\ln(-\mu^2x^2+i\epsilon) \quad , \qquad (A.26d)$$

where $\mu = e^{\gamma}\lambda$ is an infrared regulating "mass" parameter. The corresponding representation for the massless fermionic two-point functions are:

$$S^{(+)}(x) = i\partial\!\!\!/ D^{(+)}(x) = -\frac{i}{2\pi}\frac{\gamma^\mu x_\mu}{x^2 - i\epsilon x^0} = -\frac{i}{2\pi}\begin{pmatrix} 0 & \frac{1}{x^- - i\epsilon} \\ \frac{1}{x^+ - i\epsilon} & 0 \end{pmatrix} \qquad (A.27a)$$

$$S_F(x) = -i\partial\!\!\!/ D_F(x) = -\frac{1}{2\pi}\frac{\gamma^\mu x_\mu}{x^2 - i\epsilon} = -\frac{1}{2\pi}\begin{pmatrix} 0 & \frac{x^+}{x^2 - i\epsilon} \\ \frac{x^-}{x^2 - i\epsilon} & 0 \end{pmatrix} \quad . \qquad (A.27b)$$

We define the field $\vec{\phi}(x)$ "dual" to $\phi(x)$ by

$$\vec{\phi}(x) = \int_{x^1}^{\infty} dz^1 \partial_0\phi(x^0, z^1) \quad . \qquad (A.28)$$

A.3 *Zero-mass free field*

For ϕ satisfying $\Box\phi = 0$ it is convenient to define the linear combination

$$\phi_{\substack{R\\L}}(x) = (\phi(x) \pm \vec{\phi}(x)) \quad . \tag{A.29}$$

Since in this case $\partial_\mu \vec{\phi}(x) = \tilde{\partial}_\mu \phi(x)$, one has,

$$\partial_0 \phi_{\substack{R\\L}}(x) = \partial_\mp \phi(x) \quad , \qquad \partial_1 \phi_{\substack{R\\L}}(x) = \mp\partial_\mp \phi(x) \quad . \tag{A.30}$$

This implies $\partial_+\phi_R = 0$, $\partial_-\phi_L = 0$. Hence $\phi_R(\phi_L)$ represent right (left) moving waves. Although $\vec{\phi}$, defined by (A.28), and $\tilde{\phi}$, defined in chapter 2, both satisfy the differential equations (2.20) and (2.21), their commutators differ. Indeed, one finds

$$\begin{aligned}
[\vec{\phi}^{(+)}(x), \vec{\phi}^{(-)}(0)] &= D^{(+)}(x)\\
[\phi^{(+)}(x), \vec{\phi}^{(-)}(0)] &= \tilde{D}^{(+)}(x) + \frac{i}{4}\\
[\vec{\phi}^{(+)}(x), \phi^{(-)}(0)] &= \tilde{D}^{(+)}(x) - \frac{i}{4}\\
[\phi_R^{(+)}(x), \phi_R^{(-)}(0)] &= 2(D^{(+)}(x) + \tilde{D}^{(+)}(x))\\
[\phi_L^{(+)}(x), \phi_L^{(-)}(0)] &= 2(D^{(+)}(x) - \tilde{D}^{(+)}(x))\\
[\phi_R^{(+)}(x), \phi_L^{(-)}(0)] &= 2(D^{(+)}(x) - \tilde{D}^{(+)}(x)) - \frac{i}{2}\\
[\phi_L^{(+)}(x), \phi_R^{(-)}(0)] &= 2(D^{(+)}(x) - \tilde{D}^{(+)}(x)) + \frac{i}{2}
\end{aligned} \tag{A.31}$$

where

$$\tilde{D}^{(+)}(x) = \frac{1}{4\pi}\ln\left(\frac{i\mu x^+ + \epsilon}{i\mu x^- + \epsilon}\right), \quad D^{(+)}(x) = -\frac{1}{4\pi}\ln((i\mu x^+ + \epsilon)(i\mu x^- + \epsilon)) \quad . \tag{A.32a}$$

In particular

$$\begin{aligned}
D^{(+)}(x) + \tilde{D}^{(+)}(x) \quad &= -\frac{1}{2\pi}\ln(i\mu x^- + \epsilon)\\
D^{(+)}(x) - \tilde{D}^{(+)}(x) \quad &= -\frac{1}{2\pi}\ln(i\mu x^+ + \epsilon)\\
i\tilde{D}(x) = \tilde{D}^{(+)}(x) - \tilde{D}^{(+)}(-x) &= \frac{i}{2}\theta(-x^2)\epsilon(x^1) \quad .
\end{aligned} \tag{A.32b}$$

Hence

$$\begin{aligned}
[\phi_R^{(+)}(x), \phi_L^{(-)}(0)] &= -\frac{1}{\pi}\ln(i\mu x^- + \epsilon) - \frac{i}{2}\\
[\phi_L^{(+)}(x), \phi_R^{(-)}(0)] &= -\frac{1}{\pi}\ln(i\mu x^+ + \epsilon) + \frac{i}{2} \quad .
\end{aligned} \tag{A.33}$$

One checks that $\partial_\mu \tilde{D}(x) = \tilde{\partial}_\mu D(x)$, etc.

The additive, purely imaginary constants appearing in the commutation relations (A.31) are the reason for having to introduce the phase-factor $(-i\pi\gamma^5/4)$ in the Mandelstam bosonization formula (10.38a) for the free fermion field. They also insure anticommutativity of the free fermion field in terms of its Mandelstam representation (10.38).

For the equal-timecommutators one has

$$\begin{aligned}
[\phi^{(+)}(x), \partial_0 \phi^{(-)}(y)]_{ET} &= \frac{i}{2}\delta(x^1 - y^1) \\
[\tilde{\phi}^{(+)}(x), \partial_0 \tilde{\phi}^{(-)}(y)]_{ET} &= \frac{i}{2}\delta(x^1 - y^1) \\
[\tilde{\phi}^{(+)}(x), \phi^{(-)}(y)]_{ET} &= -\frac{i}{2}\theta(y^1 - x^1) \\
[\phi^{(+)}(x), \tilde{\phi}^{(-)}(y)]_{ET} &= \frac{i}{2}\theta(y^1 - x^1) \quad .
\end{aligned} \tag{A.34}$$

These properties follow from $\partial_0 D^{(+)}(x)\Big|_{x^0=0} = -\frac{i}{2}\delta(x^1)$, and

$$\tilde{D}^{(+)}(x)\Big|_{x^0=0} = \frac{1}{4\pi}\ln\left(\frac{ix^1 + \epsilon}{-ix^1 + \epsilon}\right) = \frac{i}{4}\epsilon(x^1) \tag{A.35}$$

where $\epsilon(x) = \theta(x) - \theta(-x)$.

Appendix B

Notation (Euclidean Space)

In this Appendix we summarize our notation and conventions in *euclidian* space. We use only lower indices running from 1 to 2, where the index "2" now replaces the Minkowski index "0".

Kinematics

The metric is the Kronecker delta, $\delta_{\mu\nu}$. The Levi-Civita tensor is taken to be

$$\epsilon_{12} = -\epsilon_{21} = 1, \quad \epsilon_{\mu\nu} = -\epsilon_{\nu\mu} \quad . \tag{B.1}$$

We use the compact notation,

$$\tilde{a}_\mu = \epsilon_{\mu\nu} a_\nu \quad , \quad a_\mu = -\epsilon_{\mu\nu} \tilde{a}_\nu \quad . \tag{B.2}$$

In particular

$$\tilde{\partial}_\mu = \epsilon_{\mu\nu} \partial_\nu, \ \partial_\mu = -\epsilon_{\mu\nu} \tilde{\partial}_\nu \tag{B.3}$$

with $\partial_\mu = \frac{\partial}{\partial x_\mu}$. One has the useful relations,

$$\begin{aligned} a_\mu a_\nu + \tilde{a}_\mu \tilde{a}_\nu &= a^2 \delta_{\mu\nu} \\ \tilde{a}_\mu a_\nu - a_\mu \tilde{a}_\nu &= a^2 \epsilon_{\mu\nu} \quad . \end{aligned} \tag{B.4}$$

The Laplacian Δ is given by

$$\Delta = \partial_1^2 + \partial_2^2 \quad . \tag{B.5}$$

From here it follows $\tilde{\partial}_\mu \tilde{\partial}_\mu = \Delta$, $\tilde{\partial}_\mu \partial_\mu = 0$. The combinations $(\partial_\mu \pm i\tilde{\partial}_\mu)$ are light-like vectors,

$$(\partial_\mu \pm i\tilde{\partial}_\mu)(\partial_\mu \pm i\tilde{\partial}_\mu) = 0 \quad . \tag{B.6}$$

The tensor

$$P^{(\pm)}_{\mu\nu} = \frac{1}{2}(g_{\mu\nu} \pm i\epsilon_{\mu\nu}) \tag{B.7}$$

is a projector:

$$P^{(\pm)}_{\mu\lambda} P^{(\pm)}_{\lambda\nu} = \delta_{\mu\nu}, \quad P^{(\pm)}_{\mu\lambda} P^{(\mp)}_{\lambda\nu} = 0 \quad . \tag{B.8}$$

With the definitions

$$\begin{aligned} a_\pm &= a_1 \pm i a_2 \quad , \\ \tilde{a}_\pm &= \tilde{a}_1 \pm i \tilde{a}_2 \quad , \end{aligned} \tag{B.9}$$

one has $\tilde{a}_\pm = \mp i a_\pm$, and

$$\begin{aligned} a_+b_- &= a_\mu(\delta_{\mu\nu} - i\epsilon_{\mu\nu})b_\nu \\ a_-b_+ &= a_\mu(\delta_{\mu\nu} + i\epsilon_{\mu\nu})b_\nu \end{aligned} \tag{B.10}$$

as well as the inverse relation,

$$\begin{aligned} a_\mu b_\mu &= \frac{1}{2}(a_+b_- + a_-b_+) \\ \epsilon_{\mu\nu}a_\mu b_\nu &= \frac{i}{2}(a_+b_- - a_-b_+) \quad . \end{aligned} \tag{B.11}$$

We work in the Weyl representation, where

$$\gamma_5 = -i\gamma_1\gamma_2 = \sigma_3 \quad , \tag{B.12}$$

and

$$\gamma_1 = \sigma_2 = \begin{pmatrix} & -i \\ i & \end{pmatrix}, \quad \gamma_2 = -\sigma_1 = \begin{pmatrix} & -1 \\ -1 & \end{pmatrix} \quad , \tag{B.13}$$

i.e. $\gamma_\mu = \epsilon_{\mu\nu}\sigma_\nu$. In particular,

$$\gamma_+ = \begin{pmatrix} 0 & -2i \\ 0 & 0 \end{pmatrix}, \quad \gamma_- = \begin{pmatrix} 0 & 0 \\ 2i & 0 \end{pmatrix} \quad , \tag{B.14a}$$

$$(\gamma_+\gamma_-)^n = 2^{2n-2}\gamma_+\gamma_- \, , \, \gamma_+^2 = \gamma_-^2 = 0 \quad . \tag{B.14b}$$

The γ-matrices have the properties

$$\begin{aligned} i\gamma_\mu\gamma_5 &= \epsilon_{\mu\nu}\gamma_\nu \\ \{\gamma_\mu\gamma_\nu\} &= 2\delta_{\mu\nu} \\ [\gamma_\mu\gamma_\nu] &= 2i\epsilon_{\mu\nu}\gamma_5 \quad , \end{aligned} \tag{B.15}$$

from where one has

$$\gamma_\mu\gamma_\nu = \delta_{\mu\nu} + i\epsilon_{\mu\nu}\gamma_5 \quad . \tag{B.16}$$

The axial vector current is defined by

$$\tilde{j}_\mu = \bar{\psi} i\gamma_\mu\gamma_5\psi \tag{B.17}$$

which preserves the relation (B.2) $\tilde{j}_\mu = \epsilon_{\mu\nu}j_\nu$. The projector on right (left) handed fermions is given by $P_+(P_-)$, where

$$P_\pm = \frac{1}{2}(1 \pm \gamma^5) \quad . \tag{B.18}$$

$P_\pm$ satisfies the relations

$$\gamma_\mu P_\pm = (\delta_{\mu\nu} \mp i\epsilon_{\mu\nu})\gamma_\nu \quad . \tag{B.19}$$

Two-point functions

The boson- and fermion- two-point functions satisfy the equations

$$(-\Delta + m^2)\Delta_E(x-y) = \delta^2(x-y) \tag{B.20a}$$

$$(i\partial\!\!\!/ + im)S_E(x-y) = \delta^2(x-y) \quad . \tag{B.20b}$$

From here we obtain the Fourier representations

$$\Delta_E(x) = \int \frac{d^2q}{(2\pi)^2}\frac{e^{iq\cdot x}}{q^2+m^2} \quad , \tag{B.21a}$$

$$S_E(x) = -\int \frac{d^2q}{(2\pi)^2}\frac{e^{ip\cdot x}}{p\!\!\!/ - im} \quad . \tag{B.21b}$$

We evidently have

$$S_E(x) = (i\partial\!\!\!/ - im)\Delta_E(x) \quad . \tag{B.22}$$

Since $\Delta_E(x_1,x_2) = -i\Delta_F(-ix_2,x_1)$, we have from (A.24)

$$\Delta_E(x) = \frac{1}{2\pi}K_0(m\sqrt{x^2}) \quad , \tag{B.23}$$

or for small argument

$$\Delta_E(x) \simeq \frac{-1}{4\pi}\ln \tilde{m}^2x^2 \tag{B.24}$$

with $\tilde{m}$ defined in (A.25). For the zero-mass two-point functions we have correspondingly

$$D_E(x) = \frac{-1}{4\pi}\ln \mu^2x^2 \tag{B.25a}$$

$$S_E(x) = -\frac{i}{2\pi}\frac{\gamma_\mu x_\mu}{x^2} \quad . \tag{B.25b}$$

Appendix C

Further Conventions

In this Appendix we list some conventions, definitions, and formulae not contained in Appendices A and B.

a) SU(N) generators

The antihermitian $SU(N)$ generators are normalized as follows:

$$[t^a, t^b] = f_{abc}t^c, \quad tr(t^a t^b) = -\delta^{ab} \tag{C.1a}$$

$$[T^a, T^b] = f_{abc}T^c, \quad tr(T^a T^b) = -\delta^{ab} \quad . \tag{C.1b}$$

Here $t^a(T^a)$ denote the generators in the fundamental (adjoint) representation of $SU(N)$. We have the useful relations

$$\sum_{a=1}^{N^2-1} t^a_{ij} t^a_{kl} = -(\delta_{il}\delta_{jk} - \frac{1}{N}\delta_{ij}\delta_{kl}) \tag{C.2}$$

$$(T^c)_{ab} = f_{acb} \quad . \tag{C.3}$$

b) Covariant derivatives

We denote by

$$D_\mu = \partial_\mu + e\ A_\mu \tag{C.4a}$$

and

$$\mathcal{D}_\mu = \partial_\mu + e\mathcal{A}_\mu \tag{C.4b}$$

the covariant derivatives in the fundamental and adjoint representation, respectively. Here

$$A_\mu \equiv \sum_a t^a A^a_\mu \tag{C.5a}$$

$$\mathcal{A}_\mu \equiv \sum_a T^a A^a_\mu \quad . \tag{C.5b}$$

Explicitly, we have

$$\mathcal{D}^{ab}_\mu = \delta^{ab}\partial_\mu + e f_{acb} A^c_\mu \quad . \tag{C.6}$$

The field-strength tensor is defined by

$$F^a_{\mu\nu} = \partial_\mu A^a_\nu - \partial_\nu A^a_\mu + e f_{abc} A^b_\mu A^c_\nu \tag{C.7a}$$

or

$$F_{\mu\nu} = \partial_\mu A_\nu - \partial_\nu A_\mu + e[A_\mu, A_\nu] \qquad (C.7b)$$

The covariant derivative D_μ acts by simple matrix multiplication from the left on ***group-valued*** elements U, and as a commutator on the *Lie-algebra* valued elements, both taken in the fundamental representation:

$$D_\mu U = \partial_\mu U + eA_\mu U \qquad (C.8a)$$

$$D_\mu M = \partial_\mu M + e[A_\mu, M] \quad . \qquad (C.8b)$$

Except for chapter 6, we have found it more convenient to work with *hermitian* Lie algebra valued fields defined by

$$A_\mu = \sum \tau^a A^a_\mu \qquad (C.9)$$

where $\tau^a = it^a$. Thus

$$[\tau^a, \tau^b] - if_{abc}\tau^c \quad , \mathrm{tr}(\tau^a\tau^b) = \delta^{ab} \quad . \qquad (C.10)$$

Correspondingly eqs.(C.4a), (C.7b) and (C.8) are then replaced by

$$\begin{aligned} D_\mu &= \partial_\mu - ieA_\mu \quad , \\ F_{\mu\nu} &= \partial_\mu A_\nu - \partial_\nu A_\mu - ie[A_\mu, A_\nu] \quad , \\ D_\mu U &= \partial_\mu U - ieA_\mu U \quad , \\ D_\mu M &= \partial M - ie[A_\mu, M] \quad . \end{aligned} \qquad (C.11)$$

c) Operator identities

If $[A, B] = c - number$, the following relations hold:

$$\begin{aligned} e^A e^B &= e^{[A,B]} e^B e^A \\ &= e^{\frac{1}{2}[A,B]} e^{A+B} \quad , \end{aligned} \qquad (C.12)$$

and

$$[A, e^B] = [A, B]e^B \quad , \qquad (C.13a)$$

or equivalently

$$e^{-B} A e^B = A + [A, B] \quad . \qquad (C.13b)$$

If $[B, A] = \lambda A$ then

$$e^B A e^{-B} = e^\lambda A \quad . \qquad (C.14)$$

d) Berezin integration rules

For c and $\bar{c}$ Grassman valued, one has

$$\begin{aligned} c\bar{c} + \bar{c}c &= 0 \\ c^2 = \bar{c}^2 &= 0 \end{aligned} \tag{C.15}$$

$$\begin{aligned} &\int d\bar{c}\bar{c} = 1, \quad \int dcc = 1 \\ &\int dc \int d\bar{c}\bar{c}c = \int d\bar{c} \int dcc\bar{c} = 1 \\ &\int dc = \int d\bar{c} = 0 \quad . \end{aligned} \tag{C.16}$$

If

$$P(\bar{c}, c) = p_0 + p_1 c + p_2 \bar{c} + p_{12} c\bar{c} \tag{C.17}$$

then

$$\begin{aligned} \int dcP &= p_1 = \frac{\partial}{\partial c} P \\ \int d\bar{c}P &= p_2 = \frac{\partial}{\partial \bar{c}} P \\ \int d\bar{c}\, dc &= p_{12} = \frac{\partial}{\partial \bar{c}} \frac{\partial}{\partial c} P \quad . \end{aligned} \tag{C.18}$$

e) Analytic continuation from $\Re^2$ to M^2

The analytic continuation from Euclidean to Minkowski space is realized by the substitutions

$$\delta_{\mu\nu} \to g_{\mu\nu} \tag{C.19}$$

$$\begin{aligned} x_1 \to x^1 \quad &, \quad x_2 \to ix^0 \quad , \\ \partial_1 \to \partial_1 \quad &, \quad \partial_2 \to -i\partial_0 \quad , \\ \tilde{\partial}_1 \to -i\partial^0 \quad &, \quad \tilde{\partial}_2 \to -\partial_1 \quad , \\ A_1 \to A_1 \quad &, \quad A_2 \to -iA_0 \quad . \end{aligned} \tag{C.20}$$

Hence

$$\begin{aligned} \partial_+ \to \partial_+ \quad &, \quad \partial_- \to -\partial_- \quad , \\ A_+ \to A_+ \quad &, \quad A_- \to -A_- \quad . \end{aligned} \tag{C.21}$$

For scalar products this implies

$$\begin{aligned} \partial_\mu a \partial_\mu b &\to -\partial_\mu a \partial^\mu b \quad , \\ \epsilon_{\mu\nu} \partial_\mu a \partial_\mu b &\to -i\epsilon^{\mu\nu} \partial_\mu a \partial_\nu b \quad , \end{aligned} \tag{C.22}$$

as well as

$$
\begin{aligned}
A_\mu \partial_\mu &\to -A^\mu \partial_\mu \quad , \\
\epsilon_{\mu\nu} A_\mu \partial_\nu &\to -i\epsilon^{\mu\nu} A_\mu \partial_\nu \quad ,
\end{aligned}
\tag{C.23}
$$

or equivalently

$$
\begin{aligned}
A_+ \partial_- &\to -A_+ \partial_- \quad , \\
A_- \partial_+ &\to -A_- \partial_+ \quad .
\end{aligned}
\tag{C.24}
$$

For the action and effective action the relations are, respectively,

$$
\begin{aligned}
-S_E &\to iS \quad , \\
-W_E &\to iW \quad .
\end{aligned}
\tag{C.25}
$$

Appendix D

Functional Bosonization of the Massive Thirring Model

Although operator methods have been shown to be extremely powerful, it is clear from the usual experience in quantum field theory, that functional methods can be dealt with much bigger insight in order to develop new ideas. This is our aim, namely to "rediscover" abelian bosonization by functional methods, the generalization to the non-abelian case having been treated in detail in chapter 11. With this aim in mind, consider the massive Thirring model, given by the Lagrangian density

$$\mathcal{L} = \overline{\psi} i \not\partial \psi + M\overline{\psi}\psi + \frac{g}{2}(\overline{\psi}\gamma^\mu\psi)^2 \quad , \tag{D.1}$$

with the partition function $\mathcal{Z} = \int D\overline{\psi}D\psi e^{i\int d^2x\mathcal{L}}$. We use the identity

$$e^{i\int d^2x \frac{g}{2}(\overline{\psi}\gamma^\mu\psi)^2} = \int DA_\mu e^{i\int d^2x(-\frac{1}{2g}A_\mu^2 + A^\mu\overline{\psi}\gamma_\mu\psi)} \quad , \tag{D.2}$$

in order to replace (D.1) by a Lagrangian quadratic in the fermions fields. A vector field in two dimensions can be written as

$$A_\mu = \frac{1}{\sqrt{\pi}}\epsilon_{\mu\nu}\partial^\nu\varphi + \frac{1}{\sqrt{\pi}}\partial_\mu\eta \quad , \tag{D.3}$$

with a trivial (constant) Jacobian for the change of integration measure

$$DA_\mu = JD\varphi D\eta \quad . \tag{D.4}$$

Therefore, we obtain

$$\mathcal{L} = \overline{\psi} i \not\partial \psi + M\overline{\psi}\psi + \frac{1}{\sqrt{\pi}}\overline{\psi}\gamma^\mu(\epsilon_{\mu\nu}\partial^\nu\varphi + \partial_\mu\eta)\psi + \frac{1}{2g\pi}\left[(\partial_\mu\varphi)^2 - (\partial_\mu\eta)^2\right] \quad . \tag{D.5}$$

The η field decouples from the theory. This may be verified by a simple change of variables. Define a new fermionic field by

$$\psi(x) = e^{\frac{-i}{\sqrt{\pi}}\gamma_5\varphi + \frac{i}{\sqrt{\pi}}\eta}\xi(x). \tag{D.6}$$

In terms of $\xi(x)$, the Lagrangian (D.5) reads

$$\mathcal{L} = \overline{\xi} i \not\partial \xi + M\overline{\xi}e^{\frac{-2i}{\sqrt{\pi}}\gamma_5\varphi}\xi + \frac{1}{2g\pi}\left[(\partial_\mu\varphi)^2 - (\partial_\mu\eta)^2\right] \quad . \tag{D.7}$$

The Jacobian corresponding to the above change of variable (D.6), $J = \frac{\det(i\partial\!\!\!/+A\!\!\!/)}{\det(i\partial\!\!\!/)}$, is given by (12.2$b$), continued to Minkowski space. Hence

$$D\overline{\psi}D\psi = D\overline{\xi}D\xi e^{-\frac{i}{2\pi^2}\int d^2x(\partial_\mu\varphi)^2} . \tag{D.8}$$

The full quantum Lagrangian is thus given by

$$\mathcal{L} = \overline{\xi}i\,\partial\!\!\!/\xi + M\overline{\xi}e^{\frac{-2i}{\sqrt{\pi}}\gamma_5\varphi}\xi - \frac{1}{2g\pi}(\partial_\mu\eta)^2 + \frac{1}{2\pi}\left(\frac{1}{g} - \frac{1}{\pi}\right)(\partial_\mu\varphi)^2 \quad , \tag{D.9}$$

showing that the η field decouples completely from the theory. The remaining problem is related to the mass term, which now describes an interaction. The standard way to treat this term, is by expanding in the mass, which leaves us with the partition function:

$$\mathcal{Z} = \int D\varphi D\overline{\xi}D\xi e^{i\int d^2x\{\overline{\xi}i\partial\!\!\!/\xi + \frac{1}{2\frac{g\pi}{1-\frac{g}{\pi}}}(\partial_\mu\varphi)^2\}} \times$$
$$\sum_{n=0} \frac{(iM)^n}{n!}\prod_1^n \int d^2x_j\overline{\xi}(x_j)e^{\frac{-2i}{\sqrt{\pi}}\gamma_5\varphi(x_j)}\xi(x_j) \quad . \tag{D.10}$$

Vacuum to vacuum amplitudes may thus be computed as a power series in M from the generating functional

$$\begin{aligned}\mathcal{Z}[J,\overline{J}] &= \int D\varphi D\overline{\xi}D\xi e^{i\int d^2x\left[\overline{\xi}(i\partial\!\!\!/+e^{\frac{-2i}{\sqrt{\pi}}\gamma_5\varphi})\xi + \overline{J}e^{\frac{-i}{\sqrt{\pi}}\gamma_5\varphi}\xi + \overline{\xi}e^{\frac{-i}{\sqrt{\pi}}\gamma_5\varphi}J + \frac{1}{2\frac{\pi g}{1-\frac{g}{\pi}}}(\partial_\mu\varphi)^2\right]} \\ &= \int D\varphi e^{\mathrm{tr}\,\ln(i\partial\!\!\!/+Me^{\frac{-2i}{\sqrt{\pi}}\gamma_5\varphi})} \times \\ &\qquad e^{i\int d^2x\left[\left(\frac{1+\frac{g}{\pi}}{\pi g}\right)\frac{1}{2}(\partial_\mu\varphi)^2 + \overline{J}e^{\frac{-i}{\sqrt{\pi}}\gamma_5\varphi}(i\partial\!\!\!/+e^{\frac{-2i}{\sqrt{\pi}}\gamma_5\varphi})^{-1}e^{\frac{-i}{\sqrt{\pi}}\gamma_5\varphi}J\right]} \quad ,\end{aligned} \tag{D.11}$$

where J and $\overline{J}$ are external Grassmann valued sources. From this generating functional, the correlation functions of the massive Thirring model in the bosonic formulation can be computed. We now show that the partition function (D.10) is equivalent to that of the sine-Gordon theory. To this end we first make the decomposition

$$\overline{\xi}e^{-\frac{2i}{\sqrt{\pi}}\gamma_5\varphi}\xi = e^{-\frac{2i}{\sqrt{\pi}}\varphi}\overline{\xi}\frac{1+\gamma_5}{2}\xi + e^{\frac{2i}{\sqrt{\pi}}\varphi}\overline{\xi}\frac{1-\gamma_5}{2}\xi \quad . \tag{D.12}$$

From (D.10) we see that ξ is a free, canonically normalized massless fermion field, whereas φ is a free, zero mass bosonic field with the non-canonically normalized two point function

$$\langle 0 \mid \varphi(x)\varphi(y) \mid 0\rangle = \frac{\pi g}{1-\frac{g}{\pi}} D^{(+)}(x-y) \quad . \tag{D.13}$$

The chiral selection rule carried by zero mass fermionic fields ξ allows us to write (D.10) in the form

$$\sum_{n=0}^{\infty} \frac{(iM)^{2n}}{n!} \prod_{i=1}^{n} d^2x_i d^2y_i \left\langle e^{-\frac{2i}{\sqrt{\pi}}\sum_{j=1}^{n}[\varphi(x_i)-\varphi(y_j)]} \right\rangle_{bos.vac.} \times \left\langle \prod_{i=1}^{n} \bar{\xi}(x_i)\frac{1+\gamma_5}{2}\xi(x_i)\bar{\xi}(y_j)\frac{1-\gamma_5}{2}\xi(y_j) \right\rangle_{ferm.vac.} \quad . \tag{D.14}$$

The exponential in the φ fields satisfies the superselection rule required to have a non-vanishing expectation value (see chapter 2). Using (D.13) and the corresponding expression for the fermion two-point function, Eq. (A.27*a*), one obtains (see Eq. (12.11))

$$\mathcal{Z} = \sum_{n=0}^{\infty} \left(\frac{M}{2\pi}\right)^{2n} \int d^2x_i d^2y_i \frac{\prod_{i>j}^{n} \left([-\mu^2(x_i-x_j)^2]\,[-\mu^2(y_i-y_j)^2]\right)^{\frac{1}{1-\frac{g}{\pi}}}}{\prod_{i,j}^{n} [-\mu^2(x_i-y_j)^2]^{\frac{1}{1-\frac{g}{\pi}}}} \tag{D.15}$$

It is easy to check[565] that the same expansion is obtained for the partition function of the sine-Gordon theory defined by the Lagrangian (3.33), with

$$\frac{\tilde{\beta}^2}{4\pi} = \frac{1}{1-\frac{g}{\pi}} \tag{D.16}$$

and a suitable identification of the mass M.

Appendix E

Bosonization of the Kinetic Term

We prove here the bosonization formula (3.29) for the kinetic term $\overline{\psi^{(0)}} i\gamma_1\partial_1\psi^{(0)}$ of a Dirac field $\psi^{(0)}$, by showing that the corresponding Wilson short distance expansion has the form

$$\frac{1}{2}\left(\overline{\psi}^{(0)}(x+\epsilon)i\gamma_1\partial_1\psi^{(0)}(x)+\text{h.c.}\right)-V.E.V.=\frac{1}{2}\left[:(\partial_0\phi)^2:+(\partial_1\phi)^2\right]+\mathcal{O}(\epsilon)\quad , \tag{E.1}$$

where $V.E.V.$ stands for "vacuum expectation value", and where ϕ is the "potential" of the free fermionic current.

It is convenient to write this operator product in terms of the left (L) and right (R) moving Dirac fields:

$$\begin{aligned}\overline{\psi}^{(0)}(x+\epsilon)i\gamma_1\partial_1\psi^{(0)}(x)&=-i\psi^{(0)\dagger}(x+\epsilon)\gamma^5\partial_1\psi(x)\\&=-i\left(\psi^{(0)\dagger}_R(x+\epsilon)\partial_1\psi_R(x)-\psi^{(0)\dagger}_L(x+\epsilon)\partial_1\psi_L(x)\right)\quad .\end{aligned}\tag{E.2}$$

$\psi^{(0)}_{\substack{R\\L}}(x)$ has the following bosonic representation:

$$\psi^{(0)}_{\substack{R\\L}}(x)=\left(\frac{\mu}{2\pi}\right)^{1/2}:e^{i\sqrt{\pi}\phi_{\substack{R\\L}}(x)}:\quad , \tag{E.3}$$

where $\phi_R(x)\,(\phi_L(x))$ are right (left) moving *chiral* bosonic, free zero mass fields satisfying

$$\partial_+\phi_R(x)=0\quad ,\quad \partial_-\phi_L(x)=0\quad . \tag{E.4a}$$

The corresponding commutators and two-point functions are listed in Appendix A. The following property will be useful:

$$\epsilon^\mu\partial_\mu\phi_{\substack{R\\L}}(x)=\mp\epsilon_\mp\partial_\pm\phi(x)\quad , \tag{E.4b}$$

where $\phi(x)$ is the usual bosonic field satisfying $\Box\phi=0$, and where our convention are $\epsilon_\pm=\epsilon_0\pm\epsilon_1$, $\partial_\pm=\partial_0\pm\partial_1$.

Normal ordering the exponentials, one has, using (C.12),

$$\begin{aligned}-i\psi^{(0)\dagger}_{\substack{L\\R}}(x+\epsilon)\partial_1\psi^{(0)}_{\substack{L\\R}}(x)&=\left(\frac{\mu}{2\pi}\right)e^{\pi D^{(+)}_{\substack{L\\R}}(\epsilon)}e^{-i\sqrt{\pi}(\phi^{(-)}_{\substack{L\\R}}(x+\epsilon)-\phi^{(-)}_{\substack{L\\R}}(x))}\\&\quad\times e^{-i\sqrt{\pi}\phi^{(-)}_{\substack{L\\R}}(x+\epsilon)}\sqrt{\pi}\partial_1\phi_{\substack{L\\R}}(x)e^{i\sqrt{\pi}\phi^{(+)}_{\substack{L\\R}}(x)}\end{aligned}\tag{E.5}$$

Unlike the case of the chiral densities $\overline{\psi}\frac{1\pm\gamma^5}{2}\psi$ and current $\overline{\psi}\gamma^\mu\psi$, we need to retain terms up to second order in ϵ in the exponentials. Noting that

$$e^{\pi D^{(+)}_{\substack{L\\R}}(\epsilon)} = \frac{1}{i\mu\epsilon_\mp} \qquad (E.6)$$

and (recall (C.13))

$$e^{-i\sqrt{\pi}\phi^{(+)}_{\substack{L\\R}}(x+\epsilon)} \sqrt{\pi}\partial_1\phi_{\substack{L\\R}}(x) e^{i\sqrt{\pi}\phi^{(+)}_{\substack{L\\R}}(x)} =$$

$$\left(\sqrt{\pi}\partial_1\phi_{\substack{L\\R}}(x) \mp \frac{i}{\epsilon_\mp}\right) e^{i\sqrt{\pi}(\phi^{(+)}_{\substack{L\\R}}(x+\epsilon)-\phi^{(+)}_{\substack{L\\R}}(x))} \qquad (E.7)$$

we find, upon expanding in powers of ϵ, that the first order terms in ϵ cancel identically and we are left with the result,

$$\psi^{(0)\dagger}_{\substack{L\\R}}(x+\epsilon)\partial_1\psi^{(0)}_{\substack{L\\R}}(x) = \left\{\mp\frac{1}{2\pi\epsilon^2_\mp} \mp \frac{1}{4}{:}(\partial_\pm\phi)^2{:} \pm \frac{i}{4\sqrt{\pi}}\partial^2_\pm\phi\right\} + \mathcal{O}(\epsilon) \quad . \qquad (E.8)$$

The occurrence of the purely imaginary terms in (E.8) is evidently a consequence of the non-hermiticity of the left hand side. Adding the hermitian conjugate it drops out. Combining the contributions from left- and handed right-fermions as in (E.2), we arrive at the result (E.1).

Appendix F

Classical Integrability in the Massive Thirring Model

The integrability of the massive Thirring model is a consequence of the fact that the equations

$$\epsilon 2\partial_+\chi - \chi + 2i\epsilon g\chi\psi_1^+\psi_1 + 4i\epsilon g\psi_1\chi^+\chi + \psi_1 = 0 \tag{F.1}$$

$$\partial_-\chi + m^2\epsilon\chi + 2ig\chi\psi_2^+\psi_2 + 4\epsilon gm\psi_2\chi^+\chi + im\psi_2 = 0 \tag{F.2}$$

are compatible. Indeed, one computes

$$\epsilon\partial_+(\chi^+\chi) = 2\chi^+\chi - (\chi^+\psi_1 + \psi_1^+\chi) \tag{F.3}$$

$$\partial_-(\chi^+\chi) = -2m^2\epsilon\chi^+\chi + im(\chi^+\psi_2 - \psi_2^+\chi) \quad . \tag{F.4}$$

We shall also use the field equations

$$\begin{aligned} i\partial_-\psi_1 &= m\psi_2 + 2g\psi_1\psi_2^+\psi_2 \\ i\partial_+\psi_2 &= m\psi_1 + 2g\psi_2\psi_1^+\psi_1 \\ i\partial_-(\psi_1^+\psi_1) &= m(\psi_1^+\psi_2 - \psi_2^+\psi_1 \\ i\partial_+(\psi_2^+\psi_2) &= m(\psi_2^+\psi_1 - \psi_1^+\psi_2 \quad . \end{aligned} \tag{F.5}$$

We have, from ∂_- applied on Eq.(F.2):

$$\begin{aligned} &\epsilon\partial_+\partial_-\chi = \\ &\left(-m^2\epsilon\chi - 2ig\chi\psi_2^+\psi_2 - 4\epsilon gm\psi_2\chi^+\chi - im\psi_2\right) \\ &- 2i\epsilon g\psi_1^+\psi_1\left(-m^2\epsilon\chi - 2ig\chi\psi_2^+\psi_2 - 4\epsilon gm\psi_2\chi^+\chi - im\psi_2\right) \\ &- 4ig\epsilon\psi_1\left(-2m^2\epsilon\chi^+\chi - im(\chi^+\psi_2 - \psi_2^+\chi)\right) \\ &- 2ig\chi m\left(\psi_1^+\psi_2 - \psi_2^+\psi_1\right) - 4\epsilon g(m\psi_2 + 2g\psi_1\psi_2^+\psi_2)\chi^+\chi \\ &+ i(m\psi_2 + 2g\psi_1\psi_2^+\psi_2) \end{aligned} \tag{F.6}$$

and from ∂_+ on Eq. (F.3):

$$\begin{aligned} &\partial_+\partial_-\chi = \\ &- \frac{m^2\epsilon}{\epsilon}\left(\chi - 2i\epsilon g\chi\psi_1^+\psi_1 - 4i\epsilon g\psi_1\chi^+\chi - \psi_1\right) \\ &- \frac{2ig}{\epsilon}\left(\chi - 2i\epsilon g\chi\psi_1^+\psi_1 - 4i\epsilon g\psi_1\chi^+\chi - \psi_1\right)\psi_2^+\psi_2 \\ &- 4\epsilon gm\frac{1}{i}(m\psi_1 + 2g\psi_2\psi_1\psi_1)\chi^+\chi \\ &- 4\epsilon gm\psi_2\left(\frac{2}{\epsilon}\chi^+\chi - \frac{1}{\epsilon}(\chi^+\psi_1 + \psi_1^+\chi)\right) \\ &- 2ig\chi\frac{m}{i}(\psi_2^+\psi_1 - \psi_1^+\psi_2) - \frac{im}{i}(m\psi_1 + 2g\psi_2\psi_1^+\psi_1) \quad . \end{aligned} \tag{F.7}$$

Comparing Eqs. (F.6) and (F.7), we immediately obtain the desired compatibility. In order to solve (F.2), we use the series expansion $\chi = \sum \epsilon^n \chi^{(n)}$, obtaining

$$\begin{aligned}
\chi^{(0)} &= \psi_1 \\
\chi^{(1)} &= \partial_+ \psi_1 \\
\chi^{(2)} &= \partial_+^2 \psi_1 + 2ig\partial_+\psi_1\psi_1^+\psi_1 + 4ig\psi_1\psi_1^+\partial_+\psi_1 \\
&= \partial_+^2\psi_1 - 2ig\psi_1^+\psi_1\partial_+\psi_1 \quad .
\end{aligned} \tag{F.8}$$

Define the currents

$$J_+ = i\psi_1^+\chi - i\chi^+\psi_1 \tag{F.9a}$$

$$J_- = \epsilon m(\psi_2^+\chi + \chi^+\psi_2) \quad . \tag{F.9b}$$

They form a conserved current. Indeed

$$\begin{aligned}
\partial_+ J_- + \partial_- J_+ = {} & \left(m\psi_2^+ + 2g\psi_2^+\psi_2\psi_1^+\right)\chi - \chi^+\left(m\psi_2 + 2g\psi_1\psi_2^+\psi_2\right) \\
& + m\psi_2^+\left(\chi - 2i\epsilon g\chi\psi_1^+\psi_1 - 4i\epsilon g\psi_1\chi^+\chi - \psi_1\right) \\
& + m\left(\chi^+ + 2i\epsilon g\psi_1^+\psi_1\chi^+ + 4i\epsilon g\chi^+\chi\psi_1^+ - \psi_1^+\right)\psi_2 - \frac{\epsilon m}{i}\left(m\psi_1^+ + 2g\psi_1^+\psi_1\psi_2^+\right)\chi \\
& + \frac{\epsilon m}{i\sqrt{2}}\chi^+\left(m\psi_1 + 2g\psi_2\psi_1^+\psi_1\right) + i\psi_1^+\left(-m^2\epsilon\chi - 2ig\chi\psi_2^+\psi_2 - 4\epsilon gm\psi_2\chi^+\chi - im\psi_2\right) \\
& - i\left(-m^2\epsilon\chi^+ + 2ig\psi_2^+\psi_2\chi^+ - 4\epsilon gm\chi^+\chi\psi_2^+ + im\psi_2^+\right)\psi_1 = 0 \quad .
\end{aligned} \tag{F.10}$$

From Eqs. (F.8) we obtain for the first four terms in the expansion of the currents (F.9)

$$J_+^{(0)} = 0 \tag{F.11a}$$

$$J_-^{(0)} = 0 \tag{F.11b}$$

$$J_+^{(1)} = i(\psi_1^+\partial_+\psi_1 - \partial_+\psi_1^+\psi_1) \tag{F.12a}$$

$$J_-^{(1)} = m(\psi_2^+\psi_1 + \psi_1^+\psi_2) \tag{F.12b}$$

$$J_+^{(2)} = i(\psi_1^+\partial_+^2\psi_1 - \partial_+\psi_1^+\psi_1) \tag{F.13a}$$

$$J_-^{(2)} = m(\psi_2^+\partial_+\psi_1 + \partial_+\psi_1\psi_2) \tag{F.13b}$$

$$J_+^{(3)} = i(\psi_1^+\partial_+^3\psi_1 - \partial_+^3\psi_1^+\psi_1) - 12g\psi_1^+\psi_1\partial_+\psi_1^+\partial_+\psi_1 \tag{F.14a}$$

$$J_-^{(3)} = m\left((\psi_2^+\partial_+^2\psi_1 + \partial_+^2\psi_1^+\psi_2) - 2ig\psi_1^+\psi_1(\psi_2^+\partial_+\psi_1 - \partial_+\psi_1^+\psi_2)\right) \tag{F.14b}$$

Appendix G

Quantum Non-Local Charge: Action on Asymptotic States

We have seen that the integrability equation obeyed by the conserved current in integrable models, leads to the conservation of the nonlocal charge

$$Q^{(2)} = \int dy_1 dy_2\, \epsilon(y_1 - y_2) j_0(t, y_1) j_0(t, y_2) + \int dy\, j_1(t, y) \quad . \tag{G.1}$$

We have also seen that the conservation of this charge leads to severe constraints in the dynamics of the model. In this appendix we shall see how to obtain the action of the quantized version of this charge on asymptotic states, as well as some further information concerning the quantized version of the above charge. The pedestrian way to get the action of the quantum charge on asymptotic states, is to use the fact that the nonlocal charge is conserved, and to write the asymptotic charges Q_{out} and Q_{in} in terms of the asymptotic currents. This procedure determines the action of the quantum nonlocal charge on the states, as well as the renormalization constant Z_δ. We have

$$Q^{(2)}_{quantum} = \lim_{\delta \to 0} \left\{ \int_{|y_1 - y_2 < \delta} dy_1 dy_2\, \epsilon(y_1 - y_2) j_0(t, y_1) j_0(t, y_2) + Z_\delta \int dy\, j_1(t, y) \right\} \tag{G.2}$$

However, this determination leads to inconsistency, since the mass parameter used to define the above renormalization constant is renormalization group invariant. Using the free field representation, it is possible to obtain also the mass gap[153,403,85] exactly. On the other hand, a renormalization group invariant mass should have the form

$$\mu \sim (b_0 f)^{-\frac{b_1}{b_0^2}} e^{-\frac{1}{b_0 f}} [1 + \mathcal{O}(f)] \quad , \tag{G.3}$$

where b_0 and b_1 and the first two coefficients of the expansion of the β function. Since the coefficient b_1 does not vanish, there is a non perturbative contribution to the mass, and $1/N$ perturbation theory as we have studied, fails to describe it. Therefore, the free field calculation of the nonlocal action must be at least incomplete. Nevertheless, we have shown that the main results follow from rather general arguments, and we do not need the perturbative result for the renormalization constant. The main ingredient is the Lorentz transformation property of the nonlocal charge. One considers the Lorentz transformations

$$\Lambda^\mu{}_\nu = \begin{pmatrix} \cosh\chi & \sinh\chi \\ \sinh\chi & \cosh\chi \end{pmatrix} \quad . \tag{G.4}$$

The corresponding unitary operator $U(\Lambda)$, acts on currents as

$$U(\Lambda)j_\mu(x)U(\Lambda)^{-1} = j_\nu(\Lambda x)\Lambda^\nu{}_\mu \quad . \tag{G.5}$$

The corresponding generator of infinitesimal transformations is defined by

$$T = \frac{d}{d\chi}U(\Lambda)|_{\chi=0} \quad , \tag{G.6}$$

Its action on the current is given by

$$[T, j_\mu(x)] = j_\nu(x)t^\nu{}_\mu + t^\nu{}_\rho x^\rho \partial_\nu j_\mu(x) \quad . \tag{G.7}$$

where $t_{\mu\nu} = \epsilon_{\mu\nu}$; it is not difficult to compute the action of the Lorentz generator on the nonlocal charge itself. We have

$$[T, Q^{ab}] = \lambda \int_{-\infty}^{\infty} dy j_0^{ab}(0,y) \quad , \tag{G.8}$$

where the constant λ is given by

$$\lambda = -\frac{1}{n}\lim_{\delta\to 0}\left\{\delta^4 C_2(-\delta^2) + 2\delta^2 D_1(-\delta^2) + 2\delta^2 D_2(-\delta^2)\delta^2 D_3(-\delta^2)\right\} \quad . \tag{G.9}$$

Using the normalization of the current, we obtain

$$[T, Q^{ab}] = -\frac{x}{\pi n} I^{ab} \quad , \tag{G.10}$$

where I^{ab} is the isospin generator. The computation of the action of the nonlocal charge on the vacuum and on one particle states is now very simple. We have to take into account whether the representation is real (and hence the indices a and b are equivalent under the action of the isospin generator) or complex (hence a and b are inequivalent). Using Eq.(G.8), the realization $T = \frac{\partial}{\partial\theta}$ on asymptotic states with rapidity θ, and the fact that Q^{ab} is a second order isospin-tensor, we find

$$Q^{ab}|\theta\,\rangle = -i\frac{n-2}{\pi n}\theta\left\{\delta^{bc}|\theta a\,\rangle - \delta^{ac}|\theta b\,\rangle\right\} \tag{G.11}$$

for $O(n)$ symmetry, and

$$Q^{ab}|\theta\,\rangle = -i\frac{1}{\pi}\theta\delta^{bc}|\theta a\,\rangle \tag{G.12}$$

for an $SU(n)$ symmetry. Further cases can be treated analogously. In order to deal with the action on two particle states, we need further information. It has been shown[195] that one can deal with a charge Q, once it exhibits the following properties:

i. it commutes with time translations,

ii. there is an integer N, called the "genus", such that for any collection of local fields $\phi_{\alpha_1} \cdots \phi_{\alpha_{N+1}}$, we have

$$\left[\cdots\left[Q, \phi_{\alpha_1}(x_1)\right], \cdots \phi_{\alpha_{N+1}}(x_{N+1})\right] = 0 \quad , \tag{G.13}$$

iii. Q is a hermitian operator.

The above requirements are obeyed by (G.2); moreover it has genus 2. It has been shown[195] that a charge of genus N changes the asymptotic particle number by not more than $N-2$ unities. Therefore, in the present case, the particle number is conserved. This is enough to assure that the S matrix is factorizable. The action on asymptotic two particle states of the type $\phi_1 \times^{in} \phi_2$, with non overlapping momentum distributions can be obtained using techniques from collision theory[195,204]. One has

$$\begin{aligned}\langle \phi_1 \times \phi_2 | Q | \phi_1 \times \phi_2 \rangle = \lim_{t \to \infty} \int dx dy \epsilon(x-y) \langle \phi_1 | j_0(t,x) | \phi_1 \rangle \langle \phi_2 | j_0(t,y) | \phi_2 \rangle \\ + \langle \phi_1 | Q | \phi_1 \rangle + (1 \to 2) \quad .\end{aligned} \tag{G.14}$$

We see that the first term is completely determined by the local properties of the charge. The action of the charge on a one particle state is used in order to determine the second term. The result agrees completely with the naive calculation using the asymptotic form of the charge, if we disregard the computation of the renormalization constant Z_δ, which in any case, turns out to be wrong, as we already remarked.

Appendix H

Non Linear Sigma Models

Given a natural invariant scalar product, we are able to write down the Lagrangian, corresponding to the simplest scalar. In order to express these facts in terms of mathematically well defined objects, we state some results. Let E be a vector bundle over $\mathcal{X}$, the basis space, which is either R^2 or R^1, and let $\Lambda^r T^* X \times E$ be the vector bundle of E-valued r-forms on $\mathcal{X}$. We recall that a bundle is a generalization of the cartesian product; locally they are, indeed, the same object. In the latter case, $\Lambda^r T^* X \times E$ is the set of r-forms depending on a variable $x \in \mathcal{X}$. This vector bundle will be useful to define a general space in which the derivative of the field lives. If there is a pseudo-Riemannian (in the real case) or hermitian (in the complex case) fibre metric, then E is a pseudo-Riemannian (hermitian complex) vector bundle. In general E is denoted by q^*TM. This means that E will be the space of tangent tensors to a given manifold, on which the σ model is going to be defined.

The fibre metric can be used to construct the Lagrangian. In terms of differential forms, the Lagrangian may be defined as

$$S = \int d^2x\mathcal{L} = \frac{1}{2}\int (dq, dq) \quad , \tag{H.1}$$

where dq is the differential of the field q. In terms of a coordinate system, Eq. (H.1) is equal to (6.95). Equation (H.1) implies the field equation

$$\delta d\varphi = *d * d\varphi = 0 \quad .$$

In terms of q, the field equation must contain the constraint, by means of a Lagrange multiplier.

The requirement that sigma models do not present accidental degeneracy takes us in a natural way to homogeneous spaces, which are cosets of the form G/H, that is, the vacuum has a degeneracy described by the gauge group H. It is supposed that G acts transitively on M. This means that given two points, there always exists a $g \in G$, which transforms one point into the other. This implies also that the manifold cannot be divided into disjoint orbits. Moreover, if $H \subset G$ is the subgroup leaving a fixed point $p_0 \in M$ invariant, a homeomorphism of G/H into M must be a diffeomorfism. Thus M, which is a homogeneous space, is also a coset space, if the above conditions are fulfilled. The target space is a coset $M = G/H$. This means that the field configurations q are maps from $\mathcal{X}$, which is either $\Re^2$, or $\mathcal{M}^2$, to $M = G/H$. G is a connected Lie group with Lie algebra $\mathcal{G}$ and $H \subset G$ is a closed subgroup with Lie algebra $\mathcal{H} \subset \mathcal{G}$ such that the group $Ad_H\mathcal{G}$ of linear transformations on $\mathcal{G}$, is compact. Actually H will be realized as a gauge group, for nonlinear sigma models. This means that the coset space G/H is realized by a G valued field, which we call $g(x)$, $g(x) \in G$, such that, when it is (locally) transformed by an element $h(x) \in H$,

$$g^h(x) = g(x)h(x) \quad , \tag{H.2}$$

the original field $g(x)$ and the transformed one $g^h(x)$ are identified. This means that a field configuration corresponds to a class of gauge equivalent fields $g(x)$. In other words, they are considered to be different representatives of the same element of G/H. There exists an invariant inner product on G with respect to $Ad_G H$, which means, that it is gauge invariant. As we shall see, this enables us to define a gauge invariant action. If $M = G/H$ is a homogeneous space of left cosets gH, there is a natural projection,

$$\Pi : G \to G/H \quad , \tag{H.3}$$

which can be considered as a fibration, defining a principal H-bundle, where G acts from the left. Thus, locally we write G as a product of a basis space $M = G/H$, and H, which is the fiber. Therefore, each fibre contains all gauge equivalent representative of a single point in G/H. This is essential in the construction, because a local symmetry is naturally defined. The field $g(x)$ is gauge transformed inside the fibre, which thus defines an equivalence class. The gauge field we already mentioned several times is by now a very natural element of the whole construction. The inner product, which is nothing but the Killing form, is left invariant under G and is right invariant under H. Notice that in the present construction, global transformations $g_o \in G$ act on $g(x)$ by left multiplication

$$g(x) \to g_{g_o}(x) = g_o g(x) \tag{H.4}$$

while $h(x)$, the local gauge transformation, acts by right multiplication as explained above. The inner product (.,.) is left-invariant under G and right-invariant under H. Therefore the Lagrangian is globally, as well as locally invariant.

One can also define a left-invariant and a right-invariant Maurer-Cartan 1-form

$$g^{-1}dg = g^{-1}\partial_\mu g dx^\mu \tag{H.5a}$$

and

$$dgg^{-1} = \partial_\mu g g^{-1} dx^\mu \quad , \tag{H.5b}$$

respectively.

In terms of forms, one naturally defines the gauge field as the left-invariant connection, by simply taking the decomposition of (H.5a) with respect to (6.83), and retaining the (gauge) part; in formulae

$$A = \Pi(g^{-1}dg) \tag{H.6}$$

where Π is the projection given by the fibration defined above in equation (H.3); in components we have

$$A_\mu = \left(g^{-1}\partial_\mu g\right)_{\mathcal{H}} \quad , \qquad (H.7)$$

defining the gauge field in terms of the algebra decomposition.

The above gauge potential is obviously not an independent field. Its gauge potential character becomes explicit when considering its behaviour under a gauge transformation:

$$A_\mu^h(x) = \left[h^{-1}g^{-1}\partial_\mu(gh)\right]_{\mathcal{H}} = \left[h^{-1}(g^{-1}\partial_\mu g)h\right]_{\mathcal{H}} + h^{-1}\partial_\mu h \quad . \qquad (H.8)$$

Due to the fact that h acts on each fibre (Eq. (H.3), or equivalently due to the fact that the decomposition (6.83) is invariant under $\mathcal{H}$, we have

$$A_\mu^h = h^{-1}(g^{-1}\partial_\mu g)_{\mathcal{H}} h + h^{-1}(A_\mu + \partial_\mu)h \qquad (H.9)$$

which is the usual rule for a gauge field transformation.

Appendix I

Conservation Laws in NLS Models

We here show that this is a transformation of solutions into solutions if and only if we are in a symmetric space, and $\mathcal{U}^{(\gamma)}$ obeys the set of differential equations given by (6.116), which can be conveniently summarized in the form,

$$\partial_\mu \mathcal{U}^{(\gamma)} = \mathcal{U}^{(\gamma)}\Big\{\big(1 - \frac{\gamma^{-1}+\gamma}{2}\big)j_\mu + \frac{\gamma^{-1}-\gamma}{2}\epsilon_{\mu\nu}j^\nu\Big\} \quad . \tag{I.1}$$

Proof: Suppose $g^{(\gamma)}$ satisfies the equation of motion (6.96), i.e.

$$D^\mu D_\mu g^{(\gamma)} = D_\mu g^{(\gamma)} g^{(\gamma)^{-1}} D^\mu g^{(\gamma)} \quad . \tag{I.2}$$

Using (6.119) we obtain

$$D^\mu D_\mu g^{(\gamma)} = \partial^\mu \partial_\mu \mathcal{U}^{(\gamma)} g + 2\partial^\mu \mathcal{U}^{(\gamma)} D_\mu g + \mathcal{U}^{(\gamma)} D^\mu g g^{-1} D_\mu g \quad , \tag{I.3L}$$

for the left hand side, and

$$\begin{aligned} D_\mu g^{(\gamma)} g^{(\gamma)^{-1}} D^\mu g^{(\gamma)} = {} & \partial^\mu \mathcal{U}^{(\gamma)} \mathcal{U}^{(\gamma)^{-1}} \partial^\mu \mathcal{U}^{(\gamma)} g + \partial^\mu \mathcal{U}^{(\gamma)} D_\mu g \\ & + \mathcal{U}^{(\gamma)} D_\mu g g^{-1} \mathcal{U}^{(\gamma)^{-1}} \partial^\mu \mathcal{U}^{(\gamma)} g + \mathcal{U}^{(\gamma)} D_\mu g g^{-1} D_\mu g \quad , \end{aligned} \tag{I.3R}$$

for the right hand side. We make the Ansatz

$$\partial_\mu \mathcal{U}^{(\gamma)} = \mathcal{U}^{(\gamma)} M_\mu \tag{I.4}$$

which gives $D_\mu g^{(\gamma)} = \mathcal{U}^{(\gamma)}(D_\mu + M_\mu)g$. Substituting this result back into (I.3L) and (I.3R) and identifying them, we have:

$$\partial^\mu M_\mu = D^\mu g g^{-1} M_\mu - M^\mu D_\mu g g^{-1} \quad , \tag{I.5}$$

or, in terms of $\mathcal{M}_\mu = g^{-1} M_\mu g$, we get

$$D^\mu \mathcal{M}_\mu = 0 \quad . \tag{I.6}$$

Notice that M_μ is gauge invariant and globally covariant; hence, $\mathcal{M}_\mu$ is gauge covariant and globally invariant. The only gauge covariant and globally invariant vector in the sigma model is k_μ, equation (6.91). Therefore $\mathcal{M}_\mu = \alpha k_\mu + \beta\epsilon_{\mu\nu}k^\nu$, which satisfies $D^\mu \mathcal{M}_\mu = 0$ if and only if

$$D^\mu k_\mu = 0 \quad , \tag{I.7a}$$

which is equivalent to current conservation upon use of Eq. (6.97), and

$$D_\mu k_\nu - D_\nu k_\mu = 0 \quad . \tag{I.7b}$$

This last equation implies that the theory is defined on a symmetric space, once we compare with Eq. (6.99).

Appendix J

S-Matrices

We shall write down now the Ansätze for the S matrices of σ models for the BD, C, AI and AII series.

For the BD and C cases, the indices $\{k\}$ transform under the $SO(N)_L$ $(Sp(N)_L)$ and the indices $\{l\}$ under $SO(N)_R$ $(Sp(N)_R)$ representations and can not contract. However, the representations N_L and N_R are equivalent, with the matrix I as the intertwining operator. Thus we have

$$
\begin{aligned}
{}_{out}\langle\theta_1' k_1' l_1', \theta_2' k_2' l_2' | \theta_1 k_1 l_1, \theta_2 k_2 l_2\rangle_{in} &= 16\pi^2 \delta(\theta_1' - \theta_1)\delta(\theta_2' - \theta_2)\times \\
&\Big\{ u_1(\theta)\delta^{k_1'k_1}\delta^{k_2'k_2}\delta^{l_1'l_1}\delta^{l_2'l_2} + u_2(\theta)\delta^{k_1'k_2}\delta^{k_2'k_1}\delta^{l_1'l_1}\delta^{l_2'l_2} + u_3(\theta)\delta^{k_1'k_2'}\delta^{k_2k_1}\delta^{l_1'l_1}\delta^{l_2'l_2} \\
&+u_4(\theta)\delta^{k_1'k_1}\delta^{k_2'k_2}\delta^{l_1'l_2}\delta^{l_2'l_1} + u_5(\theta)\delta^{k_1'k_2}\delta^{k_2'k_1}\delta^{l_1'l_2}\delta^{l_2'l_1} + u_6(\theta)\delta^{k_1'k_2'}\delta^{k_2k_1}\delta^{l_1'l_2}\delta^{l_2'l_1} \\
&+u_7(\theta)\delta^{k_1'k_1}\delta^{k_2'k_2}I^{l_1'l_2'}I^{l_1l_2} + u_8(\theta)\delta^{k_1'k_2}\delta^{k_2'k_1}I^{l_1'l_2'}I^{l_1l_2} + u_9(\theta)I^{k_1'k_2'}I^{k_1k_2}I^{l_1'l_2'}I^{l_1l_2}\Big\} \\
&+16\pi^2\delta(\theta_1' - \theta_2\delta(\theta_2' - \theta_1)\times \\
&\Big\{ u_5(\theta)\delta^{k_1'k_1}\delta^{k_2'k_2}\delta^{l_1'l_1}\delta^{l_2'l_2} + u_4(\theta)\delta^{k_1'k_2}\delta^{k_2'k_1}\delta^{l_1'l_1}\delta^{l_2'l_2} \pm u_6(\theta)\delta^{k_1'k_2'}\delta^{k_2k_1}\delta^{l_1'l_1}\delta^{l_2'l_2} \\
&+u_2(\theta)\delta^{k_1'k_1}\delta^{k_2'k_2}\delta^{l_1'l_2}\delta^{l_2'l_1} + u_1(\theta)\delta^{k_1'k_2}\delta^{k_2'k_1}\delta^{l_1'l_2}\delta^{l_2'l_1} \pm u_3(\theta)\delta^{k_1'k_2'}\delta^{k_2k_1}\delta^{l_1'l_2}\delta^{l_2'l_1} \\
&\pm u_8(\theta)\delta^{k_1'k_1}\delta^{k_2'k_2}I^{l_1'l_2'}I^{l_1l_2} \pm u_7(\theta)\delta^{k_1'k_2}\delta^{k_2'k_1}I^{l_1'l_2'}I^{l_1l_2} + u_9(\theta)I^{k_1'k_2'}I^{k_1k_2}I^{l_1'l_2'}I^{l_1l_2}\Big\}
\end{aligned}
\tag{J.1}
$$

The discrete symmetry under (8.140) implies that $u_2 = u_4$, $u_3 = u_7$, and $u_6 = u_8$.

For the case of quotient spaces AI and AII, we have, respectively, symmetry and antisymmetry under exchanges $k \leftrightarrow l$. Moreover, the N and N^* representations are inequivalent. Thus we obtain

$$
\begin{aligned}
{}_{out}\langle\theta_1' k_1' l_1', \theta_2' k_2' l_2' | \theta_1 k_1 l_1, \theta_2 k_2 l_2\rangle_{in} &= 16\pi^2 \delta(\theta_1' - \theta_1)\delta(\theta_2' - \theta_2)\times \\
&\Big\{ u_1(\theta)\big(\delta^{k_1'k_1}\delta^{k_2'k_2}\delta^{l_1'l_1}\delta^{l_2'l_2} \pm k_1 \leftrightarrow l_1 , \quad k_2 \leftrightarrow l_2\big) \\
+u_2(\theta)\big(\delta^{k_1'k_2}\delta^{k_2'k_1}\delta^{l_1'l_1}\delta^{l_2'l_2} \pm k_1 \leftrightarrow l_1 , \quad & k_2 \leftrightarrow l_2 , \quad k_1' \leftrightarrow l_1' , \quad k_2' \leftrightarrow l_2'\big)
\end{aligned}
$$

$$+u_3(\theta)\big(\delta^{k_1'k_2}\delta^{k_2k_1'}\delta^{l_1'l_2}\delta^{l_2'l_1} \pm k_1 \leftrightarrow l_1\ ,\quad k_2 \leftrightarrow l_2\big)\Big\}$$

$$+16\pi^2\delta(\theta_1' - \theta_2)\delta(\theta_2' - \theta_1) \times \Big\{u_3(\theta)\big(\delta^{k_1'k_1}\delta^{k_2'k_2}\delta^{l_1'l_1}\delta^{l_2'l_2} \pm k_1 \leftrightarrow l_1\ ,\quad k_2 \leftrightarrow l_2\big)$$

$$+u_2(\theta)\big(\delta^{k_1'k_2}\delta^{k_2'k_1}\delta^{l_1'l_1}\delta^{l_2'l_2} \pm k_1 \leftrightarrow l_1\ ,\quad k_2 \leftrightarrow l_2 k_1' \leftrightarrow l_1'\ ,\quad k_2' \leftrightarrow l_2'\big)$$

$$u_1(\theta)\big(\delta^{k_1'k_2}\delta^{k_2k_1'}\delta^{l_1'l_2}\delta^{l_2'l_1} \pm k_1 \leftrightarrow l_1,\quad k_2 \leftrightarrow l_2\big)\Big\} \tag{J.2}$$

$${}_{out}\langle\overline{\theta_1' k_1' l_1'}, \theta_2' k_2' l_2' | \overline{\theta_1 k_1 l_1}, \theta_2 k_2 l_2\rangle_{in} = 16\pi^2\delta(\theta_1' - \theta_1)\delta(\theta_2' - \theta_2)\times$$

$$\Big\{t_1(\theta)\left(\delta^{k_1'k_1}\delta^{k_2'k_2}\delta^{l_1'l_1}\delta^{l_2'l_2} \pm k_1 \leftrightarrow l_1\ ,\quad k_2 \leftrightarrow l_2\right)$$

$$+t_2(\theta)\left(\delta^{k_1'k_2}\delta^{k_2'k_1}\delta^{l_1'l_1}\delta^{l_2'l_2} \pm k_1 \leftrightarrow l_1\ ,\quad k_2 \leftrightarrow l_2 k_1' \leftrightarrow l_1'\ ,\quad k_2' \leftrightarrow l_2'\right)$$

$$t_3(\theta)\left(\delta^{k_1'k_2}\delta^{k_2k_1'}\delta^{l_1'l_2}\delta^{l_2'l_1} \pm k_1 \leftrightarrow l_1\ ,\quad k_2 \leftrightarrow l_2\right)\Big\}$$

$$+16\pi^2\delta(\theta_1' - \theta_2)\delta(\theta_2' - \theta_1) \times \quad \Big\{r_3(\theta)\left(\delta^{k_1'k_1}\delta^{k_2'k_2}\delta^{l_1'l_1}\delta^{l_2'l_2} \pm k_1 \leftrightarrow l_1\ ,\quad k_2 \leftrightarrow l_2\right)$$

$$+r_2(\theta)\left(\delta^{k_1'k_2}\delta^{k_2'k_1}\delta^{l_1'l_1}\delta^{l_2'l_2} \pm k_1 \leftrightarrow l_1\ ,\quad k_2 \leftrightarrow l_2 k_1' \leftrightarrow l_1'\ ,\quad k_2' \leftrightarrow l_2'\right)$$

$$r_1(\theta)\left(\delta^{k_1'k_2}\delta^{k_2k_1'}\delta^{l_1'l_2}\delta^{l_2'l_1} \pm k_1 \leftrightarrow l_1\ ,\quad k_2 \leftrightarrow l_2\right)\Big\} \tag{J.3}$$

The S-matrices of further non linear sigma models may be obtained by these methods, such as the S-matrix of the Grassmannian non linear sigma models with fermions[310,311], describing the scattering of bound states constructed from the gauge invariant objects in that theory, and the factorizable S-matrix for $SO(N)/SO(2) \otimes SO(N-2)$ non linear σ models with fermions, describing the scattering of fundamental particles. For the case of Grassmannian models with fermions, one has the Lagrangian

$$\mathcal{L} = \overline{D^\mu z} D_\mu z + \overline{\psi} \not{D} \psi \quad . \tag{J.4}$$

If we make the Ansatz that the bound state

$$\Big|B_{ab}(\frac{1}{2}(\theta_1 + \theta_2))\Big\rangle = 2\pi^2 \int d\mu(k_1) d\mu(k_2) \delta(k - k_1 - k_2) d^{\dagger a}(k_1) d^{\dagger b}(k_2) \big|0\Big\rangle \tag{J.5}$$

for the bound state, we have the following Ansatz for the S-matrix

$$
\begin{aligned}
&\langle B_{e_1e_2}(\theta)B_{f_1f_2}(\theta)|B_{c_1c_2}(\theta)B_{d_1d_2}(\theta)\rangle \\
&= [\ \sigma_1(\theta)\delta_{c_1c_2}\delta_{d_1d_2}\delta_{e_1e_2}\delta_{f_1f_2} + \sigma_2(\theta)\delta_{c_1d_2}\delta_{d_1c_2}\delta_{e_1f_2}\delta_{f_1e_2} + \sigma_3(\theta)\delta_{c_1c_2}\delta_{d_1d_2}\delta_{e_1f_2}\delta_{f_1e_2} \\
&+\sigma_4(\theta)\delta_{c_1d_2}\delta_{d_1c_2}\delta_{e_1e_2}\delta_{f_1f_2} + \sigma_5(\theta)\delta_{c_1c_2}\delta_{d_1f_1}\delta_{d_2f_2}\delta_{e_1e_2} + \sigma_6(\theta)\delta_{c_1e_1}\delta_{d_1d_2}\delta_{c_2e_2}\delta_{f_1f_2} \\
&\sigma_7(\theta)\delta_{c_1c_2}\delta_{d_1e_1}\delta_{d_2e_2}\delta_{f_1f_2} + \sigma_8(\theta)\delta_{c_1f_1}\delta_{d_1d_2}\delta_{e_1e_2}\delta_{f_2c_2} + \sigma_9(\theta)\delta_{c_1c_2}\delta_{d_1e_1}\delta_{d_2f_2}\delta_{f_1e_2} \\
&+\sigma_{10}(\theta)\delta_{c_1e_1}\delta_{d_1d_2}\delta_{f_1e_2}\delta_{c_2f_2} + \sigma_{11}(\theta)\delta_{c_1c_2}\delta_{d_1f_1}\delta_{d_2e_2}\delta_{e_1f_2} + \sigma_{12}(\theta)\delta_{c_1f_1}\delta_{d_1d_2}\delta_{c_2e_2}\delta_{e_1f_2} \\
&+\sigma_{13}(\theta)\delta_{c_1f_1}\delta_{d_1c_2}\delta_{e_1e_2}\delta_{d_2f_2} + \sigma_{14}(\theta)\delta_{c_1e_1}\delta_{d_1c_2}\delta_{d_2e_2}\delta_{f_1f_2} + \sigma_{15}(\theta)\delta_{c_1d_2}\delta_{d_1f_1}\delta_{e_1e_2}\delta_{f_1d_1} \\
&+\sigma_{16}(\theta)\delta_{c_1d_2}\delta_{d_1e_1}\delta_{c_2e_2}\delta_{f_1f_2} + \sigma_{17}(\theta)\delta_{c_1e_1}\delta_{d_1c_2}\delta_{f_1e_2}\delta_{d_2f_2} + \sigma_{18}(\theta)\delta_{c_1d_2}\delta_{d_1f_1}\delta_{e_1f_2}\delta_{c_2e_2} \\
&+\sigma_{19}(\theta)\delta_{c_1f_1}\delta_{d_1c_2}\delta_{d_2e_2}\delta_{e_1f_2} + \sigma_{20}(\theta)\delta_{c_1d_2}\delta_{d_1e_1}\delta_{f_1e_2}\delta_{c_2f_2} + \sigma_{21}(\theta)\delta_{c_1e_1}\delta_{d_1f_1}\delta_{c_2e_2}\delta_{d_2f_2} \\
&+\sigma_{22}(\theta)\delta_{c_1f_1}\delta_{d_1e_1}\delta_{d_2e_2}\delta_{c_2f_2} + \sigma_{23}(\theta)\delta_{c_1e_1}\delta_{d_1f_1}\delta_{d_2e_2}\delta_{c_2f_2}\ ] \times \\
&\delta(\theta_1 - \theta_1')\delta(\theta_2 - \theta_2') \quad . \qquad (J.6)
\end{aligned}
$$

The factorization equations determine all coefficients in terms of the amplitude σ_{21}, and the bound state position α, which in this approach remains undetermined. A cross check may be obtained considering the 4 into 4 scattering, and subsequently going to the respective bound state pole. This requires an enormous algebraic effort, which eventually leads to the same solution. Further details may be found in [310,311].

Appendix K

Complete S-matrix of the Gross-Neveu Model

The $O(2N)$ Gross-Neveu model has a rich spectrum of solutions. The semi classical analysis reveals a rich bound state spectrum. The scattering of these objects can be also computed in general. Indeed, the so called kinks are isospinors[566]. However, computation of the S matrix based on the methods developed in chapter 8 will be difficult to perform. However, there are new constraints arising from the fact that the particles are bound state of each other; as it turns out, the "elementary" fermions are kink-kink bound states. A very exhaustive derivation of the properties of the S matrix for the $O(2N)$ symmetric Gross Neveu model has been given by Karowski and Thun, which we follow. They use some group theoretical facts peculiar of $O(2N)$. This group has rank N, and the Dynkin diagram is displayed in figure 15.1. There are two spinor representations $\rho_\pm$, a vector representation ρ_1, and $n-3$ representations by antissymmetric tensors ρ_r.

The representations $\rho_{\pm,1}$ are called elementary representations. Any other representation is contained in the tensor product of appropriately chosen elementary representation. Indeed, we have the following statements:

$$\begin{aligned} \rho_r \in (\rho_1)^r \quad &, \quad \rho_{n-2k} \in \rho_\pm \otimes \rho_\pm \\ \rho_{n-2k-1} \in \rho_\pm \otimes \rho_\mp \quad & \quad \rho_\pm \in \rho_m p \otimes \rho_1 \quad . \end{aligned} \tag{K.1}$$

We can also make a general identification as follows. First notice that for even n the spinor representations are real, while for odd n, $\rho_\pm \sim \rho^*_\mp$. Therefore, we obtain, in general

$$\rho_{2k} \in \rho_\pm \otimes \rho^*_\pm, \qquad \rho_{2k+1} \in \rho_\pm \otimes \rho^*_\mp, \qquad \rho_1 \in \rho_\pm \otimes \rho^*_\pm \quad . \tag{K.2}$$

We therefore arrive at the following set of useful properties to be used in order to obtain the exact total S matrix of this model:[566]

1. Tensor states are bound states of elementary fermions.
2. Elementary fermions are bound states of kinks.
3. A kink is a bound state of a kink with opposite chirality, and an elementary fermion.

As is by now well known (see chapter 5) the above state bound structure implies that the fields obey a generalized statistics; we define a generalized relative spin s_{12} for the fields φ_1 and φ_2 as

$$\varphi_1(t,x)\varphi_2(t,y) = \varphi_2(t,y)\varphi_1(t,x)e^{2i\pi s_{12}\epsilon(x-y)} \quad . \tag{K.3}$$

In order to obtain the values of the relative spins above, we notice that the kink-kink bound states corresponds to the fundamental representations ρ_r; moreover, for odd n they are fermions. The cases $n = 2, 3, 4$ can be explicitly computed from the isomorphisms of the groups $O(4) \approx SU(2) \times SU(2)$, $O(6) \approx SU(4)$, and by the triality of $O(8)$, which relates the spinors and vector representations. The Ansatz

$$s_{++} = s_{--} = \frac{n}{8}, \qquad s_{+-} = \frac{n}{8} + \frac{1}{4} \tag{K.4}$$

satisfies these requirements, and have a simple behavior as a function of n. The final check is the consistency with the S matrix.

The starting point of the construction of the complete S matrix is the kink-kink S matrix. The two particle scattering of kinks is described by the following Ansatz in terms of invariant amplitudes

$$_{\gamma\delta}S_{\alpha\beta}(\theta) = 2^{-N} \sum_{r=0}^{2N} \frac{1}{r!} u_r(\theta) \sigma^{(r)}_{\gamma\beta} \sigma^{(r)}_{\delta\alpha} \quad , \tag{K.5a}$$

where

$$\sigma^{(r)}_{\alpha\beta} \equiv \sigma^{a_1 \cdots a_r}_{\alpha\beta} = \frac{1}{r!} \sum_{\pi \in S_r} \mathrm{sign}(\pi) (\gamma^{a_{\pi_1}} \cdots \gamma^{a_{\pi_r}})_{\alpha\beta} \tag{K.5b}$$

and γ^a is a matrix representation of the $O(2N)$ Clifford algebra, which may be defined in terms of the Pauli matrices as

$$\gamma_{2j-1} = \prod_{k=1}^{j-1} \sigma^3_k \sigma^1_j, \qquad \gamma_{2j} = \prod_{k=1}^{j-1} \sigma^3_k \sigma^2_j \quad . \tag{K.6}$$

We can also define a matrix anticommuting with all generators of the Clifford algebra, and with square equal to one, as

$$\Gamma \equiv (-i)^N \prod_{a=1}^{2n} \gamma_a = \prod_{j=1}^{n} \sigma^3_j \quad . \tag{K.7}$$

Moreover, the amplitude $u_r(\theta)$ in (K.5a) has been found to be

$$u_r(\theta) = -u^{SG}(\theta) \times e^{\int_0^\infty \frac{dx}{x} \left[\frac{e^{-x(1+\frac{1}{N-1})}}{\sinh \frac{x}{N-1}} \left(1 - e^{x\frac{\xi}{\lambda}}\right) + \frac{e^{-x(1+\frac{1}{2N-2})} \sinh\left[\frac{x}{2}(1+\frac{1}{N-1})\right]}{\cosh \frac{x}{2} \sinh \frac{x}{N-1}} \right] \sinh\left[x(1 - \frac{\theta}{i\pi})\right]} \tag{K.8}$$

with $u^{SG}(\theta)$ given by the sine-Gordon scattering amplitude

$$u^{SG}(\theta) = e^{\int_0^\infty \frac{dx}{x} \frac{\sinh\left[\frac{x}{2}(1-\frac{1}{N-1})\right]}{\cosh\frac{x}{2}\sinh\frac{x}{2N-2}} \sinh\left[x\frac{\theta}{i\pi}\right]} \quad . \tag{K.9}$$

On the other hand, the fermion-fermion amplitude is known (see chapter 8). We can compute this amplitude from the above expression by using the fusion rules. It can be shown that the result is given by the result (8.90,93), proving the correctness of the above Ansatz. Moreover, using the same fusion rules, we find also the fermion-kink amplitude, which has the factorized form

$${}_{b\beta}S_{a\alpha}(\theta) = (\Gamma^n)_{\beta\alpha}\,\delta^{ba} t_1(\theta) + \left(\Gamma^n \sigma^{ba}\right)_{\beta\alpha} t_2(\theta) \tag{K.10}$$

where

$$t_1(\theta) = \frac{2}{n-2}\left(\frac{\theta}{i\pi} - \frac{1}{2}\right) t_2(\theta) \tag{K.11}$$

and

$$t_1(\theta) - t_2(\theta) = \frac{\theta - i\frac{\pi}{2}\left(1 + \frac{1}{n-1}\right)}{\theta - i\frac{\pi}{2}\left(1 - \frac{1}{n-1}\right)} S_2(\theta) \tag{K.12}$$

$$S_2(\theta) = -\varepsilon_{1,n} S^{SG}_{sol,b}(\theta) e^{\int_0^\infty \frac{dx}{x} \frac{e^{-x(1+\frac{1}{2(n-1)})}}{\cosh\frac{x}{2}} \sinh\frac{x\theta}{i\pi}} \tag{K.13}$$

with the sine Gordon scattering amplitude for soliton breather given by

$$S^{SG}_{sol,b}(\theta) + \frac{\sinh\theta + i\sin\left(\frac{\pi}{2}(1+\frac{1}{n-1})\right)}{\sinh\theta - i\sin\left(\frac{\pi}{2}(1+\frac{1}{n-1})\right)} \quad . \tag{K.14}$$

These expressions can be checked in the cases $n = 2, 3, 4$. Higher bound state amplitudes can be computed as well.

Appendix L

Poisson Brackets and Commutators

a) Poisson Brackets

The following Poisson brackets are required in the computations of chapter 10. $\hat{\mathcal{P}}$ stands for $\hat{\Pi}$ (eq.(13.191)) or $\hat{P}$ (eq.(13.200)) with $\epsilon = +1$ and $\epsilon = -1$, respectively.

$$\begin{aligned}
i\{\mathrm{tr}(\hat{\mathcal{P}}^T g\tau^a)(x), \mathrm{tr}(\hat{\mathcal{P}}^T g\tau^b)(y)\} &= f_{abc}[\frac{\epsilon}{4\pi}\mathrm{tr}(g^{-1}\partial_1 g\tau^c) + \mathrm{tr}(\hat{\mathcal{P}}^T g\tau^c)]\delta(x^1 - y^1) \\
i\{\mathrm{tr}(g\hat{\mathcal{P}}^T \tau^a)(x), \mathrm{tr}(g\hat{\mathcal{P}}^T \tau^b)(y)\} &= -f_{abc}[\frac{\epsilon}{4\pi}\mathrm{tr}(g\partial_1 g^{-1}\tau^c) + \mathrm{tr}(g\hat{\mathcal{P}}^T \tau^c)]\delta(x^1 - y^1) \\
i\{\mathrm{tr}(\hat{\mathcal{P}}^T g\tau^a)(x), \mathrm{tr}(g\hat{\mathcal{P}}^T \tau^b)(y)\} &= -\frac{i\epsilon}{4\pi}\partial_1 \mathrm{tr}(\tau^a g^{-1}\tau^b g)\delta(x^1 - y^1) \\
i\{\mathrm{tr}(g\hat{\mathcal{P}}^T \tau^a)(x), \mathrm{tr}(\hat{\mathcal{P}}^T g\tau^b)(y)\} &= \frac{i\epsilon}{4\pi}\partial_1 \mathrm{tr}(\tau^a g\tau^b g^{-1})\delta(x^1 - y^1) \\
i\{\mathrm{tr}(\hat{\mathcal{P}}^T g\tau^a)(x), \mathrm{tr}(\hat{\mathcal{P}}^T g\hat{\mathcal{P}}^T g)(y)\} &= \frac{\epsilon\delta(x^1 - y^1)}{2\pi}\left[\mathrm{tr}(g^{-1}\partial_1 g\hat{\mathcal{P}}^T g\tau^a) - \mathrm{tr}(\partial_1 g\tau^a \hat{\mathcal{P}}^T)\right]
\end{aligned} \tag{L.1}$$

$$\{\mathrm{tr}(\hat{\mathcal{P}}^T g(x)\tau^a), \mathrm{tr}(\hat{\mathcal{P}}^T g\hat{\mathcal{P}} g(y))\} = \frac{\epsilon}{2\pi}[\mathrm{tr}(g^{-1}\partial_1 g\hat{\mathcal{P}}^T g\tau^a) - \mathrm{tr}(\partial_1 g\tau^a\hat{\mathcal{P}}^T)]\delta(x^1 - y^1) \tag{L.2}$$

$$\begin{aligned}
i\{\mathrm{tr}(\hat{\mathcal{P}}^T g\tau^a)(x), \mathrm{tr}(g\partial_1 g^{-1}\tau^b)(y)\} &= i(\tau^a g^{-1}(y)\tau^b g(y))\partial_1\delta(x^1 - y^1) \\
i\{\mathrm{tr}(\hat{\mathcal{P}}^T g\tau^a)(x), \mathrm{tr}(g^{-1}\partial_1 g\tau^b)(y)\} &= i\delta^{ab}\delta'(x^1 - y^1) + f_{abc}\,\mathrm{tr}(g^{-1}\partial_1 g\tau^c)\delta(x^1 - y^1) \\
i\{\mathrm{tr}(g\hat{\mathcal{P}}^T \tau^a)(x), \mathrm{tr}(g\partial_1 g^{-1}\tau^b)(y)\} &= -i\delta^{ab}\partial_1\delta(x - y) - f_{abc}\,\mathrm{tr}(g\partial_1 g^{-1}\tau^c)\delta(x^1 - y^1) \\
i\{\mathrm{tr}(g\hat{\mathcal{P}}^T \tau^a)(x), \mathrm{tr}(g^{-1}\partial_1 g\tau^b)(y)\} &= i\delta(x^1 - y^1)\partial_1 \mathrm{tr}(\tau^a g\tau^b g^{-1}) \\
&\quad + i\mathrm{tr}(\tau^a g(x)\tau^b g^{-1}(x))\partial_1\delta(x^1 - y^1) \quad .
\end{aligned} \tag{L.3}$$

$$\begin{aligned}
i\{\mathrm{tr}(g^{-1}\partial_1 g(x)\tau^a), \mathrm{tr}(\hat{\mathcal{P}}^T g\hat{\mathcal{P}}^T g(y))\} &= 2\mathrm{tr}\left([g^{-1}\partial_1 g(x), \hat{\mathcal{P}}^T g(x)]\tau^a\right)\delta(x^1 - y^1) \\
&\quad + 2\mathrm{tr}(\hat{\mathcal{P}}^T g(y)\tau^a)\partial_1\delta(x^1 - y^1)
\end{aligned} \tag{L.4}$$

$$i\{\mathrm{tr}(\hat{\mathcal{P}}^T g(x)\tau^a), \mathrm{tr}(\partial_1 g \partial_1 g^{-1}(y))\} = -2i\mathrm{tr}(g^{-1}(y)\partial_1 g(y)\tau^a)\partial_1\delta(x^1-y^1) \quad . \quad (L.5)$$

$$i\{\mathrm{tr}(\hat{\mathcal{P}}^T g(x)\tau^a), \mathrm{tr}(g\tau^c g^{-1}(y)\tau^b)\} = f_{acd}\mathrm{tr}(g\tau^d g^{-1}(x)\tau^b)\delta(x^1-y^1) \quad . \quad (L.6)$$

b) Commutators

The commutators of chiral QCD_2 follow from the correspondence $i\{A,B\}_D \to [A,B]$. The Dirac brackets involving only g and $\hat{\mathcal{P}}$, are constructed in the standard way and formally coincide with the above Poisson brackets. We therefore list only those involving $A_0^a(x)$, which are of non-canonical form:

$$[A_0^a(x), \mathrm{tr}(g\tau^b g^{-1}(y)\tau^c)]_{ET} = \frac{4\pi i}{(1-a)e} f_{abd}\mathrm{tr}(g\tau^d g^{-1}(x)\tau^c)\delta(x^1-y^1)$$

$$[A_0^a(x), \mathrm{tr}(g\hat{\mathcal{P}}^T(y)\tau^b)]_{ET} = \frac{1}{(1-a)e}\mathrm{tr}(g(y)\tau^a g^{-1}(y)\tau^b)\delta'(x^1-y^1)$$

$$[A_0^a(x), \mathrm{tr}(g\partial_1 g^{-1}(y)\tau^b)]_{ET} = \frac{4\pi}{(1-a)e}\mathrm{tr}(g(y)\tau^a g^{-1}(y)\tau^b)\delta'(x^1-y^1)$$

$$[A_0^a(x), \mathrm{tr}(\hat{\mathcal{P}}^T g(y)\tau^b)]_{ET} = \frac{1}{(1-a)e}\left[\delta^{ab}\delta'(x^1-y^1) + 4\pi i f_{abc}\mathrm{tr}(\hat{\mathcal{P}}^T g(x)\tau^c)\delta(x^1-y^1)\right]$$

$$[A_0^a(x), \mathrm{tr}(g^{-1}\partial_1 g(y)\tau^b)]_{ET} = \frac{-4\pi}{(1-a)e}\left[\delta^{ab}\delta'(x^1-y^1) - i f_{abc}\mathrm{tr}(g^{-1}\partial_1 g(x)\tau^c)\delta(x^1-y^1)\right]$$

$$[A_-^a(x), A_-^b(y)]_{ET} = \frac{8\pi i}{(1-a)e^2}\mathcal{D}_1^{ab}\delta(x^1-y^1) + i\left(\frac{4\pi}{(1-a)e^2}\right)^2 e f_{abc}\left[\Omega_2^c(x) + \frac{e^2}{4\pi}(A_1^c(x) + (1-a)A_0^c(x))\right]\delta(x^1-y^1) \quad (L.7)$$

where Ω_2^c is the Gauss operator (13.208).

Appendix M

Chiral Bosons

Self dual fields are important ingredients in field theory; case examples are $4-D$ self dual Yang Mills theory, which is an integrable theory, and right (left) moving bosons in the construction of the heterotic string. The latter are the issue of this section.

Chiral bosons may be defined by the Siegel action[448], for left moving fields

$$S = \frac{1}{2}\int d^2x \left[\partial_+\varphi\partial_-\varphi + \lambda(\partial_-\varphi)^2\right] \quad , \tag{M.1}$$

where λ is an auxiliary field constraining the bosonic field to obey

$$\partial_-\varphi = 0 \quad . \tag{M.2}$$

A comment is necessary at this point. The above action is quadratic in the constraint $\partial_-\varphi \approx 0$, otherwise the field λ will acquire dynamics, due to its role as an external force.

The above classical action is invariant under the transformation

$$\delta\varphi = \frac{1}{2}\epsilon\partial_-\varphi \tag{M.3a}$$

$$\delta\lambda = -\partial_+\epsilon + \frac{1}{2}\epsilon\partial_-\lambda - \frac{1}{2}\lambda\partial_-\epsilon \quad , \tag{M.3b}$$

which is the "Siegel symmetry"[448]; it is reminiscent of the general coordinate transformation, since we can obtain the above action from the interaction of the bosonic field φ with gravity, if we restrict the metric to be

$$g^{\mu\nu} = \begin{pmatrix} g^{++} & g^{+-} \\ g^{-+} & g^{--} \end{pmatrix} = \begin{pmatrix} 0 & 1/2 \\ 1/2 & \frac{1}{2}\lambda \end{pmatrix} \quad . \tag{M.4}$$

Indeed, the action

$$S = \frac{1}{2}\int d^2x\sqrt{-g}g^{\mu\nu}\partial_\mu\varphi\partial_\nu\varphi \tag{M.5}$$

is invariant under general coordinate transformations

$$\delta x^\pm = \epsilon^\pm \tag{M.6a}$$

$$\delta\varphi = \frac{1}{2}(\epsilon^+\partial_+ + \epsilon^-\partial_-)\varphi \tag{M.6b}$$

$$\delta g^{\mu\nu} = (-g^{\mu\kappa}\partial_\kappa\delta^\nu_\rho - g^{\kappa\nu}\partial_\kappa\delta^\mu_\rho + \partial_\rho g^{\mu\nu})\epsilon^\rho \quad , \tag{M.6c}$$

and Weyl transformations

$$\delta g^{\mu\nu} = \Lambda g^{\mu\nu} \quad , \tag{M.7}$$

which leave the gauge choice (M.4) invariant if

$$\partial_- \epsilon^+ = 0 \tag{M.8a}$$

and

$$2\Lambda = \partial_- \epsilon^- + \partial_+ \epsilon^+ \quad . \tag{M.8b}$$

If we take $\epsilon^+ = 0$, we obtain the Siegel symmetry (M.3).

The Hamiltonian treatment of the Siegel action is rather simple, using the Dirac method.The canonical momenta are

$$\pi_\varphi = (\lambda + 1)\dot{\varphi} - \lambda \varphi' \quad , \tag{M.9a}$$

$$\pi_\lambda = 0 \quad . \tag{M.9b}$$

For general λ, we have

$$\dot{\varphi} = \frac{\pi_\varphi + \lambda \varphi'}{\lambda + 1} \quad , \tag{M.10}$$

and the primary constraint (M.9*b*).The canonical Hamiltonian is given by

$$\begin{aligned} H_c &= \pi_\varphi \dot{\varphi} + \pi_\lambda \dot{\lambda} - \mathcal{L} \\ &= \pi_\varphi \varphi' + \frac{1}{2(\lambda+1)} (\pi_\varphi - \varphi')^2 \quad . \end{aligned} \tag{M.11}$$

The time derivative of (M.9*b*) is now easily computed using (M.11), and we obtain

$$\dot{\Omega} \equiv \dot{\pi}_\lambda = \{\pi_\lambda, H\} = \frac{(\pi_\varphi - \varphi')^2}{2(\lambda+1)^2} \quad , \tag{M.12}$$

which implies

$$\Omega_2 = \pi_\varphi - \varphi' \approx 0 \quad . \tag{M.13}$$

The constraint Ω_2 is constant in time. Indeed

$$\dot{\Omega}_2 = \{\Omega_2, H\} = \Omega_2' \quad . \tag{M.14}$$

The constraint Ω_2 is second class

$$\{\Omega_2(x), \Omega_2(y)\} = -2\delta'(x - y) \quad , \tag{M.15}$$

while Ω_1 is first class, and generates the Siegel gauge symmetry, by the Anderson-Bergman algorithm. We can fix the gauge by requiring

$$\Omega_3 = \lambda - f(x) \approx 0 \quad , \tag{M.16}$$

where $f(x)$ is a given function.

The Dirac algebra may be computed straightforwardly, and we obtain

$$\{\varphi(x), \varphi(y)\} = -\frac{1}{4}\epsilon(x-y) \tag{M.17a}$$

$$\{\pi_\varphi(x), \pi_\varphi(y)\} = \frac{1}{2}\delta'(x-y) \tag{M.17b}$$

$$\{\varphi(x), \pi_\varphi(y)\} = \frac{1}{2}\delta(x-y) \tag{M.17c}$$

We may obtain other forms of the Lagrangian in such a way that (M.2) is one of the equations of motion. Thus, consider the generating functional

$$Z[J] = \int \mathcal{D}\varphi \mathcal{D}\pi_\varphi \mathcal{D}\lambda \mathcal{D}\pi_\lambda \delta(\pi_\lambda)\delta(\lambda - f)\delta(\pi_\varphi - \varphi')e^{iS[J]} \quad , \tag{M.18}$$

where $Q_{\alpha\beta}$ is the Dirac matrix

$$Q_{\alpha\beta} = \{\Omega_\alpha, \Omega_\beta\} = \begin{pmatrix} 0 & -\delta(x-y) & 0 \\ \delta(x-y) & 0 & 0 \\ 0 & 0 & -2\delta'(x-y) \end{pmatrix} \quad . \tag{M.19}$$

with Ω_i, $i = 1, 2, 3$ the constraints (M.9*b*), (M.13) and (M.16), and

$$S[J] = \int d^2x \left(\pi_\varphi\dot{\varphi} + \pi_\lambda\dot{\lambda} - \pi_\varphi\varphi' - \dot{f}\pi_\lambda - J\varphi\right) \tag{M.20}$$

the phase space expression for the Siegel action in the presence of an external source. Noticing that det$Q$ is a constant,we integrate over the fields π_λ, λ and π_φ obtaining

$$Z[J] = N^{-1} \int \mathcal{D}\varphi e^{i\int d^2x(\varphi'\dot{\varphi} - \varphi'^2 - J\varphi)} \quad , \tag{M.21}$$

from which we read off the new Lagrangian

$$\mathcal{L} = \varphi'\dot{\varphi} - \varphi'^2 \quad , \tag{M.22}$$

obtained by Floreanini and Jackiw[448]. Its equation of motion is the constraint (M.2).

The canonical quantization of this theory leads to the previous results given in Eq. (M.17). It is a theory with a local Lagrangian, and a non-local commutation relation (M.17*a*). We can also obtain a non-local Lagrangian with local commutation relations, defining a new field χ by

$$\varphi(x) = \frac{1}{2}\int dy\epsilon(x-y)\chi(y) \quad . \tag{M.23}$$

One thus obtains the non-local action

$$S=\int dtdxdy\frac{1}{2}\chi(x)\epsilon(x-y)\chi(y)-\int dxdt\chi(x) \qquad (M.24a)$$

and the local commutation relations

$$\{\chi(x),\chi(y)\}=-2\delta'(x-y) \quad . \qquad (M.24b)$$

The equation of motion of χ is

$$\chi(x)=\frac{1}{2}\int dy\epsilon(x-y)\dot{\chi}(y) \quad , \qquad (M.25)$$

and implies the boundary condition $\chi(\infty)=-\chi(-\infty)$, showing that it describes a "kink"-like solution.

Actually, the Siegel Lagrangian describes the interaction of a scalar field with gravity, which is anomalous (see chapter 15). Indeed, consider the Siegel Lagrangian (M.1). The (-) component of the corresponding energy momentum tensor

$$\theta_{--}=\frac{1}{4}(\partial_-\varphi)^2 \quad , \qquad (M.26)$$

is seen to coincide with the constraint (M.12). In the gauge $\lambda=0$ it takes the form

$$\theta_{--}=\frac{1}{4}(\pi_\varphi-\varphi')^2=\frac{1}{4}\Omega_S \quad . \qquad (M.27)$$

Its Poisson algebra is computed to be

$$\{\Omega_S(x),\Omega_S(y)\}=-4(\Omega_S(x)+\Omega_S(y))\delta'(x-y) \quad . \qquad (M.28)$$

Therefore, the energy momentum tensor obeys a classical Virasoro algebra.In the quantum theory, we must use the normal order prescription in order to define the square of $\pi_\varphi-\varphi'$. In this case, the quantum algebra obeyed by θ_{--} is found to have a central term

$$\{\theta_{--}(x),\theta_{--}(y)\}=-(\theta_{--}(x)+\theta_{--}(y))\delta'(x-y)+c\delta'''(x-y) \quad , \qquad (M.29)$$

with $c=\frac{1}{24\pi}$, and we have an anomaly in the Siegel symmetry[448,567].

The Lagrangian should be modified to include a Wess-Zumino term. Imbimbo and Schwimmer considered the Lagrangian[567]

$$\mathcal{L}=\mathcal{L}_S+\mathcal{L}_g+\mathcal{L}_{WZ} \qquad (M.30a)$$

$$\mathcal{L}_S=\partial_+\varphi\partial_-\varphi+\lambda(\partial_-\varphi)^2 \qquad (M.30b)$$

$$\mathcal{L}_{gh}=-\partial_+\bar{c}c+\lambda\left[-\frac{3}{2}\partial_-(\bar{c}c)+\frac{1}{2}(\bar{c}\partial_-c-\partial_-\bar{c}c)\right] \qquad (M.30c)$$

$$\mathcal{L}_{WZ}=\alpha\lambda\partial_-^2\varphi \quad , \quad \alpha=\sqrt{\frac{25}{6\pi}} \qquad (M.30d)$$

where $\mathcal{L}_{gh}$ is the ghost Lagrangian associated with the gauge fixing condition (M.16). The energy-momentum tensor of the matter and the ghost field are given, respectively, by (see section (15.3))

$$\theta_{--} = \frac{1}{2}\left[(\partial_-\varphi)^2 + \alpha\partial_-^2\varphi\right] \tag{M.31a}$$

$$\theta_{--}^{gh} = \frac{1}{4}\left[-3\partial_-(\bar{b}b) + (\bar{b}\partial_-b - \partial_-\bar{b}b)\right] \quad . \tag{M.31b}$$

The choice (M.30*b*) for α corresponds to an anomaly free Virasoro algebra for the total energy momentum tensor.

The Siegel Lagrangian with a WZ term has a gauge symmetry at the quantum level. Indeed, $\mathcal{L}_S + \mathcal{L}_{WZ}$ is invariant, at the quantum level, under the gauge transformation

$$\delta\varphi = \frac{1}{2}\epsilon\partial_-\varphi - \frac{\alpha}{4}\partial_-\epsilon \tag{M.32a}$$

$$\delta\lambda = -\partial_+\epsilon + \frac{1}{2}\epsilon\partial_-\lambda - \frac{1}{2}\lambda\partial_-\epsilon \quad , \tag{M.32b}$$

the action S_{WZ} associated with $\mathcal{L}_{WZ}$ transforming as

$$\delta S = \int d^2x \frac{\alpha^2}{8}\epsilon\partial_-^3\lambda \quad . \tag{M.33}$$

The variation (M.33) just cancels the quantum anomaly (see chapter 15 for the problem of two dimensional gravity, whose action arises from an anomaly as well).

We can also treat the Lagrangian $\mathcal{L}_S + \mathcal{L}_{WZ}$ by the Hamiltonian formalism if we take the above information concerning the quantum gauge symmetry as an input. The canonical momenta are

$$\pi_\varphi = \dot{\varphi} + \lambda(\dot{\varphi} - \varphi') - \frac{\alpha}{2}(\dot{\lambda} - \lambda') \tag{M.34a}$$

$$\pi_\lambda = -\frac{\alpha}{2}(\dot{\varphi} - \varphi') \quad , \tag{M.34b}$$

implying the canonical Hamiltonian

$$H = \left(\pi_\varphi + \frac{2}{\alpha}\pi_\lambda\right)\varphi' - \frac{2}{\alpha}\pi_\varphi\pi_\lambda + \pi_\lambda\lambda' + \frac{2}{\alpha^2}(1+\lambda)\pi_\lambda^2 \quad . \tag{M.35}$$

There are no constraints. However, at the one loop level we have the gauge symmetry (M.32), which can again be fixed by imposing a subsidiary condition of the form (M.16). In fact, only after quantization can the above be interpreted as a choice of gauge; choosing $f = -1$, constancy in time of the subsidiary condition requires

$$\pi_\varphi - \varphi' \approx 0 \quad , \qquad (M.36a)$$

which in turns implies

$$\pi'_\lambda \approx 0 \quad , \qquad (M.36b)$$

implying that we have chiral bosons again.

Comparing with (M.9*b*), (M.13) and (M.16) we see that we have essentially recovered the constraints of the original formulation[568].

The methods just presented allow a study of the interaction of chiral bosons with gauge fields. Non abelian chiral bosons may be studied in an analogous[569] way. The fact that a chiral boson describes a reparametrization invariant theory, permits also some nice new descriptions generalizing the previous one, and including spin fields. Classes of reparametrization invariant theories may be obtained, in terms of an arbitrary parameter, with the previous formulation being a particular case[568].

Appendix N

Teichmüller Parameters

In this Appendix we address some details concerning the integration over geometries. This is a very refined computation. The quantum sum over geometries is invariant under the group of reparametrizations. Therefore, we have to divide the integration measure by the volume of the diffeomorphism group. We thus go to the space of metrics, and define an integration measure by

$$\|\delta g\|^2 = \int d^2z\sqrt{-g}\left(g^{abcd} + ug^{ab}g^{cd}\right)\delta g_{ab}\delta g_{cd} \quad , \tag{N.1}$$

where ${g_{ab}}^{cd} = \frac{1}{2}\left(\delta_a^c\delta_b^d + \delta_a^d\delta_b^c - g_{ab}g^{cd}\right)$ is the projector on the space of symmetric tensors. We can decompose the variation of the metric into a symmetric traceless piece δh_{ab}, and a trace part, such that a variation of the metric is given by

$$\delta g_{ab} = \delta h_{ab} + 2g_{ab}\delta\tau \quad . \tag{N.2}$$

Moreover, we can also rewrite the variation in terms of differential operators acting on vector fields defined on the manifold in question. We have

$$\delta g_{ab} = \nabla_a\delta V_b + \nabla_b\delta V_a + 2g_{ab}\delta\sigma \quad , \tag{N.3}$$

in such a way that the variations in the symmetric traceless piece, and that of the conformal piece are given by

$$\delta h_{ab} = 2{G_{ab}}^{cd}\nabla_c\delta V_d + 2g_{ab}\delta\sigma \equiv [P\delta V]_{ab} \tag{N.4a}$$

$$\delta\tau = \delta\sigma + \frac{1}{2}g^{cd}\nabla_c\delta V_d \quad . \tag{N.4b}$$

The Jacobian of the transformation from $d\tau dh$ to $d\sigma dV$ is

$$J = \left|\det\begin{pmatrix}1 & \text{anything}\\ 0 & P\end{pmatrix}\right| = |\det P| = |\det P^\dagger P|^{\frac{1}{2}} \quad , \tag{N.5}$$

with, as we read from (N.4a), $[P^\dagger\delta h]^a = -\nabla_b\delta h^{ba}$. However, this procedure does not eliminate all problems, since we have also to consider the problem of zero modes of the operators P and $P^\dagger$. The fact that $P^\dagger$ has zero modes implies that the conformal gauge cannot be achieved (see Polyakov's book[534] for a very simple discussion of this fact). On the other hand, zero modes of P, means that diffeomorphisms generated by the Killing equation, i.e. by the zero modes, is equivalent to a change in the conformal factor. Therefore, these changes are being overcounted. Fortunately, they are degrees of freedom with a finite number of parameters, and the number of these parameters is -3χ, where χ is the genus of the manifold. For the plane, we have no parameters. These parameters characterize inequivalent n-tori, and since the theory is conformal invariant, and conformal transformations modify the size of the torus, it is not difficult to see, using a "paper" model of the n-torus, that a finite number of parameters is left invariant by these transformations. In the case of the one torus, we have the ratio of the lengths of the sides, and the angle (after we cut off the torus – notice that we have, for $\chi = 0$, two parameters, and not -3χ, which is valid for $\chi < -1$; notice also that $\chi = 2 - 2h$, where h is the number of handles, for a closed surface.)

One way of computing the remaining determinant, is in terms of the complex structure of the manifold. In a complex manifold we have n tensors defined by the transformation under the conformal group as

$$T' = \left(\frac{\partial z'}{\partial z}\right)^n T \tag{N.6}$$

for positive or negative values of n, since in the case of an index $\bar{z}$ we can use the metric $g_{\bar{z}z}$ in order to change a $\bar{z}$ index into a z index[1]. We define the differential operator ∇_n^z, mapping n tensors into $n+1$ tensors, by

$$\nabla_n^z T = g^{z\bar{z}} \partial_{\bar{z}} T \quad ; \tag{N.7}$$

and ∇_z^n, mapping n tensors into $n-1$ tensors, by

$$\nabla_z^n T = \left(g^{z\bar{z}}\right)^n \partial_z \left((g_{z\bar{z}})^n T\right) \quad . \tag{N.8}$$

The operator P acts as a first order (respectively "minus" first order) operator in z as

$$PV = \nabla_1^z V^z, \quad \nabla_z^{-1} V_z \qquad P^\dagger h = \nabla_z^2 h^{zz}, \quad \nabla_{-2}^z h_{zz} \quad . \tag{N.9}$$

The previously mentioned fact that the kernel of the operator $P^\dagger$, which are the so called Teichmüller deformations, is related to the remaining (finite) residual symmetry, given by the conformal Killing vectors, is realized by the famous Riemann-Roch theorem, which states that the number of conformal Killing vectors minus the number of Teichmüller parameters is $3\chi = 6 - 6h$. The computation of the relevant determinants is now a technical question, and we shall refer to the methods developed in chapter 4, in general, or to the paper of O. Alvarez[154], in this more specific case.

---

1) Fermions can also be defined, if we allow n to be half integer.

Appendix O

Dirac Quantization

This Appendix is meant as an elementary introduction to the Hamiltonian quantization of constraints systems. There exists an extensive literature on the subject, which the reader may consult[338]. Our discussion will refer to quantum mechanical systems with a finite number of degrees of freedom. The generalization to quantum field theory will be obvious from the context.

Consider a quantum mechanical system whose dynamics is described by a Lagrange function[1)]

$$L(q,\dot{q}) \equiv L(\{q_i\},\{\dot{q}_i\}) \quad , \quad i=1,\cdots,n \tag{O.1}$$

Canonical quantization of such a system involves the transition from the $2n$ dimensional space of coordenates $\{q_i\}$ and velocities $\{\dot{q}_i\}$ to *phase space* spanned by the coordinates and their canonically conjugate momenta $\{p_i\}$, defined by

$$p_i = \frac{\partial L(q,\dot{q})}{\partial \dot{q}_i} \quad . \tag{O.2}$$

Correspondingly the Lagrangian (O.1) is replaced by the Hamiltonian H_c (the subscript "c" stands for "canonical") defined as the Legendre-transform of L,

$$H_c = \sum_i p_i \dot{q}_i - L(q,\dot{q}) \quad . \tag{O.3}$$

The velocities $\dot{q}_i$ in (O.4) should be expressed in terms of the phase-space variables q_i, p_i, by solving Eqs. (O.2) for the $\dot{q}_i$'s. This is however only possible provided that

$$\det W \neq 0 \quad , \quad W_{ij} = \frac{\partial^2 L}{\partial \dot{q}_i \partial \dot{q}_j} \quad . \tag{O.4}$$

The crucial role the matrix W plays is given by noting that the Euler-Lagrange equations $\frac{d}{dt}\frac{\partial L}{\partial \dot{q}_i} - \frac{\partial L}{\partial q_i} = 0$ can also be written in the form

$$\sum_i W_{ij}(q,\dot{q})\ddot{q}_j = \frac{\partial L}{\partial q_i} - \frac{\partial^2 L}{\partial \dot{q}_i \partial q_j}\dot{q}_j \quad . \tag{O.5}$$

This shows that, if $\det W = 0$, the dynamics of the system is not uniquely determined by the specification of the coordinates and velocities at some initial time. In that case one says that the Lagrangian is *singular*.

[1)] We consider only Lagrangians which do not depend explicitly on time.

As seen in the course of this book, it is exactly the singular Lagrangians which play an important role in physics. One must therefore extend the "text-book" canonical quantization procedures to include the case of singular Lagrangians. This was done by Dirac·· in his now classic book on quantization of constrained systems.

Let us suppose that the rank of the matrix W is $R_W < n$. In this case we may solve (O.2) for R_W of the velocities as a function of the n coordenates and the $n - R_W$ remaining velocities,

$$\dot{q}_a = f_a(\{q_i\}, \{p_b\}, \{\dot{q}_\rho\}) \quad . \tag{O.6}$$

Here the indices take the values $i = 1, \cdots, n \quad ; \quad a, b = 1, \cdots, R_W \quad ; \quad \rho = R_W + 1, \cdots, n$.

Substituting (O.6) into (O.2) we formally obtain

$$p_i = g_i(q, \{p_b\}, \{\dot{q}_\rho\}) \quad . \tag{O.7}$$

It is clear that for $i = a \in [1, R_W]$, those equations reduce to an identity. On the other hand, we have for $i = \rho \in [R_W + 1, n]$,

$$p_\rho = g_\rho(q, \{p_b\}) \quad , \quad \rho = R_W + 1, \cdots, n \tag{O.8}$$

where g_ρ cannot depend on the velocities $\{\dot{q}_\rho\}$, since we could otherwise eliminate further velocities, contrary to our hypothesis.

The equations (O.8) represent $m = n - R_W$ *constraints* following alone from the definition (O.2) of the momenta p_i; they are thus called *primary* constraints. They imply that the dynamics in phase-space Γ evolves on the m dimensional subspace $\Gamma^{(p)}$ (superscript "p" for primary) defined by

$$\Gamma^{(p)}: \phi_\rho = 0 \quad , \quad \rho = R_W + 1, \cdots, n \tag{O.9a}$$

with

$$\phi_\rho(q, p) := p_\rho - g_\rho(q, \{p_a\}) \quad . \tag{O.9b}$$

We come now to the first important statement:

Theorem: The Euler-Lagrange equations imply on $\Gamma^{(p)}$ the (modified) Hamilton equations

$$\dot{q}_a = \frac{\partial H_c^{(p)}}{\partial p_a} - \sum \dot{q}_\rho \frac{\partial g_\rho}{\partial p_a} \quad , \quad a = 1, \cdots, R_W \tag{O.10a}$$

$$\dot{p}_i = -\frac{\partial H^{(p)}}{\partial q_i} + \sum \dot{q}_\rho \frac{\partial g_\rho}{\partial q_i} \quad , i = 1, \cdots, n \tag{O.10b}$$

where $H_c^{(p)}$ stands for H_c evaluated on the constraint surface $\Gamma^{(p)}$.

The proof of this theorem departs from the explicit computation of the derivatives $\partial H_c^{(p)}/\partial q_i$ and $\partial H_c^{(p)}/\partial p_a$, and uses the fact that H_c does not depend on the velocities $\{\dot{q}_\rho\}$ as a consequence of the definition (O.2) of the canonical momenta:

$$H_c^{(p)} = H_c^{(p)}(q, \{p_a\}) \quad . \tag{O.11}$$

The "velocities" $\dot{q}_\rho$ appearing in (O.10) are so far undetermined functions of time. The degree to which their dynamics is fixed by the Euler-Lagrange equations characterizes the type of system one is dealing with (first class, second class, or mixed). Thus in general the equations (O.10) alone are not equivalent to the Euler-Lagrange equations. We return to this point later on.

The $n + R_W$ Hamilton equations (O.10) treat the phase-space variables asymmetrically. This asymmetry is removed by the following

Corollary 1: Eqs. (O.10a, b) can be written in the form

$$\dot{q}_i = \frac{\partial H_c^{(p)}}{\partial p_i} + \sum_\rho \frac{\partial \phi_\rho}{\partial p_i}\dot{q}_\rho \tag{O.12a}$$

$$\dot{p}_i = -\frac{\partial H_c^{(p)}}{\partial q_i} - \sum_\rho \frac{\partial \phi_\rho}{\partial q_i}\dot{q}_\rho \quad . \tag{O.12b}$$

The proof of this corollary follows from the identities

$$\frac{\partial g_\rho}{\partial q_i} = -\frac{\partial \phi_\rho}{\partial q_i} \quad , \quad \frac{\partial g_\rho}{\partial p_a} = -\frac{\partial \phi_\rho}{\partial p_a} \quad , \quad \frac{\partial \phi_\rho}{\partial p_\sigma} = \delta_{\rho\sigma} \tag{O.13}$$

and the observation, that for $i = R_W + 1, \cdots, n$, Eqs. (O.10a) are *identically* satisfied, since $\frac{\partial H_c^{(p)}}{\partial p_\rho} = 0$.

Often is not possible or desirable to solve Eq.(O.2) explicitly for the velocities $\dot{q}_a$. It is thus desirable to free the Hamiltonian in (O.12) from the restriction to $\Gamma^{(p)}$. To this end it is useful to introduce the notation of "weak-equality", denoted by $\approx$, defined as follows:

Definition: Two functions f and h on Γ are said to be *weakly* equal, if they coincide on the subspace Γ' defined by the constraints. We have the following

Proposition: Let f and h be two functions defined on Γ. Then, the equality of f and h on $\Gamma^{(p)}$: $\{\phi_\beta = 0\}$, that is, $f \approx h$, implies

$$\frac{\partial}{\partial q_i}\left(f - \sum_\rho \phi_\rho \frac{\partial f}{\partial p_\rho}\right) \approx \frac{\partial}{\partial q_i}\left(h - \sum_\rho \phi_\rho \frac{\partial h}{\partial p_\rho}\right) \tag{O.14a}$$

$$\frac{\partial}{\partial p_i}\left(f - \sum_\rho \phi_\rho \frac{\partial f}{\partial p_\rho}\right) \approx \frac{\partial}{\partial p_i}\left(h - \sum_\rho \phi_\rho \frac{\partial h}{\partial p_\rho}\right) \quad . \tag{O.14b}$$

The proof of this proposition follows from the identities (O.13). Note that the left and right hand side of (O.14b), separately, vanish weakly for $i = R_W + 1, \cdots, n$.

Let now $H'(q,p)$ be a function with the property

$$H'(q_i, p_i) \approx H_c^{(p)}(q, \{p_a\}) \quad . \tag{O.15}$$

Choosing $f = H_c^{(p)}$ and $h = H'$ in Eqs.(O.14), and recalling (O.11), we obtain,

$$\frac{\partial H_c^{(p)}}{\partial q_i} \approx \frac{\partial}{\partial q_i}\left(H' - \sum_\rho \phi_\rho \frac{\partial H'}{\partial p_\rho}\right) \tag{O.16a}$$

$$\frac{\partial H_c^{(p)}}{\partial p_i} \approx \frac{\partial}{\partial p_i}\left(H' - \sum_\rho \phi_\rho \frac{\partial H'}{\partial p_\rho}\right) \quad . \tag{O.16b}$$

Substitution of (O.16) into (O.12) results in the following

Corollary 2:

$$\dot{q}_i \approx \frac{\partial H}{\partial p_i} + \sum_\rho v_\rho \frac{\partial \phi_\rho}{\partial p_i} \quad , \tag{O.17a}$$

$$\dot{p}_i \approx -\frac{\partial H}{\partial q_i} - \sum_\rho v_\rho \frac{\partial \phi_\rho}{\partial q_i} \quad , \tag{O.17b}$$

where

$$H := H' - \sum_\rho \phi_\rho \frac{\partial H'}{\partial p_\rho} \approx H_c^{(p)} \tag{O.17c}$$

and $v_\rho := \dot{q}_\rho$ are arbitrary functions (velocities) of time.

Finally, defining the "total" Hamiltonian H_T by

$$H_T = H + \sum_{\rho=1}^{m} v_\rho \phi_\rho \quad , \tag{O.18}$$

we may write Eqs. (O.16) also in the form,

$$\begin{aligned} \dot{q}_i &\approx \{q_i, H\} + \sum_\rho v_\rho \{q_i, \phi_\rho\} \approx \{q_i, H_T\} \quad , \\ \dot{p}_i &\approx \{p_i, H\} + \sum_\rho v_\rho \{p_i, \phi_\rho\} \approx \{p_i, H_T\} \quad , \end{aligned} \tag{O.19}$$

where $\{\, ,\, \}$ denotes the Poisson bracket, defined for functions on Γ by[2],

$$\{F,G\} = \sum_{i=1}^{n} \left(\frac{\partial F}{\partial q_i} \frac{\partial G}{\partial p_i} - \frac{\partial G}{\partial q_i} \frac{\partial F}{\partial p_i} \right) \quad . \tag{O.20}$$

Using the familiar properties of the Poisson brackets, Eqs. (O.19) imply,

$$\dot{F}(q,p) \approx \{F(q,p), H_T\} \quad . \tag{O.21}$$

It is important to realize, that the weak equality in (O.17) means that the primary constraints $\{\phi_\rho = 0\}$ are to be imposed only *after* all the differentiations have been performed.

The Dirac algorithm

Eqs.(O.17) are in general not equivalent to the Lagrange equations. This is clear once one recognizes that these equation do in general not guarantee that the primary constraints $\{\phi_\rho = 0\}$ be satisfied at all times. In addition to (O.19), we must thus require

$$\dot{\phi}_\rho \approx \{\phi_\rho, H_T\} \approx 0 \quad , \quad \rho = R_W + 1, \cdots, n \tag{O.22}$$

or equivalently

$$\{\phi_\rho, H\} + \sum_\sigma v_\sigma \{\phi_\rho, \phi_\sigma\} \approx 0 \quad . \tag{O.23}$$

In the following it will be convenient to relabel the primary constraints so that the index ρ runs from 1 to $m = n - R_W$.

Let R be the rank of the matrix $\|\{\phi_\rho, \phi_\sigma\}\|$, $\rho, \sigma = 1, \cdots, m = n - R_W$. If $R = m$, we may solve (O.23), for all the velocities, and H_T is completely determined. If $R < m$, we may choose linear combinations $\phi'_\rho = \sum a_{\rho\sigma}\phi_\sigma$ of the primary constraints, such that, after suitable relabelling,

$$\det \|\{\phi'_\rho, \phi'_\sigma\}\| \not\approx 0 \quad , \quad \rho, \sigma = 1, \cdots, R < m \tag{O.24a}$$

---

[2] The form (O.20) only applies to commuting phase-space coordinates. If anticommuting (Grassmann valued) degrees of freedom are also involved, the definition must be generalized to

$$\{A,B\} = \sum (-1)^{gh(q)} \left(\frac{\partial^{(r)} A}{\partial q_i} \frac{\partial^{(l)} B}{\partial p_i} - (-1)^{\epsilon(A)\epsilon(B)} \frac{\partial^{(r)} B}{\partial q_i} \frac{\partial^{(l)} A}{\partial p_i} \right)$$

where $\epsilon(A) = 0(1)$ for A a bosonic (fermionic) function, and $gh(q_i)$ is the "ghost number" of the coordinate q_i : $gh(q_i) = 1(0)$ for a q_i bosonic (Grassmann valued). $\partial^{(l)}$ and $\partial^{(r)}$ stands for "left" and "right" derivative, respectively. They are conveniently defined by $df = dq \frac{\partial^{(l)}}{\partial q} f = \frac{\partial^{(r)}}{\partial q} f dq$.

while

$$\{\phi'_\rho, \phi'_\sigma\} \approx 0 \quad , \quad \rho = R+1, \cdots, m \qquad\qquad (O.24b)$$
$$\sigma = 1, \cdots, m \quad .$$

Hence Eq. (O.23) now reads

$$\{\phi'_\rho, H\} + \sum_{\sigma=1}^{R} \{\phi'_\rho, \phi'_\sigma\} v'_\sigma \approx 0 \quad , \quad \rho = 1, \cdots, R \qquad (O.25a)$$

$$\{\phi'_\rho, H\} \approx 0 \quad , \quad \rho = R+1, \cdots, m \quad . \qquad (O.25b)$$

Because of Eq. (O.24*a*) we may solve (O.25*a*) for the velocities v'_ρ, $\rho = 1, \cdots, R$, which are thus determined as functions of q and p. Unless all equations (O.25*b*) are identically satisfied, they imply further, *secondary*, constraints φ_s, $s = 1, \cdots, S \leq m - R$, which must again be satisfied for all times. Repetitive application of the above procedure will thus generally lead to several "generations" of second class constraints. The algorithm ends, if no further secondary constraints are generated. At this point, some, or all of the velocities v_ρ have been determined; if all have been determined, one speaks of a second-class Hamiltonian system; if l velocities remain undetermined, the Hamiltonian system in question possesses a local symmetry associated with an l-parameter group of transformations, as we shall see below.

Equations of Motion

We now collect all primary and secondary constraints into a vector $(\Omega_A) = (\phi_1, \cdots, \phi_m, \varphi_{m+1}, \cdots, \varphi_M)$. The corresponding subspace of Γ we call Γ_Ω. The equations of motion then read,

$$\{F(q,p), H_T\} \approx 0 \qquad (O.26a)$$
$$\{\Omega_A, H_T\} \approx 0 \quad , \quad A = 1, \cdots, M \qquad (O.26b)$$

where the sign of weak equality now means "evaluation on Γ_Ω".

It is convenient to introduce at this point the notion of first- and second class quantities. A function $F(q,p)$ on Γ is called ***first-class***, if its Poisson brackets with *all* constraints vanish on Γ_Ω:

$$\{F, \Omega_A\} \approx 0 \quad , \quad A = 1, \cdots, M \quad . \qquad (O.27a)$$

This means that $\{F, \Omega_A\}$ has the property

$$\{F, \Omega_A\} = \sum_B c_{AB}(q,p) \Omega_B \quad . \qquad (O.27b)$$

Quantities not having this property are called ***second class***.

We separate the constraints $\{\Omega_A\}$ generated by the above algorithm into first (second) class constraints $\Omega_a^{(1)}$, $a = 1, \cdots, M^{(1)}$ ($\Omega_\alpha^{(2)}$, $\alpha = 1, \cdots, M^{(2)}$), with $M^{(1)} + M^{(2)} = M$. Correspondingly we write $\phi_a^{(1)}$, $a = 1, \cdots, m^{(1)}$ ($\phi_\alpha^{(2)}$, $\alpha = 1, \cdots, m^{(2)}$) for the primary constraints, with $m^{(1)} + m^{(2)} = m$. We have

$$\left\{\Omega_a^{(1)}, \Omega_b^{(1)}\right\} \approx 0 \quad , \quad \left\{\Omega_a^{(1)}, \Omega_\alpha^{(2)}\right\} \approx 0 \quad ,$$

and

$$\det \|\{\Omega_\alpha^{(2)}, \Omega_\beta^{(2)}\}\| \not\approx 0 \quad . \tag{O.28}$$

It is thus convenient to define a matrix Q with elements

$$Q_{\alpha\beta} := \{\Omega_\alpha^{(2)}, \Omega_\beta^{(2)}\} \quad . \tag{O.29}$$

Because of (O.28), Q^{-1} exists.

The above compact notation now allows us to streamline the calculation. Conservation of the constraints in time requires (recall (O.17c))

$$\{\Omega_a^{(1)}, H\} \approx 0 \quad , \tag{O.30a}$$

$$\{\Omega_\alpha^{(2)}, H\} + \sum_{\beta=1}^{m^{(2)}} \{\Omega_\alpha^{(2)}, \phi_\beta^{(2)}\} v_\beta^{(2)} \approx 0 \quad , \tag{O.30b}$$

where the sum extends over the $m^{(2)}$ second class primary constraints. Our algorithm ensures consistency of these equations.

Defining

$$H^{(1)} = H + \sum_{\beta=1}^{m^{(2)}} \phi_\beta^{(2)} v_\beta^{(2)} \tag{O.31}$$

we have for the total Hamiltonian,

$$H_T = H^{(1)} + \sum_{a=1}^{m^{(1)}} \phi_a^{(1)} v_a^{(1)} \tag{O.32}$$

where the sum extends over the $m^{(1)}$ first-class primary constraints. Since by construction

$$\{\Omega_A, H_T\} \approx 0 \quad , \quad A = 1, \cdots, M \quad , \tag{O.33}$$

we also have

$$\{\Omega_A, H^{(1)}\} \approx 0 \quad , \quad A = 1, \cdots, M \quad . \tag{O.34}$$

Thus H_T and $H^{(1)}$ are both first class quantities (hence the superscript of $H^{(1)}$).

From Eq. (O.30b) we obtain

$$v_\alpha^{(2)} \approx -\sum_{\beta=1}^{M^{(2)}} R_{\alpha\beta}\{\Omega_\beta^{(2)}, H\} \quad , \quad \alpha = 1, \cdots, m^{(2)} \tag{O.35a}$$

and

$$0 \approx -\sum_{\beta=1}^{M^{(2)}} R_{\alpha\beta}\{\Omega_\beta^{(2)}, H\} \quad , \quad \alpha = m^{(2)} + 1, \cdots, M^{(2)} \tag{O.35b}$$

where

$$R = Q^{-1} \quad . \tag{O.35c}$$

Hence we may write the equations of motion (O.26a) in the compact form

$$\dot{F} \approx \{F, H\} - \sum \{F, \Omega_\alpha^{(2)}\} R_{\alpha\beta}\{\Omega_\beta^{(2)}, H\} + \sum_{a=1}^{m^{(1)}} \{F, \phi_a^{(1)}\} v_a^{(1)} \tag{O.36}$$

Define the *Dirac* bracket of two functions F and G on Γ by

$$\{F, G\}_D \equiv \{F, G\} - \sum_{\alpha,\beta=1}^{M^{(2)}} \{F, \Omega_\alpha^{(2)}\} R_{\alpha\beta}\{\Omega_\beta^{(2)}, G\} \quad . \tag{O.37}$$

The equation of motion (O.36) then reads

$$\dot{F} \approx \{F, H\}_D + \sum_{b=1}^{m^{(1)}} \{F, \phi_b^{(1)}\} v_b^{(1)} \quad . \tag{O.38}$$

If the system is second class, the second term in this equation will be absent. In this case the transition to the quantum theory is achieved by the substitution $i\{A, B\}_D \to [\hat{A}, \hat{B}]$, with $\hat{A}$ and $\hat{B}$ now operators, and the constraints are realized strongly (as operator equations) as a result of the identity $\{\Omega_i, \Omega_j\}_D \approx 0$. On the other hand, if the system is first class, we have two methods available for quantizing the system:

i) The first-class constraints are realized as conditions on the *physical* states. For first class constraints this can consistently be done.

ii) We impose subsidiary (gauge) conditions, in order to fix the remaining parameters; this then allows for the introduction of Dirac brackets in the reduced phase space obtained by adding to the constraints $\Omega_a \approx 0$ the subsidiary conditions.

First class constraints and gauge transformations

The freedom in the choice of the velocities $v_a^{(1)}$ corresponds to an invariance under local (gauge) transformations in the Lagrangian formalism. In the Hamiltonian formalism these transformations are generated by

$$G(t) = \sum_{a=1}^{M^{(1)}} \epsilon_a(t)\Omega_a^{(1)}(q,p) \quad . \tag{O.39}$$

The corresponding infinitesimal transformation of a function $F(q,p)$ is that given by

$$\delta_G F = \{F, G\} \quad . \tag{O.40}$$

Here a strong equality is meant, since Eq.(O.40) is supposed to describe symmetry transformations of L, and is thus not limited to the subspace of solutions to the equations of motion.

In order to leave the corresponding action invariant, the parameters ϵ_a can however not be chosen independently. Indeed, consider

$$S = \int dt \left(\sum_{i=1}^{n} p_i \dot{q}_i - H_T \right) \quad . \tag{O.41}$$

Under the infinitesimal transformation (O.40)

$$\delta_G(p\dot{q}) = \frac{d}{dt}\left(p\frac{\partial G}{\partial p} - G \right) + \sum_{a=1}^{M^{(1)}} \dot{\epsilon}_a \Omega_a^{(1)} \quad . \tag{O.42}$$

Now, since $H^{(1)}$ in (O.32) is first class, it has the property,

$$\{H^{(1)}, \Omega_a^{(1)}\} = \sum_b h_{ab}\Omega_b^{(1)} \quad . \tag{O.43a}$$

Likewise, we must have the involution algebra

$$\{\Omega_a^{(1)}, \Omega_b^{(1)}\} = \sum_c c_{abc}\Omega_c^{(1)} \quad , \tag{O.43b}$$

where c_{ijk} are called "structure functions". They may or may not depend on the phase-space variables[446].

Now (summation over repeated indices is implied)

$$\delta_G H_T = \left[\epsilon_a h_{ac} + v_a^{(1)} \epsilon_b c_{abc} \right] \Omega_c^{(1)} + \delta_G v_a^{(1)} \phi_a^{(1)} \quad . \tag{O.44}$$

The requirement $\delta_G S = 0$ thus leads to

$$\delta_G v_a^{(1)} = \dot{\epsilon}_a - \epsilon_b(h_{ba} + v_c^{(1)} c_{cba}) \,,\; a = 1, \cdots, m^{(1)} \tag{O.45}$$

$$0 = \dot{\epsilon}_a - \epsilon_b(h_{ba} + v_c^{(1)} c_{cba}) \,,\; a = m^{(1)} + 1, \cdots, M^{(1)} \tag{O.46}$$

where we have assumed that the first term in (O.42) gives a vanishing contribution. See ref [446] for a discussion of this point.

Eq.(O.45) states how the parameters must transform, whereas Eq. (O.46) in general implies $M^{(1)} - m^{(1)}$ restrictions on the parameters ϵ_a, corresponding to the $M^{(1)} - m^{(1)}$ first-class secondary constraints.

The fact, that the restrictions on ϵ_a are in general incompatible with the choice $\epsilon_a = 0 \,,\; a = m^{(1)}+1, \cdots, M^{(1)}$, demonstrates the necessity of including also the first-class *secondary* constraints in (O.39). This also follows from the observation that observables should be invariant under single, as well as repeated infinitesimal transformations generated by only the first-class primary constraints. If we denote the infinitesimal parameters of two such transformations by ϵ_a and ϵ'_b, $(a, b = 1, \cdots, m^{(1)})$ and take the difference between the results of two consecutive transformations performed in either order, one obtains, making use of the Jacobi identity for Poisson bracket,

$$\Delta F = \sum \epsilon_a \epsilon'_b \{F, \{\phi_a^{(1)}, \phi_b^{(1)}\}\} \quad . \tag{O.47}$$

Since in general

$$\{\phi_a^{(1)}, \phi_b^{(1)}\} = \sum_{c=1}^{M^{(1)}} f_{abc} \Omega_c^{(1)} \quad , \quad a, b = 1, \cdots, m^{(1)} \quad , \tag{O.48}$$

we are led from the requirement $\{F, \phi_a^{(1)}\} \approx 0$ for an observable F, to the stronger requirement $\{F, \Omega_a^{(1)}\} \approx 0$. This is the *Dirac conjecture*[468]: "Observables" and states of a physical system are invariant under infinitesimal transformations generated by the primary and secondary, first-class constraints, acting *independently*.

The implications of Dirac's conjecture are, that the dynamics of the *observables* should be equivalently described by working with either H_T, or the extended Hamiltonian

$$H_E = H^{(1)} + \sum_{a=1}^{M^{(1)}} \xi_a \Omega_a^{(1)}$$

with ξ_j a correspondingly extended set of Lagrange multipliers. The restrictions on ϵ_a implied by (O.46) are thereby lifted, Eq. (O.46) being replaced by

$$\delta \xi_a = \dot{\epsilon}_a - \epsilon_b(h_{ba} + \xi_c c_{cba}) \quad . \tag{O.49}$$

Dirac's conjecture has been criticized by many authors[570], since, whereas H_T leads to equations of motion which are equivalent to the Euler-Lagrange equations of motion, H_E leads in general to different equations of motion. Dirac's conjecture however applies only to gauge-invariant quantities. For such "observables", the equivalence of the dynamics generated by H_T and H_E has been proven for the case of systems with only first-class constraints[468].

Implementation of gauge conditions

With the transition $H_T \to H_E$ one now has available a natural procedure for quantizing the theory, with all constraints imposed ***strongly***. The equation of motion (O.38) is replaced by

$$\dot{F} \approx \{F, H\}_D + \sum_a \left\{F, \Omega_a^{(1)}\right\}_D \xi_a \quad . \tag{O.50}$$

We impose as many subsidiary conditions

$$\chi_a \approx 0 \quad , \quad a = 1, \cdots, M^{(1)} \quad , \tag{O.51}$$

as there are first-class primary and secondary constraints. In order that all Lagrange parameters ξ_a be determined, we require that

$$\det \|\{\chi_a, \Omega_b^{(1)}\}_D\| \not\approx 0 \quad , \tag{O.52}$$

where the sign of weak equality now refers to $\Gamma^* \subset \Gamma_\Omega \subset \Gamma$, defined by the constraints ***and*** subsidiary conditions. The conditions $\dot{\chi}_a \approx 0$ now read,

$$\{\chi_a, H\}_D + \sum_{b=1}^{M^{(1)}} \{\chi_a, \Omega_b^{(1)}\}_D \xi_b \approx 0 \quad . \tag{O.53}$$

These equations determine the Lagrange multipliers $\{\xi_a\}$ on Γ^* uniquely.

In order to streamline the notation it is convenient to collect the first class constraints and subsidiary conditions in one vector $(\Phi_r) = (\Omega^{(1)}, \chi)$, and to define the matrix

$$\overline{Q} := \begin{pmatrix} \overline{Q}^{11} & \overline{Q}^{12} \\ \overline{Q}^{2} & \overline{Q}^{22} \end{pmatrix} = \begin{pmatrix} \{\Omega^{(1)}, \Omega^{(1)}\}_D & \{\Omega^{(1)}, \chi\}_D \\ \{\chi, \Omega^{(1)}\}_D & \{\chi, \chi\}_D \end{pmatrix} \quad , \tag{O.54a}$$

with the inverse $\overline{R} \equiv \overline{Q}^{-1}$. We then have from (O.51)

$$\xi_a \approx -\overline{R}_{ab}^{12} \{\chi_b, H\}_D \quad . \tag{O.55}$$

In order to make the transition to the quantum theory we must impose the conditions $\Omega_a^{(1)} \approx 0$, strongly. This leads us to define the Dirac bracket on Γ^* by

$$\{A,B\}_{D^*} = \{A,B\}_D - \sum_{r,s=1}^{2M^{(1)}} \{A,\Phi_r\}_D \overline{R}_{rs} \{\Phi_s,B\}_D \quad . \tag{O.56}$$

We evidently have,

$$\{\Phi_r,B\}_{D^*} = 0 \quad , \quad r = 1,\cdots,2M^{(1)} \tag{O.57a}$$

$$\{\Omega_\alpha^{(2)},B\}_{D^*} = 0 \quad , \quad \alpha = 1,\cdots,M^{(2)} \quad . \tag{O.57b}$$

The new Dirac bracket thus implements strongly the first and second class constraints as well as subsidiary conditions. Furthermore,

$$\dot{A} \approx \{A,H\}_{D^*} = \{A,H\}_D - \sum_{r,s} \{A,\Phi_r\}_D \overline{R}_{rs} \{\Phi_s,H\}_D \quad . \tag{O.58}$$

Indeed, noting that (O.30a) also implies $\{\Omega_a^{(1)},H\}_D \approx 0$, and noting further that $\overline{Q}^{11} \approx 0$ implies $\overline{R}^{22} \approx 0$, we have

$$\dot{A} \approx \{A,H\}_D - \sum \left\{A,\Omega_a^{(1)}\right\}_D \overline{R}_{ab}^{12} \{\chi_b,H_c\}_D \quad . \tag{O.59}$$

Recalling (O.55), this proves our claim.

The above procedure is only applicable to the case, where the subsidiary conditions (O.51) do not involve the Lagrange multipliers ξ_a. It therefore excludes the case of the so-called covariant α-gauges, as well as the temporal gauge $A^0 = 0$[3] in gauge-theories, where A^0 plays the role of Lagrange multiplier. Methods for coping with such gauges have been developed in a series of papers by Fradkin and collaborators[445]; in this book we use alternative and simpler approaches to deal with this case (see chapter 13 and 14).

---

[3] Axial gauges not involving A^0, such as $A_3 = 0$ can be treated by the above methods, but are nevertheless problematic (see ref. [455]).

REFERENCES

[1] P. A. M. Dirac, *Proc. Roy. Soc.* **117** (1928) 610; **118** (1928) 351.

[2] R. P. Feynman, *Phys. Rev.* **76** (1949) 769; **80** (1950) 440.

[3] F. J. Dyson, *Phys. Rev.* **75** (1949) 1736; **85** (1952) 631.
A. Salam, *Phys. Rev.* **82** (1951) 217.
J. C. Ward, *Phys. Rev.* **78** (1950) 182; *Proc. Phys. Soc.* **A64** (1951) 54.

[4] N. N. Bogoliubov and O. S. Parasiuk, *Acta Math.* **97** (1957) 227.
K. Hepp, *Commun. Math. Phys.* **2** (1966) 301.

[5] W. Zimmermann, in *Lectures on Elementary Particles and Quantum Field Theory*, vol 1, 1970, Brandeis Univ. Summer Institut.

[6] G. 't Hooft and M. Veltman, *Nucl. Phys.* **B44** (1972) 189.
G. Cicuta and E. Montaldi, *Nuovo Cimento Letters* **4** (1972) 329.
T. Gustafson, *Arkiv fõr Matematik, Astronomi och Fysik* **34A** (1947) 1.
E. Speer, *J. Math. Phys.* **9** (1968) 1404.
C. G. Bollini, J. J. Giambiaggi and A. G. Dominguez, *Nuovo Cimento* **31** (1964) 550.

[7] H. Lehmann, K. Symanzik and W. Zimmermann, *Nuovo Cimento* **1** (1955) 205.

[8] A. S. Wightman, *Phys. Rev.* **101** (1956) 860.

[9] R. Haag, *Dan. Mat. Fys. Medd.* **29** (1955) 12.
R. Haag and D. Kastler, *J. Math. Phys.* **5** (1964) 848.

[10] G. Källen, *Quantum Electrodynamics*, Springer Verlag, Berlin 1972.

[11] J. Schwinger, *Phys. Rev. Lett.* **3** (1959) 296.

[12] N. N. Bogoliubov and D. V. Shirkov, *Introduction to the theory of quantized fields.* Interscience New York, 1959.

[13] R. Haag, *Phys. Rev.* **112** (1958) 69.
D. Ruelle, *Helv. Phys. Acta* **35** (1962) 147.
K. Hepp, *Commun. Math. Phys.* **1** (1965) 95.

[14] R. F. Streater and A. S. Wightman, *PCT, Spin Statistics and all that*, W. A. Benjamin, N.Y. (1964).

[15] M. Jacob, ed. *Dual theory, Physics Reports Reprint Volume I*, Noth Holland, Amsterdam, 1974.

[16] R. Dolen, D. Horn and C. Schmid, *Phys. Rev. Lett.* **19** (1967) 402; *Phys. Rev.* **166** (1968) 1768.

[17] G. Veneziano, *Nuovo Cimento* **57A** (1968) 190.

[18] J. Scherk, *Rev. Mod. Phys.* **47** (1975) 123.

[19] S. Weinberg, *Phys. Rev. Lett.* **19** (1967) 1264.
A. Salam, in *Elementary Particle Theory*, ed. N. Svartholm, Almquist and Wiksells, Stockholm 1968.

[20] M. Gell Mann and Y.Ne'eman, in *The eightfold way*, Benjamin, New York, 1964.

[21] W. Marciano and H. Pagels, *Phys. Rep.* **C36** (1978) 137.

[22] C. G. Callan, *Phys. Rev.* **D2** (1970) 1541.
K. Symanzik, *Commun. Math. Phys.* **18** (1970) 227; *Springer Tracts in Mod. Phys.* **57** (1971) 222.
D. J. Gross and F. A. Wilczek, *Phys. Rev. Lett.* **30** (1973) 1346.
H. D. Polizer, *Phys. Rev. Lett.* **30** (1973) 1346.
H. D. Politzer, *Phys. Rep.* **14** (1974) 129.

[23] S. Coleman, *Aspects of symmetry,* Cambridge University Press, 1985.
S. Coleman, *The magnetic monopole fifty years later*, Erice, Italy, 1981.

[24] W. Thirring, *Ann. of Phys.* **3** (1958) 91.

[25] B. Klaiber, in *Lectures in Theoretical Physics*, Boulder 1967, Gordon and Breach, New York, 1968.

[26] A. S. Wightman, in *Cargese Lectures of Theoretical Physics* , ed. Gordon and Breach, New York, 1964.

[27] J. Schwinger, *Phys. Rev.* **128** (1962) 2425.

[28] J. Lowenstein and J. A. Swieca, *Annals of Phys.* **68** (1971) 172.

[29] A. Casher, J. Kogut and L. Susskind, *Phys. Rev. Lett.* **31** (1973) 31; *Phys. Rev.* **D10** (1974) 732.

[30] H. J. Rothe, K. D. Rothe and J. A. Swieca, *Phys. Rev.* **D15** (1977) 1675.

[31] K. Wilson, *Phys. Rev.* **179** (1969) 1499.

[32] S. Coleman and D. Gross, *Phys. Rev. Lett.* **31** (1973) 851.

[33] R. Rajaraman, *Phys. Rep.* **5** (1975) 227.

[34] K. Pohlmeyer, *Commun. Math. Phys.* **46** (1976) 207.

[35] M. Lüscher, *Nucl. Phys.* **B135** (1978) 1.

[36] A. B. Zamolodchikov and Al. B. Zamolodchikov, *Annals of Phys.* **120** (1979) 253.

[37] M. Karowski, *Phys. Rep.* **49** (1979) 229.

[38] E. Abdalla, *Lec. Notes in Phys.* **226** (1984) 140 ed. N. Sanchez and H. J. de Vega.

[39] H. J. de Vega, *Phys. Lett.* **87B** (1979) 233;
H. Eichenherr, H. J. de Vega and J. M. Maillet, *Proc. Meudon and Paris VI*, 83/84,171;
H. J. de Vega, H. Eichenherr and J. M. Maillet, *Commun. Math. Phys.* **92** (1984) 507;
H. J. de Vega, H. Eichenherr and J. M. Maillet, *Nucl. Phys.* **B240** (1984) 377;
H. J. de Vega, H. Eichenherr and J. M. Maillet, *Phys. Lett.* **132B** (1983) 337.

[40] M. Karowski, Proceedings - *String theory, Quantum cosmology and quantum gravity: integral and conformal invariant theories*, Paris 1986; *Nucl. Phys.* **B300** (1988) 473;

[41] M. K. Prasad, A. Sinha and L. L. Chau Wang, *Phys. Rev. Lett.* **43** (1979) 750; *Phys. Lett.* **B87** (1979) 237;

L. L. Chau, Proceedings - *Swansea 1988 in Mathematical Physics*, p. 346;
I. Ya Aref'eva and I. V. Volovich, *Phys. Lett.* **149B** (1984) 131;
L. L. Chau, M. K. Prasad and A. Sinha, *Phys. Rev.* **D23** (1981) 2321; **D24** (1981) 1574;
L. L. Chau, M. L. Ge and Z. Popowicz, *Phys. Rev. Lett.* **52** (1984) 1940;
E. Witten, *Phys. Lett.* **77B** (1978) 394; *Nucl. Phys.* **B266** (1986) 245;
E. Abdalla, M. Forger and M. Jacques, *Nucl. Phys.* **B307** (1988) 198;
J. Harnad, J. Hurtubise, M. Légaré and S. Shnider, *Nucl. Phys.* **B256** (1985) 609;
J. Avan, H. J. de Vega and J. M. Maillet, *Phys. Lett.* **171B** (1986) 255.

[42] M. Karowski and P. Weisz, *Nucl. Phys.* **B139** (1978) 455.

[43] M. Karowski, *Int. Colloq. on complex analysis, Microlocal calculus and relativistic quantum theory*, Les Houches, 1979; *Nucl. Phys.* **B153** (1979) 244.

[44] T. H. R. Skyrme, *Proc. Roy. Soc.* **A260** (1961) 127.

[45] S. Coleman, *Phys. Rev.* **D11** (1975) 2088.

[46] B. Schroer, Berlin-Prep. 75/5 unpublished.

[47] S. Mandelstam, *Phys. Rev.* **D11** (1975) 3026.

[48] P. Weisz, *Nucl. Phys.* **B122** (1977) 1.

[49] J. A. Swieca, *Fortschritte der Physik* **25** (1977) 303.

[50] A. M. Polyakov and P. B. Wiegman, *Phys. Lett.* **131B** (1983) 121;**141B** (1984) 223.

[51] E. Witten, *Commun. Math. Phys.* **92** (1984) 455.

[52] J. Wess and B. Zumino, *Phys. Lett.* **37B** (1971) 95.

[53] E. Abdalla and M. C. B. Abdalla, *Nucl. Phys.* **B255** (1985) 392.

[54] H. Eichenherr, *Nucl. Phys.* **B146** (1978) 215; E **B155** (1979) 544.

[55] H. Eichenherr and M.Forger, *Nucl. Phys.* **B155** (1979) 381.

[56] A. D'Adda, P. di Vecchia and M. Lüscher, *Nucl. Phys.* **B146** (1978) 63.

[57] A. A. Migdal, *Soviet Phys. Jetp* **42** (1976) 413,742.

[58] A. M. Polyakov, *Phys. Lett.* **59B** (1975) 79.

[59] A. Chakrabarti, *Introduction to classical solutions of Yang-Mills equations*, Rencontre de Rabat, 1978;
A. Actor, *Rev. Mod. Phys.* **51** (1979) 461.

[60] A. M. Din, W. J. Zakrzewski, *Nucl. Phys.* **B174** (1980) 397; *Phys. Lett.* **95B** (1980) 419;
A. M. Din, Z. Horvath and W. J. Zakrzewski, *Nucl. Phys.* **B233** (1984) 269.

[61] C. G. Callan, R. Dashen and D. J. Gross, *Phys. Rev. Lett.* **44** (1980) 435.

[62] E. Abdalla, M. Forger and M. Gomes, *Nucl. Phys.* **B210** (1982) 181.

[63] A. D'Adda and A. C. Davis, *Phys. Lett.* **101B** (1981) 85.

[64] A. D'Adda, A. C. Davis and P. di Vecchia, *Nucl. Phys.* **B218** (1983) 191.

[65] A. D'Adda, P. di Vecchia and M. Lüscher, *Phys. Rep.* **49** (1979) 239.

[66] M. Forger, Proceedings - *Clausthal*, Germany 1981 ed. S.I. Andersson, H.D. Doebner; *Lect. Notes in Math.* 1037, 1983.
[67] A. A. Tseytlin, *Int. J. Mod. Phys.* **A5** (1990) 589.
[68] C. G. Callan, E. J. Martinec, M. J. Perry and D. Friedan; *Nucl. Phys.* **B262** (1985) 593.
[69] C. G. Callan, I. R. Klebanov and M. J. Perry, *Nucl. Phys.* **B278** (1986) 78.
[70] D. J. Gross and E. Witten, *Nucl. Phys.* **B277** (1986) 1.
[71] R. Jackiw and R. Rajaraman, *Phys. Rev. Lett.* **54** (1985) 1219.
[72] L. D. Faddeev, *Phys. Lett.* **145B** (1984) 81
L. D. Faddeev, S. L. Shatashivili, *Phys. Lett.* **167B** (1986) 225.
S. L. Shatashvili, *Teor. Mat. Fiz.* **71** (1987) 40.
[73] H. O. Giroti, H. J. Rothe and K. D. Rothe, *Phys. Rev.* **D33** (1986) 514.
[74] B. Zumino, *Les Houches*, August 1983, Notes by K. Sibold.
B. Zumino, *Nucl. Phys.***B253** (1985) 477;
B. Zumino, Wu Young Shi and A. Zee, *Nucl. Phys.* **B239** (1984) 477;
L. Baulieu,*Nucl. Phys.* **B241** (1984) 577;
R. Stora, in *New Developments in Quantum Field Theory and Statistical Mechanics*, eds. M. Levy and P. Mitter, Plenum, New York, 1977.
[75] E. Abdalla, M. C. B. Abdalla and M. Gomes, *Phys. Rev.* **D23** (1981) 1800.
[76] E. Witten and Y. Y. Goldschmidt, *Phys. Lett.* **91B** (1980) 392.
[77] E. Abdalla, M. C. B. Abdalla and M. Gomes, *Phys. Rev.* **D25** (1982) 452.
[78] V. Kurak and R. Köberle, *Phys. Rev.* **D36** (1987) 627.
[79] A. A. Belavin, A. M. Polyakov and A. B. Zamolodchikov, *Nucl. Phys.* **B241** (1984) 333.
[80] A. Capelli, C. Itzykson and J. B. Zuber, *Commun. Math. Phys.* **113** (1987) 1.
[81] K. H. Rehren and B. Schroer, *Phys. Lett.* **198B** (1987) 480.
J. Fröhlich, *Proceedings of the Cargese School*, 1987.
[82] A. M. Polyakov, *Mod. Phys. Lett.* **A2** (1987) 893.
[83] A. M. Polyakov, *Les Houches*, 1988.
E. Abdalla, M. C. B. Abdalla and A. Zadra, Trieste-Preprint IC/89/56 unpublished.
[84] A. M. Polyakov, A. B. Zamolodchikov, *Mod. Phys. Lett.* **A3** (1988) 1213.
E. Abdalla, M. C. B. Abdalla and A. Zadra, *Mod. Phys. Lett.* **A4** (1989) 849.
[85] J. L. Cardy, in *Phase transitions and critical phenomena vol.11*, ed. by C. Domb, J.L. Lebowitz, Acad. Press, 1987.
[86] A. Rocha Caridi, in *Vertex Operator in Mathematics and Physics*, Berkeley, 1983, Springer Verlag, p. 451.
[87] V. G. Kac, *Lecture Notes in Physics* **94** (1979) 441.
[88] E. Witten, *Commun. Math. Phys.* **121** (1989) 351.
[89] E. Witten, *Commun. Math. Phys.* **117** (1988) 353.

E. Witten and J. H. Horne, *Phys. Rev. Lett.* **62** (1989) 501.
S. Deser, R. Jackiw and S. Templeton, *Phys. Rev. Lett.* **48** (1982) 975.
[90] A. M. Polyakov, *Mod. Phys. Lett.* **A3** (1988) 325.
A. Coste and M. Lüscher, *Nucl. Phys.* **B323** (1989) 631.
M. Lüscher, *Nucl. Phys.* **B326** (1989) 557.
[91] M. B. Green, J. H. Schwarz and E. Witten, *Superstring Theory.* Cambridge, Univ. Press, 1987.
[92] B. Schroer, *Fortschritte der Physik* **11** (1963) 1.
[93] S. Coleman, *Commun. Math. Phys.* **31** (1973) 259.
N. D. Mermin and H. Wagner, *Phys. Rev. Lett.* **17** (1966) 1133.
[94] L. Schwarz, *Théorie des Distribuitions*, Paris, Hermann, 1951.
I. M. Gel'fand and N. YD. Vilenkin, *Generalized Functions*, N.Y., Academic Press. 1964.
[95] L. Gårding and J. L. Lions, *Nuovo Cimento* (supp.) **14** (1959) 45.
[96] A. Jaffe, Ann. Phys. **32** (1965) 127.
[97] P. Jordan and E. P. Wigner, *Zeitsch. für Phys.* **47** (1928) 631.
[98] B. Schroer, J. A. Swieca and A. H. Völkel, *Phys. Rev.* **D11** (1975) 1509.
J. Kupsch, W. Rühl and B. C. Yunn, *Ann. of Phys.* (NY) **89** (1975) 115.
[99] Vl. S. Dotsenko and V. A. Fateev, *Nucl. Phys.* **B240** (1984) 312; **B251** (1985) 691.
[100] K. D. Rothe and I. O. Stamatescu, *Ann. of Phys.* (NY) **95** (1975) 202.
R. Banerjee *Z. Phys.* **C25** (1984) 251.
[101] L. Gårding and A. S. Wightman, *Arkiv Fysik* **28** (1964) 129.
[102] H. Epstein, *Nuovo Cimento* **27** (1963) 886.
[103] P. Jordan, *Zeitsch. für Phys.* **93** (1935) 464;
A. Sokolow, *Phys. Z. der Sowj.* **12** (1937) 148;
M. Born and N. S. Nagendra Nath, Proceedings *Ind. Acad. Sci.* **3** (1936) 318.
[104] W. Glaser, *Nuovo Cimento* **9** (1958) 990.
[105] K. Johnson, *Nuovo Cimento* **20** (1961) 773.
[106] F. Scarf and J. Wess, *Nuovo Cimento* **26** (1962) 150.
[107] B. Klaiber, *Helv. Phys. Acta* **37** (1964) 554.
[108] D. Finkelstein and J. Rubinstein, *J. Math. Phys.* **9** (1968) 1762.
[109] P. P. Kulisch and E. R. Nissimov, *Pisma Zh. Eksp. Teor. Fiz.* **24** (1976) 244.
R. Flume, *Phys. Lett.* **62B** (1976) 93;
I. Ya. Aref'eva and V. E. Korepin, *JETP Lett.* **20** (1974) 321;
R. Dashen, B. Hasslacher and A. Neveu, *Phys. Rev.* **D11** (1975) 3424;
B. Berg, M. Karowski and H. J. Thun, *Phys. Lett.* **62B** (1976) 187;
B. Yoon, *Phys. Rev.* **D132** (1976) 3440;
B. Berg, M. Karowski and H. J. Thun, *Phys. Lett.* **62B** (1976) 63;
R. Flume, P. K. Mitter and N. Papanicolaou, *Phys. Lett.* **64B** (1976) 289.
[110] J. Lowenstein, *Seminars on Renormalization Theory*, Maryland 1972.

[111] P. Lam, *Phys. Rev.* **D6** (1972) 2154.
[112] M. Karowski and H. J. Thun, *Nuovo Cimento* **38A** (1977) 11.
R. Flume and S. Meyer, *Lett. Nuovo Cimento* **18** (1977) 238.
[113] M. D. Kruskal and D. Wiley, American Math. Society, *Summer Seminar on Nonlinear Wave motion*, ed. by A.C. Newell (Potsdam NY 1972).
[114] V. E. Korepin and L. D. Fadeev, *Theor. Math. Phys.* **25** (1975) 1039;
A. B. Zamolodchikov, *Pisma v JETP* **25** (1977) 499.
[115] J. H. Lowenstein and W. Zimmerman, *Nucl. Phys.* **B86** (1985) 77.
[116] E. Abdalla, M. Gomes and R. Köberle, *Phys. Rev.* **D18** (1978) 3634.
[117] B. Berg, *Nuovo Cimento* **41A** (1977) 58.
[118] B. Berg, M. Karowski and H. J. Thun, *Phys. Lett.* **64B** (1976) 286;
B. Berg, M. Karowski and H. J. Thun, unpublished.
[119] G. Källen and J. Toll, *J. Math. Phys.* **6** (1965) 299.
[120] S. Weinberg, *Phys. Rev.* **118** (1960) 838.
[121] H. Neuberger, Thesis, Tel-Aviv Univ. 1976;
M. Bander, Univ. Calif. Irvine, Preprint (1975);
M. Halpern, *Phys. Rev.* **D12** (1975) 1684;
H. Lehman and J. Stehr, Hamburg, Desy-Preprint, 1976/29.
[122] B. Schroer and T. Truong, *Phys. Rev.* **D15** (1977) 1684.
[123] L. P. Kadanoff and H. Ceva, *Phys. Rev.* **B3** (1971) 3918).
[124] R. Köberle and E. C. Marino, *Phys. Lett.* **126B** (1983) 475.
[125] E. Fradkin and L. Susskind, *Phys. Rev.* **D17** (1978) 2637.
[126] J. Kogut, *Rev. Mod. Phys.* **51** (1979) 659.
[127] E. C. Marino, *Nucl. Phys.* **B217** (1983) 413; *Nucl. Phys.* **B230** (1984) 149;
A. A. S. de Macedo and E. C. Marino, *Phys. Rev.* **D40** (1989) 1360.
[128] E. C. Marino, B. Schroer and J. A. Swieca, *Nucl. Phys.* **B200** (1982) 473.
[129] P. Furlan, G. M. Sotkov anf I. T. Todorov, *Rivista del Nuovo Cimento* **12** (1989) 1.
[130] L. Heck, *Fortschrit der Physik* **27** (1979) 169.
[131] M. Gomes, J. Lowenstein, *Nucl. Phys.* **B45** (1972) 252.
[132] A. C. Scott, F. Y. F. Chu and D. W. Mc. Laughlin, Proceedings of the *IEEE* **61** (1973) 1443.
[133] L. A. Takhtadzhyan and L. D. Fadeev, *Theor. Math. Phys.* **21** (1975) 1046;
M. S. Alblowitz, O. J. Kaup, A. C. Newell and N. Segur, *Phys. Rev. Lett.* **31** (1973) 125.
[134] L. D. Fadeev and L. A. Takhtadjan, *Hamiltonian Methods in the Theory of Solitons*, Springer, 1987;
S. P. Novikov, *Theory of Solitons: the inverse scattering method* , New York, Contemporary Soviet Mathematics, 1984;
R. Rajaraman, *An Introduction to Solitons and Instantons in Quantum Field Theory*, North Holland, 1982.

[135] L. D. Fadeev and V. N. Popov, *Phys. Lett.* **25B** (1967) 29.
[136] P. T. Mathews and A. Salam, *Phys. Rev.* **94** (1954) 185.
[137] F. Berezin, *The Method of Second Quantization*, Academic Press, New York, London (1966).
[138] R. T. Seeley, *Am. Math. Soc. Proc. Symp. Pure Math.* **10** (1967) 288.
[139] P. B. Gilkey, *Proc. Symp. Pure Math.* **27** (1973); *J. Diff. Geom.* **10** (1975) 601.
[140] S. W. Hawking, *Commun. Math. Phys.* **55** (1977) 133.
[141] K. D. Rothe and B. Schroer, *Field Theoretical Methods in Particle Physics*, ed. W. Rühl, Plenum, N.Y., 1980.
[142] R. E. Gamboa-Saravi, M. A. Muschietti and J. E. Solomin, *Commun. Math. Phys.* **89** (1983) 363; *ibid* **93** (1984) 407.
[143] P. A. M. Dirac, *Ann. Math.* **36** (1935) 657.
[144] S. L. Adler, *Phys. Rev.* **D6** (1972) 3445;
R. Jackiw and C. Rebbi, *Phys. Rev.* **D14** (1976) 517.
[145] J. Kiskies, *Phys. Rev.* **D16** (1977) 2535.
[146] N. K. Nielsen and B. Schroer, *Nucl. Phys.* **B127** (1977) 493.
[147] J. W. Milnor and J. D. Stasheff, *Characteristic Classes*, *Annals of Mathematics Studies* **76**, Princeton University Press 1974.
[148] R. E. Gamboa-Saravi, M. A. Muschietti, F. A. Schaposnik and J. E. Solomin, *Ann. of Phys.* **157** (1984) 360.
[149] J. Singe, in *Relativity: The General Theory*, North-Holland, Amsterdam, 1960.
[150] B. De Witt, *Dinamical Theory of Groups and Fields*, Gordon and Breach 1965;
N. K. Nielsen, Nordita 78/24.
[151] J. S. Dowker and R. Critchley, *Phys. Rev.* **D13** (1976) 3224;
[152] J. S. Dowker, *J. Phys.* **A11** (1978) 347;
N. K. Nielsen, H. Römer and B. Schroer, *Nucl. Phys.* **B136** (1978) 445.
[153] J. Schwinger, *Theor. Phys.* (IAEA) Vienna (1963) 88.
[154] O. Alvarez, *Nucl. Phys.* **B216** (1983) 125; *ibid* **B238** (1984) 61.
[155] K. D. Rothe, *Nucl. Phys.* **B269** (1986) 269.
[156] K. Fujikawa, *Phys. Rev. Lett.* **42** (1979) 1195; *Phys. Rev.* **D21** (1980) 2848; *Phys. Rev.* **D22** (1980) 1499(E).
[157] K. Fujikawa, *Phys. Rev. Lett.* **44** (1980) 1733; *Phys. Rev.* **D23** (1981) 2262.
[158] K. Fujikawa, Proceedings of Nato Advanced Research Workshop *Super Field theories*, Simon Fraser University, Burnaby, B.C. Canada, 1986; *Proceedings of the Kyoto Summer Institute*, 1985, ed. T. Inami (World Scientific, Singapore, 1986).
[159] C. Itzykson and J. B. Zuber, *Quantum Field Theory* , Mc Graw Hill, 1980.
[160] G. A. Christos, I. C. Trieste preprint 82/97, 1982.
[161] K. Fujikawa, Proceedings of the 1988 Stefan Banach Workshop, *Gauge Theories of Fundamental Interactions*, Warsaw, Poland.

[162] M. F. Atiyah and I. M. Singer, *Amer. Math. Soc.* **69** (1963)422; *Ann. Math.* **87** (1968) 596.
[163] R. Banerjee, *Phys. Lett.* **B174** (1986) 313.
[164] K. Fujikawa, *Phys. Rev.* **D31** (1985) 341.
[165] L. Alvarez-Gaume and P. Ginsparg, *Nucl. Phys.* **B243** (1984) 449.
[166] H. Banerjee, R. Banerjee and P. Mitra, *Zeitsch. für Phys.* **C32** (1986) 445.
[167] M. A. Atiyah, R. Bott and V. K. Patodi, *Inv Math.* **19** (1973) 279; P. B. Gilkey, *The Index Theorem and the Heat Equation*, Math. Lect. Series, Publish on Perish Inc. 1974.
[168] A. S. Schwarz, *Commun. Math. Phys.* **64** (1979) 233.
[169] S. Coleman, unpublished.
[170] B. Schroer, *Acta Physica Austriaca*, suppl. XIX (1978) 155.
[171] J. Kiskies, *Phys. Rev.* **D15** (1977) 2329.
[172] L. S. Brown, R. D. Carlitz and C. Lee, (1977).
[173] R. Jackiw and C. Rebbi, *Phys. Rev.* **D16** (1977) 1052; *Phys. Rev.* **D13** (1976) 339.
[174] M. Ansourian, *Phys. Lett.* **70B** (1977) 301.
[175] M.F. Atiyah, V. G. Drinfeld, N. J. Hitchin and Yu. I. Manin *Phys. Lett.* **65A** (1978) 185
[176] F.A. Shaposnik, *Quantum Mechanics of Fundamental Systems II*, Ed. C. Teitelboim and J. Zanelli, Plenum, 1989.
[177] Y. Nambu and G. Jona-Lasinio, *Phys. Rev.* **122** (1961) 345.
[178] D. Gross and A. Neveu, *Phys. Rev.* **D10** (1974) 3235.
[179] A. Neveu and N. Papanicolau, *Commun. Math. Phys.* **58** (1978) 31.
[180] A. B. Zamolodchikov and Al. B. Zamolodchikov, *Phys. Lett.* **B72** (1978) 481.
[181] R. Dashen, B. Hasslacher and A. Neveu, *Phys. Rev.* **D12** (1975) 2443.
[182] T. L. Curthright and C. Zachos, *Phys. Rev.* **D24** (1981) 2661.
[183] C. Brézin, C. Itzykson, J. Zinn-Justin and J. B. Zuber, *Phys. Lett.* **82B** (1979) 442.
[184] S. Coleman and E. Weinberg, *Phys. Rev.* **D7** (1973) 1888; **D11** (1975) 3040.
[185] K. Symanzik, *Renormalization of theories with broken symmetry*, Cargese, 1970.
[186] H. Fleming and K. Furuya, *Nuovo Cimento* **49A** (1979) 101.
[187] H. D. I. Abarbanel, *Nucl. Phys.* **130** (1977) 29.
[188] H. Fleming, K. Furuya and J. L. de Lyra, *Nuovo Cimento* **53A** (1979) 405.
[189] I. S. Gradshteyn and M. Ryzhik, *Table of integrals, series and products*, Academic Press, 1980.
[190] D. Iagolnitzer, *Phys. Rev.* **D18** (1978) 1275.
[191] B. Berg, M. Karowski, V. Kurak and P. Weisz, *Phys. Lett.* **76B** (1978) 502.
[192] G. Källen and J. Toll, *J. Math. Phys.* **6** (1965) 299.
[193] E. Abdalla, A. Lima-Santos, *Rev. Bras. Fis.* **12** (1982) 293.

[194] M. Lüscher. Erratum of *Nucl. Phys.* **B135** (1978) 1, unpublished.
[195] D. Buchholtz and J. T. Lopuszanski, *Lett. Math. Phys.* **3** (1979) 175.
[196] B. Berg and P. Weisz, *Nucl. Phys.* **B145** (1978) 205.
[197] E. Witten, *Nucl. Phys.* **B145** (1978) 110.
[198] E. Abdalla, B. Berg and P. Weisz, *Nucl. Phys.* **B157** (1979) 387.
[199] R. Köberle, V. Kurak and J. A. Swieca, *Phys. Rev.* **D20** (1979) 897; E **D20** (1979) 2638.
[200] W. A. Bardeen, *Phys. Rev.* **184** (1969) 1848;
J. Bell and R. Jackiw, *Nuovo Cimento* **60A** (1969) 47.
[201] S. Adler, *Phys. Rev.* **177** (1969) 2426; in *Lectures on Elementary Particle Physics and Quantum Field Theory*, ed. S. Deser et al., vol. I, MITPress (1970).
[202] A. D'Adda, A. Davis, P. di Vecchia and P. Salomonson, *Nucl. Phys.* **B222** (1983) 45.
[203] E. Abdalla and M. C. B. Abdalla, *Nuovo Cimento* **57A** (1980) 334.
[204] H. Araki and R. Haag, *Commun. Math. Phys.* **4** (1967) 77.
[205] A. J. da Silva, M. Gomes and R. Köberle, *Phys. Rev.* **20** (1979) 895.
[206] E. Abdalla and Lima-Santos, *Phys. Rev.* **D29** (1984) 1851.
[207] J. Schwinger, *Ann. Phys.* **2** (1958) 407.
[208] M. Gell-Mann and M. Levy, *Nuovo Cimento* **16** (1960) 705.
[209] B. W. Lee, *Chiral Dynamics*, Gordon and Breach, Cargese, 1970.
[210] S. B. Treiman, R. Jackiw, B. Zumino and E. Witten, *Current algebra and anomalies*, World Scientific, 1985.
[211] J. Honerkamp, *Nucl. Phys.* **B36** (1972) 130.
[212] F. Nicodemi, P. di Vecchia, P. Rossi, R. Musto and R. Petorino, *Nucl. Phys.* **B235** (1984)478.
E. Abdalla, M. C. B. Abdalla and N. Kawamoto, *Phys. Rev.* **D31** (1985) 3213.
[213] E. Cremmer, J. Scherk, *Phys. Lett.* **74B** (1978) 341; *Nucl. Phys.* **B118** (1977) 61.
[214] J. Goldstone, *Nuovo Cimento* **19** (1961) 154.
Y. Nambu, *Phys. Rev. Lett.* **4** (1960) 380.
[215] D. J. Gross, *Applications of the renormalization group to high energy physics*, Les Houches, 1975.
[216] J. M. Maillet, *Renormalisation et Theories Completement Integrables* , "These 3° cycle", Univ. Paris Sud, 1982.
[217] E. Witten, *Phys. Rev.* **D16** (1977) 2991.
[218] H. Eichenherr, Ph D Thesis, Heidelberg, 1978.
[219] M. Forger, Ph D Thesis, Berlin, 1980.
[220] H. Eichenherr and M. Forger, *Nucl. Phys.* **164** (1980) 528; **B282** (1987) 745.
[221] H. Eichenherr and M. Forger, *Commun. Math. Phys.* **82** (1981) 227.

[222] I. Ya. Aref'eva, S. I. Azakov, *Nucl. Phys.* **B162** (1980) 298.
[223] W. Buchmüller, *Acta Physica Austr. Suppl.* **27** (1985) 517.
[224] L. Alvarez Gaumé, J. Bost, G. Moore, P. Nelson and C. Vafa, *Phys. Lett.* **B178** (1986) 41.
[225] J. F. Nash, *Ann. Math.* **63** (1956) 20.
[226] S. Kobayashi and K. Nomizu, *Foundations of Differential Geometry* , Interscience, New York, 1963/1969.
[227] E. Calabi, in *Algebraic Geometry and Topology: A Symposium in Honor of S. Lefschetz*, (Princeton Univ. Press, 1955).
S. T. Yau, *Proc. Natl. Acad. Sci.* **74** (1977) 1798.
[228] P. Candelas, G. Horowitz, A. Strominger and E. Witten, *Nucl. Phys.* **B258** (1985) 46.
[229] S. Helgason, *Differential geometry, Lie groups, and symmetric spaces.* New York: Academic Press 1978.
[230] V. E. Zakharov, A. V. Mikhailov, *Sov. Phys. JETP* **47** (1978) 1017.
[231] M. Lüscher and K. Pohlmeyer, *Nucl. Phys.* **B137** (1978) 46.
[232] E. Abdalla, M. Forger and A. Lima Santos, *Nucl. Phys.* **B256** (1985) 145.
[233] E. Abdalla and M. Forger, *Commun. Math. Phys.* **104** (1986) 123.
[234] G. Moore and P. Nelson, *Phys. Rev. Lett.* **53** (1984) 1519.
[235] N. E.Steenrod, *The Topology of fibre bundles*, Princeton Univ. Press, 1951.
[236] M. Forger, Private communication
[237] T. Curtright and C. Zachos, *Phys. Rev.* **D21** (1980) 411; *Phys. Lett.* **88B** (1979) 276; Proceedings *Supergravity*, Stony Brook, 1979.
[238] B. Zumino, *Phys. Lett.* **87B** (1979) 203.
[239] A. D'Adda, P. di Vecchia and M. Lüscher, *Nucl. Phys.* **B152** (1979) 125.
[240] J. Schonfeld, *Nucl. Phys.* **B169** (1980) 49.
S. Aoyama, *Nuovo Cimento* **57A** (1980) 176;
D. Z. Freedman and P. K. Townsend, *Nucl. Phys.* **B177** (1981) 282.
[241] L. Alvarez-Gaumé and D. Z. Freedman, *Commun. Math. Phys.* **80** (1981) 443.
[242] M. Dubois-Viollete and Y. Georgelin, *Phys. Lett.* **82B** (1979) 251.
B. Berg and M. Lüscher, *Commun. Math. Phys.* **69** (1979) 57; *Nucl. Phys.* **B160** (1979) 281; *Nucl. Phys.* **B190** (1981) 412.
M. Lüscher, *Nucl. Phys.* **B200** (1982) 61; *Commun. Math. Phys.* **85** (1982) 39; *Nucl. Phys.* **B205** (1982) 483.
A. M. Polyakov, *Phys. Lett.* **82B** (1979) 247; *Phys. Lett.* **72B** (1977) 224; *Nucl. Phys.* **B164** (1980) 171.
D. Maison, Max Planck Institute Prep.PAE/PTh 52/79 (unpublished).
Yu Makeenko and A. A. Migdal, *Phys. Lett.* **88B** (1979) 135.
[243] M. Forger, in *Proceedings, Clausthal*, ed. H. Doebner *Lec. Notes in Physics* **139**, 1978.

[244] V. Golo and A. Perelomov, *Phys. Lett.* **79B** (1978) 112.
[245] W. Buchmüller, R. D. Peccei and T. Yanagida, *Nucl. Phys.* **B227** (1983) 503; *ibid* **B231** (1984) 63.
[246] W. Buchmüller, R. D. Peccei and T. Yanagida, *Nucl. Phys.* **B244** (1984) 186.
[247] M. T. Grisaru, A. van de Ven and D. Zanon, *Phys. Lett.* **173B** (1986) 423.
[248] A. Borel, *Kählerian Coset Spaces of Semisimple Lie Groups*, Proc. Acad. Sci. USA **40** (1954) 1147.
[249] M. Bordemann, M. Forger and H. Römer, *Commun. Math. Phys.* **102** (1986) 605.
[250] M. Forger, *Scheveningen Lectures* (Lecture Notes in Physics 311).
[251] R. Gilmore, *Lie Algebras, and Some of their Applications* , John Wiley and Sons, 1974.
[252] W. Lerche, *Nucl. Phys.* **B238** (1984) 582.
[253] W. Buchmüller and W. Lerche, *Ann. of Phys.* **175** (1987) 159.
[254] J. Lowenstein and E. R. Speer, *Nucl. Phys.* **B158** (1979) 397;
W. Bardeen, B. Lee and R. Schrock, *Phys. Rev.* **D14** (1976) 985.
[255] E. Brezin, S. Hikami, J. Zinn Justin, *Nucl. Phys.* **B165** (1980) 528.
[256] F. Delduc and G. Valent, *Phys. Lett.* **148B** (1984) 124;
C. Becchi, A. Blasi, G. Bonneau, R. Collina and F. Delduc, *Commun. Math. Phys.* **120** (1988) 121.
[257] I. Ya Aref'eva, *Theor. Math. Phys.* **36** (1979) 573.
[258] L. Alvarez Gaumé and D. Z. Freedman, *Phys. Rev.* **D22** (1980) 846.
[259] L. Alvarez Gaumé, D. Z. Freedman and S. Mukhi, *Annals of Phys.* **134** (1981) 85.
[260] M. T. Grisaru, A. E. M. van de Ven and D. Zanon, *Nucl. Phys.* **B277** (1986) 388,409; *Nucl. Phys.* **B277** (1986) 388.
M. T. Grisaru, D. I. Kazakov and D. Zanon, *Nucl. Phys.* **B287** (1987) 189.
[261] E. Braaten, T. L. Curthright and C. K. Zachos, *Nucl. Phys.* **B260** (1985) 630.
[262] B. Cobra and M. Gomes, *Rev. Bras. Fis.* **12** (1982) 693.
E. Abdalla. (unpublished)
[263] S. Weinberg, *Phys. Rev.* **D11** (1975) 3583.
[264] G. 't Hooft, *Phys. Rev. Lett.* **37** (1976) 8; *Phys. Rev.* **D14** (1976) 3432.
[265] R. J. Crewther, *Phys. Lett.* **70B** (1977) 349; *"Field Theoretical Methods in Particle Physics"*, ed. W. Rühl, Plenum, N.Y., 1980.
[266] P. Di Vecchia, *Phys. Lett.* **85B** (1979) 357.
[267] E. Witten, *Nucl. Phys.* **B149** (1979) 285.
[268] C. G. Callan, R. Dashen and D. Gross, *Phys. Lett.* **63B** (1976) 334.
[269] R. Jackiw and C. Rebbi, *Phys. Rev. Lett.* **37** (1976) 172.
[270] J. Kogut and L. Susskind, *Phys. Rev.* **D11** (1975) 3594.

[271] V. L. Berezinski, *Sov. Phys. Jetp.* **32** (1970) 493.
J. M. Kosterlitz and D. J. Thouless, *J. Phys.* **C6** (1973) 1181.
[272] E. Abdalla, M. C. B. Abdalla and M. Gomes, *Phys. Rev.* **D27** (1983) 825.
[273] F. Gliozzi, *Phys. Lett.* **153B** (1985) 403.
[274] R. Köberle and V. Kurak, *Phys. Rev. Lett.* **58** (1987) 627.
[275] J. Schwinger, *Phys. Rev.* **82** (1951) 664.
[276] G. 't Hooft, and M. Veltman, *Annales de L'Inst. H. Poincaré*, XX (1974) 69.
B. de Witt, *Phys. Rev.* **162** (1967) 1195.
[277] I. A. Batalin, S. G. Matinyan and G. K. Savoidi, *J. Nucl. Phys.* (USSR) **26** (1977) 407.
[278] B. S. de Witt, *Phys. Rep.* **19C** (1975) 295.
[279] I. Jack and H. Osborn, *Nucl. Phys.* **B207** (1982) 474.
M. T. Grisaru, N. K. Nielsen and D. Zanon, *Nucl. Phys.* **B247** (1984) 157.
E. Abdalla and M. C. B. Abdalla, *Nucl. Phys.* **B255** (1985) 392.
L. F. Abbott, *Nucl. Phys.* **185** (1981) 189.
[280] J. Hadamard, *Leçons sur la propagation des ondes, et les equation de l' hydrodynamique*, Hermann, Paris, 1903.
L. P. Eisenhart, *Riemannian Geometry*, Princeton Univ. Press, 1926.
A. Z. Petrov, *Einstein Spaces* , Pergamon Press, 1969.
[281] D. Friedan, *Ann. Phys.* **163** (1985) 318.
[282] S. J. Gates, M. T. Grisaru, M. Roček and W. Siegel, *Superspace*, Frontiers in Physics, Lecture Note series 58, Benjamin Cummings Publ., 1983.
[283] L. Alvarez-Gaumé, S. Coleman and P. Ginsparg, *Commun. Math. Phys.* **103** (1986) 423.
[284] G. Curci and G. Paffuti, *Nucl. Phys.* **B286** (1987) 399.
[285] G. Moore and P. Nelson, *Commun. Math. Phys.* **100** (1985) 83.
[286] G. 't Hooft, *Naturalness, chiral symmetry and spontaneous symmetry breaking in recent developments in gauge theories*, ed. G. 't Hooft et al, Plenum, N.Y. 1980.
[287] P. di Vecchia, S. Ferrara and L. Girardello, *Phys. Lett.* **151B** (1985) 199.
[288] S. Coleman and J. Mandula, *Phys. Rev.* **159** (1967) 1251.
[289] R. Haag, J. Lopuszanki and M. Sohnius, *Nucl. Phys.* **B88** (1975) 257.
[290] R. Shankar and E. Witten, *Phys. Rev.* **D17** (1978) 2134.
[291] D. Iagolnitzer, *The S matrix* , North Holland, 1978.
[292] L. Castillejo, R. H. Dalitz and F. J. Dyson, *Phys. Rev.* **101** (1956) 453.
[293] B. Schroer, T. T. Truong and P. Weisz. *Phys. Lett.* **63B** (1976) 422.
[294] B. Berg, M. Karowski, W. R. Theis and H. J. Thun, *Phys. Rev.* **D17** (1978) 1172.
[295] H. J. de Vega, in *Two dimensional models and string theories*, ed. E. Abdalla and M. C. B. Abdalla, World Scientific, 1988;
H. B. Thacker, *Rev. Mod. Phys.* **53** (1981) 253;

H. J. de Vega and J. M. Maillet, *Phys. Lett.* **101B** (1981) 382;
O. Babelon, H. J. de Vega and C. M. Viallet, *Nucl. Phys.* **B200** (1982) 266.

[296] M. Karowski, *Nucl. Phys.* **B153** (1979) 244.

[297] B. Berg, M. Karowski, V. Kurak and P. Weisz, *Nucl. Phys.* **B134** (1978) 125.

[298] I. Ya Aref'eva and V. E. Korepin, *Pisma Zh. Eksp. Teor. Piz.* **21** (1975) 302.

[299] A. Luther, *Phys. Rev.* **B14** (1976) 2153.
S. Nussimov, *Phys. Rev.* **D14** (1976) 647.

[300] A. B. Zamolodchikov, *Commun. Math. Phys.* **55** (1977) 183.
M. Karowski, H. J. Thun, T. T. Truong and P. H. Weisz, *Phys. Lett.* **67B** (1977) 321.

[301] A. B. Zamolodchikov and Al. B. Zamolodchikov, *Nucl. Phys.* **B133** (1978) 525.

[302] R. Köberle and J. A. Swieca, *Phys. Lett.* **83B** (1979) 209.

[303] B. Mc. Coy, C. A. Tracy and T. T. Wu, *Phys. Rev. Lett.* **38** (1977) 1793;
B. Schroer, T. T. Truong, *Phys. Lett.* **72B** (1978) 371; **73B** (1978) 149;
M. Sato, T. Miwa and M. Jimbo, RIMS Kyoto Preprint 207 (1976).

[304] E. Abdalla and M. C. B. Abdalla, *Nuovo Cimento* **53A** (1980) 334.

[305] E. Witten, *Nucl. Phys.* **B142** (1978) 285.

[306] M. Karowski and H.J. Thun, *Nucl. Phys.* **B130** (1977) 224.

[307] B. Berg and P. Weisz, *Commun. Math. Phys.* **67** (1979) 241.

[308] T. Banks and A. Zaks, *Nucl. Phys.* **B128** (1977) 333.

[309] E. Brézin and J. Zinn Justin, *Phys. Rev.* **B14** (1976) 3110.

[310] E. Abdalla and A. Lima-Santos, *Mod. Phys. Lett.* **A3** (1988) 311.

[311] E. Abdalla and A. Lima Santos, *Phys. Lett.* **B206** (1988) 281.

[312] M. C. B. Abdalla and A. Lima Santos, *Acta Phys. Pol.* **B15** (1984) 813.

[313] M. Karowski, V. Kurak and B. Schroer, *Phys. Lett.* **81B** (1979) 200.

[314] E. Abdalla, M. C. B. Abdalla, N. A. Alves and C. E. I. Carneiro, *Phys. Rev.* **D41** (1990) 571.

[315] G. 't Hooft, *Nucl. Phys.* **B72** (1974) 461.

[316] E. Abdalla, M. C. B. Abdalla and A. Lima Santos, *Phys. Lett.* **140B** (1984) 71; **146B** (1984) 457.

[317] P. Wiegman, *Phys. Lett.* **142B** (1984) 173; *ibid* **141B** (1984) 217.

[318] M. C. B. Abdalla, *Phys. Lett.* **164B** (1985) 71.

[319] E. Abdalla, M. C. B. Abdalla and M. Forger, *Nucl. Phys.* **B297** (1988) 374.

[320] E. Ogievetsky, N. Reshetikhin and P. Wiegman, *Nucl. Phys.* **B280** (1987) 45.

[321] C. Orzalezi, *Rev. Mod. Phys.* **42** (1970) 381.

[322] J. Lowenstein, *Les Houches*, 1982.

[323] L. D. Fadeev, *Sov. Sci. Rev.* **C1** (1980) 107; *Les Houches*, 1982.

[324] H. J. de Vega, *Int. J. Mod. Phys.* **A4** (1989) 2371.
[325] O. Babelon, H. J. de Vega and C. M. Viallet, *Nucl. Phys.* **B190** (1981) 542.
[326] I. V. Cherednik, *Theor. Math. Phys.* **47** (1981) 225.
[327] V. N. Dutyshev, *Jetp Lett.* **51** (1980) 671.
[328] A. P. Buthvostov, L. N. Lipatov, *Nucl. Phys.* **B180** (1981) 116.
[329] B. J. Zuber, *Lectures at J.A. Swieca Summer School*, 1991.
[330] O. Babelon and C. M. Viallet, Sissa preprint 54 EP (May 89).
[331] E. Witten, *Nucl. Phys.* **B223** (1983) 422.
[332] P. di Vecchia, Proc. *Schladming Inst. Acta Phys. Austr. Suppl.* XXII (1980) 341.
[333] A. P. Balachandran, V. P. Nair, S. G. Rajeev and A. Stern, *Phys. Rev. Lett.* **49** (1982) 1124.
[334] E. Witten, *Nucl. Phys.* **B223** (1983) 433.
[335] G. S. Adkins, C. R. Nappi and E. Witten, *Nucl. Phys.* **B228** (1983) 552.
[336] S. P. Novikov, *Usp. Mat. Nauk.* **37** (1982) 3.
[337] E. Abdalla, K. D. Rothe, *Phys. Rev.* **D36** (1987) 3190.
[338] P. A. M. Dirac, *Lectures on Quantum Mechanics*, Belfer Graduate School of Science, Yeshiva Univ., 1964;
P. A. M. Dirac *Can. J. Math.* **2** (1950) 120; *Proc. Roy. Soc.* **1246** (1958) 326;
A. Hanson, T. Regge and C. Teitelboim, *Accademia Nazionale dei Lincei*, Rome, 1976;
E. Sudarshan and N. Mukunda, *Classical Dynamics: A Modern Perspective*, Wiley N.Y., 1974;
K. Sundermeyer, *Constraint Systems*, Lecture Notes in Physics **169**, Springer, 1982.
[339] E. Majorana, *Nuovo Cimento*, XIII (1937) 171.
[340] H. Sugawara, *Phys. Rev.* **170** (1968) 1659.
[341] C. G. Callan, R. Dashen, D. H. Sharp. *Phys. Rev.* **165** (1968) 1883.
[342] S. Coleman, D. Gross and R. Jackiw, *Phys. Rev.* **180** (1969) 1359.
[343] J. L. Petersen, *Acta Phys. Polon.* **B16** (1985) 271.
P. di Vecchia, B. Durhuus and J. L. Petersen, *Phys. Lett.* **B144** (1984) 245.
D. Gonzales and A. N. Redlich, *Phys. Lett.* **B147** (1984) 150; *Nucl. Phys.* **B256** (1985) 621.
[344] M. C. B. Abdalla, *Phys. Lett.* **B152** (1985) 215.
[345] H. J. de Vega, *Phys. Lett.* **87B** (1979) 233.
[346] E. Abdalla and M. C. B. Abdalla, *Phys. Lett.* **152B** (1985) 59.
[347] P. di Vecchia, V. G. Knizhnik, J. L. Petersen and P. Rossi, *Nucl. Phys.* **B253** (1985) 701.
[348] L. S. Brown, *Nuovo Cimento* **29** (1963) 617;

B. Zumino, *Phys. Lett.* **10** (1964) 224; *Acta Phys. Austriaca* Supl. **2** (1966) 212;
C. R. Hagen, *Nuovo Cimento* **B51** (1967) 169.
[349] L. V. Belvedere, K. D. Rothe, B. Schroer and J. A. Swieca, *Nucl. Phys.* **112** (1979) 112.
[350] K. D. Rothe and B. Schroer, *Nucl. Phys.* **B185** (1981) 429.
P. J. Steinhardt, Harvard preprint HUTP–80/A022, unpublished.
[351] H. J. Rothe, K. D. Rothe and J. A. Swieca, *Phys. Rev.* **D19** (1979) 3020.
[352] S. Coleman, R. Jackiw and L. Susskind, *Ann. Phys.* (N.Y.) **13** (1975) 267.
[353] S. Coleman, *Ann. Phys.* (N.Y.) **101** (1976) 239.
[354] V. Kurak, B. Schroer and J. A. Swieca, *Nucl. Phys.* **B134** (1978) 61.
[355] N. K. Nielsen and B. Schroer *Nucl. Phys.* **B120** (1977) 62.
[356] R. J. Crewther, Sun–Sheng Shei and Tung–Mow Yan, *Phys. Rev.* **D8** (1973) 1730.
[357] N. Nakanishi, *Progr. Theor. Phys.* **57** (1977) 580; *ibid.* **57** (1977) 1025; *ibid* **58** (1977) 1007.
[358] C. R. Hagen, *Nucl. Phys.* **B85** (1975) 477.
[359] H. J. Rothe, K. D. Rothe and I. Stamatescu, *Ann. Phys.* (N.Y.) **105** (1977) 63.
[360] H. Georgi and J. M. Rawls *Phys. Rev.* **D3** (1971) 874.
[361] G. F. Dell'Antonio, Y. Frishman and D. Zwanziger, *Phys. Rev.* **D6** (1972) 988.
[362] C. Morchio, D. Pierotti and F. Strocchi, *Ann. Phys.* (1988) 188.
[363] N. Nakanishi, *Progr. Theor. Phys.* **58** (1977) 1927.
[364] G. A. Christos, *Phys. Rep.* **116** (1984) 251.
G. 't Hooft, *Phys. Rep.* **142** (1986) 357;
S. Weinberg, in *"Proceedings of the XVII International Conference on High Energy Physics"* London 1974, Didcot, Berkshire, England, 1974.
[365] J. Bernstein, *Rev. Mod. Phys.* **46** (1974) 1.
[366] G. S. Guralnik, C. R. Hagen and T. W. Kibble, *Phys. Rev. Lett.* **13** (1964) 585;
R. Brout and F. Englert, *Phys. Rev. Lett.* **13** (1964) 321;
G. S. Guralnik and C. R. Hagen, *Nuovo Cim.* **43A** (1966) 1;
P. W. Higgs, *Phys. Rev.* **145** (1966) 1156;
T. W. Kibble, *Phys. Rev.* **155** (1967) 1554.
[367] K. D. Rothe and J. A. Swieca, *Phys. Rev.* **D15** (1977) 541.
[368] C. Callan, R. Dashen and D. Gross, *Phys. Lett.* **66B** (1977) 375.
[369] B. Schroer and T. T. Truong, *Nucl. Phys.* **B144** (1978) 80.
[370] R. Haag and D. Kastler, *J. Math. Phys.* **5** (1964) 848.
[371] J. A. Swieca, *Phys. Rev.* **D13** (1976) 312.
[372] K. Wilson, *Phys. Rev.* **D10** (1974) 2445.

[373] G. Segré and W. Weisberger, *Phys. Rev.* **D10** (1974) 1767;
M. B. Halpern, *Phys. Rev.* **D13** (1976) 337.
[374] A. Patrascioiu, *Phys. Rev.* **D15** (1977) 3592.
[375] P. Mitra and P. Roy, *Phys. Lett.* **79B** (1978) 469.
[376] K. D. Rothe and J. A. Swieca, *Phys. Rev.* **D15** (1977) 1675.
[377] I. Bender, H. J. Rothe, and K. D. Rothe, *Nucl. Phys.* **B251** (1985) 745.
[378] B. Schroer and J. A. Swieca, *Nucl. Phys.* **121B** (1977) 505.
[379] E. C. Marino and J. A. Swieca, *Nucl. Phys.* **B170** (1980) 175.
[380] W. E. Thirring and J. E. Wess, *Ann. Phys.* (N.Y.) **27** (1964) 331.
[381] N. Nakanishi, *Progr. Theor. Phys.* **58** (1977) 1580.
[382] K. D. Rothe and J. A. Swieca, *Nucl. Phys.* **B168** (1980) 454.
[383] G. 't Hooft, *Nucl. Phys.* **B75** (1974) 461.
[384] C. Callan, N. Coote and D. J. Gross, *Phys. Rev.* **D13** (1975) 1649.
[385] T. T. Wu, *Phys. Rev. Lett.* **71B** (1977) 142.
[386] Y. Frishman, C. T. Sachrajda, H. Abarbanel and R. Blankenbecler, *Phys. Rev.* **D15** (1977) 2275.
[387] E. E. Salpeter and H. Bethe, *Phys. Rev.* **84** (1951) 1232.
[388] I. O. Stamatescu and T. T. Wu, *Nucl. Phys.* **B143** (1978) 503.
[389] V. Baluni, *Phys. Lett.* **90B** (1980) 407.
[390] P. J. Steinhardt, *Nucl. Phys.* **B176** (1980) 100.
[391] D. Gepner, *Nucl. Phys.* **B252** (1985) 481.
[392] M. Hortaçsu, K. D. Rothe and B. Schroer, *Nucl. Phys.* **B164** (1980) 333.
[393] E. Brézin, C. Itzykson, G. Parisi and J. Zuber, *Comm. Math. Phys.* **59** (1978) 35;
C. Itzykson and J. B. Zuber, *J. Math. Phys.* **21** (1980) 411.
A. B. Zamolodchikov, *Phys. Lett.* **B117** (1982) 87.
[394] A. Jevicki and N. Papanicolasu, *Nucl. Phys.* **B171** (1980) 362;
A. Jevicki and B. Sakita, *Nucl. Phys.* **B185** (1981) 89.
[395] N. K. Nielsen, K. D. Rothe and B. Schroer, *Nucl. Phys.* **B160** (1979) 330.
[396] R. E. Gamboa–Saravi, F. A. Schaposnik and J. E. Solomin, *Phys. Rev.* **D30** (1984) 1353;
L. C. L. Botelho and M. A. R. Monteiro, *Phys. Rev.* **D30** (1984) 2242;
R. Roskies and F. Schaposnik, *Phys. Rev.* **D23** (1981) 558;
R. Roskies, *"Festschrift for Feza Gursey's 60'th birthday"*, Symmetries in Particle Physics, eds. I. Bars., A. Chodos and C.–H. Tze, Plenum, New York, 1984.
[397] Y. Frishman, *Nucl. Phys.* **B148** (1979) 74, 93.
[398] H. Partovi, *Phys. Lett.* **80B** (1978) 377.
C. Sorensen and G. H. Thomas, *Phys. Rev.* **D21** (1980) 1625.
[399] M. Forger, private communication.
[400] M. Makowka, G. Wanders, *Helv. Phys. Acta* **59** (1986) 1366.

[401] H. Falomir and E. M. Santangelo, *Phys. Rev. Lett.* **56** (1986) 1659.
[402] D. A. Aragão de Carvalho and K. D. Rothe, *Phys. Rev.* **D36** (1987) 1897.
[403] M. B. Einhorn, *Phys. Rev.* **D14** (1976) 3451.
[404] S. J. Brodsky and G. R. Farrar, *Phys. Rev. Lett.* **31** (1973) 1153; *Phys. Rev.* **D11** (1975) 1309.
[405] S. D. Drell and T. M. Yan, *Phys. Rev. Lett.* **24** (1970) 181;
G. B. West, *Phys. Rev. Lett.* **24** (1970) 1206.
[406] E. D. Bloom and F. J. Gilman, *Phys. Rev. Lett.* **25** (1970) 1140; *Phys. Rev.* **D4** (1971) 2901.
[407] A. A. Belavin, A. M. Polyakov, A. S. Schwartz, and Yu. S. Tyupkin, *Phys. Lett.* **59B** (1975) 85.
[408] N. K. Nielsen and B. Schroer, *Phys. Lett.* **66B** (1977) 373, 475.
[409] K.D. Rothe and J. A. Swieca, *Ann. Phys.* **117** (1979) 382.
[410] H. J. Rothe and J. A. Swieca, *Nucl. Phys.* **B149** (1979) 237.
[411] K. D. Rothe and J. A. Swieca, *Nucl. Phys.* **B138** (1978) 26.
[412] N. S. Manton, *Ann. of Phys.* **159** (1985) 220.
[413] J. von Neumann, *Mathematical foundations of quantum mechanics* , Princeton University Press, 1955.
V. Bargmann, *Ann. Math.* **59** (1954) 1.
[414] M. Hortaçsu, K. D. Rothe, and B. Schroer, *Phys. Rev.* **D20** (1979) 3203; *Phys. Rev.* **D22** (1980) 3145.
[415] M. Hortaçsu, K. D. Rothe and B. Schroer, *Nucl. Phys.* **B171** (1980) 530.
[416] A. Patrascioiu, *Phys. Rev.* **D20** (1979) 491.
[417] M. F. Atiyah, V. K. Patodi and I. M. Singer, *Math. Proc. Camb. Phil. Soc.* **77** (1975) 43.
[418] V. Rubakov, *Zh. Eksp. Teor. Fiz. Pis'ma Red.* **33** (1981) 658; *JETP Lett.* **33** (1981) 644; *Nucl. Phys.* **B203** (1982) 311;
C. Callan, *Phys. Rev.* **D25** (1982) 2141; **D26** (1982) 2058;
V. Rubakov and M. Sevebryakov, *Nucl. Phys.* **B218** (1983) 240.
[419] *Theory and Detection of Magnetic Monopoles in Gauge Theories*, ed. N. Craigie, World Scientific, 1986.
[420] D. Sutherland, *Nucl. Phys.* **B2** (1967) 433;
M. Veltman, *Proc. Roy. Soc.* **A301** (1967) 107.
[421] L. D. Faddeev and S. L. Shatashvili, *Theor. Math. Phys.* **60** (1984) 206.
[422] H. J. Rothe and D. K. Rothe, *Phys. Rev.* **D40** (1989) 545.
[423] W. T. Kim and B. H. Cho, *Mod. Phys. Lett.* **A4** (1989) 1645.
[424] S. -G. Jo, *Phys. Lett.* **124** (1985) 353; *Nucl. Phys.* **B259** (1985) 616.
[425] M. Kobayashi, K. Seo and A. Sugamoto, *Nucl. Phys.* **B273** (1986) 607.
[426] K. Harada and I. Tsutsui, *Phys. Lett.* **B183** (1987) 311.
O. Babelon, F. A. Shaposnik and C. M. Viallet, *Phys. Lett.* **B177** (1986) 385;
A. V. Kulikov, IHEP preprint 86-83 (Serpukov) unpublished.

[427] R. Jackiw, CECS Scientific Series, C. Teitelboim, Eds., Summer Mtg. on *Quantum Mechanics of Fundamental Systems*, (CECS), Santiago, Chile, Dec. 1985, Plenum 1987.
[428] H. Cartan and S. Eilenberg, *Homological Algebra*, Princeton University Press, 1956.
[429] C. Becchi, A. Rouet, and R. Stora, *Commun. Math. Phys.* **42** (1975) 127; *Ann. Phys.* **98** (1976) 287.
[430] W. A. Bardeen and B. Zumino, *Nucl. Phys.* **B244** (1984) 421.
[431] R. Banerjee, preprint HD-THEP-1990-10;
H. Banerjee and R. Banerjee, *Phys. Lett.* **174B** (1986) 313.
[432] H. Banerjee, R. Banerjee and P. Mitra, *Z. Phys.* **C32** (1986) 445.
[433] E. D'Hoker and E. Farhi, *Nucl. Phys.* **B248** (1984) 59;77.
[434] H. O. Girotti and K. D. Rothe, *J. Mod. Phys.* **A4** (1989) 3041.
[435] K. Harada and H. Mukaido, *Z. Phys.* **C48** (1990) 151.
[436] K. Seo *Phys. Rev.* **D42** (1990) 2819.
[437] L. Baulieu, *Phys. Rep.* **129** (1985) 1.
[438] C. Aragão de Carvalho, C. A. Linhares, H. J. Rothe and K. D. Rothe, *Phys. Lett.* **B194** (1987) 539.
[439] R. Percacci and R. Rajaraman, *Phys. Lett.* **B201** (1988) 256;
N. K. Pak and P. Rossi, *Nucl. Phys.* **B250** (1985) 279.
[440] K. Harada, *Nucl. Phys.* **B329** (1990) 723.
[441] N. K. Falck and G. Kramer, *Phys. Lett.* **B193** (1987) 257; *Ann. Phys.* **174** (1987) 330.
[442] Ken-ichi Shizuja, *Phys. Rev.* **D40** (1989) 1260; *Phys. Lett.* **B213** (1988) 298.
[443] R. Rajaraman, *Phys. Lett.* **165B** (1985) 148.
[444] K. Fujikawa, *Phys. Lett.* **B171** (1986) 424.
M. Testa and K. Yoshida, *Phys. Lett.* **167B** (1986) 83.
[445] E. S. Fradkin and G. A. Vilkovisky, *CERN report* TH 2332-CERN;
E. S. Fradkin and G. A.Vilkovisky, *Phys. Lett.* **55B** (1975) 224;
E. S. Fradkin and T. E. Fradkina, *Phys. Lett.* **72B** (1978) 343;
I. A. Batalin and G. A.Vilkovisky, *Phys. Lett.* **69B** (1977) 309.
[446] M. Henneaux, *Phys. Rep.* **126** (1985) 1.
[447] T. Kugo and I. Ojima, *Phys. Lett.* **73B** (1978) 459; *Progr. Theor. Phys.* **60** (1978) 1869;
G.Curci and R. Ferrari, *Nuovo Cimento* **35A** (1976) 273.
[448] W. Siegel *Nucl. Phys.* **B238** (1984) 307.
R. Floreanini, R. Jackiw *Phys. Rev. Lett.* **59** (1987) 1873.
[449] K. Harada, *Int. J. Mod. Phys.* **A5** (1990) 4469.
[450] K. Harada and K. D. Rothe, *Phys. Lett.* **B237** (1990) 495.

[451] W. T. Kim, W. H. Kye, B. H. Cho and D. K. Park, *Phys. Rev.* **D42** (1990) 2748.

[452] T. D. Bjorken, *Phys. Rev.* **148** (1966) 1467;
K. Johnson and F. E. Low, *Progr. Theor. Phys. Suppl.* 37-38 (1966) 74.

[453] S.-G. Jo, *Phys. Lett.* **163B** (1985) 353.

[454] A. V. Ramallo, *Z. Phys.* **C42** (1989) 125;
C. E. M. Wagner, *Phys. Lett.* **209B** (1988) 300;
M. Moshe and Y. Oz, *Phys. Lett.* **224B** (1989)145.

[455] H. J. Rothe and K. D. Rothe, *Il Nuovo Cimento* **74A** (1983) 129.

[456] M. B. Halpern, *Phys. Rev.* **19** (1979) 517.

[457] F. Otto, H. J. Rothe and K. D. Rothe, *Phys. Lett.* **B231** (1989) 299.

[458] F. S. Otto and K. D. Rothe, *Phys. Rev.* **D42** (1990) 2829.

[459] F. S. Otto, *Phys. Rev.* **D43** (1991) 548; PhD thesis, Heidelberg, 1991.

[460] R. Banerjee, H. J. Rothe and K. D. Rothe, *Mod. Phys. Lett.* **A5** (1990) 2147;
R. Banerjee and H. J. Rothe, to be published in *Int. J. Mod. Phys.* (1991)

[461] P. Mitra, *Phys. Rev. Lett.* **60** (1988) 265; *ibid* **61** (1988) 2005.

[462] I. G. Halliday, E. Rabinovici, A. Schwimmer and M. Chanowitz, *Nucl. Phys.* **B268** (1986) 413;
M. S. Chanowitz, *Phys. Lett.* **B171** (1986) 280;
R. Link, A. Roberge and G. W. Semenoff, *Z. Phys.* **C39** (1988) 269;
A. J. Niemi and G. W. Semenoff, *Phys. Lett.* **B175** (1986) 439.

[463] R. Rajaraman, *Phys. Lett.* **154B** (1985) 305.

[464] H. O. Girotti, H. J. Rothe and K. D. Rothe, *Phys. Rev.* **D34** (1986) 592.

[465] R. Banerjee, *Phys. Rev. Lett.* **56** (1986) 1889.

[466] D. Boyanovsky, *Nucl. Phys.* **B294** (1987) 223;
D. Boyanovsky, I. Schmidt and M. F. L. Golterman, *Ann. Phys.* (1988) 111.

[467] P. Mitra and R. Rajaraman, *Phys. Rev.* **D37** (1988) 448.

[468] M. E. V. Costa, H. O. Girotti and T. J. M. Simoes, *Phys. Rev.* **D32** (1985) 405.

[469] K. D. Rothe and H. O. Girotti, *Lett. Nuovo Cimento* **31** (1981) 545.

[470] K. Harada and I. Tsutsui, *Z. Phys.* **C39** (1988) 137.

[471] K. Harada, *Phys. Rev.* **D42** (1990) 4170.

[472] S. Miyake and Ken–ichi–Shizuya, *Phys. Rev.* **D36** (1987) 3781; *Phys. Rev.* **D37** (1988) 2282.

[473] T. Berger, *Z. Phys.* **C47** (1990) 297.

[474] C. A. Linhares, H. J. Rothe and K. D. Rothe, *Phys. Rev.* **D35** (1987) 2501.

[475] C. A. Linhares, H. J. Rothe and K. D. Rothe, *Phys. Rev.* **D37** (1988) 427.

[476] T. Berger, N. K. Falck and G. Kramer, *Int. J. Mod. Phys.* **A4** (1989) 427;
N. K. Falck and G. Kramer, *Z. Phys.* **C37** (1988) 321.

[477] A. Recknagel, Master Thesis, Heidelberg University, 1990.

[478] A. Das, *Phys. Rev. Lett.* **55** (1985) 2126.

[479] F. S. Otto, H. J. Rothe and A. Recknagel, *Phys. Rev.* **D42** (1990) 1203.

[480] K. Harada, *Phys. Rev.* Lett. **64** (1990) 139.

[481] G. L. Delmarco and R. C. Trinchero, *Phys. Rev.* **D38** (1988) 427.

[482] J. E. Wess, *Nuovo Cimento* , **18** (1960) 1086.
H. A. Kastrup, *Ann. Physik* **7** (1962) 388; *Phys. Rev.* **140** (1965) 183; **142** (1966) 1060.

[483] B. Schroer, *Nuovo Cimento Letters* **2** (1971) 867.

[484] A. A. Migdal, *Phys. Lett.* **37B** (1971) 98; *ibid* **37B** (1971) 386.
G. Mack and K. Symanzik, *Commun. Math. Phys.* **27** (1972) 247.
S. Ferrara, R. Gatto and A. Grillo, *Springer Tracts in Modern Physics* 67, *Conformal Algebra in Space Time*, Springer 1973.
G. Mack and A. Salam, *Ann. Phys.* (NY)**53** (1969) 174.

[485] M. Hortaçsu, R. Seiler and B. Schroer, *Phys. Rev.* **D5** (1972) 2519.
J. A. Swieca and A. H. Völkel, *Commun. Math. Phys.* **29** (1973) 319.
S. Cicciarello and G. Sartori, *Nuovo Cimento* **19A** (1974) 470.

[486] B. Schroer and J. A. Swieca, *Phys. Rev.* **D10** (1974) 480.
B. Schroer, *Cargese Lectures*, 1987.

[487] A. Polyakov, *Jetp. Lett.* **12** (1970) 381; *Sov. Phys. Jetp.* **39** (1974) 10.
A. A. Migdal, *Phys. Lett.* **44B** (1972) 112.

[488] K. Wilson, *Phys. Rev.* **D2** (1970) 1473; *Rev Mod. Phys.* **47** (1975) 773.

[489] B. Zumino, *Brandeis Lectures on Theoretical Physics*, 1970.

[490] I. T. Todorov, M. C. Mintcheva and V. B. Petkova, *Conformal Invariance in Quantum Field Theory*, Pisa, Classe di Scienze 1978.
T. Eguchi, *Lectures Academia Sinica*, 1986.

[491] F. Jegerlehner and B. Schroer, *Nucl. Phys.* **B68** (1974) 461.

[492] P. A. M. Dirac, *Rev. Mod. Phys.* **21** (1949) 392;
S. Fubini, A. J. Hansen and R. Jackiw, *Phys. Rev.* **D7** (1973) 1732.

[493] P. Ginsparg, *Applied Conformal Field Theory*, Les Houches, 1988.

[494] A. N. Turin, *Usp. Math. Nauk.* **33** (1978) 149.

[495] V. G. Kac, *Infinite dimensional Lie algebras, Progress in Math.*, vol. 44 Birkhänser, 1984.

[496] B. L. Feigin and D. B. Fucks, *Funct. Anal. and Appl.* **16** (1982) 114.

[497] D. Friedan, Z. Qiu and S. Shenker, in *Vertex Operators in Mathematics and Physics*, Berkeley, Ed. J. Lepowski, S. Mandelstam, I. M. Singer, 1983.

[498] D. Friedan, Z. Qiu and S. Shenker, *Phys. Rev. Lett.* **52** (1984) 1575.

[499] V. Kazakov, *Mod. Phys. Lett.* **A4** (1989) 2125.

[500] E. Verlinde, *Nucl. Phys.* **B300** (1988) 360.

[501] P. M. Morse and H. Feshbach, *Methods of Theoretical Physics*, Mac Graw Hill Book Comp. Inc., 1953.

[502] V. G. Knizhnik and A. B. Zamolodchikov *Nucl. Phys.* **B247** (1984) 83.

[503] D. Gepner and E. Witten, *Nucl. Phys.* **B278** (1986) 493.
[504] V. A. Fateev and A. B. Zamolodchikov, *Sov. J. Nucl. Phys.* **43** (1986) 657; *Sov. Phys. JETP* **62** (1985) 215.
[505] R. Dashen and Y. Frishman, *Phys. Rev.* **D11** (1975) 2781.
[506] C. M. Sommerfield, *Phys. Rev.* **176** (1968) 2019.
[507] M. Gomes, V. Kurak, A. J. da Silva, *Nucl. Phys.* **B295** (1988) 139.
[508] E. Abdalla and M. C. B. Abdalla, *Rev. Bras. Fis.* **19** (1989) 166.
E. Abdalla, in *Quantum Mechanics of Fundamental Systems II*, ed. C. Teitelboim and J. Zanelli, Plenum Press, 1989.
[509] C. Destri, H. J. de Vega, *Phys. Lett.* **208B** (1988) 255.
[510] M. Bershadskii and h. Ooguri, *Commun. Math. Phys.* **126** (1989) 49.
[511] Vl. Dotsenko, *Nucl. Phys.* **B338** (1990) 747.
[512] J. Bagger, D. Nemeshanski, N. Seiberg and S. Yankielowicz, *Nucl. Phys.* **B289** (1987) 53.
[513] V. G. Kac, *Math. USSR Izv.* **2** (1968) 127;
[514] R. V. Moody, *J. Algebra* **10** (1968) 2.
[515] I. B. Frenkel and V. G. Kac, *Inv. Math.* **62** (1980) 23.
[516] P. Goddard and D. Olive, in *Vertex Operators in Mathematics and Physics* , ed. J. Lepowsky, S. Mandelstam and I. M. Singer, Berkeley, Springer Verlag, 1983.
[517] P. Goddard and D. Olive, *Int. J. Mod. Phys.* **A1** (1986) 303.
[518] E. del Giudice, P. di Vecchia and S. Fubini, *Ann. Phys.* **70** (1972) 378.
[519] J. P. Serre, *A course in Arithmetic*, Springer Verlag, New York, 1973.
[520] J. Schwarz, *Superstrings - The first 15 years*, World Scientific, 1985.
[521] A. Neveu and J. Schwarz, *Nucl. Phys.* **B31** (1971) 86.
[522] P. Ramond, *Phys. Rev.* **D3** (1971) 2415.
[523] K. S. Narain, M. H. Samardi and E. Witten, *Nucl. Phys.* **B279** (1986) 369.
[524] Th. Kaluza, *Sitz. Preuss. Akad. Wiss.* **K1** (1921) 966.
O. Klein, *Z. Phys.* **37** (1926) 895.
[525] P. Fayet and S. Ferrara, *Phys. Rep.* **32C** (1977) 250.
[526] P. van Niewvenhuizen, *Phys. Rep.* **68C** (1981) 189.
[527] J. Schwarz, *Phys. Rep.* **89** (1982) 223.
[528] D. Friedan, E. Martinec, S. Shenker, *Nucl. Phys.* **B271** (1986) 93.
[529] S. Deser and B. Zumino, *Phys. Lett.* **65B** (1976) 369.
[530] M. A. Bershadsky, V. G. Knizhnik and M. Teitelman, *Phys. Lett.* **151B** (1985) 31.
[531] W. Greub, S. Halperin and R. Vanstone, *Connections, Curvature and Cohomology vol.II* , Academic Press, New York and London, 1973.
[532] L. Alvarez-Gaumé, C. Gomez and C. Reina, *Proc. Trieste Spring School on Superstrings*, 1987.

[533] L. Alvarez-Gaumé, J. B. Bost, G. Moore, P. Nelson and C. Vafa, *Commun. Math. Phys.* **112** (87) 503.
[534] A. M. Polyakov, *Gauge Fields and Strings*, Harwood Academic Publishers, 1987.
[535] K. Isler and J. M. Lina, *Canonical quantization of the induced* $2-D$ *gravity*, UDEM-LPN-TH-12, 1990.
[536] V. G. Knizhnik, A. M. Polyakov and A. B. Zamolodchikov, *Mod. Phys. Lett.* **A3** (1988) 819.
[537] I. Fukai, *Nucl. Phys.* **B299** (1988) 346.
[538] E. Abdalla and R. S. Jasinschi, *Nucl. Phys.* **B232** (1984) 426.
[539] D. J. Gross and N. Miljković, *Phys. Lett.* **238B** (1990) 217.
[540] F. David, *Nucl. Phys.* **257** (1985) 45,543.
[541] H. B. Nielsen, (unpublished).
[542] V. A. Kazakov, *Phys. Lett.* **119A** (1986) 140;
D. Boulatov and V. A. Kazakov, *Phys. Lett.* **186B** (1987) 379.
[543] D. Gross and A. A. Migdal, *Phys. Rev. Lett.* **64** (1990) 717.
[544] V. A. Kazakov, *Nucl. Phys.* B(FS4) (1988) 93.
[545] B. Duplantier and I. K. Kostov, *Phys. Rev. Lett.* **61** (1988) 1433.
[546] I. K. Kostov, *Mod. Phys. Lett.* **A4** (1989) 217.
[547] T. D. Lee and C. N. Yang, *Phys. Rev.* **87** (1952) 404,410.
[548] M. Staudacher, *Nucl. Phys.* **B336** (1990) 349.
[549] D. J. Gross and I. Klebanov. *Nucl. Phys.* **B344** (1990) 475.
[550] M. R. Douglas, *Phys. Lett.* B238 (1990) 176.
[551] T. Banks, M. Douglas, N. Seiberg and S. Shenker, *Phys. Lett.* **238B** (1990) 279.
E. Brézin, M. Douglas, V. Kazakov and S. Shenker, *Phys. Lett.* **237B** (1990) 43.
[552] M. A. Awada, Z. Qiu, *Phys. Lett.* **B245** (1990) 359.
[553] V. A. Kazakov and A. A. Migdal, *Nucl. Phys.* **B311** (1988) 171.
[554] E. Witten, *Nucl. Phys.* **B340** (1990) 281.
[555] J. Distler and H. Kawai, *Nucl. Phys.* **B321** (1988) 171.
[556] Weingarten, *Phys. Lett.* **B90** (1980) 28.
T. Eguchi, H. Kawai, *Phys. Lett.* **B114** (1982) 247.
[557] S. Chaudhuri, H. Kawai and S.-H. Henry Tye, *Phys. Rev.* **D36** (1986) 1148.
[558] D. Boulatov, V. A. Kazakov, I. K. Kostov and A. A. Migdal, *Nucl. Phys.* **B275** (1986) 641.
[559] I. K. Kostov, M. Mehta, *Phys. Lett.* **B189** (1987) 118.
F. David, *Mod. Phys. Lett.* **A3** (1988) 1651.
[560] D. Gross and A. A. Migdal, *Nucl. Phys.* **B340** (1990) 333.
[561] J. H. Schwarz, *Lecture*, Trieste, 4th july,1990.

[562] F. Gliozzi, J. Scherk and D. Olive, *Nucl. Phys.* **B122** (1977) 253.
[563] M. B. Green and J. H. Schwarz, *Phys. Lett.* **149B** (1984) 117; **151B** (1985) 21.
[564] D. J. Gross, J. A. Harvey, E. Martinec and R. Rohm, *Nucl. Phys.* **256** (1985) 253.
[565] C. M. Naon, *Phys. Rev.* **D31** (1985) 2035.
L. C. Botelho, *Phys. Rev.* **D33** (1986) 1195.
[566] E. Witten, *Nucl. Phys.* **B142** (1978) 285;
R. Shankar, E. Witten, *Nucl. Phys.* **B141** 91978) 349;
M. Karowski and H. J. Thun, *Nucl. Phys.* **B190** (1981) 61.
[567] M. Bernstein and J. Sonnenschein, *Phys. Rev. Lett.* **60** (1988) 1772;
C. Imbimbo and A. Schwimmer, *Phys. Lett.* **B193** (1987) 455;
J. Sonnenschein, *Nucl. Phys.* **B309** (1988) 752;
E. Abdalla, M. C. B. Abdalla, F. P. Devecchi and A. Zadra, *Phys. Lett.* **B215** (1988) 564.
[568] H. O. Girotti et al., *Phys. Rev. Lett.* **60** (1988) 1913;
H. O. Girotti, M. Gomes, V. Kurak, V. O. Rivelles and A. J. da Silva, *Phys. Rev. Lett.* **19** (1988) 1913;
M. N. Sanielevici and G. W. Semenoff, *Nucl. Phys.* **B312** (1989) 197;
L. D. Faddeev and R. Jackiw, *Phys. Rev. Lett.* **60** (1988) 1692;
P. Solomonson and B.-S. Skagerstam, *Phys. Rev. Lett.* **62** (1989) 1817.
[569] E. Abdalla and M. C. B. Abdalla, *Phys. Rev.* **D40** (1989) 491.
[570] R. Cawley, *Phys. Rev. Lett.* **42** (1979) 413; *Phys. Rev.* **D21** (1980) 2988;
A. Frenkel, *Phys. Rev.* **D21** (1980) 2986;
N. Mukunda, *Phys. Scr.* **21** (1980) 783;
L. Castellani, *Ann. Phys.* (N.Y.)**143** (1982) 357;
R. Sugano and H. Kamo, *Prog. Theor. Phys.* **67** (1972) 1966;
R. Sugano and T. Kimura, *Prog. Theor. Phys.* **69** (1983) 252.

INDEX

www.ingramcontent.com/pod-product-compliance
Ingram Content Group UK Ltd.
Pitfield, Milton Keynes, MK11 3LW, UK
UKHW020405150926
8624IPUK00004B/363

9 789810 204624